BOUNDARY ELEMENT METHODS IN MECHANICS

MECHANICS AND MATHEMATICAL METHODS

A SERIES OF HANDBOOKS

General Editor

J.D. ACHENBACH

Northwestern University, Evanston, Illinois, U.S.A.

First Series

COMPUTATIONAL METHODS IN MECHANICS

Editors

T. BELYTSCHKO

Northwestern University, Evanston, Illinois, U.S.A.

K.J. BATHE

Massachusetts Institute of Technology, Cambridge, Massachusetts, U.S.A.

NORTH-HOLLAND
AMSTERDAM · NEW YORK · OXFORD · TOKYO

BOUNDARY ELEMENT METHODS IN MECHANICS

Volume 3 in
Computational Methods in Mechanics

Edited by

Dimitri E. BESKOS

University of Patras
Patras, Greece

1987

NORTH-HOLLAND
AMSTERDAM · NEW YORK · OXFORD · TOKYO

ISBN 0 444 87990 0

PUBLISHERS:
ELSEVIER SCIENCE PUBLISHERS B.V.
P.O. Box 1991
1000 BZ Amsterdam
The Netherlands

SOLE DISTRIBUTORS FOR THE USA AND CANADA:
ELSEVIER SCIENCE PUBLISHING COMPANY, INC.
52 Vanderbilt Avenue
New York, N.Y. 10017,
U.S.A.

PRINTED IN THE NETHERLANDS

Preface

The study of the various complicated contemporary phenomena encountered in engineering mechanics is usually accomplished in two steps. The first step deals with the construction of the pertinent mathematical model describing the phenomenon, usually in the form of a set of governing differential equations accompanied by appropriate boundary and initial conditions, while the second step deals with the solution of this set of equations. In most practical situations the great complexities of the problem preclude the use of analytical methods of solution and render the employment of numerical methods of solution imperative.

The advent and widespread use of the electronic computer during the last 30 years created the impetus for the improvement and efficient use of already known numerical methods, such as the Finite Difference Method (FDM) and the development of new powerful numerical methods, such as the Finite Element Method (FEM) which appears to be the most widely known and used numerical method of solution among scientists and engineers.

The last 15 years, however, witnessed the emergence and proliferation of another powerful numerical method, the Boundary Element Method (BEM). This method appears to be superior to the FEM for quite a number of classes of problems and inferior or equally efficient for some other. Combinations of the BEM with the FDM and especially the FEM in a number of problems have also resulted in very efficient computational schemes. The BEM has been successfully used to a great variety of problems in engineering science, such as potential theory, elastostatics, elastodynamics, elastoplasticity, fracture, fluid mechanics, heat conduction, acoustics, electromagnetism and soil– or fluid–structure interaction.

There are presently available in the literature two major texts, a number of texts dealing with specialized subjects, two series of volumes coming out periodically and describing the new developments in the field and a great number of journal and conference proceedings articles on the BEM. The present book entitled "Boundary Element Methods in Mechanics" represents a collection of the most important topics in the field written by some of the most well known researchers in boundary elements. It is really a handbook characterized by a combination of tutorial and state-of-the-art aspects which attempts to provide a more comprehensive treatment of the various subjects than in the two existing series of new developments in the field and in a more authoritative way than in a text through the use of specialists.

The book is primarily intended for scientists and engineers with some background in engineering mechanics and consists of twelve chapters each one accompanied by numerous references. The first chapter provides an introduction to the fundamentals of the BEM, its history, advantages and disadvantages and future developments. In the second chapter the potential theory is used as a vehicle to illustrate the mathematical and numerical aspects of the method. Further illustration of the method is provided by the third chapter which deals with two- and three-dimensional elastostatics. Chapters 4 and 5 treat two- and three-dimensional elastodynamics including viscoelasticity from a general and a special point of view, respectively. Nonlinear solid mechanics including material and geometric nonlinearities is taken up in the sixth chapter, while two- and three-dimensional fracture analysis is treated in the seventh chapter. Chapter 8 is devoted to fluid mechanics and in particular to potential, viscous and ground water flow and water-waves, while Chapter 9 concerns itself with acoustics. Chapter 10 discusses heat conduction and the mathematically related phenomena of transient thermoelasticity and soil-consolidation. The last two chapters deal with two important interaction phenomena. Thus, Chapter 11 covers dynamic soil-structure interaction, while Chapter 12 treats the phenomenon of fluid–structure interaction.

Unavoidably, certain areas of application of the method are not covered in the book. In many cases, however, a brief treatment of these subjects is provided together with a number of appropriate references.

Closing this preface, the editor would like to express his appreciation to all those who helped him in one way or another to make this book become a reality: to Professor T. Belytschko for his encouragement and valuable assistance, to all the authors of the book for their efforts and cooperation, to his wife Zetta for her patience and understanding and to the publishing house of North-Holland for their prompt cooperation.

Dimitri E. Beskos

Contents

CHAPTER 1

Introduction to Boundary Element Methods

Dimitri E. BESKOS

Department of Civil Engineering
University of Patras
Patras 26110, Greece

Boundary Element Methods in Mechanics
Edited by D.E. Beskos

Introduction

Realistic problems of engineering mechanics, usually formulated in terms of a set of governing differential equations together with appropriate boundary and initial conditions, are characterized by great complexities in geometry and material constitution. As a result of that, analytical solutions of these problems are practically impossible to obtain and resort has to be made to numerical methods for the determination of an approximate solution. Numerical solutions of very complicated problems are feasible today due to the availability of powerful digital computers. A numerical method is of the differential or the integral type depending on whether the numerical analysis precedes or follows the integration of the governing equations.

The most widely known and used numerical methods of solution are the Finite Difference Method (FDM) and the Finite Element Method (FEM), both of the differential type. The FDM replaces the differential equations by algebraic ones, valid at a set of nodes within the domain, through the approximation of derivatives by finite differences, while the FEM replaces the domain itself by a set of finite subdomains or elements connected through their nodes and approximately reproducing the behavior of the subdomain that represent. The FEM presents some very distinct advantages over the FDM such as better conformity to the domain geometry, much easier handling of the boundary conditions and easier construction of variable-size meshes. These advantages have made the FEM the most popular numerical method among scientists and engineers. Even the FEM, however, presents some disadvantages such as rather high costs for the preparation and input of data and time-consuming solutions of three-dimensional problems, especially those with distant boundaries.

The Boundary Element Method (BEM), which is based on an integral formulation of the problem, has emerged during the last 15 years as a new powerful computational tool. This method usually requires only a surface discretization and not a discretization of both the interior and the surface of the domain of interest as it is the case with 'domain-type' techniques, such as the FDM and the FEM. This fact makes the BEM more efficient than the FEM for quite a number of classes of problems. For some other problems the BEM might be equally good or inferior to the FEM, while in certain cases a combination of the two methods usually creates the optimum numerical scheme. The BEM has been successfully used to a great variety of problems in engineering science, such as potential theory, elastostatics, elastodynamics, elastoplasticity, viscoelasticity, viscoplasticity, fracture analysis, fluid mechanics, acoustics, heat conduction, electromagnetism, soil–structure interaction and fluid–structure interaction.

This chapter provides a brief introduction to the BEM consisting of some historical notes in section 1, the fundamental aspects of the method in section 2, a list of its advantages and disadvantages in section 3, and a discussion on its future developments in section 4.

1. Historical notes

The BEM considered as an integral equation method has a long history that can go back to 1903 when Fredholm [45] published his rigorous work on integral equations encountered in potential theory. The classical works of Kellogg [64] in 1929 on potential theory and especially of Muskhelishvili [79, 80] in 1953 and Kupradze [66] in 1965 on elastostatics represent applications of integral equation techniques. It should be mentioned, however, that Muskhelishvili's [79, 80] approach utilizes complex variables which preclude its extension to three-dimensional problems. The notable works of the Russian authors Mikhlin [74–76] in 1957 and 1965, Smirnov [98] in 1964, Gakhov [47] in 1966, and Ivanov [58] in 1976 provided a wealth of information on integral equations, their properties and their approximate solution. However, no general numerical solution capabilities emerged during the period before the advent of the digital computer.

The methods developed by Trefftz [107] and Prager [85] in 1917 and 1928, respectively, for solving integral equations in potential fluid flow theory might be considered the precursors of modern boundary integral techniques, even though such methods are really impractical without the use of a computer. Integral equation methods as practical, efficient and general computational tools started appearing during the 1960s, a period characterized by the widespread use of the computer, showed a considerable expansion and development during the 1970s, in step with the rapid improvements in computers, and are approaching a level of maturity during the first half of the 1980s.

The term BEM, which first appeared in the literature in 1977 in the works of Banerjee and Butterfield [6] and Brebbia and Dominguez [15] to indicate the surface discretization character of the method, finally prevailed over the term BIEM (Boundary Integral Equation Method) first introduced by Cruse [40] in 1973 in conjunction with the direct version of the method, which nevertheless is still being used by some authors. It was first proven in 1977–78 by Brebbia et al. [14, 15, 18] that the BEM, like the FEM and many other numerical methods, can be obtained as a special case of the general weighted residuals statement, thereby establishing a connection between the various existing numerical techniques and providing an additional justification for prefering the term BEM instead of BIEM.

During the same period, Zienkiewicz et al. [112, 113] and Atluri and Grannell [1] proved the same thing on the basis of variational methods and described ways of coupling together the FEM and the BEM for the efficient solution of various complex problems.

There are basically two kinds of BEMs, the indirect and the direct ones. In the indirect approach the discretized integral equations are first solved for the density of the singular solutions over the boundary surface and then the remaining boundary quantities are computed in terms of these densities which have no physical significance. In the direct approach the discretized integral equations are formulated with the help of certain fundamental integral theorems and connect directly the unknown with the known boundary quantities. Even though it has been shown by Brebbia and Butterfield [16] in 1978 that the indirect and direct BEMs are formally equivalent, more emphasis is usually given to direct BEMs because they are more appealing to scientists and engineers. In addition, direct BEMs appear to be, at least so far, more easily amenable to improvements of the FEM type for the development of advanced BEMs than indirect ones. The term semi-direct BEM has also been used for some time in the past to indicate the treatment of some potential and elastostatic problems by a direct formulation with unknowns functions analogous to stream or stress functions from which the physical quantities of the problem can be obtained by differentiation after the solution of the integral equations [59, 102, 60, 89, 63]. This BEM category, however, is very limited and of a special character to be an independent one and for this reason it is usually considered as part of either the direct or, more frequently, the indirect BEM.

Among those who first developed indirect boundary element methodologies one can mention here Smith and Pierce [99], Hess [51], Hess and Smith [52], Jaswon [59], Symm [102], Jaswon and Ponter [60], Harrington et al. [48], McDonald and Wexler [71] and Silvester and Hsieh [97] for their work on potential theory and Massonnet [72], Rim and Henry [89], Jaswon et al. [61], Oliveira [83], Butterfield and Banerjee [30, 31], Benjumea and Sikarskie [13], Banerjee [5], Banerjee and Butterfield [6] and Crouch [33, 34] for their work in elastostatics.

The direct BEM was first introduced in an explicit and general way by Jaswon [59] in 1963 in connection with the potential theory and by Rizzo [90] in 1967 in connection with elastostatics, even though one can observe here that the works of Friedman and Shaw [46, 94] on acoustics and Banaugh and Goldsmith [3, 4] on acoustics and steady elastic waves are essentially based on the idea of the direct BEM. Other early notable works employing the direct BEM are those of Cruse and Rizzo [36, 37] on transient elastodynamics and Cruse [38] on three-dimensional elastostatics.

Applications of the BEM abound in many areas of engineering mechanics such as general potential theory [59, 102], potential fluid flow [99, 51, 52], acoustics [46, 94, 3, 95], torsion of shafts [60, 72, 73], electric and magnetic field theory

[48, 71, 97], elastostatics [72, 61, 89, 90, 83, 38, 30, 31, 13, 33, 40, 5, 34], elastodynamics [4, 36, 37], plates and shells [62, 81, 44], transient heat conduction [91], viscoelasticity [92], fracture [39, 43, 100, 105, 106], plasticity [101, 88, 73, 65, 7, 77], water waves [59, 96, 68], viscous fluid flow [110, 111], ground water flow [82, 69] and thermoelasticity [93]. This list of references is by no means complete and it just contains titles of some of the early papers in each area. The chapters which follow will provide almost complete reference lists for the aforementioned areas of applications as well as for other areas such as soil– and fluid–structure interaction.

The first book on the BEM and its various applications appeared in 1975 as a collection of review-type articles edited by Cruse and Rizzo [41]. Jaswon and Symm [63] in 1977 published the first and Brebbia [18] in 1978 the second textbook on the method as applied to potential theory and elastostatics, while Brebbia and Walker [19] in 1980 revised [18] and expanded it by adding a chapter on time-dependent and nonlinear problems.

The textbook of Banerjee and Butterfield [9], published in 1981, represents the first comprehensive work on the BEM and its applications in the various fields of engineering science. The more recent (1984) textbook of Brebbia, Telles and Wrobel [26] is another comprehensive and general work on the same subject. The boundary elements literature is recently becoming richer and richer in textbooks devoted to specialized subjects, such as those of Mukherjee [78] on creep and fracture, Parton and Perlin [84] and Crouch and Starfield [35] on solid mechanics, Liggett and Liu [70] on porous media flow, Telles [104] on inelastic problems, Venturini [109] on geomechanics and Hromadka [53] and Ingham and Kelmanson [57] on potential theory.

In an effort to provide information to scientists and engineers on the most recent developments in the field of boundary elements in the form of review-type articles written by experts, publication of two series of volumes has started and is being continued under the editorship of Banerjee et al. [8, 10–12] and Brebbia [22, 24, 28, 29]. Other sources of information on the BEM and its various applications, besides the numerous scientific and technical journal articles, are the proceedings of a number of international conferences that have been edited by Brebbia [18, 20, 21, 23, 25, 27].

2. Fundamentals of the method

Since the purpose of this section is to provide a brief introduction to the fundamental analytical and numerical aspects of the BEM in an easy to understand manner, the two-dimensional Laplace's equation is selected as the ideal vehicle for accomplishing this task. This equation is characterized by enough simplicity and yet it can provide the ideal setting for clearly illustrating all the essential aspects of the method. A detailed treatment of Laplace's equation in both two and three dimensions with applications will be given in the next chapter of this book in the framework of the discussion on potential theory.

2.1. Integral formulation

Consider a smooth, simply connected and two-dimensional domain V with a boundary surface S. It is required to approximately determine the function $u(\boldsymbol{x})$, which satisfies Laplace's equation

$$\nabla^2 u(\boldsymbol{x}) = 0\,, \qquad \boldsymbol{x} \in V\,, \tag{2.1}$$

in the domain V and the boundary conditions

$$u(\boldsymbol{x}) = \bar{u}(\boldsymbol{x})\,, \qquad \boldsymbol{x} \in S_1 \tag{2.2}$$

$$v(\boldsymbol{x}) = \frac{\partial u(\boldsymbol{x})}{\partial n} = \bar{v}(\boldsymbol{x})\,, \qquad \boldsymbol{x} \in S_2 \tag{2.3}$$

on the surface $S = S_1 \cup S_2$, where $\boldsymbol{n}$ is the unit outward normal vector to that surface.

The fundamental solution of (2.1) is the two point function $u^*(\boldsymbol{x}, \boldsymbol{\xi})$ that satisfies

$$\nabla^2 u^*(\boldsymbol{x}, \boldsymbol{\xi}) = -\delta(\boldsymbol{x}, \boldsymbol{\xi}) \tag{2.4}$$

where $\delta(\boldsymbol{x}, \boldsymbol{\xi})$ is the Dirac delta function characterized by the properties

$$\delta(\boldsymbol{x}, \boldsymbol{\xi}) = 0 \qquad \text{for } \boldsymbol{x} \neq \boldsymbol{\xi} \tag{2.5}$$

$$\int_V u(\boldsymbol{x})\delta(\boldsymbol{x}, \boldsymbol{\xi})\, \mathrm{d}V(\boldsymbol{x}) = u(\boldsymbol{\xi}) \tag{2.6}$$

Thus, the symmetric singular function $u^*(\boldsymbol{x}, \boldsymbol{\xi})$ represents the value of the function $u(\boldsymbol{x})$ at $\boldsymbol{x}$ generated by a unit point source at $\boldsymbol{\xi}$. It can be proven that the solution of (2.4) is [64]

$$u^*(\boldsymbol{x}, \boldsymbol{\xi}) = \frac{1}{2\pi} \ln\left(\frac{1}{r}\right) \tag{2.7}$$

where

$$\boldsymbol{r} = \boldsymbol{x} - \boldsymbol{\xi} \tag{2.8}$$

According to the indirect approach [9], the function $u(\boldsymbol{x})$ can be obtained as the superposition of unit singular solutions $u^*(\boldsymbol{x}, \boldsymbol{\xi})$ with amplitude (density or source surface density) $\sigma(\boldsymbol{\xi})$ distributed over the surface S, i.e.

$$u(\boldsymbol{x}) = \int_S u^*(\boldsymbol{x}, \boldsymbol{\xi})\sigma(\boldsymbol{\xi})\, \mathrm{d}S(\boldsymbol{\xi}) + C\,, \qquad \boldsymbol{x} \in V \tag{2.9}$$

where the constant C will be determined with the aid of the condition

$$\int_S \sigma(\boldsymbol{\xi})\, \mathrm{d}S(\boldsymbol{\xi}) = 0 \tag{2.10}$$

which enforces uniqueness of the solution [9]. Similarly, $v(\boldsymbol{x})$ can be expressed as

$$v(\boldsymbol{x}) = \int_S u_n^*(\boldsymbol{x}, \boldsymbol{\xi})\sigma(\boldsymbol{\xi})\, \mathrm{d}S(\boldsymbol{\xi})\,, \qquad \boldsymbol{x} \in V \tag{2.11}$$

where

$$u_n^*(\boldsymbol{x}, \boldsymbol{\xi}) = \frac{\partial u^*(\boldsymbol{x}, \boldsymbol{\xi})}{\partial n(\boldsymbol{x})}\,. \tag{2.12}$$

A limiting process that brings $\boldsymbol{x}$ on the boundary S reduces (2.9) and (2.11) to [9]

$$u(\boldsymbol{x}) = \int_S u^*(\boldsymbol{x}, \boldsymbol{\xi})\sigma(\boldsymbol{\xi})\, \mathrm{d}S(\boldsymbol{\xi}) + C\,, \qquad \boldsymbol{x} \in S \tag{2.13}$$

$$v(\boldsymbol{x}) = -\tfrac{1}{2}\sigma(\boldsymbol{x}) + \int_S u_n^*(\boldsymbol{x}, \boldsymbol{\xi})\sigma(\boldsymbol{\xi})\, \mathrm{d}S(\boldsymbol{\xi})\,, \qquad \boldsymbol{x} \in S \tag{2.14}$$

which represent the two boundary integral equations of the problem. These are supplemented by Eq. (2.10). Thus, solution of the system of equations (2.13), (2.14) and (2.10) subject to the boundary conditions (2.2) and (2.3) provides the values of $\sigma(\boldsymbol{\xi})$ over the boundary S and then $u(\boldsymbol{x})$ and $v(\boldsymbol{x})$ can be determined at any interior point $\boldsymbol{x}$ from (2.9) and (2.11), while $u(\boldsymbol{x})$ on S_2 and $v(\boldsymbol{x})$ as S_1 can be obtained from (2.13) and (2.14) with $\boldsymbol{x} \in S_2$ and $\boldsymbol{x} \in S_1$, respectively.

According to the direct approach [63], the starting point is Green's third theorem or identity

$$u(\boldsymbol{\xi}) = \int_S [v(\boldsymbol{x})u^*(\boldsymbol{x}, \boldsymbol{\xi}) - u(\boldsymbol{x})u_n^*(\boldsymbol{x}, \boldsymbol{\xi})]\, \mathrm{d}S(\boldsymbol{x})\,, \qquad \boldsymbol{\xi} \in V \tag{2.15}$$

which is obtained from Green's second theorem

$$\int_V (u\nabla^2 w - w\nabla^2 u)\, \mathrm{d}V = \int_S \left(u\, \frac{\partial w}{\partial n} - w\, \frac{\partial u}{\partial n}\right) \mathrm{d}S \tag{2.16}$$

for $w = u^*$ and with the aid of Eqs. (2.4) and (2.6). Differentiation of (2.15) in accordance with the definition of $v(\xi) = \partial u(\xi)/\partial n(\xi)$ with $\xi \in V$ produces

$$v(\xi) = \int_S \left[v(x) u_n^*(x, \xi) - u(x) \frac{\partial}{\partial n} u_n^*(x, \xi) \right] \mathrm{d}S(x) , \qquad \xi \in V . \tag{2.17}$$

A limiting process that brings ξ on the boundary S reduces Eq. (2.15) to the boundary integral equation

$$\tfrac{1}{2} u(\xi) = \int_S [v(x) u^*(x, \xi) - u(x) u_n^*(x, \xi)] \, \mathrm{d}S(x) , \qquad \xi \in S \tag{2.18}$$

The solution of (2.18) subject to the boundary conditions (2.2) and (2.3) consists of $u(\xi)$ on S_2 and $v(\xi)$ on S_1 and thus knowledge of the whole boundary data permits the determination of $u(\xi)$ and $v(\xi)$ in V from Eqs. (2.15) and (2.17), respectively.

2.2. Numerical solution

For the numerical solution of Eqs. (2.13), (2.14) and (2.10) the boundary surface S is discretized into a finite number N of linear segments – boundary elements of length S_j ($j = 1, 2, \ldots, N$) characterized by their midpoints and a constant variation of the density $\sigma(\xi)$ over them. Thus, Eqs. (2.13), (2.14) and (2.10) in conjunction with the boundary conditions (2.2) and (2.3) attain the discretized form

$$\bar{u}_i = \sum_{j=1}^{N} \sigma_j \int_{S_j} u^* \, \mathrm{d}S + C , \qquad i = 1, 2, \ldots N_1 \tag{2.19}$$

$$\bar{v}_k = -\frac{1}{2} \sigma_k + \sum_{j=1}^{N} \sigma_j \int_{S_j} u_n^* \, \mathrm{d}S , \qquad k = N_1 + 1, N_1 + 2, \ldots , N \tag{2.20}$$

$$\sum_{j=1}^{N} \sigma_j S_j = 0 \tag{2.21}$$

where the N_1 elements correspond to the surface S_1 and the remaining ones to the surface S_2. Writing the above equations for all the N points results in the matrix form

$$\begin{bmatrix} [G_{ij}] & [I] \\ [G_{kj}^n] & 0 \\ \{S_j\}^T & 0 \end{bmatrix} \begin{Bmatrix} \{\sigma_j\} \\ C \end{Bmatrix} = \begin{Bmatrix} \{\bar{u}_i\} \\ \{\bar{v}_k\} \\ 0 \end{Bmatrix} \tag{2.22}$$

where

$$G_{ij} = \int_{S_j} u^* \, \mathrm{d}S \,, \qquad \bar{G}^n_{kj} = \int_{S_j} u^*_n \, \mathrm{d}S \,, \qquad G^n_{kj} = -\tfrac{1}{2}\delta_{kj} + \bar{G}^n_{kj} \tag{2.23}$$

δ_{kj} is the Kronecker's delta and $[I]$ is the identity matrix. Solution of (2.22) provides $\{\sigma_j\}$ and C so that (2.9) and (2.11) in their discretized form

$$u_m = \sum_{j=1}^{N} G_{mj}\sigma_j + C \tag{2.24}$$

$$v_m = \sum_{j=1}^{N} \bar{G}^n_{mj}\sigma_j \tag{2.25}$$

serve to obtain u and v at any point $\boldsymbol{x}_m$ in V. Values of u and v on S_2 and S_1, respectively, can finally be obtained from (2.19) and (2.20) with $i = N_1 + 1, N_1 + 2, \ldots, N$ and $k = 1, 2, \ldots, N_1$, respectively.

For the numerical solution of (2.18) a discretization of the boundary surface S into a finite number N of linear boundary elements of length S_j ($j = 1, 2, \ldots, N$) characterized by their midpoints and a constant variation of the functions $u(\boldsymbol{x})$ and $v(\boldsymbol{x})$ is made. This process enables one to write (2.18) in the discretized form

$$\tfrac{1}{2}u_i = \sum_{j=1}^{N} v_j \int_{S_j} u^* \, \mathrm{d}S - \sum_{j=1}^{N} u_j \int_{S_j} u^*_n \, \mathrm{d}S \,. \tag{2.26}$$

Employment of Eq. (2.26) for all the points N of the boundary and use of the boundary conditions (2.2) and (2.3) results in the matrix equation

$$\begin{bmatrix} [G^n_{ij}]_1 & [\bar{G}^n_{ij}]_2 \\ [\bar{G}^n_{kj}]_1 & [G^n_{kj}]_2 \end{bmatrix} \begin{Bmatrix} \{\bar{u}_j\}_1 \\ \{u_j\}_2 \end{Bmatrix} = \begin{bmatrix} [G_{ij}]_1 & [G_{ij}]_2 \\ [G_{kj}]_1 & [G_{kj}]_2 \end{bmatrix} \begin{Bmatrix} \{v_j\}_1 \\ \{\bar{v}_j\}_2 \end{Bmatrix} \tag{2.27}$$

where $i = 1, 2, \ldots, N_1$; $k = N_1 + 1, N_1 + 2, \ldots, N$, the subscripts 1 and 2 indicate that $j = i$ and $j = k$, respectively, and G_{ij}, $\bar{G}^n_{ij}$ and G^n_{ij} are given by expressions analogous to those in (2.23). Reordering of Eq. (2.27) produces the matrix equation

$$[A]\{z\} = [B]\{\bar{z}\} = \{b\} \tag{2.28}$$

which can be solved for the unknown vector $\{z\}$ consisting of values of u and v on S_2 and S_1, respectively. Thus, it is apparent that, even though the boundary matrix equation (2.28) for the direct approach is more involved than the corresponding one (2.22) for the indirect approach, the former provides all the necessary boundary data while the latter furnishes only the values of the density function which can be subsequently used to determine the unknown boundary data. Values of u and v at any point $\boldsymbol{\xi}_m$ in V can finally be obtained from the discretized versions of Eqs. (2.15) and (2.17) which read

$$u_m = \sum_{j=1}^{N} G_{mj} v_j - \sum_{j=1}^{N} \bar{G}^{n}_{mj} u_j \tag{2.29}$$

$$v_m = \sum_{j=1}^{N} \bar{G}^{n}_{mj} v_j - \sum_{j=1}^{N} \bar{G}^{nn}_{mj} u_j \tag{2.30}$$

where

$$\bar{G}^{nn}_{mj} = \int_{S_j} \frac{\partial u_n^*}{\partial n} \, \mathrm{d}S \, . \tag{2.31}$$

More details about the numerical implementation of the direct BEM can be found in the next chapter.

Closing this subsection, the following important remarks can be made:

(a) The presence of a known function $\psi = \psi(\boldsymbol{x})$ in the right-hand side of (2.1) results in the appearance of the integrals

$$I_1 = \int_V u^*(\boldsymbol{x}, z)\psi(z) \, \mathrm{d}V(z) \, , \qquad I_2 = \int_V u_n^*(\boldsymbol{x}, z)\psi(z) \, \mathrm{d}V(z) \, , \quad z \in V \tag{2.32}$$

in the right-hand side of Eqs. (2.13), (2.14) and (2.18), (2.17), respectively. The numerical evaluation of these integrals is accomplished by dividing the interior of the domain V into a number M of cells V_k ($k = 1, 2, \ldots, M$) over which $\psi(\boldsymbol{x})$ is assumed to be constant and writing them in the form

$$\begin{aligned} I_1 &= \sum_{k=1}^{M} \psi(z^k) \int_{V_k} u^*(\boldsymbol{x}^i, z^k) \, \mathrm{d}V(z^k) \, , \\ I_2 &= \sum_{k=1}^{M} \psi(z^k) \int_{V_k} u_n^*(\boldsymbol{x}^i, z^k) \, \mathrm{d}V(z^k) \, . \end{aligned} \qquad z \in V \tag{2.33}$$

It is very important to observe that the order of the influence matrices involving elements of the form G_{ij} and G^n_{ij} in Eqs. (2.22) and (2.28) is just $N \times N$, where N is the number of boundary elements having nothing to do with the number M of the internal cells of the domain.

(b) The evaluation of the boundary integrals of the form

$$G_{ij} = \int_{S_j} u^*(\boldsymbol{x}^j, \boldsymbol{\xi}^i) \, \mathrm{d}S(\boldsymbol{x}^j) \, , \qquad \bar{G}_{ij} = \int_{S_j} u_n^*(\boldsymbol{x}^j, \boldsymbol{\xi}^i) \, \mathrm{d}S(\boldsymbol{x}^j) \tag{2.34}$$

over line segments can be easily accomplished analytically due to their simple integrand expressions. The integrals G_{ij} and $\bar{G}_{ij}$ become singular for $\boldsymbol{x}^j = \boldsymbol{\xi}^i$, i.e. for $\boldsymbol{r} = 0$, due to the presence of the $\ln r$ and $1/r$ terms in their integrands, respectively. These singular integrals can be treated also analytically.

(c) Instead of assuming constant variation of the unknown or known functions over boundary elements and internal cells one can assume higher-order parametric representation of functions and the element geometry, thereby considerably increasing the accuracy of the method. In this case the various boundary and domain integrals have to be evaluated numerically by Gaussian quadrature formulae, except at singular points where an analytic treatment is usually done.

More on these three items will be presented in the next chapter.

2.3. *Connection with other methods*

Even though at a first glance the BEM seems to be partially or completely unrelated to other known numerical methods and based on not so general principles for its formulation, it can be obtained, like many other numerical methods, as a special case of the general weighted residuals statement [18, 26, 2].

With reference the boundary value problem described by Eqs. (2.1)–(2.3), the general weighted residuals statement requires the satisfaction of

$$\int_V (\nabla^2 u) w \, \mathrm{d}V + \int_{S_1} (u - \bar{u}) w_n \, \mathrm{d}S - \int_{S_2} (v - \bar{v}) w \, \mathrm{d}S = 0 \tag{2.35}$$

where u is the approximating or trial function, w is a weighting or test function with continuous first derivatives and $w_n = \partial w / \partial n$. Integration of (2.35) by parts once yields

$$\int_V \frac{\partial u}{\partial x_i} \frac{\partial w}{\partial x_i} \, \mathrm{d}V - \int_{S_1} v w \, \mathrm{d}S - \int_{S_2} \bar{v} w \, \mathrm{d}S - \int_{S_1} (u - \bar{u}) w_n \, \mathrm{d}S = 0 \tag{2.36}$$

with summation convention implied over repeated indices, while one more integration results in

$$\int_V (\nabla^2 w) u \, \mathrm{d}V + \int_S v w \, \mathrm{d}S - \int_S u w_n \, \mathrm{d}S = 0 \tag{2.37}$$

It can then be easily proven [18, 26, 2] that:

(a) if the functions u and w are the same, then:

- (i) Eq. (2.36) can lead to the Galerkin finite element formulation [103, 55].
- (ii) Eq. (2.37) in conjunction with the requirement that u satisfies the governing equation (2.1) can lead to Trefftz type of boundary methods [108, 49, 50].

(b) if the functions u and w are different, then:

(i) Eq. (2.35) in conjunction with the requirements that the boundary conditions (2.2) and (2.3) are satisfied, w is replaced by a Dirac delta function and the derivatives in the ∇^2 operator are replaced by finite difference expressions, can lead to the classical finite difference method. An extension of this idea forms the basis of the cell collocation method [67].

(ii) Eq. (2.35) in conjunction with the requirement that u satisfies the governing equation (2.1) can lead to a variety of boundary weighted residuals methods, such as the edge function method [86, 87, 1, 2] and the various continuous (integral) and discrete (point-matching) least-squares methods [32, 54, 42].

(iii) Eq. (2.37) in conjunction with the requirement that w is the fundamental solution of the problem, i.e. $w = u^*$ and satisfies Eq. (2.4) leads to the direct boundary element formulation as it was demonstrated in subsection 2.1.

3. Advantages and disadvantages of the method

A brief description of the history and areas of application as well as the fundamental analytical and numerical aspects of the BEM were given in the previous two sections. This section deals with a systematic enumeration of the advantages and disadvantages of the BEM as compared with the FEM, which is presently considered the most popular and widely used numerical method in engineering mechanics.

The BEM presents some distinct advantages, such as

(1) It is a general numerical method with a wide range of applications in engineering mechanics. It enjoys a firm mathematical basis and constitutes a practical and efficient computational tool. It is applicable to both linear and nonlinear problems, provided that the latter ones are or can be approximated as incrementally linear.

(2) It usually requires only a surface discretization and not a discretization of both the interior and the surface of the domain of interest as in 'domain' type techniques like the FEM. This reduction of the spatial dimensions of the problem by one greatly facilitates the data preparation job, permits an easy conduction of mesh refinement studies and leads to a system of algebraic equations much smaller than the one encountered in the FEM.

(3) The facts that the approximations involved in the method are confined to the surface of the domain and that the influence matrices associated with the BEM consist of dominant elements on or near their main diagonal due to the singular nature of the employed Green's functions are the reasons for obtaining results of higher accuracy than by the FEM, especially in problems involving

sharp gradients of the unknown functions. This accuracy of the method can be considerably increased by employing sophisticated elements and highly accurate numerical integration techniques.

(4) The employment of the Green's function or fundamental singular solution associated with the problem of interest makes possible the easy treatment of infinite or semi-infinite domains without artificial "box-type" discretizations as it is the case with the FEM.

(5) The above advantages of the BEM over the FEM are greatly pronounced in linear problems characterized by three-dimensional (especially infinite or semi-infinite), homogeneous and isotropic domains.

(6) The method is capable of providing values of the unknowns in the interior of the domain in a pointwise fashion so that no inter-element continuity problems can arise as in the FEM. In addition the computation is restricted to points of interest and does not involve all the mesh points as in the FEM.

(7) The BEM can be used for the construction of stiffness matrices for homogeneous very big finite elements (super-elements) or infinite elements which can be used in a finite element formulation characterized by a drastic reduction in the number of the unknowns of the system.

As any other approximate method, the BEM is also characterized by a number of disadvantages, such as

(1) It is associated with non-symmetric and fully populated influence matrices in contrast to the FEM, which involves symmetric and sparse matrices that require less computational effort for their inversion. The order of the former matrices is, however, much lower than that of the latter and their numerical treatment can be done efficiently through special schemes. Besides, the former matrices are numerically well conditioned.

(2) It is very difficult or practically impossible to obtain the fundamental solutions for some types of problems, especially for those involving strong anisotropies and inhomogeneities, while sometimes the expressions for these solutions are too complicated or known in discrete form leading to inefficient schemes. Of course, approximate fundamental solutions in conjunction with an iterative approach can be employed at an increased computational cost. The FEM is a better choice for problems characterized by rapidly changing physical properties in the domain.

(3) The BEM is inefficient as compared with the FEM for problems with a geometry for their mathematical model characterized by one or two spatial dimensions disproportionally small with respect to the others but dimensionally effective, such as those involving the analysis of moderately thick plates and shells or thin narrow strips. The BEM is also inefficient for one-dimensional problems, such as those involving the analysis of beams and frameworks.

(4) The presence of known distributed body forces in linear problems or pseudo-incremental body forces in nonlinear problems creates volume integrals which require an internal discretization of the domain. However, this internal

discretization is much simpler than that of the FEM, can be restricted to a small portion of the domain in some cases and most importantly, does not lead in any increase in the order of the final system of algebraic equations.

(5) There are very few special and general-purpose BEM computer programs presently available as compared with the great number of corresponding FEM programs. This is because the BEM is still in the stage of development, competes against a very efficient and well established technique, the FEM and the architecture of a BEM computer program is not conceptually as easy as the one for the FEM.

Although at a first glance the BEM appears to have few similarities with the FEM, the two methods share a common basis as they can be derived from the general weighted residuals statement which provides a connection between the BEM and quite a number of other known numerical methods, as it was shown in the previous section. Thus, from the viewpoint of their common origin and the fact that many advantages of one method correspond to disadvantages for the other method, one can consider the BEM and the FEM not as antagonistic techniques but rather as techniques that can complement each other. The BEM indeed has been successfully combined with the FEM (as well as with some other numerical methods) for the solution of a variety of problems in mechanics thereby creating a hybrid type of method that combines the advantages of the two methods and reduces or eliminates their disadvantages.

4. Future developments

As it was mentioned in the first section of this chapter, the BEM is presently approaching a level of maturity. However, many further developments of a special or general character are still needed. This section presents an attempt to list these future needs in the field of the BEM and its applications in engineering mechanics with the hope that they will provide the directions for future research and development and further demonstrate the great potential of the method. No claim is made about the completeness of this list of future developments which primarily reflects the personal bias of the author and by no means exhausts the subject. The list of these developments consists of four categories as follows:

4.1. Software developments

In this category the following developments are mentioned:

(1) Use of computer symbolic manipulation for the automatic generation and synthesis of influence matrices in the most efficient way.

(2) Use of adaptive mesh refinement, i.e. refinement of the h or p type or their combination in a way that each step depends on the error information provided by the previous ones.

(3) Development of comprehensive general-purpose computer programs and more special-purpose ones to cover all the existing areas of application of the BEM both equipped with increased user-oriented features and graphic display capabilities.

(4) Development of computer programs that can take advantage of the special features of the new super-computers (e.g. Cray-1, CDC Cyber 205), such as vectorization and parallel processing, for the more efficient solution of large order problems.

4.2. Methodology developments

This category enumerates the following developments in the area of constructing accurate and efficient methodologies to cope with the various old and new applications of the BEM:

(1) Efficient treatment of linear transient analysis techniques by especially following time-domain formulations thereby creating the basis for the solution of nonlinear problems.

(2) Efficient treatment of the problems of combining the BEM with other numerical methods, especially the FEM and the construction of finite super-elements with the aid of the BEM. There is also a need to resolve the efficiency controversy between the BEM and the FEM by a series of detailed comparison studies for various classes of problems in order to be able to combine them appropriately on the basis of their particular advantages and disadvantages.

(3) Construction of exact and approximate Green's functions for certain classes of problems and evaluation of the effectiveness of the latter functions or development of non-conventional BEM's without the use of Green's functions with the aid of the weighted residuals formulation.

(4) Development of efficient techniques for the treatment of geometric and material nonlinearities in problems dealing with, e.g., plasticity or viscoplasticity under small or large deformations, postbuckling behavior and inelastic fracture.

(5) Development of dynamic nonlinear analysis techniques for problems, such as dynamic plasticity, dynamic inelastic stability, inelastic creep, fatigue and damage.

(6) Development of efficient methodologies for the treatment of large-scale nonlinear static or dynamic problems by domain decomposition techniques.

4.3. Physics developments

This category includes various physical problems which have not been solved (completely or efficiently) before by other techniques and can be efficiently solved by the BEM itself or in conjunction with other techniques.

(1) Three-dimensional geotechnical problems, linear or nonlinear, under static or dynamic loading, such as those encountered in engineering seismology, soil

consolidation, excavation technology, geothermal energy, hydraulic fracture, and slope stability.

(2) Three-dimensional dynamic soil–structure interaction problems of a linear or a nonlinear character such as those dealing with the dynamics of foundations, the behavior of underground structures and the isolation of structures from surface waves.

(3) Three-dimensional dynamic fluid–structure interaction problems dealing with, e.g., liquid storage tanks or vessels, submerged or floating structures, aerospace structures and buildings under wind forces.

(4) Complicated problems characterized by a combination of many of the previously mentioned physical aspects, such as those encountered in off-shore platforms or modern space stations.

(5) Problems involving structural optimization, optimal structural control and identification studies including the development of nondestructive testing techniques.

(6) Various other special problems such as linear or nonlinear contact problems under static or dynamic conditions.

4.4. Educational Developments

This fourth category describes the educational developments that are needed for the method to become more widely known among scientists and engineers, i.e.

(1) Development and introduction of new graduate level courses in engineering schools on the BEM and its applications.

(2) Offering of various short courses on the method and its applications.

(3) Production of some application-oriented books on the BEM for the sake of practising engineers.

(4) Continuation of special conferences and seminars devoted to the BEM exclusively or in conjunction with other numerical methods.

Acknowledgments

The author would like to express his gratitude to the National Science Foundation for supporting his work on Boundary Element Methods through the grants NSF/CEE-8024725 and NSF/CEE-8109723 to the University of Minnesota. During the final stages of the preparation of this chapter the author was staying as a Visiting Professor at Colorado State University. Professor C.N. Papadakis, Head of the Civil Engineering Department and the faculty of the Geotechnical Engineering Program of that University are gratefully acknowledged for their hospitality and technical assistance. Thanks are also due to Dr. T.A. Cruse and Professor S.N. Atluri for their technical assistance.

References

[1] S.N. Atluri and J.J. Grannell, Boundary Element Methods (BEM) and Combination of BEM-FEM, Report No. GIT-ESM-SA-78-16, Center for the Advancement of Computational Mechanics, Georgia Institute of Technology, Atlanta, Georgia (1978).

[2] S.N. Atluri, Computational Solid Mechanics (Finite Elements and Boundary Elements): Present Status and Future Directions, in: *Proceedings of the 4th International Conference on Applied Numerical Modelling*, H.M. Hsia et al., Eds (Tainan, Taiwan, 1984) pp. 19–37.

[3] R.P. Banaugh and W. Goldsmith, Diffraction of Steady Acoustic Waves by Surfaces of Arbitrary Shape, *Journal of the Acoustical Society of America* **35**, 1590-1601 (1963).

[4] R.P. Banaugh and W. Goldsmith, Diffraction of Steady Elastic Waves by Surfaces of Arbitrary Shape, *Journal of Applied Mechanics* **30**, 589–597 (1963).

[5] P.K. Banerjee, Integral Equation Methods for Analysis of Piecewise Nonhomogeneous Three-Dimensional Elastic Solids of Arbitrary Shape, *International Journal of Mechanical Sciences* **18**, 293–303 (1976).

[6] P.K. Banerjee and R. Butterfield, Boundary Element Methods in Geomechanics, in: *Finite Elements in Geomechanics*, G. Gudehus, Ed. (John Wiley and Sons, London, 1977) pp. 529–570.

[7] P.K. Banerjee and G.C.W. Mustoe, The Boundary Element Method for Two-Dimensional Problems of Elastoplasticity, in: *Recent Advances in Boundary Element Methods*, C.A. Brebbia, Ed. (Pentech Press, London, 1978) pp. 283–300.

[8] P.K. Banerjee and R. Butterfield, Eds., *Developments in Boundary Element Methods*–1 (Applied Science, London, 1979).

[9] P.K. Banerjee and R. Butterfield, *Boundary Element Methods in Engineering Science* (McGraw-Hill, London, 1981).

[10] P.K. Banerjee and R.P. Shaw, Eds., *Developments in Boundary Element Methods*–2 (Applied Science, London, 1982).

[11] P.K. Banerjee and S. Mukherjee, Eds., *Developments in Boundary Element Methods*–3 (Elsevier Applied Science, London, 1984).

[12] P.K. Banerjee and J.O. Watson, Eds., *Developments in Boundary Element Methods*–4 (Elsevier Applied Science, London, 1986).

[13] R. Benjumea and D.L. Sikarskie, On the Solutions of Plane, Orthotropic Elasticity Problems by an Integral Equation Method, *Journal of Applied Mechanics* **39**, 801–808 (1972).

[14] C.A. Brebbia, Approximate Methods in Mathematical Modelling, in: *Proceedings of the 1st International Conference on Mathematical Modelling*, X.J.R. Avula, Ed. (Department of Engineering Mechanics, University of Missouri-Rolla, 1977) pp. 43–57.

[15] C.A. Brebbia and J. Dominguez, Boundary Element Methods for Potential Problems, *Applied Mathematical Modelling* **1**, 372–378 (1977).

[16] C.A. Brebbia and R. Butterfield, Formal Equivalence of Direct and Indirect Boundary Element Methods, *Applied Mathematical Modelling* **2** 132–134 (1978).

[17] C.A. Brebbia, Ed., *Recent Advances in Boundary Element Methods* (Pentech Press, London, 1978).

[18] C.A. Brebbia, *The Boundary Element Method for Engineers* (Pentech Press, London, 1978).

[19] C.A. Brebbia and S. Walker, *Boundary Element Techniques in Engineering* (Newnes–Butterworths, London, 1980).

[20] C.A. Brebbia, Ed., *New Developments in Boundary Element Methods* (CML Publication, Southampton, 1980).

[21] C.A. Brebbia, Ed., *Boundary Element Methods* (Springer-Verlag, Berlin, 1981).

[22] C.A. Brebbia, Ed., *Progress in Boundary Element Methods*, Vol. 1 (Pentech Press, London, 1981).

[23] C.A. Brebbia, Ed., *Boundary Element Methods in Engineering* (Springer-Verlag, Berlin, 1982).

[24] C.A. Brebbia, Ed., *Progress in Boundary Element Methods*, Vol. 2 (Pentech Press, London, 1983).

[25] C.A. Brebbia, T. Futagami and M. Tanaka, Eds., *Boundary Elements* (Springer-Verlag, Berlin, 1983).

[26] C.A. Brebbia, J.C.F. Telles and L.C. Wrobel, *Boundary Element Techniques* (Springer-Verlag, Berlin, 1984).
[27] C.A. Brebbia, Ed., *Boundary Elements* VI (Springer-Verlag, Berlin, 1984).
[28] C.A. Brebbia, Ed., *Topics in Boundary Element Research*, Vol. 1 (Springer-Verlag, Berlin, 1984).
[29] C.A. Brebbia, Ed., *Topics in Boundary Element Research*, Vol. 2 (Springer-Verlag, Berlin, 1985).
[30] R. Butterfield and P.K. Banerjee, The Problem of Pile Reinforced Half Space, *Geotechnique* **20**, 100–103 (1970).
[31] R. Butterfield and P.K. Banerjee, The Elastic Analysis of Compressible Piles and Pile Groups, *Géotechnique* **21**, 43–60 (1971).
[32] H.D. Conway and A.W. Leissa, Application of the Point-Matching Method to Shallow-Spherical Shell Theory, *Journal of Applied Mechanics* **29**, 745–747 (1962).
[33] S.L. Crouch, Two-Dimensional Analysis of Near-Surface Single Seam Extraction, *International Journal of Rock Mechanics and Mining Sciences & Geomechanics Abstracts* **10**, 85–96 (1973).
[34] S.L. Crouch, Solution of Plane Elasticity Problems by the Displacement Discontinuity Method, I and II, *Internation Journal for Numerical Methods in Engineering* **10**, 301–343 (1976).
[35] S.L. Crouch and A.M. Starfield, *Boundary Element Methods in Solid Mechanics* (George Allen and Unwin, London, 1983).
[36] T.A. Cruse and F.J. Rizzo, A Direct Formulation and Numerical Solution of the General Transient Elastodynamic Problem, I, *Journal of Mathematical Analysis and Applications* **22**, 244–259 (1968).
[37] T.A. Cruse, A Direct Formulation and Numerical solution of the General Transient Elastodynamic Problem, II, *Journal of Mathematical Analysis and Applications* **22**, 341–355 (1968).
[38] T.A. Cruse, Numerical Solutions in Three-Dimensional Elastostatics, *International Journal of Solids and Structures* **5**, 1259–1274 (1969).
[39] T.A. Cruse and W. Van Buren, Three-Dimensional Elastic Stress Analysis of a Fractured Specimen with an Edge Crack, *International Journal of Fracture Mechanics* **7**, 1–15 (1971).
[40] T.A. Cruse, Application of the Boundary Integral Equation Method to Three-Dimensional Stress Analysis, *Computers and Structures* **3**, 509–527 (1973).
[41] T.A. Cruse and F.J. Rizzo, Eds., Boundary Integral Equation Method: Computational Applications in Applied Mechanics (AMD-11, American Society of Mechanical Engineers, New York, 1975).
[42] E.D. Eason, A Review of Least-Squares Methods for Solving Partial Differential Equations, *International Journal for Numerical Methods in Engineering* **10**, 1021–1046 (1976).
[43] F. Erdogan, G.D. Gupta and T.S. Cook, Numerical Solution of Integral Equations, in: *Mechanics of Fracture*, Vol. 1, *Methods of Analysis and Solutions of Crack Problems*, G.C. Sih, Ed. (Noordhoff Publishing Co., Leyden, 1973) pp. 368-425.
[44] D.J. Forbes and A.R. Robinson, Numerical Analysis of Elastic Plates and Shallow Shells by an Integral Equation Method, *Structural Research Series, Report No. 345*, University of Illinois, Urbana, Illinois (1969).
[45] I. Fredholm, Sur une classe d'Equations Fonctionelles, *Acta Mathematica, Sweden* **27**, 365–390 (1903).
[46] M.B. Friedman and R. Shaw, Diffraction of Pulses by Cylindrical Obstacles of Arbitrary Cross Section, *Journal of Applied Mechanics* **29**, 40–46 (1962).
[47] F.D. Gakhov, *Boundary Value Problems* (Pergamon Press, Oxford, 1966).
[48] R.F. Harrington, K. Kontoppidan, P. Abrahamsen and N.C. Albertsen, Computation of Laplacian Potentials by an Equivalent-Source Method, *Proceedings of the Institution of Electrical Engineers* **116**, 1715–1720, (1969).
[49] I. Herrera, Trefftz Method, in: *Topics in Boundary Element Research*, Vol. 1, C.A. Brebbia, Ed. (Springer-Verlag, Berlin, 1984), pp. 225–253.
[50] I. Herrera, *Boundary Methods: An Algebraic Theory* (Pitman Publishing Co., London, 1984).
[51] J.L. Hess, Calculation of Potential Flow About Bodies of Revolution Having Axes Perpendicular to the Free Stream Direction, *Journal of the Aerospace Sciences* **29**, 726–742 (1962).
[52] J.L. Hess and A.M.O. Smith, Calculation of Nonlifting Potential Flow About Arbitrary Three-Dimensional Bodies, *Journal of Ship Research* **8**, 22–44 (1964).

[53] T.V. Hromadka II, *The Complex Variable Boundary Element Method*, (Springer-Verlag, Berlin, 1984).
[54] L.E. Hulbert and F.W. Niedenfuhr, Accurate Calculation of Stress Distribution in Multiholed Plates, *Journal of Engineering for Industry*, *Transactions of the ASME* **87**, 331–336 (1965).
[55] S.G. Hutton and D.L. Anderson, Finite Element Method: A Galerkin Approach, *Journal of the Engineering Mechanics Division*, *Proceedings of the ASCE* **97** (EM 5), 1503–1520 (1971).
[56] L.S. Hwang and E.O. Tuck, On the Oscillations of Harbors of Arbitrary Shape, *Journal of Fluid Mechanics* **42**, 447–464 (1970).
[57] D.B. Ingham and M.A. Kelmanson, *Boundary Integral Equation Analysis of Singular Potential and Biharmonic Problems* (Springer-Verlag, Berlin, 1984).
[58] V.V. Ivanov, *The Theory of Approximate Methods and Their Application to the Numerical Solution of Singular Integral Equations* (Noordhoff Publishing Co., Leyden, 1976).
[59] M.A. Jaswon, Integral Equation Methods in Potential Theory: I, *Proceedings of the Royal Society of London* **275** (A), 23–32 (1963).
[60] M.A. Jaswon and A.R.S. Ponter, An Integral Equation Solution of the Torsion Problem, *Proceedings of the Royal Society of London* **273** (A), 237–246 (1963).
[61] M.A. Jaswon, M. Maiti and G.T. Symm, Numerical Biharmonic Analysis and Some Applications, *International Journal of Solids and Structures* **3**, 309–332 (1967).
[62] M.A. Jaswon and M. Maiti, An Integral Equation Formulation of Plate Bending Problems, *Journal of Engineering Mathematics* **2**, 83–93 (1968).
[63] M.A. Jaswon and G.T. Symm, *Integral Equation Methods in Potential Theory and Elastostatics* (Academic Press, London, 1977).
[64] O.D. Kellogg, *Foundations of Potential Theory* (J. Springer, Berlin, 1929; also published by Dover, New York, 1953).
[65] V. Kumar and S. Mukherjee, A Boundary Integral Equation Formulation for Time-Dependent Inelastic Deformation in Metals, *International Journal of Mechanical Sciences* **19**, 713–724 (1977).
[66] V.D. Kupradze, *Potential Methods in the Theory of Elasticity* (Israel Program for Scientific Translations, Jerusalem, 1965).
[67] P.C.M. Lau and C.A. Brebbia, The Cell Collocation Method in Continuum Mechanics, *International Journal of Mechanical Sciences* **20**, 83–95 (1978).
[68] J.J. Lee, Wave Induced Oscillation in Harbors of Arbitrary Geometry, *Journal of Fluid Mechanics* **45**, 375–394 (1971).
[69] J.A. Liggett, Locations of Free Surface in Porous Media, *Journal of the Hydraulics Division*, *Proceedings of the ASCE* **103**, (HY 4), 353–365 (1977)
[70] J.A. Liggett and P.L.-F. Liu, *The Boundary Integral Equation Method for Porous Media Flow* (George Allen and Unwin, London, 1983).
[71] B.H. McDonald and A. Wexler, Finite-Element Solution of Unbounded Field Problems, *IEEE Transactions on Microwave Theory and Techniques* **MTT-20**, 841–847 (1972).
[72] C.E. Massonnet, Numerical Use of Integral Procedures, in: *Stress Analysis*, O.C. Zienkiewicz and G.S. Holister, Eds. (John Wiley and Sons, London, 1965) pp. 198–235.
[73] A. Mendelson, Boundary-Integral Methods in Elasticity and Plasticity, *NASA Technical Note* TN D-7418, Washington, D.C. (1973).
[74] S.G. Mikhlin, *Integral Equations* (Pergamon Press, Oxford, 1957).
[75] S.G. Mikhlin, *Approximate Solutions of Differential and Integral Equations* (Pergamon Press, Oxford, 1965).
[76] S.G. Mikhlin, *Multidimensional Singular Integrals and Integral Equations* (Pergamon Press, Oxford, 1965).
[77] S. Mukherjee and V. Kumar, Numerical Analysis of Time-Dependent Inelastic Deformation in Metallic Media Using the Boundary Integral Equation Method, *Journal of Applied Mechanics* **45**, 785–790 (1978).
[78] S. Mukherjee, *Boundary Element Methods in Creep and Fracture* (Applied Science, London, 1982).
[79] N.I. Muskhelishvili, *Singular Integral Equations* (Noordhoff Publishing Co., Groningen, 1953).
[80] N.I. Muskhelishvili, *Some Basic Problems of the Mathematical Theory of Elasticity* (Nordhoff Publishing Co., Groningen, 1953).

[81] D.A. Newton and H. Tottenham, Boundary Value Problems in Thin Shallow Shells of Arbitrary Plan Form, *Journal of Engineering Mathematics* **2**, 211–224 (1968).
[82] Y. Niwa, S. Kobayashi and T. Fukui, An Application of the Integral Equation Method to Seepage Problems, *Theoretical and Applied Mechanics* **24**, 479–486 (1976).
[83] E.R.A. Oliveira, Plane Stress Analysis by a General Integral Method, *Journal of the Engineering Mechanics Division, Proceedings of the ASCE* **94**, (EM 1), 79–101 (1968).
[84] V.Z. Parton and P.I. Perlin, *Integral Equations in Elasticity* (Mir Publishers, Moscow, 1982).
[85] W. Prager, Die Druckverteilung an Körpern in Ebener Potentialströmung, *Physikalische Zeitschrift* **29**, 865–869 (1928).
[86] P.M. Quinlan, The Torsion of an Irregular Polygon, *Proceedings of the Royal Society of London* **282** (A), 208–227 (1964).
[87] P.M. Quinlan, J.E. Fitzgerald and S.N. Atluri, The Edge Function Method, in: *Proceedings of the International Symposium on Innovative Numerical Analysis in Applied Engineering Science*, T.A. Cruse et al., Eds., (CETIM, Cenlins, France, 1977) pp. 15–18.
[88] P. Riccardella, An Implementation of the Boundary Integral Technique for Plane Problems of Elasticity and Elastoplasticity, Ph.D. Thesis, Carnegie Mellon University, Pittsburgh, Pennsylvania (1973).
[89] K. Rim and A.S. Henry, An Integral Equation Method in Plane Elasticity, NASA Contractor Report CR-779 (1967).
[90] F.J. Rizzo, An Integral Equation Approach to Boundary Value Problems of Classical Elastostatics, *Quarterly of Applied Mathematics* **25**, 83–95 (1967).
[91] F.J. Rizzo and D.J. Shippy, A Method of Solution for Certain Problems of Transient Heat Conduction, *American Institute of Aeronautics and Astronautics Journal* **8**, 2004–2009 (1970).
[92] F.J. Rizzo and D.J. Shippy, An Application of the Correspondence Principle of Linear Viscoelasticity Theory, *SIAM Journal of Applied Mathematics* **21**, 321–330 (1971).
[93] F.J. Rizzo and D.J. Shippy, An Advanced Boundary Integral Equation Method for Three-Dimensional Thermoelasticity, *International Journal for Numerical Methods in Engineering* **11**, 1753–1768 (1977).
[94] R.P. Shaw and M.B. Friedman, Diffraction of Pulses by Deformable Cylindrical Obstacles of Arbitrary Cross Section, in: *Proceedings of the 4th US National Congress of Applied Mechanics*, R.M. Rosenberg, Ed. (American Society of Mechanical Engineers, New York, 1962) pp. 371–379.
[95] R.P. Shaw, Diffraction of Acoustic Pulses by Obstacles of Arbitrary Shape with a Robin Boundary Condition–Part A, *Journal of the Acoustical Society of America* **41**, 855–859 (1962).
[96] R.P. Shaw, Methods of Solution for Water Wave Scattering Problems, in: *Topics in Ocean Engineering* II, C. Bretschneider, Ed. (Gulf Publishing Co., Houston, 1970) pp. 180–198.
[97] P. Silvester and M.S. Hsieh, Finite Element Solution of 2-Dimensional Exterior Field Problems, *Proceedings of the Institution of Electrical Engineers* **118**, 1743–1747 (1971).
[98] V.I. Smirnov, *Integral Equations and Partial Differential Equations*, in: *A Course of Higher Mathematics*, Vol. 4 (Pergamon Press, Oxford, 1964).
[99] A.M.O. Smith and J. Pierce, Exact Solution of the Neumann Problem, Calculation of Non-circulatory Plane and Axially Symmetric Flows About or Within Arbitrary Boundaries, in: *Proceedings of the 3rd US National Congress of Applied Mechanics*, R.M. Haythornthwaite, Ed. (The American Society of Mechanical Engineers, New York, 1958) pp. 807–815.
[100] M.D. Snyder and T.A. Cruse, Boundary Integral Equation Analysis of Cracked Anisotropic Plates, *International Journal of Fracture* **11**, 315–328 (1975).
[101] J.L. Swedlow and T.A. Cruse, Formulation of Boundary Integral Equations for Three-Dimensional Elastoplastic Flow, *International Journal of Solids and Structures* **7**, 1673–1681 (1971).
[102] G.T. Symm, Integral Equation Methods in Potential Theory: II, *Proceedings of the Royal Society of London* **275** (A), 33–46 (1963).
[103] B.A. Szabo and G.C. Lee, Derivation of Stiffness Matrices for Problems in Plane Elasticity by Galerkin's Method, *International Journal for Numerical Methods in Engineering* **1**, 301–310 (1969).
[104] J.C.F. Telles, *The Boundary Element Method Applied to Inelastic Problems* (Springer-Verlag, Berlin, 1983).

[105] P.S. Theocaris and N.I. Ioakimidis, Numerical Integration Methods for the Solution of Singular Integral Equations, *Quarterly of Applied Mathematics* **35**, 173–183 (1977).
[106] P.S. Theocaris, Numerical Solution of Singular Integral Equations: Parts I & II, *Journal of the Engineering Mechanics Division, Proceedings of the ASCE* **107** (EM 5), 733–752 and 753–771 (1981).
[107] E. Trefftz, Über die Kontraktion Kreisförmiger Flüssigkeitsstrahlen, *Zeitschrift für Mathematik und Physik* **64**, 34–61 (1917).
[108] E. Trefftz, Ein Gegenstück zum Ritzchen Verfahren, in: *Proceedings of the 2nd International Congress of Applied Mechanics*, E. Meissner, Ed. (Orell Füssli Verlag, Zurich, 1927) pp. 131–137.
[109] W.S. Verturini, *Boundary Element Method in Geomechanics* (Springer-Verlag, Berlin, 1983).
[110] J.C. Wu and J.F. Thompson, Numerical Solution of Time-Dependent Incompressible Navier–Stokes Equations Using an Integrodifferential Formulation, *Computers and Fluids* **1**, 197–215 (1973).
[111] J.C. Wu, A.H. Spring and N.L. Sankar, A Flow Field Segmentation Method for Numerical Solution of Viscous Flows Problems, in: *Lecture Notes in Physics*, Vol. 35 (Springer-Verlag, Berlin, 1974) pp. 452–457.
[112] O.C. Zienkiewicz, D.W. Kelly and P. Bettess, The Coupling of the Finite Element Method and Boundary Solution Procedures, *International Journal for Numerical Methods in Engineering* **11**, 355–375 (1977).
[113] O.C. Zienkiewicz, *The Finite Element Method*, 3rd Edition (McGraw-Hill, London, 1977).

CHAPTER 2

Potential Theory

Dimitri E. BESKOS

Department of Civil Engineering
University of Patras
Patras 26110, Greece

Boundary Element Methods in Mechanics
Edited by D.E. Beskos

Introduction

As it was mentioned in the previous introductory chapter, the Boundary Element Method (BEM) represents today one of the most widely used numerical methods for the solution of a great variety of problems in engineering mechanics.

This chapter deals with the application of the BEM to the solution of steady-state potential flow problems, i.e. problems which are governed by the Laplace or Poisson equations. This very rich class of potential problems encompasses, e.g., potential fluid flow, flow in porous media, heat flow, electric and magnetic flow and torsion of shafts problems. Potential problems are treated first in this handbook for the following three important reasons: (a) they are described by very simple governing equations; (b) they were historically the first to be treated by the BEM, and (c) they represent a large number of applications. Thus, the present chapter, taking advantage of the simplicity of potential problems and its leading position among the other chapters in this handbook, becomes the right vehicle to introduce the BEM and illustrate in detail the mathematical and numerical aspects of the method. A large number of applications is also discussed in this chapter to demonstrate the versatility of the method and its advantages.

The BEM, as applied to potential problems, basically consists of formulating the problem in terms of an integral equation relating only boundary values and determining its solution numerically. If interior domain values are needed, they can be subsequently calculated from the boundary data. Thus, the dimensionality of the problem is reduced by one and the system of equations resulting from the discretization of the boundary integral formulation is smaller in size than the one encountered in either the Finite Element Method (FEM) or the Finite Difference Method (FDM), which are both based on a differential equation formulation.

There are basically two kinds of boundary element methodologies, the indirect and the direct one. In the indirect method, also known as source method, the boundary integral formulation is accomplished in terms of fictitious source densities which after their computation lead to the physical quantities of the problem. In the direct method, however, the boundary integral formulation deals with quantities, such as potentials and fluxes, which have a physical significance and are computed directly.

Historically, the employment of integral equations for the solution of potential problems can be traced back to 1903 when Fredholm [58] established the existence of solution to such problems on the basis of his limiting discretization procedure. Unfortunately, neither Fredholm [58] nor his immediate followers were able to visualize that approximate solutions could actually be obtained by discretization and due to the difficulty involved in obtaining analytical solutions, the use of integral equations had for many years been restricted to problems of existence and uniqueness of solution.

The first application of the direct BEM for the numerical solution of potential problems can apparently be attributed to Trefftz [181] who in 1917 solved the fluid mechanics problem of the axisymmetric jet by employing a method of successive approximations to satisfy his integral equation. Prager [153] in 1928, utilizing a kind of direct BEM formulation, was able to numerically determine the doubly symmetric potential flow about an elliptic cylinder by dividing the boundary into elements and reducing the integral equations into a system of algebraic equations, thereby really introducing the BEM as it is known today. Lotz [123] in 1931 developed a kind of indirect BEM, the so-called surface-source method, for studying the potential flow about axisymmetric bodies and in 1951 Vandrey [184] was able to apply it to certain bodies by means of hand computations.

It appears that boundary integral techniques were well known during the 1950s but they were enjoying a limited popularity due to the computational effort involved in their use [185, 162]. The advent of the electronic computer and its widespread use during the middle 1950s gave a tremendous boost to the development and popularity of the BEM. The first use of the electronic computer for solving problems by the BEM was probably done by Smith and Pierce [168] who in 1958 presented results for axisymmetric potential fluid flow problems. The following year Martensen [131] was able to extend Prager's method [153] to two-dimensional potential fluid flow problems encountered in turbo machines and numerically solve them also with the aid of an electronic computer. Smith and Pierce [168] were followed in the succeeding years by Hess [73], Davenport [52] and Hess and Smith [74] who employed the indirect BEM for solving complex three-dimensional potential fluid flow problems encountered in the aerospace industry. The numerical method of solving the boundary integral equations was based on the discretization of the body surface into flat quadrilateral elements–panels with constant potential and flux values. For this reason their approach is also known as the panel method. A comprehensive account of this method and its applications can be found in the 1967 review paper of Hess and Smith [75].

In 1963, Jaswon [92] and Symm [174] presented a numerical technique for solving Fredholm integral equations of the first kind resulting from the indirect BEM of two-dimensional potential problems. In those same papers they also presented a kind of BEM based on Green's third identity which is capable of directly relating in integral form the physical quantities of the general potential problem characterized by mixed boundary conditions. Thus, they first introduced the direct BEM, which because of its conceptual simplicity and engineering appeal enjoys nowadays much greater popularity than the indirect one, even though, as it was proven by Brebbia and Butterfield [27] and Jaswon [95], these two approaches are essentially equivalent. Jaswon and Ponter [93] utilized the direct BEM to study problems of torsion of shafts which are governed by potential type of equations. At this point it should be mentioned that, as it was demonstrated in 1977 by Brebbia et al. [26, 25], the direct BEM formulation can be easily obtained in a general manner by employing the method of weighted residuals, which is also capable of constructing formulations for quite a number of numerical methods, such as the FEM and the FDM, thereby indicating that all of these methods share a common basis.

The introduction of higher-order elements in the BEM significantly improved its accuracy and efficiency characteristics. Among those who first utilized during the early 1970s higher-order elements in potential problems one can mention here Silvester and Hsieh [165] and McDonald and Wexler [125] in connection with their method of coupling boundary elements with finite elements and Argyris and Scharpf [8] and Hess [76], and Wu et al. [191] in connection with the indirect and direct BEM, respectively. An attempt was made in 1978 by Clements and Rizzo [43] to generalize the BEM and apply it to boundary-value problems governed by second-order elliptic systems.

Both the indirect and especially the direct BEM have been successfully applied to a great variety of problems in engineering science governed by Laplace or Poisson equations. A comprehensive account of these applications will be given in the subsequent sections of this chapter. The interested reader can consult quite a number of textbooks treating the subject of potential theory for more details, such as those of Jaswon and Symm [94], Brebbia [28], Brebbia and Walker [30], Banerjee and Butterfield [20], Liggett and Liu [119], Brebbia, Telles and Wrobel [31], Ingham and Kelmanson [89] and Alarcón and Reyero [4] in Spanish. The book by Hromadka [82] also treats potential problems with a BEM based on complex variables. Use of complex variables, however, limits the applicability of the method to two dimensions, and for this reason this version of the BEM is not discussed here. Another source of comprehensive information on the subject is the availability of quite a number of review-type articles such as those by Rizzo [159], Wu [190], Cruse [48], Christiansen and Hansen [42], Paris et al. [145], Jaswon [96] and Liggett and Liu [120].

The present chapter has been divided into eight sections. The first section provides a brief summary of those parts of classical potential theory that are needed in the subsequent developments as well as a listing of some of the various physical problems governed by Laplace or Poisson equations. The second and third sections describe in detail the formulation and numerical solution of the Laplace equation by both the indirect and the direct BEM for two and three dimensions, respectively. Only constant elements are utilized in sections 2 and 3. Higher-order boundary elements are taken up in section 4. Some special problems, such as internal sources and the Poisson equation, infinite and semi-infinite regions, inhomogeneity, anisotropy, complex geometry and boundary conditions and modern software developments are treated in section 5. Section 6 deals with problems involving nonlinear material and boundary conditions, while section 7 discusses methods for coupling the BEM with the FEM so as to optimize the solution procedure. Finally, section 8 briefly describes two promising variations of the BEM, namely the edge function method of Quinlan [154] and the boundary method of Herrera [71]. A large number of examples and applications are provided throughout the chapter in order to illustrate the method and demonstrate its merits and very comprehensive lists of references are given in every section for further study.

1. Potential theory and its applications

1.1 Theoretical fundamentals

This subsection briefly describes fundamental concepts and results from classical potential theory relevant to subsequent developments in this chapter and it is based on the work of Cruse [48], which has been written with the BEM in mind. For more details, the interested reader can consult the classical books, e.g. Kellogg [101], Sternberg and Smith [171] and Courant and Hilbert [44].

1.1.1. Harmonic functions and potentials

Potential theory is the theory dealing with the solution of the Laplace and Poisson equations. In this subsection discussion is restricted to the Laplace equation and the study of the Poisson equation is treated as a special case in section 5. Consider a C^2 function $p = p(\boldsymbol{x})$ that satisfies the Laplace equation

$$\nabla^2 p(\boldsymbol{x}) = 0 \tag{1.1}$$

in a three-dimensional and simply connected domain V bounded by a closed smooth surface S and subjected to boundary conditions

$$p(\boldsymbol{x}) = \bar{p}(\boldsymbol{x}) \quad \text{on } S_1 , \tag{1.2}$$

$$f(\boldsymbol{x}) = \frac{\partial p(\boldsymbol{x})}{\partial n} = \nabla p(\boldsymbol{x}) \cdot \boldsymbol{n} - f(\boldsymbol{x}) \cdot \boldsymbol{n} = \bar{f}(\boldsymbol{x}) \quad \text{on } S_2 , \tag{1.3}$$

where $\boldsymbol{n}$ is the unit outward normal vector to the surface $S = S_1 \cup S_2$ and $\bar{p}(\boldsymbol{x})$ and $\bar{f}(\boldsymbol{x})$ are prescribed boundary values, as shown in Fig. 1. In this general case the boundary conditions are of mixed or Cauchy type, in that they combine Dirichlet and Neumann type boundary conditions given by (1.2) and (1.3), respectively. Other more complicated boundary conditions are also possible and in fact they are discussed later in section 6 of this chapter. A function $p(\boldsymbol{x})$ is

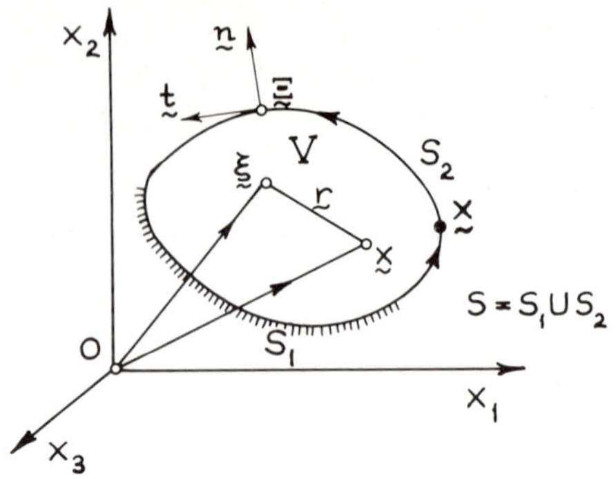

Fig. 1. Domain V with surface $S = S_1 \cup S_2$, interior ($\boldsymbol{x}$, $\boldsymbol{\xi}$) and surface ($\boldsymbol{X}$, $\boldsymbol{\Xi}$) points and tangential ($\boldsymbol{t}$) and normal ($\boldsymbol{n}$) vectors to S.

called harmonic within the domain V bounded by a closed surface S, if it is continuous in V and S, is differentiable to at least the second degree in V and satisfies the Laplace equation in V. In the following, the concept of potential is defined and its properties and relation with harmonic functions are briefly described.

The potential of a conservative force field vector $\boldsymbol{f} = \boldsymbol{f}(\boldsymbol{x})$ is defined as the scalar function

$$p(\boldsymbol{x}) = -\int_{\boldsymbol{\xi}}^{\boldsymbol{x}} \boldsymbol{f}(\boldsymbol{x}) \cdot \mathrm{d}\boldsymbol{r} , \tag{1.4}$$

which expresses the negative of the work done by the force $\boldsymbol{f}$ as it moves a particle along the path from the fixed point $\boldsymbol{\xi}$ to the point $\boldsymbol{x}$ with $\mathrm{d}\boldsymbol{r}$ being the differential motion on the path. Because the force field is conservative, the integral in (1.4) is path-independent. When the force field is gravitational the potential is a Newtonian one, which for point masses m_1 and m_2 is given by

$$p(\boldsymbol{x}) = \int_{\boldsymbol{x}}^{\boldsymbol{\xi}} Gm_1m_2\nabla\left(\frac{1}{r}\right) \cdot \mathrm{d}\boldsymbol{r} = -Gm_1m_2\left(\frac{1}{r}\right) + \text{const.} , \tag{1.5}$$

where $r = r(\boldsymbol{x}, \boldsymbol{\xi})$ is the distance between the two particles located at the space points $\boldsymbol{x}$ and $\boldsymbol{\xi}$. In the following, the symbols $\boldsymbol{x}$ and $\boldsymbol{\xi}$ will represent field points in a volume V, while the symbols $\boldsymbol{X}$ and $\boldsymbol{\Xi}$ will represent field points on a surface S.

For a distribution of mass with density $\rho(\boldsymbol{x})$ in a domain V, the attractive forces create a volume potential

$$p(\boldsymbol{x}) = \int_V \rho(\boldsymbol{\xi}) \frac{1}{r(\boldsymbol{x}, \boldsymbol{\xi})} \, \mathrm{d}V(\boldsymbol{\xi}) , \tag{1.6}$$

while for a mass distribution of density $\mu(\Xi)$ on a surface S, the attractive forces create a surface potential

$$p(x) = \int_S \mu(\Xi) \frac{1}{r(x, \Xi)} \, \mathrm{d}S(\Xi) , \tag{1.7}$$

which is actually called a single-layer potential. The potential obtained as the limit of the potential of two single layers of opposite signs that approach coincidence is called a double-layer potential and is of the form

$$p(x) = \int_S \lambda(\Xi) \frac{d}{\mathrm{d}n(\Xi)} \left(\frac{1}{r(x, \Xi)}\right) \mathrm{d}S(\Xi) , \tag{1.8}$$

where $\lambda(\Xi)$ is the surface density of this potential.

All the above potential functions are defined in a three-dimensional space. In two dimensions the function $\ln(1/r)$ replaces $1/r$ and V and S stand for a plane and a curve, respectively, for all the above equations. It is then easy to verify that the Newtonian potentials $1/r$ and $\ln(1/r)$ satisfy the respective Laplace equation (1.1) identically for $r \neq 0$. Furthermore, it can also be proven [48] that any harmonic function can be expressed as a potential distribution and every potential function is a harmonic function. This statement establishes the relation between potential and harmonic function.

The potentials in Eqs. (1.7) and (1.8) contain singularities in their integrands as $r(x, \Xi)$ approaches zero and for this reason an investigation of their behavior near the singularity, i.e. as $r(x, \Xi) \to 0$ for $x \to X$ from inside the domain V to its surface S in the direction of the vector n at X, is necessary. Through a limiting process one can prove that [48]:

(a) The single-layer potential in both three and two dimensions is continuous as the field point passes through the surface for a density σ which is bounded at all surface points.

(b) The double-layer potential in both three and two dimensions experiences a discontinuity or jump as the point x passes from inside to outside the domain. Indeed as $x \to X$ from the inside

$$p^+(X) = -2\pi\lambda(X) + \int_S \lambda(\Xi) \frac{\mathrm{d}}{\mathrm{d}n(\Xi)} \left(\frac{1}{r(X, \Xi)}\right) \mathrm{d}S(\Xi) \tag{1.9}$$

and from the outside

$$p^-(X) = 2\pi\lambda(X) + \int_S \lambda(\Xi) \frac{\mathrm{d}}{\mathrm{d}n(\Xi)} \left(\frac{1}{r(X, \Xi)}\right) \mathrm{d}S(\Xi) \tag{1.10}$$

so that the jump is

$$p^{+}(X) - p^{-}(X) = -4\pi\lambda(X) \tag{1.11}$$

for three dimensions, provided that the density λ is Holder continuous, i.e. satisfies the relation

$$|\lambda(\Xi) - \lambda(X)| \leq Ar(X, \Xi)^{q}\,; \qquad A, q > 0\,. \tag{1.12}$$

In two dimensions and under the same (1.12) condition, the potential jump is

$$p^{+}(X) - p^{-}(X) = -\pi\lambda(X) - [\pi\lambda(X)] = -2\pi\lambda(X)\,. \tag{1.13}$$

Finally it should be emphasized that integrals like those in Eqs. (1.9) and (1.10) are interpreted in the Cauchy principal-value sense.

1.1.2. Fundamental solutions

The fundamental solution $u^{*}(\boldsymbol{x}, \boldsymbol{\xi})$ of the Laplace's equation (1.1) is defined as the one that satisfies the equation

$$\nabla^{2}u^{*}(\boldsymbol{x}, \boldsymbol{\xi}) = -\delta(\boldsymbol{x}, \boldsymbol{\xi})\,, \tag{1.14}$$

where $\delta(\boldsymbol{x}, \boldsymbol{\xi})$ is the Dirac delta function for which

$$\begin{aligned} &\delta(\boldsymbol{x}, \boldsymbol{\xi}) = 0 \qquad \text{for } \boldsymbol{x} \neq \boldsymbol{\xi}\,, \\ &\int_{V} \rho(\boldsymbol{\xi})\delta(\boldsymbol{x}, \boldsymbol{\xi})\,\mathrm{d}V(\boldsymbol{\xi}) = \rho(\boldsymbol{x})\,. \end{aligned} \tag{1.15}$$

It can be proven [48] that the volume potential (1.6) satisfies the equation

$$\nabla^{2}\Big(\int_{V} \rho(\Xi)\,\frac{1}{r(\boldsymbol{x}, \boldsymbol{\xi})}\,\mathrm{d}V(\Xi)\Big) = -4\pi\rho(\boldsymbol{x})\,. \tag{1.16}$$

A comparison of Eqs. (1.15) and (1.16) reveals that

$$-\nabla^{2}\Big(\frac{1}{4\pi r(\boldsymbol{x}, \boldsymbol{\xi})}\Big) = \delta(\boldsymbol{x}, \boldsymbol{\xi})\,, \tag{1.17}$$

indicating that, on account of (1.14), the fundamental solution $u^{*}(\boldsymbol{x}, \boldsymbol{\xi})$ is given in three dimensions by

$$4\pi u^{*}(\boldsymbol{x}, \boldsymbol{\xi}) = \frac{1}{r(\boldsymbol{x}, \boldsymbol{\xi})}\,. \tag{1.18}$$

The two-dimensional fundamental solution, by following the same procedure, takes the form

$$2\pi u^*(x, \xi) = \ln\left(\frac{1}{r(x, \xi)}\right). \tag{1.19}$$

It is important to notice that for both dimensions $u^*(x, \xi) = u^*(\xi, x)$. Both fundamental solutions exhibit a singular behavior for $r \to 0$, i.e. they become of $O(1/r)$ and $O(\ln r)$ for three and two dimensions, respectively, and for this reason are also called singular solutions. They are also known as free-space Green's functions and play a very important role in the integral formulation of potential problems.

1.1.3. Green's second theorem

Green is well known, among other things, for his two theorems connecting volume to surface integrals involving two C^2 functions and defined over a finite and simply connected domain V with a regular surface S. His second theorem reads

$$\int_V (u\nabla^2 v - v\nabla^2 u)\, \mathrm{d}V = \int_S \left(u\frac{\mathrm{d}v}{\mathrm{d}n} - v\frac{\mathrm{d}u}{\mathrm{d}n}\right) \mathrm{d}S\,. \tag{1.20}$$

Eq. (1.20) is a reciprocal relationship between two functions and as it will be seen in the next section, forms the basis of the direct BEM, as the single- and double-layer potentials form the basis of the indirect BEM.

1.1.4. Fredholm integral equations

The boundary integral equations resulting from the integral formulation of a potential problem with the aid of the indirect or direct BEM, as it will be seen in the next section, are of the form

$$\int_S u(\Xi)K(X, \Xi)\, \mathrm{d}S(\Xi)\, \mathrm{d}S(\Xi) = v(X)\,, \tag{1.21}$$

$$u(X) - \int_S u(\Xi)K(X, \Xi)\, \mathrm{d}S(\Xi) = v(X)\,. \tag{1.22}$$

Eqs. (1.21) and (1.22) with u and v being the unknown and known function, respectively and K being a known kernel are recognized as the Fredholm integral equations of the first and second kind, respectively. This subsection deals with questions of existence and uniqueness of the above two kinds of equations.

For historical reasons Eq. (1.22) is treated first. Answers to the questions of existence and uniqueness of solution of Eq. (1.22) are provided by the Fredholm alternatives in the form of the following theorems [48]:

Theorem 1. Either Eq. (1.22) has one and only one solution for every function

v, or the corresponding homogeneous equation ($v = 0$) has at least one nontrivial solution $u \neq 0$.

Theorem 2. If the first alternative occurs, then the alternative also holds for the adjoint equation

$$\bar{u}(X) - \int_S \bar{u}(\Xi) K(\Xi, X) \, \mathrm{d}S(\Xi) = v(X) . \tag{1.23}$$

For either alternative, the corresponding homogeneous integral equation and its adjoint have the same finite number of linearly independent solutions.

Theorem 3. In the case of the second alternative, a necessary and sufficient condition for the existence of a solution of Eq. (1.22) is that

$$\int_S v(\Xi) \bar{u}(\Xi) \, \mathrm{d}S(\Xi) = 0 \tag{1.24}$$

where $\bar{u}(\Xi)$ is any solution of Eq. (1.23).

The above three theorems can be utilized to prove existence and uniqueness of solution for the classical Dirichlet and Neumann problems discussed in the next section.

In spite of the fact that Eq. (1.21) appears to be simpler mathematically and more illuminating physically in potential problems than Eq. (1.22), it only recently attracted attention and some theoretical background on existence and uniqueness was established. Muskhelishvili [139] and Jaswon [92] have proved for the two-dimensional case of (1.21) the

Theorem: Either the equation

$$\int_S u(\Xi) \ln|X - \Xi| \, \mathrm{d}S(\Xi) = 1 . \tag{1.25}$$

or the equation

$$\int_S u(\Xi) \ln|X - \Xi| \, \mathrm{d}S(\Xi) = 0 \tag{1.26}$$

has a solution for any smooth simple contour S. This theorem can be used to establish existence and uniqueness for Eq. (1.21) in two dimensions as it will be shown in the next section.

And this subsection closes with a brief discussion on kernels $K(X, \Xi)$ encountered in Fredholm integral equations. A kernel $K(X, \Xi)$ is singular if

$$\lim_{\Xi \to X} r^{\alpha}(X, \Xi) K(X, \Xi) \neq 0 . \tag{1.27}$$

As it will be seen in the next section, the most singular kernel in boundary integral equations of potential theory is

$$K = \frac{\mathrm{d}u^*}{\mathrm{d}n} = \frac{\mathrm{d}u^*}{\mathrm{d}r}\frac{\mathrm{d}r}{\mathrm{d}n} = \frac{\mathrm{d}(1/r)}{\mathrm{d}r}\frac{\mathrm{d}r}{\mathrm{d}n} = r^{-2}\frac{\mathrm{d}r}{\mathrm{d}n}, \tag{1.28}$$

corresponding to the three-dimensional case. For this K and for $\alpha = 2$, Eq. (1.27) reduces to

$$\lim_{\Xi \to X} \frac{\mathrm{d}r(X, \Xi)}{\mathrm{d}n(\Xi)} \tag{1.29}$$

which approaches uniformly zero in the limit. Thus the kernels u^* and $\mathrm{d}u^*/\mathrm{d}n$ in both three- and two-dimensional potential problems are non-singular and the Fredholm theorems are applicable to both Dirichlet and Neumann problems.

1.2. Applications

It was mentioned in the Introduction that the Laplace (or Poisson) equation is the governing equation for a variety of problems in engineering mechanics. This can be easily realized by briefly describing here the most important of those problems, identifying the physical quantities involved and relating them, wherever possible, with the functions $p(\boldsymbol{x})$ and $\boldsymbol{f}(\boldsymbol{x})$ of Eqs. (1.1)–(1.3) which stand in general terms for potential and flux vector, respectively. The following engineering mechanics problems are considered:

1.2.1. Gravitational force field [101]

In this case, as it was seen in subsection 1.1., the flux vector $\boldsymbol{f}$ is the force field vector $\boldsymbol{f}$, which being conservative can be written as

$$\boldsymbol{f} = \nabla p, \tag{1.30}$$

where p is the gravitational potential. The gravitational force $\boldsymbol{f}$, however, is given by Newton as

$$\boldsymbol{f} = G\frac{m_1 m_2}{r^3}\boldsymbol{r} \tag{1.31}$$

and thus satisfies the equation

$$\nabla \cdot \boldsymbol{f} = 0. \tag{1.32}$$

Use of (1.30) into (1.32) yields

$$\nabla^2 p = 0, \tag{1.33}$$

indicating that the gravitational potential satisfies Laplace's equation.

1.2.2. Potential fluid flow [101, 161]

Consider the steady-state irrotational flow of an ideal, i.e. inviscid and incompressible, fluid. The role of the flux vector $\boldsymbol{f}$ is played here by the velocity vector $\boldsymbol{v}$ which, because of the irrotational nature of the flow, can be written as

$$\boldsymbol{v} = \nabla\phi \, , \tag{1.34}$$

where ϕ is the velocity potential. On the other hand, the equation of continuity for the steady-state flow of an ideal fluid takes the form

$$\nabla \cdot \boldsymbol{v} = 0 \tag{1.35}$$

Combination of Eqs. (1.34) and (1.35) yields the Laplace's equation

$$\nabla^2\phi = 0 \tag{1.36}$$

for the velocity potential.

If the potential fluid motion is two-dimensional, an additional formulation of the problem is also possible. In two dimensions Eq. (1.35) is written as

$$\frac{\partial v_1}{\partial x_1} + \frac{\partial v_2}{\partial x_2} = 0 \, , \tag{1.37}$$

where v_1 and v_2 are the velocity $\boldsymbol{v}$ components along the x_1 and x_2 directions, respectively. Eq. (1.37) is automatically satisfied through the introduction of a new function $\psi = \psi(x_1, x_2)$, the so-called stream function, defined by

$$v_1 = \frac{\partial \psi}{\partial x_2} \, , \qquad v_2 = -\frac{\partial \psi}{\partial x_1} \, , \tag{1.38}$$

On the other hand, the irrotationality of the flow is expressed by the relation $\nabla \times \boldsymbol{v} = 0$, or

$$\frac{\partial v_2}{\partial x_1} = \frac{\partial v_1}{\partial x_2} \, , \tag{1.39}$$

which actually implies the validity of (1.34) in two dimensions. Substitution of v_1 and v_2 in (1.39) by their expressions in (1.38) yields

$$\nabla^2\psi = 0 \, , \tag{1.40}$$

which is again a Laplace equation.

1.2.3. Groundwater flow [119]

The steady-state flow of an incompressible fluid (water) through a rigid porous medium (soil) under conditions of isotropy and homogeneity is characterized by Darcy's law

$$\boldsymbol{v} = -k\nabla p \tag{1.41}$$

where $\boldsymbol{v}$ is the fluid velocity playing again the role of the flux vector $\boldsymbol{f}$, k is the permeability and p is the piezometric head, and the equation of continuity

$$\nabla \cdot \boldsymbol{v} = 0 \tag{1.42}$$

Combining of Eqs. (1.41) and (1.42) one obtains the Laplace equation

$$\nabla^2 p = 0\,, \tag{1.43}$$

for the piezometric head-potential p.

1.2.4. Heat conducting flow [35]

The steady-state heat conducting flow, under conditions of isotropy and homogeneity, is governed by Fourier's law

$$\boldsymbol{h} = c\nabla T\,, \tag{1.44}$$

where $\boldsymbol{h}$ is the heat flux vector, c is the thermal conductivity and T is the temperature, and the equation of continuity

$$\nabla \cdot \boldsymbol{h} = 0 \tag{1.45}$$

Combination of Eqs. (1.44) and (1.45) leads to the Laplace equation

$$\nabla^2 T = 0 \tag{1.46}$$

for the temperature (potential).

1.2.5. Flow of electric current [101, 170, 166]

The steady-state flow of electric current in conductors, under conditions of isotropy and homogeneity, is governed by Ohm's law

$$\boldsymbol{i} = -\sigma\nabla U \tag{1.47}$$

where $\boldsymbol{i}$ is the electrical current flux vector, σ is the electrical conductivity and U is the electrical potential, and the equation of continuity

$$\nabla \cdot \boldsymbol{i} = 0 \tag{1.48}$$

Combination of Eqs. (1.47) and (1.48) again produces the Laplace equation

$$\nabla^2 U = 0 \tag{1.49}$$

1.2.6. Magnetic field induced by current flow [170, 166]

The magnetic field induction by a steady-state electric current flow in a homogeneous and isotropic medium is governed by Ampère's law

$$\nabla \times \boldsymbol{B} = \mu \boldsymbol{i} , \tag{1.50}$$

where $\boldsymbol{B}$ is the magnetic induction field vector, μ is the magnetic permeability and $\boldsymbol{i}$ is the electric current density. The introduction of the magnetic vector potential $\boldsymbol{A}$ such that

$$\boldsymbol{B} = \nabla \times \boldsymbol{A}, \qquad \nabla \cdot \boldsymbol{A} = 0 , \tag{1.51}$$

transforms (1.50) into the vectorial Poisson's equation

$$\nabla^2 \boldsymbol{A} = -\mu \boldsymbol{i} . \tag{1.52}$$

1.2.7. Torsion of shafts [169]

According to the St. Venant's theory of torsion of elastic prismatic bars of arbitrary but simply connected cross-section R with a boundary C, the displacement field is assumed to be

$$u_1 = -\theta x_2 x_3 , \qquad u_2 = \theta x_1 x_3 , \qquad u_3 = \theta \phi(x_1, x_2) , \tag{1.53}$$

where x_3 is the longitudinal axis of the bar, θ is the angle of twist per unit length of the bar and $\phi(x_1, x_2)$ is the warping function to be determined. Indeed use of the strain–displacement and stress–strain relations (Hooke's law) result in the expressions

$$\tau_{13} = G\theta \left(\frac{\partial \phi}{\partial x_1} - x_2 \right) , \qquad \tau_{23} = G\theta \left(\frac{\partial \phi}{\partial x_2} + x_1 \right) \tag{1.54}$$

for the only nonzero stresses, where G is the shear modulus. Substitution of (1.54) into the stress equilibrium equations provides the governing equation

$$\nabla^2 \phi(x_1, x_2) = 0 \qquad \text{in } R \tag{1.55}$$

which is a Laplace equation. The warping function ϕ is subject to the boundary condition

$$\frac{d\phi(s)}{dn} = x_1 \frac{dx_1}{ds} + x_2 \frac{dx_2}{ds} \qquad \text{on } C \tag{1.56}$$

indicating that the boundary is stress-free, where $\boldsymbol{n}$ is the unit outward normal vector on C and s is a coordinate along C. Once ϕ is determined from (1.55) and (1.56), stresses and displacements can be computed from (1.54) and (1.53), respectively.

Two more formulations of the torsion problem are also possible. Introduction of the conjugate of the warping function $\psi(x_1, x_2)$ through the Cauchy–Riemann equations

$$\frac{\partial \psi}{\partial x_2} = \frac{\partial \phi}{\partial x_1}, \qquad \frac{\partial \psi}{\partial x_1} = -\frac{\partial \phi}{\partial x_2} \tag{1.57}$$

leads to the formulation

$$\begin{aligned} &\nabla^2 \psi(x_1, x_2) = 0 && \text{in } R \\ &\psi = \tfrac{1}{2}(x_1^2 + x_2^2) && \text{on } C \end{aligned} \tag{1.58}$$

involving again a Laplace equation. Finally, introduction of the Prandtl stress function $\omega(x_1, x_2)$ such that

$$\tau_{13} = G\theta \frac{\partial \omega}{\partial x_2}, \qquad \tau_{23} = -G\theta \frac{\partial \omega}{\partial x_1} \tag{1.59}$$

produces the formulation

$$\begin{aligned} &\nabla^2 \omega(x_1, x_2) = -2 && \text{in } R \\ &\omega = 0 && \text{on } C \end{aligned} \tag{1.60}$$

which is characterized by a Poisson equation.

2. Two-dimensional formulation and solution

In this section the formulation and solution of two-dimensional potential problems are described in detail for both the indirect and the direct BEM. The indirect formulation is accomplished with the aid of single- and double-layer potentials, while the direct one with the aid of Green's second identity or the general method of weighted residuals. The numerical solution of the boundary integral equations is obtained with the employment of constant boundary elements, i.e. straight segments with a constant variation of the potential and the flux values. This is done primarily for tutorial reasons in order to avoid obscuring the basic steps of the methodology with additional complexities, usually accompanying higher-order elements which are treated in detail in a separate section. Use of constant elements permits an analytic computation of all the regular and singular integrals appearing as elements in the influence matrices resulting from the discretization of the boundary integral equation. Explicit expressions of these integrals are provided. A numerical computation of the integrals is also possible and is also described in this section. The use of special algorithms for the accurate and efficient solution of systems of algebraic equations characterized by unsymmetric and fully populated matrices is additionally discussed. Finally, some examples are briefly presented to illustrate the method and stress its merits and many references are provided for further study.

2.1. *Indirect BEM formulation*

According to the results of the previous section, every harmonic function can be expressed as a single- or double-layer potential of an unknown source density $\sigma(\Xi)$, where the term 'source' has been adopted here instead of 'mass' for generalization reasons.

Thus for the Dirichlet problem defined by (1.1) and (1.2), one can assume, following the classical approach, that $p(\boldsymbol{x})$ can be expressed solely as a double-layer potential of unknown source density σ of the form

$$p(\boldsymbol{x}) = \int_S \sigma(\Xi) \frac{\mathrm{d}u^*(\boldsymbol{x}, \Xi)}{\mathrm{d}n(\Xi)} \, \mathrm{d}S(\Xi) , \tag{2.1}$$

where $u^*(\boldsymbol{x}, \Xi)$ is the fundamental solution (1.19). Having in mind (1.2) and (1.13), the limit of (2.1) as $\boldsymbol{x} \to \boldsymbol{X}$ becomes

$$\bar{p}(\boldsymbol{X}) = -\tfrac{1}{2}\, \sigma(\boldsymbol{X}) + \int_S \sigma(\Xi) \frac{\mathrm{d}u^*(\boldsymbol{X}, \Xi)}{\mathrm{d}n(\Xi)} \, \mathrm{d}S(\Xi) , \tag{2.2}$$

which is a Fredholm integral equation of the second kind for the unknown source density $\sigma(\boldsymbol{X})$ and where the integral in (2.2) is interpreted in the Cauchy principal-value sense. It can be proven on the basis of the Fredholm alternative theorems of the previous section, that the solution to the Dirichlet problem (2.2) exists and is unique [48].

An alternative approach to obtain an integral equation for the Dirichlet problem is to assume that the function $p(\boldsymbol{x})$ can be expressed solely as a single-layer potential with unknown density σ of the form

$$p(\boldsymbol{x}) = \int_S \sigma(\Xi) u^*(\boldsymbol{x}, \Xi) \, \mathrm{d}S(\Xi) . \tag{2.3}$$

According to the previous section, there is no jump here as $\boldsymbol{x} \to \boldsymbol{X}$ and (2.3), in view of (1.2), becomes at the boundary

$$\bar{p}(\boldsymbol{X}) = \int_S \sigma(\boldsymbol{\Xi}) u^*(\boldsymbol{X}, \boldsymbol{\Xi})\, \mathrm{d}S(\boldsymbol{\Xi})\,. \tag{2.4}$$

which is a Fredholm integral equation of the first kind. In spite of the simplicity of Eq. (2.4) as compared with (2.2), more people prefer to use (2.2) rather than (2.4) because (2.4) is not supported by theorems on existence and uniqueness of solution as (2.2) is. In fact, due to the logarithmic singularity at infinity exhibited by $u^*(\boldsymbol{X}, \boldsymbol{\Xi})$, the resulting influence matrix from the discretization of (2.4) may be ill-conditioned leading to non-uniqueness of solution. This non-uniqueness, on the basis of the theorem expressed by (1.25) and (1.26), has been proven by Jaswon [92] to occur for any given shape of the boundary curve S which is characterized by a specific curve-magnitude. Analogous observations have also been made by Hayes and Kellner [67]. In both Refs. [92] and [67] nothing is said, however, about the value of the critical magnitude forf a given shape of the curve S and it is simply suggested that when ill-conditioning is detected to change the scale of the given geometry so that the curve is away from the critical magnitude. By applying Green's third identity, described in the next subsection, Christiansen [39] was able to derive the supplementary integral condition

$$\int_S \sigma(\boldsymbol{\Xi})\, \mathrm{d}S(\boldsymbol{\Xi}) = 0\,, \tag{2.5}$$

which enforces uniqueness of solution. Jaswon and Symm [94] and Banerjee and Butterfield [20] propose the same condition and rewrite Eq. (2.4) in the form

$$\bar{p}(\boldsymbol{X}) = \int_S \sigma(\boldsymbol{\Xi}) u^*(\boldsymbol{X}, \boldsymbol{\Xi})\, \mathrm{d}S(\boldsymbol{\Xi}) + C\,, \tag{2.6}$$

where the additional unknown C will be determined from (2.5). It should be mentioned that this non-uniqueness problem is absent in the three-dimensional case which is characterized by a kernel $u^*(\boldsymbol{X}, \boldsymbol{\Xi})$ which tends to zero for r approaching infinity.

In the Neumann problem defined by Eqs. (1.1) and (1.3) it is assumed that the function $p(\boldsymbol{x})$ can be expressed solely as a single-layer potential with unknown source density σ of the form

$$p(\boldsymbol{x}) = \int_S \sigma(\boldsymbol{\Xi}) u^*(\boldsymbol{x}, \boldsymbol{\Xi})\, \mathrm{d}S(\boldsymbol{\Xi}) \tag{2.7}$$

The directional derivative of $p(\boldsymbol{x})$ along the direction $\boldsymbol{n}$ in an interior point $\boldsymbol{x}$ is then

$$\frac{\mathrm{d}p(x)}{\mathrm{d}n} = \int_S \sigma(\Xi)\, \frac{\mathrm{d}u^*(x, \Xi)}{\mathrm{d}n}\, \mathrm{d}S(\Xi)\,. \tag{2.8}$$

For $\boldsymbol{n}$ becoming the outer normal vector on the surface S as $\boldsymbol{x} \to \boldsymbol{X}$, Eq. (2.8), on account of (1.3) and (1.13), takes the form

$$\bar{f}(\boldsymbol{X}) = -\tfrac{1}{2}\sigma(\boldsymbol{X}) + \int_S \sigma(\Xi)\, \frac{\mathrm{d}u^*(\boldsymbol{X}, \Xi)}{\mathrm{d}n}\, \mathrm{d}S(\Xi)\,, \tag{2.9}$$

which is a Fredholm integral equation of the second kind. It can be proven on the basis of the Fredholm theory of the previous section that a necessary and sufficient condition for the existence of solution of (2.9) is [48]

$$\int_S \bar{f}(\Xi)\, \mathrm{d}S(\Xi) = 0 \tag{2.10}$$

which is known as the Gauss condition. The solution to (2.9) is unique to within an arbitrary additive constant. A unique solution of (2.9) can be obtained by a normalizing condition quite similar to the one previously imposed in the Dirichlet problem [94].

For Cauchy type of problems involving mixed boundary conditions defined by Eqs. (1.1)–(1.3) one can go beyond the classical results and following Ref. [20] express $p(\boldsymbol{x})$ as a single-layer potential of the form

$$p(\boldsymbol{x}) + \int_S \sigma(\Xi) u^*(\boldsymbol{x}, \Xi)\, \mathrm{d}S(\Xi) + C \tag{2.11}$$

and obtain by a directional differentiation

$$f(\boldsymbol{x}) = \int_S \sigma(\Xi)\, \frac{\mathrm{d}u^*(\boldsymbol{x}, \Xi)}{\mathrm{d}n}\, \mathrm{d}S(\Xi) \tag{2.12}$$

with the constant C being in Eq. (2.11) to ensure uniqueness of solution. The limiting expressions of (2.11) and (2.12) as $\boldsymbol{x} \to \boldsymbol{X}$, in view of (1.2), (1.3) and (1.13), take the form

$$\bar{p}(\boldsymbol{X}) = \int_S \sigma(\Xi) u^*(\boldsymbol{X}, \Xi)\, \mathrm{d}S(\Xi) + C\,; \qquad \boldsymbol{X} \in S_1 \tag{2.13}$$

$$\bar{f}(\boldsymbol{X}) = -\tfrac{1}{2}\, \sigma(\boldsymbol{X}) + \int_S \sigma(\Xi)\, \frac{\mathrm{d}u^*(\boldsymbol{X}, \Xi)}{\mathrm{d}n}\, \mathrm{d}S(\Xi)\,; \qquad \boldsymbol{X} \in S_2\,. \tag{2.14}$$

Thus, according to the indirect BEM the unknown function is always expressed as a surface potential with an unknown source density and the resulting boundary integral equations are solved for this source density which is subsequently used for the computation of the unknown function and its normal derivative both on the boundary and inside the domain. For example, in the mixed boundary-value problem defined by Eqs. (1.1)–(1.3) and having the integral formulation described by Eqs. (2.13) and (2.14), one first solves Eqs. (2.13) and (2.14) for $\sigma(\boldsymbol{X})$ and then obtains $p(\boldsymbol{x})$ and $f(\boldsymbol{x})$ from (2.11) and (2.12), respectively and $p(\boldsymbol{X})$ on S_2 and $f(\boldsymbol{X})$ on S_1 from

$$p(\boldsymbol{X}) = \int_S \sigma(\boldsymbol{\Xi})u^*(\boldsymbol{X}, \boldsymbol{\Xi})\,\mathrm{d}S(\boldsymbol{\Xi}) + C\,; \qquad \boldsymbol{X} \in S_2 \tag{2.15}$$

$$f(\boldsymbol{X}) = -\tfrac{1}{2}\,\sigma(\boldsymbol{X}) + \int_S \sigma(\boldsymbol{\Xi})\,\frac{\mathrm{d}u^*(\boldsymbol{X}, \boldsymbol{\Xi})}{\mathrm{d}n}\,\mathrm{d}S(\boldsymbol{\Xi})\,; \qquad \boldsymbol{X} \in S_1\,. \tag{2.16}$$

2.2. Direct BEM formulation

The main advantage of the direct BEM, also known as Boundary Integral Equation Method (BIEM), is that it employs physical rather than fictitious quantities in the formulation of the problem. Thus, it is conceptually advantageous to directly deal with potentials and fluxes rather than source densities which bear no physical relation to the problem. The direct BEM formulation can be accomplished either by employing Green's second or third theorem [92] or through the general method of weighted residuals [26, 25]. Both approaches will be presented here.

Use of Green's second theorem (1.20) with $u = u^*(\boldsymbol{x}, \boldsymbol{\xi})$, $\partial u/\partial n = \partial u^*(\boldsymbol{x}, \boldsymbol{\xi})/\partial n(\boldsymbol{x}) = u_n^*(\boldsymbol{x}, \boldsymbol{\xi})$, $v = p(\boldsymbol{x})$ and $\partial v/\partial n = \partial p(\boldsymbol{x})/\partial n(\boldsymbol{x}) = f(\boldsymbol{x})$ in conjunction with the definitions (1.14), (1.19) and (1.15) for the fundamental solution u^* and the delta function δ, respectively, results in the integral equation

$$p(\boldsymbol{\xi}) + \int_S p(\boldsymbol{X})u_n^*(\boldsymbol{\xi}, \boldsymbol{X})\,\mathrm{d}S(\boldsymbol{X}) = \int_S f(\boldsymbol{X})u^*(\boldsymbol{\xi}, \boldsymbol{X})\,\mathrm{d}S(\boldsymbol{X})\,, \tag{2.17}$$

also known as Green's third theorem or identity. It should be noticed here that if $\boldsymbol{\xi}$ lies outside the domain V then $p(\boldsymbol{\xi}) = 0$ in (2.17). Having in mind the results of the continuity study on single- and double-layer potentials presented in the previous section, one can take the limit of (2.17) as $\boldsymbol{\xi} \to \boldsymbol{\Xi}$ and obtain the boundary integral equation

$$c(\boldsymbol{\Xi})p(\boldsymbol{\Xi}) + \int_S p(\boldsymbol{X})u_n^*(\boldsymbol{\Xi}, \boldsymbol{X})\,\mathrm{d}S(\boldsymbol{X}) = \int_S f(\boldsymbol{X})u^*(\boldsymbol{\Xi}, \boldsymbol{X})\,\mathrm{d}S(\boldsymbol{X})\,, \tag{2.18}$$

where the coefficient $c(\boldsymbol{\Xi})$ has the value of $\frac{1}{2}$ for a smooth boundary point or the value of $\theta = \pi + \alpha_1 - \alpha_2$ in radians for a corner boundary point, where θ is the internal angle of the boundary at the point $\boldsymbol{\Xi}$ as shown in Fig. 2. In other words,

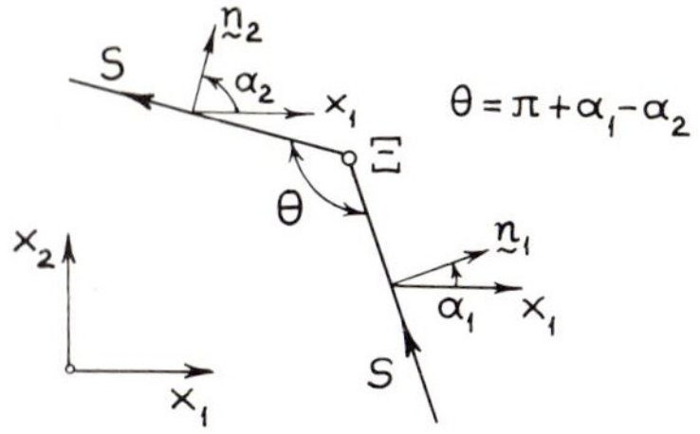

Fig. 2. Geometry of a corner surface point Ξ for two-dimensions.

the direct BEM has the additional advantage over the indirect one of dealing with more general (regular) surfaces allowing corners or edges to be present and not just with smooth ones as it is the case with the classical indirect BEM.

Equation (2.18) can be viewed as a constraint equation between $p(X)$ and $f(X)$ over the surface S thereby ensuring compatibility of the boundary data. In the case of a Dirichlet problem Eq. (2.18) becomes the Fredholm integral equation of the first kind

$$c(\Xi)\bar{p}(\Xi) + \int_S \bar{p}(X)u_n^*(\Xi, X)\,\mathrm{d}S(X) = \int_S f(X)u^*(\Xi, X)\,\mathrm{d}S(X)\,. \tag{2.19}$$

Thus (2.19) is solved for f on the boundary and then (2.17) is used to compute p inside the domain. In the case of a Neumann problem, Eq. (2.18) becomes the Fredholm integral equation of the second kind

$$c(\Xi)p(\Xi) + \int_S p(X)u_n^*(\Xi, X)\,\mathrm{d}S(X) = \int_S \bar{f}(X)u^*(\Xi, X)\,\mathrm{d}S(X)\,. \tag{2.20}$$

Thus (2.20) is solved for p on the boundary and then (2.17) is used to compute p inside the domain. For the solution of a Cauchy problem, Eq. (2.18) takes the form

$$\begin{aligned} c(\Xi)\bar{p}(\Xi) &+ \int_{S_1} \bar{p}(X)u_n^*(\Xi, X)\,\mathrm{d}S(X) + \int_{S_2} p(X)u_n^*(\Xi, X)\,\mathrm{d}S(X) \\ &= \int_{S_1} f(X)u^*(\Xi, X)\,\mathrm{d}S(X) + \int_{S_2} \bar{f}(X)u^*(\Xi, X)\,\mathrm{d}S(X)\,; \qquad \Xi \in S_1 \end{aligned} \tag{2.21}$$

$$\begin{aligned} c(\Xi)p(\Xi) &+ \int_{S_1} \bar{p}(X)u_n^*(\Xi, X)\,\mathrm{d}S(X) + \int_{S_2} p(X)u_n^*(\Xi, X)\,\mathrm{d}S(X) \\ &= \int_{S_1} f(X)u^*(\Xi, X)\,\mathrm{d}S(X) + \int_{S_2} \bar{f}(X)u^*(\Xi, X)\,\mathrm{d}S(X)\,; \qquad \Xi \in S_2 \end{aligned} \tag{2.22}$$

from which the boundary values of f on S_1 and p on S_2 are obtained. The complete boundary data is then utilized in (2.17) to determine p in the interior of the domain. For all the above three cases the values of the directional derivative $f = \mathrm{d}p/\mathrm{d}n$ in the interior of the domain can be obtained, if desired, from the equation

$$f(\xi) = -\int_S p(X) \frac{\mathrm{d}u_n^*(\xi, X)}{\mathrm{d}n} \,\mathrm{d}S(X) + \int_S f(X) \frac{\mathrm{d}u^*(\xi, X)}{\mathrm{d}n} \,\mathrm{d}S(X) \qquad (2.23)$$

which is obtained from (2.17) by differentiation and where

$$f(\xi) = \frac{\mathrm{d}p(\xi)}{\mathrm{d}n} = \frac{\partial p(\xi)}{\partial \xi_i} n_i(\xi), \qquad (2.24)$$

with summation convention assumed over repeated indices. It should be also noticed that in the direct BEM, problems of non-uniqueness associated with Fredholm integral equations of the first kind also exist and their treatment is the same as the one described in the indirect BEM.

The derivation of the integral equation (2.17) from Green's second theorem represents a way of deriving direct integral equation formulations with the aid of known theorems or principles. Thus, for example, in elastostatics the role of Green's second theorem is played by Betti's reciprocal theorem [158]. There is, however, a much more general way of deriving direct integral equation formulations in a systematic and conceptually simple manner and this is through the method of weighted residuals first introduced by Brebbia et al. [26, 25]. According to this method, if the unknown potential and flux functions in (1.1)–(1.3) are replaced by their approximate expressions p and f, the error thus introduced can be minimized by writing the weighted residual statement

$$\int_V \nabla^2 p(x) u^*(\xi, x) \,\mathrm{d}V(x) = \int_{S_2} [f(X) - \bar{f}(X)] u^*(\xi, X) \,\mathrm{d}S(X) - \int_{S_1} [p(X) - \bar{p}(X)] u_n^*(\xi, X) \,\mathrm{d}S(X), \qquad (2.25)$$

where u^* is the weighting function and $u_n^* = \partial u^*/\partial n$. Integration of the volume integral of (2.25) by parts twice reduces (2.25) to

$$\int_V \nabla^2 u^*(\xi, x) p(x) \,\mathrm{d}V(x) = -\int_{S_1} f(X) u^*(\xi, X) \,\mathrm{d}S(X) - \int_{S_2} \bar{f}(X) u^*(\xi, X) \,\mathrm{d}S(X) + \int_{S_2} p(X) u_n^*(\xi, X) \,\mathrm{d}S(X) + \int_{S_1} \bar{p}(X) u_n^*(\xi, X) \,\mathrm{d}S(X) \qquad (2.26)$$

or more generally to

$$\int_V \nabla^2 u^*(\boldsymbol{\xi}, \boldsymbol{x}) p(\boldsymbol{x})\, \mathrm{d}V(\boldsymbol{x}) = -\int_S f(\boldsymbol{X}) u^*(\boldsymbol{\xi}, \boldsymbol{X})\, \mathrm{d}S(\boldsymbol{X}) + \int_S p(\boldsymbol{X}) u_n^*(\boldsymbol{\xi}, \boldsymbol{X})\, \mathrm{d}S(\boldsymbol{X})\,. \tag{2.27}$$

Identifying u^* with the fundamental solution (1.14) and taking into account Eqs. (1.15) defining the delta function, one is able to reduce (2.27) to the form (2.17) previously obtained with the aid of Green's second theorem.

2.3. Numerical implementation and solution

The present discussion on the numerical implementation of the BEM will be restricted to the direct approach because of the many similarities of these two approaches and the fact that the direct BEM is more popular and more widely used. A brief description of the numerical treatment of the indirect BEM as applied to the Cauchy problem can be found in the first chapter of this book.

The goal is here to numerically solve the boundary integral equation (2.18). To this end, the boundary S of the two-dimensional body V shown in Fig. 3 is divided into a finite number N of straight segments or "boundary elements" of variable, in general, length S_j ($j = 1, 2, \ldots, N$) and the assumption is made that the values of p and f remain constant on every such element and are assigned at points called "nodes" at the middle of every element. These are the so-called "constant elements" which, because of their simplicity, permit a better introductory presentation of the numerical implementation procedure. Section 4 of this chapter discusses the use of higher-order elements such as linear, quadratic, etc., which are capable of achieving a better representation of the functional behavior of the functions p and f and the geometry of the boundary surface S.

Equation (2.18) can, in general, be written in the discretized form

$$c_i p_i + \sum_{j=1}^{N} \int_{S_j} p u_n^* \,\mathrm{d}S = \sum_{j=1}^{N} \int_{S_j} f u^* \,\mathrm{d}S \tag{2.28}$$

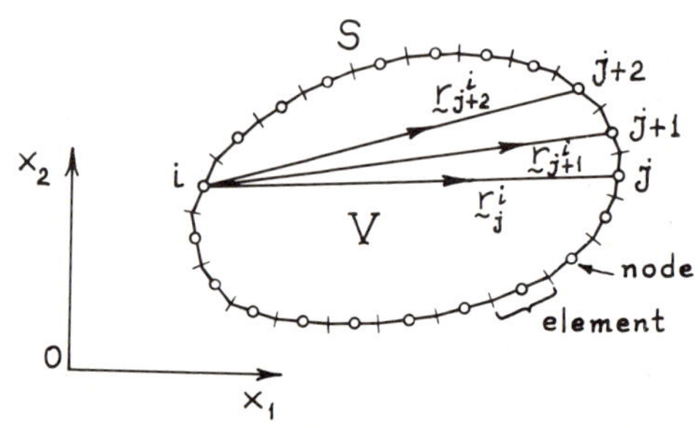

Fig. 3. Discretization of a two-dimensional domain into constant boundary elements.

which actually represents a relation between the node i, which has replaced point Ξ, and all the j elements, including the one $i = j$, as shown in Fig. 3.

The constant element type of discretization always creates a smooth boundary and thus the c_i coefficients in (2.28) become equal to $\frac{1}{2}$. Furthermore, the p and f functions remain constant in every element and hence they can be taken out of the integrals over the various elements. Thus, equation (2.28) becomes

$$\sum_{j=1}^{N} G^{n}_{ij} p_j = \sum_{j=1}^{N} G_{ij} f_j \tag{2.29}$$

where

$$\bar{G}^{n}_{ij} = \int_{S_j} u^*_n \, \mathrm{d}S \,, \qquad G_{ij} = \int_{S_j} u^* \, \mathrm{d}S$$
$$G^{n}_{ij} = \bar{G}^{n}_{ij} + \tfrac{1}{2}\delta_{ij} \,, \quad u^*_n = u^*_n(r^i_j) \,, \quad u^* = u^*(r^i_j) \tag{2.30}$$

with δ_{ij} being the Kronecker delta. The whole set of equations is then obtained by letting i in (2.29) take the values $i = 1, 2, \ldots, N$ thereby creating the matrix equation

$$[G^n]\{p\} = [G]\{f\} \tag{2.31}$$

where $[G^n]$ and $[G]$ are $N \times N$ influence matrices consisting of elements G^n_{ij} and G_{ij}, respectively and $\{p\}$ and $\{f\}$ are $N \times 1$ vectors with elements p_j and f_j, respectively

Equation (2.31) can be used directly for the solution of Dirichlet or Neumann boundary value problems. For instance, for the Dirichlet problem one has

$$[G]\{f\} = [G^n]\{\bar{p}\} = \{a\} \tag{2.32}$$

where $\{a\}$ is a known vector and [G] is a fully populated and unsymmetric $N \times N$ influence matrix, and he can easily solve (2.32) for the flux vector $\{f\}$.

In a Cauchy mixed boundary-value problem, N_1 values of p and N_2 values of f, where $N_1 + N_2 = N$, are known on S and hence Eq. (2.31) has to be partitioned and appropriately reordered before it is ready to be solved. To this end consider Eqs. (2.21) and (2.22) in their discretized form

$$\tfrac{1}{2}\bar{p}_i = \sum_{j=1}^{N_1} f_j G_{ij} + \sum_{j=N_1+1}^{N} \bar{f}_j G_{ij} - \sum_{j=1}^{N_1} \bar{p}_j \bar{G}^n_{ij} - \sum_{j=N_1+1}^{N} p_j \bar{G}^n_{ij} \tag{2.33}$$

$$\tfrac{1}{2}p_k = \sum_{j=1}^{N_1} f_j G_{kj} + \sum_{j=N_1+1}^{N} \bar{f}_j G_{kj} - \sum_{j=1}^{N_1} \bar{p}_j \bar{G}^n_{kj} - \sum_{j=N_1+1}^{N} p_j \bar{G}^n_{kj} \tag{2.34}$$

where $i = 1, 2, \ldots, N_1$; $k = N_1 + 1, N_1 + 2, \ldots, N$ and G_{ij} and $\bar{G}^n_{ij}$ are given by (2.30). Writing the above equations for all the N points on S and assembling them in matrix form one receives

$$\begin{bmatrix} [G^n_{ij}]_1 & [\bar{G}^n_{ij}]_2 \\ [\bar{G}^n_{ij}]_1 & [G^n_{ij}]_2 \end{bmatrix} \begin{Bmatrix} \{\bar{p}_j\}_1 \\ \{p_j\}_2 \end{Bmatrix} = \begin{bmatrix} [G_{ij}]_1 & [G_{ij}]_2 \\ [G_{ij}]_1 & [G_{ij}]_2 \end{bmatrix} \begin{Bmatrix} \{f_j\}_1 \\ \{\bar{f}_j\}_2 \end{Bmatrix} \tag{2.35}$$

where the subscripts 1 and 2 indicate that $j = i$ and $j = k$, respectively. Reordering of Eq. (2.35) so that known vectors are separated from unknown ones results in a matrix equation of the form

$$[K]\{z\} = [B]\{\bar{z}\} = \{b\} \tag{2.36}$$

where $\{z\}$ and $\{\bar{z}\}$ are the vectors of the unknown and known, respectively, values of p and f and hence $\{b\}$ is a known vector and $[K]$ is a fully populated and unsymmetric $N \times N$ influence matrix. In a computer program G_{ij} and G^n_{ij} are assembled directly into $[K]$, without using (2.35), for increased efficiency. Solution of (2.36) for $\{z\}$ leads to knowledge of all the boundary data. Then p and f can be computed pointwise at any interior point l by utilizing Eqs. (2.17) and (2.23) in their discretized forms

$$p_l = \sum_{j=1}^{N} G_{lj} f_j - \sum_{j=1}^{N} \bar{G}^n_{lj} p_j \,, \tag{2.37}$$

$$f_l = \sum_{j=1}^{N} \bar{G}^n_{lj} f_j - \sum_{j=1}^{N} H_{lj} p_j \,, \tag{2.38}$$

where

$$H_{lj} = \int_{S_j} \frac{\mathrm{d}u^*_n}{\mathrm{d}\boldsymbol{n}} \, \mathrm{d}S \,. \tag{2.39}$$

Other problems characterized by more complicated boundary conditions can also be solved by the direct BEM. For example, the solution of the Robin type of problem, characterized by the boundary condition

$$\alpha p + \beta f = \gamma \qquad \text{on } S \,, \tag{2.40}$$

where α, β and γ are constants, can be found in Symm [176, 178], while the solution of some special electrical and thermal problems, characterized by the boundary condition

$$\nabla p \cdot \boldsymbol{n} + \beta \nabla p \cdot \boldsymbol{t} = 0 \qquad \text{on } S \tag{2.41}$$

where $\boldsymbol{n}$ and $\boldsymbol{t}$ are the unit normal and tangential vectors on S, respectively and β is a constant, can be found in De Mey [55] and Anza et al. [7].

2.4. Computational aspects

This subsection discusses the problems of computing the influence matrices G_{ij}, $\bar{G}^n_{ij}$ and H_{ij} and efficiently solving the matrix Eqs. (2.32) and (2.36).

The integrals G_{ij}, $\bar{G}^n_{ij}$ and H_{ij}, because of their relatively simple integrands for the case of constant elements, can be computed analytically and expressed in closed form for both the regular or non-singular case for which $i \neq j$ or $r \neq 0$ and the singular one for which $i = j$ or $r = 0$. Integrals involving $\ln r$ singularity are weakly singular and can be integrated in the normal sense, while integrals involving $1/r$ singularity exist only in the Cauchy principle value sense as $r \to 0$. Integrals with $1/r^2$ singularity, however, do not exist at all as $r \to 0$ [20]. In the following, a list of closed-form expressions is provided for these integrals which has been taken from Banerjee and Butterfield [20] where one can also find the relevant derivations. Consider first the explicit expressions for the fundamental solution u^* and its derivatives

$$
\begin{aligned}
u^*(\boldsymbol{x}, \boldsymbol{\xi}) &= -\frac{1}{2\pi}\ln r\,; \qquad r^2 = y_i y_i = (x-\xi)_i (x-\xi)_i \\
u^*_n(\boldsymbol{x}, \boldsymbol{\xi}) &= \frac{\partial u^*(\boldsymbol{x}, \boldsymbol{\xi})}{\partial x_i}\, n_i(\boldsymbol{x}) = -\frac{n_i(\boldsymbol{x})(x-\xi)_i}{2\pi r^2} \\
\frac{\partial u^*_n(\boldsymbol{x}, \boldsymbol{\xi})}{\partial n} &= \frac{\partial u^*_n(\boldsymbol{x}, \boldsymbol{\xi})}{\partial \xi_j}\, n_j(\boldsymbol{\xi}) = \frac{1}{2\pi r^2}\left(\frac{2 y_i y_i}{r^2} - \delta_{ij}\right) n_i(\boldsymbol{x}) n_j(\boldsymbol{\xi})
\end{aligned}
\tag{2.42}
$$

with summation convention assumed over repeated indices. The above expressions as $r \to 0$ exhibit a singular behavior symbolically described by

$$
u^* = \mathrm{O}(\ln r)\,, \qquad u^*_n = \mathrm{O}(1/r), \ \frac{\partial u^*_n}{\partial n} = \mathrm{O}(1/r^2)\,. \tag{2.43}
$$

The various integrals, involving the functions of Eq. (2.42) with reference to Fig. 4 are given as

$$
G_{ij} = \int_{S_j} u^*(\boldsymbol{x}, \boldsymbol{\xi})\, \mathrm{d}S = -\frac{1}{2\pi}\left[r \sin\theta(\ln r - 1) + \theta h\right]_{\theta_A, r_A}^{\theta_B, r_B} \tag{2.44}
$$

$$
G_{ii} = \lim_{\boldsymbol{x}\to\boldsymbol{\xi}} \int_{S_j} u^*(\boldsymbol{x}, \boldsymbol{\xi})\, \mathrm{d}S = -\frac{S_j}{2\pi}\left[\ln\left(\frac{S_j}{2}\right) - 1\right] \tag{2.45}
$$

$$
\bar{G}^n_{ij} = \int_{S_j} u^*_n(\boldsymbol{x}, \boldsymbol{\xi})\, \mathrm{d}S = \frac{1}{2\pi}\left[n_1(\boldsymbol{x})\theta - n_2(\boldsymbol{x}) \ln\left|\frac{\mathrm{h}}{\mathrm{r}}\right|\right]_{\theta_A, r_A}^{\theta_B, r_B} \tag{2.46}
$$

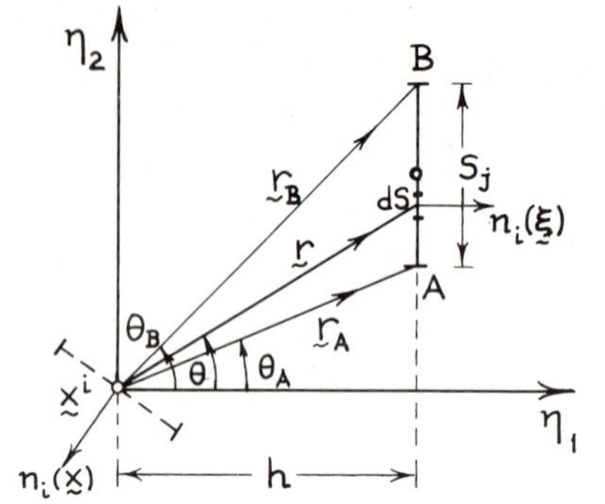

Fig. 4. Line constant boundary element geometry for the analytic evaluation of boundary integrals.

$$\bar{G}^n_{ii} = \lim_{x \to \xi} \int_{S_j} u^*_n(x, \xi)\, dS = \tfrac{1}{2} \tag{2.47}$$

$$H_{ij} = \int_{S_j} \frac{\partial u^*_n(x, \xi)}{\partial n}\, dS = \frac{1}{8\pi h}\left[n_1(x)(2\theta - \sin 2\theta) + n_2(x)\cos 2\theta\right]^{\theta_B, r_B}_{\theta_A, r_A} \tag{2.48}$$

As it is evident from (2.43), the $\partial u^*_n/\partial n$ exhibits a very strong singularity and thus H_{ii} cannot be evaluated for $r \to 0$. Fortunately, H_{ij} is used for interior points and only near the boundary one might experience difficulties.

The regular integrals G_{ij}, $\bar{G}^n_{ij}$ and H_{ij} can also be evaluated numerically by using simple Gauss quadrature rules of the form

$$L_{ij} = \int_{S_j} l(s)\, dS \simeq \frac{S_j}{2} \sum_{k=1}^{K} l(s_k) w_k\,, \tag{2.49}$$

where L_{ij} and l stand for G_{ij}, $\bar{G}^n_{ij}$, H_{ij} and u^*, u^*_n, $\partial u^*_n/\partial n$, respectively and w_k is the weight corresponding to the numerical integration point k where the function l is evaluated. Tables for the integration points and the corresponding weights can be found, e.g. in Stroud and Secrest [173]. For two-dimensional problems four integration points usually provide sufficient accuracy. It should be also noticed that the G_{ii} integrals can also be evaluated numerically with the aid of special quadrature formulae of the form [173]

$$\int_0^1 \ln(1/x) l(x)\, dx \simeq \sum_{k=1}^{K} l(x_k) w_k\,. \tag{2.50}$$

Kutt [107–109] has shown that integrals involving higher-order singularities of the form $1/r^m$ $(m \geq 1)$ can be numerically evaluated by finite-part integration for which he constructed special quadrature formulae.

In an effort to avoid the integral singularities altogether, Patterson and Sheikh [146, 147] consider an auxiliary surface S' completely surrounding S and being close to it and assume that point Ξ in a direct BEM lies on the surface S' rather than S, thereby ensuring that r is always different from zero. In addition, because Ξ lies outside the surface S enclosing the domain V, $c(\Xi)$ in Eq. (2.18) becomes zero, further simplifying the problem. They call this method the regular direct BEM. Similar techniques have also been employed in connection with the indirecrt BEM by, among others, De Mey [54] and Johnston et al. [98]. The main disadvantage of this approximate technique is that there is not a rational way of defining the auxiliary surface S'. Recently, however, Johnston et al. [98] developed an adaptive indirect BEM which determines automatically the location of S' via a least-squares procedure.

Another important computational problem is the efficient solution of matrix equation (2.32) which involves the fully populated unsymmetric matrix $[K]$. Tomlin [180], Lachat and Watson [110], Das [51], Hofmeister [81] and Crotty [47] have developed out-of-core block elimination techniques for this purpose.

2.5. Examples and applications

Both the indirect and direct BEM have been used for the solution of a large variety of potential problems. In this section two examples are briefly described and a comprehensive list of references dealing with various applications of the BEM in potential problems is provided.

2.5.1. Torsion of a prismatic bar

Jaswon and Ponter [93] considered the torsion of a prismatic bar with an elliptical cross-section with semi-axes a and b such that $a = 2b$ and solved that problem formulated as a Neumann potential problem given by Eqs. (1.55) and (1.56) by the direct BEM. Their results in terms of the torsional rigidity

$$D = G \int_V \left(x_1^2 + x_2^2 + x_1 \frac{\partial \phi}{\partial x_2} - x_2 \frac{\partial \phi}{\partial x_1} \right) \mathrm{d}x_1 \, \mathrm{d}x_2$$

and the maximum shear stress $\tau_{\max}$ are given in Table 1 for various discretizations

Table 1
Computer data for the torsion of the elliptical cross-section of Example 2.5.1.

n	D/b^4	$\tau_{\max}/b$
8	4.985	2.134
16	5.087	1.645
32	5.033	1.603
analytic	5.026	1.600

together with the analytic solution for comparison purposes. The index n indicates the number of constant elements of the discretization, even though only $n/4$ of the elements were really used because of symmetry. The convergence of the method to the exact analytic solution for increasing values of n is apparent.

2.5.2. Potential fluid flow

Brebbia and Dominguez [25] studied the potential fluid flow around a cylinder between parallel plates as shown in Fig. 5(a) by employing the direct BEM with constant and linear (to be discussed in section 4) elements. They also compared their results against solutions obtained by the finite element and the finite difference methods. Figures 5(b) and (c) show the constant boundary element and the finite element discretizations, respectively, while Table 2 provides values of the stream function ψ at selected internal points of the finite element mesh of Fig. 5(c) as obtained by the finite and boundary element methods as well as by a curvilinear finite difference method. Values of ψ obtained by the BEMs agree well with those obtained by the FEM and the FDM. No exact solution of this problem is known for a more detailed comparison study. The advantage of the BEM over the FEM, as far as discretization simplicity is concerned, is clearly demonstrated in Figs. 5(b,c).

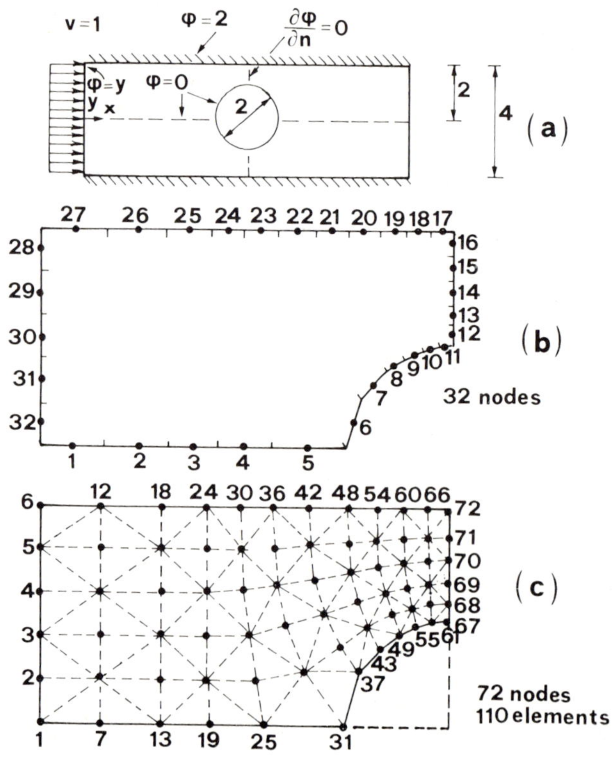

Fig. 5. Potential fluid flow around a cylinder between parallel plates: (a) geometry of the problem, (b) BEM discretization, (c) FEM discretization.

Table 2
Comparison of values of ψ for some mesh points of Fig. 5(c)

Node	FEM	BEM^c	BEM^l	FDM
16	1.154	1.227	1.227	1.230
27	0.801	0.824	0.822	0.829
44	0.404	0.426	0.420	0.416
46	1.205	1.217	1.215	1.210
58	1.205	1.214	1.211	1.210
63	0.768	0.776	0.773	0.771

^{c} = constant and ^{l} = linear boundary elements.

2.5.3. Various application

The BEM in its direct or indirect form has been successfully used for the solution of various potential problems dealing with torsion of prismatic bars [93, 133, 100, 122, 136, 22, 105, 25, 94, 2, 41, 9, 160, 163, 10, 49, 59], potential fluid flow [131, 90, 91, 188, 64, 119, 120], groundwater flow [32, 180, 33, 141, 34], heat conduction [38, 97] and various electrical engineering applications [66, 165, 125, 53, 40, 113, 183, 177, 138, 50].

3. Three-dimensional formulation and solution

In this section the formulation and solution of truly three-dimensional as well as axisymmetric three-dimensional potential problems are described in detail for the BEM. Only constant boundary elements of flat triangular or quadrilateral geometry are used for the boundary surface representation of truly three-dimensional problems. Higher-order curved surface elements are discussed in the next section. Three-dimensional potential problems exhibiting axisymmetry can be effectively reduced to two-dimensional ones which can then be solved by utilizing the machinery that has already been developed in the previous section.

3.1. Three-dimensional case

The formulation and solution procedure of three-dimensional potential problems is exactly the same as the one described in the previous section for two-dimensional problems. The surface S, however, is now discretized into a finite number of triangular or quadrilateral boundary elements, which in this section are assumed to be flat and constant and the fundamental solution and its derivatives present singularities of one degree higher than in the two-dimensional case. As a result of this increased complexity in geometry and singular behavior the computation of the elements of the influence matrices is more involved in this case. Nevertheless, closed-form expressions for these elements can still be obtained. In the following the three-dimensional BEM formulation and solution procedure is illustrated on the basis of the direct approach. Quite analogous results are valid for the indirect BEM.

Consider the direct boundary integral equation (2.18) with the understanding that S is now a surface and not a curve and u^* is the three-dimensional fundamental solution (1.18). The surface S is now discretized into a finite number N of flat triangular or quadrilateral elements S_j ($j = 1, 2, \ldots, N$) with constant values of the functions p and f. These constant values are assigned to one nodal point on the element which is taken to be its centroid C, as shown in Fig. 6. Under these assumptions Eq. (2.18) takes the discretized form

$$c_i p_i + \sum_{j=1}^{N} \left(\int_{S_j} u_n^* \, dS \right) p_j = \sum_{j=1}^{N} \left(\int_{S_j} u^* \, dS \right) f_j \, . \tag{3.1}$$

Consider the case of using constant flat triangular boundary elements. Because it is more convenient to perform the element integrations indicated in parentheses in (3.1) in terms of the local coordinates η_1, η_2, η_3, rather than the global Cartesian ones x_1, x_2, x_3, as shown in Fig. 6(a), the coordinate transformation

$$dS = |J| \, d\eta_1 \, d\eta_2 \tag{3.2}$$

is utilized where the Jacobian of the transformation $|J|$ equals twice the area of the triangle. More details about $|J|$ and its explicit expression are given in the next section. Thus, the integrals to be computed are of the form [31]

$$\bar{G}_{ij}^n = \int_{S_j} u_n^* \, dS = |J| \int_0^1 \left[\int_0^{1-\eta_2} u_n^*(\eta) \, d\eta_1 \right] d\eta_2 \, , \tag{3.3}$$

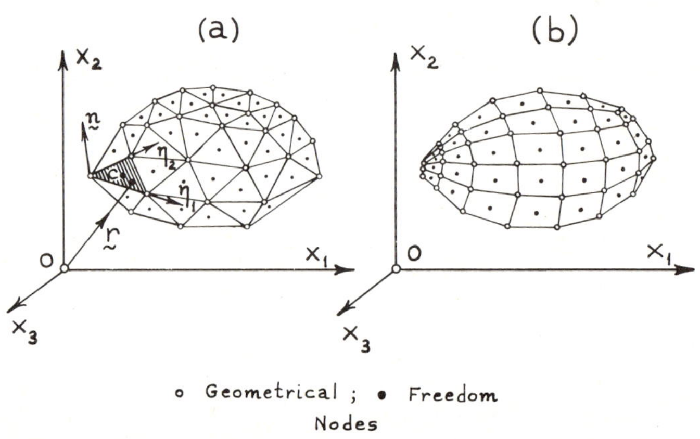

Fig. 6. Surface discretization of a three-dimensional domain into: (a) flat triangular and, (b) flat quadrilateral elements.

$$G_{ij} = \int_{S_j} u^* \, dS = |J| \int_0^1 \left[\int_0^{1-\eta_2} u^*(\eta) \, d\eta_1 \right] d\eta_2 \, . \tag{3.4}$$

When the node i does not belong to the element j these integrals are regular and can be evaluated numerically by the quadrature scheme of the form [25]

$$\bar{G}^n_{ij} = 2A_j d_{ij} \sum_{k=1}^{K} \frac{1}{r^3_{ik}} w_k \, , \tag{3.5}$$

$$G_{ij} = 2A_j \sum_{k=1}^{K} \frac{1}{r_{ik}} w_k \, , \tag{3.6}$$

where A_j is the area of the element j, d_{ij} is the perpendicular distance of point i from the plane of element j and the integration points k and the corresponding weights w_k are given in tabular form in Hammer et al. [65].

When the node i falls on the element j the integrals (3.3) and (3.4) are singular because for $r \rightarrow 0$, $u^* = O(1/r)$ and $u^*_n = O(1/r^2)$ and have to be evaluated analytically. The results of this evaluation with reference Fig. 7 are [31]

$$\bar{G}^n_{ii} = \frac{1}{2} \tag{3.7}$$

$$G_{ii} = \frac{2A_i}{3} \left[\frac{1}{2r_{23}} \ln \left(\frac{\tan[(\theta_1 + \alpha_2)/2]}{\tan(\alpha_2/2)} \right) + \frac{1}{r_{31}} \ln \left(\frac{\tan[(\theta_2 + \alpha_3)/2]}{\tan(\alpha_3/2)} \right) + \frac{1}{r_{12}} \ln \left(\frac{\tan[(\theta_3 + \alpha_1)/2]}{\tan(\alpha_1/2)} \right) \right] . \tag{3.8}$$

Exact analytic expressions for $\bar{G}^n_{ij}$ and G_{ij} for both the regular and singular cases for the case of flat constant quadrilateral elements can be found in Hess and Smith [74, 75]. Integrals with $1/r$ singularity are weakly singular and hence integrable in the normal sense, while those will $1/r^2$ singularity exist only in the

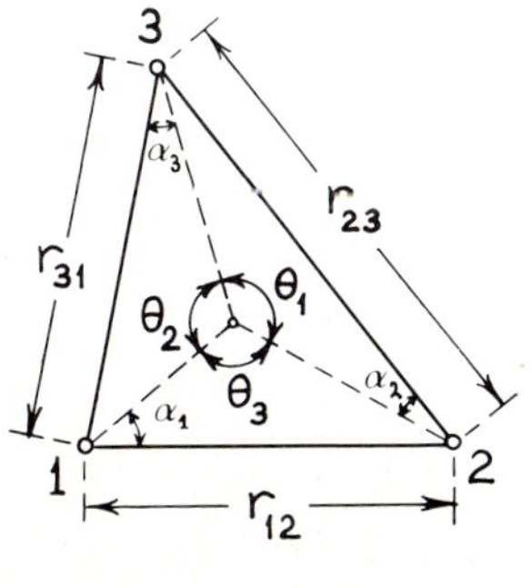

Fig. 7. Surface constant triangular boundary element geometry for the analytic evaluation of boundary integrals.

Cauchy principal value sense as $r \to 0$. Integrals with $1/r^3$ singularity do not exist at all as $r \to 0$ [20].

3.2. Axisymmetric case

In many practical problems the domain of interest as well as the boundary conditions are axisymmetric. In these cases the three-dimensional problem can be essentially reduced to a two-dimensional one by taking appropriate advantage of the existing symmetry about an axis. Even problems geometrically axisymmetric with arbitrary boundary conditions can be reduced to completely axisymmetric ones. Thus, axisymmetric analysis of potential problems is very important in that it considerably reduces the computational effort involved in a truly three-dimensional analysis.

Under the assumption of full axisymmetry, i.e. with respect to both geometry and boundary conditions, use of polar coordinates R, Θ, Z defined as shown in Fig. 8 and with $\mathrm{d}S(x_1, x_2, x_3) = R\,\mathrm{d}\Theta\,\mathrm{d}\bar{S}(R, Z)$, transforms Eq. (2.18) into

$$c(\Xi)p(\Xi) + \int_{\bar{S}} p(X) \int_0^{2\pi} u_n^*(\Xi, X)\,\mathrm{d}\Theta(X) R(X)\,\mathrm{d}\bar{S}(X)$$

$$= \int_{\bar{S}} f(X) \int_0^{2\pi} u^*(\Xi, X)\,\mathrm{d}\Theta(X) R(X)\,\mathrm{d}\bar{S}(X)\,, \tag{3.9}$$

where the three-dimensional fundamental solution in polar coordinates is

$$u^*(\Xi, X) = 1/r(\Xi, X) = 1/[R^2(\Xi) + R^2(X) - 2R(\Xi)R(X) \times \cos(\Theta(\Xi) - \Theta(X)) + (Z(\Xi) - Z(X))^2]^{1/2}\,. \tag{3.10}$$

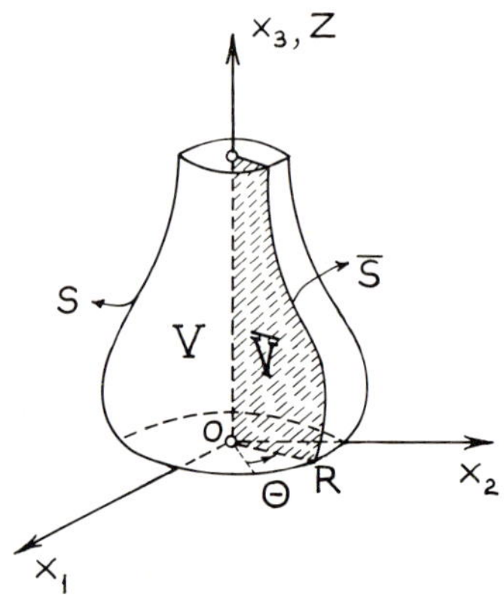

Fig. 8. Geometry of an axisymmetric domain.

Thus, the axisymmetric fundamental solution takes the form

$$\bar{u}^*(\Xi, X) = \int_0^{2\pi} u^*(\Xi, X)\,\mathrm{d}\Theta(X) = \frac{4K(m)}{(a+b)^{1/2}}\,, \tag{3.11}$$

where

$$\begin{aligned} &m = 2b/(a+b)\,; \qquad 0 \leq m \leq 1\,,\\ &a = R^2(\Xi) + R^2(X) + (Z(\Xi) - Z(X))^2\,,\\ &b = 2R(\Xi)R(X)\,, \end{aligned} \tag{3.12}$$

and $K(m)$ represents the complete elliptic integral of the first kind which has a logarithmic singularity as $m \to 1$. Similarly one has that

$$\begin{aligned} \bar{u}_n^*(\Xi, X) &= \int_0^{2\pi} u_n^*(\Xi, X)\,\mathrm{d}\Theta(X)\\ &= \frac{4}{(a+b)^{1/2}}\left[\frac{1}{2R(X)}\left(\frac{R^2(\Xi) - R^2(X) + (Z(\Xi) - Z(X))^2}{(a-b)}\,E(m) - K(m)\right)n_R(X)\right.\\ &\quad\left. + \frac{Z(\Xi) - Z(X)}{(a-b)}\,E(m)n_Z(X)\right], \end{aligned} \tag{3.13}$$

where $E(m)$ is the complete elliptic integral of the second kind which also has a logarithmic singularity as $m \to 1$.

In view of Eqs. (3.11) and (3.13), Eq. (3.9) becomes

$$c(\Xi)p(\Xi) + \int_{\bar{S}} p(X)\bar{u}_n^*(\Xi, X)R(X)\,\mathrm{d}\bar{S}(X) = \int_{\bar{S}} f(X)\bar{u}^*(\Xi, X)R(X)\,\mathrm{d}\bar{S}(X)\,, \tag{3.14}$$

which is now a boundary integral equation for the two-dimensional shaded domain $\bar{V}$ with boundary $\bar{S}$ shown in Fig. 8 and can be treated with the aid of the methodology developed in the previous section. After discretization, both the regular and the singular integrals $\bar{G}_{ij}''$ and G_{ij} are evaluated numerically. This is facilitated by approximating the elliptic integrals $K(m)$ and $E(m)$ by polynomials as [1, 119]

$$\begin{aligned} &K(m) = (a_0 + a_1 m_1 + \cdots + a_4 m_1^4) + (b_0 + b_1 m_1 + \cdots + b_4 m_1^4)\ln(1/m_1) + \varepsilon(m)\,,\\ &E(m) = (1 + c_1 m_1 + \cdots + c_4 m_1^4) + (d_1 m_1 + \cdots + d_4 m_1^4)\ln(1/m_1) + \varepsilon(m)\\ &m_1 = 1 - m\,, \qquad |\varepsilon(m)| \leq 2 \times 10^{-8}\,. \end{aligned} \tag{3.15}$$

Using the above expressions for $K(m)$ and $E(m)$ one can very easily evaluate $\bar{G}_{ij}^n$ and G_{ij} by the regular Gauss quadrature (2.49) when $i \neq j$ and by the special Gauss quadrature (2.50) when $i = j$ in conjunction with the integration point and weight tables of Ref. [173].

The case of a geometrically axisymmetric domain with arbitrary boundary conditions cannot be treated directly as an axisymmetric one due to the dependence of all functions involved on the polar angle Θ. The usual method of treating these problems is by appropriately expanding all the problem variables in Fourier series thereby reducing the problem into a sequence of uncoupled two-dimensional problems which are independent of the angle Θ [164].

3.3. *Examples and applications*

Two examples are discussed in this subsection and a list of references dealing with various applications of three-dimensional potential theory is also provided.

3.3.1. *Non-lifting potential fluid flow around bodies*

This example is a truly three-dimensional one and deals with the determination of the flow velocity distribution on the surface of an ellipsoid with semi-axes 1, 2, $\frac{1}{2}$ in the x, y, z directions, respectively, which is subjected to a potential fluid onset flow in turn along the x, y and z axes with velocity $\boldsymbol{v}_0$ as shown in Fig. 9(a). The velocity field $\boldsymbol{v}$ in the fluid is equal to $\boldsymbol{v} = \boldsymbol{v}_0 + \boldsymbol{v}_d$, where $\boldsymbol{v}_d$ is the disturbance velocity due to the presence of the ellipsoid such that $\boldsymbol{v}_d = -\nabla\phi$. The condition of incompressibility then implies that $\nabla^2\phi = 0$, while the requirement of zero total normal velocity on the surface of the ellipsoid leads to the boundary condition $\partial\phi/\partial n = \nabla\phi \cdot \boldsymbol{n} = \boldsymbol{v}_0 \cdot \boldsymbol{n}$. Thus, the problem is essentially an exterior Neumann problem which, because of the infinite extent of its domain, has to satisfy some regularity conditions at infinity described in subsection 5.1 of this chapter. This problem was first solved by the indirect BEM by Hess and Smith [74, 75] who employed constant flat quadrilateral elements and took advantage of the existing symmetry. The same problem was also later solved by Grodtkjaer [63] by the direct BEM in a more efficient manner (406 versus 4320 effective elements) which utilizes, however, an analytic representation of the ellipsoid to calculate the derivatives of the surface and not a numerical representation as in [74, 75] which can thus handle more complicated geometries. The results from both works are in excellent agreement with the analytic ones as Fig. 9(b) clearly indicates for the case of the flow in the xz plane.

3.3.2. *Temperature distribution in a hollow cylinder*

Consider the steady-state heat conduction problem for the hollow cylinder of Fig. 10(a) which is subjected to the boundary conditions

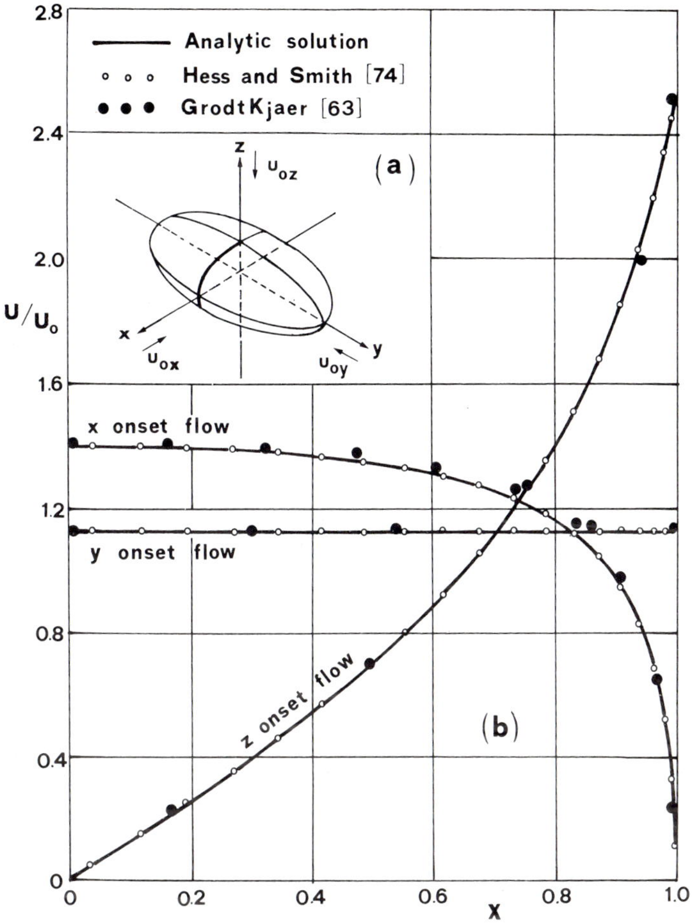

Fig. 9. Fluid velocity distributions in the xz plane on an ellipsoid with axes ratios $1:2:\frac{1}{2}$ for onset flow along the x, y, z directions.

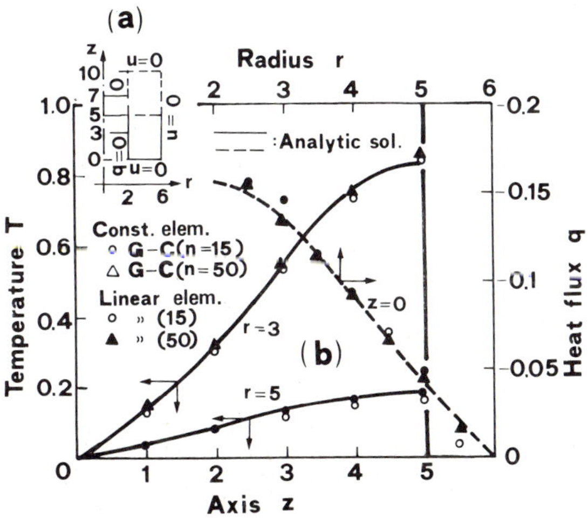

Fig. 10. Temperature and heat flux profiles in a hollow cylinder.

$T = 0$ at $r = 6$, $z = 0$ and 10

$q = 0$ at $r = 2$, $0 < z < 3$ and $7 < r < 10$

$q = 1$ at $r = 2$, $3 < z < 7$

for the temperature T and the heat flux q. This axisymmetric potential problem was solved by Yoshikawa and Tanaka [192] who employed the direct BEM in an axisymmetric formulation but with the axisymmetric fundamental solution derived numerically rather than analytically. Because of the symmetry with respect to the plane $z = 5$, half a cross-section of the cylinder below the plane $z = 5$ is discretized into 18 equal constant elements. Linear elements are also used and in that case two nodal points are placed near the corners of the domain and the point on the inner surface $r = 2$ at which the boundary conditions change. The numerical results for the temperature and heat flux profiles obtained by both constant and linear elements are in very good agreement with the analytical solution as Fig. 10(b) clearly demonstrates.

3.3.3. Various applications

Most of the truly three-dimensional BEM applications of potential theory come from potential fluid flow [73–75, 63, 77, 78, 80, 84, 64]. Among the other applications one can mention geoelectrical [142], groundwater flow [116, 119] and general potential [175, 11, 12] three-dimensional problems.

Axisymmetric potential problems have also been solved by the BEM. These include fluid flow problems [168, 83,79], torsion problems [103, 104, 135, 137], groundwater flow problems [114, 115, 119], heat conduction problems [7, 192] and general potential problems [134, 24, 164, 189, 18].

4. Higher-order boundary elements

In the previous two sections constant boundary elements were utilized. These elements are characterized by the simplest possible geometry and functional behavior. Thus, they are just straight segments and flat triangles or quadrilaterals for two- and three-dimensional problems, respectively and are characterized by constant values of p and f over them. In general, however, p and f can have a linear or higher-order variation and the shape of the elements can be a curved one, thereby achieving a better functional and geometrical representation. This means that fewer higher-order elements will be required for obtaining a solution of the same degree of accuracy as the one obtained by employing constant elements.

Even though it appears to be possible to select different orders of variation for the functions p and f as well as between these functions and the geometry of the elements, in practice it is computationally simpler to take p, f and the element geometry to be the same order of variation, thereby creating the so-called isoparametric elements. Elements with different orders in p and f are called mixed elements. Martin et al. [132] have constructed mixed line elements for two-dimensional potential problems by assuming a linear variation for p and a constant one for f and have been able to avoid problems of overdetermination of f at corners, which usually appear when boundary elements with a linear variation in both p and f are used.

In this section, which is mainly based on Ref. [31], the case of higher-order line elements is illustrated by the quadratic line element and the linear line element is subsequently presented as a special case, while the case of higher-order surface elements is illustrated by the quadratic triangular and quadrilateral surface elements. Other special types of elements, such as the singular or discontinuous elements are discussed in the next section dealing with special problems. Two examples are presented at the end of this section for illustration purposes and a list of relevant references is also given.

4.1. Line higher-order elements

Consider the quadratic line element consisting of three nodes as shown in Fig. 11(a) over which p and f vary quadratically and attain specific values at the nodes. The functions p and f can then be expressed in terms of their nodal values as

$$
\begin{aligned}
p(\eta) &= \phi_1 p_1 + \phi_2 p_2 + \phi_3 p_3 = \lfloor \phi_1 \phi_2 \phi_3 \rfloor \begin{Bmatrix} p_1 \\ p_2 \\ p_3 \end{Bmatrix} = \boldsymbol{\phi}^T \boldsymbol{p} \\
f(\eta) &= \phi_1 f_1 + \phi_2 f_2 + \phi_3 f_3 = \lfloor \phi_1 \phi_2 \phi_3 \rfloor \begin{Bmatrix} f_1 \\ f_2 \\ f_3 \end{Bmatrix} = \boldsymbol{\phi}^T \boldsymbol{f}
\end{aligned}
\tag{4.1}
$$

where the shape functions ϕ_1, ϕ_2 and ϕ_3 given by

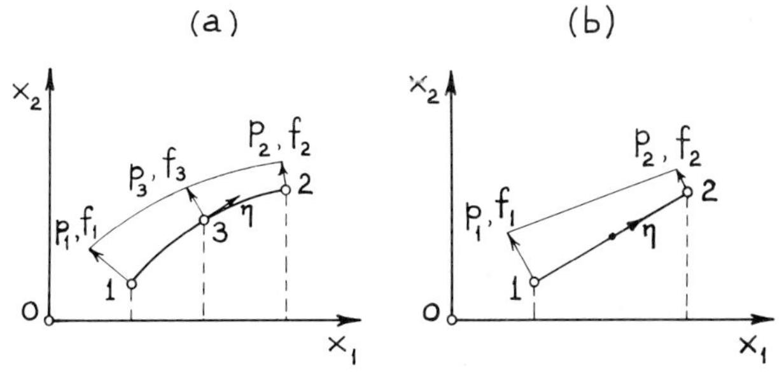

Fig. 11. Line higher-order boundary elements: (a) quadratic, and (b) linear element.

$$\phi_1 = \tfrac{1}{2}\eta(\eta - 1)\,, \qquad \phi_2 = \tfrac{1}{2}\eta(1 + \eta)\,, \qquad \phi_3 = (1 - \eta)(1 + \eta) \tag{4.2}$$

vary quadratically with respect to the coordinate η. Introduction of the expressions (4.1) for p and f into the integrals of Eq. (2.28) results in

$$\int_{S_j} p u_n^* \, \mathrm{d}S = \int_{S_j} \lfloor \phi_1 \phi_2 \phi_3 \rfloor u_n^* \, \mathrm{d}S \begin{Bmatrix} p_1 \\ p_2 \\ p_3 \end{Bmatrix} = \lfloor \bar{G}_{ij}^{n1} \bar{G}_{ij}^{n2} \bar{G}_{ij}^{n3} \rfloor \begin{Bmatrix} p_1 \\ p_2 \\ p_3 \end{Bmatrix} \tag{4.3}$$

$$\int_{S_j} f u^* \, \mathrm{d}S = \int_{S_j} \lfloor \phi_1 \phi_2 \phi_3 \rfloor u^* \, \mathrm{d}S \begin{Bmatrix} f_1 \\ f_2 \\ f_3 \end{Bmatrix} = \lfloor G_{ij}^1 G_{ij}^2 G_{ij}^3 \rfloor \begin{Bmatrix} f_1 \\ f_2 \\ f_3 \end{Bmatrix} \tag{4.4}$$

where

$$\bar{G}_{ij}^{nk} = \int_{S_j} \phi_k u_n^* \, \mathrm{d}S\,, \qquad G_{ij}^k = \int_{S_j} \phi_k u^* \, \mathrm{d}S\,; \qquad k = 1, 2, 3\,. \tag{4.5}$$

The boundary integral equation (2.28) then takes the form

$$c_i p_i + \sum_{j=1}^{N} \bar{G}_{ij}^n p_j = \sum_{j=1}^{N} G_{ij} f_j\,, \tag{4.6}$$

where this equation for a node i common to two adjoining elements $j - 1$ and j contains contributions from both elements and where each $\bar{G}_{ij}^n(G_{ij})$ term is equal to the $\bar{G}_{ij}^{n2}(G_{ij}^2)$ term of element $j - 1$ plus the $\bar{G}_{ij}^{n1}(G_{ij}^1)$ term of element j for a counterclockwise numbering system. Equation (4.6) can be written in the simple compact matrix form

$$[G^n]\{p\} = [G]\{f\}\,, \tag{4.7}$$

where

$$\begin{aligned} G_{ij}^n &= \bar{G}_{ij}^n && \text{for } i \neq j \\ G_{ij}^n &= \bar{G}_{ij}^n + c_i && \text{for } i = j\,. \end{aligned} \tag{4.8}$$

The evaluation of the integrals (4.5) requires the use of the Jacobian of the transformation $|\bar{J}|$ from (x_1, x_2) to η, which reads

$$|\bar{J}| = \left[\left(\frac{\mathrm{d}x_1}{\mathrm{d}\eta} \right)^2 + \left(\frac{\mathrm{d}x_2}{\mathrm{d}\eta} \right)^2 \right]^{1/2} = \frac{\mathrm{d}S}{\mathrm{d}\eta}\,. \tag{4.9}$$

Thus the integrals of (4.5) become

$$\bar{G}_{ij}^{nk} = \int_{(1)}^{(2)} \phi_k(\eta) u_n^* |\bar{J}| \, \mathrm{d}\eta \qquad G_{ij}^{k} = \int_{(1)}^{(2)} \phi_k(\eta) u^* |\bar{J}| \, \mathrm{d}\eta \tag{4.10}$$

and for the non-singular case can be evaluated numerically. Evaluation of (4.10), however, requires knowledge of the element geometry as a function of η. It is assumed that the geometry varies in the same way as the functions p and f, i.e.

$$\begin{aligned} x_1 &= \phi_1 x_1^1 + \phi_2 x_1^2 + \phi_3 x_1^3 \\ x_2 &= \phi_1 x_2^1 + \phi_2 x_2^2 + \phi_3 x_2^3 \end{aligned} \tag{4.11}$$

thereby adopting the notion of the isoparametric element. It can be shown [31] that the reproduction of constant-potential type of conditions requires that the order of the function p be at least the same as the one used to describe the geometry of the domain.

The coefficient c_i in Eq. (4.6) is $\frac{1}{2}$ when the surface is smooth at the point i. When this is not the case the diagonal elements of $[G^n]$ in (4.7) can easily be computed in terms of the off-diagonal ones if one takes into account that when a uniform potential is applied on a closed domain the f's must be zero, i.e.

$$[G^n]\{I\} = \{0\} \tag{4.12}$$

where $\{I\}$ is the identity vector. Equation (4.12) indicates that

$$G_{ii}^n = -\sum_{\substack{j=1 \\ j\neq i}}^{N} G_{ij}^n \,, \qquad i = 1, 2, \ldots, N \,. \tag{4.13}$$

Computation of the G_{ii}^n and hence of the G_{ii} integrals can be done either analytically or better numerically since there exist special quadrature rules for integrands possessing a logarithmic singularity as the one here [173]. Once all the elements of the matrices $[G^n]$ and $[G]$ have been computed, Eq. (4.7) can be solved as it was described in section 2.

Increasing the number of intermediate nodes and following the same procedure one can construct and employ in a BEM formulation even higher-order line elements than quadratic ones. For more details the reader is advised to consult Refs. [20, 31].

The case of linear line elements, as the one shown in Fig. 11(b), can be considered as a special case of the quadratic element. Here the shape functions

$$\phi_1 = \tfrac{1}{2}(1 - \eta) \,, \qquad \phi_2 = \tfrac{1}{2}(1 + \eta) \tag{4.14}$$

are linear in η and so is the geometry of the element. Due to the simplicity of the functions ϕ_1 and ϕ_2 the terms G_{ii} can be computed in this case analytically and with the aid of Fig. 11(b) to yield [31]

$$G_{ii} = \tfrac{1}{2} l_{i-1}\left[\tfrac{3}{2} - \ln(l_{i-1})\right] + \tfrac{1}{2} l_i\left[\tfrac{3}{2} - \ln(l_i)\right]. \tag{4.15}$$

A comprehensive treatment of linear elements in potential theory and elastostatics including special problems due to discontinuities in domain geometry and boundary conditions can be found in Alarcón et al. [3] as well as in the next chapter of this book on Elastostatics.

4.2. Surface higher-order elements

The boundary discretization of three-dimensional bodies is usually accomplished by quadrilateral or triangular elements. Flat constant surface elements were discussed in section 3. This subsection deals with higher-order surface quadrilateral and triangular elements. The study of these elements requires a knowledge of the connection between the global (x_1, x_2, x_3) system and the local (η_1, η_2, η_3) system defined over the surface of the domain, as shown in Fig. 12. The transformation of coordinates between the two systems as referred to a function p reads

$$\begin{Bmatrix} \dfrac{\partial p}{\partial \eta_1} \\[2mm] \dfrac{\partial p}{\partial \eta_2} \\[2mm] \dfrac{\partial p}{\partial \eta_3} \end{Bmatrix} = \begin{bmatrix} \dfrac{\partial x_1}{\partial \eta_1} & \dfrac{\partial x_2}{\partial \eta_1} & \dfrac{\partial x_3}{\partial \eta_1} \\[2mm] \dfrac{\partial x_1}{\partial \eta_2} & \dfrac{\partial x_2}{\partial \eta_2} & \dfrac{\partial x_3}{\partial \eta_2} \\[2mm] \dfrac{\partial x_1}{\partial \eta_3} & \dfrac{\partial x_2}{\partial \eta_3} & \dfrac{\partial x_3}{\partial \eta_3} \end{bmatrix} \begin{Bmatrix} \dfrac{\partial p}{\partial x_1} \\[2mm] \dfrac{\partial p}{\partial x_2} \\[2mm] \dfrac{\partial p}{\partial x_3} \end{Bmatrix} \tag{4.16}$$

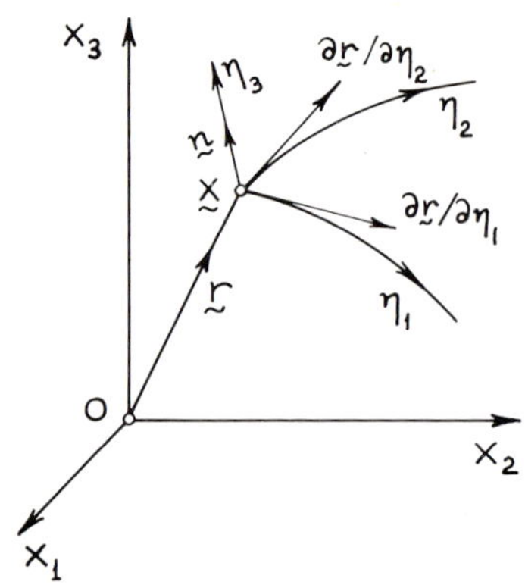

Fig. 12. Coordinate systems for the surface of a three-dimensional domain.

or in compact form

$$\left\{\frac{\partial p}{\partial \eta_i}\right\} = [J]\left\{\frac{\partial p}{\partial x_i}\right\}, \tag{4.17}$$

where $[J]$ is the Jacobian matrix. From (4.17) one has

$$\left\{\frac{\partial p}{\partial x_i}\right\} = [J]^{-1}\left\{\frac{\partial p}{\partial \eta_i}\right\}. \tag{4.18}$$

Thus, the differentials of volume dV and surface dS can now be written as

$$\mathrm{d}V = \left|\left(\frac{\partial \boldsymbol{r}}{\partial \eta_1} \times \frac{\partial \boldsymbol{r}}{\partial \eta_2}\right) \cdot \frac{\partial \boldsymbol{r}}{\partial \eta_3}\right| \mathrm{d}\eta_1\, \mathrm{d}\eta_2\, \mathrm{d}\eta_3 = |J|\, \mathrm{d}\eta_1\, \mathrm{d}\eta_2\, \mathrm{d}\eta_3\,, \tag{4.19}$$

$$\mathrm{d}S = \left|\frac{\partial \boldsymbol{r}}{\partial \eta_1} \times \frac{\partial \boldsymbol{r}}{\partial \eta_2}\right| \mathrm{d}\eta_1\, \mathrm{d}\eta_2 = |\bar{J}|\, \mathrm{d}\eta_1\, \mathrm{d}\eta_2\,, \tag{4.20}$$

where

$$|\bar{J}| = (J_1^2 + J_2^2 + J_3^2)^{1/2}$$

$$J_1 = \frac{\partial x_2}{\partial \eta_1}\frac{\partial x_3}{\partial \eta_2} - \frac{\partial x_2}{\partial \eta_2}\frac{\partial x_3}{\partial \eta_1}, \qquad J_2 = \frac{\partial x_3}{\partial \eta_1}\frac{\partial x_1}{\partial \eta_2} - \frac{\partial x_1}{\partial \eta_1}\frac{\partial x_3}{\partial \eta_2} \tag{4.21}$$

$$J_3 = \frac{\partial x_1}{\partial \eta_1}\frac{\partial x_2}{\partial \eta_2} - \frac{\partial x_2}{\partial \eta_1}\frac{\partial x_1}{\partial \eta_2}.$$

Knowledge of the expression (4.29) for the differential area dS permits one to write the integrals

$$\int_{S_j} p u_n^*\, \mathrm{d}S = \int_{S_j} p u_n^* |\bar{J}|\, \mathrm{d}\eta_1\, \mathrm{d}\eta_2 \tag{4.22}$$

$$\int_{S_j} f u^*\, \mathrm{d}S = \int_{S_j} f u^* |\bar{J}|\, \mathrm{d}\eta_1\, \mathrm{d}\eta_2 \tag{4.23}$$

and subsequently evaluate them numerically for the case of $r \neq 0$ [45, 65] and analytically–numerically for the case of $r \to 0$ as described in Ref. [20]. Cristescu and Loubignac [46] and Pina et al. [151] have presented numerical quadrature formulae for integration over triangles and squares for the case of $1/r$ singularity. Of course, the singular integrals of the form (4.22) augmented with the term $c_i p_i$ can also be obtained indirectly with the aid of the notion of uniform potential as discussed in the previous subsection.

Consider first the simple quadratic quadrilateral element of Fig. 13(a) with four corner nodal points for which p and f can be expressed as

$$p = \phi_1 p_1 + \phi_2 p_2 + \phi_3 p_3 + \phi_4 p_4$$
$$f = \phi_1 f_1 + \phi_2 f_2 + \phi_3 f_3 + \phi_4 f_4 \tag{4.24}$$

where the shape functions ϕ_i $(i = 1, 2, 3, 4)$ are given by

$$\phi_1 = \tfrac{1}{4}(1 - \eta_1)(1 - \eta_2)\,, \qquad \phi_2 = \tfrac{1}{4}(1 + \eta_1)(1 - \eta_2)\,,$$
$$\phi_3 = \tfrac{1}{4}(1 + \eta_1)(1 + \eta_2)\,, \qquad \phi_4 = \tfrac{1}{4}(1 - \eta_1)(1 + \eta_2)\,, \tag{4.25}$$

p_i and f_i are the values of p and f at the nodes $i = 1, 2, 3, 4$ and η_1 and η_2 are dimensionless coordinates shown in Fig. 13(a). The above shape functions can also be used to describe the geometry of the element as

$$x_i = \phi_1 x_i^1 + \phi_2 x_i^2 + \phi_3 x_i^3 + \phi_4 x_i^4\,; \qquad i = 1, 2, 3, 4 \tag{4.26}$$

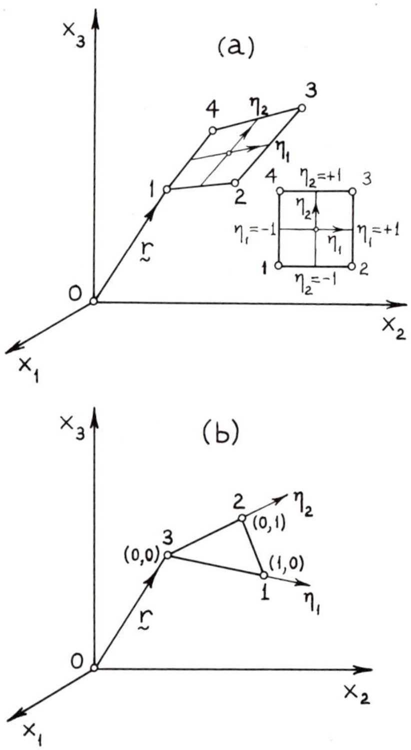

Fig. 13. Surface higher-order boundary elements: (a) quadratic quadrilateral, and (b) linear triangular element.

making it an isoparametric one. Knowledge of the relations (4.26) permits one to compute the values of $|\bar{J}|$ and $|J|$ from (4.21) and (4.17), respectively. Quadrilaterals of higher order than quadratic can also be constructed by introducing additional nodes along the sides of the element. The interested reader can consult Refs. [20, 31] for more details.

Now consider the simple linear triangular element of Fig. 13(b) with three corner nodal points for which p and f can be expressed as

$$p = p_1\eta_1 + p_2\eta_2 + p_3\eta_3 = \sum_{i=1}^{3} p_i\phi_i\,,$$
$$f = f_1\eta_1 + f_2\eta_2 + f_3\eta_3 = \sum_{i=1}^{3} f_i\phi_i\,, \tag{4.27}$$

where p_i and f_i are the values of p and f at the nodes $= 1, 2, 3$ and $\eta_1, \eta_2, \eta_3 = 1 - \eta_1 - \eta_2$ are the triangular coordinates of Fig. 13(b) which are used here also as shape functions. The geometry of the element is also described by the same shape functions, i.e.

$$x_i = x_i^1\eta_1 + x_i^2\eta_2 + x_i^3\eta_3\,; \qquad i = 1, 2, 3\,, \tag{4.28}$$

rendering it isoparametric. Introduction of additional nodes along the sides of the triangular element results in higher order than linear triangular elements which sometimes require the establishment of one interior node at their centroid for maintaining the polynomial completeness and assuring even treatment of all directions. More on the subject of higher-order triangular elements can be found in Refs. [20, 31].

4.3. Examples and applications

Two examples dealing with the use of higher-order elements in potential theory are presented in this subsection together with a list of references describing various applications on the subject.

4.3.1. The magnetic field in a transformer core

This example taken from Wu et al. [191] deals with the determination of the magnetic field in a transformer core shown in Fig. 14(a). The current density i within a coil is a source of magnetic potential A, which satisfies the Poisson equation (1.52), i.e. $\nabla^2 A = -\mu_0 i$ throughout the coil area, where μ_0 is the permebility of free space. In the air surrounding the coils, $\nabla^2 A = 0$. The core permeability is assumed to be infinite. Because of the symmetry only one quadrant of the transformer has to be considered, which is shown in detail in Fig. 14(b), where α_1 and α_2 represent coil areas with current densities i_1 and i_2, while

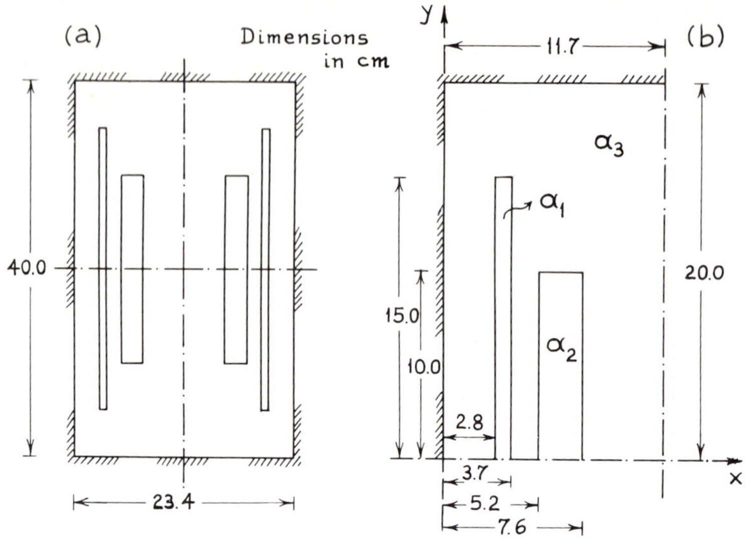

Fig. 14. Magnetic field in a transformer core: (a) complete transformer window, and (b) region of interest.

α_3 stands for the air area. Because the net current flow is zero the condition $i_1\alpha_1 + i_2\alpha_2 = 0$ has to be satisfied. The boundary-value problem that was solved in [191] is then stated as follows: A function is sought which satisfies $\nabla^2 A = 0$ in α_3, $\nabla^2 A = -1$ in α_1 and $\nabla^2 A = \alpha_1/\alpha_2$ in α_2 and the boundary condition $\partial A/\partial n = 0$ on the outermost rectangular boundary. Across the interfaces between the regions the usual continuity of A and $\partial A/\partial n$ must be enforced. The last two governing equations are of the Poisson type but they can be easily treated, as explained in the next section, in a manner analogous to that used for the Laplace equation because their right-hand side is a constant which can be considered a degenerate harmonic function. This problem was solved in [191] by three different direct BEM discretizations. BEM-1 and BEM-2 had quadratic boundary and interface elements with 26 and 38 nodes, respectively, while the BEM-3 discretization had constant elements with 47 nodes. Table 3 provides the maximum and average percentage errors in A at the nodes as well as the x and y components of the force F acting on each conductor computed by the three BEMs together with their

Table 3
Error analysis and comparison of results for the electromagnetic problem of Example 4.3.1.

	BEM-1	BEM-2	BEM-3	Analytical
max. error (%)	13.7	2.4	4.5	–
aver. error (%)	5.5	0.7	2.0	–
F_x in α_1 (N/m)	4.83	4.43	4.46	4.46
F_y in α_1 (N/m)	1.17	1.41	1.34	1.42
F_x in α_2 (N/m)	5.50	4.84	4.71	4.76
F_y in α_2 (N/m)	2.29	2.55	2.48	2.54

exact analytical values. The errors reported are relative errors based upon the maximum value of A within the entire region. The averge error is an average of error magnitudes. The forces $(F_x)_k$ and $(F_y)_k$ for the coil area α_k are computed from the formula $(F_i)_k = i_k \int_{S_k} An_i \, ds$, where S_k is the boundary of α_k. It is apparent from Table 3 that the values of the force F for all three BEMs are in very good agreement with the analytical results. It is interesting to observe that the potential results of the BEM-2 discretization employing quadratic elements enjoy a higher degree of accuracy than those of the constant element BEM-3 discretization, even though the results for the forces appear to be comparable due to the fact that the integration process required to obtain the forces from the potential smooths out the error.

4.3.2. Axisymmetric potential fluid flow around bodies

Hess [79] studied the effectiveness of curved surface elements with variable source densities over them in connection with the problem of the axisymmetric potential fluid flow around various axisymmetric bodies in an indirect BEM framework. This is an exterior Neumann problem involving an infinite domain for which a discussion is given in the next section. Part of his results is shown in Fig. 15, which gives the error in calculated surface velocity on an 8-to-1 prolate spheroid in a uniform onset flow parallel to the x-axis for two discretizations. The first discretization employs constant surface elements, while the second higher-order one uses parabolic surface elements with a parabolic variation of the source over them. Actually, owing to the axisymmetry, these curved elements are essentially line elements. Fig. 15 clearly demonstrates the fact that a higher degree of accuracy can be achieved by fewer higher-order elements in comparison with constant elements.

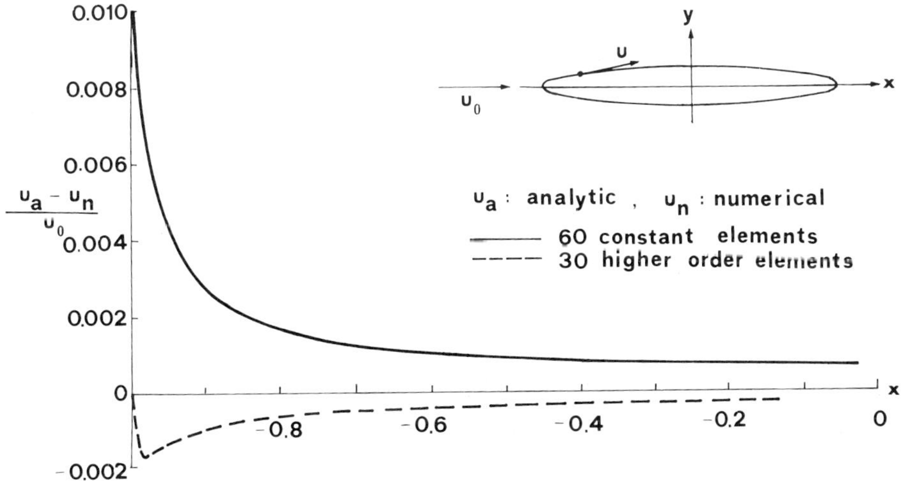

Fig. 15. Errors in calculated surface velocity on an 8-to-1 prolate spheroid in a uniform onset flow parallel to the x axis for two discretizations.

4.3.3. Various applications

Among the many references employing higher-order elements in analyzing potential problems by the BEM one can mention here Refs. [8, 63, 76, 79, 25, 2] in connection with potential fluid flow, Refs. [165, 125, 140, 191, 113] in connection with electrical engineering problems, Refs. [114, 115, 119] in connection with groundwater flow and Refs. [2, 57, 86, 192, 15, 16] in connection with various other applications.

5. Special problems

The BEM presented in the previous sections for two- and three-dimensional potential problems is subject to a number of important restrictions, such as absence of internal sources, closed, homogeneous, isotropic and simply connected domains, and simple geometry and boundary conditions. All these restrictive assumptions are relaxed in this section and the necessary modifications and/or additions to the BEM required for handling these special problems are reviewed. In addition, the last subsection discusses modern software developments and emphasizes the importance of adaptive procedures.

5.1. Internal sources

The presence of known, continuous or discrete sources $g(\boldsymbol{x})$ in the interior of the domain V essentially leads to a non-homogeneous Laplace equation or, as it is usually known, to a Poisson equation of the form

$$\nabla^2 p(\boldsymbol{x}) = g(\boldsymbol{x}) \tag{5.1}$$

One way of solving Eq. (5.1) is by reducing it to a Laplace equation through a subtraction of a particular solution independent of the boundary conditions. For example, consider the formulation (1.60) of the torsion problem involving Poisson's equation. Assuming its solution to be

$$\omega = \omega_1 + \omega_2 , \tag{5.2}$$

where ω_1 is a particular solution of $(1.60)_1$ of the form

$$\omega_1 = -\tfrac{1}{2}(x_1^2 + x_2^2) , \tag{5.3}$$

the problem is then reduced to the determination of the complementary solution ω_2 which satisfies the system

$$\begin{aligned} &\nabla^2 \omega_2 = 0 && \text{in } R , \\ &\omega_2 = -\omega_1 = \tfrac{1}{2}(x_1^2 + x_2^2) && \text{on } C , \end{aligned} \tag{5.4}$$

which is recognized to be identical to (1.58) and involves the Laplace equation $(5.4)_1$.

This approach, however, is difficult to apply in cases where $g(\boldsymbol{x})$ is defined pointwise, or is given by complicated continuous expressions, even though a numerical evaluation of particular solutions is also possible [14]. The general approach is to develop a boundary integral equation directly on the basis of (5.1). Indeed application of the weighted residual statement on (5.1) supplemented by the boundary conditions (1.2) and (1.3) yields

$$\begin{aligned}&\int_V (\nabla^2 p(\boldsymbol{x}) - g(\boldsymbol{x}))u^*(\boldsymbol{\xi}, \boldsymbol{x})\,\mathrm{d}V(\boldsymbol{x})\\ &\quad = \int_{S_2} [f(\boldsymbol{X}) - \bar{f}(\boldsymbol{X})]u^*(\boldsymbol{\xi}, \boldsymbol{X})\,\mathrm{d}S(\boldsymbol{X}) - \int_{S_1} [p(\boldsymbol{X}) - \bar{p}(\boldsymbol{X})]u_n^*(\boldsymbol{\xi}, \boldsymbol{X})\,\mathrm{d}S(\boldsymbol{X}),\end{aligned} \tag{5.5}$$

which after two integrations by parts and taking $\boldsymbol{\xi}$ to the boundary reduces to

$$\begin{aligned}&c(\boldsymbol{\Xi})p(\boldsymbol{\Xi}) + \int_S p(\boldsymbol{X})u_n^*(\boldsymbol{\Xi}, \boldsymbol{X})\,\mathrm{d}S(\boldsymbol{X}) + \int_V g(\boldsymbol{x})u^*(\boldsymbol{\Xi}, \boldsymbol{x})\,\mathrm{d}V(\boldsymbol{x})\\ &\quad = \int_S f(\boldsymbol{X})u^*(\boldsymbol{\Xi}, \boldsymbol{X})\,\mathrm{d}S(\boldsymbol{X}),\end{aligned} \tag{5.6}$$

which is the same with (2.18) apart from the volume integral

$$I_\mathrm{i} = \int_V g(\boldsymbol{x})u^*(\boldsymbol{\Xi}, \boldsymbol{x})\,\mathrm{d}V(\boldsymbol{x}). \tag{5.7}$$

This integral, however, involves only known functions in its integrand and its evaluation presents no problems. For the purpose of numerically evaluating I_i the interior domain V is discretized into a finite number N_c of internal cells, triangular and/or quadrilateral for two dimensions and tetrahedral and/or hexahedral for three dimensions, as shown in Fig. 16, and regular numerical quadrature is

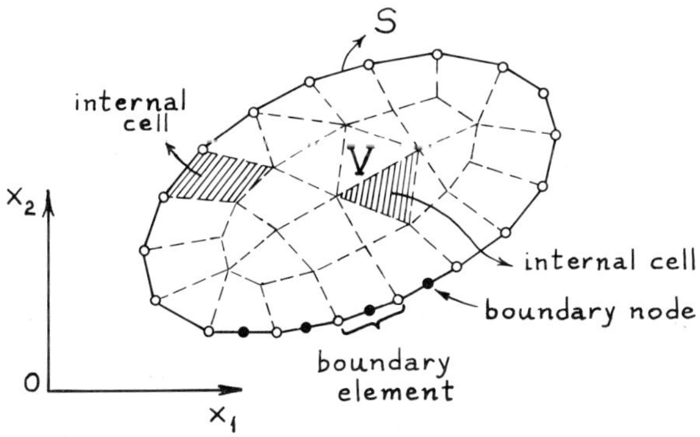

Fig. 16. Discretization of the interior of a domain V into a number of internal cells.

employed. Thus, for two dimensions I_i takes the form

$$I_i = \int_V gu^* \, dV = \sum_{l=1}^{N_c} \left(\sum_{k=1}^{K} (gu^*)_k w_k \right) A_l , \tag{5.8}$$

where w_k is the weight at the kth integration point and A_l is the area of the lth cell. A more general way for computing I_i, which is the rule in three dimensions, is to assign nodes to the cells, express the behaviour of the cells by higher-order interpolation functions defined in terms of local special coordinates and perform the integrations as described in section 4. More on this subject can be found in Ref. [20]. For the computation of the function f, which can be obained by differentiation of Eq. (5.6) before it is brought to the boundary, the evaluation of the integral

$$\bar{I}_i = \int_V g(\boldsymbol{x}) u_n^*(\boldsymbol{\xi}, \boldsymbol{x}) \, dV \tag{5.9}$$

is also necessary and can be accomplished in the same manner as that for I_i.

In the special case of M discrete internal sources of the form

$$g_m = G_m \delta_m ; \qquad m = 1, 2, \ldots, M \tag{5.10}$$

where δ_m is the Dirac delta function one simply has

$$I_i = \sum_{m=1}^{M} (G_m u_m^*) . \tag{5.11}$$

Another special case discussed in Ref. [57] is the one for which $g(\boldsymbol{x})$ is a harmonic function, i.e.

$$\nabla^2 g(\boldsymbol{x}) = 0 . \tag{5.12}$$

Then, if a function v^* can be computed such that

$$\nabla^2 v^* = u^* , \tag{5.13}$$

Green's second theorem (1.20) with $u = g$ and $v = v^*$ yields

$$\int_V gu^* \, dV = \int_S (g v_n^* - v^* g_n) \, dS , \tag{5.14}$$

thereby reducing the volume into a surface integral. One function v^* that satisfies (5.13) is the [57]

$$v^* = (r^2/8\pi)[\ln(1/r) + 1] . \tag{5.15}$$

5.2. Infinite and semi-infinite regions

The Green's second theorem (1.20) and the boundary integral equation (2.18) have been derived on the assumption that the region V is a finite simply-connected region bounded by a regular surface S (interior problem case). Here these fundamental equations are extended to infinite regions $\bar{V}$ external to the regular surface S (exterior problem case).

Thus it can be proven [48] that for an infinite region $\bar{V}$ external to the regular surface S Green's second theorem takes the form

$$\int_{\bar{V}} (u\nabla^2 v - v\nabla^2 u)\,\mathrm{d}\bar{V} = \int_S \left(u \frac{\mathrm{d}v}{\mathrm{d}n} - v \frac{\mathrm{d}u}{\mathrm{d}n}\right) \mathrm{d}S \tag{5.16}$$

provided that the functions u and v are regular ones, i.e. they satisfy the regularity conditions at infinity $u, v = \mathrm{O}(1/R)$, meaning that they asymptotically behave as $\mathrm{O}(1/R)$ for $R \to \infty$.

Regarding the boundary integral equation (2.18) one can prove [31] that the same equation holds true on the surface S internal of an infinite regular region $\bar{V}$, provided that $p(\boldsymbol{x}) = \mathrm{O}(1/R)$ for three dimensions and $p(\boldsymbol{x}) = \mathrm{O}(\ln R)$ for two dimensions as $R \to \infty$. The same is also true for the integral equation (2.17) written for points inside the infinite region $\bar{V}$. Consider now the Neumann problem defined over the infinite domain $\bar{V}$. It can be proven [31] that the Gauss condition (2.10) is satisfied if and only if $p(\boldsymbol{x}) = \mathrm{O}(R^{-2})$ for three dimensions and $p(\boldsymbol{x}) = \mathrm{O}(R^{-1})$ for two dimensions as $R \to \infty$.

It should be finally mentioned that for infinite regions it is also possible to evaluate the diagonal singular coefficients of the matrix $[G^n]$ of Eq. (4.7) with the aid of the relation (4.12) without forgetting that the regularity conditions at infinity are violated as the function p is now assumed to be constant in the whole region V. In this case it can be shown [31] that

$$G^n_{ii} = -\sum_{\substack{j=1\\ i\neq j}}^{N} G^n_{ij} + 1\,; \qquad i = 1, 2, \ldots, N\,. \tag{5.17}$$

The fundamental solution utilized in the BEM formulation is usually assumed to be Green's function for the infinite two- or three-dimensional space, i.e. the function $u^*(\boldsymbol{x}, \boldsymbol{\xi})$ satisfying Eq. (1.14) and subject to no boundary conditions. Thus, in an actual problem to be solved by the BEM the satisfaction of the boundary conditions requires discretization of the boundary. In those portions of the boundary surface, however, where the Green's function satisfies such boundary conditions that can result in the elimination of the surface integrals containing unknown boundary data in the boundary integral equation (2.18), no discretization is required. If such a Green's function can be constructed for the whole boundary, then that approach becomes the classical Green's function

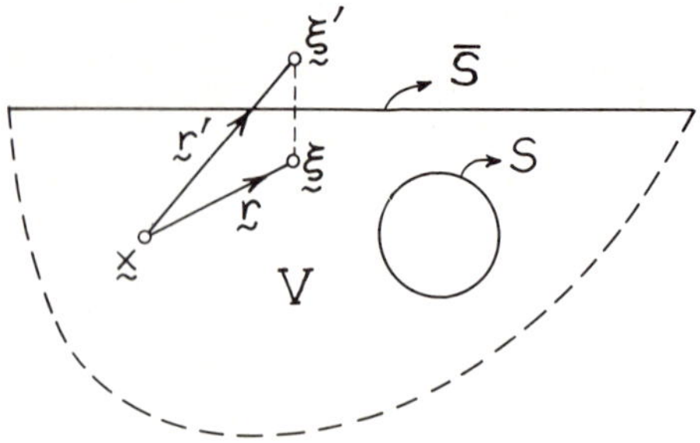

Fig. 17. Geometry of a problem involving a half-plane domain.

method [48, 62], which unfortunately is only applicable to very simple geometries such as spheres, cylinders and half-planes. In many potential problems dealing with a semi-infinite region V with a surface $S \cup \bar{S}$, such as the one depicted in Fig. 17, use of a Green's function u^* which becomes zero on $\bar{S}$ does not require any discretization of the infinite surface $\bar{S}$ and the discretization is restricted to the finite surface S. The Green's functions satisfying the conditions $u^* = 0$ or $u_n^* = 0$ on the surface of a semi-infinite region can be easily obtained by the method of images for two- and three-dimensional problems [62, 31]. For example, the Green's functions satisfying $u^* = 0$ and $u_n^* = 0$ on the boundary of the half-plane have, respectively, the form [31]

$$u^*(\boldsymbol{x}, \boldsymbol{\xi}) = (1/2\pi)\ln(r'/r) , \tag{5.18}$$

$$u^*(\boldsymbol{x}, \boldsymbol{\xi}) = (1/2\pi)\ln(rr') , \tag{5.19}$$

where $\boldsymbol{r} = \boldsymbol{\xi} - \boldsymbol{x}$, $\boldsymbol{r}' = \boldsymbol{\xi}' - \boldsymbol{x}$ and $\boldsymbol{\xi}'$ is symmetric to $\boldsymbol{\xi}$ with respect to $\bar{S}$, as shown in Fig. 17.

5.3. Anisotropic regions

In all the previous developments, the assumption has been made that the domain of definition V of the problem was isotropic. There are problems, however, characterized by domain anisotropy and this subsection briefly describes the treatment of these problems by the BEM. The governing equation for the steady-state potential flow of a three-dimensional anisotropic medium has the form

$$k_{ij} \frac{\partial^2 p(\boldsymbol{x})}{\partial x_i \partial x_j} = 0 ; \qquad i, j = 1, 2, 3 \tag{5.20}$$

where the summation convention is assumed over repeated indices and where k_{ij} is a second-rank tensor independent of $\boldsymbol{x}$ with a physical interpretation depending on the particular problem. For example, in groundwater flow k_{ij} is the permeability tensor.

Equation (5.20) can, of course, be treated by the general method of Clements and Rizzo [43] for second-order elliptic systems. In this subsection, however, a much simpler and more elegant way will be described for treating (5.20). The idea is to reduce the anisotropic case of Eq. (5.20) to an isotropic one by appropriate transformations of coordinates. Indeed, expressing Eq. (5.20) in terms of a new coordinate system with its axes y_i $(i = 1, 2, 3)$ directed along the principal axes or axes of anisotropy k_1, k_2 and k_3 of k_{ij} yields

$$k_1 \frac{\partial^2 p}{\partial y_1^2} + k_2 \frac{\partial^2 p}{\partial y_2^2} + k_3 \frac{\partial^2 p}{\partial y_3^2} = 0 . \tag{5.21}$$

Introduction of new scaled coordinates z_i such that

$$z_1 = \frac{y_1}{\sqrt{k_1}} , \qquad z_2 = \frac{y_2}{\sqrt{k_2}} , \qquad z_3 = \frac{y_3}{\sqrt{k_3}} , \tag{5.22}$$

finally reduces (5.21) to

$$\frac{\partial^2 p}{\partial z_1^2} + \frac{\partial^2 p}{\partial z_2^2} + \frac{\partial^2 p}{\partial z_3^2} = \nabla^2 p = 0 , \tag{5.23}$$

which is the Laplace equation.

The above transformation of coordinates approach can alternatively be used to analytically determine the Green's function for Eq. (5.20) in anisotropic media. Thus one has [38]

$$u^*(\boldsymbol{x}, \boldsymbol{\xi}) = \frac{1}{|k_{ij}|^{1/2}} \ln \frac{1}{r(\boldsymbol{x}, \boldsymbol{\xi})} \tag{5.24}$$

$$u^*(\boldsymbol{x}, \boldsymbol{\xi}) = \frac{1}{|k_{ij}|^{1/2}} \frac{1}{r(\boldsymbol{x}, \boldsymbol{\xi})} , \tag{5.25}$$

for two and three dimensions, respectively, where $|k_{ij}|$ is the determinant of the tensor k_{ij} and

$$r(\boldsymbol{x}, \boldsymbol{\xi}) = \left[\frac{1}{k_{ij}} (x_i - \xi_i)(x_j \quad \zeta_j) \right]^{1/2} , \tag{5.26}$$

with $i, j = 1, 2$ for two and $i, j = 1, 2, 3$ for three dimensions, respectively, and where the summation convention over repeated indices is assumed in (5.26). The function f for anisotropic media has the form

$$f(\boldsymbol{x}) = \frac{\partial p(\boldsymbol{x})}{\partial n} = k_{ij} \cos(\boldsymbol{n}, x_i) \frac{\partial p(\boldsymbol{x})}{\partial x_j} , \tag{5.27}$$

with the same as in (5.26) definition for the indices i and j. A complete formulation and solution of the anisotropic problem based on Eqs. (5.24)–(5.27) can be found in Chang et al. [38].

5.4. *Inhomogeneous regions*

When the physical properties of a medium, as for example the permeabilities in a groundwater problem, change with position, the medium is called inhomogeneous and the previously defined methodology for homogeneous media cannot be applied without some modifications. In general, one has to deal with continuous and discrete or piecewise inhomogeneous domains. In the first case the physical properties vary continuously from point to point in the medium, while in the second case the medium consists of a finite number of homogeneous subregions simply or multiply connected to each other, as shown in Fig. 18.

Potential problems for continuously inhomogeneous domains are very difficult to solve because of the lack of knowledge of the associated inhomogeneous Green's functions. Butterfield [34] has shown that employment of the corresponding homogeneous Green's function in a direct BEM formulation of the continuously inhomogeneous problem can lead to its solution with the aid of an iterative procedure. Of course, a continuously inhomogeneous domain can also be treated in an approximate way as a piecewise inhomogeneous one.

A piecewise inhomogeneous domain can easily be treated by the methodology developed in the previous sections which now has to be applied separately on every homogeneous subregion. The resulting equations for every subregion are appropriately combined through equilibrium and compatibility at their interfaces to form the equations for the whole inhomogeneous domain.

This procedure is illustrated by the two-dimensional problem of Fig. 18(c)

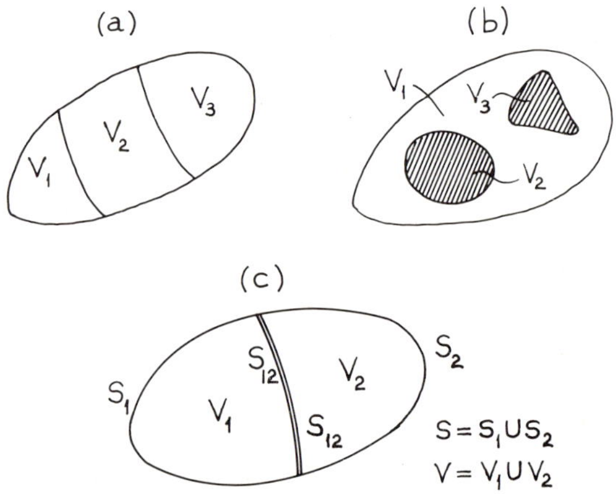

Fig. 18. Piecewise inhomogeneous domains: (a) simply connected and, (b) multiply connected domains; (c) domain consisting of two subdomains V_1 and V_2.

concerning a region V with surface $S = S_1 \cup S_2$ consisting of two subregions V_1 and V_2 with corresponding surfaces $S_1 \cup S_{12}$ and $S_2 \cup S_{12}$, where S_{12} is their interface. The boundary element equation (2.31) is utilized for both regions V_1 and V_2 to produce the equations

$$[\boldsymbol{G}^{n1} \quad \boldsymbol{G}_{12}^{n1}]\begin{Bmatrix} \boldsymbol{p}^1 \\ \boldsymbol{p}_{12}^1 \end{Bmatrix} = [\boldsymbol{G}^1 \quad \boldsymbol{G}_{12}^1]\begin{Bmatrix} \boldsymbol{f}^1 \\ \boldsymbol{f}_{12}^1 \end{Bmatrix}, \tag{5.28}$$

$$[\boldsymbol{G}^{n2} \quad \boldsymbol{G}_{12}^{n2}]\begin{Bmatrix} \boldsymbol{p}^2 \\ \boldsymbol{p}_{12}^2 \end{Bmatrix} = [\boldsymbol{G}^2 \quad \boldsymbol{G}_{12}^2]\begin{Bmatrix} \boldsymbol{f}^2 \\ \boldsymbol{f}_{12}^2 \end{Bmatrix}, \tag{5.29}$$

where superscripts 1 and 2 refer to S_1 and S_2, respectively, while subscript 12 refers to the interface S_{12}. The equations of compatibility and equilibrium at the interface S_{12}

$$\boldsymbol{p}_{12}^1 = \boldsymbol{p}_{12}^2 = \boldsymbol{p}_{12}\,, \tag{5.30}$$

$$\boldsymbol{f}_{12}^1 = -\boldsymbol{f}_{12}^2 = \boldsymbol{f}_{12}\,, \tag{5.31}$$

are then introduced into Eqs. (5.28) and (5.29) which are subsequently combined and take the form [29, 145, 31]

$$\begin{bmatrix} \boldsymbol{G}^{n1} & \boldsymbol{G}_{12}^{n1} & 0 \\ 0 & \boldsymbol{G}_{12}^{n2} & \boldsymbol{G}^{n2} \end{bmatrix}\begin{Bmatrix} \boldsymbol{p}^1 \\ \boldsymbol{p}_{12} \\ \boldsymbol{p}^2 \end{Bmatrix} = \begin{bmatrix} \boldsymbol{G}^1 & \boldsymbol{G}_{12}^1 & 0 \\ 0 & -\boldsymbol{G}_{12}^2 & \boldsymbol{G}^2 \end{bmatrix}\begin{Bmatrix} \boldsymbol{f}^1 \\ \boldsymbol{f}_{12} \\ \boldsymbol{f}^2 \end{Bmatrix} \tag{5.32}$$

which is a matrix equation involving banded matrices and not fully populated ones as in (5.28) and (5.29). Depending on the boundary conditions of the problem, Eq. (5.32) is appropriately reordered in order to separate known from unknown values of the potential $\boldsymbol{p}$ and the flux $\boldsymbol{f}$ vectors. Thus, if the potential problem for the composite region V of Fig. 18(c) is a Neumann one, Eq. (5.32) is rearranged as

$$\begin{bmatrix} \boldsymbol{G}^{n1} & \boldsymbol{G}_{12}^{n1} & -\boldsymbol{G}_{12}^1 & 0 \\ 0 & \boldsymbol{G}_{12}^{n2} & \boldsymbol{G}_{12}^2 & \boldsymbol{G}^{n2} \end{bmatrix}\begin{Bmatrix} \boldsymbol{p}_1 \\ \boldsymbol{p}_{12} \\ \boldsymbol{f}_{12} \\ \boldsymbol{p}^2 \end{Bmatrix} = \begin{bmatrix} \boldsymbol{G}^1 & 0 \\ 0 & \boldsymbol{G}^2 \end{bmatrix}\begin{Bmatrix} \boldsymbol{f}^1 \\ \boldsymbol{f}^2 \end{Bmatrix}, \tag{5.33}$$

i.e. with the known $\boldsymbol{f}^1$ and $\boldsymbol{f}^2$ vectors on one side and Eq. (5.33) is solved for the potential vectors $\boldsymbol{p}^1$ and $\boldsymbol{p}^2$ as well as the interface potential and flux vectors $\boldsymbol{p}_{12}$ and $\boldsymbol{f}_{12}$, respectively.

Another approach to coupling two subregions is by transforming Eqs. (5.28) and (5.29) into equivalent finite element type ones and combining them as in the FEM [180, 29, 20]. This is described in detail in section 7.

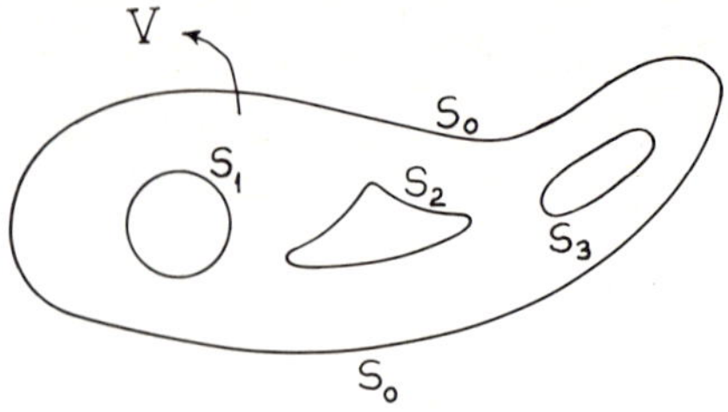

Fig. 19. Multiply connected domain.

The treatment of multiply connected regions like the one of Fig. 19 presents no difficulties. The basic boundary integral equation (2.18) applies here with $S = S_0 \cup S_1 \cup S_2 \cup S_3$. Care should be exercised, however, in ensuring existence and uniqueness of solution for the given boundary-value problem through imposition of appropriate conditions. Thus, for instance, in a Neumann problem the conditions

$$\int_S \bar{f}(X)\,\mathrm{d}S(X) = 0\,; \qquad \int_{S_0} p(X)\,\mathrm{d}S(X) = 0 \tag{5.34}$$

have to be satisfied [178]. For more information on the subject as well as applications the reader can consult Refs. [92, 174, 152, 94, 21, 178, 36, 37, 14].

5.5. Complex geometry and boundary conditions

During the solution of some potential problems special problems might arise due to the complex nature of the surface geometry and/or the boundary conditions. This subsection briefly discusses the treatment of problems of surface geometric discontinuities due to corners or data discontinuities due to abrupt changes in boundary conditions. The present discussion is limited to two dimensions but extension to three dimensions is straightforward.

At a corner point, as the one of Fig. 20(a), even though there is continuity of potential, there is discontinuity of the flux vector, which implies that an additional unknown is created without a corresponding increase in the number of equations. One easy way out of this difficulty is to round-off the corner, as shown in Fig.

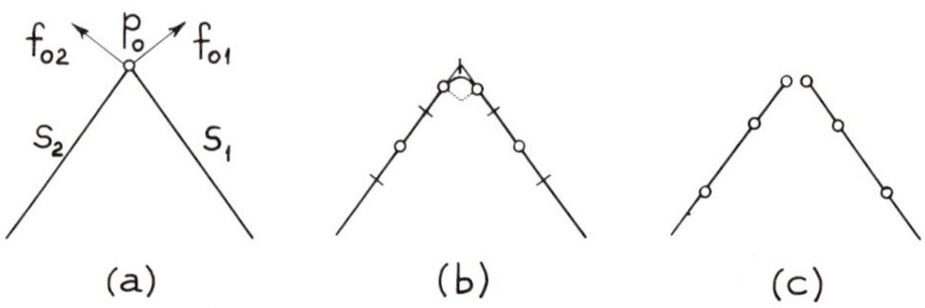

Fig. 20. Discontinuity of flux vector at a corner point of plane domain (a) and treatment by rounding-off the corner (b) or introducing two nodes close to each other (c).

20(b), by introducing two nodal points close to the corner and expect that this approximation will not influence considerably the results at points away from the corner [94]. There are many problems, however, such as those involving re-entrant corners, for which information near the corner is of great importance. Another simple solution to the corner problem is, of course, to assume, with reference Fig. 20(a), that $f_{01}=f_{02}=f_0$. This single-node representation first introduced by Lachet and Watson [111] in elastostatics is again only good for points away from the corner.

Riccardella [157], working in elastostatics, in an effort to resolve this problem of lack of single-valueness at a corner, introduced two nodes close to each other, as shown in Fig. 20(c). The disadvantage of this multiple independent-node representation is that unless a sufficient gap is left between the corner nodes, instabilities occur in the equations produced by the direct BEM. A multiple-node representation supplemented with auxiliary relations was proposed by Alarcón et al. [2, 145]. According to this approach, at a corner node of a Dirichlet problem where the evolution of p is known along the boundary one can approximately compute its corner derivatives $(\partial p/\partial s)_i$, $i=1,2$, before and after the node by finite differences. Projecting these values on the normals before and after the node produces the values $f_i = f\cos\alpha_i$ of the two fluxes f_1 and f_2 at the corner node where α_i is the angle between f_i and the respective normal n_i at the boundary. That way the only unknown quantity at the corner node is f. Another similar technique is the one based on the concept of the crossing element, i.e. a pair of boundary elements intersecting each other at the corner, supplemented by the auxiliary condition $(\partial p/\partial s)_1 + (\partial p/\partial s)_2 = 0$ at the corner in order to take care of the additional unknown introduced by the crossing element [59]. Use of mixed elements with p having a linear variation and f having a constant variation can also solve this overdetermination problem at a corner [132].

In the presence of boundary singularities, i.e. singular behavior of p resulting in infinite values for f due to discontinuous boundary conditions or re-entrant corners different remedies can be utilized in order to obtain good results in the neighborhood of these singular points. Among the existing methods one can mention simple or logarithmic mesh refinement around the singular point [172], introduction of an auxiliary harmonic function [94, 105, 144], asymptotic analysis by extrapolation of results near the singular node [179] and special shape functions including the singularity for the construction of special singular elements [61, 145]. According to the method of Papamichael and Symm [144] an auxiliary harmonic function v is introduced such that $p=p_s+v$, where p_s is a known harmonic function which has the discontinuity properties of the required solution p. Then the BEM is applied to v taking any point of discontinuity in the boundary conditions to be a point of the boundary subdivision. After the determination of v, p can be obtained by adding p_s and v. For example, for the case of a re-entrant corner with interior angle $\alpha > \pi$, as shown in Fig. 21(a), one has

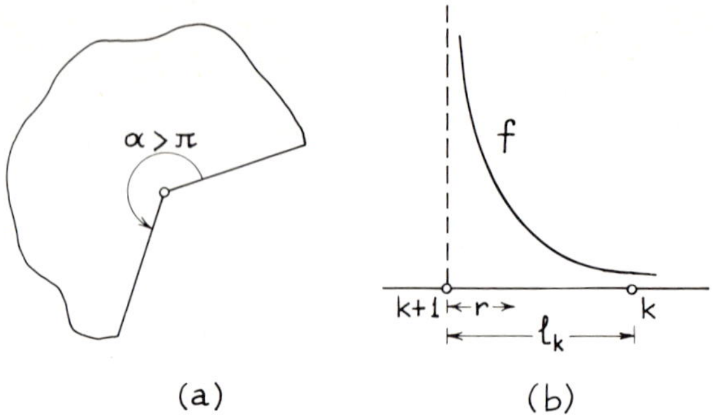

Fig. 21. Singularities due to domain geometry or boundary conditions: (a) re-entrant corner and (b) singular behavior of flux vector at a point with boundary condition discontinuity.

$$\begin{aligned}
p_s(r,\theta) &= A_0 + A_1 r^{\pi/\alpha}\cos(\pi\theta/\alpha) + A_2 r^{2\pi/\alpha}\cos(2\pi\theta/\alpha) + \cdots \\
f_s(r,\theta) &= \frac{1}{r}\frac{\partial p_s}{\partial \theta} \\
&= -A_1(\pi/\alpha) r^{(\pi/\alpha)-1}\sin(\pi\theta/\alpha) - A_2(2\pi/\alpha) r^{(2\pi/\alpha)-1}\sin(2\pi\theta/\alpha) \\
&\quad + \cdots
\end{aligned} \tag{5.35}$$

clearly indicating that $\lim_{r\to 0} f_s \to \infty$. For more details on this method one should look at [144, 86]. The construction of special singular elements for singularity of the type $1/\sqrt{r}$ as shown in Fig. 21(b) and encountered, for example, in cases with discontinuous boundary conditions [61] is accomplished by interpolating $f\sqrt{r}$ in place of f near the singularity, thereby obtaining, e.g., for the case of linear interpolation

$$f_s = [N_1'/\sqrt{r}, N_2/\sqrt{r}]\left\{\begin{matrix} f_k \\ f'_{k+1} \end{matrix}\right\}, \tag{5.36}$$

where N_1 and N_2 are the shape functions without singularity and $N_1' = N_1\sqrt{l_k}$ and $f'_k = l_k f_k$, with r, k, $k+1$ and l_k as shown in Fig. 21(b). More details on this subject of singular elements can be found in Gomez Lera et al. [61] and Paris et al. [145].

The introduction of the so-called discontinuous or non-conforming boundary elements by Patterson and his co-workers [146, 148–150] for two- and three-dimensional potential and elastostatic problems permits one to avoid singularity problems, like the ones mentioned in this subsection. This is possible due to the fact that these higher-order elements with internal nodes, as shown in Fig. 22(a), do not enforce interelement continuity, a property necessary in the FEM but not necessary in the BEM as the success of the constant element demonstrates. Discontinuous elements, however, lead to an increased number of nodal points due to the fact that at any interface of two adjacent elements twice as many nodes

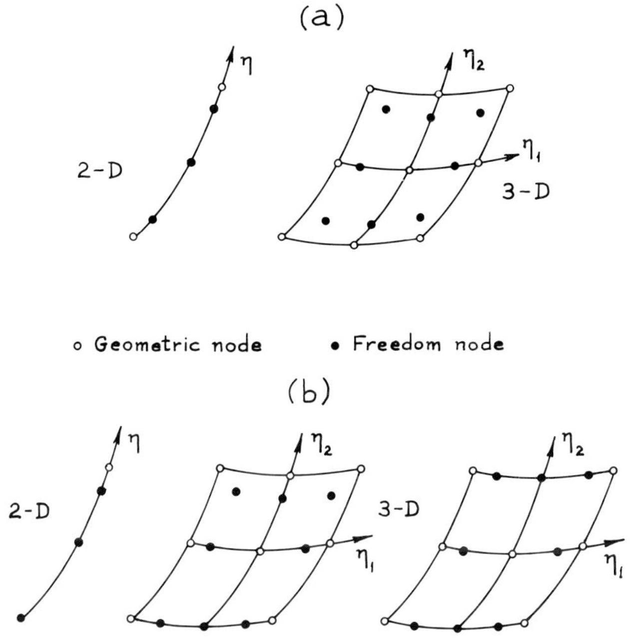

Fig. 22. Line and surface discontinuous elements: (a) completely, and (b) partially discontinuous elements.

are introduced than if these elements were satisfying interelement continuity. For this reason the partially discontinuous elements shown in Fig. 22(b) have been also developed by Patterson et al. [149, 150] so that interelement discontinuity is retained only where it is desirable, e.g. at corners or points where the boundary data is discontinuous.

5.6. Modern software developments

Boundary element techniques are approaching maturity and the emphasis from fundamental developments is nowadays being shifted towards the development of better software in order to achieve higher accuracy and efficiency in treating complex problems. This effort is presently directed towards the development of general-purpose, user-oriented computer programs and adaptive mesh refinement procedures based on error estimation and control. Adaptive procedures in the BEM have recently been introduced in connection with potential problems by Alarcón et al. [5, 6] following an intensive effort undertaken by the FEM people. More details on this can be found in the next chapter of this book. At this point the theoretical studies of Wendland et al. [186, 187, 112] on the asymptotic convergence of solution of the BEM should be also mentioned.

Presently there is a number of simple and more advanced computer programs available for the solution of potential problems [28, 2, 4, 50, 119]. With the appearance of the new supercomputers (e.g. Cray 1, CDC Cyber 205) an effort is under way by people in the FEM community to improve old algorithms by vectorizing them and utilize substructuring in a parallel fashion so as to take full advantage of the new features of the supercomputers. This effort is expected to be shared by people being involved with the BEM.

5.7. Examples and applications

This subsection presents two examples and in addition provides a number of references dealing with the subjects treated in section 5.

5.7.1. Heat conduction in an anisotropic body

This example which is taken from Chang et al. [38] deals with the determination of the steady-state temperature distribution on the cross-section of a hollow anisotropic cylinder which is subjected to constant temperatures at the inner and outer surfaces, as shown in Fig. 23(a). The results of this problem are given by Fig. 23(b) in the form of isotherms for various values of the elements k_{ij} of the thermal conductivity tensor and the polar angles θ which clearly demonstrate the effect of anisotropy.

5.7.2. Heat conduction in a fibrous composite body

In this example taken from Jaworski [97] the steady-state temperature distribution on the cross-section of a unidirectional fibrous composite, as depicted in

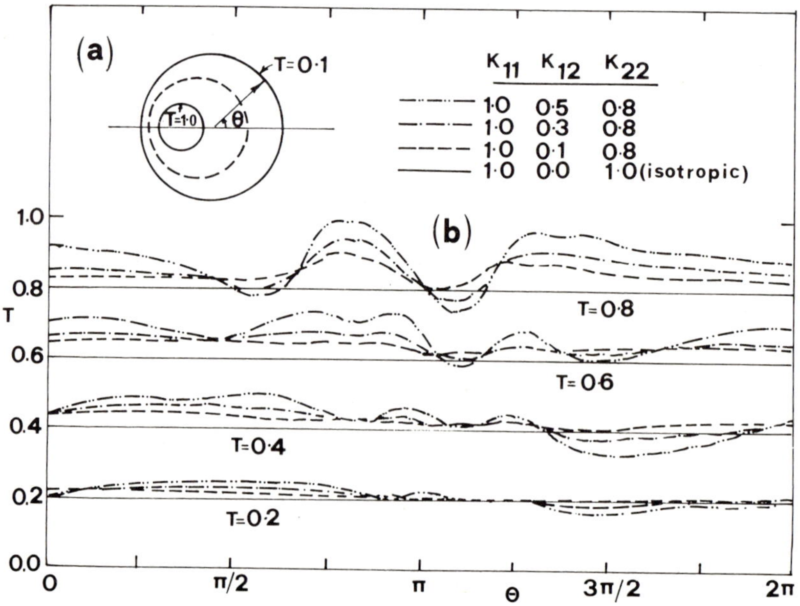

Fig. 23. Isotherms for a hollow anisotropic cylinder for various degrees of anisotropy and values of the polar angle θ.

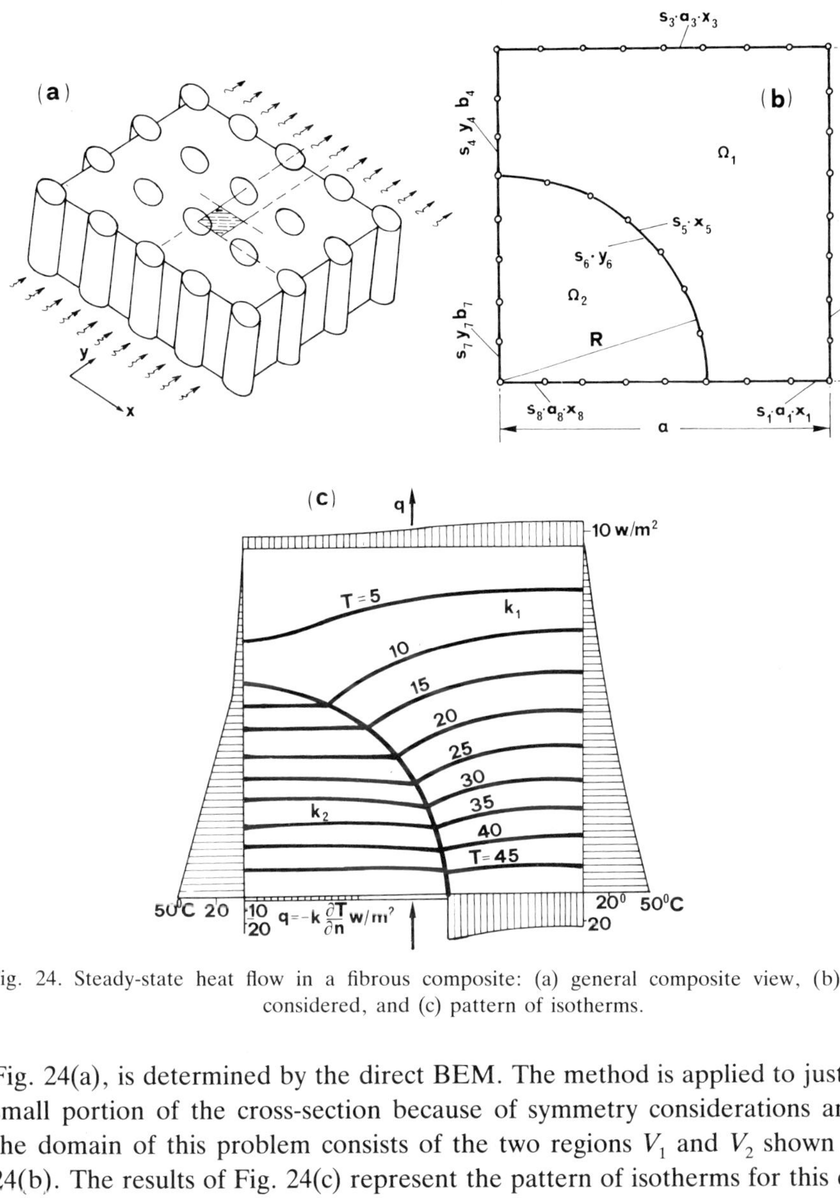

Fig. 24. Steady-state heat flow in a fibrous composite: (a) general composite view, (b) domain considered, and (c) pattern of isotherms.

Fig. 24(a), is determined by the direct BEM. The method is applied to just a very small portion of the cross-section because of symmetry considerations and thus the domain of this problem consists of the two regions V_1 and V_2 shown in Fig. 24(b). The results of Fig. 24(c) represent the pattern of isotherms for this domain when its lower and upper sides are kept at constant temperatures of 50°C and 0°C, respectively, while the lateral sides are thermally insulated and $R/\alpha = 0.6$, $\kappa = k_1/k_2 = 10$, where k_1 and k_2 are the thermal conductivities of regions V_1 and V_2, respectively.

5.7.3. Various applications

There is a considerable body of literature describing applications related to the first five topics considered in section 5. Among those one can mention Refs.

[53, 94, 191, 138, 59, 14] for the problem of internal sources, Refs. [168, 73–79, 191, 80, 84, 142, 128] for the problem of infinite and semi-infinite regions, Refs. [32, 180, 33, 38, 141] for the problem of anisotropic regions, Refs. [92, 174, 152, 32, 180, 33, 94, 24, 34, 21, 178, 36, 37] for the problem of inhomogeneous regions and Refs. [157, 144, 105, 111, 94, 179, 2, 24, 132, 145, 146, 61, 148, 59, 149, 150, 172, 13] for the problem of the complex geometry and boundary conditions. In addition to those references one can also look at the texts [20, 31] and especially [119] for more applications.

6. Nonlinear problems

In many practical potential problems one has to solve the equation

$$\frac{\partial}{\partial x_i}\left(k\frac{\partial p}{\partial x_i}\right) = q \qquad \text{in } V \tag{6.1}$$

subject to boundary conditions

$$p(x_i) = \bar{p}(x_i) \qquad \text{on } S_1 \tag{6.2}$$

$$f(x_i) = -k\frac{\partial p}{\partial x_i} n_i = \bar{f}(x_i) \qquad \text{on } S_2 \tag{6.3}$$

where $S = S_1 \cup S_2$ is the total boundary surface of the domain V, q is the internal source term and summation convention is assumed over repeated indices. The solution of (6.1) subject to the conditions (6.2) and (6.3) under assumptions of linearity can be easily obtained by employing the methodology developed in previous sections. Here attention is focused on the following two kinds of nonlinearities that may arise in potential problems:

(a) The material property k (permeability or conductivity, etc.) and/or the source term q may depend on p or f; this is the case of material nonlinearity.

(b) A part of the boundary S may not be known initially and/or the boundary conditions themselves may be nonlinear; this is the case of nonlinearity due to the boundary conditions.

Review type of articles dealing with the above two types of nonlinearities in potential problems have been prepared by Banerjee [19] and Skerget and Brebbia [167]. These two cases of nonlinearities are discussed in the following two subsections.

6.1. *Material nonlinearities*

Consider the problem defined by Eqs. (6.1)–(6.3) and

$$k = k(p, f), \qquad q = q(p, f) \tag{6.4}$$

and try to find its solution by following [19]. Expressing as

$$k = k_0 + k_n , \tag{6.5}$$

where k_0 is the constant part and $k_n = k_n(p, f)$ one can rewrite Eq. (6.1) in the form

$$k_0 \frac{\partial^2 p}{\partial x_i \partial x_i} = q^* \tag{6.6}$$

where

$$q^* = q - \frac{\partial}{\partial x_i}\left(k_n \frac{\partial p}{\partial x_i}\right) . \tag{6.7}$$

Boundary condition (6.2) remains unchanged while (6.3) changes according to

$$f_0 = -k_0 \frac{\partial p}{\partial x_i} n_i = f + k_n \frac{\partial p}{\partial x_i} n_i . \tag{6.8}$$

Application of the direct BEM on the problem governed by Eq. (6.6) and satisfying the boundary conditions (6.2) and (6.8) yields

$$cp = \int_S (f_0 G - pG'')\,\mathrm{d}S - \int_V \frac{\partial}{\partial x_i}\left(k_n \frac{\partial p}{\partial x_i}\right) G\,\mathrm{d}V + \int_V qG\,\mathrm{d}V . \tag{6.9}$$

Use of the divergence theorem in the first volume integral of (6.9) reduces it to

$$cp = \int_S (fG - pG'')\,\mathrm{d}S + \int_V k_n \frac{\partial p}{\partial x_i}\frac{\partial G}{\partial x_i}\,\mathrm{d}V + \int_V qG\,\mathrm{d}V . \tag{6.10}$$

Discretization of the boundary into a number of boundary elements and the interior of the domain into a number of interior cells permits one to write (6.10) in the matrix form

$$[K]\{z\} = \{b\} - [A]\{Q\} , \tag{6.11}$$

where $\{Q\}$ is a function of p and f and hence $\{z\}$. Solution of Eq. (6.11) is accomplished iteratively by starting initially with $k_n = 0$ and $q = 0$. The solution of (6.11) then provides $\{z\}$ and hence p and f from which k_n and q can be computed in the domain cells and produce values for $\{Q\}$ to be used for the second step solution of (6.11) for $\{z\}$, etc. Treatment of potential problems with material nonlinearities of the form $k = k(p)$ and $q = 0$ has been reported by Khader and Hanna [106], Ikeuchi and Onishi [85] and Ingham and Kelmanson [88, 89]. Problems of compressible flow [124], elastohydrodynamics of lubrication [68] and torsion of an elasto-plastic shaft [136] with $k = k_0$ and $q = q(p)$ have also been formulated and solved by the above iterative method.

Another way of treating material nonlinearities, which is especially recommended for cases where k changes very quickly with p, say exponentially, is through a transformation of coordinates known as Kirchhoff's transformation which succeeds in reducing the problem to a linear one. Indeed consider the case of

$$k = k(p) \tag{6.12}$$

and introduce a new variable $\phi(p)$ such that

$$\frac{\partial \phi}{\partial x_i} = -\frac{\mathrm{d}\phi}{\mathrm{d}p}\frac{\partial p}{\partial x_i} = -k(p)\frac{\partial p}{\partial x_i} \tag{6.13}$$

or, in integral form

$$\phi = -\int_{p_0}^{p} k(p)\,\mathrm{d}p\,, \tag{6.14}$$

where p_0 is a reference value. Equation (6.14) is called the Kirchhoff's transformation and succeeds in transformingf Eqs. (6.1)–(6.3) into the linear equation

$$\nabla^2 \phi = -q \tag{6.15}$$

subject to the boundary conditions

$$\phi = \bar{\phi} = -\int_{p_0}^{\bar{p}} k\,\mathrm{d}p \qquad \text{on } S_1 \tag{6.16}$$

$$\frac{\partial \phi}{\partial n} = -k\frac{\partial p}{\partial n} = \bar{f} \qquad \text{on } S_2\,. \tag{6.17}$$

The above method was first introduced by Bialecki and Nowak [23] who employed it for solving some nonlinear heat conduction problems. Figure 25(a) depicts a square domain with prescribed boundary conditions in terms of the temperature T and its normal derivatives; $T_0 = 300$ K, $T_f = 500$ K, $h = 10$ W/m^{2}K and $k = 1(1 + \alpha T)$ W/mK. The steady-state temperature distribution for this domain, discretized as shown in Fig. 25(a), has been computed in [23] and is portrayed in Figs. 25(b,c) for the linear ($\alpha = 0$) and nonlinear ($\alpha \neq 0$) cases.

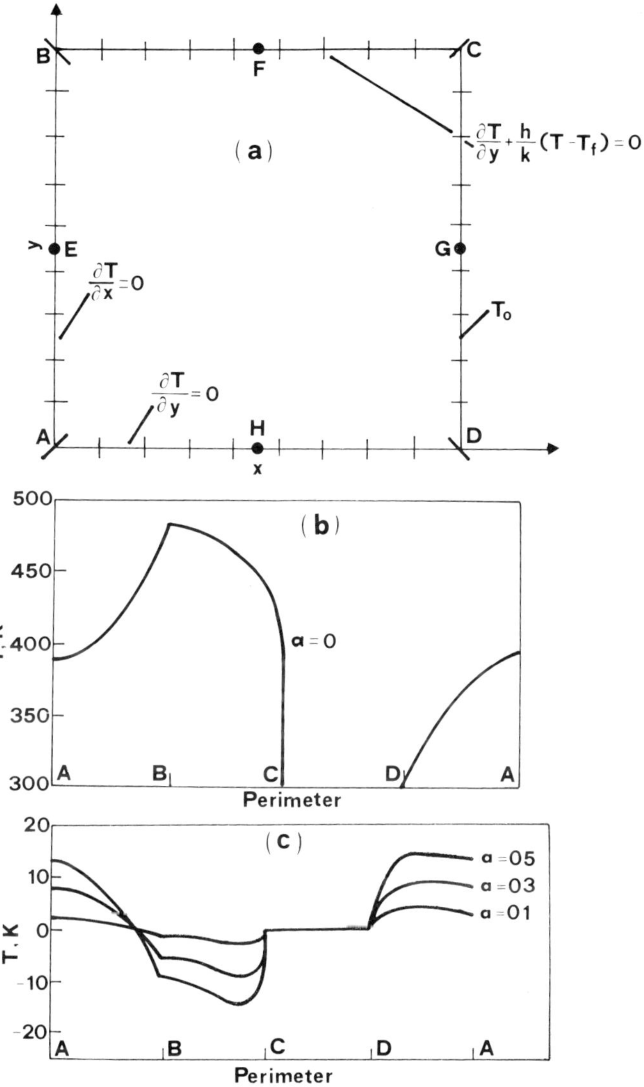

Fig. 25. Steady-state heat conduction in a square domain with temperature varying conductivity: (a) domain discretization and boundary conditions, (b) temperature profiles for various degrees of the material nonlinearity.

6.2. *Nonlinearities due to boundary conditions*

Nonlinearities due to the presence of nonlinear boundary conditions like

$$f = \alpha(p_0 - p) - \beta(p^4 - p_s^4) \tag{6.18}$$

occurring during heat radiation and where α, β, p_0 and p_s are constants, can be treated as explained in [23] by writing (6.18) in the form

$$f = f_0 + f_n\,, \tag{6.19}$$

where f_0 is the linear and f_n is the nonlinear part of f and formulating the problem in terms of the direct BEM as

$$[K]\{z\} = \{b\} - [A]\{Q\} \tag{6.20}$$

where the vector $\{Q\}$ is now a function of p and unlike $\{Q\}$ in Eq. (6.11) contains only surface integrals meaning that no interior cells are necessary. Equation (6.20) is, in general, solved by iteration as previously described, even though it has been observed in [23] that incremental procedures should be preferred for problems with strong nonlinearities since iterative methods often fail to converge in those cases. Similar work on the subject of nonlinear boundary conditions has been reported by Ingham et al. [87], Khader and Hanna [106] and Ingham and Kelmanson [88, 89].

In various problems of fluid flow, such as seepage through porous media with a free surface or flow over a spillway crest, the free surface of the flow is not known initially. Thus, even though these problems are governed by linear equations, they are essentially nonlinear due to their boundary conditions. This kind of problem will be illustrated on the basis of the two-dimensional sluice gate steady-state flow problem analysed in Liu [121]. The problem is described in Fig. 26 and consists of four fixed (bottom FE, sluice gate BC and two lateral boundaries AF and DE) and two initially unknown (free surfaces AB and CD) boundaries. The governing equation of the problem in terms of the stream function reads

$$\nabla^2\psi = 0\,. \tag{6.21}$$

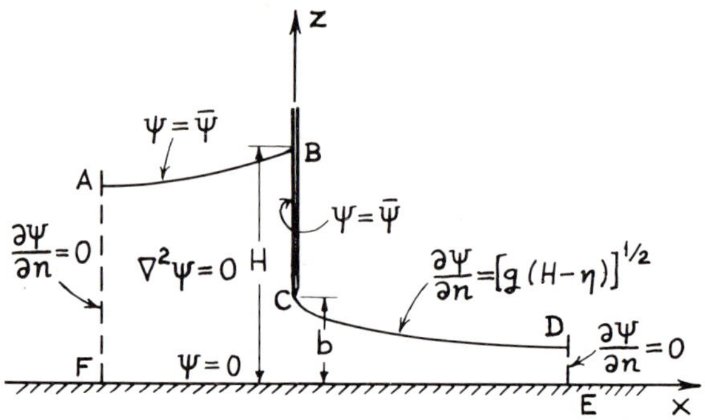

Fig. 26. Geometry and boundry conditions for the two-dimensional sluice gate steady-state flow problem.

The boundary conditions of the problem are

$$\partial\psi/\partial n = 0 \qquad \text{on AF and DE}\,, \tag{6.22}$$

$$\psi = 0 \qquad \text{on EF}\,, \tag{6.23}$$

$$\psi = \bar{\psi} \qquad \text{on AB, BC and CD}\,, \tag{6.24}$$

where $\bar{\psi}$ is the rate of volume discharge of the problem. Along the free surface AB and CD Bernoulli's equation in the form

$$\partial\psi/\partial n = [2g(H-\eta)]^{1/2}\,, \qquad z = \eta(x) \tag{6.25}$$

is satisfied. Clearly this problem is nonlinear because the free surface elevation $z = \eta(x)$ is not known a-priori and $\bar{\psi}$ is also not known. Thus the problem is solved iteratively.

Initially the assumption is made that the free surface AB is the flat one $z = H$, while CD is represented by a quarter of an ellipse with minor axis $0.4b$ and major axis $2H$. Because $\bar{\psi}$ is also not known initially a guess is made for $\partial\psi/\partial n$ along BC. The free surface elevation d_2 far downstream from the gate is given to be $0.6b$. Application of the BEM then produces ψ along the free surface and the sluice gate from which a starting trial value for $\bar{\psi}$ can be obtained. A new application of the BEM succeeds in providing ψ and $\partial\psi/\partial n$ and hence H and $\bar{\psi}$ on the boundary, which of course are different from the ones assumed. For this reason an adjustment of upstream and downstream free surface locations is made through the equations

$$\eta_i^{k+1} = \eta_i^k + (H - H_i^k)\,, \tag{6.25}$$

$$\eta_{i+1}^{k+1} = \eta_i^k \bar{\psi}^k / \psi_i^k\,, \tag{6.27}$$

respectively, where k denotes the number of iterations and i denotes a point P_i on the free surface. Using the updated geometry and ψ the computational procedure continues until after a number of iterations the values of H and $\bar{\psi}$ converge.

The first applications of the BEM in free surface flow problems with the aid of iterations were done by Niwa et al. [141] and Liggett [117] both in connection with groundwater flow phenomena. More information on this subject can be found in Liggett [118, 120], the text by Liggett and Liu [119] and the chapter on Fluid Mechanics in this book.

7. Coupling with other methods

As it has been mentioned in the first introductory chapter of this book the BEM has its advantages and disadvantages. This is also true for any other numerical method, such as the FEM and the FDM. Thus, it seems natural to attempt to solve a potential problem by combining methods in such a way as to fully exploit the advantages and reduce or completely eliminate the disadvantages of these methods. The BEM presents definite advantages for problems involving infinite or semi-infinite domains and high flux concentrations, while the FEM or the FDM represent the right choice for problems involving finite, inhomogeneous and anisotropic domains as well as material and/or geometric nonlinearities. Therefore, in a problem involving a bounded domain, characterized by inhomogeneities, anisotropy and/or linearities and governed by the Laplace equation or by some other complex equation, which is embedded or is in contact with an infinitely or semi-infinitely extended homogeneous, isotropic domain without nonlinearities in which the Laplace equation is to be satisfied, the FEM or the FDM should be used for the interior and the BEM for the exterior domain. Problems of this type are encountered, for example, in flow through porous media, electrical engineering or in fluid–structure interaction. Among the various types of articles on the subject one can mention those of Lean et al. [113], Silvester and Hsieh [165], McDonald and Wexler [125–127], Zienkiewicz et al. [193, 102], Atluri and Grannell [17], Margulies [129, 130], Kagawa et al. [99], Onuki et al. [143] and Dunbar [56]. General treatment of the coupling problem can also be found in the texts [194, 20, 119, 31] as well as in the chapters on Elastostatics, Elastodynamics, Fluid Mechanics, Soil–Structure Interaction and Fluid-Structure Interaction of this book.

The discussion in this section will be restricted to the coupling between the BEM and the FEM only, as the latter method is more widely used than the FDM. In principle, coupling between the two methods does not present great difficulties. Boundary and finite element matrix equations are written down for the two respective domains and their coupling is achieved by appropriately enforcing equilibrium and compatibility at the interface of the two domains.

Consider a potential problem for the composite domain V consisting of two subdomains V_1 and V_2 to be analysed by the FEM and the BEM, respectively, as shown in Fig. 27. The FEM and the BEM matrix equations describing the potential behavior of domains V_1 and V_2, respectively, are of the form

$$[K^1]\{p^1\} = \{F^1\}\,, \tag{7.1}$$

$$[G^n]\{p^2\} = [G]\{f^2\}\,. \tag{7.2}$$

In Eq. (7.1) $[K^1]$ is the finite element matrix, $\{p^1\}$ is the vector of nodal potential values and $\{F^1\}$ is the vector of the equivalent nodal forces in V_1. The

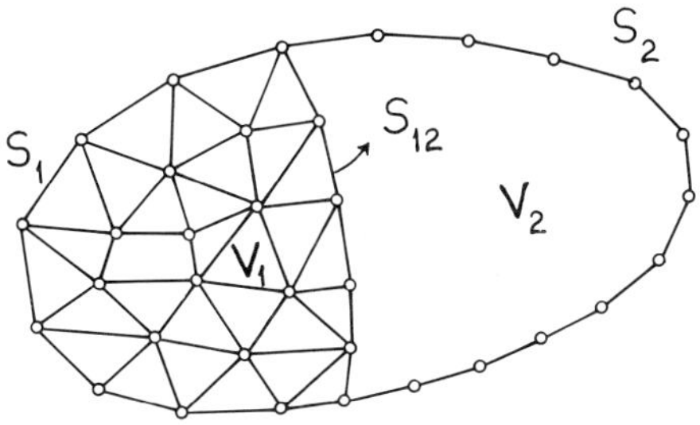

Fig. 27. Coupling of the BEM and the FEM.

quantities in Eq. (7.2) have the meaning given to them in previous sections. For a given flux vector $\{f^i\}$, a force vector $\{F^i\}$ can be constructed of the form

$$\{F^i\} = [N^i]\{f^i\} ; \quad i = 1, 2 , \tag{7.3}$$

where $[N^i]$ is a matrix of shape functions. Equation (7.2) can also be written as

$$[G]^{-1}[G^n]\{p^2\} = \{f^2\} , \tag{7.4}$$

or, after premultiplication by $[N^2]$ of Eq. (7.3), as

$$[K^2]\{p^2\} = \{F^2\} , \tag{7.5}$$

where

$$[K^2] = [N^2][G]^{-1}[G^n] , \tag{7.6}$$

$$\{F^2\} = [N^2]\{f^2\} . \tag{7.7}$$

Equation (7.5), which is of the same form as (7.1), clearly indicates that the subdomain V_2 has been transformed into an equivalent finite element. The matrix $[K^2]$, however, unlike $[K^1]$, is not symmetric. This asymmetry is due to the kind of discretization and collocation procedures as well as the special weighting functions – fundamental solutions employed by the BEM. For reasons of computational efficiency it is desirable to render the matrix $[K^2]$ symmetric.

One way of symmetrizing this matrix due to Georgiou et al. [60, 29] is by defining an error in the nondiagonal terms of $[K^2]$. The error for any coefficient K_{ij}^{2s} of the symmetric to be constructed matrix $[K^{2s}]$ is

$$\varepsilon_{ij} = \tfrac{1}{2}\,[(K_{ij}^{2s} - K_{ij}^2) + (K_{ij}^{2s} - K_{ji}^2)] . \tag{7.8}$$

Minimizing the square of ε_{ij} with respect to K_{ij}^{2s} yields

$$\frac{\partial(\varepsilon_{ij})^2}{\partial K_{ij}^{2s}} = 2K_{ij}^{2s} - K_{ij}^2 - K_{ji}^2 = 0 \tag{7.9}$$

and hence

$$K_{ij}^{2s} = \tfrac{1}{2}[K_{ij}^2 + K_{ji}^2] . \tag{7.10}$$

Thus, the equivalent symmetric stiffness matrix for V_2 takes the form

$$[K^{2s}] = \tfrac{1}{2}[[K^2] + [K^2]^{\mathrm{T}}] \tag{7.11}$$

and Eq. (7.5) becomes

$$[K^{2s}]\{p^2\} = \{F^2\} \tag{7.12}$$

Coupling of the two Eqs. (7.1) and (7.5) or (7.12) for the domains V_1 and V_2, respectively, is now achieved by enforcing compatibility and equilibrium at their interface S_{12}, which read

$$\{p_{12}^1\} = \{p_{12}^2\} , \tag{7.13}$$

$$\{f_{12}^1\} = -\{f_{12}^2\} , \tag{7.14}$$

and utilize these equations in the partitioned forms of Eqs. (7.1) and (7.5) or (7.12) which separate the interface nodal values from the rest. Of course, one should have the parts of $[N^1]$ and $[N^2]$ that correspond to S_{12} equal in order that (7.14) implies $\{F_{12}^1\} = -\{F_{12}^2\}$. Another approach can couple Eqs. (7.1) and (7.2) according to the manner described in subsection 5.4 of this chapter and with the aid of Eq. (7.3) for $i = 1$.

Energy approaches have also been used to symmetrize the matrix $[K^2]$ produced from a boundary element formulation. Margulies [130] and Brebbia et al. [31] describe two such approaches which are similar and their only difference is that, while the first one due to Zienkiewicz et al. [193, 102] uses free-space Green's functions, the second one due to Margulies [129, 130] uses bounded Green's functions. The starting point of the first approach is the potential energy expression for the BEM domain of the form

$$\Pi = \tfrac{1}{2} \int_V f_i \frac{\partial p}{\partial x_i}\, \mathrm{d}V - \int_{S_2} \bar{f} p \, \mathrm{d}S , \tag{7.15}$$

where the values p and f are assumed to be those obtained after the BEM solution of the potential problem. Use of the divergence theorem and the fact that p

satisfies Laplace's equation converts (7.15) into

$$\Pi = \tfrac{1}{2}\int_{S} fp\,\mathrm{d}S - \int_{S_2} \bar{f}p\,\mathrm{d}S\,. \tag{7.16}$$

The stationary condition $\delta\Pi = 0$ then implies

$$\tfrac{1}{2}\int_{S} (f\,\delta p + \delta f\,p)\,\mathrm{d}S - \int_{S_2} \bar{f}\delta p\,\mathrm{d}S = 0\,. \tag{7.17}$$

Assuming that p and f are given in the interpolatory form

$$p = \boldsymbol{\phi}^{\mathrm{T}}\boldsymbol{p}^{n}\,, \qquad f = \boldsymbol{\phi}^{\mathrm{T}}\boldsymbol{f}^{n} \tag{7.18}$$

one can write (7.17) in matrix form as

$$\tfrac{1}{2}[\delta\{p\}^{\mathrm{T}}[M]\{f\} + \delta\{f\}^{\mathrm{T}}[M]^{\mathrm{T}}\{p\}] - \delta\{p\}^{\mathrm{T}}\{F\} = 0 \tag{7.19}$$

where the matrix $[M]$ consists of integrals of the shape functions $\boldsymbol{\phi}$. Substitution of $\{f\}$ taken from (7.2) in the form

$$\{f\} = [G]^{-1}[G^{n}]\{p\} \tag{7.20}$$

into (7.19) finally results in the equation

$$[K]\{p\} = \{F\} \tag{7.21}$$

where the matrix $[K]$ given by

$$[K] = \tfrac{1}{2}\{[M][G]^{-1}[G^{n}] + [[M][G]^{-1}[G^{n}]]^{\mathrm{T}}\}\,, \tag{7.22}$$

is obviously symmetric and essentially of the same form as the one in (7.11).

It has been observed, however, by Margulies [129, 130] and Brebbia et al. [31] that the above deduction is not rigorously correct from the mathematical viewpoint exactly because the substitution of $\{f\}$ of (7.20) into (7.19) creates small errors due to the fact that $\{f\}$ is approximate as coming from the numerical solution of the boundary integral equation (7.2) and not from the potential field corresponding to the fundamental solution. As a result of that an imperfect matching between the solutions in the domains V_1 and V_2 is created which may lead to detectable errors. This non-conformity between solutions at the interface S_{12} may be reduced by Lagrange multipliers [193, 194, 102] but never completely eliminated. Margulies [129, 130] was able to achieve a perfect matching of the solutions and construct symmetric matrices through energy approaches by using bounded Green's functions in the BEM domain. He first observed that the surface S_{12} can always be selected to have a very simple shape, e.g. circle, straight line, square, so that a Green's function satisfying the condition

$$u^*(\boldsymbol{x}, \boldsymbol{\xi}) = 0 \qquad \text{on } S \tag{7.23}$$

can be easily found or constructed. In view of (7.23) the boundary integral equation (2.18) becomes

$$cp + \int_S u_n^* p \, \mathrm{d}S = 0 \,. \tag{7.24}$$

Use of (7.24) in the energy expression (7.16) yields

$$\Pi = \tfrac{1}{2} \int_S f(\boldsymbol{x}) \Big[- \int_S u_n^*(\boldsymbol{x}, \boldsymbol{\xi}) p(\boldsymbol{\xi}) \, \mathrm{d}S(\boldsymbol{\xi}) \Big] \mathrm{d}S(\boldsymbol{x}) - \int_{S_2} \bar{f}(\boldsymbol{x}) p(\boldsymbol{x}) \, \mathrm{d}S(\boldsymbol{x}) \,. \tag{7.25}$$

Introduction as before of the interpolatory expressions (7.18) for p and f and use of the stationary condition $\delta\Pi = 0$ on (7.25) produce the relation

$$[K]\{p\} = \{F\} \,, \tag{7.26}$$

$$[K] = \Big[\int_S \Big[\phi(\boldsymbol{x}) \int_S u_n^*(\boldsymbol{x}, \boldsymbol{\xi}) [\phi(\boldsymbol{\xi})]^{\mathrm{T}} \, \mathrm{d}S(\boldsymbol{\xi}) \, \mathrm{d}S(\boldsymbol{x}) \Big] \Big] \,, \tag{7.27}$$

$$\{F\} = - \Big\{ \int_{S_2} [\phi(\boldsymbol{x})] \{ \bar{f}(\boldsymbol{x}) \} \, \mathrm{d}S(\boldsymbol{x}) \Big\} \,. \tag{7.28}$$

It is now obvious that matrix $[K]$ of (7.27) is symmetric.

A similar approach has been used in many electrical engineering potential problems [165, 125, 113, 127] in connection with the indirect BEM. Starting with expressions

$$p = \int_S \sigma u^* \, \mathrm{d}S \,, \qquad f = \int_S \sigma u_n^* \, \mathrm{d}S \tag{7.29}$$

$$\Pi = -\tfrac{1}{2} \int_S \sigma f \, \mathrm{d}S \tag{7.30}$$

for the potential p the flux f and the potential energy Π without the S_2 term one can express Π as

$$\Pi = -\tfrac{1}{2} \int_S \sigma(\boldsymbol{x}) \Big[\int_S u_n^*(\boldsymbol{x}, \boldsymbol{\xi}) \sigma(\boldsymbol{\xi}) \, \mathrm{d}S(\boldsymbol{\xi}) \Big] \mathrm{d}S(\boldsymbol{x}) \,, \tag{7.31}$$

which has a form analogous to that in (7.25) and can thus lead to a symmetric matrix $[K]$ of exactly the same form as in (7.27) if one assumes that the source

$$\sigma = \boldsymbol{\phi}^{\mathrm{T}} \boldsymbol{\sigma}^{n} . \tag{7.32}$$

Very recently Dunbar [56] proposed a different method of combining the FEM and BEM regions according to which these two regions are treated separately and at their interface potentials and fluxes are matched in an iterative manner. In this way matrix inversion may be avoided and there is no restriction on the type of solution used for the BEM region implying that an analytic solution could very well be used.

It should be finally mentioned that for problems with a BEM region spherically

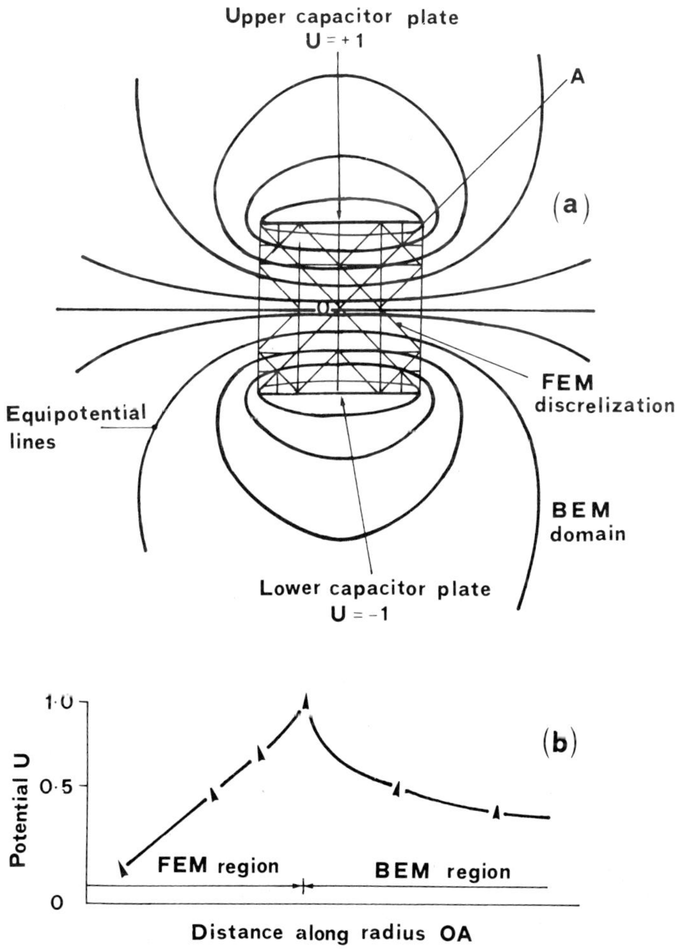

Fig. 28. Electrical field surrounding a parallel plate capacitor: (a) geometry and discretization and (b) electrical potential profile along the radius 0A.

surrounding the FEM region and extending to infinity one can construct approximate boundary elements which are described by the equation

$$r\frac{\partial p}{\partial r} + p = \text{constant} \tag{7.33}$$

on the spherical interface of the two regions [31]. Equation (7.33) can be applied at each node of the interface independently of the others but its neighbors, thereby creating a banded system of equations.

7.1. Numerical example

In order to demonstrate the applicability of the coupling procedures between the FEM and the BEM a numerical example taken from Zienkiewicz et al. [193] is presented. Coupling of the two methods is used in this example to compute the electrical field surrounding a two-dimensional charged parallel place capacitor as shown in Fig. 28(a). The same figure also shows the surface S_{12} separating the interior and exterior regions of the problem. Combination of the two methods was implemented with the aid of the free-space Green's function direct BEM. The results in the form of the potential U versus the radial distance r are shown by Fig. 28(b).

8. Variations of the BEM

The conventional BEM in its indirect or direct form that was presented in detail in the previous sections of this chapter is associated with the employment of singular fundamental solutions or Green's functions. The presence of Green's functions in the BEM creates in some cases problems in connection with the construction of these functions and always computational difficulties concerning the evaluation of singular integrals.

This section briefly describes two similar boundary methods, which, instead of employing Green's functions, express the solution of a potential problem in terms of a series of complete harmonic functions with constant coefficients to be numerically determined through utilization of the boundary conditions. Thus, both of these methods can be thought of as improved and systematic versions of Trefftz's method [182]. The first of these boundary methods is the so-called Edge-Function Method (EFM) developed by Quinlan [154] in connection with the torsion problem of an irregular polygonal section and subsequently extended to a great variety of problems in elastostatics by Quinlan and his co-workers [155, 17, 156]. The second method is the boundary method developed by Herrera [69–72] on the basis of his studies on variational methods, the theory of connectivity and his T-complete systems of functions, which has been so far applied to problems governed by the Laplace–Poisson and Helmholtz's equations.

According to the EFM the solution of the Laplace equation (1.1) is assumed to be of the form

$$p = \sum_{n=1}^{N} a_n \phi_n + \sum_{m=1}^{M} b_m \psi_m \,, \tag{8.1}$$

where the functions ϕ_n and ψ_m are harmonic in the domain V and such that ϕ_n are regular polynomial solutions with origin at the center of the domain, while ψ_m are asymptotic solutions in the neighborhood of the boundary. The functions ψ_m are constructed by considering the asymptotic solution of the Laplace equation near plane boundary segments, boundary edges and vertices, as it is assumed that the domain V is bounded by a surface consisting of plane segments. The boundary conditions are then imposed for the determination of the coefficients a_n and b_m by harmonic matching or fitting, which are essentially equivalent to continuous or discrete least-squares fitting, respectively, or, more recently, by boundary variational techniques. The boundary variational approach of Ref. [17] used in elastostatics is applied here to potential problems. The starting point is the expression for the potential energy Π which reads

$$\Pi = \tfrac{1}{2} \int_V f_i \frac{\partial p}{\partial x_i} \, \mathrm{d}V - \int_{S_2} \bar{f} p \, \mathrm{d}S - \int_{S_1} f(p - \bar{p}) \, \mathrm{d}S \,. \tag{8.2}$$

Application of the divergence theorem on the volume integral and use of the fact that p satisfies the Laplace equation enable one to write (8.2) exclusively in the form of boundary integrals as

$$\Pi = \int_{S_2} (\tfrac{1}{2} f - \bar{f}) p \, \mathrm{d}S - \int_{S_1} (\tfrac{1}{2} p - \bar{p}) f \, \mathrm{d}S \,. \tag{8.3}$$

The stationary condition $\delta\Pi = 0$ on (8.3) yields

$$\int_{S_2} [(\tfrac{1}{2} f - \bar{f}) \delta p + \tfrac{1}{2} p \delta f] \, \mathrm{d}S - \int_{S_1} [(\tfrac{1}{2} p - \bar{p}) \delta f + \tfrac{1}{2} f \delta p] \, \mathrm{d}S = 0 \,. \tag{8.4}$$

Assuming for simplicity and in order to conform with the choice taken in [154] that

$$p = \sum_{m=1}^{M} b_m \psi_m + \phi_0 \,, \tag{8.5}$$

$$f = \sum_{m=1}^{M} b_m \psi_m^* + \phi_0^* \,, \tag{8.6}$$

where

$$\psi_m^* = \partial\psi_m/\partial n\,, \qquad \phi_0^* = \partial\phi_0/\partial n\,, \tag{8.7}$$

and substituting (8.5) and (8.6) into (8.4) one can obtain the matrix equation

$$[A_{ij}]\{b_j\} = \{c_i\}\,; \qquad i, j = 1, 2, \ldots, M \tag{8.8}$$

where

$$A_{ij} = \int_{S_2} \tfrac{1}{2}(\psi_j^*\psi_i + \psi_j\psi_i^*)\,\mathrm{d}S - \int_{S_1} \tfrac{1}{2}(\psi_j^*\psi_i + \psi_j\psi_i^*)\,\mathrm{d}S \tag{8.9}$$

$$c_i = \int_{S_2} [\tfrac{1}{2}(\phi_0^*\psi_i + \phi_0\psi_i^*)\,\mathrm{d}S - \bar{f}\psi_i]\,\mathrm{d}S - \int_{S_1} [(\phi_0^*\psi_i + \phi_0\psi_i^*) - \bar{p}\psi_i^*]\,\mathrm{d}S\,. \tag{8.10}$$

It is evident from (8.9) that the matrix A_{ij} is symmetric but, in general, fully populated.

The boundary method of Herrera [71, 72] will be now illustrated for the simple case of the Dirichlet problem described by Eqs. (1.1) and (1.2). The solution is assumed to be given in the form

$$p = \sum_{n=1}^{N} a_n w_n\,, \tag{8.11}$$

where a_n are coefficients to be determined from the boundary conditions through a least-squares fitting and w_n is a T-complete (previously called by Herrera C-complete) set of functions which has been determined by Herrera [71, 72] for two and three dimensions and interior and exterior Laplace's problems. For example, for the two-dimensional interior problem

$$w_n = \{1, r^n \cos n\theta, r^n \sin n\theta\}\,. \tag{8.12}$$

The coefficients a_n are chosen so that the square of the error ε

$$\varepsilon^2 = \left|\bar{p} - \sum_{n=1}^{N} a_n w_n\right|^2 \tag{8.13}$$

is minimized on the boundary S. This requires that the condition

$$\left\langle \bar{p} - \sum_{n=1}^{N} a_n w_n, w_m \right\rangle = 0\,, \qquad m = 1, 2, \ldots, N \tag{8.14}$$

is satisfied at the boundary S, where for two functions p_1 and p_2 the symbol $\langle p_1, p_2 \rangle$ means

$$\langle p_1, p_2 \rangle = \int_S p_1 p_2 \, \mathrm{d}S \, . \tag{8.15}$$

Expansion of (8.14) results in the maxrix equation

$$[A_{nm}]\{a_n\} = \{d_m\} \, , \tag{8.16}$$

where

$$A_{nm} = \int_S w_n w_m \, \mathrm{d}S \, , \tag{8.17}$$

$$d_m = \int_S \bar{p} w_m \, \mathrm{d}S \, . \tag{8.18}$$

If the computation of f is subsequently desired on the boundary, a convenient way is to utilize the relation resulting from Green's second theorem (1.20) with u and v harmonic functions such that $u = p$ and $v = w_m$, i.e. the relation

$$\int_S \left(\bar{p} \frac{\partial w_m}{\partial n} - w_m f \right) \mathrm{d}S = 0 \, . \tag{8.19}$$

The assumption is made again that

$$f = \sum_{n=1}^{N} b_n w_n \tag{8.20}$$

and a least-squares approach leads to

$$[B_{nm}]\{b_n\} = \{e_m\} \, , \tag{8.21}$$

where

$$B_{nm} = \int_S w_n w_m \, \mathrm{d}S \, ; \qquad n, m = 1, 2, \ldots, N \tag{8.22}$$

and

$$e_m = \int_S f w_m \, \mathrm{d}S = \int_S \bar{p} \frac{\partial w_m}{\partial n} \, \mathrm{d}S \tag{8.23}$$

due to the validity of (8.19). It is again evident from (8.17) and (8.22) that matrices A_{nm} and B_{nm} are symmetric but, in general, fully populated.

Both of the aforementioned boundary methods have some very attractive features and it is expected to see more extensive application of these techniques in the near future. Apparently the EFM appears to be more efficient than Herrera's boundary method due to the presence of those asymptotic boundary functions ψ_m which, because of their accounting of complex field behavior, permit one to have a smaller number of coefficients required for a satisfactory solution. Of course, some preparatory theoretical work is necessary for the computation of ψ_m but this is done only once.

Acknowledgements

The author would like to express his gratitude to the National Science Foundation for supporting his work on Boundary Element Methods through the grants NSF/CEE-80 24727 and NSF/CEE-81 09723 to the University of Minnesota. During the final stages of the preparation of this chapter the author was staying as a Visiting Professor at Colorado State University. Professor C.N. Papadakis, Head of the Civil Engineering Department, and the faculty of the Geotechnical Engineering Program of that University are gratefully acknowledged for their hospitality and technical assistance. Thanks are also due to Professor G.D. Manolis for reading the manuscript and offering helpful suggestions and to Professor F.J. Rizzo, Professor S.N. Atluri and Dr. T.A. Cruse for their technical assistance.

References

[1] M. Abramowitz and I.A. Stegun, *Handbook of Mathematical Functions* (Dover, New York, 1974).

[2] E. Alarcón, A. Martin and F. Paris, Improved Boundary Elements in Torsion Problems, in: *Recent Advances in Boundary Element Methods*, C.A. Brebbia, Ed. (Pentech Press, London, 1978) pp. 149–165.

[3] E. Alarcón, A. Martin and F. Paris, Boundary Elements in Potential and Elasticity Theory, *Computers and Structures* **10**, 341–362 (1979).

[4] E. Alarcón and P. Reyero, *Guia de Introduccion al Metodo des los Elementos de Contorno* (Impreso por E.T.S. de Ingenieros Industriales, Madrid, 1983).

[5] E. Alarcón, L. Abia and A. Reverter, On the Possibility of Adaptive Boundary Elements, in: *Proceedings of the International Conference on Accuracy Estimates and Adaptive Refinements in Finite Element Computations*, I. Babuska, E.R.A. Oliveira and O.C. Zienkiewicz, Eds. (Technical University of Lisbon, Lisbon, 1984) pp. 25–34.

[6] E. Alarcón, A. Reverter and J. Molina, Hierarchical Boundar Elements, *Computers and Structures* **20**, 151–156 (1985).

[7] J.J. Anza, E. Ahedo, I. Da Riva and E. Alarcón, A New Boundary Condition Solved with BIEM, in: *Boundary Element Methods in Engineering*, C.A. Brebbia, Ed. (Springer-Verlag, Berlin, 1982) pp. 607–617.

[8] J.H. Argyris and D.W. Scharpf, Two and Three-Dimensional Potential Flow by the Method of Singularities, *Aeronautical Journal of the Royal Aeronautical Society* **73**, 959–961 (1969).

[9] G. Athanasiadis, Torsion Prismatischer Stäbe Nach der Singularitätenethode, *Ingenieur Archiv* **49**, 89–96 (1980).

[10] G. Athanasiadis and A. Mitakidis, Die Untersuchung Einiger Integralgleichungen des St. Venantschen Torsionsproblems, *Ingenieur Archiv* **53**, 303–316 (1983).

[11] K.E. Atkinson, The Numerical Solution of Laplace's Equation in Three-Dimensions – II, in:

Numerical Treatment of Integral Equations, J. Albrecht and L. Collatz, Eds. (Birkhàuser Verlag, Basel, 1980) pp. 1–23.
[12] K.E. Atkinson, The Numerical Solution of Laplace's Equation in Three-Dimensions, *SIAM Journal on Numerical Analysis* **19**, 263–274 (1982).
[13] K. Atkinson and F. De Hoog, The Numerical Solution of Laplace's Equation on a Wedge, *IMA Journal of Numerical Analysis* **4**, 19–41 (1984).
[14] K.E. Atkinson, The Numerical Evaluation of Particular Solutions for Poisson's Equation, *IMA Journal of Numerical Analysis* **5**, 319–338 (1985).
[15] K.E. Atkinson, Piecewise Polynomial Collocation for Integral Equations on Surfaces in Three-Dimensions, *Journal of Integral Equations* **9** (Supplementary Issue), 25–48 (1985).
[16] K.E. Atkinson, Solving Integral Equations on Surfaces in Space, in: *Constructive Methods for the Practical Treatment of Integral Equations*, G. Hämmerlin and K. Hoffman, Eds. (Birkhäuser Verlag, Basel, 1985) pp. 20–43.
[17] S.N. Atluri and J.J. Grannell, Boundary Element Methods (BEM) and Combination of BEM-FEM, Report No. GIT-ESM-SA-78-16, Center for the Advancement of Computational Mechanics, Georgia Institute of Technology, Atlanta, Georgia, 1978.
[18] M.C. Au and C.A. Brebbia, A Three Dimensional Boundary Element Approach for Rotationally Symmetrical Problems, in: *Boundary Elements*, C.A. Brebbia, T. Futagami and M. Tanaka, Eds. (Springer-Verlag, Berlin, 1983) pp. 915–928.
[19] P.K. Barnerjee, Non-Linear Problems of Potential Flow, in: *Developments in Boundary Elements Methods* – 1, P.K. Banerjee and R. Butterfield, Eds. (Applied Science Publishers, London, 1979) pp. 21–30.
[20] P.K. Banerjee and R. Butterfield, *Boundary Element Methods in Engineering Science* (McGraw-Hill, London, 1981).
[21] M.R. Barone and D.A. Caulk, Special Boundary Integral Equations for Approximate Solution of Laplace's Equation in Two-Dimensional Regions with Circular Holes, *Quarterly Journal of Mechanics and Applied Mathematics* **34**, 265–286 (1981).
[22] H. Battenbo and B.H. Baines, A Boundary Integral Solution to Torsion of Cylinders and Solids of Revolution, *International Journal for Numerical Methods in Engineering* **9**, 461–476 (1975).
[23] R. Bialecki and A.J. Nowak, Boundary Value Problems in Heat Conduction with Nonlinear Material and Nonlinear Boundary Conditions, *Applied Mathematical Modelling* **5**, 417–421 (1981).
[24] J.L. Blue, Boundary Integral Solutions of Laplace's Equation, *The Bell System Technical Journal* **57**, 2797 2822 (1978).
[25] C.A. Brebbia and J. Dominguez, Boundary Element Methods for Potential Problems, *Applied Mathematical Modelling* **1**, 372–378 (1977).
[26] C.A. Brebbia, Approximate Methods in Mathematical Modelling, in: *Proceeding of 1st International Conference on Mathematical Modeling*, X.J.R. Avula, Ed. (Department of Engineering Mechanics, University of Missouri-Rolla, 1977) pp. 43–57.
[27] C.A. Brebbia and R. Butterfield, Formal Equivalence of Direct and Indirect Boundary Element Methods, *Applied Mathematical Modelling* **2**, 132–134 (1978).
[28] C.A. Brebbia, *The Boundary Element Method for Engineers* (Pentech Press, London, 1978).
[29] C.A. Brebbia and P. Georgiou, Combination of Boundary and Finite Elements for Elastostatics, *Applied Mathematical Modelling* **3**, 212–220 (1979).
[30] C.A. Brebbia and S Walker, *Boundary Element Techniques in Engineering* (Newnes–Butterworths, London, 1980).
[31] C.A. Brebbia, J.C.F. Telles and L.C. Wrobel, *Boundary Element Techniques* (Springer-Verlag, Berlin, 1984).
[32] R. Butterfield, The Application of Integral Equation Methods to Continuum Problems in Soil-Mechanics, in: *Stress Strain Behaviour of Soils*, R.H.G. Parry, Ed. (G.T. Foulis and Co., London, 1972) pp. 573–587.
[33] R. Butterfield and G.R. Tomlin, Integral Techniques for Solving Zoned Anisotropic Continuum Problems, in: *Variational Methods in Engineering*, C.A. Brebbia and H. Tottenham, Eds. (University of Southampton Press, Southampton, 1973) pp. 9/30–9/50.
[34] R. Butterfield, An Application of the Boundary Element Method to Potential Flow Problems in Generally Inhomogeneous Bodies, in: *Recent Advances in Boundary Element Methods*, C.A. Brebbia, Ed. (Pentech Press, London, 1978) pp. 123–135.

[35] H.S. Carslaw and J.C. Jaeger, *Conduction of Heat in Solids* (Clarendon Press, Oxford, 1959).
[36] D.A. Caulk, Analysis of Elastic Torsion in a Bar with Circular Holes by a Special Boundary Integral Method, *Journal of Applied Mechanics* **50**, 101–108 (1983).
[37] D.A. Caulk, Special Boundary Integral Equations for Potential Problems in Regions with Circular Holes, *Journal of Applied Mechanics* **51**, 713–716 (1984).
[38] Y.P. Chang, C.S. Kang and D.J. Chen, The Use of Fundamental Green's Functions for the Solution of Problems of Heat Conduction in Anisotropic Media, *International Journal of Heat and Mass Transfer* **16**, 1905–1918 (1973).
[39] S. Christiansen, Integral Equations Without a Unique Solution Can be Made Useful for Solving Some Plane Harmonic Problems, *Journal of the Institute of Mathematics and its Applications* **16**, 143–159 (1975).
[40] S. Christiansen and H. Rasmussen, Numerical Solutions for Two-Dimensional Annular Electrochemical Machining Problems, *Journal of the Institute of Mathematics and its Applications* **18**, 295–307 (1976).
[41] S. Christiansen, A Review of Some Integral Equations for Solving the St. Venant Torsion Problem, *Journal of Elasticity* **8**, 1–20 (1978).
[42] S. Christiansen and E.B. Hansen, Numerical Solution of Boundary Value Problems Through Integral Equations, *Zeitschrift für Angewandte Mathematik und Mechanik* **58**, T14–T25 (1978).
[43] D.L. Clements and F.J. Rizzo, A Method for the Numerical Solution of Boundary Value Problems Governed by Second-Order Elliptic Systems, *Journal of the Institute of Mathematics and its Applications* **22**, 197–202 (1978).
[44] R. Courant and D. Hilbert, *Methods of Mathematical Physics*, Vol. 2 (Interscience Publishers, New York, 1962).
[45] G.R. Cowper, Gaussian Quadrature Formulas for Triangles, *International Journal for NumericaL Methods in Engineering* **7**, 405–408 (1973).
[46] M. Cristescu and G. Loubignac, Gaussian Quadrature Formulas for Functions with Singularities in $1/R$ Over Triangles and Quadrangles, in: *Recent Advances in Boundary Element Methods*, C.A. Brebbia, Ed. (Pentech Press, London, 1978) pp. 375–390.
[47] J.M. Crotty, A Block Equation Solver for Large Unsymmetric Matrices Arising in the Boundary Integral Equation Method, *International Journal for Numerical Methods in Engineering* **18**, 997–1017 (1982).
[48] T.A. Cruse, *Mathematical Foundations of the Boundary Integral Equation Method in Solid Mechanics*, Air Force Office of Scientific Research Technical Report AFOSR-TR-77-1002 (1977).
[49] D.J. Danson, Using BEASY to Solve Torsion Problems, in: *Boundary Elements*, C.A. Brebbia, T. Futagami and M. Tanaka, Eds. (Springer-Verlag, Berlin, 1983) pp. 821–834.
[50] D.J. Danson and C.A. Brebbia, Further Engineering Applications of BEASY, in: *Boundary Elements*, C.A. Brebbia, T. Futagami and M. Tanaka, Eds. (Springer-Verlag, Berlin, 1983) pp. 857–880.
[51] P.C. Das, A Disc Based Block Elimination Technique Used for the Solution of Non-symmetrical Fully Populated Matrix Systems Encountered in the Boundary Element Method, in: *Recent Advances in Boundary Element Methods*, C.A. Brebbia, Ed. (Pentech Press, London, 1978) pp. 391–404.
[52] F.J. Davenport, Singularity Solutions to General Potential Flow Airfoil Problem, Report D6-7207, Boeing Airplane Company, Seattle, Washington (1963).
[53] G. De Mey, Determination of the Electric Field in a Hall Generator Under the Influence of an Alternating Magnetic Field, *Solid State Electronics* **17**, 977–979 (1974).
[54] G. De Mey, Integral Equations for Potential Problems with the Source Function not Located on the Boundary, *Computers and Structures* **8**, 113–115 (1978).
[55] G. De Mey, Improved Boundary Element Methods for Solving the Laplace Equation in Two Dimensions, in: *New Developments in Boundary Element Methods*, C.A. Brebbia, Ed. (CML Publications, Southampton, 1980) pp. 90–100.
[56] W.S. Dunbar, Finite Element – Boundary Integral Linkage Algorithms, in: *Engineering Mechanics in Civil Engineering*, A.P. Boresi and K.P. Chong, Eds. (American Society of Civil Engineers, New York, 1984) pp. 128–131.

[57] G. Fairweather, F.J. Rizzo, D.J. Shippy and Y.S. Wu, On the Numerical Solution of Two-Dimensional Potential Problems by an Improved Boundary Integral Equation Method, *Journal of Computational Physics* **31**, 96–112 (1979).
[58] I. Fredholm, Sur une Classe d'Equations Fonctionelles, *Acta Mathematica, Sweden* **27**, 365–390 (1903).
[59] T. Fukui, On Corner Solutions by Indirect B.I.E.M., in: *Boundary Elements*, C.A. Brebbia, T. Futagami and M. Tanaka, Eds. (Springer-Verlag, Berlin, 1983) pp. 929–938.
[60] P. Georgiou, The Coupling of the Direct Boundary Element Method with the Finite Element Displacement Technique in Elastostatics, Ph.D. Thesis, Southampton University, Southampton, England (1981).
[61] S. Gomez-Lera, E. Paris and E. Alarcón, Treatment of Singularities in 2-D Domains Using BIEM, *Applied Mathematical Modelling* **6**, 111–118 (1982).
[62] M.D. Greenberg, *Application of Green's Functions in Science and Engineering* (Prentice Hall, Englewood Cliffs, N.J., 1971).
[63] E. Grodtkjaer, A. Direct Integral Equation Method for the Potential Flow About Arbitrary Bodies, *International Journal for Numerical Methods in Engineering*, **6**, 253–264 (1973).
[64] P.H.L. Groenenboom, The Application of Boundary Elements to Steady and Unsteady Potential Fluid Flow Problems in Two and Three Dimensions, in: *Boundary Element Methods*, C.A. Brebbia, Ed. (Springer-Verlag, Berlin, 1981) pp. 37–52.
[65] P.C. Hammer, O.J. Marlowe and A.H. Stroud, Numerical Integration over Simplexes and Cones, *Mathematics of Computation* **10**, 130–137 (1956).
[66] R.F. Harrington, K. Pontoppidan, P. Abrahamsen and N.C. Albertsen, Computation of Laplacian Potentials by an Equivalent-Source Method, *Proceedings of the Institution of Electrical Engineers* **116**, 1715–1720 (1969).
[67] J. Hayes and R. Kellner, The Eigenvalue Problem for a Pair of Coupled Integral Equations Arising in the Numerical Solution of Laplace's Equation, *SIAM Journal on Applied Mathematics* **22**, 503–513 (1972).
[68] K. Herrebrugh, Solving the Incompressible and Isothermal Problem in Elastohydrodynamic Lubrication Through an Integral Equation, *Journal of Lubrication Technology, Transactions of the ASME* **90**, 262–268 (1968).
[69] I. Herrera, Theory of Connectivity: A Systematic Formulation of Boundary Element Methods, *Applied Mathematical Modelling* **3**, 151–156 (1979).
[70] I. Herrera, Boundary Methods: Development of Complete Systems of Solutions, in: *Finite Element Flow Analysis*, T. Kawai, Ed. (University of Tokyo Press, Tokyo, 1982) pp. 897–906.
[71] I. Herrera, Trefftz Method, in: *Topics in Boundary Element Research*, Vol. 1, C.A. Brebbia, Ed. (Springer-Verlag, Berlin, 1984) pp. 225–253.
[72] I. Herrera, *Boundary Methods, An Algebraic Theory* (Pitman Publishing Co., London, 1984).
[73] J.L. Hess, Calculation of Potential Flow About Bodies of Revolution Having Axes Perpendicular to the Free-Stream Direction, *Journal of the Aerospace Sciences* **29**, 726–742 (1962).
[74] J.L. Hess and A.M.O. Smith, Calculation of Nonlifting Potential Flow About Arbitrary Three-Dimensional Bodies, *Journal of Ship Research* **8**, 22–44 (1964).
[75] J.L. Hess and A.M.O. Smith, Calculation of Potential Flow About Arbitrary Three-Dimensional Bodies, in: *Progress in Aeronautical Sciences*, Vol. 8, D. Küchemann, Ed. (Pergamon Press, New York, 1967) pp. 1–138.
[76] J.L. Hess, Higher Order Numerical Solution of the Integral Equation for the Two-Dimensional Neumann Problem, *Computer Methods in Applied Mechanics and Engineering* **2**, 1–15 (1973).
[77] J.L. Hess, The Problem of Three-Dimensional Lifting Potential Flow and its Solution by Means of Surface Singularity Distribution, *Computer Methods in Applied Mechanics and Engineering* **4**, 283–319 (1974).
[78] J.L. Hess, Review of Integral-Equation Techniques for Solving Potental-Flow Problems with Emphasis on the Surface–Source Method, *Computer Methods in Applied Mechanics and Engineering* **5**, 145–196 (1975).
[79] J.L. Hess, Improved Solution for Potential Flow About Arbitrary Axisymmetric Bodies by the Use of a Higher Order Surface Source Method, *Computer Methods in Applied Mechanics and Engineering* **5**, 297–308 (1975).

[80] J.L. Hess, Review of Integral-Equation Techniques for Solving Potential Flow Problems with Complicated Boundaries, in: *Innovative Numerical Analysis for the Engineering Sciences*, R. Shaw, et al., Eds. (University Press of Virginia, Charlottesville, 1980) pp. 131–143.

[81] L.D. Hofmeister, An Out-of-Core Equation Solver for Full, Unsymmetric Matrices, *International Journal for Numerical Methods in Engineering* **12**, 721–731 (1978).

[82] T.V. Hromadka II, *The Complex Variable Boundary Element Method* (Springer-Verlag, Berlin, 1984).

[83] B.W. Hunt, Numerical Solutions of an Integral Equation for Flow from a Circular Orifice, *Journal of Fluid Mechanics* **31**, 361–377 (1968).

[84] B. Hunt, The Mathematical Basis and Numerical Principles of the Boundary Integral Method for Incompressible Potential Flow Over 3-D Aerodynamic Configurations, in: *Numerical Methods in Applied Fluid Dynamics*, B. Hunt, Ed. (Academic Press, London, 1980) pp. 49–135.

[85] M. Ikeuchi and K. Onishi, Boundary Element Solutions to Steady Convective Diffusion Equations, *Applied Mathematical Modelling* **7**, 115–118 (1983).

[86] D.B. Ingham, P.J. Heggs and M. Manzoor, The Numerical Solution of Plane Potential Problems by Improved Boundary Integral Equation Methods, *Journal of Computational Physics* **42**, 77–98 (1981).

[87] D.B. Ingham, P.J. Heggs and M. Manzoor, Boundary Integral Equation Solution of Nonlinear Plane Potential Problems, *IMA Journal of Numerical Analysis* **1**, 416–426 (1981).

[88] D.B. Ingham and M.A. Kelmanson, Solution of Nonlinear Elliptic Equations with Boundary Singularities by an Integral Equation Method, in: *Boundary Elements*, C.A. Brebbia, T. Futagami and M. Tanaka, Eds. (Springer-Verlag, Berlin, 1983) pp. 61–71.

[89] D.B. Ingham and M.A. Kelmanson, *Boundary Integral Equation Analysis of Singular Potential and Biharmonic Problems* (Springer-Verlag, Berlin, 1984).

[90] K. Jakob, Berechnung der Inkompressiblen Strömung um Dicke Tragflügel-und Gitterprofile mit Eckiger Hinterkante, *Zeitschrift fuer Flugwissenschaften* **15**, 341–346 (1967).

[91] K. Jakob, Berechung der Abgelösten Inkompressiblen Strömung um Tragflügelprofile und Bestimmung des Maximalen Auftriebs, *Zeitschrift fuer Flugwissenschaften* **17**, 221–230 (1969).

[92] M.A. Jaswon, Integral Equation Methods in Potential Theory, I, *Proceedings of the Royal Society of London* **275** (A), 23–32 (1963).

[93] M.A. Jaswon and A.R.S. Ponter, An Integral Equation Solution of the Torsion Problem, *Proceedings of the Royal Society of London* **273** (A), 237–246 (1963).

[94] M.A. Jaswon and G.T. Symm, *Integral Equation Methods in Potential Theory and Elastostatics* (Academic Press, London, 1977).

[95] M.A. Jaswon, Some Theoretical Aspects of Boundary Integral Equations, *Applied Mathematical Modelling* **5**, 409–413 (1981).

[96] M.A. Jaswon, A Review of the Theory, in: *Topics in Boundary Element Research*, Vol. 1, C.A. Brebbia, Ed. (Springer-Verlag, Berlin, 1984) pp. 13–40.

[97] A.R. Jaworski, Boundary Elements for Heat Conduction in Composite Media, *Applied Mathematical Modelling* **5**, 45–48 (1981).

[98] R.L. Johnston, G. Fairweather and P.S. Han, The Method of Fundamental Solutions. An Adaptive Boundary Element Method for Problems in Potential Flow and Solid Mechanics, in: *Engineering Mechanics in Civil Engineering*, A.P. Boresi and K.P. Chong, Eds. (American Society of Civil Engineers, New York, 1984) pp. 140–143.

[99] Y. Kagawa, T. Yamabuchi and S. Kitagami, Infinite Boundary Element and its Application to a Combined Finite-Boundary Element Technique for Unbounded Field Problems, in: *Boundary Elements*, C.A. Brebbia, T. Futagami and M. Tanaka, Eds. (Springer-Verlag, Berlin, 1983) pp. 1017–1026.

[100] P. Kandler, Numerische Behandlung des Saint-Venantschen Torsionsproblems bei Beliebigen Querschnittsbereichen mit Hilfe von Integralgleichungen, Mitteilungen aus Mathematik, *Naturwissenschaft, Medizin und Technik* **9**, 637–649 (1967).

[101] O.D. Kellogg, *Foundations of Potential Theory* (J. Springer, Berlin, 1929; also published by Dover, New York, 1953).

[102] D.W. Kelly, G.G. Mustoe and O.C. Zienkiewicz, Coupling of Boundary Element Methods with Other Numerical Methods, in: *Developments in Boundary Element Methods* – 1, P.K. Banerjee and R. Butterfield, Eds. (Applied Science, London, 1979) pp. 251–285.

[103] T. Kermanidis, Eine Integralgleichungs Methode zur Lösung des Torsionsproblems des Umdrehungs Körpers, *Acta Mechanica* **16**, 175–181 (1973).

[104] T. Kermanidis, A Numerical Solution for Axially Symmetrical Elasticity Problems, *International Journal of Solids and Structures* **11**, 493–500 (1975).

[105] T. Kermanidis, Kupradze's Functional Equation for the Torsion Problem of Prismatic Bars, Pts. I & II, *Computer Methods in Applied Mechanics and Engineering* **7**, 39–46 & 249–259 (1976).

[106] M.S. Khader and M.C. Hanna, An Iterative Boundary Numerical Solution for General Steady Heat Conduction Problems, *Journal of Heat Transfer, Transactions of the ASME* **103**, 26–31 (1981).

[107] H.R. Kutt, The Numerical Evaluation of Principal Value Integrals by Finite-Part Integration, *Numerische Mathematik* **24**, 205–210 (1975).

[108] H.R. Kutt, Quadrature Formulae for Finite Part Integrals, *Report WISK* 178, *The National Research Institute of Mathematical Sciences, Pretoria* (1975).

[109] H.R. Kutt, On the Numerical Evaluation of Finite Part Integrals Involving an Algebraic Singularity, *Report WISK* 179, The National Research Institute for Mathematical Sciences, Pretoria (1975).

[110] J.C. Lachat and J.O. Watson, Progress in the Use of Boundary Integral Equations Illustrated by Examples, *Computer Methods in Applied Mechanics and Engineering* **10**, 273–289 (1975).

[111] J.C. Lachat and J.O. Watson, Effective Numerical Treatment of Boundary Integral Equations: A Formulation for Three-Dimensional Elastostatics, *International Journal for Numerical Methods in Engineering* **10**, 991–1005 (1976).

[112] U. Lamp, T. Schleicher, E. Stephan and W. Wendland, Theoretical and Experimental Asymptotic Convergence of the Boundary Value Problem, in: *Boundary Element Methods in Engineering*, C.A. Brebbia, Ed. (Springer-Verlag, Berlin, 1982) pp. 3–17.

[113] M.H. Lean, M. Friedman and A. Wexler, Application of the Boundary Element Method in Electrical Engineering Problems, in: *Developments in Boundary Element Methods* – 1, P.K. Banerjee and R. Butterfield, Eds. (Applied Science Publishers, London, 1979) pp. 207–250.

[114] G.P. Lennon, P.L-F. Liu and J.A. Liggett, Boundary Integral Equation Solution to Axisymmetric Potential Flows: 1, Basic Formulation, *Water Resources Research* **15**, 1102–1106 (1979).

[115] G.P. Lennon, P.L-F. Liu and J.A. Liggett, Boundary Integral Equation Solution to Axisymmetric Potential Flows: 2, Recharge and Well Problems in Porous Media, *Water Resources Research* **15**, 1107–1115 (1979).

[116] G.P. Lennon, P.L-F. Liu and J.A. Liggett, Boundary Integral Solutions to Three-Dimensional Unconfined Darcy's Flow, *Water Resources Research* **16**, 651–658 (1980).

[117] J.A. Liggett, Location of Free Surface in Porous Media, *Journal of Hydraulics Division, ASCE* **103**, (HY4), 353–365 (1977).

[118] J.A. Liggett, Hydrodynamics Calculations Using Boundary Elements, in: *Finite Element Flow Analysis*, T. Kawai, Ed. (University of Tokyo Press, Tokyo, 1982) pp. 889–896.

[119] J.A. Liggett and P.L-F. Liu, *The Boundary Integral Equation Method for Porous Media Flow* (G. Allen and Unwin, London, 1983).

[120] J.A. Liggett and P.L-F. Liu, Applications of Boundary Element Methods to Fluid Mechanics, in: *Topics in Boundary Element Research*, Vol. 1, C.A. Brebbia, Ed. (Springer-Verlag, Berlin, 1984) pp. 78–96.

[121] P.L-F. Liu, Boundary Integral Equation Method Applied to Free Surface Flow Problems, in: *Innovative Numerical Analysis for the Engineering Sciences*, R. Shaw, et al, Eds. (University Press of Virgina, Charlottesville, 1980) pp. 179–188.

[122] C.C. Lo and F.W. Niedenfuhr, Singular Integral Equation Solution for Torsion, *Journal of the Engineering Mechancis Division, ASCE* **96**, 535–542 (1970).

[123] I. Lotz, Zur Berechnung der Potentialströmung um Quergestellte Luftschiffkörper, *Ingenieur Archiv* **2**, 507–527 (1931).

[124] T.S. Luu and G. Coulmy, Method of Calculating the Compressible Flow Around an Aerofoil or a Cascade up to the Stockfree Transonic Range, *Computers and Fluids* **5**, 261–275 (1977).

[125] B.H. McDonald and A. Wexler, Finite-Element Solution of Unbounded Field Problems, *IEEE Transactions on Microwave Theory and Techniques* **MTT-20**, 841–847 (1972).

[126] B.H. McDonald, M. Friedman, M. Decreton and A. Wexler, Integral Finite-Element Approach for Solving the Laplace Equation, *Electronics Letters* **9**, 242–244 (1973).

[127] B.H. McDonald and A. Wexler, Mutually Constrained Partial Differential and Integral Equation Field Formulations, in: *Finite Elements in Electrical and Magnetic Field Problems*, M.V.K. Chari and P.P. Silvester, Eds. (John Wiley & Sons, Chichester, 1980) pp. 161–190.

[128] D. Makepeace, M.J. Diss, N. Mullineux and J.R. Reed, Some Aspects of the Boundary Element Method, *Applied Mathematical Modelling* **6**, 229–232 (1982).

[129] M. Margulies, Exact Treatment of the Exterior Problem in the Combined FEM–BEM, in: *New Developments in Boundary Element Methods*, C.A. Brebbia, Ed. (CML Publications, Southhampton, 1980) pp. 43–64.

[130] M. Margulies, Combination of the Boundary Element and Finite Element Methods, in: *Progress in Boundary Element Methods*, Vol. 1, C.A. Brebbia, Ed. (Pentech Press, London, 1981) pp. 258–288.

[131] E. Martensen, Berechnung der Druckverteilung an Gitterprofilen in Ebener Potentialströmung mit Einer Fredholmschen Integralgleichung, *Archive for Rational Mechanics and Analysis* **3**, 235–270 (1959).

[132] A. Martin, I. Rodriguez and E. Alarcón, Mixed Elements in the Boundary Theory, in: *New Developments in Boundary Element Methods* (CML Publications, Southampton, 1980) pp. 34–42.

[133] C.E. Massonnet, Numerical Use of Integral Procedures, in: *Stress Analysis*, O.C. Zienkiewicz and G.S. Holister, Eds. (Wiley, London, 1965) pp. 198–235.

[134] J.R. Mautz and R.F. Harrington, Computation of Rotationally Symmetric Laplacian Potentials, *Proceedings of the Institution of Electrical Engineers* **117**, 850–852 (1970).

[135] M. Mayr and W. Neureiter, Ein Numerisches Verfahren zur Lösung des Axiallysymmetrischen Torsionsproblems, *Ingenieur Archiv* **46**, 137–142 (1977).

[136] A. Mendelson, Boundary-Integral Methods in Elasticity and Plasticity, *NASA Technical Note* TN D-7418 (1973).

[137] Y. Miyamoto, Analysis of Axisymmetrical Torsion Problems by the Boundary ELement Method, in: *Boundary Elements*, C.A. Brebbia, T. Futagami and M. Tanaka, Eds. (Springer-Verlag, Berlin, 1983) pp. 355–364.

[138] S. Mukherjee, M.A. Morjaria and F.C. Moon, Eddy Current Flows Around Cracks in Thin Plates for Nondestructive Testing, *Journal of Applied Mechanics* **49**, 389–395 (1982).

[139] N.I. Muskhelishvili, *Singular Integral Equations*, (Noordhoff Publishing Co., Groningen, 1953).

[140] J.C. Nedelec, Curved Finite Element Method for the Solution of Singular Integral Equations on Surfaces in R^3, *Computer Methods in Applied Mechanics and Engineering* **8**, 61–80 (1976).

[141] Y. Niwa, S. Kobayashi and T. Fukui, An Application of the Integral Equation Method to Seepage Problems, *Theoretical and Applied Mechanics* **24**, 479–486 (1976).

[142] M. Okabe, A Boundary Integral Approach in the Geoelectrical Cavity Prospecting, *Computer Methods in Applied Mechanics and Engineering* **29**, 297–311 (1981).

[143] T. Onuki, A. Ishiyama, T. Abe and Y. Murai, Hybrid Finite-Boundary Element Method Using Vector and Scalar Potentials for Open Boundary Magnetic Field Problem, in: *Boundary Elements*, C.A. Brebbia, T. Futagami and M. Tanaka, Eds. (Springer-Verlag, Berlin, 1983) pp. 113–122.

[144] N. Papamichael and G.T. Symm, Numerical Techniques for Two-Dimensional Laplacian Problems, *Computer Methods in Applied Mechanics and Engineering* **6**,175–194 (1975).

[145] F. Paris, A. Martin and E. Alarcón, Potential Theory, in: *Progress in Boundary Element Methods*, Vol. 1, C.A. Brebbia, Ed. (Halsted Press, New York, 1981) pp. 45–83.

[146] C. Patterson and M.A. Sheikh, Non-conforming Boundary Elements for Stress Analysis, in: *Boundary Element Methods*, C.A. Brebbia, Ed. (Springer-Verlag, Berlin, 1981) pp. 137–152.

[147] C. Patterson and M.A. Sheikh, A Regular Boundary Element Method for Fluid Flow, *International Journal for Numerical Methods in Fluids*, **2**, 239–251 (1982).

[148] C. Patterson and N.A.S. El-Sebai, A Regular Boundary Method Using Nonconforming Elements for Potential Problems in Three Dimensions, in: *Boundary Element Methods in Engineering*, C.A. Brebbia, Ed. (Springer-Verlag, Berlin, 1982) pp. 112–126.

[149] C. Patterson and N.A.S. El-Sebai, A Family of Partially Discontinuous Boundary Elements for Three-Dimensional Analyses, in: *Boundary Elements*, C.A. Brebbia, T. Futagami and M. Tanaka, Eds. (Springer-Verlag, Berlin, 1983) pp. 193–206.

[150] C. Patterson and M.A. Skeikh, Interelement Continuity in the Boundary Element Method, in: *Topics in Boundary Element Research*, Vol. 1, C.A. Brebbia, Ed. (Springer-Verlag, Berlin, 1984) pp. 123–141.

[151] H.L.G. Pina, J.L.M. Fernandes and C.A. Brebbia, Some Numerical Integration Formulae Over Triangles and Squares with a $1/R$ Singularity, *Applied Mathematical Modelling* **5**, 209–211 (1981).

[152] A.R.S. Ponter, An Integral Equation Solution of the Ihomogeneous Torsion Problem, *SIAM Journal on Applied Mathematics* **14**, 819–830 (1966).

[153] W. Prager, Die Druckverteilung an Körpern in Ebener Potentialströmung, *Physikalische Zeitschrift* **29**, 865–869 (1928).

[154] P.M. Quinlan, The Torsion of an Irregular Polygon, *Proceedings of the Royal Society of London* **282** (A), 208–227 (1964).

[155] P.M. Quinlan, J.E. Fitzgerald and S.N. Atluri, The Edge-Function Method, in: *Proceedings of the International Symposium on Innovative Numerical Analysis in Applied Engineering Science*, T.A. Cruse et al, Eds. (CETIM, Senlis, France, 1977) pp. 15–18.

[156] P.M. Quinlan, J.J. Grannell, S.N. Atluri and J.E. Fitzgerald, Boundary Discretization Using the Edge-Function Method in Three Dimensional Elasticity, *Applied Mathematical Modelling* **3**, 18–24 (1979).

[157] P. Riccardella, An Implementation of the Boundary Integral Techniques for Plane Problems in Elasticity and Elastoplasticity, Ph.D. Thesis, Carnegie Mellon University, Pittsburgh, Pennsylvania (1973).

[158] F.J. Rizzo, An Integral Equation Approach to Boundary Value Problems of Classical Elastostatics, *Quarterly of Applied Mathematics* **25**, 83–95 (1967).

[159] F.J. Rizzo, The Boundary-Integral Equation Method: A Modern Computational Procedure in Applied Mechanics, in: *Boundary-Integral Equation Method: Computational Applications in Applied Mechanics*, AMD – Vol. 11, T.A. Cruse and F.J. Rizzo, Eds. (ASME, New York, 1975) pp. 1–5.

[160] F.J. Rizzo, A.K. Gupta and Y. Wu, A Boundary Integral Equation Method for Torsion of Variable Diameter Circular Shafts and Related Problems, in: *Innovative Numerical Analysis for the Engineering Sciences*, R.P. Shaw et al, Eds. (University Press of Virginia, Charlottesville, 1980) pp. 373–380.

[161] H. Rouse, *Fluid Mechanics for Hydraulic Engineers* (Dover Publications, New York, 1961).

[162] M.A. Sadowsky and E. Sternberg, Elliptic Integral Representation of Axially Symmetric Flows, *Quarterly of Applied Mathematics* **8**, 113–126 (1950).

[163] E. Sauer and G. Mehlhorn, Application of the Boundary Element Method to Shear and Torsion Problems in Elastic Prismatic Members, in: *Innovative Numerical Analysis for the Engineering Sciences*, R.P. Shaw et al, Eds. (University Press of Virginia, Charlottesville, 1980) pp. 655–665.

[164] D.J. Shippy, F.J. Rizzo and A.K. Gupta, Boundary-Integral Solution of Potential Problems Involving Axisymmetric Bodies and Non-symmetric Boundary Conditions, in: *Developments in Theoretical and Applied Mechanics* Vol. 10, J.E. Stoneking, Ed. (*Proceedings of the 10th Southeastern Conference of Theoretical and Applied Mechanics*, Knoxville, Tennessee, April 17–18, 1980) pp. 189–206.

[165] P. Silvester and M.-S. Hsieh, Finite-Element Solution of 2-Dimensional Exterior-Field Problems, *Proceedings of the Institution of Electrical Engineers* **118**, 1743–1747 (1971).

[166] P.P. Silvester, The Fundamental Equations of Electric and Magnetic Fields, in: *Finite Elements in Electrical and Magnetic Field Problems*, M.V.K. Chari and P.P. Silvester, Eds. (John Wiley & Sons, Chichester, 1980) pp. 33–48.

[167] P. Skerget and C.A. Brebbia, Non-Linear Potential Problems, in: *Progress in Boundary Element Methods*, Vol. 2, C.A. Brebbia, Ed. (Pentech Press, London, 1983) pp. 1–23.

[168] A.M.O. Smith and J. Pierce, Exact Solution of the Neumann Problem. Calculation of Non-circulatory Plane and Axially Symmetric Flows About or Within Arbitrary Boundaries, in: *Proceedings of the 3rd US National Congress of Applied Mechanics*, R.M. Haythornthwaite, Ed. (The American Society of Mechanical Engineers, New York, 1958) pp. 807–815.

[169] I.S. Sokolnikoff, *Mathematical Theory of Elasticity* (McGraw-Hill, New York, 1956).

[170] L. Solymar, *Lectures on Electromagnetic Theory: A Short Course for Engineers* (Oxford University Press, London, 1976).

[171] W.J. Sternberg and T.L. Smith, *The Theory of Potential and Spherical Harmonics* (University of Toronto Press, Toronto, 1944).

[172] H. Strese, Remarks Concerning the Boundary Element Method in Potential Theory, *Applied Mathematical Modelling*, **8**, 40–44 (1984).

[173] A.H. Stroud and D. Secrest, *Gaussian Quadrature Formulas* (Prentice Hall, Englewood Cliffs, NJ, 1966).

[174] G.T. Symm, Integral Equation Methods in Potential Theory, II, *Proceedings of the Royal Society of London* **275** (A), 33–46 (1963).

[175] G.T. Symm, Potential Problems in Three-Dimensions, in: *Numerical Solution of Integral Equations*, L. Delves and J. Walsh, Eds. (Clarendon Press, Oxford, 1974) pp. 312–320.

[176] G.T. Symm, The Robin Problem for Laplace's Equation, in: *New Developments in Boundary Element Methods*, C.A. Brebbia, Ed. (CML Publications, Southampton, 1980) pp. 112–121.

[177] G.T. Symm, Two Methods for Computing the Capacitance of a Quadrilateral, in: *Boundary Element Methods*, C.A. Brebbia, Ed. (Springer-Verlag, Berlin, 1981) pp. 3–13.

[178] G.T. Symm, The Robin Problem in a Multiply-Connected Domain, in: *Boundary Element Methods in Engineering*, C.A. Brebbia, Ed. (Springer-Verlag, Berlin, 1982) pp. 89–100.

[179] J.C. Thompson and E. Post, Asymptotic Analysis Techniques for Stress-Concentration Regions in Plane Problems, *Strain* **13**, 137, 139–143 (1977).

[180] G.R. Tomlin, Numerical Analysis of Continuum Problems in Zoned Anisotropic Media, Ph.D. Thesis, Southampton University, Southampton, England, 1972.

[181] E. Trefftz, Über die Kontraktion Kreisförmiger Flüssigkeitsstrahlen, *Zeitschrift für Mathematik und Physik* **64**, 34–61 (1917).

[182] E. Trefftz, Ein Gegenstück zum Ritzschen Verfahren, in: *Proceedings of the 2nd International Congress for Applied Mechanics*, E. Meissner, Ed. (Orell Füssli Verlag, Zurich, 1927) pp. 131–137.

[183] C.W. Trowbridge, Applications of Integral Equation Methods to the Numerical Solution of Magnetostatic and Eddy-Current Problems, in: *Finite Elements in Electrical and Magnetic Field Problems*, M.V.K. Chari and P.P. Silvester, Eds. (John Wiley & Sons, Chichester, 1980) pp. 191–213.

[184] F. Vandrey, On the Calculation of the Transverse Potential Flow Past a Body of Revolution with the Aid of the Method of Mrs. Flugge-Lotz, *Astia* AD-40089 (1951).

[185] A. Weinstein, On Axially Symmetric Flows, *Quarterly of Applied Mathematics* **5**, 429–444 (1948).

[186] W.L. Wendland, On the Asymptotic Convergence of Boundary Integral Methods, in: *Boundary Element Methods*, C.A. Brebbia, Ed. (Springer-Verlag, Berlin, 1981) pp. 412–430.

[187] W.L. Wendland, Asymptotic Accuracy and Convergence, in: *Progress in Boundary Element Methods*, Vol. 1, C.A. Brebbia, Ed. (Pentech Press, London, 1981) pp. 289–313.

[188] D.H. Wilkinson, Analysis and Design of Aerofoils and Cascades in Incompressible Flow, *Journal of Science and Technology* **36**, 99–104 (1969).

[189] L.C. Wrobel and C.A. Brebbia, Axisymmetric Potential Problems, in: *New Developments in Boundary Element Methods*, C.A. Brebbia, Ed. (CML Publications, Southampton, 1980) pp. 77–89.

[190] Y.S. Wu, The Boundary-Integral Equation Method Using Various Approximation Techniques for Problems Governed by Laplace's Equation, Air Force Office of Scientific Research Technical Report AFOSR-TR-76-1313 (1976).

[191] Y.S. Wu, F.J. Rizzo, D.J. Shippy and J.A. Wagner, An Advanced Boundary Integral Equation Method for Two-Dimensional Electromagnetic Field Problems, *Electric Machines and Electromechanics* **1**, 301–313 (1977).

[192] F. Yoshikawa and M. Tanaka, Boundary Elements in Axisymmetric Potential Problems, in: *Boundary Element Methods in Engineering*, C.A. Brebbia, Ed. (Springer-Verlag, Berlin, 1982) pp. 101–111.

[193] O.C. Zienkiewicz, D.W. Kelly and P. Bettess, The Coupling of the Finite Element Method and Boundary Solution Procedures, *International Journal for Numerical Methods in Engineering* **11**, 355–375 (1977).

[194] O.C. Zienkiewicz, *The Finite Element Method* (McGraw-Hill, London, 1977).

CHAPTER 3

Elastostatics

Maria S. GÓMEZ-LERA and Enrique ALARCÓN

School of Industrial Engineering
Polytechnical University of Madrid
Madrid, Spain

Boundary Element Methods in Mechanics
Edited by D.E. Beskos

Introduction

In this chapter, we are going to describe the main features as well as the basic steps of the Boundary Element Method (BEM) as applied to elastostatic problems and to compare them with other numerical procedures. As we shall show, it is easy to appreciate the advantages of the BEM, but it is also advisable to refrain from a possible unrestrained enthusiasm, as there are also limitations to its usefulness in certain types of problems. The number of these problems, nevertheless, is sufficient to justify the interest and activity that the new procedure has aroused among researchers all over the world.

Briefly speaking, the most frequently used version of the BEM as applied to elastostatics works with the fundamental solution, i.e. the singular solution of the governing equations, as an influence function and tries to satisfy the boundary conditions of the problem with the aid of a discretization scheme which consists exclusively of boundary elements. As in other numerical methods, the BEM was developed thanks to the computational possibilities offered by modern computers on a totally "classical" basis. That is, the theoretical grounds are based on linear elasticity theory, incorporated long ago into the curricula of most engineering schools. Its delay in gaining popularity is probably due to the enormous momentum with which Finite Element Methods (FEM) penetrated the professional and academic media. Nevertheless, the fact that these methods were developed before the BEM has been beneficial because the BEM successfully uses those results and techniques studied in past decades. Some authors even consider the BEM as a particular case of the FEM, while others view both methods as special cases of the general weighted residual technique [24].

The first paper usually cited in connection with the BEM as applied to elastostatics is that of Rizzo [95], even though the works of Jaswon et al. [65], Massonnet [79] and Oliveira [86] were published at about the same time, the reason probably being the attractiveness of the "direct" approach [95] over the "indirect" one [65, 79, 86]. The work of Rizzo [95] and the subsequent work of Cruse [45] initiated a fruitful period with applications of the direct BEM to problems of elastostatics, elastodynamics, fracture, etc. The next key contribution was that of Lachat and Watson [73] incorporating all the FEM discretization philosophy in what is sometimes called the "second BEM generation". This has, no doubt, led directly to the current developments.

Among the various researchers who worked on elastostatics by employing the direct BEM, one can additionally mention Rizzo and Shippy [96–99], Cruse et al. [46–48], Lachat and Watson [74, 75], Alcarcón et al. [1–4], Brebbia et al. [26, 27], Brady et al. [22, 23], Wardle and Crotty [113], Dominguez et al. [53], Howell and Doyle [63, 64], Kuhn and Möhrmann [72] and Patterson and Sheikh [89–91], and among those who used the indirect BEM, one can additionally mention Benjumea and Sikarskie [20], Butterfield et al. [37–39], Tomlin and Butterfield [108], Banerjee et al. [13–15], Niwa et al. [85] and Altiero and Gavazza [10]. An interesting version of the indirect method, called the Displacement Discontinuity Method (DDM) has been developed by Crouch [42–44]. A comprehensive study on various special aspects of the elastostatic BEM has been done by Heisse [60–62], while review-type articles on the subject have been reported by Watson [114] and Hartmann [59].

At the present time, the method is well established and is being used for the solution of a variety of problems in engineering mechanics. Numerous introductory and advanced books have been published [66, 24, 28, 17, 44, 35] as well as research-oriented ones [16, 18, 31, 34]. In this sense, it is worth noting the series of conferences promoted by Brebbia since 1978 [25, 29, 30, 32, 33], which have provoked a continuous research effort all over the world in relation to the BEM.

In the following sections, we shall concentrate on developing the direct BEM as applied to elastostatics. Section 1 will present the fundamentals of the BEM, indicating its main features and fields of application as well as some general considerations on the different computational steps. The idea is to follow the well known path of the FEM and to comment on the major differences in a formal fashion. Section 2 contains more detailed material related to elastostatics problems, including such classical topics as the reciprocity theorem and Somigliana's identity, as well as the discretization procedure for both the direct and the indirect BEM. Special care has been taken to show the reduction to integrals on the boundary of several interesting body forces. Section 3 will present the development necessary for treating two-dimensional domains. The presentation will be done using linear elements and Kelvin-type fundamental solutions.

The method is particularly attractive in three-dimensional cases, and section 4 is dedicated to the corresponding implementation. It will be shown that owing to the properties of Kelvin's fundamental solution, the work is very simple and the treatment of conditions at infinity superior even to that of the plane case. Here the interpolation is presented with constant elements to prevent complications with corners, of the type treated in section 3, from shadowing the BEM properties. Section 5 discusses a special three-dimensional case, that of axisymmetric bodies. In this condition, the two-dimensional aspect is revised for the definition of elements although the fundamental solutions present typical features. Finally, section 6 briefly discusses the case of anisotropy, while section 7 is devoted to the special topics of bending of plates and shells, coupling of the BEM with the FEM and adaptive procedures in BEM analysis.

1. Fundamentals of the Boundary Element Method

1.1. Primary features of the method

The BEM proceeds by discretizing the boundary of the domain under study (Fig. 1) by subdivision into elements that support the zones where the data or the unknowns are supposed to vary according to a certain predefined fashion, e.g. linear, parabolic, cubic, etc. From that point of view, the BEM is similar to the FEM and the previous hypothesis would be identified with the usual "interpolation functions". A first and important difference is that of discretizing the boundary and not the domain with the logical advantages of simplification so conveyed.

Another important difference is with regard to the quantities used by the method. They correspond exactly to the field variables of the problem (stresses, displacements, temperatures, etc.) so that the concept of "equivalent nodal forces" disappears. That is essentially due to the fact that the above-mentioned interpolation is one simultaneously on the dual problem variables, i.e. stresses and displacements, temperature and fluxes, etc. It is worth noting that both interpolations can be independent because there is no relationship between their evolution along the boundary; nevertheless, it is common to apply the same degree of polynomial interpolation to both variables. In other cases, it could be advantageous to use a mixture of different degrees, generating the so-called "mixed elements" [78], or to incorporate in the polynomials the singular behavior of the solution near stress concentration or singular points [57]. In any case, it is logical to hope that the simultaneous interpolation of all field variables will produce more accurate results, as we shall see later.

It is important to point out from the beginning that the basis of the procedure hinges on the computation of the variables at the boundary. If it is desirable to compute values inside the domain, a relatively costly complementary algorithm must be established. It is well known that the FEM works precisely with values defined inside the domain and therefore, as a general conclusion, it can be said that the BEM is preferable in those cases where only the boundary values are

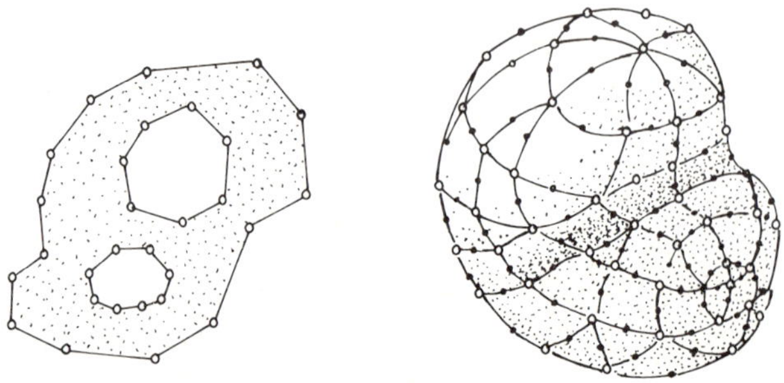

Fig. 1. Boundary element discretization of two- and three-dimensional domains.

needed, although it is important to remember the advantage of discretizing only the boundary.

Finally, a more technical peculiarity will be cited which is not obvious to the user but which hinges heavily on the programmer and on the effectiveness of the method. It is related to the "weighting functions" that, on the one hand form the core of the advantages of the method, while on the other hand burden it with their inconvenience. In short, the weighting functions are globally defined in opposition to the locally based FEM functions or to the interpolating functions. This implies two serious consequences: first, the resultant matrices are full and, second, they are not symmetric. Both conditions provoke the use of algorithms different from those of the FEM and which hinge on the storage and the methods used to solve the equations but do not substantially affect the user.

1.2. Fields of application

The areas to which the BEM can be applied are enormously varied, as can be seen by rereading the above-mentioned references. In principle, the ideal field of application includes linear problems defined over isotropic homogeneous domains, where a fundamental solution of the field problem is defined. In what follows, we shall restrict the appraisal to elastostatics.

The limitation on anisotropy is relatively easy to eliminate as is that of continuous heterogeneity, although, regarding the former, it is necessary to know the corresponding fundamental solution. The case of the piecewise heterogeneity is treated by a zoning technique (Fig. 2a), and the use for every subdomain of the

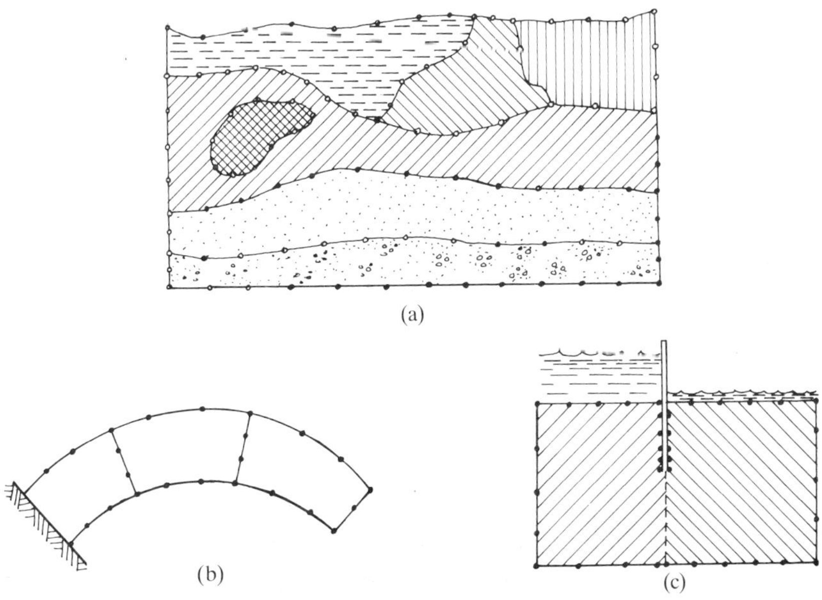

Fig. 2. Zoning technique in boundary element analysis.

corresponding fundamental solution. That is also interesting when treating aligned bodies, which would need a great number of boundary elements due to simple computational problems (Fig. 2b). A similar situation arises in cases with small or null distances (Fig. 2c), where it would be complicated to establish different conditions at the same node. Occasionally, this type of zoning is even included artificially to produce block matrices which can be favorably treated by the computer [74].

In problems with nonzero body forces or those in which an approximate fundamental solution is used, volume integrals on quantities varying in an arbitrary manner over the domain appear in the formulation. The usual technique is the definition of integration "cells" with a predefined evolution as a function of the nodal values (Fig. 3). They are not finite "elements" because they do not contribute unknowns to the equations. It is fair to recognize, nevertheless, that an important part of the spectacularity and advantages of the method are lost in those cases, because a domain mesh must be defined and values have to be computed at the mesh nodes. This is costly, although several attempts have been made to alleviate this disadvantage.

Another specific area of applicability is the treatment of areas with boundaries at infinity, as usually assumed in soil–structure interaction (Fig. 4). In those cases, the variables of interest are related to a small part of the domain (the structure) where a detailed knowledge of the field variables is important, while the other is a contact or interface boundary. A recommended procedure is then to

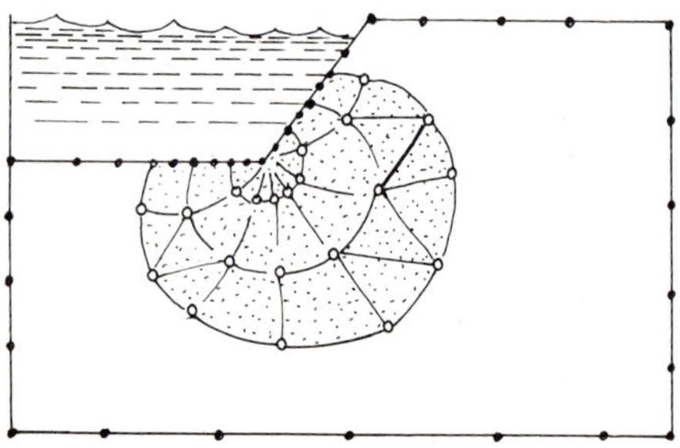

Fig. 3. Use of internal cells in the BEM.

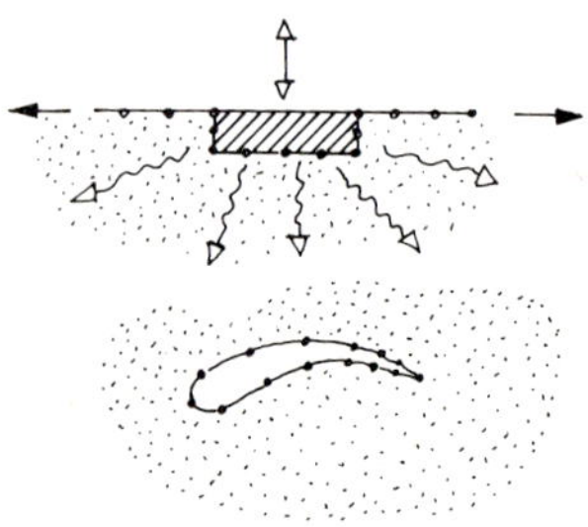

Fig. 4. Automatic treatment of domains with infinite boundaries by the BEM.

apply the FEM to the structure and to model the rest of the media by the BEM in a "marriage à la mode", as it was called by Zienckiewicz, where "the best of both worlds" is used [117].

1.3. Basic steps of the method

As in every engineering situation, the initial step is the selection of the mathematical model best suited to the properties of the problem at hand and consideration of this model from the point of view of the BEM possibilities. If the answer is affirmative, the steps are similar to those used in the FEM, i.e.:

(a) The domain is subdivided into the number of subregions necessary to reflect its peculiarities, either of heterogeneity or of a geometrical nature.

(b) The boundaries of each subregion are subdivided, using the concept of "element", i.e. the degree of interpolation on that part of the boundary is chosen. Usually, the boundary is discretized by the same shape functions (isoparametric elements), although that is not strictly necessary.

(c) In order to build the pre-established degree of interpolation, a number of nodes is defined. It is important to note that these nodes can be anywhere that they are needed (Fig. 5a). In general, it is common to work with compatible elements, in the sense that neighboring elements share nodes at their sides or vertex. This has the advantage of reducing the number of equations to be solved

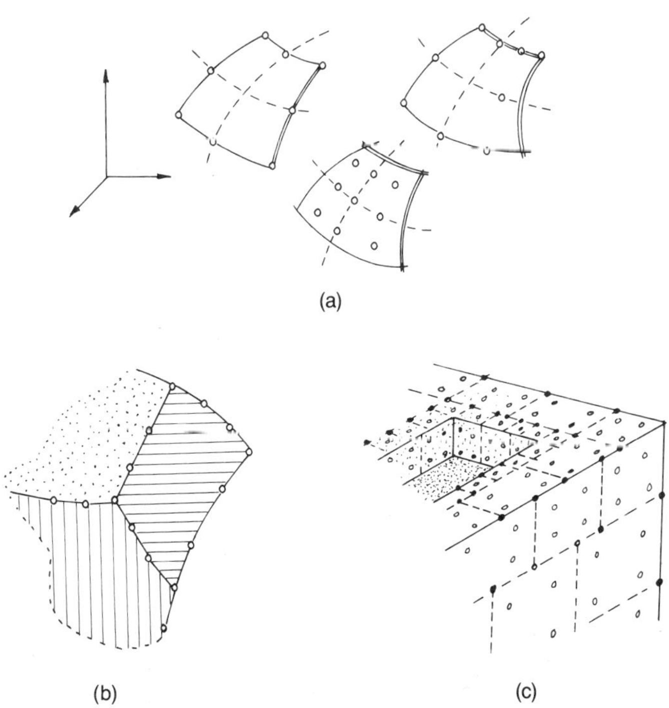

Fig. 5. Various types of higher-order boundary elements.

but may present complications with discontinuities, interfaces or singularities where it is most convenient to give the model more adaptiveness (Fig. 5b). Some authors prefer the first working possibility, especially in three-dimensional problems, where an extraordinary capacity for adaptation to the peculiarities of the problem can be achieved (Fig. 5c).

(d) For every degree of freedom (d.o.f.) at each node, an equation is built through an integral representation formula, using as weighting functions the fields of displacements and stresses generated by unit loads along the d.o.f. acting (in general) on an infinite space with the same material properties of the body under study. At this point, important computational problems appear, due to the necessity of computing singular integrals with accuracy and effectiveness. Usually, those unit loads are placed at the previously defined nodes. It is possible, nevertheless, that in some problems it might be interesting to define new nodes to which those loads would be applied. This is precisely what is done in the so-called "indirect methods" where, on occasion, the nodes are situated outside the domain, or in the "hierarchical elements" where collocation points are completely independent of the geometry.

(e) The natural and essential variables in the boundary are contained in two vectors that will be called, respectively, $\boldsymbol{t}$ and $\boldsymbol{u}$ and the previously computed factors are stored in matrices $\boldsymbol{A}$ and $\boldsymbol{B}$, forming a system

$$\boldsymbol{A u} = \boldsymbol{B t}\,, \tag{1.1}$$

where $\boldsymbol{u}$ and $\boldsymbol{t}$ contain data as well as unknowns.

(f) Once $\boldsymbol{A}$ and $\boldsymbol{B}$ have been built (or simultaneously, to gain time), the boundary conditions are established, imposing the known values of the variables on the different pieces. The problem is then reduced to the classical form

$$\boldsymbol{K x} = \boldsymbol{F}\,, \tag{1.2}$$

that is, to a system of equations where $\boldsymbol{x}$ contains all the boundary unknowns and $\boldsymbol{F}$, the product of the $\boldsymbol{A}$ and $\boldsymbol{B}$ coefficients by the corresponding data.

(g) Solution of the system (1.2) allows the determination of all the boundary information. If results at internal points are sought, an additional step is required. New $\boldsymbol{A}$ and $\boldsymbol{B}$ matrices have to be computed (one set for each internal point) and multiplied by the boundary values, as will be shown later. In this process of computing new $\boldsymbol{A}$ and $\boldsymbol{B}$ matrices, a significant amount of time is used, so that, if the number of internal points is large, the advantages of using the BEM may disappear.

1.4. *The fundamental solution*

As we mentioned in point (d) above, the construction of every equation follows the general pattern of the FEM, i.e. an integral formula relates internal and external energy of loads and stresses. There are two main differences between the method under study and the Displacements Finite Element Method. First, the basic integral formula is a reciprocity relationship among the actual problem and a series of virtual problems. That means that there are energies due to virtual stresses as well as those related to virtual displacements, so that it will be necessary to discretize both the actual stresses and the displacements. Second, the virtual state is chosen so that the field equations can be satisfied. There are several ways of doing that, but the classical BEM technique uses the so-called fundamental solution, that is, the fields generated by point loads acting on an infinite space. Thus, if $\mathscr{A}$ represents the operator of the problem, the weighting functions are the fields related to the solution of

$$\mathscr{A}\boldsymbol{u}^*(\boldsymbol{x}) = \delta(\boldsymbol{x} - \boldsymbol{x}_\mathrm{P})\boldsymbol{e} \tag{1.3}$$

where δ is the Dirac (delta) function, $\boldsymbol{x}$ represents an arbitrary point inside an infinite space, $\boldsymbol{x}_\mathrm{P}$ are the coordinates of point P (where the unit load is applied) and $\boldsymbol{e}$ is a unit vector pointing in the d.o.f. direction.

By varying the point P, it is possible to generate as many weighting functions as necessary, the simplest way being the use of the same points defined to interpolate the actual variables. The above-mentioned fundamental solution for linear elasticity is well known. It is the solution to the so-called Kelvin problem [76] and is the most appropriate for general purposes. Other fundamental solutions are related to special conditions. For instance, the Mindlin problem [35] can be used to model half-spaces, the Boussinesq problem [112] can be applied to certain soil-fundation situations, etc. Two features related to the special choice of weighting functions are worth noting. First, the reduction to only one term of the energy related to virtual loads, i.e. the work of the applied Dirac delta on the actual $u(\boldsymbol{x}_\mathrm{P})$ of point P. This implies elimination of the trouble related to computations of domain integrals. Then the "globally based" property of the weighting functions will produce interactions with all the interpolating ones, independently of their global or local definition. In other words, as we said in section 1.1, the matrices to be obtained are full, as opposed to the FEM matrices. Finally, the special character of the delta function will cause trouble in the computation, especially when trying to compute integrals relating the same collocation and interpolation points. This problem will be dealt with in subsequent sections.

2. Classical relationships in linear elasticity

In this section, the formulas controlling the response of linear elastic bodies to static loads will be established and some classical relationships directly related to boundary methods will be presented.

2.1. Elasticity problems

In what follows, the discussion will be organized around problems controlled by the Navier equation for linear isotropic and homogeneous elastostatics [76]

$$(\lambda + G)\ \mathrm{grad\,div}\ \boldsymbol{u} + G\nabla^2\boldsymbol{u} + \boldsymbol{X} = \boldsymbol{0}\,, \tag{2.1}$$

where u is the displacement vector

$$\boldsymbol{u}^{\mathrm{T}} = \{u_1, u_2, u_3\}\,, \tag{2.2}$$

$\boldsymbol{X}$ is the body force vector

$$\boldsymbol{X}^{\mathrm{T}} = \{X_1, X_2, X_3\}\,, \tag{2.3}$$

while λ and G are the classical Lame constants related to the Young modulus E and Poisson ratio ν by

$$\lambda = \frac{\nu E}{(1+\nu)(1-2\nu)}\,, \qquad G = \frac{E}{2(1+\nu)}\,. \tag{2.4}$$

The solution of Eqs. (2.1) has been treated largely from a mathematical point of view and several well known closed-form solutions constitute the experience of the academic community. Nevertheless, the solution of situations with arbitrary geometry and general conditions on the boundary can still be considered to offer computational and practical interest.

It is well known that, for the problem to be "well posed", the field equations must be completed with the specification of suitable boundary conditions. Those conditions for a body Ω with boundary $\partial\Omega$ can be of the following types:

1. Essential conditions:
   The displacements are prescribed on the boundary, i.e.

$$\boldsymbol{u}(\boldsymbol{x}) = \boldsymbol{u}_0(\boldsymbol{x})\,, \qquad \boldsymbol{x} \in \partial\Omega\,. \tag{2.5}$$

This is similar to the Dirichlet-type conditions in potential problems.

2. Natural conditions:
   The stresses are prescribed on the boundary, i.e.

$$\overset{n}{\boldsymbol{T}}(\boldsymbol{x}) = \overset{n}{\boldsymbol{T}}_0(\boldsymbol{x}) \qquad \boldsymbol{x} \in \partial\Omega \tag{2.6}$$

where $\overset{n}{\boldsymbol{T}}$ is the traction vector at a point $\boldsymbol{x}$ where the normal to the boundary is $\boldsymbol{n}$.

It is well known that

$$\overset{n}{\boldsymbol{T}} = \boldsymbol{\sigma} \cdot \boldsymbol{n} \tag{2.7}$$

where $\boldsymbol{\sigma}$ is the stress tensor, and the material law is the Hooke's relationship

$$\sigma_{ij} = \lambda \varepsilon_{mm} \delta_{ij} + 2G\varepsilon_{ij} \tag{2.8}$$

with

$$\begin{aligned} \varepsilon_{ij} &= \tfrac{1}{2}(u_{ij} + u_{j,i}) \\ \delta_{ij} &= \begin{cases} 0 & \text{if} \quad i \neq j \\ 1 & \text{if} \quad i = j \end{cases} \end{aligned} \tag{2.9}$$

a comma denoting derivative and a repeated index summation.

Using (2.7), (2.8), (2.9) and (2.6), it is possible to write the natural conditions in terms of the displacement vector as

$$\lambda \boldsymbol{n} \operatorname{div} \boldsymbol{u} + 2G \frac{\partial \boldsymbol{u}}{\partial \boldsymbol{n}} + G(\boldsymbol{n} \times \operatorname{rot} \boldsymbol{u}) = \overset{n}{\boldsymbol{T}}_0(\boldsymbol{x}) \,. \tag{2.10}$$

It is possible to see that this condition is similar to the Neumann-type conditions in potential problems, only with respect to the second term on the left-hand side of (2.10).

3. Mixed conditions:

They are a combination of the previous conditions, either on different parts of the boundary or on a fixed one, where several stresses (normal, for example) and several displacements (tangential, for example) are given.

As examples, Figs. 6 and 7 offer several familiar cases of what has been said. Thus the problem depicted by Fig. 6(a) corresponds to simple traction of a strip. Obviously, the basic variables here are the displacement vector $\boldsymbol{u}$ and its dual, the traction $\boldsymbol{T}$. Apparently the displacement is constant along BC while the tractions

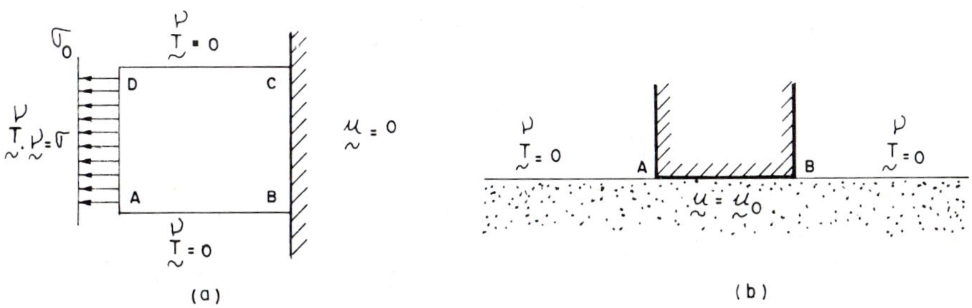

Fig. 6. Typical boundary-value problems in elastostatics with mixed boundary conditions.

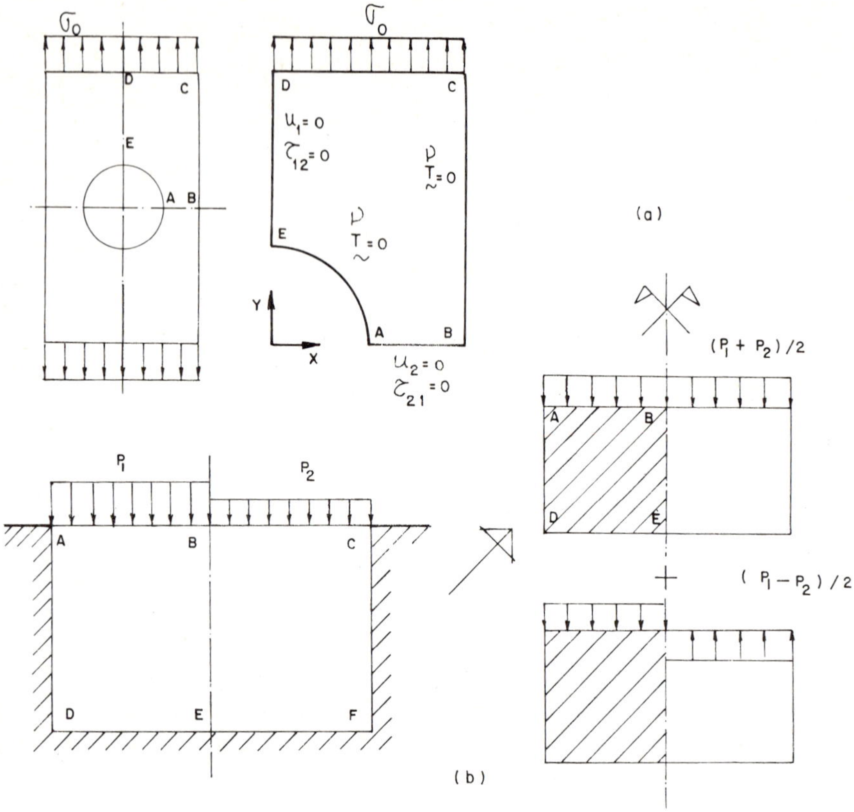

Fig. 7. Use of symmetry and antisymmetry in elastostatic boundary-value problems.

are null along AB and DC and $\boldsymbol{\sigma}_0 \cdot \boldsymbol{n}$ along DA. The problem shown in Fig. 6(b), where a rigid punch is applied to an elastic half-space, is also familiar. Along AB, the displacement is constant, while the free boundary offers no traction. Conditions at infinity are more complicated to establish. In general, it is assumed that stresses become bounded and displacements tend to zero. It is interesting to note that points A and B are singular points due to the sudden change in conditions from pure natural to pure essential. This implies the presence of infinite stresses and the need to solve an interesting numerical problem.

Occasionally, the conditions of symmetry or antisymmetry will allow a mixed problem, which is interesting for the good conditioning of matrices. In Fig. 7(a), we can see how a problem with natural boundary conditions can be transformed into a mixed one, taking advantage of the symmetry of the perforated strip and of the load. The first problem is undetermined, in the sense that a rigid-body displacement can be superimposed without affecting the state of stresses. On the other hand, after applying the conditions of symmetry, as indicated in Fig. 7(a), the displacements are relative to those specified. Problem 7(a) is also interesting because, along DE or AB, conditions are mixed in displacements and stresses.

Similarly, Fig. 7(b) presents the way to treat symmetrical geometry with antisymmetrical loads. The problem is decomposed into another two. In the first one, the interface BE cannot move horizontally, while in the second the vertical displacement is nil. The symmetry in the first case impedes the development of tangential vertical stresses, while, in the second, the antisymmetry along BE precludes the development of normal horizontal stresses. If, in addition, walls AD and CF are smooth, the first problem is a trivial one and is solved immediately. Once again a singular situation will develop at B, where the stress tensor has to produce two different traction values $\boldsymbol{p}_1$ and $\boldsymbol{p}_2$ at a point B with only one normal.

In the previous examples, it has been implicitly assumed that the medium was isotropic and homogeneous. For piecewise heterogeneities, as in Fig. 8(a), it is possible to establish a zoning scheme and, for each zone, apply the previous equations along with the specific material properties. In addition to the conditions at the exterior boundary, it is necessary now to introduce the compatibility of displacements and equilibrium of tractions along the interfaces. The zoning is also an artifice allowing the solution of asymmetric situations such as that of Fig. 8(b), by the introduction of an artificial interface.

In addition to the three classical boundary conditions, there are occasions in which the natural values are combined with the esential ones, either in a Robin–Newton mode of the form

$$a\boldsymbol{u} + b\overset{n}{\boldsymbol{T}} = \boldsymbol{c}\,, \tag{2.11}$$

or in a generalized oblicuum derivative, relating different components of $\overset{n}{\boldsymbol{T}}$, according to

$$a(\overset{n}{\boldsymbol{T}} \cdot \boldsymbol{n}) + b(\overset{n}{\boldsymbol{T}} \cdot \boldsymbol{s}) = c\,, \tag{2.12}$$

where $\boldsymbol{s}$ indicates the tangent unit vector.

An example of the first situation is a beam resting on springs (Fig. 9a), while the second one can be represented by Coulomb-like interfaces, as in Fig. 9(b).

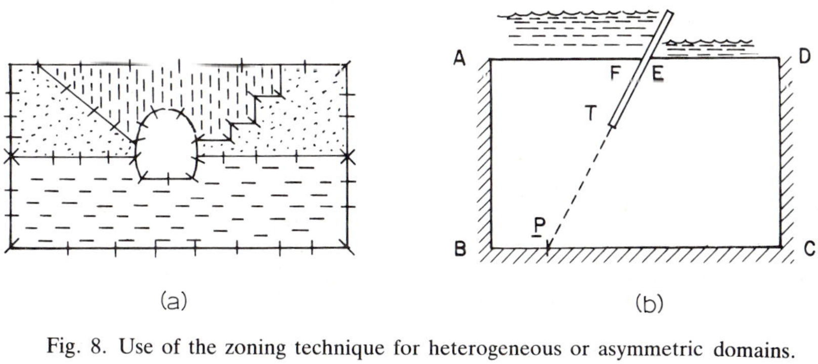

Fig. 8. Use of the zoning technique for heterogeneous or asymmetric domains.

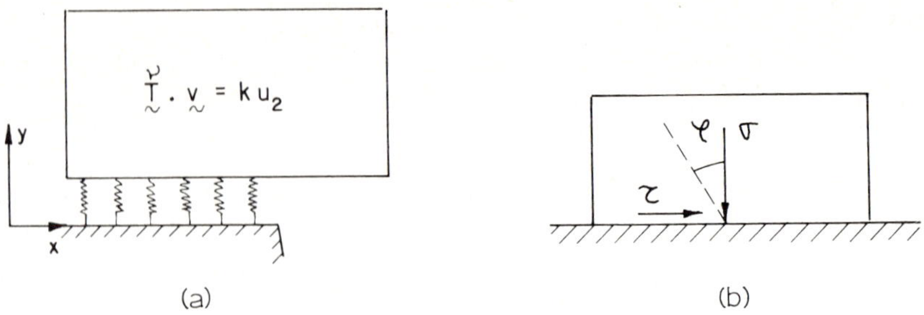

Fig. 9. Boundary-value problems with non-classical boundary conditions.

2.2. Virtual work and reciprocity theorem

It is known that the equilibrium equations inside the domain can be written as

$$\sigma_{ij,j} + X_i = 0\,.$$

If a displacement field u_i^* is imagined, then it is possible to write

$$\int_V \sigma_{ij,j} u_i^* + \int_\Omega X_i u_i^* = 0\,, \tag{2.13}$$

where Ω represents the body, and where here and throughout, the differentials of volume are omitted for the sake of convenience. On account of

$$(\sigma_{ij} u_i^*)_{,j} = \sigma_{ij,j} u_i^* + \sigma_{ij} u_{i,j}^* \tag{2.14}$$

and due to the symmetry of σ_{ij}

$$(\sigma_{ij} u_i^*)_{,j} = \sigma_{ij,j} u_i^* + \sigma_{ij} \varepsilon_{ij}^*$$

so that, application of the divergence theorem yields

$$\int_\Omega \sigma_{ij,j} u_i^* = \int_\Omega (\sigma_{ij} u_i^*)_{,j} - \int_\Omega \sigma_{ij} \varepsilon_{ij}^* = \int_{\partial\Omega} \sigma_{ij} n_j u_i^* - \int_\Omega \sigma_{ij} \varepsilon_{ij}^*\,. \tag{2.15}$$

Thus, introduction of (2.15) in (2.13) results in

$$\int_{\partial\Omega} \overset{n}{T}_i u_i^* + \int_\Omega X_i u_i^* = \int_\Omega \sigma_{ij} \varepsilon_{ij}^*\,. \tag{2.16}$$

This is the principle of virtual work, relating an equilibrium set ($\overset{n}{T}_i$, X_i, ε_{ij}) to a completely independent, compatible one (u_i^*, ε_{ij}^*).

The right-hand side of Eq. (2.16) can be rewritten by using Hooke's law as follows:

$$\int_{\Omega} \sigma_{ij}\varepsilon^*_{ij} = \lambda \int_{\Omega} \varepsilon_{mm}\delta_{ij}\varepsilon^*_{ij} + 2G\int_{\Omega} \varepsilon_{ij}\varepsilon^*_{ij}$$

$$= \lambda \int_{\Omega} \varepsilon_{mm}\varepsilon^*_{mm} + 2G\int_{\Omega} \varepsilon_{ij}\varepsilon^*_{ij} \,. \tag{2.17}$$

Due to the symmetry of the right-hand side, it is clear that

$$\int_{\Omega} \sigma_{ij}\varepsilon^*_{ij} = \int_{\Omega} \varepsilon_{ij}\sigma^*_{ij} \tag{2.18}$$

and thus, introducing (2.18) in (2.16), we obtain

$$\int_{\partial\Omega} \overset{n}{T}_i u^*_i + \int_{\Omega} X_i u^*_i = \int_{\partial\Omega} \overset{n}{T}{}^*_i u_i + \int_{\Omega} X^*_i u_i \,, \tag{2.19}$$

expressing the Maxwell–Betti reciprocity relationship between two different elastic problems established on the same domain.

2.4. *Somigliana's identity*

An interesting application of Eq. (2.19) is obtained when

$$X^*_i = \boldsymbol{e}\delta(\boldsymbol{x} - \mathrm{P}) \,,$$

where $\boldsymbol{e}$ denotes a direction and $\delta(\boldsymbol{x} - \mathrm{P})$ represents a unit value concentrated force applied at point $\boldsymbol{x} = \mathrm{P}$.

Due to the properties of the Dirac δ-function, the last term of the right-hand of Eq. (2.19) reduces to u_i and thus a representation formula can be obtained of the form

$$u_i(\mathrm{P}) = \int_{\partial\Omega} \overset{n}{T}_i(\mathrm{Q})u^*_i(\mathrm{P},\mathrm{Q}) - \int_{\partial\Omega} \overset{n}{T}{}^*_i(\mathrm{P},\mathrm{Q})u_i(\mathrm{Q}) + \int_{\Omega} X_i(\mathrm{Q})u^*_i(\mathrm{Q}) \,; \qquad \mathrm{P} \in \Omega \,. \tag{2.20}$$

This formula, along with other interesting contributions, was developed by the Italian mathematician Somigliana [101–103]. It is important to note that, on the right-hand side of (2.20), only the first two terms include unknowns. The third is completely known. Let us forget it, however, for a moment and imagine that there are no body forces.

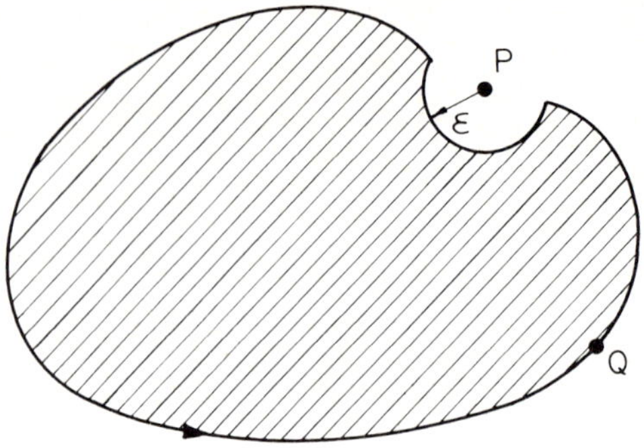

Fig. 10. Illustration of the limiting procedure as an interior point P approaches the boundary.

If P is located at the boundary, a limiting procedure (Fig. 10) allows (2.20) to be written as

$$u_i(\mathrm{P}) = \lim_{\varepsilon \to 0} \Big[\int_{\partial(\Omega - \mathscr{E})} \overset{n}{T}_i(\mathrm{Q}) u_i^*(\mathrm{P}, \mathrm{Q}) + \int_{\partial \mathscr{E}} \overset{\mathrm{n}}{\mathrm{T}}_{\mathrm{i}}(\mathrm{Q}) \mathrm{u}_{\mathrm{i}}^*(\mathrm{P}, \mathrm{Q})$$

$$- \int_{\partial(\Omega - \mathscr{E})} \overset{n}{T}{}_i^*(\mathrm{P}, \mathrm{Q}) u_i(Q) - \int_{\partial \mathscr{E}} \overset{n}{T}{}_i^*(\mathrm{P}, Q) u_i(\mathrm{Q}) \Big] . \tag{2.21}$$

The second integral on the right-hand side of (2.21) represents the work done by the resultant real force on $\partial \mathscr{E}$ on the virtual displacement $\boldsymbol{u}^*$ and tends to zero with ε. On the other hand, the last integral represents the resultant of the virtual stresses around $\partial \mathscr{E}$ on the real displacement of the point Q. When the boundary at P is smooth and based on the symmetry and antisymmetry conditions with respect to the normal and with respect to the tangent, it is clear that

$$\lim_{\varepsilon \to 0} \int_{\partial \mathscr{E}} \overset{n}{T}{}_i^*(\mathrm{P}, \mathrm{Q}) u_i(\mathrm{Q}) = -\tfrac{1}{2} u_i(\mathrm{P}) , \tag{2.22}$$

i.e. each half of the load is supported by a half-space. The minus sign is due to the change in positive directions on normals to the surface and $\partial \mathscr{E}$. This condition is proved rigorously by Rizzo [95], using the expression for the fundamental solution.

When the boundary is not smooth, the resultant of the stresses along the surface $\partial \mathscr{E}$ will have to be computed explicity by means of

$$\lim_{\varepsilon \to 0} \int_{\partial \mathscr{E}} T_{ij}^*(\mathrm{P}, \mathrm{Q}) , \tag{2.23}$$

where T_{ij}^* are the stresses $\overset{n}{T}{}_i^*$ at point Q in direction j, provoked by a unit load acting at P and directed along i.

In general, it would be possible then to write (2.20) as

$$c_{ji}(\mathrm{P})u_i(\mathrm{P}) = \int_{\partial\Omega} \overset{n}{T}_i(\mathrm{Q})U^*_{ij}(\mathrm{P},\mathrm{Q}) - \int_{\partial\Omega} T^*_{ji}(\mathrm{P},\mathrm{Q})u_i(\mathrm{Q})$$

$$+ \int_{\Omega} U^*_{ji}(\mathrm{P},\mathrm{Q})X_i(\mathrm{R})\,, \qquad \mathrm{P},\mathrm{Q} \in \partial\Omega\,; \qquad \mathrm{R} \in \Omega\,, \tag{2.24}$$

where points P and Q are on the boundary, point R is inside the domain,

$$C_{ji}(\mathrm{P}) = \delta_{ji} - \lim_{\varepsilon\to 0} \int_{\partial\mathscr{E}} T^*_{ji}(\mathrm{P},\mathrm{Q}) \tag{2.25}$$

and U^*_{ji} is the displacement at point Q in direction i when a unit load of direction j has been applied at point P. The pair of tensors U^*_{ji} and T^*_{ji} constitutes the fundamental solution of the problem.

2.4. *Boundary integral equation*

Equations (2.24) and (2.25) are ideally suited to solve the case when $\boldsymbol{X} = 0$, i.e. when there are no body forces. In this case

$$c_{ji}u_i(\mathrm{P}) = \int_{\partial\Omega} U^*_{ji}(\mathrm{P},\mathrm{Q}) - \int_{\partial\Omega} T^*_{ji}(\mathrm{P},\mathrm{Q})\,, \qquad \mathrm{P},\mathrm{Q} \in \partial\Omega\,, \tag{2.26}$$

and both P and Q are along the boundary. The representation formula (2.26) is similar to the classical one in potential theory, offering the solution at a boundary point as a contribution from single and double layer potentials.

Explicit expressions for the Kelvin's fundamental solution of elastostatics are given in three dimensions by [76, 35]

$$U^*_{ij} = \frac{1}{16\pi G(1-\nu)}\,\frac{1}{r}\,[(3-4\nu)\delta_{ij} + r_{,i}r_{,j}]$$

$$T^*_{ij} = \frac{-1}{8\pi(1-\nu)}\,\frac{1}{r^2}\left\{\frac{\partial r}{\partial n}\,[(1-2\nu)\delta_{ij} + 3r_{,i}r_{,j}] - (1-2\nu)(r_{,i}n_j - r_{,j}n_i)\right\} \tag{2.27}$$

and in two dimensions (plane strain) by

$$U^*_{ij} = \frac{1}{8\pi G(1-\nu)}\left[(3-4\nu)\ln\frac{1}{r}\,\delta_{ij} + r_{,i}r_{,j}\right],$$

$$T^*_{ij} = -\frac{1}{4\pi(1-\nu)}\,\frac{1}{r}\left\{\frac{\partial r}{\partial n}\,[(1-2\nu)\delta_{ij} + 2r_{,i}r_{,j}] - (1-2\nu)\,(r_{,i}n_j - r_{,j}n_i)\right\}, \tag{2.28}$$

where $\boldsymbol{r} = \boldsymbol{r}(\mathbf{P},\mathbf{Q})$ is the distance between the points P and Q. The singular

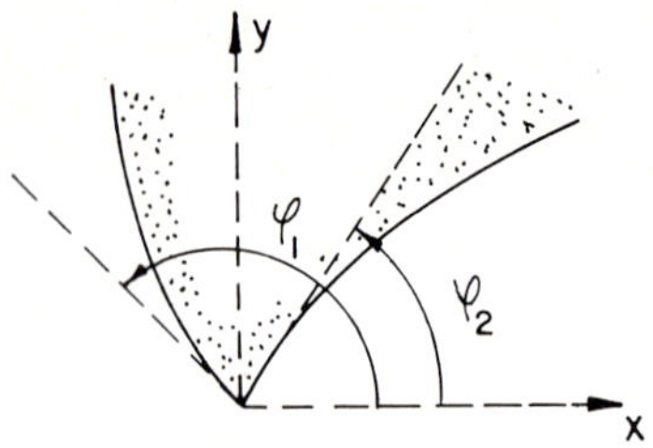

Fig. 11. Geometry of a non-smooth boundary point.

character of the above fundamental solutions is apparent. Indeed, as $r \to 0$, we have from (2.27) and (2.28) that $U^*_{ij} = O(1/r)$ and $T^*_{ij} = O(1/r^2)$ for three dimensions and $V^*_{ij} = O(\ln r)$ and $T^*_{ij} = O(1/r)$ for two dimensions.

On the other hand, for non-smooth boundary points, the tensor c_{ij} is given explicitly in two dimensions by

$$c_{ij} = \frac{1}{4\pi(1-\nu)} \times \begin{bmatrix} 2(1-\nu)(\varphi_1 - \varphi_2) + \frac{1}{2}(\sin 2\varphi_1 - \sin 2\varphi_2) & \sin^2\varphi_1 - \sin^2\varphi_2 \\ \sin^2\varphi_1 - \sin^2\varphi_2 & 2(1-\nu)(\varphi_1 - \varphi_2) - \frac{1}{2}(\sin^2\varphi_1 - \sin^2\varphi_2) \end{bmatrix}, \tag{2.29}$$

where the angles φ_1 and φ_2 are shown in Fig. 11, while for three dimensions, Hartmann [59] has presented expressions for c_{ij} for the following cases:

Quarter of a sphere:

$$c_{ij} = \frac{1}{8\pi}\begin{bmatrix} 2\pi & 0 & 0 \\ 0 & 2\pi & 2/(1-\nu) \\ 0 & 2/(1-\nu) & 2\pi \end{bmatrix}.$$

Eighth of a sphere: (2.30)

$$c_{ij} = \frac{1}{8\pi}\begin{bmatrix} \pi & 1/(1-\nu) & 1/(1-\nu) \\ 1/(1-\nu) & \pi & 1(1-\nu) \\ 1/(1-\nu) & 1/(1-\nu) & \pi \end{bmatrix}.$$

It will be shown in the following sections that the exact computation of c_{ij} is, in general, not necessary, because if the boundary is not smooth c_{ij} can be obtained through the imposition of rigid-body displacements. However, for smooth boundary points the tensor $c_{ij} = \frac{1}{2}\boldsymbol{I}$, where $\boldsymbol{I}$ is the identity matrix.

2.5. The direct method

Equation (2.26) is the basis for the BEM. The idea is to interpolate the unknowns by means of predefined shape functions in the same fashion as with the FEM, but simultaneously. That is, it is assumed that

$$\begin{aligned} \boldsymbol{u} &= \boldsymbol{N}\boldsymbol{a} \,, \\ \boldsymbol{T} &= \boldsymbol{M} \cdot \boldsymbol{b} \,, \end{aligned} \tag{2.31}$$

where $\boldsymbol{u}$ and $\boldsymbol{T}$ are the displacement and traction vectors along some piece of boundary, $\boldsymbol{N}$ and $\boldsymbol{M}$ are matrices containing the interpolating functions and $\boldsymbol{a}$ and $\boldsymbol{b}$ are vectors with the unknowns. The support of $\boldsymbol{N}$ and $\boldsymbol{M}$ define the "elements" on the boundary over which the integrals appearing in (2.26) will be calculated. $\boldsymbol{N}$ and $\boldsymbol{M}$ can be locally or globally based, they can be different and, finally, they are completely independent of the approximation used to represent the geometry.

The first idea by Rizzo [95] and Cruse [45] was the simplest one, i.e. to divide the boundary into pieces, where both $\boldsymbol{u}$ and $\boldsymbol{T}$ were assumed to remain constant. In this case, Eq. (2.26) simplifies to

$$c_{ji}u_i(\mathrm{P}) = \sum_k T_i(\mathrm{Q}) \int_{\partial\Omega_k} U^*_{ji}(\mathrm{P}, \mathrm{Q}_k) - \sum_k u_i(\mathrm{Q}) \int_{\partial\Omega_k} T^*_{ji}(\mathrm{P}, \mathrm{Q}_k) \,. \tag{2.32}$$

The pieces $\partial\Omega_k$ must still be selected, that is the boundary must be discretized and the k points P where Eq. (2.32) will be applied must be chosen. In this way, a system $(k \times k)$ given by

$$\boldsymbol{A}\boldsymbol{u} = \boldsymbol{B}\boldsymbol{t} \tag{2.33}$$

will be formed, where $\boldsymbol{A}$ contains the coefficients c_{ij} plus the last integrals of the right-hand side of Eq. (2.32) and $\boldsymbol{B}$ contains the first integrals of the right-hand side of Eq. (2.32). Several examples of this procedure will be shown in the following sections. In general, the actual geometry is substituted by plane triangular facets as in Fig. 21(a), and the collocation points are taken at the centroids of the facets. The representative values for $\boldsymbol{u}$ and $\boldsymbol{t}$ are assumed to be those occurring at the centroids.

Equation (2.33) can be transformed in a classical system of linear equations by imposing the boundary conditions at the element centroids, thereby reducing the number of unknowns. If all the displacements are known, for instance

$$\boldsymbol{t} = \boldsymbol{B}^{-1}\boldsymbol{A}\boldsymbol{u} \,, \tag{2.34}$$

while if the stresses are known (except rigid-body displacements)

$$\boldsymbol{u} = \boldsymbol{A}^{-1}\boldsymbol{B}\boldsymbol{t} \,. \tag{2.35}$$

In general, in a mixed problem, (2.33) yields

$$\begin{aligned} \boldsymbol{F} &= \boldsymbol{K}\boldsymbol{X} \\ \boldsymbol{X} &= \boldsymbol{K}^{-1}\boldsymbol{F} \end{aligned} \tag{2.36}$$

where X contains all the boundary unknowns and F all the boundary data. The problem is then formally solved.

Lachat and Watson [73] proposed to use the finite element isoparametric idea to improve the acuracy of the previous method, that is to use

$$\begin{aligned} u &= Nu^{e} , \\ T &= NT^{e} , \\ x &= Nx^{e} , \end{aligned} \tag{2.37}$$

where N is the common set of interpolatory serendipity polynomial functions and u^{e}, T^{e}, x^{e} are, respectively, the displacements, tractions and coordinates of the points chosen to discretize the boundary. In addition, the same set of points is used to locate the fundamental solution. In this way, the rigidity of the method is balanced by the simplicity of the procedure. As soon as the user selects a degree for the interpolatory polynomial, and the collocation points have been selected, everything is fixed, and the method can be completely computerized.

It is important to notice that the fundamental solution is defined everywhere so that, although functions N are locally based, the matrices A and B will be full. Nevertheless, u^{e}, T^{e} and x^{e} retain a physical significance. On the other hand, a close look at A or B will show that these matrices need not be square.

Also worth noting is the fact that, when the body force $X \neq 0$, the last integral of Eq. (2.24) must be computed. In general, X will not follow any analytical law, so that the integral will have to be computed numerically along cells, as suggested by Fig. 3. However, there are important cases where, for the sake of elegance, this volume integral can be reduced to a boundary one. Imagine, for instance, that

$$X = \nabla \Phi , \tag{2.38}$$

i.e. the body forces are derived from a potential Φ. In this case

$$I = \int_{\Omega} X u^{*} = \int_{\Omega} u^{*} \nabla \phi . \tag{2.39}$$

Use of the identity

$$\nabla(\phi u^{*}) = u^{*} \nabla \phi + \phi \nabla u^{*} ,$$

or

$$u^{*} \nabla \phi = \operatorname{div}(\phi u^{*}) - \phi \operatorname{div} u^{*} , \tag{2.40}$$

in conjunction with the divergence theorem, transforms I of (2.39) into

$$I = \int_{\Omega} \operatorname{div}(\phi u^{*}) - \int_{\Omega} \phi \operatorname{div} u^{*} = \int_{\partial\Omega} \phi u^{*} \cdot n - \int_{\Omega} \phi \operatorname{div} u^{*} . \tag{2.41}$$

On the other hand, use of the Galerkin vector G for the representation of the fundamental solution by

$$u^* = \frac{1-2\nu}{2(1-\nu)} \text{ grad div } \boldsymbol{G} - \text{rot rot } \boldsymbol{G} ,$$

or, because of

$$\nabla^2 \boldsymbol{G} = \text{grad div } \boldsymbol{G} + \text{rot rot } \boldsymbol{G}$$

by

$$\boldsymbol{u}^* = \nabla^2 \boldsymbol{G} - \frac{1}{2(1-\nu)} \text{ grad div } \boldsymbol{G} , \tag{2.42}$$

in conjunction with the fact that div(rot) = 0, leads to

$$\text{div } \boldsymbol{u}^* = \frac{1-2\nu}{2(1-\nu)} \nabla^2(\text{div } \boldsymbol{G}) . \tag{2.43}$$

If we take into account that (e.g. for three-dimensions)

$$\boldsymbol{G} = \frac{1+\nu}{4\pi E} \boldsymbol{re} , \tag{2.44}$$

and call

$$\boldsymbol{W} = \frac{(1-2\nu)(1+\nu)}{8\pi E(1-\nu)} \boldsymbol{re} , \tag{2.45}$$

then (2.43) becomes

$$\text{div } \boldsymbol{u}^* = \nabla^2(\text{div } \boldsymbol{W}) = \nabla^2 H , \tag{2.46}$$

where

$$H = \text{div } \boldsymbol{W} = \frac{(1-2\nu)(1+\nu)}{8\pi E(1-\nu)} (r_{,x} + r_{,y} + r_{,z}) . \tag{2.47}$$

Thus, the volume integral on the right-hand side of (2.41) takes the form

$$-\int_{\Omega} \phi \text{ div } \boldsymbol{u}^* = -\int_{\Omega} \phi \nabla^2 H , \tag{2.48}$$

or, upon using Green's theorem, the form

$$-\int_{\Omega} \phi \text{ div } \boldsymbol{u}^* = \int_{\partial\Omega} H \cdot \text{grad } \phi \cdot \boldsymbol{n} - \int_{\partial\Omega} \phi \text{ grad } H \cdot \boldsymbol{n} - \int_{\Omega} H \nabla^2 \phi . \tag{2.49}$$

If the function ϕ satisfies the equation

$$\nabla^2\phi = k_0 , \tag{2.50}$$

where k_0 is a constant, then

$$\int_{\Omega} H\nabla^2\phi = k_0 \int_{\Omega} \operatorname{div} \boldsymbol{W} = k_0 \int_{\partial\Omega} \frac{\partial \boldsymbol{W}}{\partial \boldsymbol{n}} \tag{2.51}$$

and in view of (2.49) and (2.51), Eq. (2.41) yields

$$I = \int_{\partial\Omega} \phi \boldsymbol{u}^* \cdot \boldsymbol{n} - \int_{\partial\Omega} \phi \operatorname{grad} H \cdot \boldsymbol{n} - k_0 \int_{\partial\Omega} \frac{\partial \boldsymbol{W}}{\partial \boldsymbol{n}} \tag{2.52}$$

as presented by Rizzo and Shippy [97]. For the case of the self-weight, for instance $k_0 = 0$, while for the case of the centrifugal force, $k_0 = \rho\omega^2$, where ρ is the material density and ω is the angular velocity of the rotating body. A review type of article on the treatment of body forces in elasticity has been written by Danson [50].

2.6. *The indirect method*

The indirect boundary element method proceeds directly from the physical meaning of the fundamental solution on a heuristic basis. Only a brief description is given here. A full treatment of the indirect method can be found in Banerjee and Butterfield [17].

A unit load applied to the boundary at point Q produces displacements $\boldsymbol{u}$ and tractions $\boldsymbol{T}$ defined by

$$\begin{aligned} u_i(\mathrm{P}) &= U^*_{ij}(\mathrm{P}, \mathrm{Q})e_j(\mathrm{Q}) , \\ T_i(\mathrm{P}) &= T^*_{ij}(\mathrm{P}, \mathrm{Q})e_j(\mathrm{Q}) , \end{aligned} \tag{2.53}$$

and, thus, a fictitious load $\boldsymbol{t}(\mathrm{Q})$ continuously distributed along the boundary will produce

$$u_i(\mathrm{P}) = \int_{\partial\Omega} U^*_{ij}(\mathrm{P}, \mathrm{Q})t_j(\mathrm{Q}) , \tag{2.54a}$$

$$T_i(\mathrm{P}) = \int_{\partial\Omega} T^*_{ij}(\mathrm{P}, \mathrm{Q})t_j(\mathrm{Q}) , \tag{2.54b}$$

$$\mathrm{P} \in \Omega ; \mathrm{Q} \in \partial\Omega ,$$

When P is moved to the boundary, Eq. (2.54b) is transformed into

$$T_i(\mathrm{P}) = \tfrac{1}{2}t_i(\mathrm{P}) + \int_{\partial\Omega} T^*_{ij}(\mathrm{P}, \mathrm{Q})t_j(\mathrm{Q}) \,, \tag{2.55}$$

while Eq. (2.54a) remains of the same form.

Provided that P is not a corner point, the two Eqs. (2.54a) and (2.55) can be used to repeat the discretization process described above for the direct BEM. In this way, the system:

$$\begin{aligned} \boldsymbol{u} &= \boldsymbol{A}\boldsymbol{t} \\ \boldsymbol{T} &= \boldsymbol{B}\boldsymbol{t} \end{aligned} \tag{2.56}$$

is applied to every collocation point around the boundary, where either $\boldsymbol{u}$ or $\boldsymbol{T}$ is known. After all conditions on $\boldsymbol{u}$ or $\boldsymbol{T}$ have been imposed, there results the system

$$\boldsymbol{F} = \boldsymbol{K}\boldsymbol{t} \tag{2.57}$$

from which the unknown fictitious distribution $\boldsymbol{t}$ can be evaluated. As soon as $\boldsymbol{t}$ is known around the boundary, Eq. (2.56) can be applied to those points where either $\boldsymbol{u}$ or $\boldsymbol{T}$ is needed.

It is possible to see that recovery of the main unknowns is done after an intermediate quantity $\boldsymbol{t}$ has been evaluated, and this is why the method is called "indirect" in opposition to the "direct" one, where the basic unknowns are directly incorporated into the system to be solved.

Of course, it is possible to use a distribution of $\boldsymbol{t}$ along a fictitious boundary, as several authors have proposed. In addition to complicating the data input, these methods seem to present several mathematical objections, the nature of which is out of context here.

It is also important to note the sensitivity of the results to sharp corners. This sensitivity has caused the trend to move towards the direct method.

2.7. The displacement discontinuity method

This interesting method is a special case of the indirect method, where the fundamental solution is substituted by an ingenious artifice, the opening or closure of fictitious cracks. The method is fully described in the works of Crouch [42–44], which the interested reader can consult for more details. Extension of the DDM to three dimensions has been reported by Dunbar and Anderson [54].

In summary, equations are established for the displacements and tractions along the given plane boundary of the form

$$\begin{aligned} u_i(\mathrm{P}) &= \sum B_{ij}D_j \,, \\ T_i(\mathrm{P}) &= \sum A_{ij}D_j \,, \end{aligned} \qquad i, j = 1, 2, \ldots, N \tag{2.58}$$

provoked by arbitrary displacements D_j given along N cracks distributed either on the boundary or outside it.

Equations (2.58) are applied at a series of N points where $\boldsymbol{u}$ or $\boldsymbol{T}$ is known, and from them, an equation of the form

$$\boldsymbol{F} = \boldsymbol{K}\boldsymbol{D} \tag{2.59}$$

can be established which, when solved, gives the values $\boldsymbol{D}$ inducing $\boldsymbol{u}$ and $\boldsymbol{T}$ along the boundary. Equation (2.58) can be applied now to the points where $\boldsymbol{u}$ and $\boldsymbol{T}$ are unknowns in order to obtain the final results.

2.8. Computational aspects

The most important computational aspects of the direct BEM are discussed in some detail in the next two sections dealing with two- and three-dimensional elastostatics. Thus, the use of linear elements and the treatment of corners are discussed in connection with two-dimensional problems, while the use of constant and higher-order elements as well as the computation of the influence matrices $\boldsymbol{A}$ and $\boldsymbol{B}$ are presented in connection with three-dimensional problems.

3. Plane problems

In this section, we shall present, from an elementary point of view, the steps needed to develop the BEM in plane-stress or plane-strain elastostatic problems. The essential boundary conditions will be on the displacements $\boldsymbol{u}$ and the natural ones on the tractions $\boldsymbol{T}$ of the two-dimensional body. After establishing the usual notions, we shall offer the computation of the influence matrices and present some simple examples.

3.1. Definitions and notations

As we indicated above, the main objective of the method is the computation of values taken by the basic variables at the boundary. The first noteworthy idea is that it is necessary to interpolate those variables both when they are data and when they are unknowns, in order to approximate their behavior by a finite number of "ad-hoc" functions.

In order to stabilize these ideas, consider the problem shown in Fig. 12 which deals with the stress analysis of the square disk ABCD. Along AB and CD, the displacements are known to be zero and the tractions are unknown, while along BC and DA, the situation is reciprocal.

In order to reduce the unknowns to a finite number, the quest for unknown values has been limited to a finite number $1, 2, \ldots, n$ of points. In addition, it is assumed that both displacements and tractions will be interpolated by "shape"

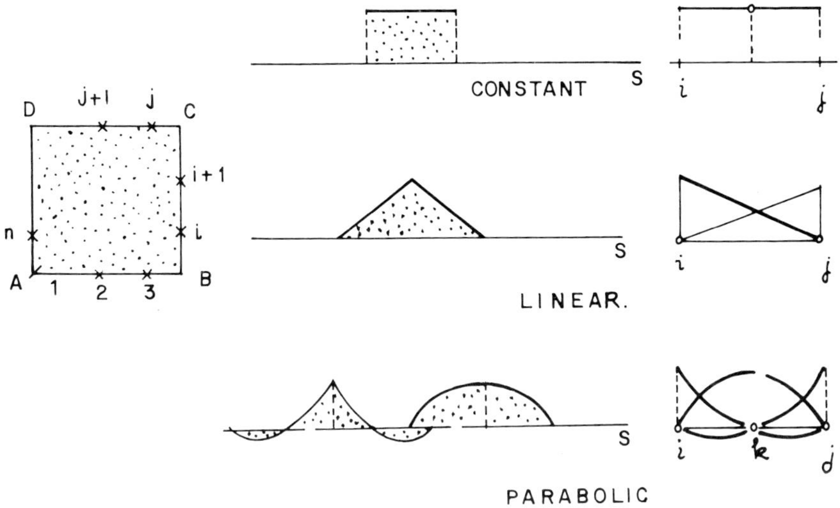

Fig. 12. Constant, linear and parabolic interpolation functions for boundary elements.

functions $N(i)$, $M(i)$, taking unit values in turn at each of the chosen points (nodes) and zero at the others, so that in an interpolation of the type

$$
\begin{aligned}
u &= a(i)\ N(i) \\
v &= b(i)\ N(i) \\
\bar{X} &= c(i)\ M(i) \\
\bar{Y} &= d(i)\ M(i) \qquad i = 1, 2, \ldots, n
\end{aligned}
\tag{3.1}
$$

the $a(i)$, $b(i)$, $c(i)$, $d(i)$ will be the values taken by the components u, v, of $\boldsymbol{u}$ and the components $\bar{X}$ and $\bar{Y}$ of $\boldsymbol{T}$, respectively, at point i.

Figure 12 shows the constant, linear and parabolic interpolations with polynomials that are zero along important parts of the boundary, simplifying the above-mentioned idea of giving physical meaning to the coefficients.

As with the FEM, it can be seen that this type of interpolation generates the concept of "element" naturally, i.e. that zone of the boundary where the approximating shape is repeated. This allows the computations (basically integrations) to be done in a local sense and in a repetitive manner, which is ideal for a treatment by the computer.

The idea of geometrical approximation of the boundary also appears by interpolating among the points chosen to define the variables. That is

$$
\begin{aligned}
x &= x(i) \cdot R(i) \\
y &= y(i) \cdot R(i)
\end{aligned}
\qquad i = 1, 2, \ldots, n
\tag{3.2}
$$

where $R(i)$ are again locally based functions.

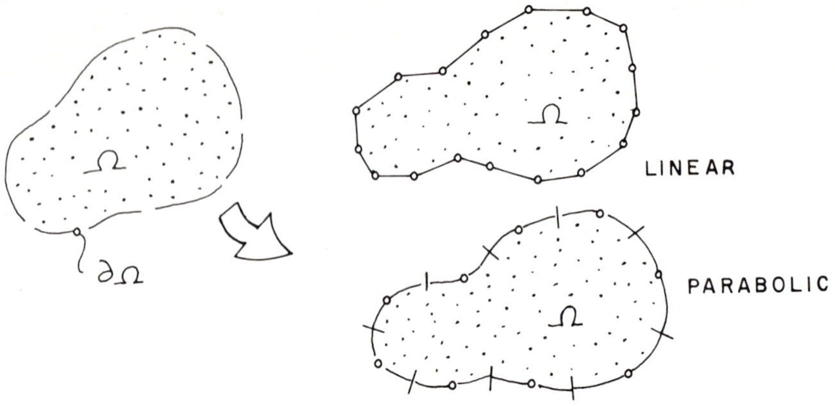

Fig. 13. Linear and parabolic isoparametric boundary element discretization of a plane domain.

In principle N, M and R are independent, but traditionally they are taken as equal, thus introducing the idea of isoparametric elements. So the analyst's first task is to discretize, that is to choose those boundary points which are going to guide the interpolation of variables and geometry (Fig. 13). After that, the method is limited to the procedure for determining the parameters $a(i)$, $b(i)$, $c(i)$, $d(i)$, which are unknown at the boundary. In general, there will be N unknowns so that N independent equations are needed. In the BEM they are obtained through the representation formula (2.26) of section 2, which must be applied as many times as necessary, as we shall see below.

3.2. *Linear elements*

The general analysis can be subdivided into the following steps:

(a) Selection of boundary elements (Fig. 14).

As usual, the mesh is refined in the areas with larger curvature where the variables could have higher gradients as shown in Fig. 14(a).

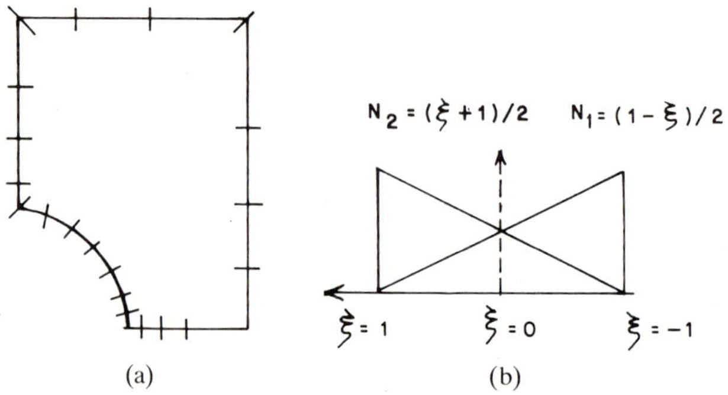

Fig. 14. Selection of size of boundary elements based on linear interpolation functions.

(b) Linear interpolation (Fig. 14).

Inside each element, both displacements and stresses are approximated by linear shape functions of the form

$$N_1(\xi) = -\tfrac{1}{2}(\xi - 1)\,; \qquad N_2(\xi) = \tfrac{1}{2}(\xi + 1)\,; \qquad \xi = s/L\,, \tag{3.3}$$

as shown in Fig. 14(b).

In general, it is assumed that the displacements are continuous along the boundary while it is necessary to allow discontinuous tractions around each sharp corner. The following analysis is based on the work of Alarcón et al. [2].

Using the notation of Fig. 15(c) and assuming that a subscript indicates "node" and a superscript "element", it is possible to write for element j

$$\boldsymbol{u} = \begin{Bmatrix} u \\ v \end{Bmatrix} = \begin{bmatrix} N_1 & N_2 & 0 & 0 \\ 0 & 0 & N_1 & N_2 \end{bmatrix} \begin{Bmatrix} u_j^j \\ u_{j+1}^j \\ v_j^j \\ v_{j+1}^j \end{Bmatrix} \tag{3.4a}$$

and

$$\boldsymbol{T} = \begin{Bmatrix} \boldsymbol{X} \\ \boldsymbol{Y} \end{Bmatrix} = \begin{bmatrix} N_1 & N_2 & 0 & 0 \\ 0 & 0 & N_1 & N_2 \end{bmatrix} \begin{Bmatrix} \boldsymbol{X}_j^j \\ \boldsymbol{X}_{j+1}^j \\ \boldsymbol{Y}_j^j \\ \boldsymbol{Y}_{j+1}^j \end{Bmatrix}. \tag{3.4b}$$

The general boundary integral equation (2.26) can be written now for every collocation point i as follows:

$$2c(i)\begin{Bmatrix} u_i \\ v_i \end{Bmatrix} + \sum_{j=1}^{N} L_j \begin{bmatrix} \int_{\partial\Omega_j} N_1 T_{11}^* & \int_{\partial\Omega_j} N_2 T_{11}^* & \int_{\partial\Omega_j} N_1 T_{12}^* & \int_{\partial\Omega_j} N_2 T_{12}^* \\ \int_{\partial\Omega_j} N_1 T_{21}^* & \int_{\partial\Omega_j} N_2 T_{21}^* & \int_{\partial\Omega_j} N_1 T_{22}^* & \int_{\partial\Omega_j} N_2 T_{22}^* \end{bmatrix} \begin{Bmatrix} u_j^j \\ u_{j+1}^j \\ v_j^j \\ v_{j+1}^j \end{Bmatrix}$$

$$= \sum_{j=1}^{N} L_j \begin{bmatrix} \int_{\partial\Omega_j} N_1 U_{11}^* & \int_{\partial\Omega_j} N_2 U_{11}^* & \int_{\partial\Omega_j} N_1 U_{12}^* & \int_{\partial\Omega_j} N_2 U_{12}^* \\ \int_{\partial\Omega_j} N_1 U_{21}^* & \int_{\partial\Omega_j} N_2 U_{21}^* & \int_{\partial\Omega_j} N_1 U_{22}^* & \int_{\partial\Omega_j} N_2 U_{22}^* \end{bmatrix} \begin{Bmatrix} \bar{X}_j^j \\ \bar{X}_{j+1}^j \\ \bar{Y}_j^j \\ \bar{Y}_{j+1}^j \end{Bmatrix} \tag{3.5}$$

where (u, v) and $(\bar{X}, \bar{Y})$ are the components of boundary displacements and

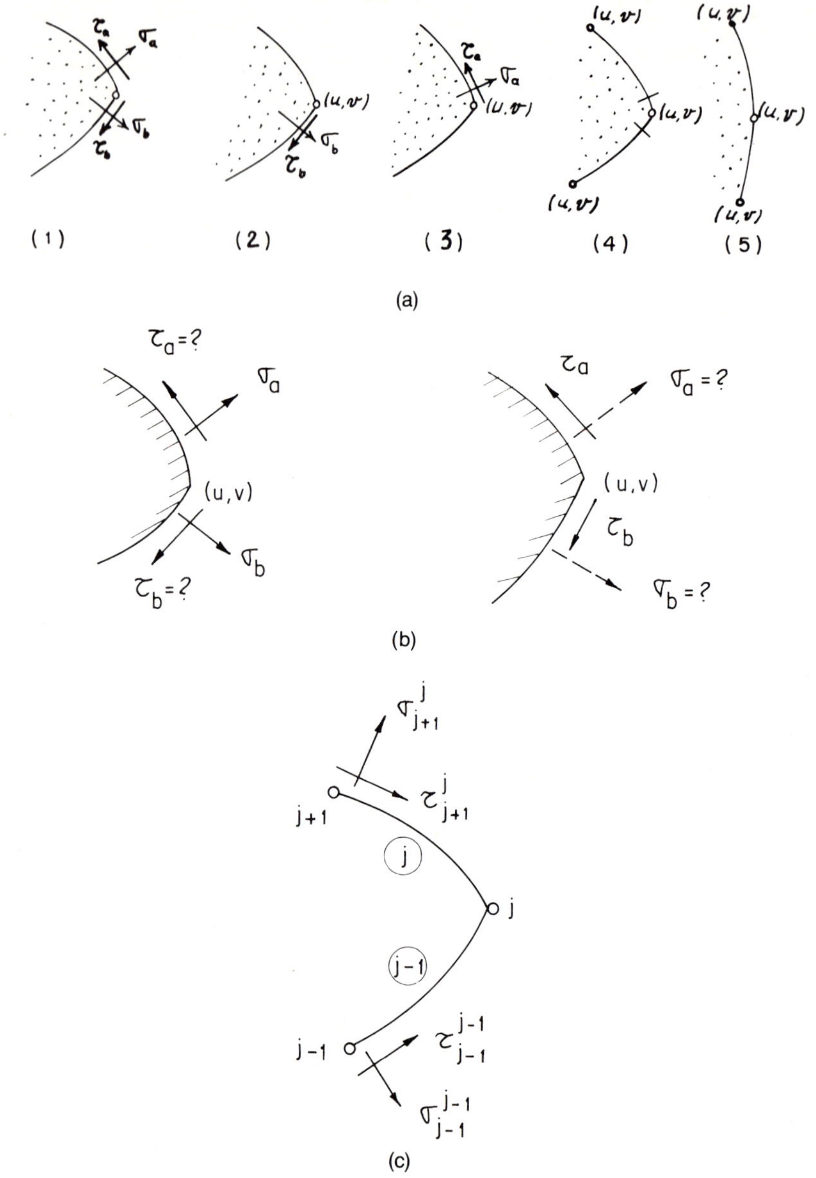

Fig. 15. Boundary condition cases at a corner.

tractions, respectively, in global coordinates and T^*_{ij}, U^*_{ij} are the components of the fundamental solution for stresses and displacements, respectively.

In general, it is more useful to express tractions as normal σ and tangential τ components. This can be done simply by using the transformation

$$\begin{Bmatrix} \sigma \\ \tau \end{Bmatrix}^{e} = \begin{bmatrix} \cos\alpha_e & \sin\alpha_e \\ -\sin\alpha_e & \cos\alpha_e \end{bmatrix} \begin{Bmatrix} \bar{X} \\ \bar{Y} \end{Bmatrix}, \tag{3.6}$$

where α_e is the angle between the x-axis and the unit normal vector to the element e.

Equation (3.5) can now be written in terms of σ and τ as

$$c(i)\begin{Bmatrix} u_i \\ v_i \end{Bmatrix} + \sum_{j=1}^{N} \begin{bmatrix} A^j_{11} & A^j_{12} & A^j_{13} & A^j_{14} \\ A^j_{21} & A^j_{22} & A^j_{23} & A^j_{24} \end{bmatrix} \begin{Bmatrix} u^j_j \\ u^j_{j+1} \\ v^j_j \\ v^j_{j+1} \end{Bmatrix}$$

$$= \sum_{j=1}^{N} \begin{bmatrix} B^j_{11} & B^j_{12} & B^j_{13} & B^j_{14} \\ B^j_{21} & B^j_{22} & B^j_{23} & B^j_{24} \end{bmatrix} \begin{Bmatrix} \sigma^j_j \\ \sigma^j_{j+1} \\ \tau^j_j \\ \tau^j_{j+1} \end{Bmatrix} \tag{3.7}$$

Varying i, one obtains from (3.7) a system of $2N$ equations of the form

$$\boldsymbol{Au} = \boldsymbol{B\sigma} \tag{3.8}$$

The scheme of contribution of each element is as follows:

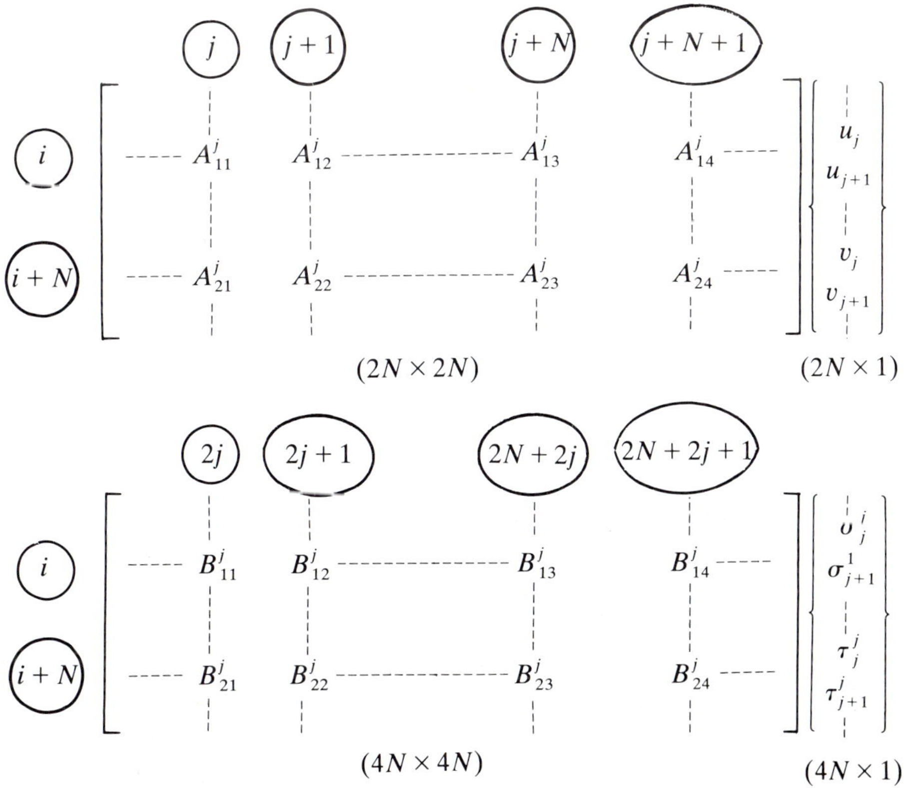

The double size of the stress vector $\boldsymbol{\sigma}$ is due to the double normal around each node. To proceed with the method, it is necessary now to establish conditions related to the continuity of the functions. This is done by identifying the different cases that can occur around a corner. The basic possibilities are those five included in Fig. 15(a) and having as follows:

(1) presents tractions which are specified at both sides of the corner with (u, v) remaining as local unknowns.

(2) presents a case where the data are the tractions before the node (σ_b, τ_b) and the node displacements (u, v). Two unknowns, the tractions after the node (σ_a, τ_a), remain.

(3) is the reciprocal of (2) with (σ_b, τ_b) as unknowns.

(4) deals with a sharp corner where there are four unknowns (σ_b, τ_b) and (σ_a, τ_a).

(5) is like (4) except that the boundary is smooth, with only one normal so that

$$\sigma_a = \sigma_b\,, \qquad \tau_a = \tau_b \tag{3.9}$$

reducing the unknowns to two.

It is also possible to imagine mixed combinations like those of Fig. 15(b) but there are always only two unknowns per node, except for case (4) above.

If there is no singularity at the sharp corner, this situation is easily solvable because knowing the derivatives of u and v along the boundary (computable since one knows the evolution of the data) allows the computation of their gradients at the corner and then the strains ε_{ij}. Using Hooke's material law, the stresses are obtained immediately so that the corner is completely solved. We are, in fact, facing a corner element in Fig. 15(c), where there are two nodes to collocate, namely $(j-1)$ and $(j+1)$ and four unknowns, namely the tractions "after" node $j-1$ and the tractions "before" node $j+1$. Another alternative can be seen in [2].

In that way, matrix $\boldsymbol{B}$ is reduced to size $2N \times 2N$ and after imposing boundary conditions, it is possible to reduce (3.8) to the system

$$\boldsymbol{KX} = \boldsymbol{F}\,. \tag{3.10}$$

For the sake of computing efficiency, the theoretical scheme described above has been slightly modified. Instead of forming $\boldsymbol{A}$ and $\boldsymbol{B}$ by rows, they are formed by columns; that is, in place of computing the integrals, viewing the elements from each node (implying the computation of normals, lengths, etc.), it is better to fix an element and to compute the contributions affecting each node. Another important aspect is the computation of the constants $c(i)$ by using a fictitious problem, i.e. by imagining rigid-body displacements which implies $\boldsymbol{\sigma} = 0$ and then imposing the condition of nullity for the sum of the row coefficients in $\boldsymbol{A}$. A more detailed description of a computer program called SERBA incorporating this approach can be seen in [2].

Figure 16 presents results obtained by Domínguez [52], using constant elements, while studying the distribution of stresses produced around rigid anchors modelled by a semirigid crack, when the anchor is near a free boundary.

An example with linear elements taken from [57] is plotted in Fig. 17, where a rectangular plate, fissured by a crack, is analyzed in order to determine the stress intensity factor. As it is shown, in spite of the fine mesh used near the singularity,

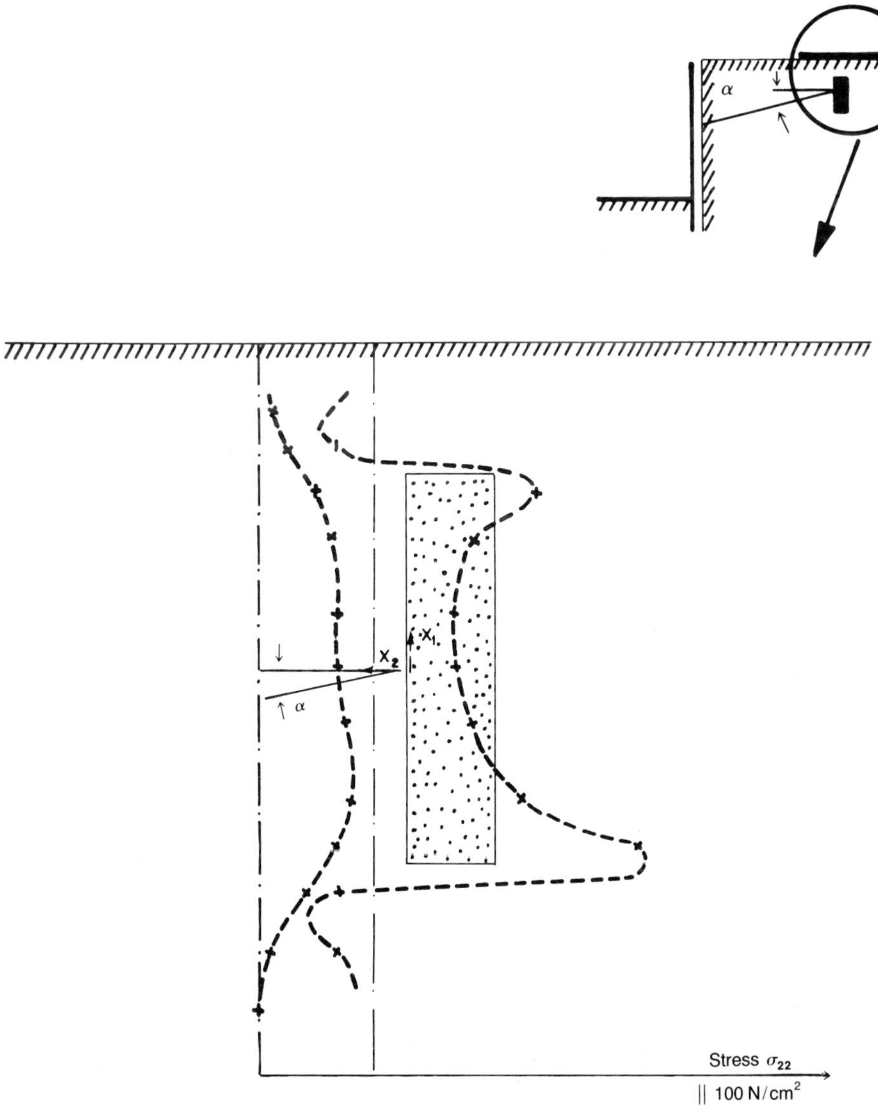

Fig. 16. Stress distribution around a rigid anchor near a free boundary.

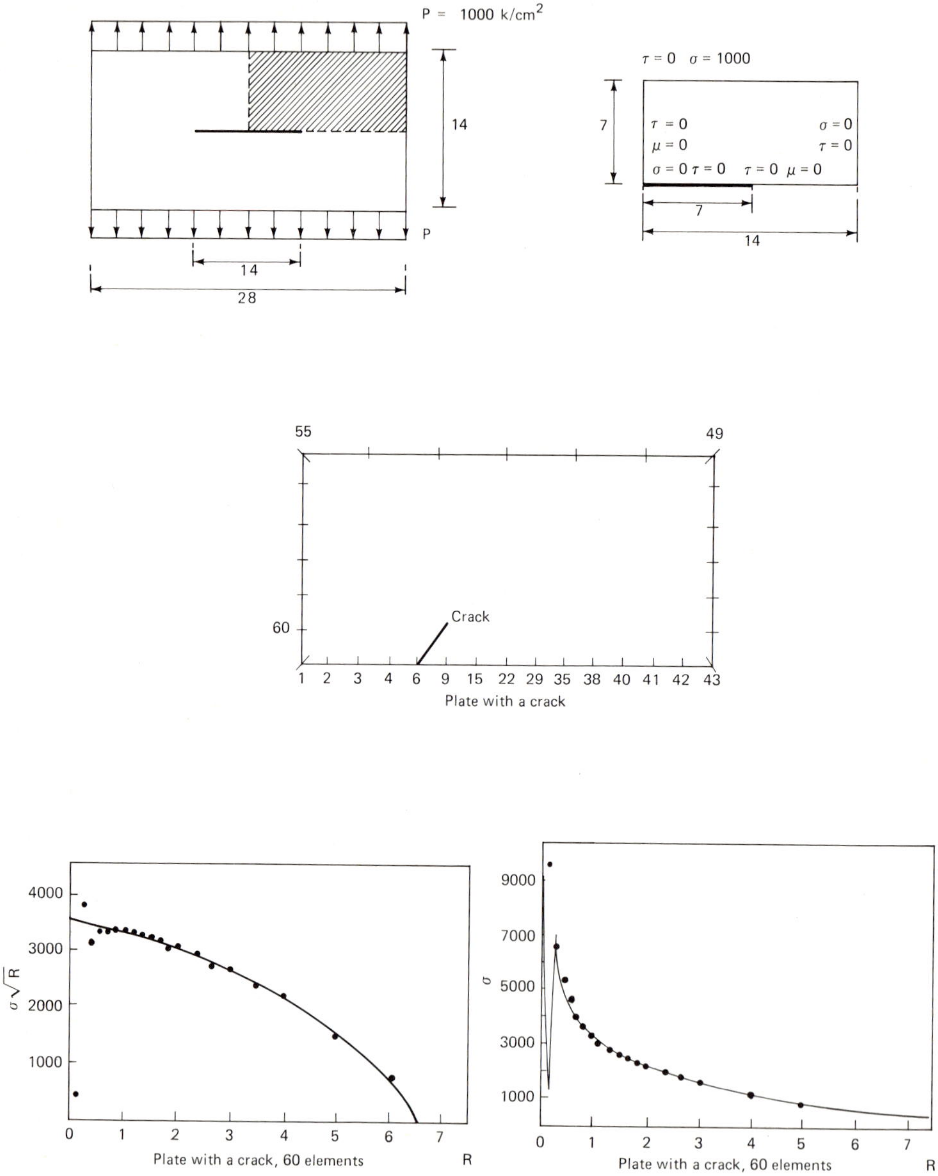

Fig. 17. Stress intensity factor for cracked rectangular plate.

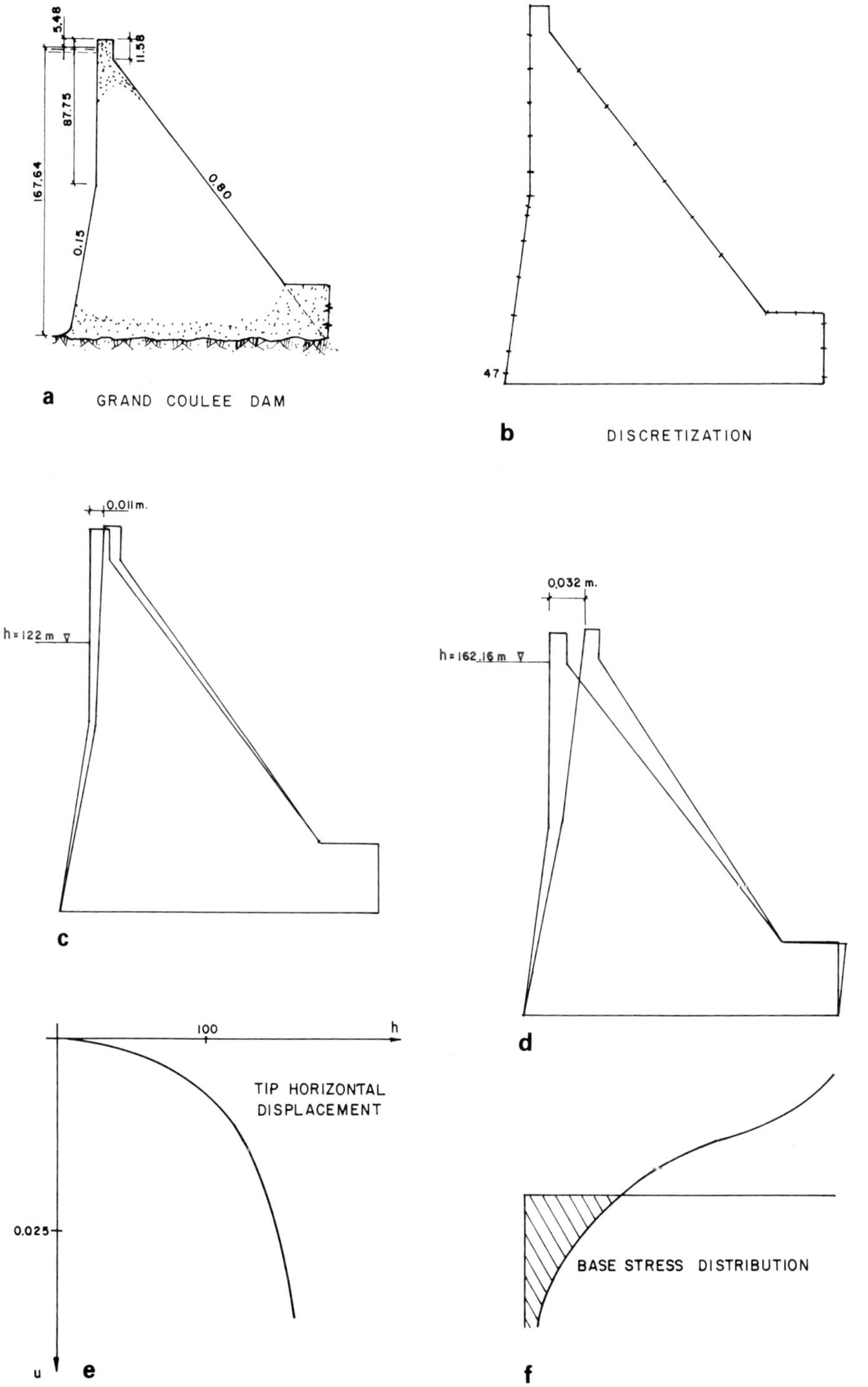

Fig. 18. Stress analysis of the Grand Coulee gravity dam.

there is contamination of results which can only be eliminated through interpolation of the results or by using singular elements.

Figure 18 is a classical example borrowed from civil engineering. It represents a simplified section of a famous gravity dam, the Grand Coulee dam. The general geometry is given in Fig. 18(a), while 18(b) presents the boundary element discretization. It is possible to observe that the mesh is finer near the sharp corner in order to take into account the higher gradients nearby. The base is assumed to be completely fixed against displacements so that, as Fig. 18(f) shows, there are singular points near the corners. Except for those points, the stresses follow the general pattern of the well known Levy's solution. The dashed zone represents tension areas. Figures 18(c) and (d) show the dam deformations for two different water levels, and Fig. 18(e) presents the evolution of the tip horizontal displacement as a function of the dam's degree of filling.

In Fig. 19(a) we have plotted the cross section of a reinforced concrete syphon

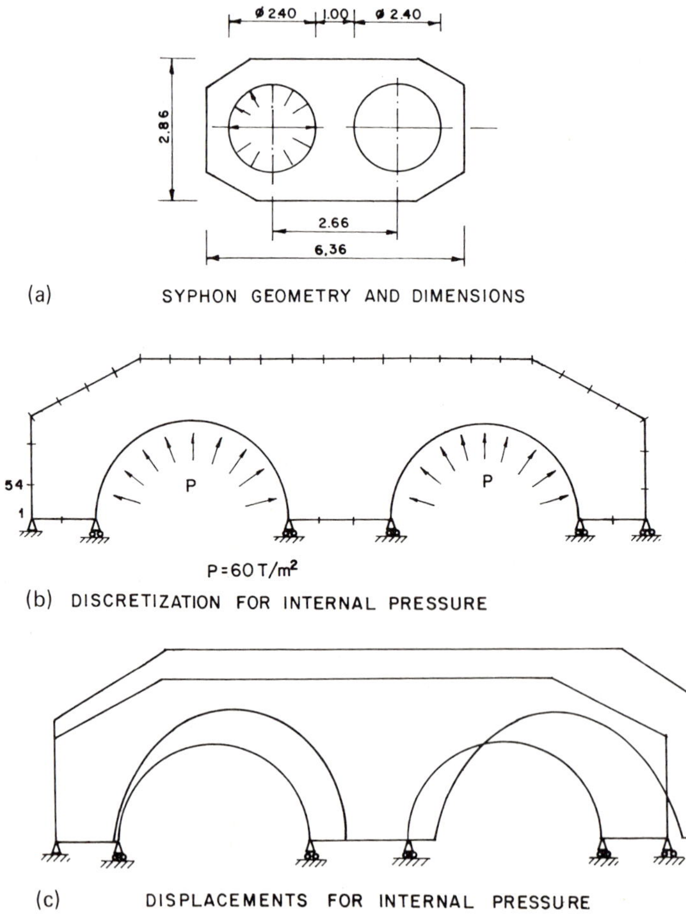

Fig. 19. Deformation analysis of a reinforced concrete syphon under external and internal loads.

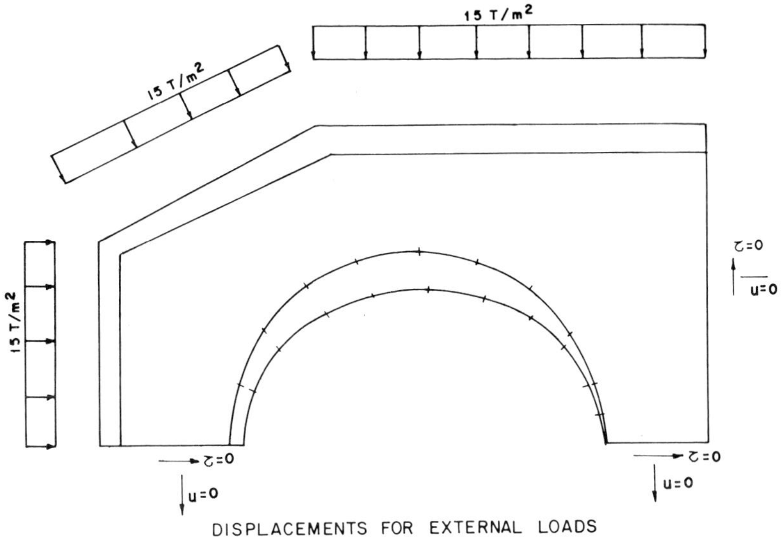

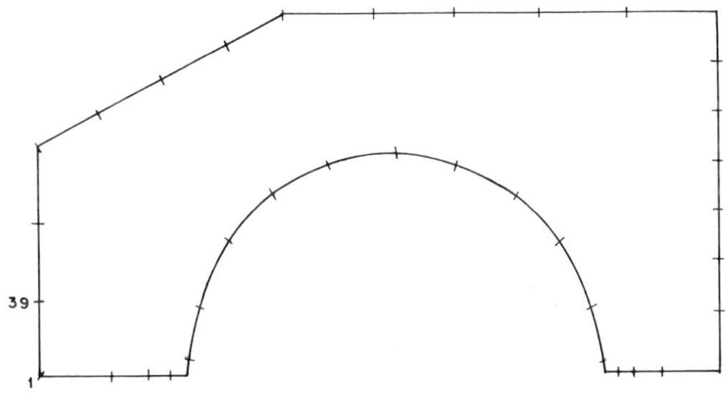

(d)

Fig. 19. (continued).

designed to sustain an interior pressure equivalent to 6 m of water and working under an exterior pressure on the order of 15 m of water. In Fig. 19(b), we show the discretization used to analyze the boundary displacements under internal pressure. Due to the symmetry with respect to the horizontal axis, it is possible to analyze only half of the section. Of course, it would also be possible to establish a vertical axis of symmetry where both cavities are equally pressurized, but it has been decided to maintain only one axis in order to study the asymmetric situation. It is worth noting, then, that the model can suffer rigid-body horizontal displacements. In order to avoid this situation, node 1 has been completely fixed and so the displacements shown in Fig. 19(c) are relative to that point. It is possible to establish that condition in several ways. What we did here was to delete the first row and column of the final matrix, substitute the diagonal elements by a 1 and changing the first element of the "force" vector to a zero [3]. As we can see, this is the classical solution used in similar situations with the FEM or the matrix analysis of beam structures. Figure 19(d) shows the discretization used for a totally symmetrical external load. In this case, there is no room for rigid-body displacements and the computation follows the usual patterns of the method.

4. Three-dimensional problems

In this section, we are going to develop some ideas related to three-dimensional geometry. If the domain presents a large ratio volume/surface, the economies that can be obtained, compared with other methods, are important.

After describing some typical problems, the simplest approach utilizing constant elements as well as some of the integration problems that arise, will be developed. We shall also indicate the treatment of higher-order elements and, finally, we shall show some examples, including the application of volume integrals to the thermoelastic case.

4.1. Typical problems

Typical problems are governed by Eq. (2.1) along with boundary conditions (2.5), (2.6), etc. In the majority of the cases, however, the situation is not strictly pure and even. Mixed conditions, can be artifically generated. In Fig. 20, for enstance, several different cases are presented for fan annular cylinder. Thus, in Fig. 20(b), the displacements are imposed along the internal and external boundaries, while in Fig. 20(c), the same happens with tractions, and Fig. 20(d) represents a mixed problem. Finally, in Figs. 20(e) and (f) we suggest the

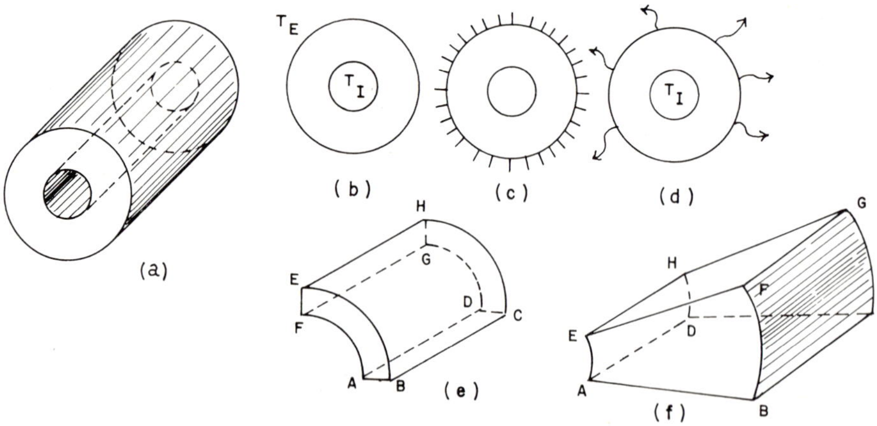

Fig. 20. Mixed boundary conditions for an annular cylinder.

possibility of using symmetry for the walls ABCD and EFGH to model mixed conditions of the type zero normal displacements plus zero tangential stresses thereby producing artificial mixed problems from the initial pure ones of Figs. 20(b, c). We shall show below that this situation can be managed in an automatic fashion while forming the matrix $\boldsymbol{K}$ of (2.36)

4.2. *Boundary elements*

Imagine for a moment that body forces are zero. The representation formula will then include variables defined only at the boundary where the discretization process has to be carried out. The "elements" are generated while interpolating with locally based functions. The simplest examples is obtained with "constant elements", i.e. plane facets where both displacements and stresses are assumed constant. In addition, the facets are taken to have triangular shapes and the representative points for displacements and stresses are taken at the centroid of every triangle as shown in Fig. 21(a).

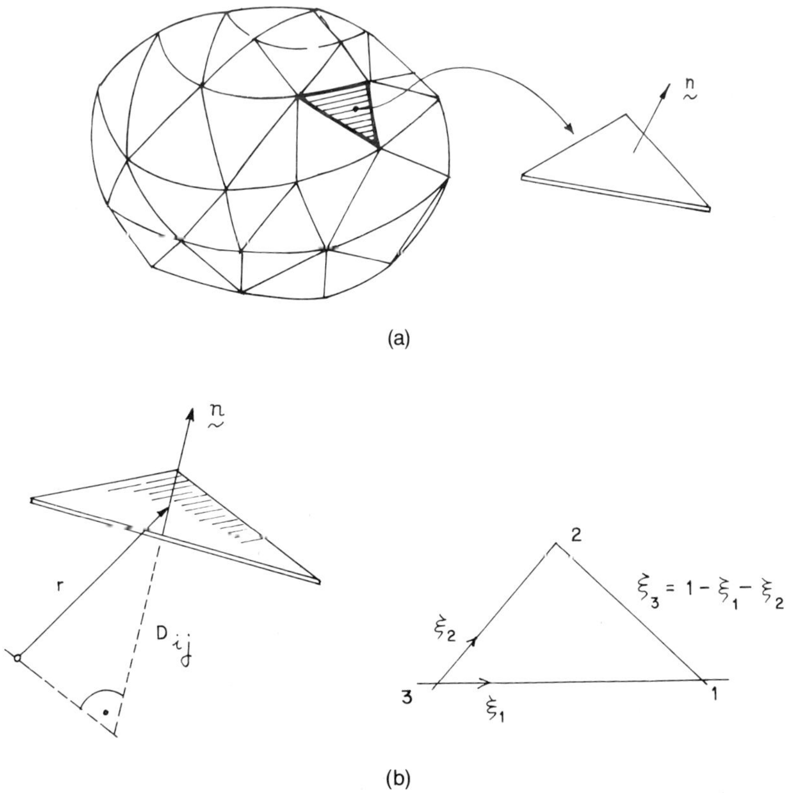

Fig. 21. Constant triangle boundary element discretization of a three-dimensional body.

The general boundary equation (2.26) then takes the form

$$\tfrac{1}{2}u_j(\boldsymbol{x}_L) + \sum_{K=1}^{N} \int_{\partial\Omega_K} T_{ji}(\boldsymbol{x}_L, \boldsymbol{x}_K) u_i(\boldsymbol{x}_K) = \sum_{K=1}^{N} \int_{\partial\Omega_K} U_{ji}(\boldsymbol{x}_L, \boldsymbol{x}_K) t_i(\boldsymbol{x}_K) , \tag{4.1}$$

where $i, j = 1, 2, 3$; $K, L = 1, 2, \ldots, N$ and where $\boldsymbol{x}_L$ represents the collocation point and $\boldsymbol{x}_K$, the running point inside the element K.

If $u_i(\boldsymbol{x}_K)$ and $t_i(\boldsymbol{x}_K)$ are constant inside the element K

$$\begin{aligned} u_i(\boldsymbol{x}_K) &= u_i(K) , \\ t_i(\boldsymbol{x}_K) &= t_i(K) , \end{aligned} \tag{4.2}$$

and the system of $3N$ equations

$$\tfrac{1}{2}u_j(\boldsymbol{x}_L) + \sum_{K=1}^{N} A_{ji}^{LK} u_i(K) = \sum_{K=1}^{N} B_{ji}^{LK} t_i(K) , \tag{4.3}$$

is obtained from (4.1) where

$$\begin{aligned} A_{ji}^{LK} &= \int_{\partial\Omega_K} T_{ji}(\boldsymbol{x}_L, \boldsymbol{x}_K) , \\ B_{ji}^{LK} &= \int_{\partial\Omega_K} U_{ji}(\boldsymbol{x}_L, \boldsymbol{x}_K) , \end{aligned} \tag{4.4}$$

and T_{ji}, U_{ji} are defined in Eq. (2.27). Equation (4.3) can be written in matrix form as

$$\boldsymbol{Au} = \boldsymbol{Bt} . \tag{4.5}$$

The integrals in (4.4) can be computed either numerically or analytically. The first procedure is implemented when the collocation point is outside the element (Fig. 21b). Change of coordinates from the global Cartesian (x_1, x_2, x_3) to the local triangular (ξ_1, ξ_2, ξ_3) ones according to the relation $\mathrm{d}x_1\,\mathrm{d}x_2\,\mathrm{d}x_3 = |J|\,\mathrm{d}\xi_1\,\mathrm{d}\xi_2\,\mathrm{d}\xi_3$ where $|J|$ is the Jacobian of the transformation, reduces the problem to the computation of the typical integral

$$I = \int_0^1 \int_0^{1-\xi_2} f(\xi_1 \xi_2 \xi_3)\,\mathrm{d}\xi_1\,\mathrm{d}\xi_2 = \sum_{i=1}^{n} w_i f(\xi_1^i \xi_2^i \xi_3^i), \tag{4.6}$$

through a quadrature scheme involving a certain number of integration points $(\xi_1^i, \xi_2^i, \xi_3^i)$ and corresponding weights w_i [106]. A typical quadrature scheme is shown in Table 1 taken from refs. [118, 17].

Table 1
Integration points and weights for a typical quadrature scheme.

n	i	ξ_1^i	ξ_2^i	ξ_3^i	$2w_i$
1	1	1/3	1/3	1/3	1
2	1	1/2	1/2	0	1/3
	2	0	1/2	1/2	1/3
	3	1/2	0	1/2	1/3
4	1	1/3	1/3	1/3	−9/16
	2	3/5	1/5	1/5	25/48
	3	1/5	3/5	1/5	25/48
	4	1/5	1/5	3/5	25/48
7	1	0.33333333	0.33333333	0.33333333	0.22500000
	2	0.79742699	0.10128651	0.10128651	0.12593918
	3	0.10128651	0.79742699	0.10128651	0.12593918
	4	0.10128651	0.10128651	0.79742699	0.12593918
	5	0.05971587	0.47014206	0.47014206	0.13239416
	6	0.47014206	0.05971587	0.47014206	0.13239416
	7	0.47014206	0.47014206	0.05971587	0.13239416

When the point is inside the element, singularities appear in the fundamental solution. Then the integration is usually done analytically by subdividing the triangle into three parts as shown in Fig. 22 and taking the (ξ_1, ξ_2, ξ_3) to be the local coordinate system such that ξ_1 is parallel to the side 2–1 and with the same orientation, ξ_3 is normal to the element and $\xi_2 = \xi_3 \times \xi_1$. If

$$\xi_i = e_{ij} X_j \tag{4.7}$$

it is possible to show that

$$\begin{aligned} A_{ij}^{LK} &= A_{ij}\big|_1^2 + A_{ij}\big|_2^3 + A_{ij}\big|_3^1 , \\ B_{ij}^{LK} &= B_{ij}\big|_1^2 + B_{ij}\big|_2^3 + B_{ij}\big|_3^1 , \end{aligned} \tag{4.8}$$

and, for instance

$$A_{ij}\big|_1^2 = \frac{1-2\nu}{8\pi(1-\nu)}\ \xi_{ijk} e_{1k} L(r+\xi_1)\big|_1^2 ,$$

$$\begin{aligned} B_{ij}\big|_1^2 = \frac{D}{16\pi G(1-\nu)} & \left\{(3-4\nu)\delta_{ij} L\left(\tan\phi + \frac{1}{\cos\phi}\right) + e_{1i}e_{1j}\right. \\ & \times\left[-\sin\phi + L\left(\tan\phi + \frac{1}{\cos\phi}\right)\right] \\ & \left. + e_{2i}e_{2j}\sin\phi - (e_{1i}e_{2j} + e_{2i}e_{1j})\cos\phi\right\}\Bigg|_{\phi_1}^{\phi_2} , \end{aligned} \tag{4.9}$$

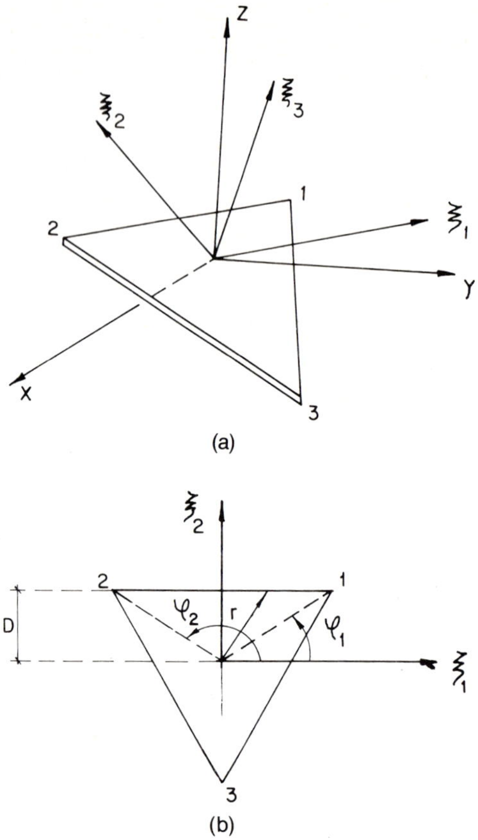

Fig. 22. Constant triangular boundary element with the collocation point inside the element.

where ξ_{ijk} is the permutation symbol.

Numerical integration in the presence of singularities is also possible [41].

4.3. Volume integrals. Thermoelasticity

As we stated in Eq. (2.38) and the following ones, in special cases it is possible to reduce volume integrals to boundary ones. An important case is thermoelasticity, where the body forces are proportional to the temperature Θ, which satisfies the steady-state heat conduction equation

$$\nabla^2\Theta = 0\,. \tag{4.10}$$

The volume integral due to these body forces becomes

$$-\gamma \int_{\partial\Omega} \Omega_j(\boldsymbol{x}_L, \boldsymbol{x}_K) \frac{\partial\Theta}{\partial n}(\boldsymbol{x}_K) + \gamma \int_{\partial\Omega} \Theta(x_K) \frac{\partial\Omega_j}{\partial n}(\boldsymbol{x}_L, \boldsymbol{x}_K)\,, \tag{4.11}$$

where $\Omega_j = W_{,j}$ and $\gamma = \alpha G(1+\nu)/(1-\nu)$ with α being the coefficient of thermal expansion.

If the same constant elements are defined for the temperature, as for the displacements and stresses, (4.11) takes the form

$$\sum_1^N \left[-\gamma \frac{\partial \Theta(K)}{\partial n} \int_{\partial\Omega_K} W_{,j}(x_L, x_K) + \gamma \Theta(K) n_i^K \int_{\partial\Omega_K} W_{,ji}(x_L, x_K) \right], \tag{4.12}$$

or

$$-\gamma \sum_{K=1}^N C_j^{LK} \frac{\partial \Theta}{\partial n}(K) + \gamma \sum_{K=1}^N D_{ji}^{LK} n_i^K \Theta(K), \tag{4.13}$$

with

$$\begin{aligned} C_j^{LK} &= \int_{\partial\Omega_K} W_{,j}(x_L, x_K), \\ D_{ji}^{LK} &= \int_{\partial\Omega_K} W_{,ji}(x_L, x_K). \end{aligned} \tag{4.14}$$

The computation of the C_j and D_{ji} elements is done analogously to that for the $\boldsymbol{A}$ and $\boldsymbol{B}$ matrices of (4.4), i.e.

$$\begin{aligned} C_j^{KK} &= C(1-2\nu)(I_{21} + I_{32} + I_{31}), \\ K_{ji}^{KK} &= C(1-2\nu)(I'_{21} + I'_{32} + I'_{31}), \end{aligned} \tag{4.15}$$

where, for instance

$$\begin{aligned} I_{12} &= D[\delta_{ij} L \tan(\phi/2) - e_{1i}e_{1j}(L \tan(\phi/2) + \cos\phi) \\ &\quad - (e_{1i}e_{2j} + e_{2i}e_{1j}) \sin\phi + e_{2i}e_{2j} \cos\phi]\big|_{\phi_1}^{\phi_2}, \\ I'_{12} &= \frac{D^3}{3}\left(-\frac{\cos\phi}{2\sin^2\phi} + \tfrac{1}{2} L \tan(\phi/2)\right)\bigg|_{\phi_1}^{\phi_2}. \end{aligned} \tag{4.16}$$

The following example was developed by Anza [11] and deals with the thermoelastic analysis of cylinders and spheres subjected to different exterior and interior temperatures, as shown in Fig. 23. Using symmetry, discretization of the sphere can be simulated, for instance, by the mesh shown in Fig. 24(b). For the evolution of temperature, elements 1 to 48 have the zero flux condition while elements 49 to 56 are supposed to be maintained at fixed temperatures. In this example, the temperature data are

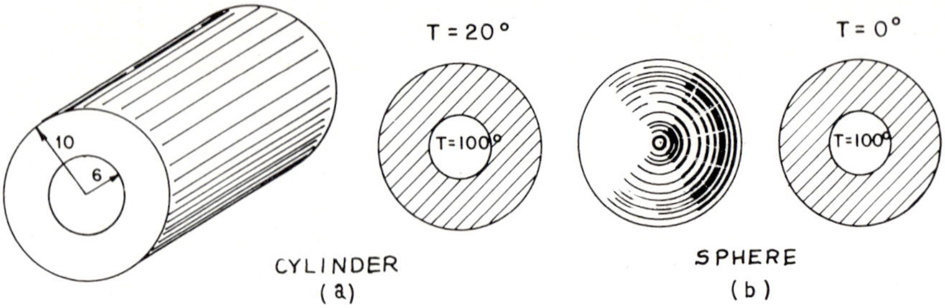

Fig. 23. Cylinder and sphere under different exterior and interior temperatures.

$$\Theta_{\text{int}} = 100° , \qquad \Theta_{\text{ext}} = 0° . \tag{4.17}$$

The elastic conditions are also imposed by the symmetry. Elements 1 to 48 present zero normal displacements and zero tangential stress, while elements 49 to 56 present known tractions (zero in this case). In addition, $\nu = 0.25$, $G = 1$, $\alpha = 1$. The closed-form solution of this problem in terms of the temperature Θ, the radial displacement u and the radial and hoop stresses σ_r and σ_θ, respectively, is given by

$$\begin{aligned} &\Theta = 1800(1/r) - 100 , \\ &u = -42.86r - 1500 - 69428.5(1/r^2) , \\ &\sigma_r = 285.72 - 600(1/r) + 277713.9(1/r^3) , \\ &\sigma_\theta = 285.71 - 3000(1/r^2) - 138857(1/r^3) , \end{aligned} \tag{4.18}$$

and it is plotted in Figs. 25(a, b and c) against the results obtained by the BEM. As we can see, the results are very good in a qualitative sense and the maximum error in displacements is of 4.7%. In general, the errors can be attributed to poor representation of the geometry. It is also important to point out that the example was chosen only to show the possibilities of the method. The symmetry conditions can be managed without elements 1 to 48 by a careful assembling of the A_{ij} and B_{ij} elements due to each element and all their symmetric ones. In this way, the number of equations could have been reduced and precision increased.

4.4. Higher-order elements

The method can easily be generalized to higher-order elements, for instance, to parabolic interpolation type of elements by shape functions which allow the adjustment by triangular or quadrilateral shells approaching complex geometry as shown in Fig. 26(a).

There are, nevertheless, problems with corner nodes that are similar to those

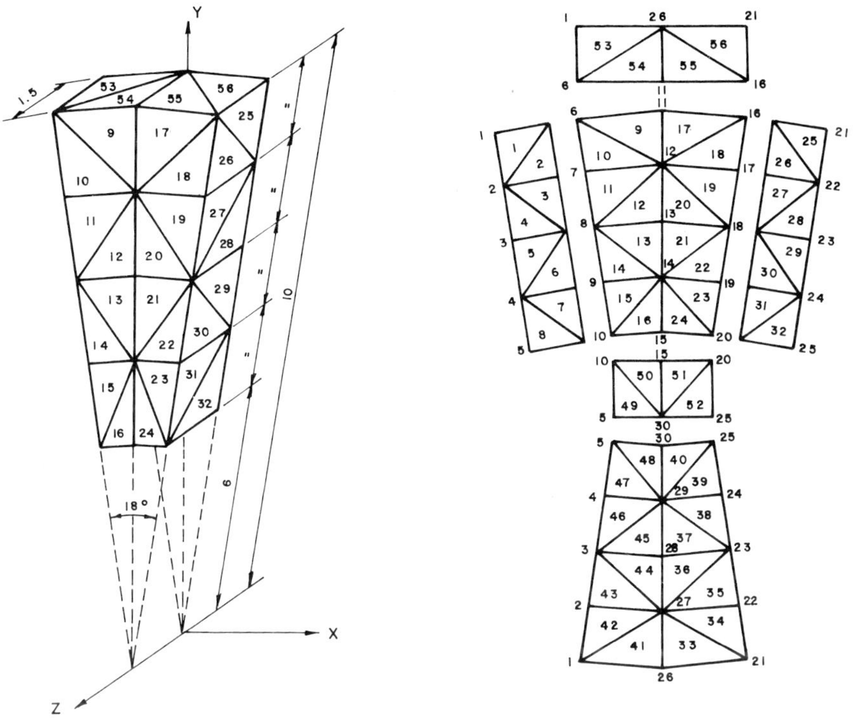

(a)

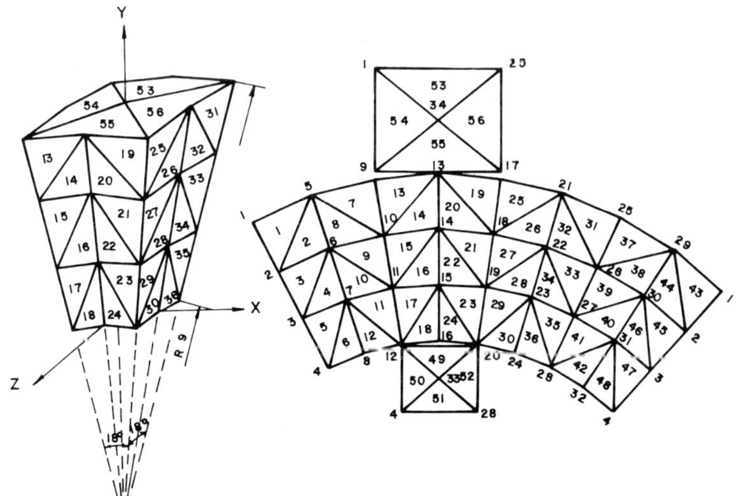

(b)

Fig. 24. Discretization of the cylinder and sphere of Fig. 23 by constant triangular boundary elements.

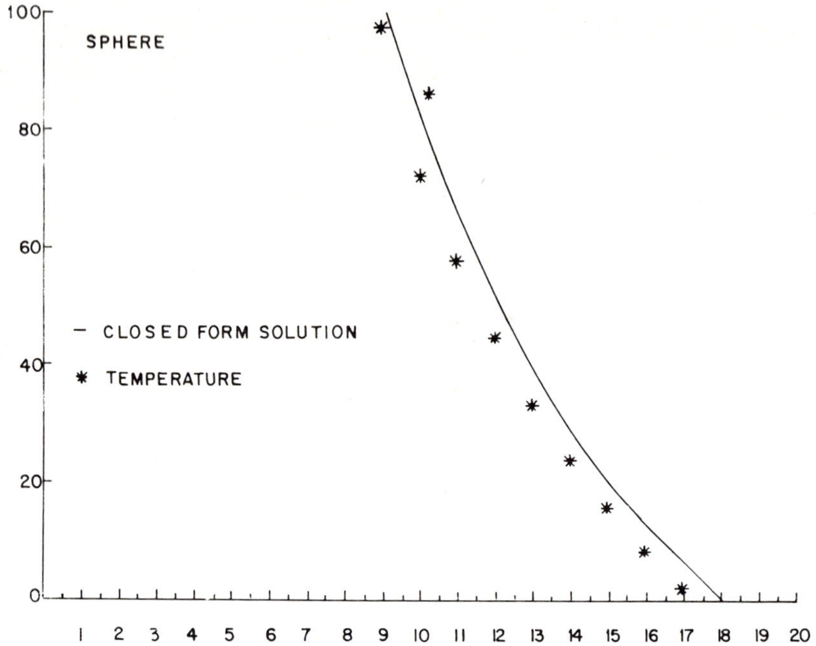

(a)

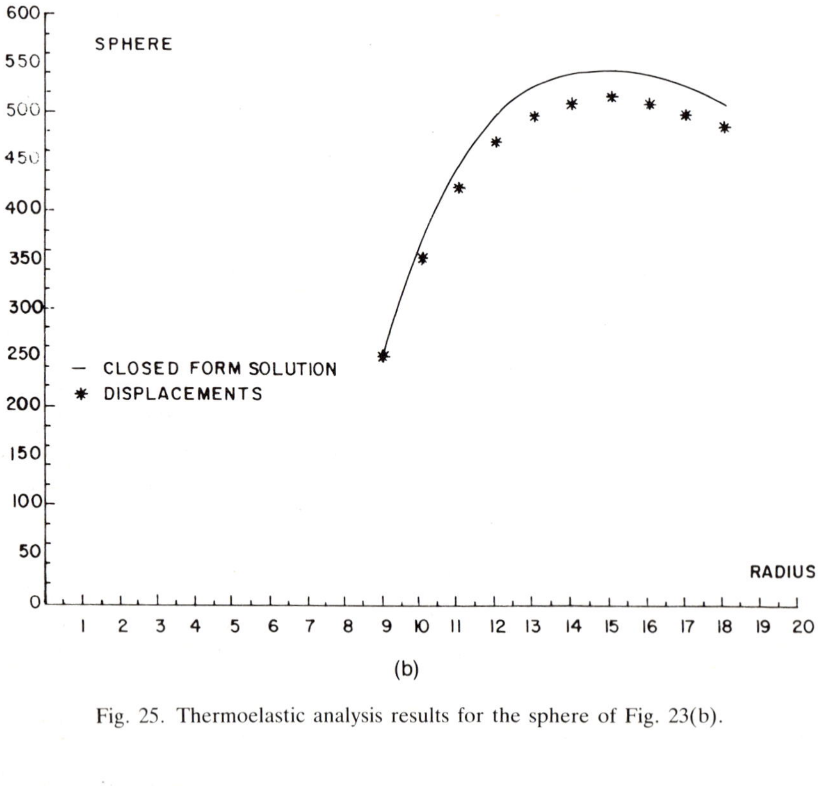

(b)

Fig. 25. Thermoelastic analysis results for the sphere of Fig. 23(b).

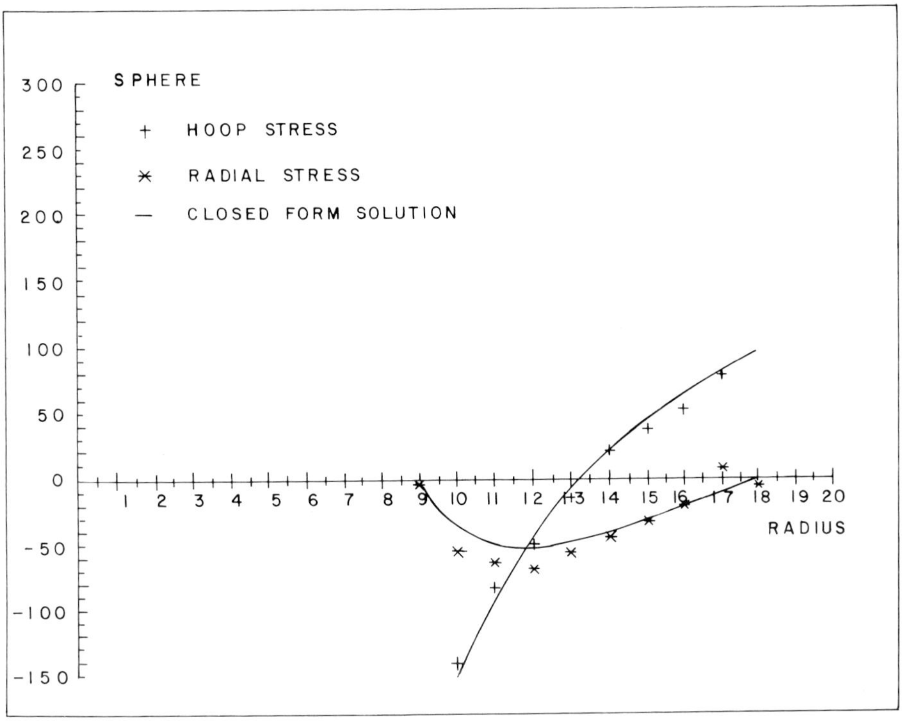

(c)

Fig. 25. (continued).

presented in plane problems. That is, for sharp corner nodes shared by different elements, there are different normals and hence different tractions which, in the assembling process, produce a rectangular ***B*** matrix. It is necessary, then, to supplement either a code or some artifice such as corner elements, master and slave degrees of freedom, etc., to eliminate the difficulty. That is why some authors prefer to define the elements by internal nodes (Fig. 4c) that eliminate the problem at the price of the size of the system of equations. The problem is really serious in piecewise heterogeneous media where it is usual to find singular behavior of fluxes or tractions.

The computation of the elements of each of the matrices ***A***, ***B***, cannot easily be done analytically because of the complexity of the Jacobian of the parabolic elements. For this reason, a numerical approach is usually followed. However, the singularity of kernels U_{ik} and T_{ik}, which tend to ∞ when r tends to zero, implies that the integration scheme needs to be done carefully while integrating over the elements near the collocation point in order to reduce the integration error. An adaptive process has been proposed by Lachat and Watson [74] which uses the standard Gaussian process, but divides each of the elements into a certain number of sub-elements and chooses the number of integration points for each of these sub-elements according to an error bound previously defined as a

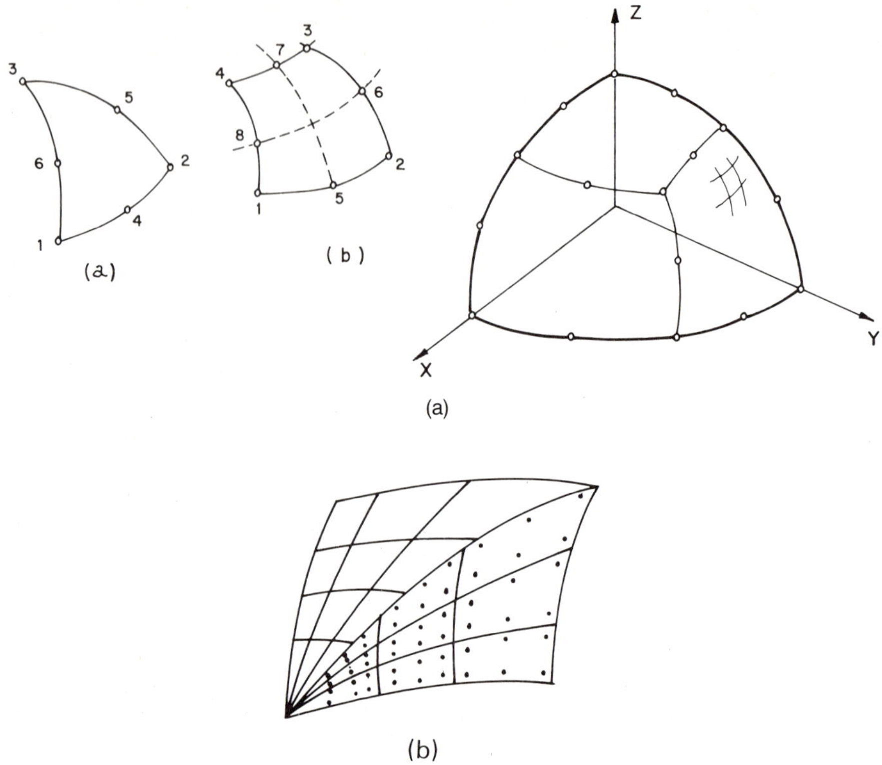

Fig. 26. Higher-order triangular and quadrilateral boundary elements.

function of some variables, as for example, the distance between the element and the singular point, the size, and the geometry of the element, etc. When the nodal point belongs to the integration element, a different approach is used in order to place a great number of integration points near the singular one. The element is divided into two or three triangles and each of them is considered as a degenerated quadrilateral sub-element. The disposition of integration points is shown in Fig. 26(b). More details can be seen in Doblaré and Alarcón [51]. Other alternatives are discussed by Watson [114] and by Rizzo and Shippy [98].

Once the equilibrium and compatibility equations have been introduced for the interface nodes, the complete linear system is reduced to a problem with $\sum_{r=1}^{R} p(r) \times 3$ equations where r is the total number of equations. However, the number of unknowns can be greater than the number of equations. This number of unknowns is $\sum_{i=1}^{I} 6p(i) + \sum_{c=1}^{R} 3p(c)$ where I is the number of interfaces and $p(i)$ and $p(c)$ are the number of nodes of each interface or subregion external boundary of the problem.

Two different problems arise during the formation of the linear system:

(a) To reduce the number of unknowns at interface nodes.

(b) To obtain a matrix structure adapted to the subsequent solving process; it is possible to choose a structure which, by a systematic process of storage, allows an appropriate situation of the zeros in order to eliminate them during the storing process and, finally, an adequate disposition for the solving process.

The first point is treated by using a new interpolation function for the stress vector in the element to which the node with more unknowns than equations belongs. Each of these functions, with a lower order than the usual ones, takes the value 1 at each node and 0 at the rest, except for the special node. After this, it is possible to extrapolate the stress vector, as Fig. 27 shows, and obtain the expressions

$$t_i = a_{1i} + a_{2i}\xi + a_{3i}\eta + a_{4i}\xi^2 + a_{5i}\eta^2 + a_{6i}\eta^2\xi + a_{7i}\xi^2\eta$$
$$t_1 = t_4 + t_2 - t_3 \,. \tag{4.19}$$

The second point is resolved by following Lachat and Watson [74, 75] and ordering the subregions as in Fig. 28, thereby obtaining a great number of grouped zeros, which are not stored, and thus saving a lot of memory.

The following three illustrative examples of three-dimensional elastic analysis by parabolic isoprametric boundary elements have been taken from [51]. The first example deals with the analysis of a sphere subject to internal pressure. This has been solved and compared with the theoretical solution given in [112]. The sphere has an internal radius of 30 and an external radius of 110, consists of a material with a Young's modulus of 2.5 and Poisson's ratio of 0.25 and is subjected to an internal pressure of 1000. The discretization of 18 elements and 36 nodes is shown in Fig. 29 and a computer memory of only 6K was used in the program for the given data. The results in terms of the radial displacement u_r and the tangential stress σ_t are shown in Table 2. The total computer time needed was 24′50″, which is very high. However, this is logical because the amount of time of input–output is very important in this case and the access time of the computer used was also very high.

Another typical test is that of a cube subjected to internal point loads. The problem consists of a simple cube supported at three different faces and subjected to two point loads of 1000 acting in the interior of the body as shown in Fig. 30(a).

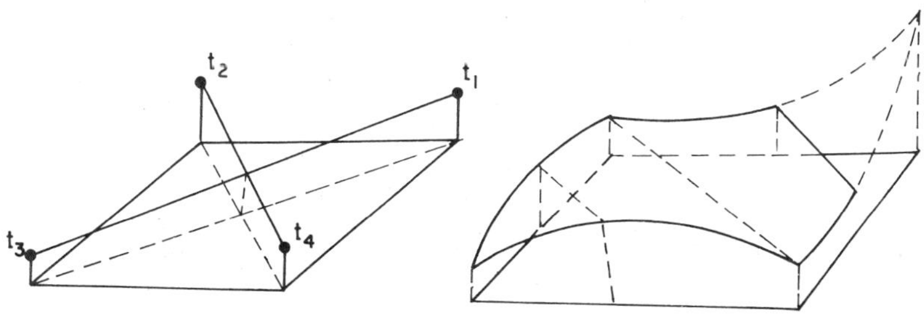

Fig. 27. Schematic illustration of a special stress vector interpolation.

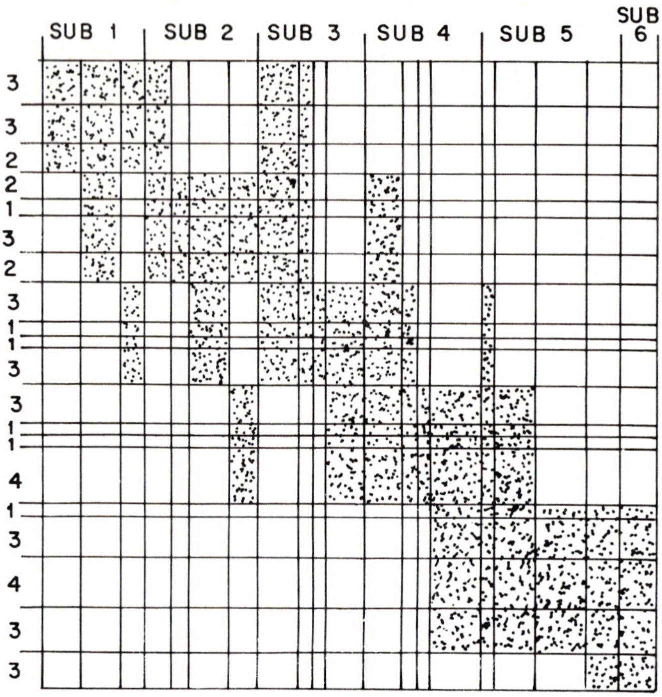

Fig. 28. Efficient ordering of subregions in the BEM.

The deformation of the cube can be seen in Fig. 30(b). Finally, as an example of the treatment of boundary conditions, the analysis of the dolos of Fig. 31(a) was also performed. The discretization of a quarter of the dolos is shown in Fig. 31(b) while the boundary conditions are such that the normal displacement restricted at the planes of symmetry, the tangential stresses are zero on them, and all the displacements for elements of the base are zero. The discretization presented was obtained using a preprocessor described in [36].

Table 2
Numerical results for the sphere of Fig. 29.

Node	Variable	Theory	BEM	Error %
1	u_r	6428.57	6401.0	−0.43
17	u_r	7164.72	7109.0	−0.78
23	u_r	8492.06	8406.0	−1.01
35	u_r	10928.57	10800.0	−1.17
41	u_r	15714.29	15590.0*l*	−0.79
1	σ_t	214.29	207.5	−3.17
17	σ_t	249.48	250.9	0.57
23	σ_t	312.17	303.9	−2.65
35	σ_t	435.43	433.5	−0.44
41	σ_t	714.29	757.0	5.98

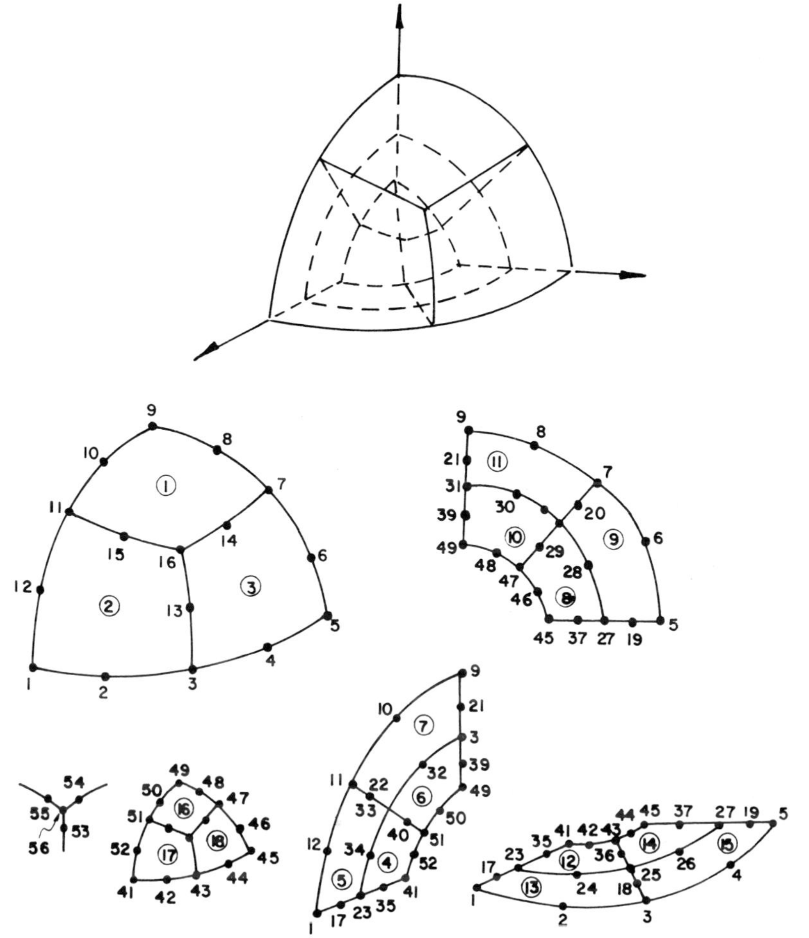

Fig. 29. Discretization of a sphere by parabolic isoparametric quadrilateral boundary elements.

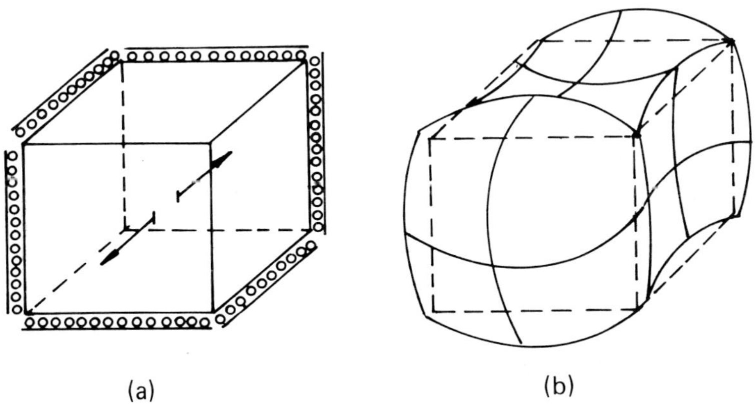

Fig. 30. Deformational configuration of a cube subjected to internal point loads.

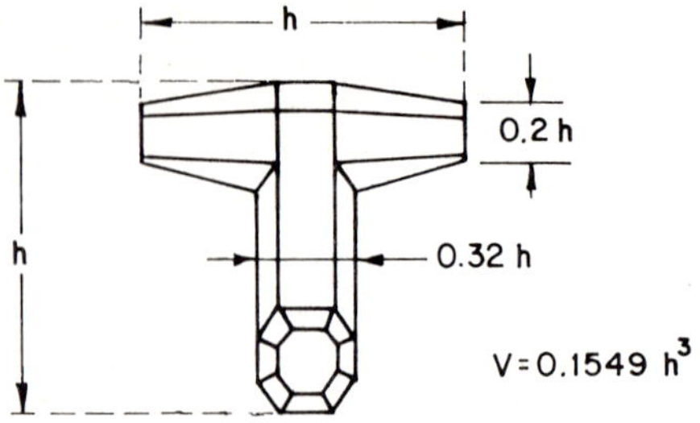

(a)

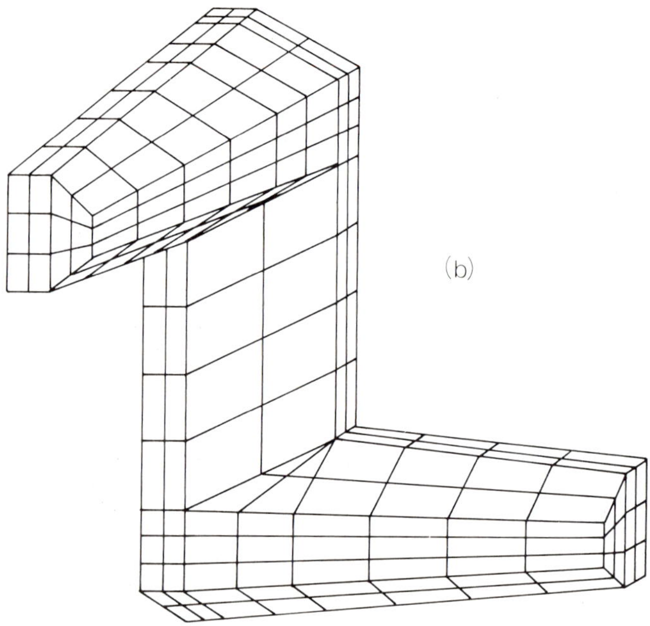

(b)

Fig. 31. Boundary element discretization of a dolos.

5. Axisymmetric problems

In this section, we shall consider those bodies whose geometry and load state have have axial symmetry (Fig. 32). The engineering applications are numerous, including, for example, rotating disks, bins, silos, reservoirs, etc.

5.1. The fundamental solution

The starting equation is Somigliana's integral equation 2.20, but now the matrices U^*_{ik}, T^*_{ik}, can be specialized by using a ring load like that in Fig. 33. A

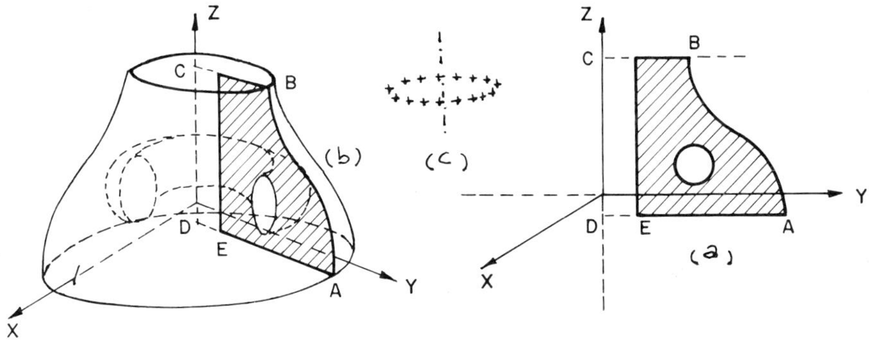

Fig. 32. Geometry of a general axisymmetric body.

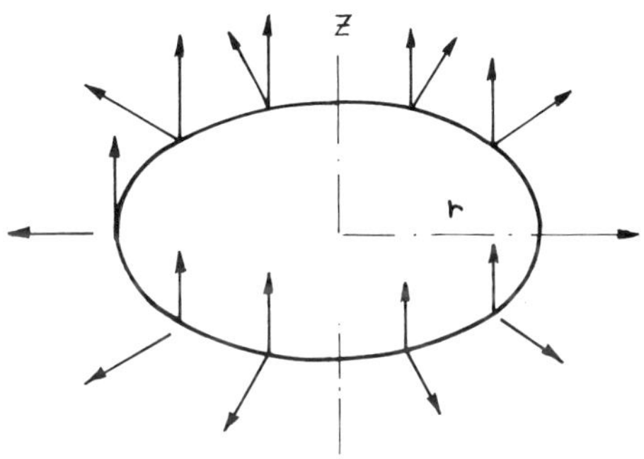

Fig. 33. Radial and axial unit concentrated loads.

current procedure for computing the fundamental solution is through the displacement function introduced by Galerkin [48, 70]. The subsequent analysis is based on the work of Cruse et al. [48]. In cylindrical coordinates, the components X_r and X_z of the Galerkin vector for a ring radial load F_r and an axial one F_z of the form

$$\begin{Bmatrix} F_r \\ F_z \end{Bmatrix} = \begin{Bmatrix} 1 \\ 1 \end{Bmatrix} \delta(R-r, Z-z)/2\pi r \tag{5.1}$$

satisfy the equations

$$\begin{gathered} \left(\nabla^2 - \frac{1}{r^2}\right)\left(\nabla^2 - \frac{1}{r^2}\right) X_r = -\frac{F_r}{G}, \\ \nabla^2\nabla^2 X_z = -\frac{F_z}{G}; \qquad \nabla^2 = \frac{\partial^2}{\partial r^2} + \frac{1}{r}\frac{\partial}{\partial r} + \frac{\partial^2}{\partial z^2}, \end{gathered} \tag{5.2}$$

The solution of (5.2) written in terms of the Legendre functions of zero order

$Q_{\pm 1/2}(\gamma)$ is

$$X_r = \frac{1}{12\pi^2 G}\sqrt{Rr}(\gamma Q_{1/2} - Q_{-1/2}),$$
$$X_z = \frac{-1}{4\pi^2 G}\sqrt{Rr}(Q_{1/2} - \gamma Q_{-1/2}), \tag{5.3}$$

where

$$\gamma = 1 + \frac{(R-r)^2 + (Z-z)^2}{2Rr}, \tag{5.4}$$

and where R, Z are the coordinates of the point of application of the load and r, z those of the running point.

The expressions for U^*_{ij} are as follows [48]:

$$2(1-\nu)U^*_{rr} = (1-2\nu)\left(\nabla^2 - \frac{1}{r^2}\right)X_r + \frac{\partial^2 X_r}{\partial z^2},$$
$$2(1-\nu)U^*_{zr} = -\frac{\partial^2 X_z}{\partial r \partial z} - \frac{1}{r}\frac{\partial X_r}{\partial z},$$
$$2(1-\nu)U^*_{rz} = -\frac{\partial^2 X_z}{\partial r \partial z}, \tag{5.5}$$
$$2(1-\nu)U^*_{zz} = (1-2\nu)\nabla^2 X_z + \frac{\partial^2 X_z}{\partial r^2} + \frac{1}{r}\frac{\partial X_z}{\partial r},$$

or in view of (5.3)

$$U^*_{rr} = \frac{1}{16\pi^2 G(1-\nu)\sqrt{Rr}}\left((3-4\nu)Q_{1/2} + \frac{(Z-z)^2}{Rr}\frac{\mathrm{d}Q_{1/2}}{\mathrm{d}\gamma}\right),$$
$$U^*_{zr} = \frac{1}{16\pi^2 G(1-\nu)}\frac{Z-z}{r\sqrt{Rr}}\left[\frac{Q_{1/2}}{2} - \left(\gamma - \frac{r}{R}\right)\frac{\mathrm{d}Q_{+1/2}}{\mathrm{d}\gamma}\right],$$
$$U^*_{rz} = \frac{-1}{16\pi^2 G(1-\nu)}\frac{Z-z}{r\sqrt{Rr}}\left[\frac{Q_{-1/2}}{2} + \left(\gamma - \frac{r}{R}\right)\frac{\mathrm{d}Q_{-1/2}}{\mathrm{d}\gamma}\right], \tag{5.6}$$
$$U^*_{zz} = \frac{1}{16\pi^2 G(1-\nu)\sqrt{Rr}}\left((3-4\nu)Q_{-1/2} - \frac{(Z-z)^2}{Rr}\frac{\mathrm{d}Q_{-1/2}}{\mathrm{d}\gamma}\right).$$

The computation of T^*_{ik} is obtained using the equilibrium equations

$$T^*_{rr} = \sigma^*_{rrr}n_r + \sigma^*_{rzr}n_z,$$
$$T^*_{zr} = \sigma^*_{rzr}n_r + \sigma^*_{zzr}n_z,$$
$$T^*_{rz} = \sigma^*_{rrz}n_r + \sigma^*_{rzz}n_z, \tag{5.7}$$
$$T^*_{zz} = \sigma^*_{rzz}n_r + \sigma^*_{zzz}n_z,$$
$$T^*_{\theta r} = 0; \qquad T^*_{\theta z} = 0,$$

where

$$\sigma^*_{rrr} = 2G\left[\frac{1-\nu}{1-2\nu}\frac{\partial U^*_{rr}}{\partial r} + \frac{\nu}{1-2\nu}\left(\frac{1}{r}U^*_{rr} + \frac{\partial U^*_{zr}}{\partial z}\right)\right],$$

$$\sigma^*_{rzr} = G\left[\frac{\partial U^*_{rr}}{\partial z} + \frac{\partial U^*_{zr}}{\partial r}\right],$$

$$\sigma^*_{zzr} = 2G\left[\frac{1-\nu}{1-2\nu}\frac{\partial U^*_{zr}}{\partial r} + \frac{\nu}{1-2\nu}\left(\frac{1}{r}U^*_{rr} + \frac{\partial U^*_{rr}}{\partial r}\right)\right],$$

$$\sigma^*_{rrz} = 2G\left[\frac{1-\nu}{1-2\nu}\frac{\partial U^*_{rz}}{\partial r} + \frac{\nu}{1-2\nu}\left(\frac{1}{r}U^*_{rz} + \frac{\partial U^*_{zz}}{\partial z}\right)\right], \tag{5.8}$$

$$\sigma^*_{rzz} = G\left(\frac{\partial U^*_{rz}}{\partial z} + \frac{\partial U^*_{zz}}{\partial r}\right),$$

$$\sigma^*_{zzz} = 2G\left[\frac{1-\nu}{1-2\nu}\frac{\partial U^*_{zz}}{\partial z} + \frac{\nu}{1-2\nu}\left(\frac{1}{r}U^*_{rz} + \frac{\partial U^*_{zz}}{\partial z}\right)\right].$$

5.2. Displacements and stresses on the boundary

The reduction to the boundary is done by transporting $x(R, Z)$ to the boundary by a standard limiting process. When $x(R, Z) \rightarrow x(r, z)$, we have that $\gamma \rightarrow 1$ and the Legendre functions can be approximated by logarithmic functions of the type

$$Q_{-1/2}(\gamma) \simeq -\tfrac{1}{2}\ln\left(\frac{\gamma-1}{32}\right), \qquad Q_{1/2}(\gamma) \simeq -\tfrac{1}{2}\ln\left(\frac{\gamma-1}{32}\right) - 2, \tag{5.9}$$

with derivatives

$$\frac{dQ_{1/2}}{d\gamma} = \frac{dQ_{-1/2}}{d\gamma} = \frac{-1}{2(\gamma-1)}, \tag{5.10}$$

and then the integrals must be computed in the sense of their Cauchy principal value. The final boundary integral equation is then

$$(\delta_{ik} - C_{ik})u_i(\boldsymbol{x}) = \int_{\partial\Omega} U^*_{ik}(\boldsymbol{x}, \boldsymbol{y})t_i(\boldsymbol{y}) - \int_{\partial\Omega} T^*_{ik}(\boldsymbol{x}, \boldsymbol{y})u_i(\boldsymbol{y}) + \int_{\Omega} U^*_{ik}(\boldsymbol{x}, \boldsymbol{z})X_i(\boldsymbol{z}),$$

$$\boldsymbol{x}, \boldsymbol{y} \in \partial\Omega;\ \boldsymbol{z} \in \Omega, \tag{5.11}$$

and the stress components can be obtained from

$$\sigma_{ij}(\boldsymbol{x}) = \int_{\partial\Omega} D^*_{ijk}t_k - \int_{\partial\Omega} S^*_{ijk}u_k + \int_{\Omega} S^*_{ijl}X_l, \tag{5.12}$$

where

$$\begin{aligned} D^*_{ijk} &= \lambda U^*_{lk,l}\delta_{ij} + G(U^*_{ij,j} + U^*_{jk,i}), \\ S^*_{ijk} &= \lambda T^*_{lk,l}\delta_{ij} + G(T^*_{ik,j} + T^*_{jk,i}). \end{aligned} \tag{5.13}$$

5.3. Discretization and treatment of the integral equation

As the discretization is limited to the meridian curve, it is possible to use constant, linear, parabolic, etc., elements as shown in Fig. 34.

If the constant element approach is used and the procedure developed in section 2.5 is applied, it is possible to write (5.11) in the form

$$\tfrac{1}{2}\delta_{ik}u_i = \sum_{i=1}^{N}\left[\int_{\partial\Omega} U^*_{ik}t_i\,\mathrm{d}s - \int_{\partial\Omega} T^*_{ij}u_i\,\mathrm{d}s + \int_{\partial\Omega} E^*_{ik}X_i\,\mathrm{d}s + \int_{\partial\Omega} F_k\,\mathrm{d}s\right], \quad (5.14)$$

where the E^*_{ik}, F_k represent the contribution of volumetric forces reducible to the boundary, or

$$\tfrac{1}{2}\begin{Bmatrix} u_r \\ u_z \end{Bmatrix} = \sum_{n=1}^{N}\left[\int_{\partial\Omega_n}\begin{bmatrix} U^*_{rr} & U^*_{zr} \\ U^*_{rz} & U^*_{zz}\end{bmatrix}\begin{Bmatrix} t_r \\ t_z \end{Bmatrix}_{(\text{centroid})_n}\right.$$

$$- \int_{\partial\Omega_n}\begin{bmatrix} T^*_{rr} & T^*_{zr} \\ T^*_{rz} & T^*_{zz}\end{bmatrix}\begin{Bmatrix} u_r \\ u_z \end{Bmatrix}_{(\text{centroid})_n}$$

$$\left. + \int_{\partial\Omega_n}\begin{bmatrix} E^*_{rr} & E^*_{zr} \\ E^*_{rz} & E^*_{zz}\end{bmatrix}\begin{Bmatrix} X_r \\ X_z \end{Bmatrix}_{(\text{centroid})_n} + \int_{\partial\Omega_n}\begin{Bmatrix} F_r \\ F_z \end{Bmatrix}\mathrm{d}s_m\right], \quad (5.15)$$

which will produce a $2N \times 2N$ system, once $2N$ boundary conditions have been fixed.

Following Eq. (5.12), the stresses are then

$$\sigma_{ij} = \sum_{n=1}^{N}\left[\int_{\partial\Omega_n} D_{ijk}t_k\,\mathrm{d}s_n - \int_{\partial\Omega_n} S_{ijk}u_k\,\mathrm{d}s_n\right.$$

$$\left. + \frac{1-2\nu}{2(1-\nu)}\int_{\partial\Omega_n}[X_kH_{ij} - H_{ijk}X_{k,k}]n_k - \int_{\partial\Omega_n} X_kI_{ijk}\,\mathrm{d}s_n\right], \quad (5.16)$$

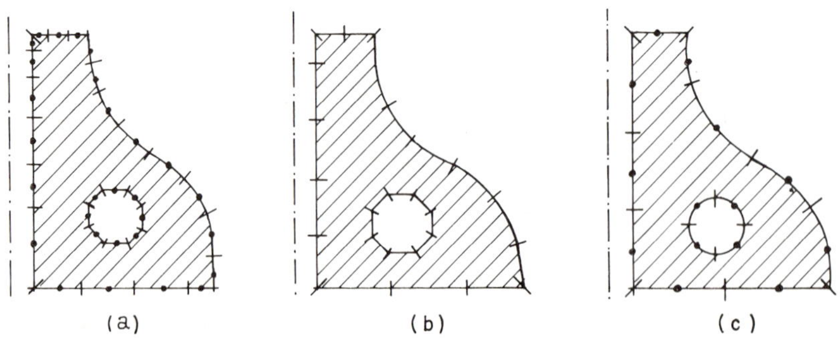

Fig. 34. Constant, linear and parabolic boundary element discretization of an axisymmetric body.

where the new tensors appear after substituting the expressions of the previously mentioned volumetric forces reducible to the boundary.

Thus, finally, Eqs. (5.15) and (5.16) can be written symbolically as

$$\begin{aligned} \boldsymbol{Bu} &= \boldsymbol{At} + \boldsymbol{P}\,, \\ \boldsymbol{\sigma} &= \boldsymbol{Dt} - \boldsymbol{Su} + \boldsymbol{P}'\,. \end{aligned} \tag{5.17}$$

For the computation of the coefficients in the influence matrices, two cases are usually distinguished:

(a) If the running point does not coincide with the collocation point, computation of the coefficients is carried out by a Gauss quadrature as

$$\int_{\partial\Omega_n} f\,\mathrm{d}S_m = \int_{-L/2}^{L/2} 2rf\,\mathrm{d}l = \int_{-1}^{1} 2\pi rfL/2\,\mathrm{d}\xi = L\sum_{k=1}^{N} w_k r(\xi_k)f(\xi_k)\,. \tag{5.18}$$

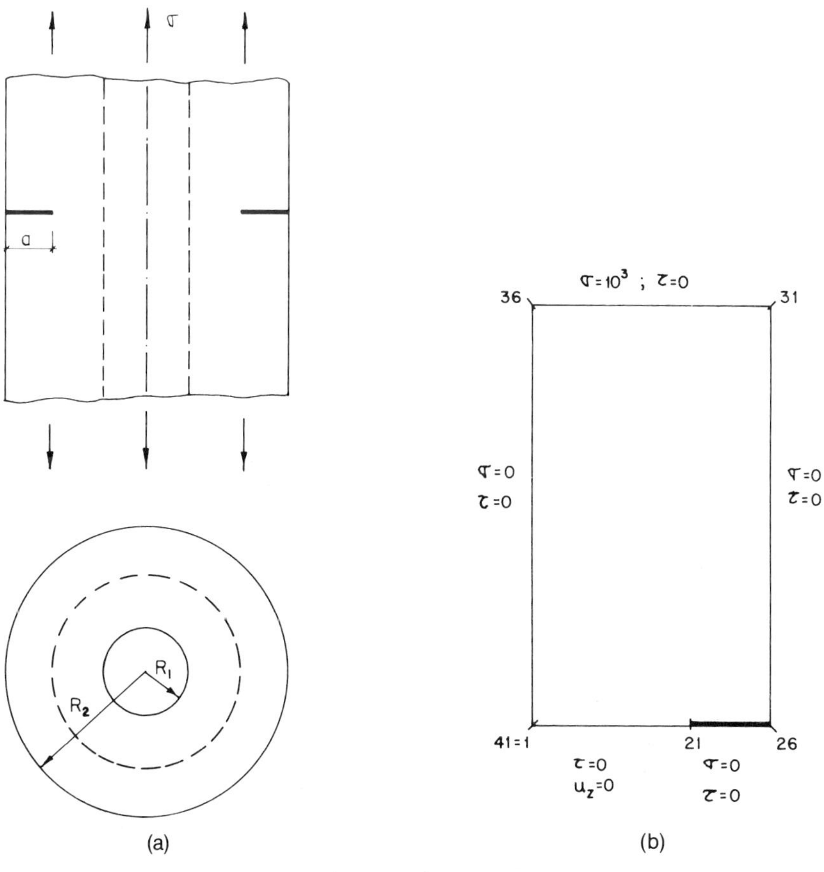

Fig. 35. Axially loaded thick walled cracked tube.

(b) If both points coincide, we use a semianalytic approach producing a closed-form solution around the singularity where the Legendre functions can be expressed by a relatively simple formula and a numerical integration of the remaining parts of the element.

5.4. Numerical examples

The first example is shown in Fig. 35(a). A thick walled tube is loaded axially and presents a crack along its exterior boundary. It is well known that the stress distribution follows the law

$$\sigma_{ij}(r, \theta) = \frac{K}{\sqrt{2\pi r}} f_{ij}(\theta) ,$$

where K is the stress intensity factor. Fig. 35(b) shows the discretization that produces the results shown in Table 3. The computed stress intensity factor is $K = 2908.372$ (Fig. 36) with an error of 4.05% against the closed-form solution in spite of the poor discretization that has been used.

Another example is presented in Fig. 37(a), where the same cylinder is

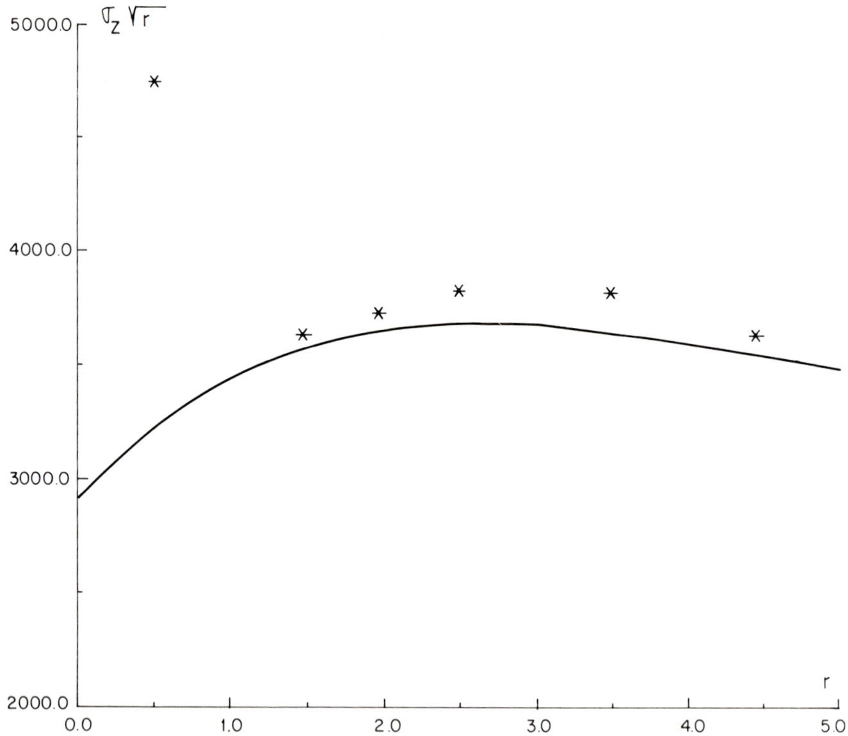

Fig. 36. Stress intensity factor for the structure of Fig. 35(a).

Table 3
Numerical results from the analysis of the thick walled tube of Fig. 35.

Node	Distance to the crack tip r (cm)	σ_z (kg/cm^2)	$\sigma_z\sqrt{r}$ ($kg \cdot cm^{-5/2}$)
20	0.5	6856.35	4848.17
19	1.5	2899	3550.54
18	2.5	2336.135	3693.75
17	3.5	2001.184	3743.87
16	4.5	1557.939	3304.89
15	5.5	1562.348	3664.03
14	6.5	1307.604	3494.37
13	7.5	1273.738	3488.28
12	8.5	1166.764	3401.67
11	9.5	1136.31	3502.67
10	10.5	908.003	3147.98

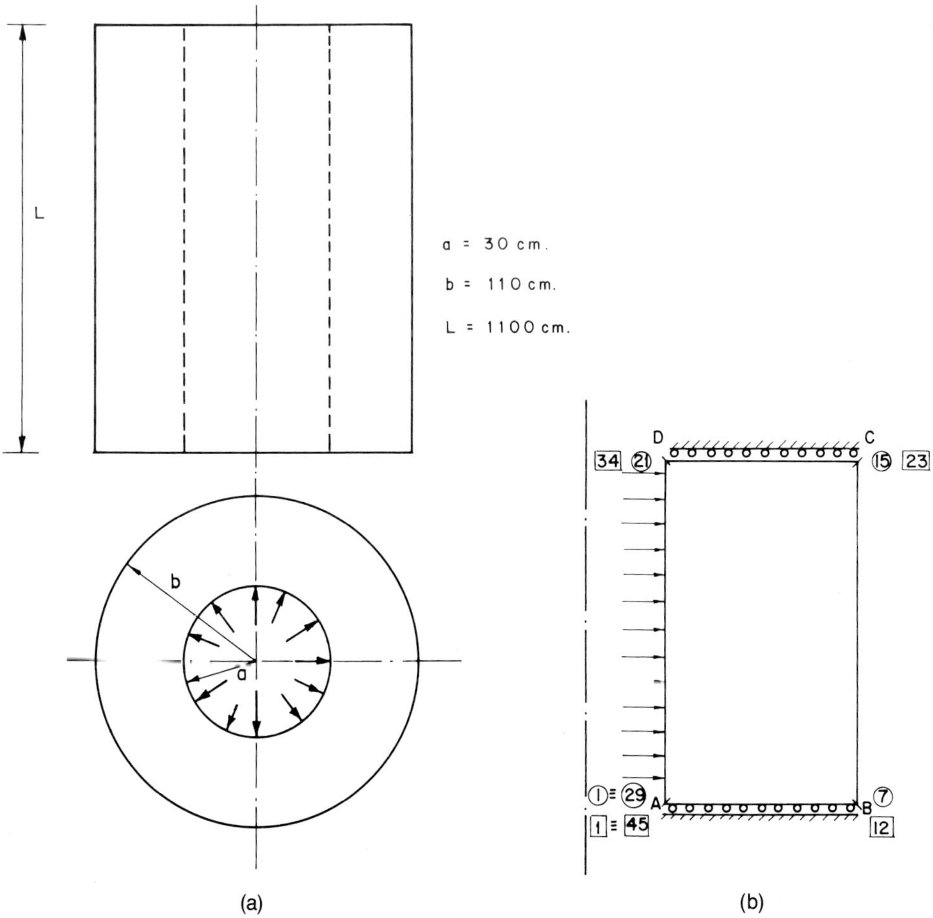

Fig. 37. Thick walled tube under internal pressure.

Table 4
Numerical results from the analysis of the cylinder of Fig. 37.

Face	Discretization	Variable	Results	Error %
AB,CD	28 elements	$\sigma_z(\mathrm{kg/cm^2})$	37.55	6.55
	44 elements		40.27	0.22
BC	28 elements	$u_r(\mathrm{cm})$	6680.75×10^{-5}	7.97
	44 elements		6627.18×10^{-5}	7.11
DA	28 elements	$u_r(\mathrm{cm})$	17362.5×10^{-5}	4.04
	44 elements		17200.91×10^{-5}	3.08

submitted to internal pressure. Nodes with the symbol ○ correspond to a 28-element discretization, while those with the □ represent a discretization with 44 elements (Fig. 37b). The results are as shown in Table 4.

5.5. Use of the Cartesian fundamental solution

As an alternative to the use of the axisymmetric fundamental solution, it is possible to treat the Cartesian equation with two changes to polar coordinates. If the field point has coordinates (ρ, θ, z) and the collocation point (ρ', θ', z'), it can be seen immediately that the boundary integral equation can be written for a smooth boundary such as

$$\tfrac{1}{2}\boldsymbol{u}^{\mathrm{p}}(\mathrm{P}) + \int_{\partial\Omega} (\boldsymbol{\Omega}'\boldsymbol{T}^{\mathrm{c}}\boldsymbol{\Omega})\boldsymbol{u}^{\mathrm{p}} = \int_{\partial\Omega} (\boldsymbol{\Omega}'\boldsymbol{U}^{\mathrm{c}}\boldsymbol{\Omega})\boldsymbol{t}^{\mathrm{p}} , \tag{5.19}$$

where superscripts c and p mean Cartesian and polar, respectively, and Ω and Ω' are defined by

$$\boldsymbol{\Omega} = \begin{pmatrix} \cos\theta & -\sin\theta & 0 \\ \sin\theta & \cos\theta & 0 \\ 0 & 0 & 1 \end{pmatrix}, \qquad \boldsymbol{\Omega}' = \begin{pmatrix} \cos\theta' & \sin\theta' & 0 \\ -\sin\theta' & \cos\theta' & 0 \\ 0 & 0 & 1 \end{pmatrix}. \tag{5.20}$$

If the collocation point P has $\theta' = 0$, then, for instance

$$\boldsymbol{T}^{\mathrm{p}} = \boldsymbol{\Omega}'\boldsymbol{T}^{\mathrm{c}}\boldsymbol{\Omega} = \begin{bmatrix} T_{11}\cos\theta + T_{12}\sin\theta & -T_{11}\sin\theta + T_{12}\cos\theta & T_{13} \\ T_{21}\cos\theta + T_{22}\sin\theta & -T_{21}\sin\theta + T_{22}\cos\theta & T_{23} \\ T_{31}\cos\theta + T_{32}\sin\theta & -T_{31}\sin\theta + T_{32}\cos\theta & T_{33} \end{bmatrix}. \tag{5.21}$$

In general, point P has to be chosen judiciously, depending on problem conditions.

A very interesting situation arises, for instance, when trying to compute the impedance of rigid circular foundations. In general, one is interested in obtaining the stresses developed under the footing when subjected to torsional, vertical, horizontal or rocking unit displacements. The distribution of displacements and stresses is given in Table 5.

Assuming constant values of u' and τ' inside concentric rings and that the collocation points P are situated at $\theta = 0$, it is easy to see that cases (a) and (b) can be solved simultaneously through the system

$$\frac{1}{2}\begin{Bmatrix} u'_\rho \\ u'_\theta \\ u'_z \end{Bmatrix} + 2\int \begin{bmatrix} T_{11}\cos\theta + T_{12}\sin\theta & 0 & T_{13} \\ 0 & -T_{21}\sin\theta + T_{22}\cos\theta & 0 \\ T_{31}\cos\theta + T_{32}\sin\theta & 0 & T_{33} \end{bmatrix} \begin{Bmatrix} u'_\rho(\mathrm{Q}) \\ u'_\theta(\mathrm{Q}) \\ u'_z(\mathrm{Q}) \end{Bmatrix}$$

$$= 2\int \begin{bmatrix} U_{11}\cos\theta + U_{12}\sin\theta & 0 & U_{13} \\ 0 & -U_{21}\sin\theta_{22} + U_{22}\cos\theta & 0 \\ U_{31}\cos\theta + U_{32}\sin\theta & 0 & U_{23} \end{bmatrix} \begin{Bmatrix} \tau'_\rho(\mathrm{Q}) \\ \tau'_\theta(\mathrm{Q}) \\ \sigma'_z(\mathrm{Q}) \end{Bmatrix} \tag{5.22}$$

where the integration is done only along the half ring, such that $0 \leqslant \theta \leqslant \pi$. For cases (c) and (d), the situation is similar although the second equation has to be computed by locating P somewhere with $\theta \neq 0$. Choosing, for instance, $\theta = -\pi/2$, the final system for every ring is

$$\frac{1}{2}\begin{Bmatrix} u'_\rho \\ u'_\theta \\ u'_z \end{Bmatrix} + 2\int \begin{bmatrix} T_{11}C^2 + T_{12}SC & T_{11}S^2 - T_{12}SC & T_{13}C \\ T^*_{11}C^2 \mid T^*_{12}SC & T^*_{11}S^2 - T^*_{12}SC & T^*_{13}C \\ T_{31}C^2 + T_{32}SC & T_{31}S^2 - T_{32}SC & T_{33}C \end{bmatrix} \begin{Bmatrix} u'_\rho \\ u'_\theta \\ u'_z \end{Bmatrix}$$

$$= 2\int \begin{bmatrix} U_{11}C^2 + U_{12}SC & U_{11}S^2 - U_{12}SC & U_{13}C \\ U^*_{11}C^2 + U^*_{12}SC & U^*_{11}S^2 - U_{12}SC & U^*_{13}C \\ U_{31}C^2 + U_{32}SC & U_{31}S^2 - U_{32}SC & U_{33}C \end{bmatrix} \begin{Bmatrix} \tau'_\rho \\ \tau'_\theta \\ \sigma'_z \end{Bmatrix} \tag{5.23}$$

where

$$C = \cos\theta,\ S = \sin\theta,\ SC = \cos\theta \cdot \sin\theta,\ C^2 = \cos^2\theta,\ S^2 = \sin^2\theta\ ,$$

and the integration for the first and third equation is carried out inside the half ring $0 \leqslant \theta \leqslant \pi$, while the second one is extended over $-\pi \leqslant \theta \leqslant \pi$.

This procedure is also useful for treating embedded foundations or piles with

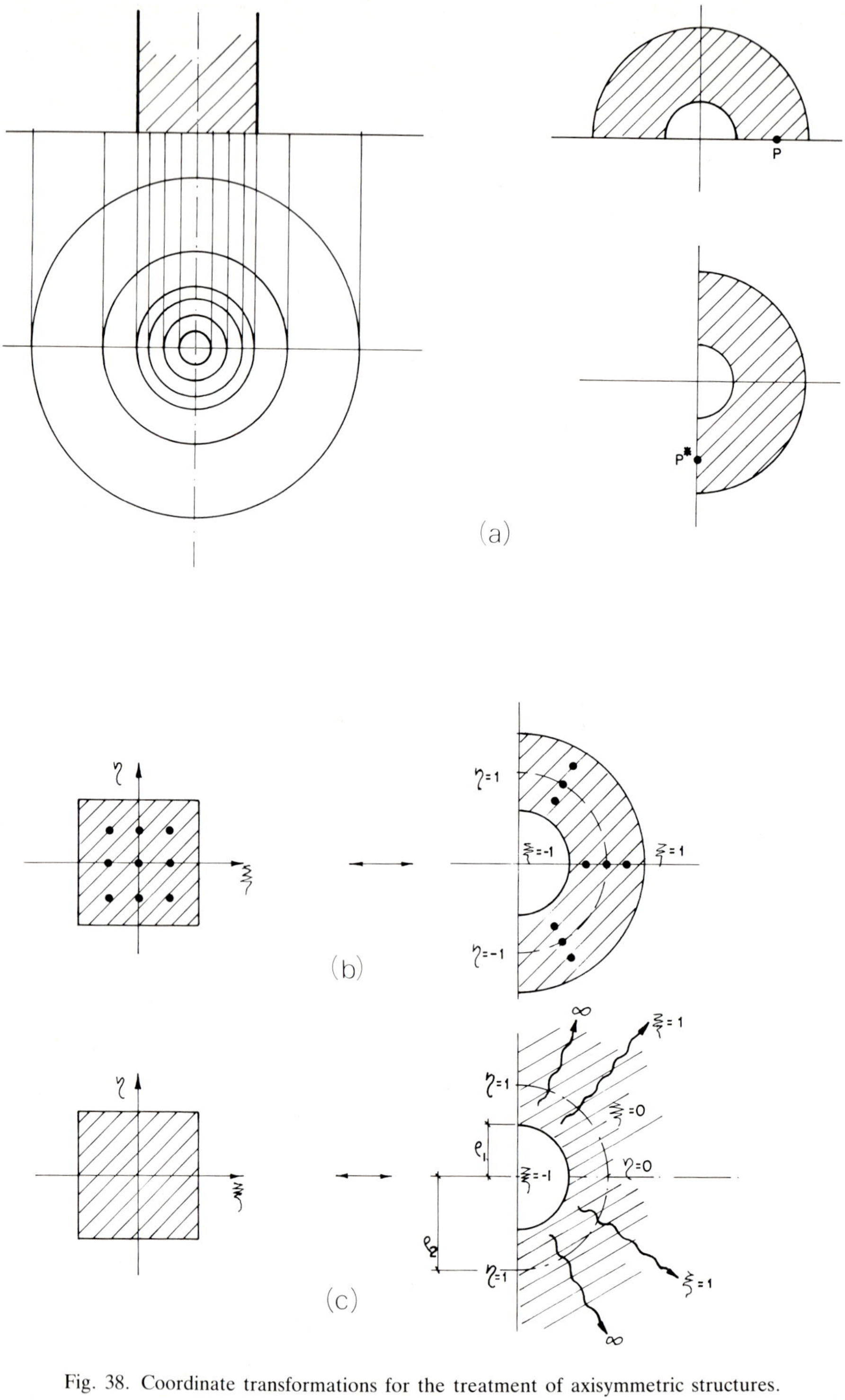

Fig. 38. Coordinate transformations for the treatment of axisymmetric structures.

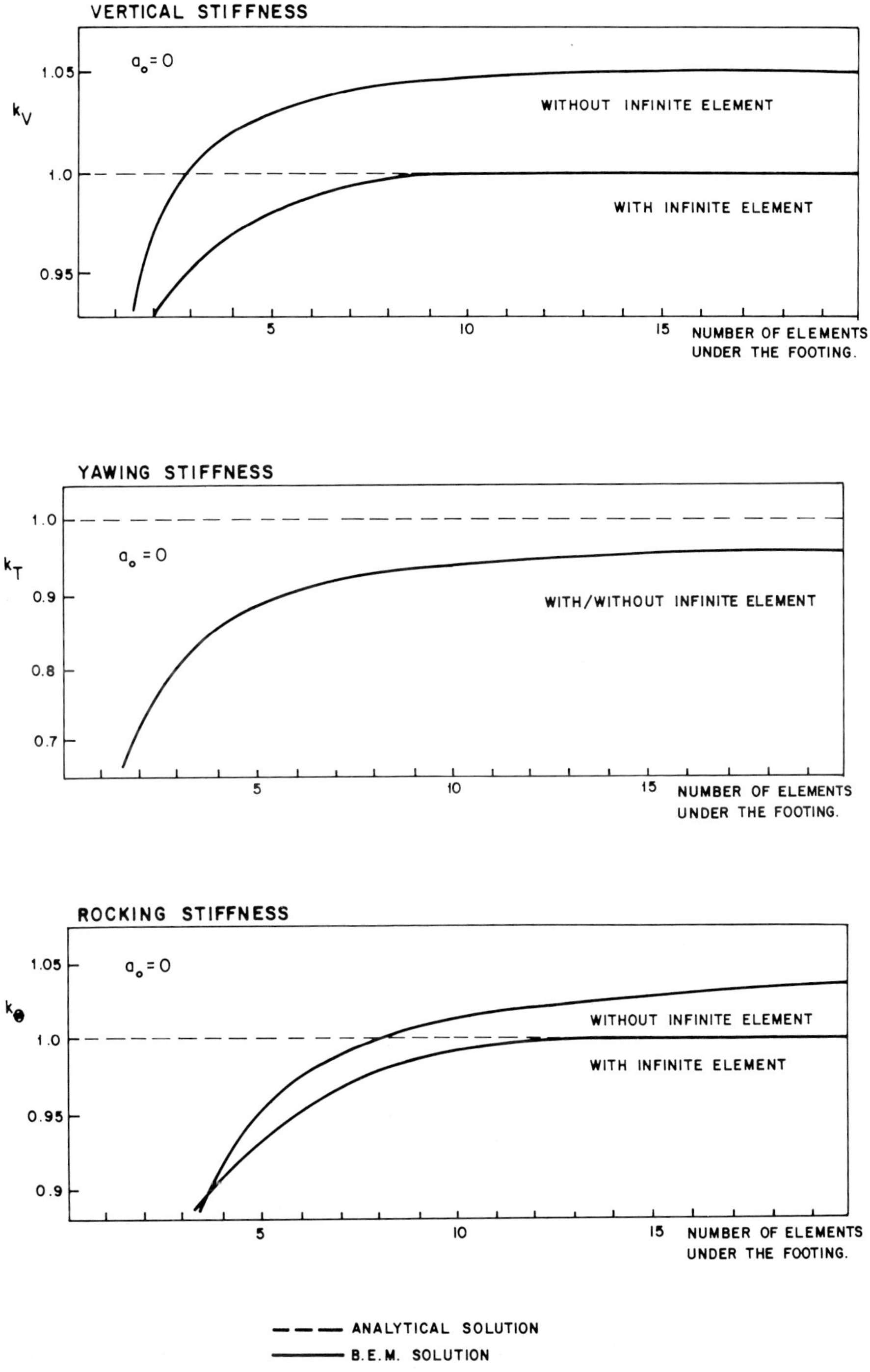

Fig. 39. Convergence of static solution discretizing only under the footing or adding an exterior infinite element.

Table 5
Stresses and displacements under a circular footing subjected to various modes of deformation.

Imposed displacement	Stresses	Displacements
(a) torsion	$\tau_\rho = 0 \quad \sigma_z = 0$ $\tau_\theta = \tau'_\theta(\rho)$	$u_\rho = 0 \quad u_z = 0$ $u_\theta = u'_\theta(\rho)$
(b) vertical	$\tau_\theta = 0 \quad \tau_\rho = \tau'_\rho(\rho)$ $\sigma_z = \sigma'_z(\rho)$	$u_\theta = 0 \quad u_\rho = u'_\rho(\rho)$ $u_z = u'_z$
(c) horizontal	$\tau_\rho = \tau'_\rho \cos\theta$ $\tau_\theta = -\tau'_\theta \sin\theta$ $\sigma_z = -\sigma'_z \cos\theta$	$u_\rho = u'_\rho \cos\theta$ $u_\theta = -u'_\theta \sin\theta$ $u_z = u'_z \cos\theta$
(d) rocking	ditto	ditto

obvious modifications. The integrals are done numerically, using the transformation indicated in Fig. 38(a, b), while sometimes it is also useful to develop infinite elements, as shown in Fig. 38(c), according to the law

$$\theta = \tfrac{1}{2}\pi\eta\,, \qquad \rho = \rho_2 + 2\,\frac{\rho_2 - \rho_1}{1 - \xi}\,\xi\,. \tag{5.24}$$

Figure 39 shows some comparisons between stiffnesses computed using elements only under the rigid footing or adding infinite elements to represent the free-field region.

Closing this section, we would like to mention that treatment of axisymmetric elastic bodies with non-axisymmetric boundary conditions is also possible by expanding the variables of the problem in a series of orthogonal functions, e.g. Fourier series, of the polar angle and reducing the problem to a sequence of two-dimensional, angularly independent problems like the ones treated in this section [99, 80, 100].

6. Anisotropy

The treatment of anisotropic elastic bodies is a field relatively scarce in publications, in comparison with the isotropic elastic ones. The difficulty lies in the fundamental solution. Only for plane problems or for certain three-dimensional cases is it possible to obtain closed-form expressions.

Rizzo and Shippy [96] presented the first application of the BEM to plane anisotropic problems, while Vogel and Rizzo [111] presented an integral representation for three-dimensional problems. Wilson and Cruse [115] used the point-load solution

$$U_{ij}(\boldsymbol{x}, \boldsymbol{y}) = \frac{1}{8\pi^2|\boldsymbol{x} - \boldsymbol{y}|} \oint_{|\boldsymbol{\xi}|=1} K_{ij}^{-1}(\boldsymbol{\xi})\,, \tag{6.1}$$

which gives the displacement at point $\boldsymbol{y}$ due to a point force applied at $\boldsymbol{x}$ in the form of a line integral taken on the unit circle in the plane passing through $\boldsymbol{x}$ and normal to $\boldsymbol{x}-\boldsymbol{y}$ with an integrand

$$K_{ij}^{-1}(\boldsymbol{\xi}) = [C_{ijkm}\xi_k\xi_m]^{-1}, \tag{6.2}$$

where C_{ijkm} represents the elastic constants of the material. This representation reduces only to closed form for the Kelvin case (isotropic material) or for transversely anisotropic materials (Pan and Chou [87] and Kobayashi and Nishimura [71]).

Wilson and Cruse [115] decided to treat what they called the modulation function

$$G_{ij}(n_1, n_2) = \int_{|\xi|=1} K_{ij}^{-1}(\xi), \tag{6.3}$$

where (n_1, n_2) is the orientation of $\boldsymbol{x}-\boldsymbol{y}$, by means of an interpolation technique, generating data bases for a variety of materials and managing them from an external computer core.

Because of the difficulties related to that presentation, Nishimura and Kobayashi [83] have recently offered an alternative, using an indirect approach where the Fourier transform is used to generate a simple layer potential approximated by piecewise-constant elements in place of the point-load solution.

Crouch [42] and Crouch and Starfield [44] have also shown how to apply the displacement discontinuity method to anisotropic elasticity. Figure 40, taken from [44], represents, for instance, their results for the tangential stress at the boundary of a circular hole in an orthotropic plate.

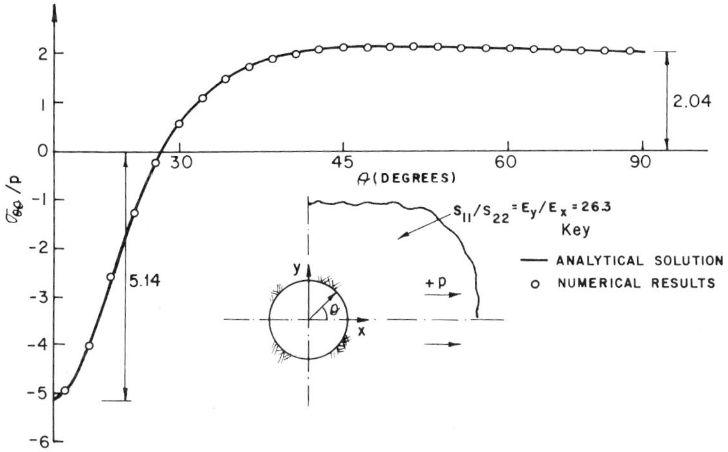

Fig. 40. Tangential stress around the boundary of a circular hole in an orthotropic plate.

7. Additional topics

Three complementary topics are included in this section. First, a discussion on the bending of thin plates and shallow shells; second, the coupling of boundary and finite elements; and finally, some hints on the possibility of developing adaptive boundary elements are briefly discussed. We think that, although some important research has already been done on the last two subjects, they deserve more attention in the future.

7.1. Bending of thin plates and shallow shells

The BEM has also been successfully applied to the analysis of thin elastic flexural plates. Both the indirect [66, 9, 84, 109] and the direct [55, 21, 104, 109] approaches have been used. The first work in this area is that of Jawson and Maiti [66] who presented an indirect approach based on the use of potentials. Comprehensive reviews of the topic can be seen in Tottenham [109] and Stern [105]. In general, the fundamental solution is based on the response to a concentrated load acting on an infinite plate, although other possibilities do exist, as shown by Altiero and Sikarskie [9]. Paris and Leon [88] have also shown the possibility of using the Marcus approach. Finally, a series of authors have presented extensions to finite deformations, buckling and shallow shell response [68].

It is well known [112] that the governing equation for thin plate flexure is

$$\nabla^2\nabla^2 w = p/D \,, \tag{7.1}$$

where w is the vertical deflection, p is the distributed loading acting on the plate, h is the plate thickness and the bending rigidity $D = Eh^3/12(1-\nu^2)$, with E and ν being the modulus of elasticity and Poisson's ratio, respectively.

Defining

$$M = D\nabla^2 w \,, \tag{7.2}$$

the governing equation can be written as

$$\nabla^2 M = p$$

and both can be treated along the same lines of potential problems to produce a pair of coupled equations

$$\begin{aligned} cM + \int_{\partial\Omega} Mq^* &= \int_{\partial\Omega} \frac{\partial M}{\partial n}\,\phi^* + \int_{\Omega} p\phi^* \,, \\ cw + \int_{\partial\Omega} wq^* &= \int_{\partial\Omega} \frac{\partial w}{\partial n}\,\phi^* + \int_{\Omega} \frac{M}{D}\,\phi^* \,, \end{aligned} \tag{7.3}$$

where ϕ^* is the fundamental solution of the Laplacian equation and M represents the value of the invariant

$$M = \frac{M_n + M_s}{1 + \nu}, \tag{7.4}$$

with M_n and M_s being the bending moments along two orthogonal directions [112].

The usual technique of discretization allows one to write

$$\begin{aligned} \boldsymbol{A}\boldsymbol{M} &= \boldsymbol{B}\boldsymbol{M}_{,n} + \mathbf{P}^0, \\ \boldsymbol{A}\boldsymbol{w} &= \boldsymbol{B}\boldsymbol{w}_{,n} + \boldsymbol{M}^0, \end{aligned} \tag{7.5}$$

where $\boldsymbol{A}$ and $\boldsymbol{B}$ are the influence matrices described in the chapter on potential problems and the comma indicates differentiation.

A very important possibility for application occurs when the boundary is simply supported and then both problems are decoupled so that it is possible to solve them in turn. In this case, it must be observed that $\boldsymbol{A}$ and $\boldsymbol{B}$ need to be computed only once.

The direct method can be obtained by considering the weak formulation of the equilibrium equation of the plate of the form

$$\frac{\partial^2 M_x}{\partial x^2} + 2\frac{\partial^2 M_{xy}}{\partial x \partial y} + \frac{\partial^2 M_y}{\partial y^2} = -p, \tag{7.6}$$

where M_x and M_y are the bending moments and M_{xy} is the twisting moment. After multiplication of both terms of (7.6) by a virtual displacement and integration by parts, it is obtained that

$$\begin{aligned} &\int_\Omega M_x \frac{\partial^2 w^*}{\partial x^2} + 2M_{xy}\frac{\partial^2 w^*}{\partial x \partial y} + M_y \frac{\partial^2 w^*}{\partial y^2} \\ &\quad = \int_\Omega p w^* - \int_{\partial\Omega} M_n \frac{\partial w^*}{\partial n} - \int_{\partial\Omega} M_{ns}\frac{\partial w^*}{\partial s} + \int_{\partial\Omega} Q w^*, \end{aligned} \tag{7.7}$$

where, as usual

$$\begin{aligned} &M_n = M_x n_1^2 + M_y n_2^2 + 2M_{xy} n_1 n_2, \\ &M_{ns} = -M_x n_1 n_2 + M_y n_1 n_2 + M_{xy} n_1^2 - M_{xy} n_2^2, \\ &n^{\mathrm{T}} = (n_1, n_2), \qquad Q = Q_x n_1 + Q_y n_2, \\ &Q_x = \frac{\partial M_x}{\partial x} + \frac{\partial M_{xy}}{\partial y}, \qquad Q_y = \frac{\partial M_y}{\partial y} + \frac{\partial M_{xy}}{\partial x}. \end{aligned} \tag{7.8}$$

It can be seen that the right-hand side of (7.7) is a bilinear symmetric form in w, w^* after the introduction of the definitions

$$M_x = -D\left(\frac{\partial^2 w}{\partial x^2} + \nu\,\frac{\partial^2 w}{\partial y^2}\right),$$

$$M_y = -D\left(\frac{\partial^2 w}{\partial y^2} + \nu\,\frac{\partial^2 w}{\partial x^2}\right),$$

$$M_{xy} = -(1-\nu)D\,\frac{\partial^2 w}{\partial x \partial y}\,.$$

Thus, a reciprocity relationship can immediately be written by interchanging w and w^* in (7.7) and subtracting the resulting equation from (7.7). This has the form

$$\int_\Omega p^* w - \int_{\partial\Omega} M_n^* \frac{\partial w}{\partial n} - \int_{\partial\Omega} M_{ns}^* \frac{\partial w}{\partial s} + \int_{\partial\Omega} Q^* w$$
$$= \int_\Omega \frac{p}{D}\, w^* - \int_{\partial\Omega} M_n \frac{\partial w^*}{\partial n} - \int_{\partial\Omega} M_{ns} \frac{\partial w^*}{\partial s} + \int_{\partial\Omega} Q w^*\,, \tag{7.9}$$

or, upon introducing

$$Q_{ef} = Q + \frac{\partial M_{ns}}{\partial s}\,, \qquad q = -\frac{\partial w}{\partial n}\,, \qquad q^* = -\frac{\partial w^*}{\partial n}$$

and applying integration by parts on the third integral on both sides, the form

$$\int_\Omega p^* w + \int_{\partial\Omega} M_n^* q + \int_{\partial\Omega} Q_{ef}^* w + \sum |M_{ns}^* w|$$
$$= \int_\Omega p w^* + \int_{\partial\Omega} M_n q^* + \int_{\partial\Omega} Q_{ef} w^* + \sum |M_{ns} w^*|\,, \tag{7.10}$$

where M_{ns} and M_{ns}^* are the possible concentrated forces in corners.

Using a fundamental solution, for instance,

$$w^* = (1/8\pi) r^2 \ln r\,, \tag{7.11}$$

we obtain the basic relations

$$cw = \int_\Omega p w^* - \int_{\partial\Omega} Q_{ef}^* w + \int_{\partial\Omega} M_n^* \frac{\partial w}{\partial n} - \int_{\partial\Omega} M_n \frac{\partial w^*}{\partial n}$$
$$+ \int_{\partial\Omega} Q_{ef} w^* + \sum_1^N |M_{ns} w^*| - \sum_1^N |M_{ns}^* w|\,, \tag{7.12}$$

$$c \frac{\partial w}{\partial n} = \int_{\Omega} p \frac{\partial w^*}{\partial n} - \int_{\partial\Omega} \frac{\partial Q^*_{ef}}{\partial n} w + \int_{\partial\Omega} \frac{\partial M^*_n}{\partial n} \frac{\partial w}{\partial n} - \int_{\partial\Omega} M_n \frac{\partial^2 w^*}{\partial n^2}$$

$$+ \int_{\partial\Omega} Q_{ef} \frac{\partial w^*}{\partial n} + \sum_{1}^{N} \left| M_{ns} \frac{\partial w^*}{\partial n} \right| - \sum_{1}^{N} \left| \frac{\partial M^*_{ns}}{\partial n} w \right| , \qquad (7.13)$$

where (7.13) was derived from (7.12) by differentiation. The pair of boundary integral equations (7.12) and (7.13) relate the variables w, $\partial w/\partial n$, M_n, Q_{ef} and is discretized along the same lines described in the previous sections, producing the set of equations

$$\begin{aligned} \boldsymbol{A}_1 \boldsymbol{w} + \boldsymbol{B}_1 \boldsymbol{w}_n + \boldsymbol{C}_1 \boldsymbol{M}_n + \boldsymbol{D}_1 \boldsymbol{Q}_{ef} &= \mathbf{P}_1 , \\ \boldsymbol{A}_2 \boldsymbol{w} + \boldsymbol{B}_2 \boldsymbol{w}_n + \boldsymbol{C}_2 \boldsymbol{M}_n + \boldsymbol{D}_2 \boldsymbol{Q}_{ef} &= \mathbf{P}_2 , \end{aligned} \qquad (7.14)$$

where $\boldsymbol{A}_i$, $\boldsymbol{B}_i$, $\boldsymbol{C}_i$ and $\boldsymbol{D}_i$ are the corresponding influence matrices. The imposition of boundary conditions allows the solution of the problem.

It is important to notice that, once again, the appearance of corners produces problems of the same type as in plane elasticity, due to the discontinuity of normals and the related variables. The solution is similar, i.e. if the w field is regular it is possible to add continuity conditions to solve the dilemma, while if there is a singularity, for instance of the re-entrant corner type, an asymptotic expansion can be used. More details can be found in Stern [104].

The treatment of shallow shells has been discussed by Forbes and Robinson [55], Newton and Tottenham [82] and Tottenham [109].

More recently Kamiya et al. [68] have presented interesting results for elastic plates experiencing finite deformations according to Berger's equation

$$\nabla^2\nabla^2 w - \kappa^2(\nabla^2 w + k_x + k_y) = p/D , \qquad (7.15)$$

where k_x, k_y are the curvatures and

$$\frac{\kappa^2 h^2}{12} = \frac{\partial u}{\partial x} + \frac{\partial v}{\partial y} + \tfrac{1}{2}\left[\left(\frac{\partial w}{\partial x}\right)^2 + \left(\frac{\partial w}{\partial y}\right)^2\right] - (k_x + k_y) w , \qquad (7.16)$$

with u and v being the in-plane displacements. It can be seen that this equation differs from Eq. (7.1) only in the second term, so that the previous scheme can be repeated, except that P_1 and P_2 in (7.14) now contain the contribution from the nonlinear terms.

If the in-plane displacement is restricted on the boundary such that

$$u = v = 0 \qquad \text{on } \partial\Omega , \qquad (7.17)$$

the magnitude of the parameter κ^2 can be estimated by integrating Eq. (7.16), producing

$$\kappa^2 = \frac{12}{h^2 A_\Omega} \int_\Omega \tfrac{1}{2}\left[\left(\frac{\partial w}{\partial x}\right)^2 + \left(\frac{\partial w}{\partial y}\right)^2\right] - (k_x + k_y)w \,, \tag{7.18}$$

where A_Ω is the area of the domain. Kamiya et al. [68] suggest the use of an iterative approach to solve the nonlinearities and extend the solution to take into account the influence of temperature, as well as the finite deflection and post buckling deformation problems.

Figure 41 shows the application of the Marcus approach to the deflection of a square plate with simply supported, fixed and free edges [89]. The domain integral has been computed using a regular mesh of 5, 9 and 25 cells, where the values are assumed as constants, while the boundary has been discretized into 16, 32 and 64 linear elements.

Figure 42 shows results for the application of the indirect BEM along with an interesting modification suggested by Fukui [56], to treat corner solutions.

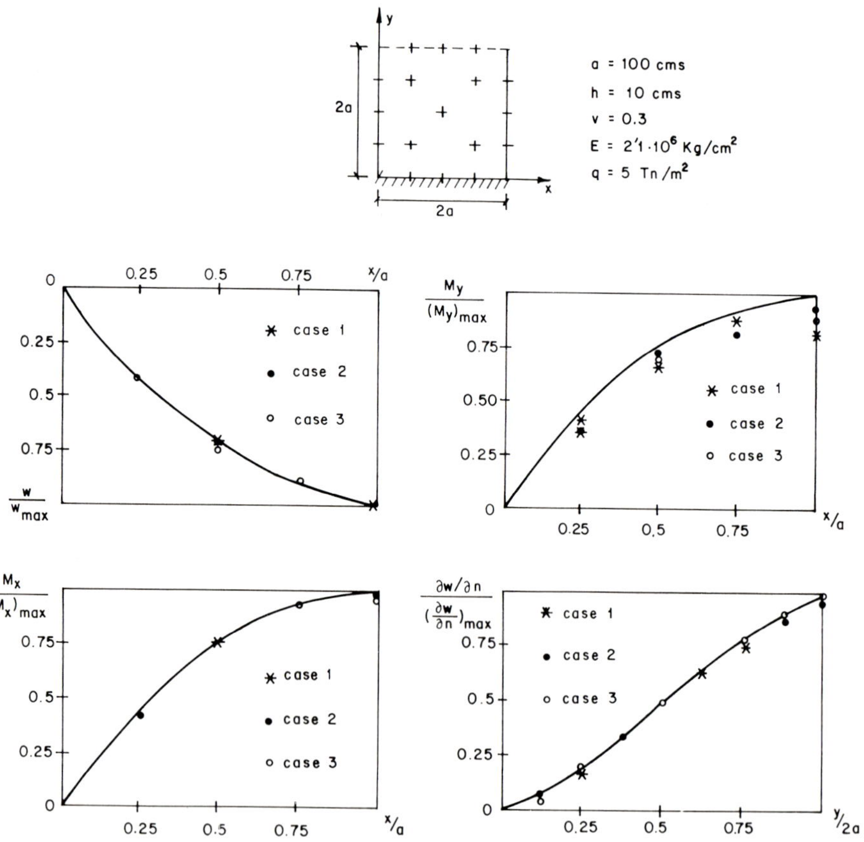

Fig. 41. Analysis results for a square plate using the BEM–Marcus approach.

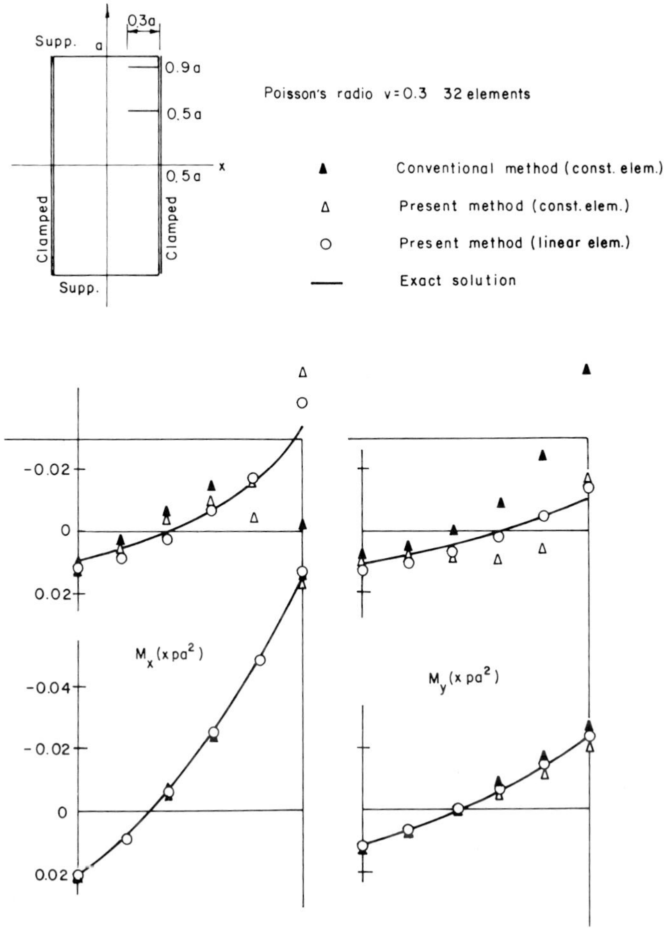

Fig. 42. Stress analysis of a rectangular plate.

7.2. BEM-FEM coupling

There are situations, such as nonlinearities, volume details, etc., where FEM versatility makes it unbeatable in treating a limited zone of the domain, while, in the remaining zones, the uniformity and extension counsel treatment with the BEM. The example in Figure 43 is typical of modelling soil–structure or fluid–structure or soil–fluid–structure interaction. The structure must be analyzed in detail and it is good to have extensive information throughout its domain, thereby requiring modelling by the FEM. On the other hand, the soil or the fluid are of interest only along the interfaces. Their size and homogeneity thus warrant the use of the BEM. Kelly et al. [69] and Margulies [77] have presented review-type articles on the subject.

The solution is a zoning, and it is important to note that the BEM works with

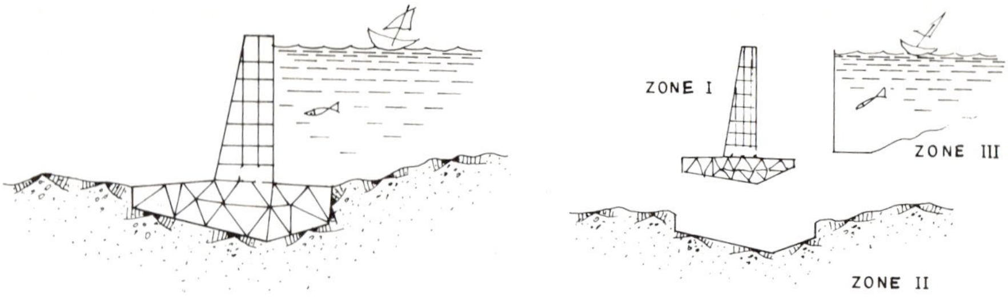

Fig. 43. The general problem of soil–structure–fluid interaction.

tractions while the FEM works with equivalent forces concentrated at the nodes. A previous integration is therefore necessary. In addition, there are problems motivated by the unsymmetric features of the BEM matrices. Thus, for each zone treated with the BEM

$$\boldsymbol{A}\boldsymbol{u} = \boldsymbol{B}\boldsymbol{t} \tag{7.19}$$

so that

$$\boldsymbol{t} = \boldsymbol{B}^{-1}\boldsymbol{A}\boldsymbol{u} \tag{7.20}$$

which will be used to obtain concentrated nodal forces. But $\boldsymbol{B}^{-1} \cdot \boldsymbol{A}$ is non-symmetric as discussed by Hartmann [58]. In particular, the non-symmetry implies that, due to the fact that the sum of the elements along columns is different from zero, the equilibrium condition is violated.

There have been several proposals to symmetrize $\boldsymbol{B}^{-1} \cdot \boldsymbol{A}$ from the simplest one

$$\tfrac{1}{2}[\boldsymbol{B}^{-1}\boldsymbol{A} + (\boldsymbol{B}^{-1} \cdot \boldsymbol{A})^{\mathrm{T}}]\,, \tag{7.21}$$

justified on variational reasons by Zienkiewickz [118, 81] or Brebbia and Walker [28] to the materialization of Hartmann [58] recommended by Tullberg and Bolteus [110]. In this sense, it seems that a good method might be the one that consists of the following steps:

(a) The elements of every column are added;

(b) the result is divided by the number of rows and the residuals are incorporated into every column element;

(c) the matrix is then symmetrized using (7.21).

Figure 44 presents a typical example of the above-mentioned procedure taken from Alarcón [4]. A lined tunnel was built after excavation of a trench which was filled a-posteriori to restore the original landscape. Due to the symmetries, it is only necessary to analyze half of the structure. The BEM was used to study the

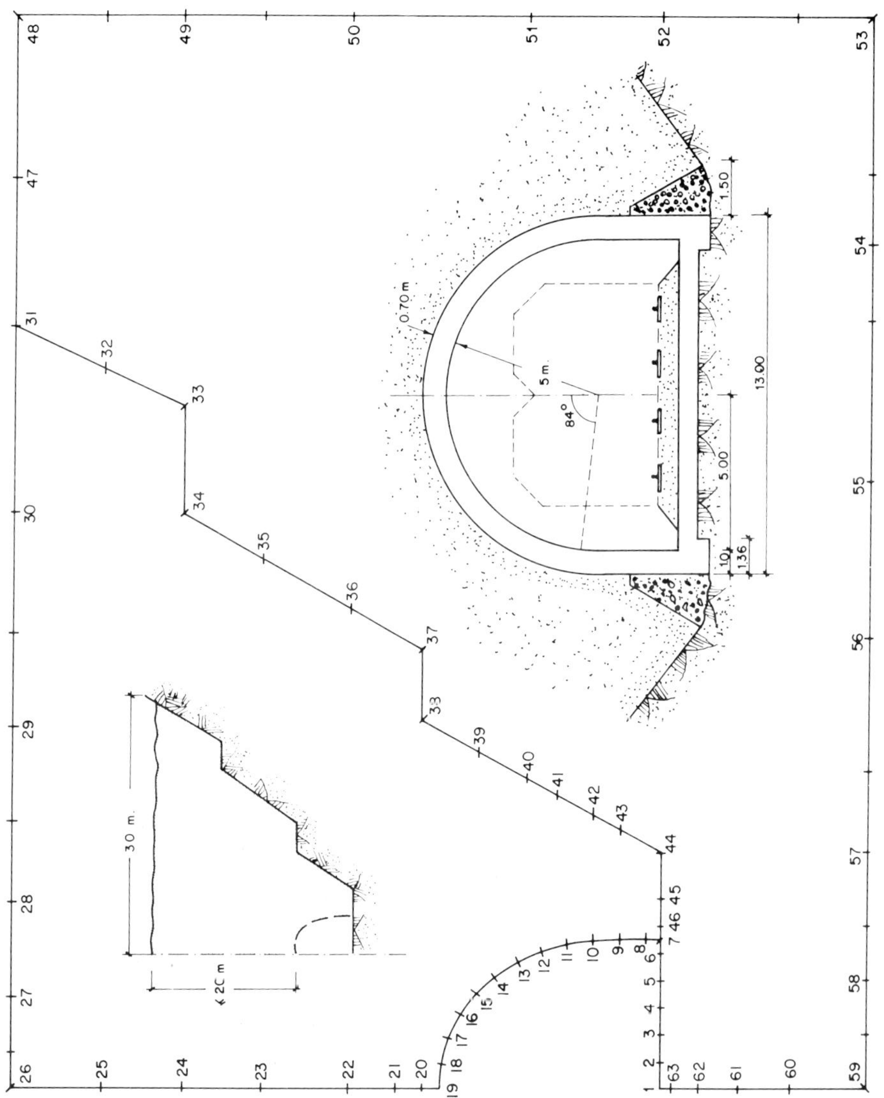

Fig. 44. Geometry and boundary discretization for the stress analysis of a lined tunnel.

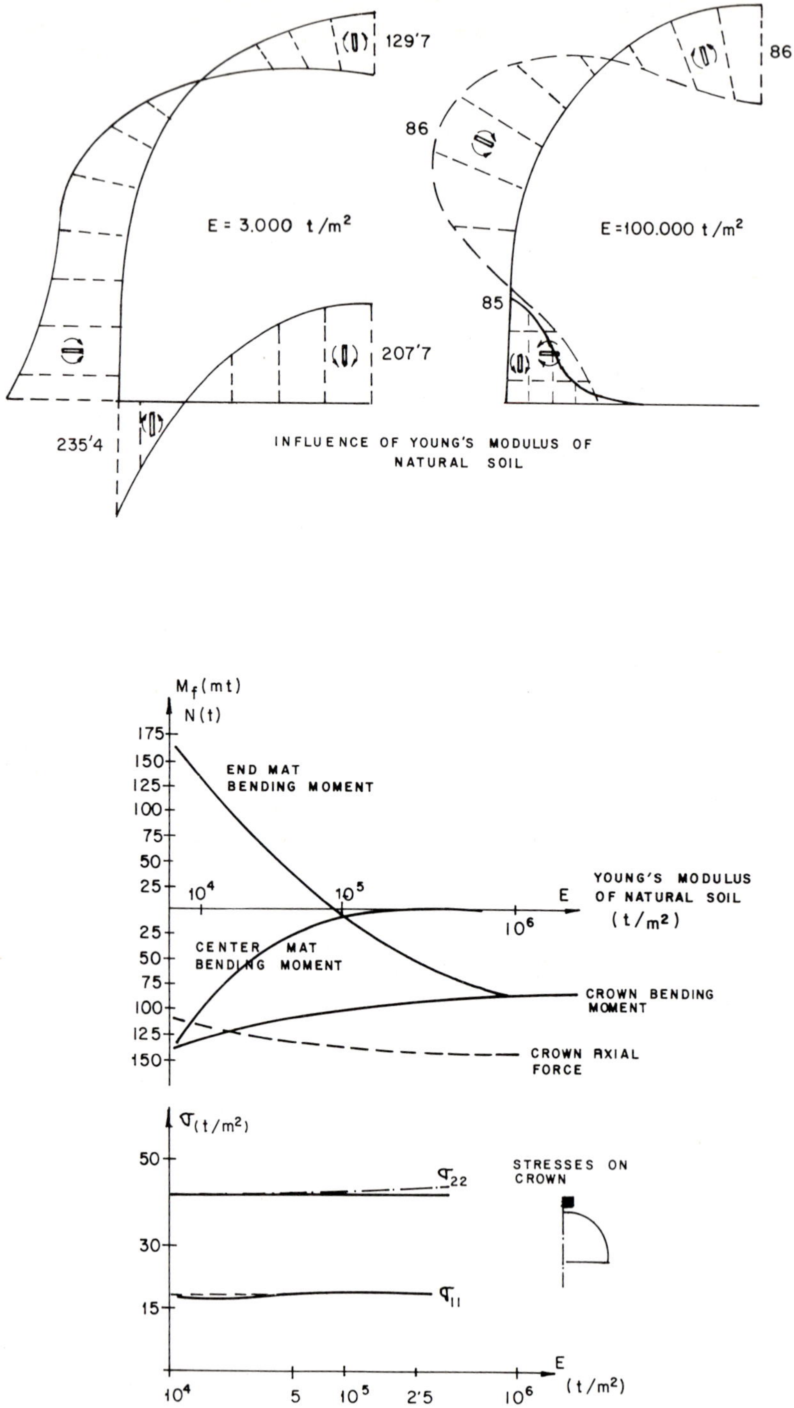

Fig. 45. Stress analysis results for the soil–structure system of Fig. 44.

earth region, while the liner was modelled through beam elements after condensation of the rotational degrees of freedom. Figure 45 shows some results of the parametric studies done with the model.

Another useful procedure akin to the BEM is sometimes used to build the stiffness matrix $\boldsymbol{K}^s$ of the soil through its definition stating that K_{ij} is the force appearing at the degree of freedom i when a unit displacement is imposed at the degree of freedom j. In this way, it is possible to treat soil–structure interaction problems as those of Figure 46 corresponding to a plate on a half-space. The interface is discretized using plane rectangular constant elements which can be thought of as boundary elements for the soil. These elements are given, in turn, unit displacements along the vertical and horizontal degrees of freedom. If only the vertical response is of concern, it suffices to give unit vertical displacements to obtain the resulting forces with the aid of the Boussinesq solution [112], in this way building a matrix $\boldsymbol{K}^s$, column by column. For its part, the plate can be discretized, using the FEM, with three degrees of freedom per node. The rotational degrees of freedom are condensed following a standard Gauss procedure and then translated along the depth of the plate to obtain its stiffness $\boldsymbol{K}^p$. The global matrix K for the soil-structure system is thus obtained as $\boldsymbol{K} = \boldsymbol{K}^s + \boldsymbol{K}^p$. After the common displacements $\boldsymbol{u}$ have been obtained, it is possible to get the contact pressures, by carrying out the product $\boldsymbol{K}^s\boldsymbol{u}$. Figures 47(a) and (b) show results obtained in this way, as compared with those presented by Cheung [40].

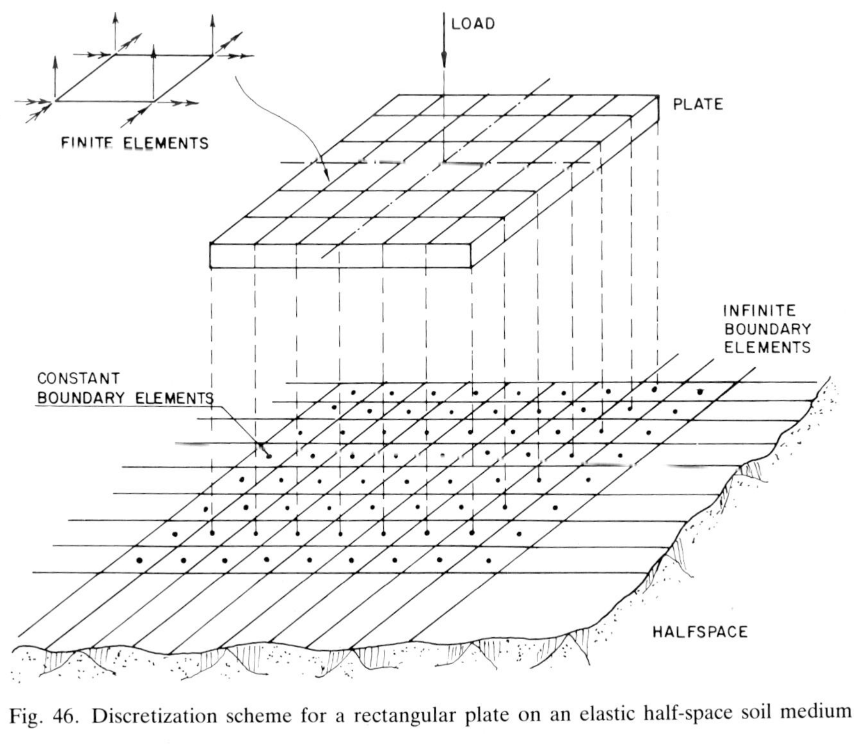

Fig. 46. Discretization scheme for a rectangular plate on an elastic half-space soil medium.

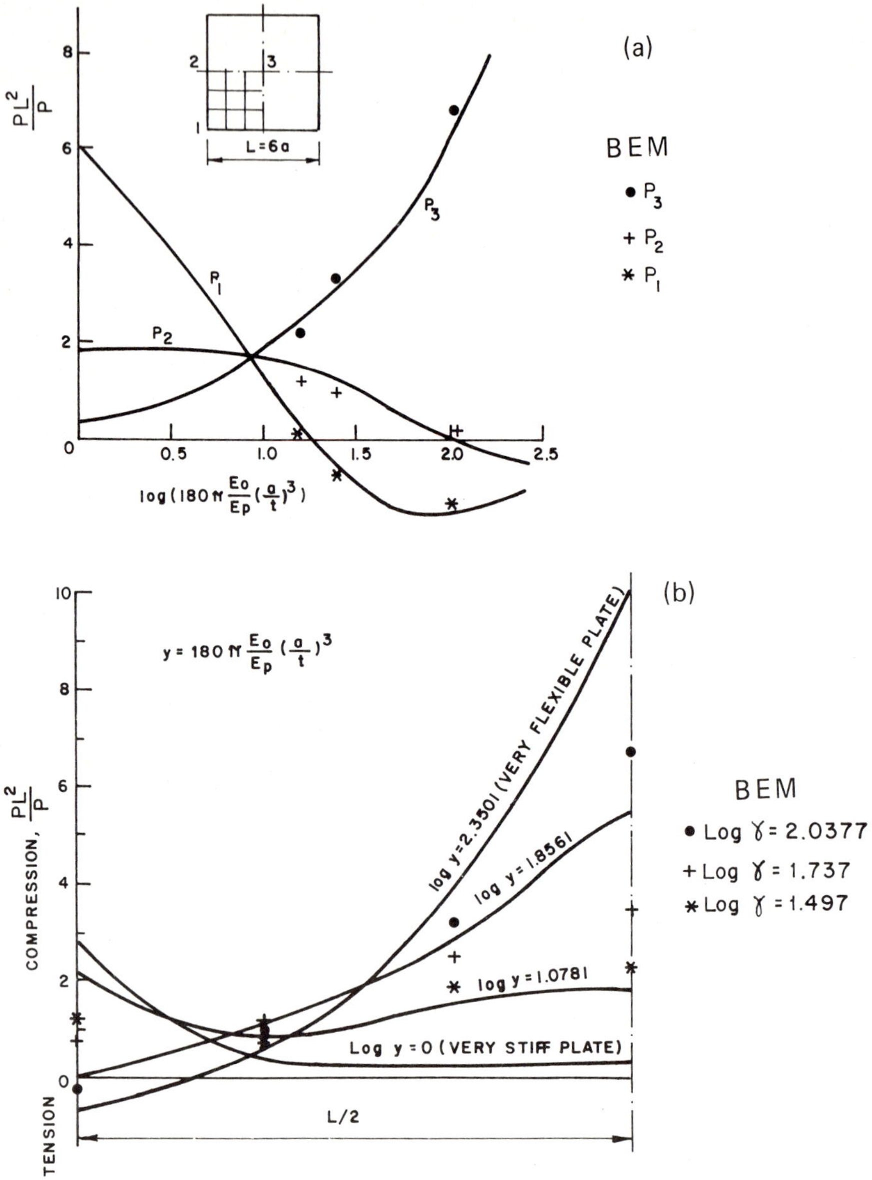

Fig. 47. Contact pressure results for the soil–foundation system of Fig. 46.

With the same procedure, using, for instance, the elements developed in section 5.5, it is possible to study soil–pile interaction problems. Another alternative has been proposed by Butterfield and Banerjee [37–39] and Banerjee et al. [13–15], which utilizes Mindlin's solution [35] for homogeneous and Kelvin's solution [76] with interface and surface discretization for layered half-spaces. Figure 48, taken from Banerjee and Butterfield [14], presents, for instance, the case of a pile group. The geometry and discretization can be seen in Fig. 48(a), while the average vertical stress over the pile cross-section is shown in Fig. 48(b).

Additional applications of the BEM–FEM coupling technique can be found in [27, 23, 19]. The coupling of the BEM with the finite difference method has also been utilized [94].

7.3. Adaptive boundary elements

It is well known from the FEM that it is common to distinguish between two types of convergence, i.e. the h-convergence and the p-convergence. The first refers to the improvement of results obtained by refining the mesh. Due to the correspondence element – nodal variables – interpolation function, that implies a progressive adjustment of the results there where the refinement has been carried

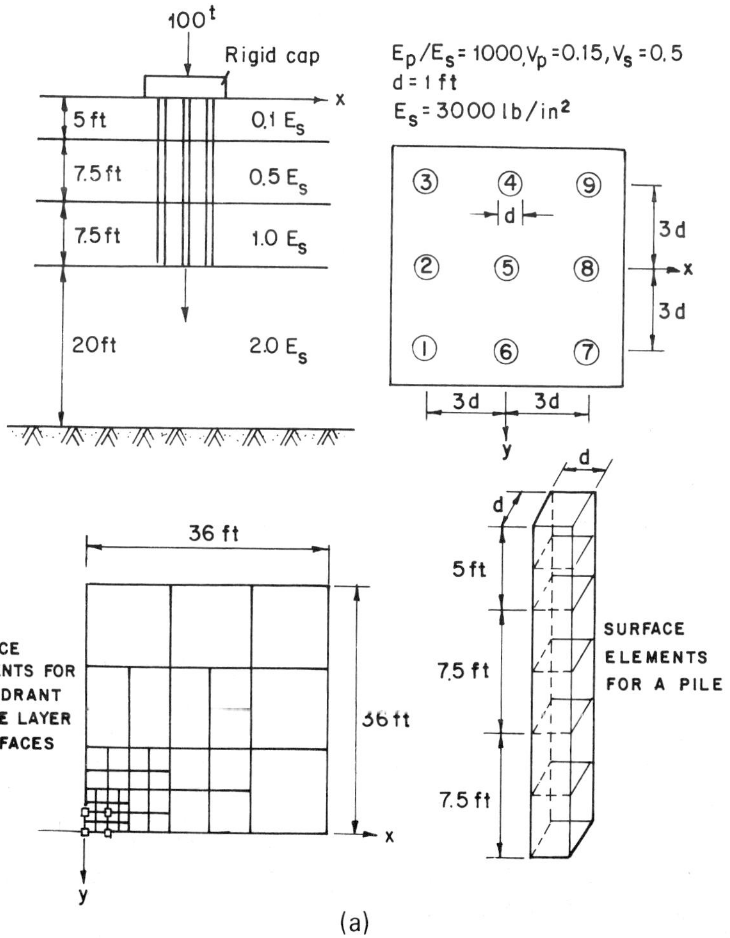

Fig. 48. Stress analysis of a vertically loaded group of piles embedded in a layered soil.

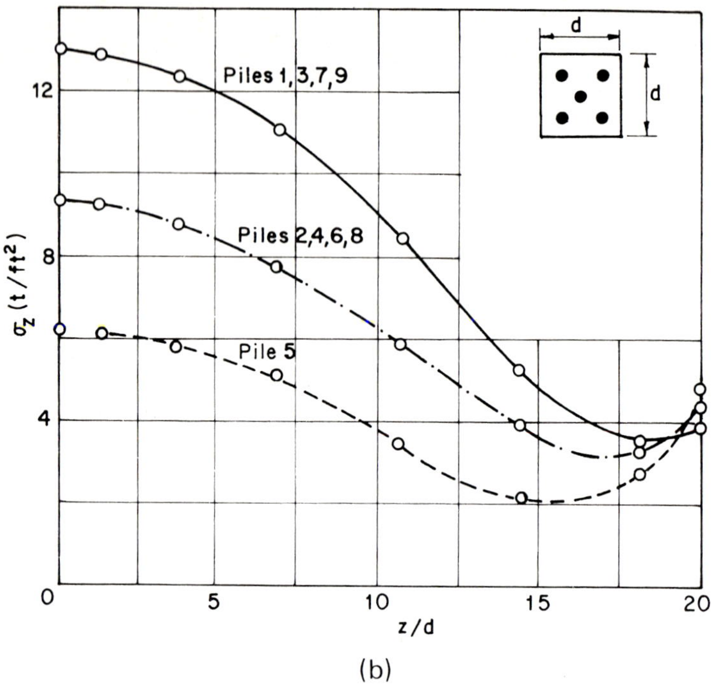

(b)

Fig. 48. (continued).

out. It is the most widely used method when, pragmatically, one wants to have an idea of the quality of the results. Its main disadvantage lies in the fact that every refinement requires the computation of different "stiffness" matrices, so that the information must be recomputed for every case, making the cost of this method quite high.

The second method analyzes p-convergence, or refinement of the approach by increasing the degree of the polynomial defined over every element. The idea is to compromise between the original Ritz idea of globally defined shape functions and the FEM of small support ones. The first step is, therefore, to define a rough mesh of elements and a hierarchy of functions inside them. With this intermediate decision between globality and locality, it is possible, on the one hand, to maintain a certain versatility for adjusting by pieces and, on the other, to refine intelligently because, as it happens with the Fourier series, the effect of adding a new term is decreasing, thus allowing truncation when a pre-established level of precision has been reached. In addition, and because every hierarchy of functions has a support fixed from each beginning, each refinement step uses all the previous computations. These ideas constitute the so-called adaptive procedures.

It seems that the first proposal was made by Zienkiewicz et al. [116] and since then, Szabo et al. [107], Peano et al. [92, 93] have contributed to the development of the method. The operating process follows these steps:

(a) establishment of a mesh on the model;

(b) definition of a hierarchy of function inside every element;

(c) establishment of an "indicator" to point to regions needing new shape functions;

(d) establishment of an "a-posteriori estimator", evaluating the error and predicting precisely when the process can be stopped.

A method following the previous steps is known as self-adaptive and is especially attractive for use with the BEM. It is well known that the BEM works with full matrices due to the globally based property of the fundamental solution, so that, except for the physical significance of the coefficients, it is not important to use locally based functions to interpolate the data or the unknowns. In order to reduce the computational effort, the idea of using hierarchical shape functions presents itself as an exciting posibility. The general isoparametric idea of the direct method, discussed in the previous sections, has been useful in the com-

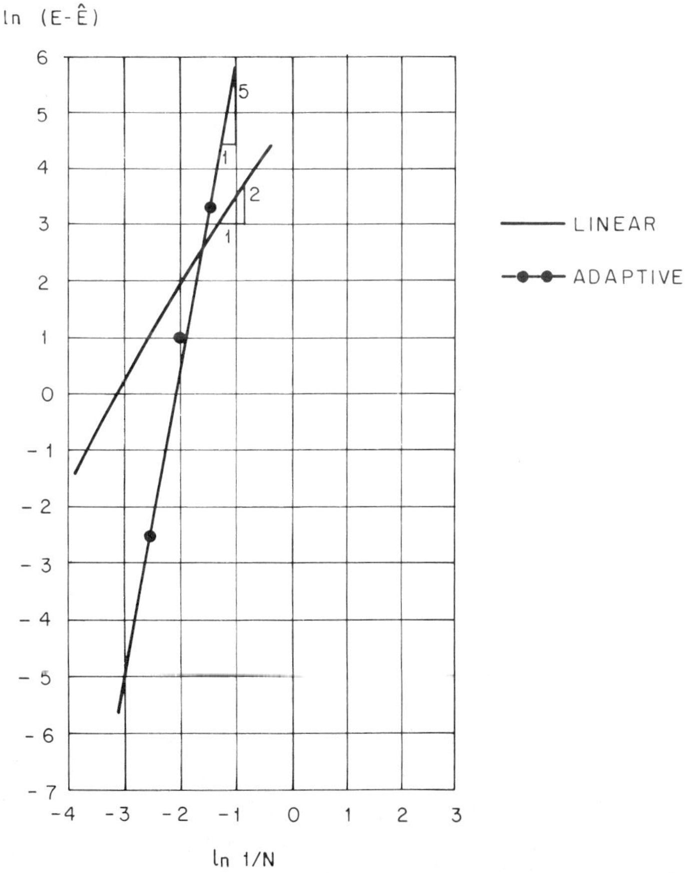

Fig. 49. Rate of convergence for adaptive boundary elements used to solve Neumann's problem in the square.

puterization of the method, but it is artificial to create elements by local shape functions when a banded structure is impossible to obtain. In addition, the rigidity introduced is enormous and as soon as the nodal points and the interpolation degree has been fixed, everything is fixed (i.e. geometry, collocation points, interpolation of data, etc.), when, usually, it is better to treat the geometry, the data and the unknowns independently.

The first attempt to adapt these ideas to the BEM as applied to potential problems was made by Alarcón [5] and subsequent advances on the subject have been presented by Alarcón et al. [6–8]. Application of adaptive procedures to elastostatics is quite straightforward.

Figure 49 shows the results (error in energy versus the inverse of the number of degrees of freedom) obtained for the classical Neumann problem in the square. As shown there and in [6–8], the increase in accuracy in comparison with the classical direct method is encouraging. In our opinion, adaptive boundary elements represent one of the more promising paths for the development of boundary elements.

Acknowledgement

The authors gratefully acknowledge Professor D. Beskos for useful comments and discussion. Financial support under Research Project 05.14 for the Comision Asesora de Investigación Cientifical y Tecnica (C.A.I.C.Y.T.) is gratefully acknowledged.

References

[1] E. Alarcón, C. Brebbia and J. Domínguez, The Boundary Element Method in Elasticity, *International Journal of Mechanical Sciences* **20**, 625–639 (1978).
[2] E. Alarcón, A. Martín and F. París, Boundary Elements in Potential and Elasticity Theory, Computers and Structures **10**, 341–362 (1979).
[3] E. Alarcón, A. Martín and F. París, Some Minor Problems with BEM, in: *Applied Numerical Modelling*, E. Alarcón and C.A. Brebbia, Eds. (Pentech Press, London, 1979) pp. 591–600.
[4] E. Alarcón, Identification of Pressures and Maintenance in Railway Tunnels, in: *Computational Methods and Experimental Measurements*, G.A. Keramidas and C.A. Brebbia, Eds. (Springer-Verlag, Berlin, 1982) pp. 516–526.
[5] E. Alarcón, Elementos de Contorno Adaptables, *Hormigón y Acero* **149**, 47–58 (1983).
[6] E. Alarcón, A. Reverter and J. Molina, Hierarchical Boundary Elements, in: *NASA Symposium on Advances and Trends in Structures and Dynamics*, A.K. Noor, Ed. (Washington D.C., 1984) pp. 441–443; also in *Computers and Structures* **20**, 151–156 (1985).
[7] E. Alarcón and A. Reverter, On the Possibility of Adaptive Boundary Elements, in: *Proceedings of the International Conference on Accuracy Estimates and Adaptive Refinements in Finite Element Computations*, I. Babuska, E.R.A. Oliveira and O.C. Zienkiewicz, Eds (Technical University of Lisbon, Lisbon, 1984) pp. 25–34.
[8] E. Alarcón and A. Reverter, P-Adaptive Boundary Elements, *International Journal for Numerical Methods in Engineering* **23**, 801–829 (1986).

[9] N. Altiero and D. Sikarskie, A. Boundary Integral Method Applied to Plates of Arbitrary Plan Form, *Computers and Structures* **9**, 163–168 (1978).

[10] N.J. Altiero and S.D. Gavazza, An Effective Boundary-Integral Approach for the Mixed Boundary-Value Problems of Linear Elastostatics, *Applied Mathematical Modelling* **3**, 99–104 (1979).

[11] J.J. Anza, El Método de los Elementos de Contorno en Problemas Termoelásticos, Ph.D. Thesis, School of Industrial Engng, Polytechnical University of Madrid, Spain (1981).

[12] I. Babuska and W.C. Rheinboldt, A Posteriori Error Estimate for the F.E.M., *International Journal for Numerical Methods in Engineering* **12**, 1597–1615 (1978).

[13] P.K. Banerjee, Integral Equation Methods for Analysis of Piece-wise Non-homogeneous Three-Dimensional Elastic Solids of Arbitrary Shape, *International Journal of Mechanical Sciences* **18**, 293–303 (1976).

[14] P.K. Banerjee and R. Butterfield, An Integral Equation Method for the Analysis of Boundary Value Problems in Irregularly Stratified Media, in: *Numerical Methods in Soil Mechanics and Rock Mechanics*, G. Borm and H. Meissner, Eds. (Karlsruhe, 1976) pp. 141–163.

[15] P.K. Banerjee and T.G. Davies, The Behaviour of Axially and Laterally Loaded Single Piles Embedded in Nonhomogeneous Soils, *Géotechnique* **28**, 309–326 (1978).

[16] P.K. Banerjee and R. Butterfield, Eds., *Developments in Boundary Element Methods*–1 (Applied Science, London, 1979).

[17] P.K. Banerjee and R. Butterfield, *Boundary Element Methods in Engineering Science* (McGraw Hill, London, 1981).

[18] P.K. Banerjee and R.P. Shaw, Eds., *Developments in Boundary Element Methods*–2 (*Applied Science, London*, 1982).

[19] G. Beer, J.L. Meek and R.G. Friday, Efficient Analysis of Problems in Geomechanics, in: *Numerical Methods in Geomechanics Edmonton*, Z. Eisenstein, Ed. (A.A. Balkema, Rotterdam, 1982) pp. 5–13.

[20] R. Benjumea and D.L. Sikarskie, On the Solutions of Plane Orthotropic Elasticity Problems by an Integral Equation Method, *Journal of Applied Mechanics* **39**, 801–808 (1972).

[21] G.P. Bézine, Boundary Integral Formulation for Plate Flexure with Arbitrary Boundary Conditions, *Mechanics Research Communications* **5**, 197–206 (1978).

[22] B.H.G. Brady, A Direct Formulation of the Boundary Element Method of Stress Analysis for Complete Plane Strain, *International Journal of Rock Mechanics and Mining Sciences & Geomechanics Abstracts* **16**, 235–244 (1979).

[23] B.H.G. Brady and A. Wassyng, A Coupled Finite Element-Boundary Element Method of Stress Analysis, *International Journal of Rock Mechanics and Mining Science & Geomechanics Abstracts* **18**, 475–485 (1981).

[24] C.A. Brebbia, *The Boundary Element Method for Engineers* (Pentech Press, London, 1978).

[25] C.A. Brebbia, Ed., *Recent Advances in Boundary Element Methods* (Pentech Press, London, 1978).

[26] C.A. Brebbia and R. Nakaguma, Boundary Elements in Stress Analysis, *Journal of the Engineering Mechanics Division of the ASCE* **105** (EM1), 55–69 (1979).

[27] C.A. Brebbia and P. Georgiou, Combination of Boundary and Finite Elements in Elastostatics, *Applied Mathematical Modelling* **3**, 212–220 (1979).

[28] C.A. Brebbia and S. Walker, *Boundary Element Techniques in Engineering* (Newnes–Butterworths, London, 1980).

[29] C.A. Brebbia, Ed., *New Developments in Boundary Element Methods* (CML Publications, Southampton, 1980).

[30] C.A. Brebbia, Ed., *Boundary Element Methods* (Springer-Verlag, Berlin, 1981).

[31] C.A. Brebbia, Ed., *Progress in Boundary Element Methods*, Vol. 1 (Pentech Press, London, 1981).

[32] C.A. Brebbia, Ed., *Boundary Element Methods in Engineering* (Springer-Verlag, Berlin, 1982).

[33] C.A. Brebbia, T. Futagami and M. Tanaka, Eds., *Boundary Elements* (Springer-Verlag, Berlin, 1983).

[34] C.A. Brebbia, Ed., *Progress in Boundary Element Methods*, Vol. 2 (Pentech Press, London, 1983).

[35] C.A. Brebbia, J.C.F. Telles and L.C. Wrobel, *Boundary Element Techniques* (Springer-Verlag, Berlin, 1984).

[36] C. Burgos, M. Doblaré, E. Alarcón, Automatic Mesh Generation Processor for 3-D Parabolic Boundary Discretization, *Advances in Engineering Software* **5**, 142–147 (1983).

[37] R. Butterfield and P.K. Banerjee, The Elastic Analysis of Compressible Piles and Pile Groups, *Géotechnique* **21**, 43–60 (1971).

[38] R. Butterfield and P. K. Banerjee, The Problem of Pile Group-Pile Cap Interaction, *Géotechnique* **21**, 135–142 (1971).

[39] R. Butterfield, The Application of Integral Equation Methods to Continuum Problems in Soil Mechanics, in: *Stress–Strain Behavior of Soils*, R.H.G. Parry, Ed., (G.T. Foulis & Co., Henley-on-Thames, Oxfordshire, 1972) pp. 573–587.

[40] Y.K. Cheung, Beams, Slabs and Pavements, in: *Numerical Methods in Geotechnical Engineering*, C.S. Desai and J.T. Christian, Eds., (McGraw-Hill, New York, 1977) pp. 176–210.

[41] M. Cristescu and G. Loubignac, Gaussian Quadrature Formulas for Functions with Singularities in $1/r$ Over Triangles, in: *Recent Advances in Boundary Element Methods*, C.A. Brebbia, Ed. (Pentech Press, London, 1978) pp. 375–390.

[42] S.L. Crouch, Solution of Plane Elasticity Problems by the Displacement Discontinuity Method: I & II, *International Journal for Numerical Methods in Engineering* **10**, 301–343 (1976).

[43] S.L. Crouch, Analysis of Stresses and Displacements Around Underground Excavations: An Application of the Displacement Discontinuity Method, Geomechanics/NSF Report, Dept. of Civil & Mineral Engng, University of Minnesota, Minneapolis (1976).

[44] S.L. Crouch and A.M. Starfield, *Boundary Element Methods in Solid Mechanics* (George Allen & Unwin, London, 1983).

[45] T.A. Cruse, Numerical Solutions in Three Dimensional Elastostatics, *International Journal of Solids and Structures* **5**, 1259–1274 (1969).

[46] T.A. Cruse, Application of the Boundary-Integral Equation Method to Three Dimensional Stress Analysis, *Computers and Structures* **3**, 509–527 (1973).

[47] T.A. Cruse, An Improved Boundary-Integral Equation Method for Three Dimensional Elastic Stress Analysis, *Computers and Structures* **4**, 741–754 (1974).

[48] T.A. Cruse, D.W. Snow and R.B. Wilson, Numerical Solutions in Axisymmetric Elasticity, *Computers and Structures* **7**, 445–451 (1971).

[49] T.A. Cruse and R.B. Wilson, Advanced Applications of Boundary-Integral Equation Methods, *Nuclear Engineering and Design* **46**, 223–234 (1978).

[50] D. Danson, Linear Isotropic Elasticity with Body Forces, in: *Progress in Boundary Element Methods*, Vol. 2, C.A. Brebbia, Ed. (Pentech Press, London, 1983) pp. 101–135.

[51] M. Doblaré and E. Alarcón, A Three-Dimensional B.I.E.M. Program, in: *Finite Element Systems: A Handbook*, C.A. Brebbia, Ed. (Springer-Verlag, Berlin, 1982) pp. 325–345.

[52] J. Domínguez, Cálculo de Tensiones en las Inmediaciones de Anclajes: Applicación del Método de los Elementos de Contorno, Ph.D. Thesis, School of Industrial Engng., University of Seville, Spain (1978).

[53] J. Domínguez, F. del Caño and E. Alarcón, B.I.E.M. in Solid Mechanics, in: *3rd Engineering Mechanics Division Speciality Conference ASCE*, Univ. of Texas at Austin, (ASCE, New York, 1979) pp. 793–796.

[54] W.S. Dunbar and D.L. Anderson, The Displacement Discontinuity Method in Three Dimensions, in: *Boundary Element Methods*, C.A. Brebbia, Ed. (Springer-Verlag, Berlin, 1981) pp. 153–165.

[55] D.J. Forbes and A.R. Robinson, Numerical Analysis of Elastic Plates and Shallow Shells by an Integral Equation Method, *Structural Research Series Report No*. 345, University of Illinois, Urbana, Illinois (1969).

[56] T. Fukui, On Corner Solutions by Indirect B.I.E.M., in: *Boundary Elements*, C.A. Brebbia, T. Futagami and M. Tanaka, Eds. (Springer-Verlag, Berlin, 1983) pp. 929–938.

[57] S. Gómez-Lera, F. París and E. Alarcón, Treatment of Singularities in 2-D Domains Using BIEM, *Applied Mathematical Modelling* **6**, 111–118 (1982).

[58] F. Hartmann, The Derivation of Stiffness Matrices from Integral Equations, *Applied Mathematical Modelling* **5**, 355–365 (1981).

[59] F. Hartmann, Elastostatics, in: *Progress in Boundary Element Methods*, Vol. 1, C.A. Brebbia, Ed. (Halsted Press, New York, 1981) pp. 84–167.

[60] U. Heisse, Comparison of Round-Off-Errors in Integral Equation Formulations of Elastostatical Boundary Value Problems, *Computer Methods in Applied Mechanics and Engineering* **28**, 145–177 (1981).

[61] U. Heisse, Removal of the Zero Eigenvalues of Integral Operator in Elastostatic Boundary Value Problems, *Acta Mechanica* **41**, 41–61 (1981).

[62] U. Heisse, Solution of Integral Equations for Plane Elastostatical Problems with Discontinuously Prescribed Boundary Values, *Journal of Elasticity* **12**, 293–312 (1982).

[63] G.C. Howell and W.S. Doyle, An Assessment of the Boundary Integral Equation Method for In-Plane Elastostatics Problems, *Applied Mathematical Modelling* **6**, 245–256 (1982).

[64] G.C. Howell and W.S. Doyle, The Plane Stress/Strain Analysis of Non-Homogeneous Continua by the Boundary Integral Equation Method, *Computers and Structures* **17**, 603–610 (1983).

[65] M.A. Jaswon, M. Maiti and G.T. Symm, Numerical Biharmonic Analysis and Some Applications, *International Journal of Solids and Structures* **3**, 309–332 (1967).

[66] M.A. Jaswon and M. Maiti, An Integral Equation Formulation of Plate Bending Problems, *Journal Engineering Mathematics* **2**, 83–93 (1968).

[67] M.A. Jaswon and G.T. Symm, *Integral Equation Methods in Potential Theory and Elastostatics* (Academic Press, New York, 1977).

[68] N. Kamiya, Y. Sawaki and Y. Nakamura, Finite and Postbuckling Deformations of Heated Plates and Shallow Shells, in: *Boundary Elements*, C.A. Brebbia, T. Futagami and M. Tanaka, Eds. (Springer-Verlag, Berlin, 1983) pp. 507–516.

[69] D.W. Kelly, G.G.W. Mustoe and O.C. Zienkiewicz, Coupling Boundary Element Methods with other Numerical Methods, in: *Developments in Boundary Element Methods* – 1, P.K. Banerjee and R. Butterfield, Eds. (Applied Science, London, 1979) pp. 251–285.

[70] T. Kermanidis, A. Numerical Solution for Axially Symmetrical Elasticity Problems, *International Journal of Solids and Structures* **11**, 493–500 (1975).

[71] S. Kobayashi and N. Nishimura, Green's Tensors for Elastic Half-Spaces: An Application of Boundary Integral Equation Method, *Memoirs Faculty Engineering, Kyoto Univ.* **42**, 228–241 (1980).

[72] G. Kuhn and W. Möhrmann, Boundary Element Method in Elastostatics: Theory and Applications, *Applied Mathematical Modelling* **7**, 97–105 (1983).

[73] J.C. Lachat and J.O. Watson, A Second Generation Integral Equation Program for Three-Dimensional Elastic Analysis, in: *Boundary-Integral Equation Method: Computational Applications in Applied Mechanics*, T.A. Cruse and F.J. Rizzo, Eds. (ASME, New York, 1975) pp. 85–100.

[74] J.C. Lachat and J.O. Watson, Effective Numerical Treatment of Boundary Integral Equations: A Formulation for Three-Dimensional Elastostatics, *International Journal for Numerical Methods in Engineering* **10**, 991–1005 (1976).

[75] J.C. Lachat and J.O. Watson, Progress in the Use of Boundary Integral Equations Illustrated by Examples, *Computer Methods in Applied Mechanics and Engineering* **10**, 273–289 (1977).

[76] A.E.H. Love, *A Treatise on the Mathematical Theory of Elasticity* (Dover Publ., New York, 1944).

[77] M. Margulies, Combination of the Boundary Element and Finite Element Methods, in: *Progress in Boundary Element Methods*, Vol. 1, C.A. Brebbia, Ed. (Halsted Press, New York, 1981) pp. 258–288.

[78] A. Martín, I. Rodríguez and E. Alarcón, Mixed Elements in the Boundary Theory, in: *New Developments in Boundary Element Methods*, C.A. Brebbia, Ed. (Comp. Mech. Centre Publ., Southampton, 1980) pp. 34–42.

[79] C.E. Massonnet, Numerical Use of Integral Procedures, in: *Stress Analysis*, O.C. Zienkiewicz and G.S. Holister, Eds. (Wiley, London, 1965) pp. 198–235.

[80] M. Mayr, W. Drexler and G. Kuhn, A Semianalytical Boundary Integral Approach for Axisymmetric Elastic Bodies with Arbitrary Boundary Conditions, *International Journal of Solids and Structures* **16**, 863–871 (1980).

[81] G.G.W. Mustoe, F. Volait and O.C. Zienkiewicz, A Symmetric Direct Boundary Integral Equation Method for Two Dimensional Elastostatics, *Res Mechanica* **4**, 57–82 (1982).

[82] D.A. Newton and H. Tottenham, Boundary Value Problems in Thin Shallow Shells of Arbitrary Plan Form, *Journal of Engineering Mathematics* **2**, 211–224 (1968).

[83] N. Nishimura and S. Kobayashi, A Boundary Integral Equation Formulation for Three Dimensional Anisotropic Elastostatics, in: *Boundary Elements*, C.A. Brebbia, T. Futagami and M. Tanaka, Eds. (Springer-Verlag, Berlin, 1983) pp. 345–354.

[84] Y. Niwa, S. Kobayashi and T. Fukui, An Application of the Integral Equation Method to Plate-Bending Problems, *Memoirs of the Faculty of Engineering, Kyoto University* **36**, 140–158.

[85] Y. Niwa, S. Kobayashi and T. Fukui, Stresses and Displacements Around an Advancing Face of a Tunnel, in: *Procedings of the 4th International Congress on Rock Mechanics*, Montreux, Switzerland (A.A. Balkema, Rotterdam, 1979) pp. 703–710.

[86] E.R.A. Oliveira, Plane Stress Analysis by a General Integral Method, *Journal of the Engineering Mechanics Division of the ASCE* **94**, (EM1), 79–101 (1968).

[87] Y.C. Pan and T.W. Chou, Green's Functional Solutions for Semi-Infinite Transverselly Isotropic Materials, *International Journal of Engineering Sciences* **17**, 545–551 (1979).

[88] F. París and S. León, Aplicación del Método de los Elementos de Contorno a Placas con Ecuaciones de Contorno Arbitrarias, *Anales de Ingenieria Mecanica, Gijon, Spain* **2**, 30–37 (1984).

[89] C. Patterson and M.A. Sheikh, Regular Boundary Integral Equations for Stress Analysis, in: *Boundary Element Methods*, C.A. Brebbia, Ed. (Springer-Verlag, Berlin, 1981) pp. 85–104.

[90] C. Patterson and M.A. Sheikh, Non-Conforming Boundary Elements for Stress Analysis, in: *Boundary Element Methods*, C.A. Brebbia, Ed. (Springer-Verlag, Berlin, 1981) pp. 137–152.

[91] C. Patterson and M.A. Sheikh, A Regular Boundary Element Method for Coupled Subdomains, in: *Proceedings of the International Conference on Numerical Methods for Coupled Problems*, R.W. Lewis, P. Bettes and E. Hinton, Eds. (Pineridge Press, Swansea, 1981) pp. 115–126.

[92] A.G. Peano, Hierarchies of Conforming Finite Elements, Ph.D. Thesis, Department of Civil Engng., Washington University, Saint Louis (1975).

[93] A. Peano, A. Pasini, R. Riccioni and L. Sardella, Adaptive Approximation in Finite Element Structure Analysis, *Computers and Structures* **10**, 333–342 (1979).

[94] R. Rangogni and M. Reali, The Coupling of the Finite Difference Method and the Boundary Element Method, *Applied Mathematical Modelling* **6**, 233–236 (1982).

[95] F.J. Rizzo, An Integral Equation Approach to Boundary Value Problems of Classical Elastostatics, *Quarterly of Applied Mathematics* **25**, 83–95 (1967).

[96] F.J. Rizzo and D.J. Shippy, A Method for Stress Determination in Plane Anisotropic Elastic Bodies, *Journal of Composite Materials* **4**, 36–61 (1970).

[97] F.J. Rizzo and D.J. Shippy, An Advanced Boundary Integral Equation Method for Three-Dimensional Thermoelasticity, *International Journal for Numerical Methods in Engineering* **11**, 1753–1768 (1977).

[98] F.J. Rizzo and D.J. Shippy, The Boundary Element Method in Thermoelasticity, in: *Developments in Boundary Element Methods*–1, P.K. Banerjee and R. Butterfield, Eds. (Applied Science, London, 1979) pp. 155–172.

[99] F.J. Rizzo and D.J. Shippy, A Boundary Integral Approach to Potential and Elasticity Problems for Axisymmetric Bodies with Arbitrary Boundary Conditions, *Mechanics Research Communications* **6**, 99–103 (1979).

[100] D.J. Shippy and F.J. Rizzo, On the Effectiveness of Three Boundary Integral Equation Formulations for Certain Axisymmetric Elastostatic Problems, *Res Mechanica* **4**, 43–56 (1982).

[101] C. Somigliana, Sopra gli Integrali delle Equazioni della Isotropia Elastica, *Il Nuovo Cimento*, Serie III, **36**, 28–39, 113–126 (1894).

[102] C. Somigliana, Sopra l'Equilibrio di un Corpo Elastico Isotropo, *Il Nuovo Cimento*, Serie III, **17**, 140–148, 272–276; **18**, 91–96, 161–166 (1885); **19**, 84–90, 278–282; **20**, 181–185 (1886).

[103] C. Somigliana, Sul Potenziale Elastico, *Annali di Mathematica Pura ed Applicata*), Serie III, **7**, 129–140 (1902).

[104] M. Stern, A General Boundary Integral Formulation for the Numerical Solution of Plate Bending Problems, *International Journal of Solids and Structures* **15**, 769–782 (1979).

[105] M. Stern, Boundary Integral Equations for Bending of Thin Plates, in: *Progress in Boundary Element Methods*, Vol. 2, C.A. Brebbia, Ed. (Pentech Press, London, 1983) pp. 158–181.
[106] A.H. Stroud and D. Secrest, *Gaussian Quadrature Formulas* (Prentice-Hall, Englewood Cliffs, N.J., 1966).
[107] B.A. Szabo and A.U. Metha, *P*-Convergent Finite Element Approximations in Fracture Mechanics, *International Journal for Numerical Methods in Engineering* **12**, 551–560 (1978).
[108] G.R. Tomlin and R. Butterfield, Elastic Analysis of Zoned Orthotropic Continua, *Journal of the Engineering Mechanics Division of the ASCE* **100** (EM3), 511–529 (1974).
[109] H. Tottenham, The Boundary Element Method for Plates and Shells, in: *Developments in Boundary Element Methods*–1, P.K. Banerjee and R. Butterfield, Eds. (Applied Science, London, 1979) pp. 173–205.
[110] O. Tullberg and L. Bolteus, A Critical Study of Different Boundary Element Stiffness Matrices, in: *Boundary Element Methods in Engineering*, C.A. Brebbia, Ed. (Springer-Verlag, Berlin, 1982) pp. 621–635.
[111] S.M. Vogel and J.F. Rizzo, An Integral Equation Formulation of Three Dimensional Anisotropic Elastostatic Boundary Value Problems, *Journal of Elasticity* **3**, 203–216 (1973).
[112] E. Volterra and J.H. Gaines, *Advanced Strength of Materials* (Prentice-Hall, Englewood Cliffs, NJ, 1971).
[113] L.J. Wardle and J.M. Crotty, Two-Dimensional Boundary Integral Equation Analysis for Non-homogeneous Mining Applications, in: *Recent Advances in Boundary Element Methods*, C.A. Brebbia, Ed. (Pentech Press, London, 1978) pp. 233–251.
[114] J.O. Watson, Advanced Implementation of the Boundary Element Method for Two- and Three-Dimensional Elastostatics, in: *Developments in Boundary Element Methods*–1, P.K. Banerjee and R. Butterfield, Eds. (Applied Science, London, 1979) pp. 31–63.
[115] R.B. Wilson and T.A. Cruse, Efficient Implementation of Anisotropic Three-Dimensional Boundary Integral Equation Stress Analysis, *International Journal for Numerical Methods in Engineering* **12**, 1383–1397 (1978).
[116] O.C. Zienkiewicz, B.M. Irons, F.E. Scott and J.S. Campbell, Three-Dimensional Stress Analysis, in: *Proceedings of the IUTAM Symposium on High Speed Computing of Elastic Structures* (University of Liege Press, Liege, 1971) pp. 413–443.
[117] O.C. Zienkiewic, Marriage a la Mode–The Best of Both Worlds (Finite Elements and Boundary Integrals), in: *Innovative Numerical Analysis in Applied Engineering Science*, T.A. Cruse et al, Eds. (CETIM, Senlis, France, 1977) pp. 19–26.
[118] O.C. Zienkiewicz, *The Finite Element Method* (McGraw-Hill, London, 1977).

CHAPTER 4

Elastodynamics

Shoichi KOBAYASHI

School of Civil Engineering
Kyoto University
Kyoto 606, Japan

Boundary Element Methods in Mechanics
Edited by D.E. Beskos

Introduction

The boundary integral equation method (BIEM) or boundary element method (BEM) is one of the most effective numerical methods for elastodynamics. It easily manages unbounded domains and reduces the number of unknowns drastically compared with domain-type techniques. Moreover, it usually provides very accurate results even if it is applied to rather complicated boundary shapes.

The use of integral equation formulations for initial-boundary-value problems has a long history. In relation to elasticity, about a century ago Betti derived the reciprocal theorem (1872), Somigliana formulated the integral representation theorem (Somigliana formula) (1885, 1886) and Kirchhoff derived an expression called the Helmholtz–Kirchhoff integral formula by the use of the retarded time potential (1883) [75].

The theory of integral equations was first studied by Fredholm [39] at the beginning of this century. Since then a great many theoretical contributions have been made by applied mathematicians, specifically of the Tbilisi school. Among them, Kupradze [70, 71] extensively discussed the integral equation method for elastodynamics by the aid of potentials.

Classical works done in the field of elastodynamics are discussed in Achenbach [2], Aki and Richards [3], Ben-Menahem and Singh [8], Eringen and Suhubi [37], Love [75], Pao and Mow [101], Pao and Varatharajulu [102] and Wheeler and Sternberg [131], among others.

Although the basic integral equation formulations for wave problems have been known for a long time, their applications in obtaining numerical solutions of initial-boundary-value problems were begun only recently. Some of the early developments in elastic wave scattering were done by Banaugh [6], Banaugh and Goldsmith [7] and Kupradze [70]. For scattering problems of the horizontally polarized shear (SH) scalar wave, similar integral formulations as in acoustics are used. Readers are advised to see Chapter 9 for its developments.

Cruse and Rizzo [23] and Cruse [24] derived the boundary integral equation formulation in conjunction with the Laplace transform in order to solve the transient wave propagation in a half-plane. A modified version of their approach was developed by Manolis [77] and Manolis and Beskos [79, 80].

The steady-state solutions and the transient responses reconstructed by the Fourier transform synthesis were obtained by Kobayashi and Nishimura [64–67], Niwa et al. [96, 97] and Rizzo et al. [106, 107].

Eigenvalue problems in elastodynamics were studied by Kitahara [57] and Niwa et al. [98]. For SH-wave problems, or for acoustics, see references in Chapter 9.

Integral formulations in the time–space domain were studied by Cole et al. [20], Friedman and Shaw [40] and Mansur and Brebbia [81] for scalar wave problems, by Manolis [77], Mansur and Brebbia [82], Niwa et al. [91] and Shaw [113] for two-dimensional elastic waves and by Karabalis and Beskos [51–53] and Karabalis et al. [54] for three-dimensional elastic waves. Recent developments in the boundary element methods for elastodynamics are also discussed in some articles [27, 61, 106].

In this chapter, first we shall discuss formulations of integral equations both in the time–space domain and the Fourier-transformed domain and then describe numerical implementation and some remarks inherent to elastodynamic problems. Finally, we shall show some application examples, which demonstrate the effectiveness and versatility of boundary integral equation methods.

1. Elastodynamics problems

1.1. Field equations

Let D be the domain in three- or two-dimensional space $\mathbb{R}^N (N = 3$ or $2)$ occupied by an elastic body, and ∂D be its boundary. The body is asumed to be in a state of motion relative to an inertial frame of reference. Also assume that the density field $\rho(\boldsymbol{x}, t)$ and the velocity field $\boldsymbol{v}(\boldsymbol{x}, t)$ are defined throughout the space–time cylinder $(\boldsymbol{x}, t) \in \mathbb{R}^N \times T$, $T = (-\infty, \infty)$.

The principle of linear momentum with the conservation of mass lead to the Euler equation of motion

$$\nabla \cdot \boldsymbol{\sigma}(\boldsymbol{x}, t) + \rho \boldsymbol{b}(\boldsymbol{x}, t) = \rho \frac{\mathrm{D}}{\mathrm{D}t} \boldsymbol{v}(\boldsymbol{x}, t) , \tag{1.1}$$

where $\boldsymbol{\sigma}$, $\boldsymbol{b}$ stand for the stress tensor and body force vector, respectively, ∇ is the del operator, $\nabla \cdot (\ )$ implies the divergence of $(\ )$, and $\mathrm{D}(\ )/\mathrm{D}t$ denotes the material derivative. The principle of the moment of momentum provides, in addition to this, the symmetry of stress $\boldsymbol{\sigma}$.

Assuming infinitesimal deformations, i.e.

$$\boldsymbol{v} = \frac{\mathrm{D}\boldsymbol{u}}{\mathrm{D}t} = \frac{\partial \boldsymbol{u}}{\partial t} + \boldsymbol{v} \cdot \nabla \boldsymbol{u} \doteqdot \frac{\partial \boldsymbol{u}}{\partial t} , \qquad \frac{\mathrm{D}\boldsymbol{v}}{\mathrm{D}t} = \frac{\partial \boldsymbol{v}}{\partial t} + \boldsymbol{v} \cdot \nabla \boldsymbol{v} \doteqdot \frac{\partial \boldsymbol{v}}{\partial t} , \tag{1.2}$$

where $\boldsymbol{u}$ stands for displacement, Eq. (1.1) reduces to the Cauchy equation of motion

$$\nabla \cdot \boldsymbol{\sigma}(\boldsymbol{x}, t) + \rho \boldsymbol{b}(\boldsymbol{x}, t) = \rho \ddot{\boldsymbol{u}}(\boldsymbol{x}, t) , \tag{1.3}$$

where $(\dot{\ }) = \partial(\ )/\partial t$, i.e. the dot denotes differentiation with respect to time.

The kinematical relation for the infinitesimal deformations is given by

$$\boldsymbol{\varepsilon}(\boldsymbol{x}, t) = \tfrac{1}{2}\{\boldsymbol{\nabla}\boldsymbol{u}(\boldsymbol{x}, t) + \boldsymbol{u}(\boldsymbol{x}, t)\boldsymbol{\nabla}\} \tag{1.4}$$

and the generalized Hooke's law for the linear elastic medium is

$$\boldsymbol{\sigma}(\boldsymbol{x}, t) = \mathbf{C} : \boldsymbol{\varepsilon}(\boldsymbol{x}, t) \, , \tag{1.5a}$$

where $\boldsymbol{\varepsilon}$, $\mathbf{C}$ stand for the strain tensor and elastic tensor, respectively, and the double dot implies the double inner product. When the medium is isotropic, we have

$$\boldsymbol{\sigma}(\boldsymbol{x}, t) = \lambda(\boldsymbol{x}) \operatorname{tr} \boldsymbol{\varepsilon}(\boldsymbol{x}, t)\mathbf{1} + 2\mu(\boldsymbol{x})\boldsymbol{\varepsilon}(\boldsymbol{x}, t) \, , \tag{1.5b}$$

where $\operatorname{tr} \boldsymbol{\varepsilon}$ means the trace of $\boldsymbol{\varepsilon}$, $\mathbf{1}$ denotes the unit tensor, and λ and μ are the Lamé constants.

Substitution of Eq. (1.4) into Eq. (1.5a) or (1.5b) leads to

$$\boldsymbol{\sigma} = \mathbf{C} : \boldsymbol{\nabla}\boldsymbol{u} \tag{1.6a}$$

or

$$\boldsymbol{\sigma} = \lambda\boldsymbol{\nabla} \cdot \boldsymbol{u}\mathbf{1} + \mu(\boldsymbol{\nabla}\boldsymbol{u} + \boldsymbol{u}\boldsymbol{\nabla}) \, . \tag{1.6b}$$

Substituting further Eq. (1.6a) into Eq. (1.3), we have the Navier–Cauchy equation of motion

$$\boldsymbol{\nabla} \cdot (\mathbf{C} : \boldsymbol{\nabla}\boldsymbol{u}) + \rho\boldsymbol{b} = \rho\ddot{\boldsymbol{u}} \, . \tag{1.7a}$$

For the isotropic medium, we have

$$\mu\nabla^2\boldsymbol{u} + (\lambda + \mu)\boldsymbol{\nabla}\boldsymbol{\nabla} \cdot \dot{\boldsymbol{u}} + (\boldsymbol{\nabla}\lambda)\boldsymbol{\nabla} \cdot \boldsymbol{u} + \boldsymbol{\nabla}\mu \cdot (\boldsymbol{\nabla}\boldsymbol{u} + \boldsymbol{u}\boldsymbol{\nabla}) + \rho\boldsymbol{b} = \rho\ddot{\boldsymbol{u}} \, . \tag{1.7b}$$

In addition, when the medium is homogeneous, Eq. (1.7b) reduces to

$$\mu\nabla^2\boldsymbol{u} + (\lambda + \mu)\boldsymbol{\nabla}\boldsymbol{\nabla} \cdot \boldsymbol{u} + \rho\boldsymbol{b} = \mathbf{E}\boldsymbol{u} + \rho\boldsymbol{b} = \rho\ddot{\boldsymbol{u}} \tag{1.8a}$$

or, equivalently

$$(\lambda + 2\mu)\boldsymbol{\nabla}\boldsymbol{\nabla} \cdot \boldsymbol{u} - \mu\boldsymbol{\nabla} \times \boldsymbol{\nabla} \times \boldsymbol{u} + \rho\boldsymbol{b} = \rho\ddot{\boldsymbol{u}} \, , \tag{1.8b}$$

where use was made of

$$\nabla^2\boldsymbol{u} = \boldsymbol{\nabla}\boldsymbol{\nabla} \cdot \boldsymbol{u} - \boldsymbol{\nabla} \times \boldsymbol{\nabla} \times \boldsymbol{u} \, .$$

In what follows, we mainly consider the homogeneous and isotropic elastic medium and sometimes disregard the body force, for simplicity. We also use the indicial notation referred to the Cartesian coordinate system $(0; x_1, x_2, x_3)$ interchangeably. In that case, the summation convention rule is implied over repeated indices. For example, $\boldsymbol{u} = u_k \boldsymbol{i}_k$, $\boldsymbol{\nabla}(\ ) = (\ )_{,k}\boldsymbol{i}_k = \partial_k(\ )\boldsymbol{i}_k = \partial(\ )/\partial x_k \boldsymbol{i}_k$ and $\boldsymbol{\nabla}\cdot\boldsymbol{\sigma}$, $\mathbf{C}:\boldsymbol{\varepsilon}$ imply $\sigma_{ij,i}$, $C_{ijkl}\varepsilon_{kl}$ in component form, where $\boldsymbol{i}_k$ is the unit base vector.

Now making use of the Stokes–Helmholtz decomposition of the displacement vector

$$\boldsymbol{u} = \boldsymbol{u}^{\mathrm{L}} + \boldsymbol{u}^{\mathrm{T}}, \qquad \boldsymbol{u}^{\mathrm{L}} = \boldsymbol{\nabla}\phi, \qquad \boldsymbol{u}^{\mathrm{T}} = \boldsymbol{\nabla}\times\boldsymbol{\psi} \tag{1.9}$$

where ϕ and $\boldsymbol{\psi}$ are scalar and vector potentials (called Lamé potentials), Eq. (1.8b) is satisfied if the following equations are fulfilled:

$$c_{\mathrm{L}}^2\nabla^2\phi = \ddot{\phi}\,, \tag{1.10}$$

$$c_{\mathrm{T}}^2\nabla^2\boldsymbol{\psi} = \ddot{\boldsymbol{\psi}} \tag{1.11}$$

where

$$c_{\mathrm{L}} = \left(\frac{\lambda + 2\mu}{\rho}\right)^{1/2}, \qquad c_{\mathrm{T}} = \left(\frac{\mu}{\rho}\right)^{1/2}. \tag{1.12}$$

Such solutions are complete [117].

By taking the divergence and curl of Eq. (1.8b), we know that the Navier–Cauchy equation of motion is satisfied whenever the following equations are assured:

$$c_{\mathrm{L}}^2\nabla^2\boldsymbol{u}^{\mathrm{L}} = \ddot{\boldsymbol{u}}^{\mathrm{L}}, \qquad \boldsymbol{\nabla}\times\boldsymbol{u}^{\mathrm{L}} = \boldsymbol{0} \tag{1.13}$$

and

$$c_{\mathrm{T}}^2\nabla^2\boldsymbol{u}^{\mathrm{T}} = \ddot{\boldsymbol{u}}^{\mathrm{T}}, \qquad \boldsymbol{\nabla}\cdot\boldsymbol{u}^{\mathrm{T}} = 0\,. \tag{1.14}$$

These equations imply that $\boldsymbol{u}^{\mathrm{L}}$ and $\boldsymbol{u}^{\mathrm{T}}$ define the dilatational (longitudinal) and the distortional (equivolumetric, or shear) waves, respectively, and hence c_{L} and c_{T} stand for the phase speeds of the respective waves.

In the two-dimensional case, when the displacement is assumed in the form

$$\boldsymbol{u}(\boldsymbol{x}, t) = u_\alpha(x_1, x_2, t)\boldsymbol{i}_\alpha + u_3(x_1, x_2, t)\boldsymbol{i}_3\,, \qquad (\alpha = 1, 2) \tag{1.15}$$

with $\boldsymbol{i}_\alpha$ being the unit base vector, we have the following equations of motion:

$$\mu u_{\alpha,\beta\beta} + (\lambda + \mu)u_{\beta,\beta\alpha} = \rho\ddot{u}_\alpha\,, \qquad (\alpha, \beta = 1, 2)\,, \tag{1.16}$$

$$\mu u_{3,\alpha\alpha} = \rho\ddot{u}_3\,, \qquad (\alpha = 1, 2)\,. \tag{1.17}$$

Equation (1.16) expresses the in-plane (x_1x_2 plane) motion and Eq. (1.17) implies the anti-plane ($\boldsymbol{i}_3$ direction) motion. The corresponding stress states are given as follows:

$$\sigma_{\alpha\beta} = \lambda u_{\gamma,\gamma}\delta_{\alpha\beta} + \mu(u_{\alpha,\beta} + u_{\beta,\alpha}) , \qquad \sigma_{33} = \lambda u_{\alpha,\alpha} , \qquad (\alpha, \beta, \gamma = 1, 2) , \tag{1.18}$$

$$\sigma_{3\alpha} = \mu u_{3,\alpha} , \qquad (\alpha = 1, 2) , \tag{1.19}$$

where $\delta_{\alpha\beta}$ denotes the Kronecker delta.

Making use of the Stokes–Helmholtz decomposition

$$\boldsymbol{u} = \boldsymbol{u}^{\mathrm{L}} + \boldsymbol{u}^{\mathrm{T}} + u_3\boldsymbol{i}_3 ,$$

we have the following wave equations from Eqs. (1.16) and (1.17):

$$\left.\begin{array}{ll} c_{\mathrm{L}}^2\nabla^2\boldsymbol{u}^{\mathrm{L}} = \ddot{\boldsymbol{u}}^{\mathrm{L}} , & \boldsymbol{\nabla} \times \boldsymbol{u}^{\mathrm{L}} = \boldsymbol{0} , \\ c_{\mathrm{T}}^2\nabla^2\boldsymbol{u}^{\mathrm{T}} = \ddot{\boldsymbol{u}}^{\mathrm{T}} , & \boldsymbol{\nabla} \cdot \boldsymbol{u}^{\mathrm{T}} = 0 , \end{array}\right\} \tag{1.20}$$

$$c_{\mathrm{T}}^2\nabla^2 u_3 = \ddot{u}_3 . \tag{1.21}$$

Equations (1.20) imply that the in-plane motion obeys the system of equations of the dilatational and distorsional waves, whereas Eq. (1.21) shows that the anti-plane motion is governed only by the distorsional wave motion. The former distorsional wave is often called the SV wave (vertically polarized shear wave), while the latter is termed the SH wave (horizontally polarized shear wave).

1.2. Elastodynamic problems

The general elastodynamic problem is defined such as to find solutions of Eq. (1.8) with initial conditions

$$\boldsymbol{u}(\boldsymbol{x}, 0) = \boldsymbol{u}_0(\boldsymbol{x}) , \qquad \dot{\boldsymbol{u}}(\boldsymbol{x}, 0) = \dot{\boldsymbol{u}}_0(\boldsymbol{x}) \tag{1.22}$$

and boundary conditions

$$\begin{array}{lll} \boldsymbol{u}(\boldsymbol{x}, t) = \bar{\boldsymbol{u}}(\boldsymbol{x}, t) , & & (\boldsymbol{x}, t) \in \partial D_1 \times T , \\ \boldsymbol{t}(\boldsymbol{x},t) = \boldsymbol{n} \cdot \boldsymbol{\sigma} = \overset{n}{\mathbf{T}}\boldsymbol{u}(\boldsymbol{x}, t) & & (\boldsymbol{x}, t) \in \partial D_2 \times T , \\ \qquad = \bar{\boldsymbol{t}}(\boldsymbol{x}, t) , & & \partial D = \partial D_1 \cup \partial D_2 , \end{array} \tag{1.23}$$

where $\boldsymbol{u}_0$, $\dot{\boldsymbol{u}}_0$ are given initial displacement and velocity, $\bar{\boldsymbol{u}}$, $\bar{\boldsymbol{t}}$ are prescribed displacement on ∂D_1 and traction on ∂D_2, and $\boldsymbol{n}$ denotes the unit outward normal vector to ∂D. The operator $\overset{n}{\mathbf{T}}$ is defined as

$$
\begin{aligned}
\overset{n}{\mathbf{T}}\boldsymbol{u} &= \boldsymbol{n}\cdot\mathbf{C}:\boldsymbol{\nabla}\boldsymbol{u} \\
&= \lambda\boldsymbol{n}(\boldsymbol{\nabla}\cdot\boldsymbol{u})\mathbf{1} + \mu\boldsymbol{n}\cdot(\boldsymbol{\nabla}\boldsymbol{u} + \boldsymbol{u}\boldsymbol{\nabla}) \\
&= \{\lambda\boldsymbol{n}\boldsymbol{\nabla}\cdot \;\; + 2\mu(\boldsymbol{n}\cdot\boldsymbol{\nabla})\mathbf{1} + \mu\boldsymbol{n}\times\boldsymbol{\nabla}\times\}\boldsymbol{u}\,.
\end{aligned}
\tag{1.24}
$$

In the two-dimensional problem, we have to solve Eq. (1.16) or (1.17) instead of Eq. (1.8) with two-dimensional initial and boundary conditions, in which the tractions are given for the inplane case by

$$
t_\beta = \overset{n}{T}_{\beta\alpha}u_\alpha = \lambda n_\beta u_{\gamma,\gamma} + \mu n_\alpha(u_{\alpha,\beta} + u_{\beta,\alpha})\,, \qquad (\alpha, \beta, \gamma = 1, 2) \tag{1.25}
$$

and for the anti-plane case by

$$
t_3 = \overset{n}{T}u_3 = \mu n_\alpha u_{3,\alpha} = \mu\,\frac{\partial u_3}{\partial n}\,, \qquad (\alpha = 1, 2)\,. \tag{1.26}
$$

For the exterior problem which includes the boundary extended to infinity, it is reasonable to require physically that the displacement at infinity must be bounded and that no wave can be reflected back from infinity. These conditions, called radiation conditions, are expressed as follows [37, 70, 114]:

$$
\begin{aligned}
&\frac{\partial\boldsymbol{u}^{\mathrm{L}}}{\partial r} + \frac{1}{c_{\mathrm{L}}}\frac{\partial\boldsymbol{u}^{\mathrm{L}}}{\partial t} = o(r^{-(N-1)/2})\,, \qquad \boldsymbol{u}^{\mathrm{L}} = o(r^{-(N-3)/2}) \\
&\frac{\partial\boldsymbol{u}^{\mathrm{T}}}{\partial r} + \frac{1}{c_{\mathrm{T}}}\frac{\partial\boldsymbol{u}^{\mathrm{T}}}{\partial t} = o(r^{-(N-1)/2})\,, \qquad \boldsymbol{u}^{\mathrm{T}} = o(r^{-(N-3)/2})
\end{aligned}
\tag{1.27}
$$

as $r = |\boldsymbol{x}| \to \infty$. In these expressions, N implies the number of dimensions ($N = 2$ or 3), and o denotes Landau's symbol, that is, $f(x) = o[g(x)]$ implies that $\lim f(x)/g(x) \to 0$ as $x \to a$, where a is a fixed value.

Radiation conditions for elastodynamics may be alternatively expressed as [28, 37, 70]:

$$
\begin{aligned}
&\overset{n}{\mathbf{T}}\boldsymbol{u}^{\mathrm{L}} + \rho c_{\mathrm{L}}\dot{\boldsymbol{u}}^{\mathrm{L}} = o(r^{-(N-1)/2})\,, \qquad \overset{n}{\mathbf{T}}\boldsymbol{u}^{\mathrm{T}} + \rho c_{\mathrm{T}}\dot{\boldsymbol{u}}^{\mathrm{T}} = o(r^{-(N-1)/2})\,, \\
&\dot{u}_{\mathrm{t}}^{\mathrm{L}} = o(r^{-(N-1)/2})\,, \qquad \dot{u}_{\mathrm{t}}^{\mathrm{T}} = o(r^{-(N-1)/2}) \\
&\dot{u}_{\mathrm{n}}^{\mathrm{L}} = o(r^{-(N-3)/2})\,, \qquad \dot{u}_{\mathrm{n}}^{\mathrm{T}} = o(r^{-(N-3)/2})\,,
\end{aligned}
\tag{1.28}
$$

where subscripts n and t mean the normal and tangential components of the velocity vector, respectively, on the infinite sphere when r approaches infinity.

The first expression, for instance for $N = 3$, may be rewritten as [28, 37]

$$\lim_{R\to\infty} \int_{S_R} |\overset{n}{\mathbf{T}}\boldsymbol{u}^{\mathrm{L}} + \rho c_{\mathrm{L}} \dot{\boldsymbol{u}}^{\mathrm{L}}|^2 \,\mathrm{d}S = 0$$

and thus

$$\lim_{R\to\infty} \int_{S_R} \overset{n}{\mathbf{T}}\boldsymbol{u}^{\mathrm{L}} \cdot \dot{\mathbf{u}}^{\mathrm{L}} \,\mathrm{d}S = -\lim_{R\to\infty} \left(\frac{1}{2\rho c_{\mathrm{L}}}\right) \int_{S_R} \{|\overset{n}{\mathbf{T}}\boldsymbol{u}^{\mathrm{L}}|^2 + \rho^2 c_{\mathrm{L}}^2 |\dot{\boldsymbol{u}}^{\mathrm{L}}|^2\} \,\mathrm{d}S < 0 .$$

Hence the efflux of energy per unit time across the surface S

$$P = -\lim_{R\to\infty} \int_{S_R} |\overset{n}{\mathbf{T}}\boldsymbol{u}^{\mathrm{L}} \cdot \dot{\boldsymbol{u}}^{\mathrm{L}} + \overset{n}{\mathbf{T}}\boldsymbol{u}^{\mathrm{T}} \cdot \dot{\boldsymbol{u}}^{\mathrm{T}}| \,\mathrm{d}S$$

is proved to be positive. Therefore, the radiation conditions of elastodynamics guarantee that an outward flow of energy is only allowed at infinity, which implies no reflection of waves from infinity.

The radiation conditions are crucial in determing the unique solution in unbounded domains, particularly in time-harmonic problems.

1.3. Fourier-transformed elastodynamic problem

It is sometimes convenient to solve elastodynamic problems with the aid of the Fourier transform with respect to time [61, 64, 66, 70]. Introducing the Fourier transform defined by

$$\hat{f}(\boldsymbol{x}; \omega) = \int_{-\infty}^{\infty} f(\boldsymbol{x}, t)\, \mathrm{e}^{\mathrm{i}\omega t} \,\mathrm{d}t , \tag{1.29}$$

we have three-dimensional elastodynamic problems expressed in the transformed domain as follows (body force is disregarded for simplicity):

$$\begin{aligned} &\mu \nabla^2 \hat{\boldsymbol{u}}(\boldsymbol{x}; \omega) + (\lambda + \mu) \boldsymbol{\nabla\nabla} \cdot \hat{\boldsymbol{u}}(\boldsymbol{x}; \omega) + \rho\omega^2 \hat{\boldsymbol{u}}(\boldsymbol{x}; \omega) \\ &\quad = (\mathbf{E} + \rho\omega^2 \mathbf{1}) \hat{\boldsymbol{u}}(\boldsymbol{x}; \omega) \equiv \mathbf{L}\hat{\boldsymbol{u}}(\boldsymbol{x}; \omega) = \mathbf{0} , \end{aligned} \tag{1.30}$$

$$\left.\begin{aligned} &\hat{\boldsymbol{u}}(\boldsymbol{x}; \omega) = \hat{\bar{\boldsymbol{u}}}(\boldsymbol{x}; \omega) , && \boldsymbol{x} \in \partial D_1 , \\ &\hat{\boldsymbol{t}}(\boldsymbol{x}; \omega) = \overset{n}{\mathbf{T}}\hat{\boldsymbol{u}}(\boldsymbol{x}; \omega) = \hat{\bar{\boldsymbol{t}}}(\boldsymbol{x}; \omega) , && \boldsymbol{x} \in \partial D_2 . \end{aligned}\right\} \tag{1.31}$$

Similar expressions are obtained for the two-dimensional problems.

The transformed problems are of a more convenient form since they are of elliptic type and similar solution techniques as for elastostatics can be available. Solutions of the original problem can be obtained by inverting the solutions obtained in the transformed domain, i.e. by

$$u(x, t) = \frac{1}{2\pi} \int_{-\infty}^{\infty} \hat{u}(x; \omega)\, e^{-i\omega t}\, d\omega\,. \tag{1.32}$$

As for the unbounded domain, the Fourier-transformed radiation conditions must be satisfied [65, 70], i.e. for example,

$$\begin{aligned} \frac{\partial \hat{u}^{L}}{\partial r} - \frac{i\omega}{c_L}\hat{u} &= o(r^{-(N-1)/2})\,, & \hat{u}^{L} &= o(r^{-(N-3)/2})\,, \\ \frac{\partial \hat{u}^{T}}{\partial r} - \frac{i\omega}{c_T}\hat{u}^{T} &= o(r^{-(N-1)/2})\,, & \hat{u}^{T} &= o(r^{-(N-3)/2})\,. \end{aligned} \tag{1.33}$$

1.4. *Fundamental solutions*

1.4.1. *General concept*

Fundamental solutions are of basic importance in the formulation of integral equations. The fundamental solution, which is sometimes called the free-space Green's function or spectral Green's tensor, is defined as the displacement field at x and at time t in the infinite domain due to an unit impulse applied at $x = y$ and $t = \tau$. If we denote the ith component of the displacement at x and at time t due to the unit impulse in the k-direction by $U_{ik}(x, t; y, \tau)$, it satisfies the equation

$$\frac{\partial}{\partial x_j}\left(C_{ijmn}\frac{\partial}{\partial x_n}U_{mk}\right) + \delta_{ik}\delta(x - y)\delta(t - \tau) = \rho \ddot{U}_{ik} \tag{1.34}$$

for the general case and

$$\mu \frac{\partial^2}{\partial x_j \partial x_j} U_{ik} + (\lambda + \mu)\frac{\partial^2}{\partial x_i \partial x_j} U_{ik} + \delta_{ik}\delta(x - y)\delta(t - \tau) = \rho \ddot{U}_{ik} \tag{1.35}$$

for the isotropic homogeneous case, where $\delta(\ )$ means the Dirac delta function. The same definition is also made for the two-dimensional case.

Hereafter we assume the validity of the causality requirement such that $U(x, t; y, \tau)$ and $\partial U(x, t; y, \tau)/\partial t$ must be zero for $t \leq \tau$ and $x \neq y$. It is easily recognized from Eq. (1.34) or (1.35) that the time origin can be shifted at will and hence

$$U(x, t; y, \tau) = U(x, t - \tau; y, 0) = U(x, -\tau; y, -t)\,, \tag{1.36}$$

which is a reciprocal relation for source and observer times. Using the elastodynamic reciprocal theorem, (see Eq. (2.2) in section 2.1), we also have the relation

$$U_{ik}(y_2, \tau + \tau_2; y_1, \tau_1) = U_{ki}(y_1, \tau - \tau_1; y_2, -\tau_2)\,, \tag{1.37a}$$

which includes, as a special case, purely spatial reciprocity

$$U_{ik}(y_2, \tau; y_1, 0) = U_{ki}(y_1, \tau; y_2, 0) \tag{1.37b}$$

and space-time reciprocity

$$U_{ik}(y_2, \tau_2; y_1, \tau_1) = U_{ki}(y_1, -\tau_1; y_2, -\tau_2) . \tag{1.37c}$$

The fundamental solution also satisfies the translation identity for time and space, i.e.

$$\left.\begin{aligned} U_{ik}(\boldsymbol{x}, t; \boldsymbol{y}, \tau) &= U_{ik}(\boldsymbol{x}-\boldsymbol{y}, t; 0, \tau) , \\ U_{ik}(\boldsymbol{x}, t; \boldsymbol{y}, \tau) &= U_{ik}(\boldsymbol{x}, t+t_0; \boldsymbol{y}, \tau+t_0) . \end{aligned}\right\} \tag{1.37d}$$

Although we know the physical meaning and properties of the fundamental solution, it is very difficult to find out general elastodynamic fundamental solutions. Hence we restrict ourselves to the homogeneous and isotropic case.

1.4.2. Fundamental solutions in the Fourier-transformed domain

Taking the Fourier transform of Eq. (1.35) with respect to time, we have for $\tau = 0$

$$\mu \frac{\partial^2}{\partial x_j \partial x_j} \hat{U}_{ik} + (\lambda + \mu) \frac{\partial^2}{\partial x_i \partial x_j} \hat{U}_{jk} + \rho\omega^2 \hat{U}_{ik} = -\delta_{ik}\delta(\boldsymbol{x}-\boldsymbol{y}) , \tag{1.38}$$

where $\hat{U}_{ik}$ is the Fourier-transformed fundamental solution.

The solution $\hat{U}_{ik}$ may be obtained as follows by the aid of the Fourier transform with respect to space variables [35, 63]: The transformed solution $\hat{\hat{U}}_{ik}$ is expressed as

$$\hat{\hat{U}}_{ik} = -\frac{1}{\mu} \frac{(k_L^2 - \xi^2)\delta_{ik} + (1 - k_L^2/k_T^2)\xi_i\xi_k}{(k_L^2 - \xi^2)(k_T^2 - \xi^2)} e^{i\xi \cdot y} , \tag{1.39a}$$

where ξ is the parameter of the Fourier transform and

$$k_L = \omega/c_L , \qquad k_T = \omega/c_T , \qquad \xi^2 = \boldsymbol{\xi} \cdot \boldsymbol{\xi} = \xi_1^2 + \xi_2^2 + \xi_3^2 . \tag{1.40}$$

Equation (1.39a) can be also expressed as

$$\hat{\hat{U}}_{ik} = \frac{1}{\mu} \left\{ \frac{\xi_i \xi_k}{k_T^2} \left(\frac{1}{\xi^2 - k_L^2} - \frac{1}{\xi^2 - k_T^2} \right) + \frac{\delta_{ik}}{\xi^2 - k_T^2} \right\} e^{i\xi \cdot y} . \tag{1.39b}$$

The inverse Fourier transform

$$I_\alpha = \frac{1}{(2\pi)^3} \int_V \frac{\exp[-i\boldsymbol{\xi} \cdot (\boldsymbol{x}-\boldsymbol{y})]}{\xi^2 - k_\alpha^2} \, d\boldsymbol{\xi} \qquad (\alpha = L, T)$$

can be evaluated by using a transformation to spherical coordinates (ρ, θ, ϕ), where the azimuthal axis is chosen in the direction of $\boldsymbol{x} - \boldsymbol{y}$, and ϕ is the azimuthal angle. Hence, $\xi = \rho$ and $\boldsymbol{\xi} \cdot (\boldsymbol{x} - \boldsymbol{y}) = \rho r \cos \phi$, where $r^2 = (x_i - y_i)(x_i - y_i)$ and $\mathrm{d}\boldsymbol{\xi} = \rho^2 \sin \phi \, \mathrm{d}\rho \, \mathrm{d}\theta \, \mathrm{d}\phi$. We have

$$I_\alpha = \frac{1}{(2\pi)^3} \int_0^\infty \int_0^\pi \int_0^{2\pi} \frac{\rho^2 \sin \phi}{\rho^2 - k_\alpha^2} \exp(-\mathrm{i}\rho r \cos \phi) \, \mathrm{d}\theta \, \mathrm{d}\phi \, \mathrm{d}\rho$$

and the integration can be reduced to

$$I_\alpha = -\frac{\mathrm{i}}{4\pi^2 r} \int_{-\infty}^{\infty} \frac{\zeta \, \mathrm{e}^{\mathrm{i}\zeta r}}{\zeta^2 - k_\alpha^2} \, \mathrm{d}\zeta = \frac{1}{4\pi r} \mathrm{e}^{\mathrm{i}k_\alpha r} ,$$

where the radiation conditions are taken into account for time dependence of the form $\exp(-\mathrm{i}\omega t)$ in evaluating the last contour integral, that is, the contour is taken in the upper half of the ζ-plane so as to contain the pole $\zeta = k_\alpha$ in it.

Using this result together with the rule that the factor $-\mathrm{i}\xi_k$ in the ξ-space corresponds to $\partial / \partial x_k$ in the x-space, we have from Eq. (1.39b)

$$\hat{U}_{ik} = \frac{1}{4\pi\mu} \left[\frac{\mathrm{e}^{\mathrm{i}k_\mathrm{T} r}}{r} \delta_{ik} + \frac{\partial^2}{\partial x_i \partial x_k} \left\{ \frac{1}{k_\mathrm{T}^2} \left(\frac{\mathrm{e}^{\mathrm{i}k_\mathrm{T} r}}{r} - \frac{\mathrm{e}^{\mathrm{i}k_\mathrm{L} r}}{r} \right) \right\} \right] .$$

For two-dimensional problems, we can easily derive fundamental solutions, which are the wave fields generated by a line load. Since a line load can be considered as a superposition of point loads, the fundamental solutions are expressed as a superposition integral of $\hat{U}_{ik}$ along, say, the x_3-axis from $-\infty$ to ∞.

Fundamental solutions in the Fourier-transformed domain are summarized as follows [61, 70, 71]:

$$\hat{U}_{jk}(\boldsymbol{x}, \boldsymbol{y}; \omega) = A(\hat{U}_1 \delta_{jk} - \hat{U}_2 r_{,j} r_{,k}) , \tag{1.40a}$$

(i) for the three-dimensional case:

$$\left. \begin{aligned} A &= \frac{1}{4\pi\mu} , \\ \hat{U}_1 &= \frac{\mathrm{e}^{\mathrm{i}k_\mathrm{T} r}}{r} + \left\{ \frac{\mathrm{i}}{k_\mathrm{T} r} - \frac{1}{(k_\mathrm{T} r)^2} \right\} \frac{\mathrm{e}^{\mathrm{i}k_\mathrm{T} r}}{r} - \left(\frac{k_\mathrm{L}}{k_\mathrm{T}} \right)^2 \left\{ \frac{\mathrm{i}}{k_\mathrm{L} r} - \frac{1}{(k_\mathrm{L} r)^2} \right\} \frac{\mathrm{e}^{\mathrm{i}k_\mathrm{L} r}}{r} , \\ \hat{U}_2 &= \left\{ 1 + \frac{3\mathrm{i}}{k_\mathrm{T} r} - \frac{3}{(k_\mathrm{T} r)^2} \right\} \frac{\mathrm{e}^{\mathrm{i}k_\mathrm{T} r}}{r} - \left(\frac{k_\mathrm{L}}{k_\mathrm{T}} \right)^2 \left\{ 1 + \frac{3\mathrm{i}}{k_\mathrm{L} r} - \frac{3}{(k_\mathrm{L} r)^2} \right\} \frac{\mathrm{e}^{\mathrm{i}k_\mathrm{L} r}}{r} , \end{aligned} \right\} \tag{1.40b}$$

(ii) for the plane-strain case:

$$\left.\begin{aligned} A &= \frac{\mathrm{i}}{4\mu}\,, \\ \hat{U}_1 &= H_0^{(1)}(k_{\mathrm{T}}r) - \frac{1}{k_{\mathrm{T}}r}\,H_1^{(1)}(k_{\mathrm{T}}r) + \left(\frac{k_{\mathrm{L}}}{k_{\mathrm{T}}}\right)^2 \frac{1}{k_{\mathrm{L}}r}\,H_1^{(1)}(k_{\mathrm{L}}r)\,, \\ \hat{U}_2 &= -H_2^{(1)}(k_{\mathrm{T}}r) + \left(\frac{k_{\mathrm{L}}}{k_{\mathrm{T}}}\right)^2 H_2^{(1)}(k_{\mathrm{L}}r)\,, \end{aligned}\right\} \tag{1.40c}$$

(iii) for the anti-plane case:

$$\hat{U}_{33}(\boldsymbol{x}, \boldsymbol{y}; \omega) = \frac{\mathrm{i}}{4\mu}\,H_0^{(1)}(k_{\mathrm{T}}r)\,, \tag{1.41}$$

where $H_n^{(1)}(\;)$ is the Hankel function of the nth order of the first kind, and $r_{,i} = \partial r/\partial x_i$.

Here we introduce a new entity $\hat{\boldsymbol{\Sigma}}$, which is obtained from $\hat{\boldsymbol{U}}$ in the same manner as we obtain the stress $\boldsymbol{\sigma}$ from the displacement $\boldsymbol{u}$. For the isotropic homogeneous case, it is defined as

$$\begin{aligned} \hat{\Sigma}_{ijk}(\hat{U}) &= \lambda \hat{U}_{mk,m}\delta_{ij} + \mu(\hat{U}_{ik,j} + \hat{U}_{jk,i}) \\ &= \rho[(c_{\mathrm{L}}^2 - 2c_{\mathrm{T}}^2)\hat{U}_{mk,m}\delta_{ij} + c_{\mathrm{T}}^2(\hat{U}_{ik,j} + \hat{U}_{jk,i})] \\ &= -\rho[(c_{\mathrm{L}}^2 - 2c_{\mathrm{T}}^2)\hat{U}_{mk.m}\delta_{ij} + c_{\mathrm{T}}^2(\hat{U}_{ik.j} + \hat{U}_{jk.i})]\,, \end{aligned} \tag{1.42}$$

where $(\;)_{,i} = \partial(\;)/\partial x_i$ and $(\;)_{.i} = \partial(\;)/\partial y_i = -\partial(\;)/\partial x_i$. Obviously, $\hat{\Sigma}_{ijk} = \hat{\Sigma}_{jik}$.

We now define the second fundamental solution by

$$\hat{W}_{jk}(\boldsymbol{x}, \boldsymbol{y}; \omega) = n_i(\boldsymbol{y})\hat{\Sigma}_{ijk} = \overset{n_y}{T}_{ki}\hat{U}_{ji}(\boldsymbol{x}, \boldsymbol{y}; \omega)\,, \tag{1.43}$$

which is explicity written as follows:

$$\begin{aligned} \hat{W}_{jk}(\boldsymbol{x}, \boldsymbol{y}; \omega) = \mu A\Bigg[&\left\{\left(\delta_{jk}\frac{\partial r}{\partial n} + n_j r_{.k}\right) + \frac{\lambda}{\mu}\,n_k r_{.j}\right\}\frac{\mathrm{d}\hat{U}_1}{\mathrm{d}r} \\ &-\left\{\left(\delta_{jk}\frac{\partial r}{\partial n} + n_j r_{.k}\right) + 2\left(n_k r_{.j} - 2r_{.j}r_{.k}\frac{\partial r}{\partial n}\right) + \alpha\,\frac{\lambda}{\mu}\,n_k r_{.j}\right\}\frac{\hat{U}_2}{r} \\ &-\left\{2r_{.j}r_{.k}\frac{\partial r}{\partial n} + \frac{\lambda}{\mu}\,n_k r_{.j}\right\}\frac{\mathrm{d}\hat{U}_2}{\mathrm{d}r}\Bigg]\,, \end{aligned} \tag{1.44a}$$

where

(i) for the three-dimensional case:

$$\alpha = 2 \text{ and } A, \hat{U}_1, \hat{U}_2 \text{ are given in Eq. (1.40b)}, \tag{1.44b}$$

(ii) for the plane-strain state:

$$\alpha = 1 \textit{ and } A, \hat{U}_1, \hat{U}_2 \text{ are given in Eq. (1.40c)}, \tag{1.44c}$$

(iii) for the anti-plane case:

$$\hat{W}_{33}(\boldsymbol{x}, \boldsymbol{y}; \omega) = -\frac{\mathrm{i}k_{\mathrm{T}}}{4} H_1^{(1)}(k_{\mathrm{T}}r)\frac{\partial r}{\partial n}, \tag{1.45}$$

with $\partial r/\partial n = (\partial r/\partial y_i)n_i$.

If we use the Laplace transform with respect to time, the corresponding fundamental solutions are easily derived by simply converting the parameter ω to $\mathrm{i}s$, where s is the parameter of the Laplace transform [4, 23, 24, 77–80].

1.4.3. Fundamental solutions in the time–space domain

The time–space fundamental solutions, i.e. $U_{ik}(\boldsymbol{x}, t; \boldsymbol{y}, \tau)$ satisfying Eq. (1.34), can be derived by inverting the Fourier-transformed solutions $\hat{U}_{ik}(\boldsymbol{x}, \boldsymbol{y}; \omega)$ with the aid of the convolution theorem. According to this theorem, if $\hat{f}(\omega)$ and $\hat{g}(\omega)$ are Fourier transforms of $f(t)$ and $g(t)$, then $\hat{f}(\omega)\hat{g}(\omega)$ is the Fourier transform of the convolution integral

$$\int_{-\infty}^{t} f(\tau)g(t-\tau)\,\mathrm{d}\tau \qquad \text{or } f(t) * g(t)$$

provided that $g(t) = 0$ for $t < 0$. Hence, for example, $\omega^{-2}\hat{f}(\omega)$ is the Fourier transform of

$$\int_{-\infty}^{t} f(\tau)(t-\tau)\,\mathrm{d}\tau$$

since $\hat{g}(\omega) = \omega^{-2}$ so that $g(t) = tH(t)$.

The inverse Fourier transform of $\hat{g}(\omega)\exp(-\mathrm{i}\omega r/c_\alpha)$ is $g(t - r/c_\alpha)$. Therefore, from Eqs. (1.40) and (1.44) we have the following expressions for $\tau = 0$ [2, 3, 8, 37, 61, 130, 131]:

$$
\begin{aligned}
&U_{jk}(\boldsymbol{x}, t; \boldsymbol{y}, 0) \\
&= \frac{1}{4\pi\rho}\left\{\frac{t}{r^2}\left(\frac{3r_j r_k}{r^3} - \frac{\delta_{jk}}{r}\right)\left[H\left(t - \frac{r}{c_\mathrm{L}}\right) - H\left(t - \frac{r}{c_\mathrm{T}}\right)\right]\right. \\
&\quad \left. + \frac{r_j r_k}{r^3}\left[\frac{1}{c_\mathrm{L}^2}\,\delta\left(t - \frac{r}{c_\mathrm{L}}\right) - \frac{1}{c_\mathrm{T}^2}\,\delta\left(t - \frac{r}{c_\mathrm{T}}\right)\right] + \frac{\delta_{jk}}{r c_\mathrm{T}^2}\,\delta\left(t - \frac{r}{c_\mathrm{T}}\right)\right\} \\
&= \frac{t}{4\pi\rho}\left\{\left(\frac{1}{r}\right)_{,jk}\left[H\left(t - \frac{r}{c_\mathrm{L}}\right) - H\left(t - \frac{r}{c_\mathrm{T}}\right)\right]\right. \\
&\quad \left. - r_{,j}\left(\frac{1}{r}\right)_{,k}\left[\frac{1}{c_\mathrm{L}}\,\delta\left(t - \frac{r}{c_\mathrm{L}}\right) - \frac{1}{c_\mathrm{T}}\,\delta\left(t - \frac{r}{c_\mathrm{T}}\right)\right] + \frac{\delta_{jk}}{r^2 c_\mathrm{T}}\,\delta\left(t - \frac{r}{c_\mathrm{T}}\right)\right\} \\
&= \frac{t}{4\pi\rho}\left\{\left(\frac{1}{r}\right)_{.jk}\left[H\left(t - \frac{r}{c_\mathrm{L}}\right) - H\left(t - \frac{r}{c_\mathrm{T}}\right)\right]\right. \\
&\quad \left. - r_{.j}\left(\frac{1}{r}\right)_{.k}\left[\frac{1}{c_\mathrm{L}}\,\delta\left(t - \frac{r}{c_\mathrm{L}}\right) - \frac{1}{c_\mathrm{T}}\,\delta\left(t - \frac{r}{c_\mathrm{T}}\right)\right] + \frac{\delta_{jk}}{r^2 c_\mathrm{T}}\,\delta\left(t - \frac{r}{c_\mathrm{T}}\right)\right\},
\end{aligned}
\tag{1.46}
$$

$$
W_{jk}(\boldsymbol{x}, t; \boldsymbol{y}, 0) = n_i(\boldsymbol{y})\Sigma_{ijk}(\boldsymbol{x}, t; \boldsymbol{y}, 0)\,, \tag{1.47}
$$

$$
\begin{aligned}
&\Sigma_{ijk}(\boldsymbol{x}, t; y, 0) \\
&= \frac{1}{4\pi}\left\{-6c_\mathrm{T}^2\left[5\,\frac{r_i r_j r_k}{r^5} - \frac{\delta_{ij} r_k + \delta_{ik} r_j + \delta_{jk} r_i}{r^3}\right]\frac{t}{r^2}\left[H\left(t - \frac{r}{c_\mathrm{L}}\right) - H\left(t - \frac{r}{c_\mathrm{T}}\right)\right]\right. \\
&\quad + 2\left[6\,\frac{r_i r_j r_k}{r^5} - \frac{\delta_{ij} r_k + \delta_{ik} r_j + \delta_{jk} r_i}{r^3}\right]\left[\delta\left(t - \frac{r}{c_\mathrm{T}}\right) - \left(\frac{c_\mathrm{T}}{c_\mathrm{L}}\right)^2\delta\left(t - \frac{r}{c_\mathrm{L}}\right)\right] \\
&\quad + 2\,\frac{r_i r_j r_k}{r^4 c_\mathrm{T}}\left[\dot{\delta}\left(t - \frac{r}{c_\mathrm{T}}\right) - \left(\frac{c_\mathrm{T}}{c_\mathrm{L}}\right)^3\dot{\delta}\left(t - \frac{r}{c_\mathrm{L}}\right)\right] \\
&\quad - \frac{r_k \delta_{ij}}{r^3}\left\{1 - 2\left(\frac{c_\mathrm{T}}{c_\mathrm{L}}\right)^2\right\}\left[\delta\left(t - \frac{r}{c_\mathrm{L}}\right) + \frac{r}{c_\mathrm{L}}\,\dot{\delta}\left(t - \frac{r}{c_\mathrm{L}}\right)\right] \\
&\quad \left. - \frac{\delta_{ik} r_j + \delta_{jk} r_i}{r^3}\left[\delta\left(t - \frac{r}{c_\mathrm{T}}\right) + \frac{r}{c_\mathrm{T}}\,\dot{\delta}\left(t - \frac{r}{c_\mathrm{T}}\right)\right]\right\}.
\end{aligned}
\tag{1.48}
$$

For the two-dimensional case, we have

$$
\begin{aligned}
V_{jk}(x_\alpha, t; y_\beta, 0) &\equiv \int_{-\infty}^{\infty} U_{jk}(\boldsymbol{x}, t; \boldsymbol{y}, 0)\,\mathrm{d}x_3 \\
S_{jk}(x_\alpha, t; y_\beta, 0) &\equiv \int_{-\infty}^{\infty} W_{jk}(\boldsymbol{x}, t; \boldsymbol{y}, 0)\,\mathrm{d}x_3
\end{aligned}
\tag{1.49}
$$

and hence in detail [37],

$$V_{\alpha\beta} = \frac{1}{2\pi\rho}\left\{\left[\frac{(2t^2 - r^2/c_L^2)}{(t^2 - r^2/c_L^2)^{1/2}} H\left(t - \frac{r}{c_L}\right) - \frac{(2t^2 - r^2/c_T^2)}{(t^2 - r^2/c_T^2)^{1/2}} H\left(t - \frac{r}{c_T}\right)\right]\frac{r_\alpha r_\beta}{r^4}\right.$$
$$-\left[H\left(t - \frac{r}{c_L}\right)(t^2 - r^2/c_L^2)^{1/2} - H(t - r/c_T)(t^2 - r^2/c_T^2)^{1/2}\right]\frac{\delta_{\alpha\beta}}{r}$$
$$\left. + \frac{H(t - r/c_T)}{c_T^2(t^2 - r^2/c_T^2)^{1/2}}\,\delta_{\alpha\beta}\right\}, \qquad (1.50)$$

$$V_{3\alpha} = 0\,,$$

$$V_{33} = \frac{H(t - r/c_T)}{2\pi\rho c_T^2(t^2 - r^2/c_T^2)^{1/2}}$$

and

$$\Sigma_{\alpha\beta\gamma} = -\rho(c_L^2 - 2c_T^2)V_{\kappa\gamma.\kappa}\delta_{\alpha\beta} - \rho c_T^2(V_{\alpha\gamma.\beta} + V_{\beta\gamma.\alpha})\,,$$
$$\Sigma_{\alpha\beta 3} = \Sigma_{\alpha 3\gamma} = 0\,,$$
$$\Sigma_{\alpha 33} = -\rho c_T^2 V_{33.\alpha}\,, \qquad \Sigma_{33\alpha} = -\rho(c_L^2 - 2c_T^2)V_{\alpha\beta.\beta} \qquad (1.51)$$
$$\Sigma_{333} = 0\,,$$

$$S_{\alpha\beta} = n_\gamma(y)\Sigma_{\gamma\alpha\beta}\,, \qquad S_{3\alpha} = S_{\alpha 3} = 0\,, \qquad S_{33} = n_\alpha(y)\Sigma_{\alpha 33}\,. \qquad (1.52)$$

2. Representation theorems

2.1. Betti's reciprocal theorem

Let $\boldsymbol{u}$ and $\boldsymbol{v}$ be two single-valued, twice continuously differentiable vector fields defined in a domain D bounded by a regular surface ∂D in the sense of Kellogg [55]. Then we have for the operator defined in Eqs. (1.8a) and (1.24),

$$\int_D [\boldsymbol{v}\cdot\mathbf{E}\boldsymbol{u} - \mathbf{E}^*\boldsymbol{v}\cdot\boldsymbol{u}]\,\mathrm{d}V = \int_{\partial D}[\boldsymbol{v}\cdot\overset{n}{\mathbf{T}}\boldsymbol{u} - \overset{n}{\mathbf{T}}{}^*\boldsymbol{v}\cdot\boldsymbol{u}]\,\mathrm{d}S\,, \qquad (2.1a)$$

where $\mathbf{E}^*$ and $\mathbf{T}^*$ are adjoint of $\mathbf{E}$ and $\mathbf{T}$, respectively, and moreover $\mathbf{E}^* = \mathbf{E}$, $\mathbf{T}^* = \mathbf{T}$ in elasticity. This identity is known as Green's second formula.

Using the indicial notation, the general form is given as

$$\int_D \{v_i\partial_j(C_{ijkl}\partial_l u_k) - u_k\partial_l(\partial_j v_i C_{ijkl})\}\,\mathrm{d}V$$
$$= \int_{\partial D}\{n_j(v_i C_{ijkl}\partial_l u_k) - n_l(\partial_j v_i C_{ijkl}u_k)\}\,\mathrm{d}S\,. \qquad (2.1b)$$

Since $\boldsymbol{u}$, $\boldsymbol{v}$ satisfy Eq. (1.8a), i.e.

$$\mathbf{E}\boldsymbol{u} = \rho\ddot{\boldsymbol{u}} - \rho\boldsymbol{b}\,, \qquad \mathbf{E}\boldsymbol{v} = \rho\ddot{\boldsymbol{v}} - \rho\boldsymbol{f}$$

we also have

$$\int_D \{\boldsymbol{v}\cdot(\ddot{\boldsymbol{u}} - \boldsymbol{b}) - (\ddot{\boldsymbol{v}} - \boldsymbol{f})\cdot\boldsymbol{u}\}\rho\,\mathrm{d}V = \int_{\partial D} (\boldsymbol{v}\cdot\overset{n}{\mathbf{T}}\boldsymbol{u} - \overset{n}{\mathbf{T}}\boldsymbol{v}\cdot\boldsymbol{u})\,\mathrm{d}S \tag{2.2a}$$

This is known as Betti's reciprocal relation.

Betti's relation does not involve initial values of $\boldsymbol{u}$ and $\boldsymbol{v}$. However, it is also valid even if $\boldsymbol{u}$ and $\boldsymbol{v}$ are respectively evaluated at different times. If we evaluate $\boldsymbol{u}$ at $t_1 = \tau$, and $\boldsymbol{v}$ at $t_2 = t - \tau$ and integrate over the temporal range 0 to t, we have

$$\begin{aligned}
&\int_0^t \{\ddot{\boldsymbol{u}}(\boldsymbol{x}, \tau)\cdot\boldsymbol{v}(\boldsymbol{x}, t-\tau) - \boldsymbol{u}(\boldsymbol{x}, \tau)\cdot\ddot{\boldsymbol{v}}(\boldsymbol{x}, t-\tau)\}\rho\,\mathrm{d}\tau \\
&\quad = \rho\int_0^t \frac{\partial}{\partial\tau}\left\{\dot{\boldsymbol{u}}(\boldsymbol{x}, \tau)\cdot\boldsymbol{v}(\boldsymbol{x}, t-\tau) + \boldsymbol{u}(\boldsymbol{x}, \tau)\cdot\dot{\boldsymbol{v}}(\boldsymbol{x}, t-\tau)\right\}\mathrm{d}\tau \\
&\quad = \rho\{\dot{\boldsymbol{u}}(\boldsymbol{x}, \tau)\cdot\boldsymbol{v}(\boldsymbol{x}, t-\tau) + \boldsymbol{u}(\boldsymbol{x}, \tau)\cdot\dot{\boldsymbol{v}}(\boldsymbol{x}, t-\tau)\}\big|_0^t\,.
\end{aligned}$$

Denoting here the convolution integral by

$$\boldsymbol{a}(\boldsymbol{x}, t) * \boldsymbol{b}(\boldsymbol{x}, t) = \begin{cases} \displaystyle\int_0^t \boldsymbol{a}(\boldsymbol{x}, \tau)\cdot\boldsymbol{b}(\boldsymbol{x}, t-\tau)\,\mathrm{d}\tau\,, & t > 0 \\ 0\,, & t \leq 0 \end{cases} \tag{2.3}$$

we have, from the Betti relation (2.2a)

$$\int_D \{(\ddot{\boldsymbol{u}} - \boldsymbol{b}) * \boldsymbol{v} - \boldsymbol{u} * (\ddot{\boldsymbol{v}} - \boldsymbol{f})\}\rho\,\mathrm{d}V = \int_{\partial D} (\overset{n}{\mathbf{T}}\boldsymbol{u} * \boldsymbol{v} - \boldsymbol{u} * \overset{n}{\mathbf{T}}\boldsymbol{v})\,\mathrm{d}S \tag{2.2b}$$

and hence [44]

$$\begin{aligned}
&\int_D \boldsymbol{u} * \boldsymbol{f}\rho\,\mathrm{d}V + \int_{\partial D} \boldsymbol{u} * \overset{n}{\mathbf{T}}\boldsymbol{v}\,\mathrm{d}S + \int_D \{\dot{\boldsymbol{u}}(\boldsymbol{x}, t)\cdot\boldsymbol{v}_0(\boldsymbol{x}) + \boldsymbol{u}(\boldsymbol{x}, t)\cdot\dot{\boldsymbol{v}}_0(\boldsymbol{x})\}\rho\,\mathrm{d}V \\
&\quad = \int_D \boldsymbol{b} * \boldsymbol{v}\rho\,\mathrm{d}V + \int_{\partial D} \mathbf{T}_n\boldsymbol{u} * \boldsymbol{v}\,\mathrm{d}S + \int_D \{\dot{v}(\boldsymbol{x}, t)\cdot\boldsymbol{u}_0(\boldsymbol{x}) + \boldsymbol{v}(\boldsymbol{x}, t)\cdot\dot{\boldsymbol{u}}_0(\boldsymbol{x})\}\rho\,\mathrm{d}V\,,
\end{aligned} \tag{2.4}$$

where $\boldsymbol{u}_0$, $\boldsymbol{v}_0$, $\dot{\boldsymbol{u}}_0$, $\dot{\boldsymbol{v}}_0$ are initial displacements and velocities, respectively.

In the special case of zero initial conditions, the third integral of both sides of Eq. (2.4) disappears. Furthermore, when the boundary conditions are homogeneous, we have

$$\int_D \boldsymbol{u} * \boldsymbol{f}\rho \, \mathrm{d}V = \int_D \boldsymbol{b} * \boldsymbol{v}\rho \, \mathrm{d}V \, . \tag{2.5}$$

In the Fourier-transformed domain, we have

$$\int_D [\hat{\boldsymbol{v}} \cdot \mathbf{L}\hat{\boldsymbol{u}} - \mathbf{L}^* \hat{\boldsymbol{v}} \cdot \hat{\boldsymbol{u}}] \, \mathrm{d}V = \int_{\partial D} [\hat{\boldsymbol{v}} \cdot \overset{n}{\mathbf{T}}\hat{\boldsymbol{u}} - \overset{n}{\mathbf{T}}{}^* \hat{\boldsymbol{v}} \cdot \hat{\boldsymbol{u}}] \, \mathrm{d}S \, , \tag{2.6}$$

where $\mathbf{L}\hat{\boldsymbol{u}}$ is defined by Eq. (1.30). This counterpart of Eq. (2.2b) is the second Green's identity and also the Fourier transform of the general Betti relation (2.2a).

2.2. *Representation theorems*

Making use of the Green's identity or the general Betti relation and the properties of the fundamental solution, we can have integral representations of the elastodynamic solution.

If we use the fundamental solution $\hat{U}_{ik}$ in place of v_i in Eq. (2.6), for example, and consider the property of $\hat{U}_{ik}$, that is

$$L^*_{ji} \hat{U}_{ik} = -\delta(\boldsymbol{x} - \boldsymbol{y}) \delta_{jk} \tag{2.7}$$

and hence

$$\int_D L^*_{ji} \hat{U}_{ik} \hat{u}_j \, \mathrm{d}V(\boldsymbol{x}) = -\int_D \hat{u}_j \delta(\boldsymbol{x} - \boldsymbol{y}) \delta_{jk} \, \mathrm{d}V(\boldsymbol{x}) = \begin{cases} -\hat{u}_k(\boldsymbol{y}) \, , & \boldsymbol{y} \in D \\ 0 \, , & \boldsymbol{y} \in D_\mathrm{c} \, , \end{cases} \tag{2.8}$$

where D_c is the complement of D, we have

$$\varepsilon(\boldsymbol{y}) \hat{u}_k(\boldsymbol{y}; \omega) = \int_{\partial D} [\hat{U}_{ik}(\boldsymbol{x}, \boldsymbol{y}; \omega) \hat{t}_i(\boldsymbol{x}; \omega) - \hat{W}_{ik}(\boldsymbol{x}, \boldsymbol{y}; \omega) \hat{u}_i(\boldsymbol{x}; \omega)] \, \mathrm{d}S(\boldsymbol{x})$$
$$+ \int_D \rho \hat{b}_i(\boldsymbol{x}; \omega) \hat{U}_{ik}(\boldsymbol{x}, \boldsymbol{y}; \omega) \, \mathrm{d}V(\boldsymbol{x}) \, , \tag{2.9}$$

where

$$\varepsilon(\boldsymbol{y}) = \begin{cases} 1 \, , & \boldsymbol{y} \in D \, , \\ 0 \, , & \boldsymbol{y} \in D_\mathrm{c} \, , \end{cases}$$

and

$$\hat{t}_i(\boldsymbol{x};\omega) = \overset{n}{T}_{ij}\hat{u}_j(\boldsymbol{x};\omega)\,, \qquad \hat{W}_{ik}(\boldsymbol{x},\boldsymbol{y};\omega) = \overset{n}{T}_{ij}\hat{U}_{jk}(\boldsymbol{x},\boldsymbol{y};\omega)\,.$$

This expression is called the representation theorem or Somigliana formula. Sometimes it may be more convenient to change integration variable from $\boldsymbol{x}$ to $\boldsymbol{y}$ and to write down Eq. (2.9) using the spatial reciprocity (see Eq. (1.37b))

$$\varepsilon(\boldsymbol{x})\hat{u}_k(\boldsymbol{x};\omega) = \int_{\partial D} [\hat{U}_{ki}(\boldsymbol{x},\boldsymbol{y};\omega)\hat{t}_i(\boldsymbol{y};\omega) - \hat{W}_{ki}(\boldsymbol{x},\boldsymbol{y};\omega)\hat{u}_i(\boldsymbol{y};\omega)]\,\mathrm{d}S(\boldsymbol{y})$$
$$+ \int_D \rho\hat{b}_i(\boldsymbol{y};\omega)\hat{U}_{ki}(\boldsymbol{x},\boldsymbol{y};\omega)\,\mathrm{d}V(\boldsymbol{y})\,, \qquad (2.10)$$

where

$$\varepsilon(\boldsymbol{x}) = \begin{cases} 1\,, & \boldsymbol{x}\in D\,,\\ 0\,, & \boldsymbol{x}\in D_{\mathrm{c}}\,,\end{cases}$$

$$\hat{t}_i(\boldsymbol{y};\omega) = \overset{n_y}{T}_{ij}\hat{u}_j(\boldsymbol{y};\omega)\,, \qquad \hat{W}_{ki}(\boldsymbol{x},\boldsymbol{y};\omega) = \overset{n_y}{T}_{ij}\hat{U}_{kj}(\boldsymbol{x},\boldsymbol{y};\omega)$$

and note that $\overset{n_y}{\boldsymbol{T}}$ operates at $\boldsymbol{y}\in\partial D$ on the boundary. This expression means that the displacement $\hat{\boldsymbol{u}}(\boldsymbol{x};\omega)$ at any point $\boldsymbol{x}$ in D can be evaluated whenever boundary displacement $\hat{\boldsymbol{u}}(\boldsymbol{y};\omega)$ on ∂D_1, and traction $\hat{\boldsymbol{t}}(\boldsymbol{y};\omega)$ on ∂D_2 are given. However, part of them, i.e. $\hat{\boldsymbol{u}}(\boldsymbol{y};\omega)$ on ∂D_2 and $\hat{\boldsymbol{t}}(\boldsymbol{y};\omega)$ on ∂D_1 are not prescribed as boundary conditions, and they must be determined beforehand in order to make use of Eq. (2.10). It is also noted here that boundary displacements and tractions must satisfy the constraint equation, that is the right-hand side of Eq. (2.10) must be zero for $\boldsymbol{x}\in D_{\mathrm{c}}$.

The time-domain representation theorem is similarly obtained from Eq. (2.4) by substituting $\delta_{ik}\delta(\boldsymbol{x}-\boldsymbol{y})\delta(t)$ for $\rho f_i(\boldsymbol{x},t)$ and $U_{ik}(\boldsymbol{x},t;\boldsymbol{y},0)$ for $v_i(\boldsymbol{x},t)$ and reads

$$\varepsilon(\boldsymbol{y})u_k(\boldsymbol{y},t) = \int_{\partial D} [t_i(\boldsymbol{x},t) * U_{ik}(\boldsymbol{x},t;\boldsymbol{y},0)$$
$$- u_i(\boldsymbol{x},t) * W_{ik}(\boldsymbol{x},t;\boldsymbol{y},0)]\,\mathrm{d}S(\boldsymbol{x})$$
$$+ \int_D \rho b_i(\boldsymbol{x},t) * U_{ik}(\boldsymbol{x},t;\boldsymbol{y},0)\,\mathrm{d}V(\boldsymbol{x})$$
$$+ \int_D \rho[\dot{U}_{ik}(\boldsymbol{x},t;\boldsymbol{y},0)u_{0i}(\boldsymbol{x}) + U_{ik}(\boldsymbol{x},t;\boldsymbol{y},0)\dot{u}_{0i}(\boldsymbol{x})]\,\mathrm{d}V(\boldsymbol{x})\,,$$
$$(2.11)$$

where

$$\varepsilon(\boldsymbol{y}) = \begin{cases} 1\,, & \boldsymbol{y} \in D \\ 0\,, & \boldsymbol{y} \in D_c \end{cases}$$

and

$$t_i(\boldsymbol{x}, t) = \overset{n}{T}_{ij} u_j(\boldsymbol{x}, t)\,, \qquad W_{ik}(\boldsymbol{x}, t; \boldsymbol{y}, 0) = \overset{n}{T}_{ij} U_{jk}(\boldsymbol{x}, t; \boldsymbol{y}, 0)\,.$$

It is helpful to rewrite Eq. (2.11) by interchanging the symbols $\boldsymbol{x}$ and $\boldsymbol{y}$, and considering the spatial reciprocity as

$$\begin{aligned} \varepsilon(\boldsymbol{x}) u_k(\boldsymbol{x}, t) = & \int_{\partial D} [t_i(\boldsymbol{y}, t) * U_{ki}(\boldsymbol{x}, t; \boldsymbol{y}, 0) - u_i(\boldsymbol{y}, t) * W_{ki}(\boldsymbol{x}, t; \boldsymbol{y}, 0)]\, \mathrm{d}S(\boldsymbol{y}) \\ & + \int_D \rho b_i(\boldsymbol{y}, t) * U_{ki}(\boldsymbol{x}, t; \boldsymbol{y}, 0)\, \mathrm{d}V(\boldsymbol{y}) \\ & + \int_D \rho[\dot{U}_{ki}(\boldsymbol{x}, t; \boldsymbol{y}, 0) u_{0i}(\boldsymbol{y}) + U_{ki}(\boldsymbol{x}, t; \boldsymbol{y}, 0) \dot{u}_{0i}(\boldsymbol{y})]\, \mathrm{d}V(\boldsymbol{y})\,, \end{aligned} \tag{2.12}$$

where

$$\varepsilon(\boldsymbol{x}) = \begin{cases} 1\,, & \boldsymbol{x} \in D \\ 0\,, & \boldsymbol{x} \in D_c \end{cases}$$

and

$$t_i(\boldsymbol{y}, t) = \overset{n_y}{T}_{ij} u_j(\boldsymbol{y}, t)\,, \qquad W_{ki}(\boldsymbol{x}, t; \boldsymbol{y}, 0) = \overset{n_y}{T}_{ij} U_{kj}(\boldsymbol{x}, t; \boldsymbol{y}, 0)\,.$$

Similar expressions are easily deduced for the two-dimensional case.
In the time–space domain, for example, we have

$$\begin{aligned} \varepsilon(\boldsymbol{x}) u_\alpha(\boldsymbol{x}, t) = & \int_{\partial D} [t_\beta(\boldsymbol{y}, t) * V_{\alpha\beta}(\boldsymbol{x}, t; \boldsymbol{y}, 0) \\ & - u_\beta(\boldsymbol{y}, t) * S_{\alpha\beta}(\boldsymbol{x}, t; \boldsymbol{y}, 0)]\, \mathrm{d}C(\boldsymbol{y}) \\ & + \int_D \rho b_\beta(\boldsymbol{y}, t) * V_{\alpha\beta}(\boldsymbol{x}, t; \boldsymbol{y}, 0)\, \mathrm{d}S(\boldsymbol{y}) \\ & + \int_D \rho[\dot{V}_{\alpha\beta}(\boldsymbol{x}, t; \boldsymbol{y}, 0) u_{0\beta}(\boldsymbol{y}) + V_{\alpha\beta}(\boldsymbol{x}, t; \boldsymbol{y}, 0) \dot{u}_{0\beta}(\boldsymbol{y})]\, \mathrm{d}S(\boldsymbol{y})\,, \end{aligned} \tag{2.13}$$

$$\varepsilon(\boldsymbol{x})u_3(\boldsymbol{x}, t) = \int_{\partial D} [t_3(\boldsymbol{y}, t) * V_{33}(\boldsymbol{x}, t; \boldsymbol{y}, 0) - u_3(\boldsymbol{y}, t) * S_{33}(\boldsymbol{x}, t; \boldsymbol{y}, 0)]\, \mathrm{d}C(\boldsymbol{y}) + \int_D \rho b_3(\boldsymbol{y}, t) * V_{33}(\boldsymbol{x}, t; \boldsymbol{y}, 0)\, \mathrm{d}S(\boldsymbol{y}) + \int_D \rho[\dot{V}_{33}(\boldsymbol{x}, t; \boldsymbol{y}, 0)u_{03}(\boldsymbol{y}) + V_{33}(\boldsymbol{x}, t; \boldsymbol{y}, 0)\dot{u}_{03}(\boldsymbol{y})]\, \mathrm{d}S(\boldsymbol{y}) , \tag{2.14}$$

where

$$\varepsilon(x) = \begin{cases} 1 , & \boldsymbol{x} \in D \\ 0 , & \boldsymbol{x} \in D_c \end{cases}$$

and

$$t_\alpha(\boldsymbol{y}, t) = \overset{n_y}{T}_{\alpha\beta} u_\beta(\boldsymbol{y}, t) , \qquad S_{\alpha\beta}(\boldsymbol{x}, t; \boldsymbol{y}, 0) = \overset{n_y}{T}_{\beta\gamma} U_{\alpha\gamma}(\boldsymbol{x}, t; \boldsymbol{y}, 0) ,$$
$$\boldsymbol{x} = (x_1, x_2) , \qquad \boldsymbol{y} = (y_1, y_2) .$$

Stresses are obtained by differentiating $\boldsymbol{u}$ with respect to $\boldsymbol{x}$ and substituting into Eq. (1.6b). For instance, from Eq. (2.10), we have

$$\hat{\sigma}_{kl}(\boldsymbol{x}; \omega) = \int_{\partial D} [\hat{t}_i(\boldsymbol{y}; \omega)\bar{\hat{U}}_{kil}(\boldsymbol{x}, \boldsymbol{y}; \omega) - \hat{u}_i(\boldsymbol{y}; \omega)\bar{\hat{W}}_{kil}(\boldsymbol{x}, \boldsymbol{y}; \omega)\, \mathrm{d}S(\boldsymbol{y}) + \int_D \rho b_i(\boldsymbol{y}; \omega)\bar{\hat{U}}_{kil}(\boldsymbol{x}, \boldsymbol{y}; \omega)\, \mathrm{d}V(\boldsymbol{y}) , \tag{2.15}$$

where

$$\{\bar{\hat{U}}_{kil}, \bar{\hat{W}}_{kil}\} = \rho(c_\mathrm{L}^2 - 2c_\mathrm{T}^2)\{\hat{U}_{kmm}, \hat{W}_{kmm}\} + \rho c_\mathrm{T}^2\{\tilde{U}_{kil} + \hat{U}_{kli}, \hat{W}_{kil} + \hat{W}_{kli}\} ,$$

$$\hat{U}_{kil} = \hat{U}_{ki,l} , \qquad \hat{W}_{kil} = \hat{W}_{ki,l} .$$

Similar expressions are easily obtained for the time–space domain formulations.

3. Integral equations

3.1. Integral equations in the Fourier-transformed domain

We know two ways of approach to time-dependent problems such as those of elastodynamics. One is to solve the problems directly in the time–space domain and the other is to treat them in conjunction with integral transforms. In what follows, we formulate integral equations of elastodynamic problems first in the Fourier-transformed domain and then in the time–space domain.

Knowing that

$$\hat{U}(\boldsymbol{x}, \boldsymbol{y}; \omega) = O(1/r)\,, \qquad \hat{W}(\boldsymbol{x}, \boldsymbol{y}; \omega) = O(1/r^2) \qquad \text{for 3D}\,,$$

and

$$\hat{U}(\boldsymbol{x}, \boldsymbol{y}; \omega) = O(\ln 1/r)\,, \qquad \hat{W}(\boldsymbol{x}, \boldsymbol{y}; \omega) = O(1/r) \qquad \text{for 2D}$$

as $r = |\boldsymbol{x} - \boldsymbol{y}| \to 0$, we can evaluate the limit value of the surface integral on the part of a sphere (or circle) $s(\boldsymbol{x}_0; \epsilon)$ with radius ϵ and centered at $\boldsymbol{x}_0$ included in D, when $\epsilon \to 0$ as shown in Fig. 1 and obtain

$$\begin{aligned} &\lim_{\epsilon\to 0} \int_{s(\boldsymbol{x}_0;\,\epsilon)\subset D} \hat{U}_{ki}(\boldsymbol{x}_0, \boldsymbol{y}; \omega) h_i(\boldsymbol{y})\, \mathrm{d}S(\boldsymbol{y}) = 0 \\ &\lim_{\epsilon\to 0} \int_{s(\boldsymbol{x}_0;\,\epsilon)\subset D} \hat{W}_{ki}(\boldsymbol{x}_0, \boldsymbol{y}; \omega) g_i(\boldsymbol{y})\, \mathrm{d}S(\boldsymbol{y}) = C^{\mathrm{c}}_{ki}(\boldsymbol{x}_0) g_i(\boldsymbol{x}_0) \end{aligned} \tag{3.1}$$

when $\boldsymbol{h}$ and $\boldsymbol{g}$ are regular in D. In these expressions, $C^{\mathrm{c}}_{ki}(\boldsymbol{x}_0)$ is δ_{ki} for $\boldsymbol{x}_0 \in D$, and $\delta_{ki}/2$ for $\boldsymbol{x}_0 \in \partial D$ when the boundary is smooth in the neighborhood of $\boldsymbol{x}_0$. However, at a corner point on ∂D, $C^{\mathrm{c}}_{ki}(\boldsymbol{x}_0)$ must be evaluated with complicated computations. As will be seen later (see section 4.3.2), since the order of singularities of $\hat{\boldsymbol{U}}$ and $\hat{\boldsymbol{W}}$ as $r \to 0$ is the same as that of the fundamental solutions for elastostatics, we can make use of C^{c}_{ki} for elastostatics [46, 106, 107].

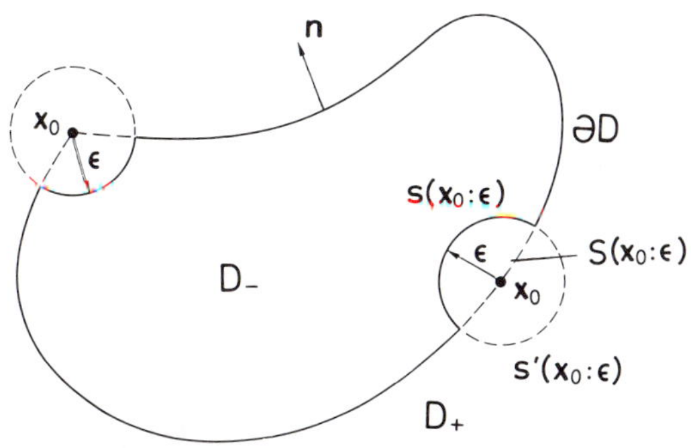

Fig. 1. Domain, boundary, normal and boundary point.

Taking the above results into consideration, we have the following integral equation as $x \in D_c \to x_0 \in \partial D$

$$C^c_{ki}(x_0)\hat{u}_i(x_0;\omega) = \int_{\partial D} \hat{U}_{ki}(x_0, y;\omega)\hat{t}_i(y;\omega)\,dS(y) - \fint_{\partial D} \hat{W}_{ki}(x_0, y;\omega)\hat{u}_i(y;\omega)\,dS(y) + \int_D \rho\hat{U}_{ki}(x_0, y;\omega)b_i(y;\omega)\,dV(y)\,, \tag{3.2}$$

where $\fint(\ )\,dS(y)$ implies the principal-value integral in Cauchy's sense.

For the two-dimensional case, we have the same expressions with the two-dimensional kernels.

It is noted that the body force is often disregarded in elastodynamics and that Eq. (3.2) then reduces to a boundary integral equation.

For unbounded exterior problems, integral equations are formulated for the scattered field $\hat{u}^S(x;\omega) = \hat{u}(x;\omega) - \hat{u}^I(x;\omega)$, where $\hat{u}^I(x;\omega)$ is the incident wave field.

However, the alternative formulation for the total field $\hat{u}(x;\omega)$ is also possible [64, 66]. Since for the scattered field Eq. (2.10) is valid for the exterior domain D_+ of the scatterer boundary ∂D

$$\fint_{\partial D} \hat{W}(x, y;\omega)\cdot\hat{u}^S(y;\omega)\,dS(y) - \int_{\partial D} \hat{U}(x, y;\omega)\cdot\hat{t}^S(y;\omega)\,dS(y) = \begin{cases} \hat{u}^S(x;\omega)\,, & x \in D_+ \\ \mathbf{0}\,, & x \in D_-\,, \end{cases} \tag{3.3}$$

where the normal on ∂D is taken positive as pointing into the exterior domain, and the incident wave field Eq. (2.10) is again true for the interior domain D_- of the scatterer

$$\int_{\partial D} \hat{U}(x, y;\omega)\cdot\hat{t}^I(y;\omega)\,dS(y) - \fint_{\partial D} \hat{W}(x, y;\omega)\cdot\hat{u}^I(y;\omega)\,dS(y) = \begin{cases} \hat{u}^I(x;\omega)\,, & x \in D_- \\ 0\,, & x \in D_+\,, \end{cases} \tag{3.4}$$

we have the following integral representation by substracting them:

$$\fint_{\partial D} \hat{W}(x, y;\omega)\cdot\hat{u}(y;\omega)\,dS(y) - \int_{\partial D} \hat{U}(x, y;\omega)\cdot\hat{t}(y;\omega)\,dS(y) = \begin{cases} \hat{u}(x;\omega) - \hat{u}^I(x;\omega)\,, & x \in D_+ \\ -\hat{u}^I(x;\omega)\,, & x \in D_-\,, \end{cases} \tag{3.5}$$

Hence we have the following boundary integral equation, noting that the incident wave field is regular:

$$\fint_{\partial D} \hat{\boldsymbol{W}}(\boldsymbol{x}, \boldsymbol{y}; \omega) \cdot \hat{\boldsymbol{u}}(\boldsymbol{y}; \omega)\, \mathrm{d}S(\boldsymbol{y}) - \int_{\partial D} \hat{\boldsymbol{U}}(\boldsymbol{x}, \boldsymbol{y}; \omega) \cdot \hat{\boldsymbol{t}}(\boldsymbol{y}; \omega)\, \mathrm{d}S(\boldsymbol{y}) + \hat{\boldsymbol{u}}^{\mathrm{I}}(\boldsymbol{x}_0; \omega)$$

$$= \boldsymbol{C}^{\mathrm{c}}(\boldsymbol{x}_0) \cdot \hat{\boldsymbol{u}}(\boldsymbol{x}_0; \omega)\,, \qquad \boldsymbol{x}_0 \in \partial D\,. \tag{3.6}$$

The boundary integral equation (3.6) thus obtained is easily applied to boundary-value-problems for multiply connected regions and for multi-foldedly connected regions as it is the case of a lined tunnel in an infinite medium. In the latter case, boundary integral equations are derived separately for each subdomain and then they are combined using the continuity conditions on the common interface of subdomains.

Sometimes, for simplicity, the Born approximation is used to find the far-field behavior in the exterior domain. This approximation is derived by substituting the incident field for the total field in the integrals on the left-hand side of Eq. (3.5) [16–18, 47].

3.2. Integral equations in time-space domain

Taking the limit of $\boldsymbol{x} \in D_{\mathrm{c}} \rightarrow \boldsymbol{x}_0 \in \partial D$ in Eqs. (2.12)–(2.14), we have integral equations expressed in the time–space domain as

$$C^{\mathrm{c}}_{ki}(\boldsymbol{x}_0) u_i(\boldsymbol{x}_0, t) = \int_{\partial D} U_{ki}(\boldsymbol{x}_0, t; \boldsymbol{y}, 0) * t_i(\boldsymbol{y}, t)\, \mathrm{d}S(\boldsymbol{y})$$

$$- \fint_{\partial D} W_{ki}(\boldsymbol{x}_0, t; \boldsymbol{y}, 0) * u_i(\boldsymbol{y}, t)\, \mathrm{d}S(\boldsymbol{y}) + \int_D \rho U_{ki}(\boldsymbol{x}_0, t; \boldsymbol{y}, 0) * b_i(\boldsymbol{y}, t)\, \mathrm{d}V(\boldsymbol{y})$$

$$+ \int_D \rho[\dot{U}_{ki}(\boldsymbol{x}_0, t; \boldsymbol{y}, 0) u_{0i}(\boldsymbol{y}) + U_{ki}(\boldsymbol{x}_0, t; \boldsymbol{y}, 0) \dot{u}_{0i}(\boldsymbol{y})]\, \mathrm{d}V(\boldsymbol{y})\,, \tag{3.7}$$

$$C^{\mathrm{c}}_{\alpha\beta}(\boldsymbol{x}_0) u_\beta(\boldsymbol{x}_0, t) = \int_{\partial D} V_{\alpha\beta}(\boldsymbol{x}_0, t; \boldsymbol{y}, 0) * t_\beta(\boldsymbol{y}, t)\, \mathrm{d}C(\boldsymbol{y})$$

$$- \fint_{\partial D} S_{\alpha\beta}(\boldsymbol{x}_0, t; \boldsymbol{y}, 0) * u_\beta(\boldsymbol{y}, t)\, \mathrm{d}C(\boldsymbol{y}) + \int_D \rho V_{\alpha\beta}(\boldsymbol{x}_0, t; \boldsymbol{y}, 0) * b_\beta(\boldsymbol{y}, t)\, \mathrm{d}S(\boldsymbol{y})$$

$$+ \int_D \rho[\dot{V}_{\alpha\beta}(\boldsymbol{x}_0, t; \boldsymbol{y}, 0) u_{0\beta}(\boldsymbol{y}) + V_{\alpha\beta}(\boldsymbol{x}_0, t; \boldsymbol{y}, 0) \dot{u}_{0\beta}(\boldsymbol{y})]\, \mathrm{d}S(\boldsymbol{y})\,, \tag{3.8}$$

$$C^{c}(\boldsymbol{x}_0)u_3(\boldsymbol{x}_0, t) = \int_{\partial D} V_{33}(\boldsymbol{x}_0, t; \boldsymbol{y}, 0) * t_3(\boldsymbol{y}, t)\, dC(\boldsymbol{y})$$
$$- \fint_{\partial D} S_{33}(\boldsymbol{x}_0, t; \boldsymbol{y}, 0) * u_3(\boldsymbol{y}, t)\, dC(\boldsymbol{y}) + \int_{D} \rho V_{33}(\boldsymbol{x}_0, t; \boldsymbol{y}, 0) * b_3(\boldsymbol{y}, t)\, dS(\boldsymbol{y})$$
$$+ \int_{D} \rho[\dot{V}_{33}(\boldsymbol{x}_0, t; \boldsymbol{y}, 0)u_{03}(\boldsymbol{y}) + V_{33}(\boldsymbol{x}_0, t; \boldsymbol{y}, 0)\dot{u}_{03}(\boldsymbol{y})]\, dS(\boldsymbol{y}) , \tag{3.9}$$

where

$$C^{c}_{ki}(\boldsymbol{x}_0)g_i(\boldsymbol{x}_0, t) = \lim_{\epsilon \to 0} \int_{s(\boldsymbol{x}_0;\epsilon)\subset D} W_{ki}(\boldsymbol{x}_0, t; \boldsymbol{y}, 0) * g_i(\boldsymbol{y}, t)\, dS(\boldsymbol{y}) \tag{3.10}$$

is similar to the previously defined Eq. $(3.1)_2$ and $\fint (\;)\, dS(\boldsymbol{y})$ implies improper integral [130, 131].

3.3. *Remarks on formulations*

3.3.1. *The indirect formulation*

In section 3.1 we have formulated integral equations by the use of the Green's identity. In the formulation, unknowns are displacements and tractions on the boundary. Hence, the formulation is called "direct", in the sense that unknowns are direct physical quantities. Alternatively, we can formulate integral equations with the aid of potential theory [48, 70, 71, 98]. In this formulation, unknowns are densities of the simple and double layer potentials, and physical quantities such as displacements and tractions can be obtained indirectly by the use of predetermined densities. Therefore, the method is sometimes called "indirect".

We denote the simple and double layer potentials by

$$S_j(\boldsymbol{x}) = \int_{\partial D} U_{jk}(\boldsymbol{x}, \boldsymbol{y})\phi_k(\boldsymbol{y})\, dS(\boldsymbol{y}) , \tag{3.11}$$

$$D_j(\boldsymbol{x}) = \int_{\partial D} U_{jk.l}(\boldsymbol{x}, \boldsymbol{y})\psi_{kl}(\boldsymbol{y})\, dS(\boldsymbol{y}) , \tag{3.12}$$

where ϕ_k and ψ_{kl} are densities distributed over the boundary ∂D. For later use, we here summarize properties of the potentials for a sufficiently smooth boundary and for continuous densities [48, 55, 57, 70, 71].

For the kernel function (fundamental solution) for elastostatics $\mathring{\boldsymbol{U}}(\boldsymbol{x}, \boldsymbol{y})$ defined by

$$E_{ij}(\partial)\mathring{U}_{jk}(\boldsymbol{x}, \boldsymbol{y}) \equiv C_{ijmn}\mathring{U}_{jk,mn}(\boldsymbol{x}, \boldsymbol{y}) = -\delta_{ik}\delta(\boldsymbol{x} - \boldsymbol{y}) , \tag{3.13}$$

(see section 4.3.2 for $\mathring{U}_{jk}(\boldsymbol{x}, \boldsymbol{y})$ for the isotropic case), the potentials have the following properties [70, 90]:

(i) Simple layer potential is continuous throughout $\mathbb{R}^N$ and

$$(\partial_i S_j)^{\pm}(\boldsymbol{x}) = \mp \tfrac{1}{2} n_i (E_{jk})^{-1}(\boldsymbol{n}) \phi_k(\boldsymbol{x}) + \int_{\partial D} \partial_i \mathring{U}_{jk}(\boldsymbol{x}, \boldsymbol{y}) \phi_k(\boldsymbol{y}) \, \mathrm{d}S(\boldsymbol{y}) , \qquad \boldsymbol{x} \in D$$

$$E_{ij}^{-1}(\boldsymbol{n}) = \frac{1}{\mu}\left(\delta_{ij} - \frac{\lambda + \mu}{\lambda + 2\mu} n_i n_j\right) \tag{3.14}$$

for the isotropic case on ∂D, where $E_{ij}^{-1}(\boldsymbol{n})$ is the inverse operator of E_{ij}. In particular,

$$(\overset{n}{\mathbf{T}}\boldsymbol{S})^{\pm}(\boldsymbol{x}) = \mp \tfrac{1}{2} \, \boldsymbol{\phi}(\boldsymbol{x}) \mathbf{1}(\boldsymbol{x}) + \int_{\partial D} \overset{n_x}{\mathbf{T}} \mathring{\boldsymbol{U}}(\boldsymbol{x}, \boldsymbol{y}) \cdot \boldsymbol{\phi}(\boldsymbol{y}) \, \mathrm{d}S(\boldsymbol{y}) . \tag{3.15}$$

(ii) Double layer potential satisfies

$$D_j^{\pm}(\boldsymbol{x}) = \pm \tfrac{1}{2} \, n_l E_{jk}^{-1}(\boldsymbol{n}) \psi_{kl}(\boldsymbol{x}) + \int_{\partial D} \mathring{U}_{jk.l}(\boldsymbol{x}, \boldsymbol{y}) \psi_{kl}(\boldsymbol{y}) \, \mathrm{d}S(\boldsymbol{y}) , \tag{3.16}$$

where $(\ )_{.l}$ implies differentiation with respect to $\boldsymbol{y}$, and its derivative is continuous throughout $\mathbb{R}^N$.

Making use of these properties and noting that the order of the singularities of the fundamental solutions for elastodynamics is the same as that of elastostatics, we can formulate indirect boundary integral equations with ease. As an example, we formulate the ones for the second exterior problem. If we assume that the displacement can be expressed in terms of the simple layer potential, we immediately have for a smooth boundary

$$-\tfrac{1}{2} \mathbf{1}(\boldsymbol{x}_0) \cdot \hat{\boldsymbol{\phi}}(\boldsymbol{x}_0; \omega) + \int_{\partial D} \overset{n_x}{\mathbf{T}} \hat{\boldsymbol{U}}(\boldsymbol{x}_0, \boldsymbol{y}; \omega) \cdot \hat{\boldsymbol{\phi}}(\boldsymbol{y}; \omega) \, \mathrm{d}S(\boldsymbol{y}) = \hat{\boldsymbol{t}}(\boldsymbol{x}_0; \omega) . \tag{3.17}$$

If we assume that the displacement can be expressed in terms of the double layer potential, we simply have

$$\hat{\boldsymbol{t}}(\boldsymbol{x}_0; \omega) = \overset{n_x}{\mathbf{T}} \int_{\partial D} \overset{n_y}{\mathbf{T}} \hat{\boldsymbol{U}}(\boldsymbol{x}, \boldsymbol{y}; \omega) \cdot \boldsymbol{\psi}(\boldsymbol{y}; \omega) \, \mathrm{d}S(\boldsymbol{y}) , \tag{3.18}$$

although the kernel of this integral equation of the first kind is strongly singular and the integral may be evaluated by regularizing the kernel. For other forms of indirect formulations, see references [57, 98].

The indirect formulations are often conveniently used for the first and second boundary value problems.

3.3.2. Integral equations for inhomogeneous media [92–95]

If the field is inhomogeneous and the elastic moduli and mass density vary as a function of position, we have to modify the treatment, since we do not know the fundamental solutions for such fields. In what follows, we formulate domain-type integral equations in the Fourier-transformed domain.

In the Fourier-transformed domain, the Navier–Cauchy equation (1.7a) is expressed as:

$$\{C_{ijkl}(\boldsymbol{x})\hat{u}_{k,l}(\boldsymbol{x};\omega)\}_{,j} + \rho(\boldsymbol{x})\hat{b}_i(\boldsymbol{x};\omega) + \rho(\boldsymbol{x})\omega^2\hat{u}_i(\boldsymbol{x};\omega) = 0\,, \tag{3.19}$$

where C_{ijkl} is, in general, the anisotropic elastic tensor.

Here we introduce for later convenience the differentiated volume potential

$$V_j(\boldsymbol{x}) = \int_D \mathring{U}_{jk.l}(\boldsymbol{x},\boldsymbol{y})\nu_{kl}(\boldsymbol{y})\,\mathrm{d}V(\boldsymbol{y})\,. \tag{3.20}$$

This potential is continuous througout $\mathbb{R}^N$ and its derivative can be written as

$$\partial_i V_j(\boldsymbol{x}) = \begin{cases} v_{ijkl}\nu_{kl}(\boldsymbol{x}) + {\displaystyle\fint_D} \partial_i\mathring{U}_{jk.l}(\boldsymbol{x},\boldsymbol{y})\nu_{kl}(\boldsymbol{y})\,\mathrm{d}V(\boldsymbol{y})\,, & \boldsymbol{x}\in D \\ {\displaystyle\int_D} \partial_i V_{jk.l}(\boldsymbol{x},\boldsymbol{y})\nu_{kl}(\boldsymbol{y})\,\mathrm{d}V(\boldsymbol{y})\,, & \boldsymbol{x}\in D_{\mathrm{c}} \end{cases} \tag{3.21}$$

where v_{ijkl} is a tensor defined by

$$\begin{aligned} v_{ijkl} &= \frac{1}{|S^N|}\int_{S^N} \xi_i E_{jk}^{-1}(\boldsymbol{\xi})\xi_l\,\mathrm{d}S(\boldsymbol{\xi}) \\ &= \frac{\{N(\lambda+2\mu)+(\lambda+3\mu)\}\delta_{il}\delta_{jk} - (\lambda+\mu)(\delta_{ij}\delta_{kl}+\delta_{ik}\delta_{jl})}{\mu(\lambda+2\mu)N(N+2)} \end{aligned} \tag{3.22}$$

for the isotropic case, where S^N is the N-dimensional unit sphere and $|S^N|$ is the area of S^N. The derivative shows a discontinuity across ∂D, and

$$\begin{aligned} (\partial_i V_j)^{\pm}(\boldsymbol{x}) &= \tfrac{1}{2}\, v_{ijkl}\nu_{kl}(\boldsymbol{x}) \mp \tfrac{1}{2} n_i E_{jk}^{-1}(\boldsymbol{n}) n_l \nu_{kl}(\boldsymbol{x}) \\ &\quad + \fint_D \partial_i\mathring{U}_{jk.l}(\boldsymbol{x},\boldsymbol{y})\nu_{kl}(\boldsymbol{y})\,\mathrm{d}V(\boldsymbol{y})\,, \qquad \boldsymbol{x}\in\partial D\,, \end{aligned} \tag{3.23}$$

where

$$\int\!\!\!\!-\,(\ )(\boldsymbol{x}, \boldsymbol{y})\,\mathrm{d}V(\boldsymbol{y}) = \lim_{\epsilon \to 0} \int_{D-B(\boldsymbol{x};\,\epsilon)} (\ )(\boldsymbol{x}, \boldsymbol{y})\,\mathrm{d}V(\boldsymbol{y})$$

is defined in the principal-value sense.

If the density ν_{ij} is a constant over D, we have

$$\int_D \mathring{U}_{ij.k}(\boldsymbol{x}, \boldsymbol{y})\nu_{jk}(\boldsymbol{y})\,\mathrm{d}V(\boldsymbol{y}) = \int_{\partial D} \mathring{U}_{ij}(\boldsymbol{x}, \boldsymbol{y})n_k \nu_{jk}(\boldsymbol{y})\,\mathrm{d}S(\boldsymbol{y})\,. \tag{3.24}$$

Now, we are ready to formulate integral equations. Multiplying Eq. (3.19) by $\mathring{U}_{im}(\boldsymbol{x}, \boldsymbol{y})$, and intergrating over the volume domain and utilizing Gauss's divergence theorem and the reciprocity of $\mathring{U}$, i.e. $\mathring{U}_{im}(\boldsymbol{x}, \boldsymbol{y}) = \mathring{U}_{mi}(\boldsymbol{y}, \boldsymbol{x})$, we have the following expression after some manipulation:

$$\begin{aligned}
&\int_{\partial D} \{\mathring{U}_{mi}(\boldsymbol{x}, \boldsymbol{y})t_i(\boldsymbol{y}) - \mathring{W}_{mi}(\boldsymbol{x}, \boldsymbol{y})u_i(\boldsymbol{y})\}\,\mathrm{d}S(\boldsymbol{y}) \\
&\qquad + \int_D \{\mathring{U}_{mi.j}(\boldsymbol{x}, \boldsymbol{y})C_{ijkl}(\boldsymbol{y})\}_{.l}u_k(\boldsymbol{y})\,\mathrm{d}V(\boldsymbol{y}) \\
&\qquad + \int_D \mathring{U}_{mi}(\boldsymbol{x}, \boldsymbol{y})\{\rho(\boldsymbol{y})b_i(\boldsymbol{y}) + \rho(\boldsymbol{y})\omega^2 u_i(\boldsymbol{y})\}\,\mathrm{d}V(\boldsymbol{y}) = 0\,,
\end{aligned} \tag{3.25}$$

where

$$\mathring{W}_{mi}(\boldsymbol{x}, \boldsymbol{y}) \equiv n_l(\boldsymbol{y})C_{kjil}(\boldsymbol{y})\mathring{U}_{mk.j}(\boldsymbol{x}, \boldsymbol{y})$$

and $(\ )_{.j} = \partial(\ )/\partial y_j$, and C_{ijkl} is assumed continuous.

Further, making use of the properties of the aforementioned potentials, we have

$$\begin{aligned}
\varepsilon_{mk}(\boldsymbol{x})u_k(\boldsymbol{x}) = &\int_{\partial D} \{\mathring{U}_{mi}(\boldsymbol{x}, \boldsymbol{y})t_i(\boldsymbol{y}) - \mathring{W}_{mi}(\boldsymbol{x}, \boldsymbol{y})u_i(\boldsymbol{y})\}\,\mathrm{d}S(\boldsymbol{y}) \\
&+ \int\!\!\!\!-_D \mathring{U}_{mi.jl}(\boldsymbol{x}, \boldsymbol{y})C_{ijkl}(\boldsymbol{y})u_k(\boldsymbol{y})\,\mathrm{d}V(\boldsymbol{y}) \\
&+ \int_D \mathring{U}_{mi.j}(\boldsymbol{x}, \boldsymbol{y})C_{ijkl.l}(\boldsymbol{y})u_k(\boldsymbol{y})\,\mathrm{d}V(\boldsymbol{y}) \\
&+ \int_D \mathring{U}_{mi}(\boldsymbol{x}, \boldsymbol{y})\{\rho(\boldsymbol{y})b_i(\boldsymbol{y}) + \rho(\boldsymbol{y})\omega^2 u_i(\boldsymbol{y})\}\,\mathrm{d}V(\boldsymbol{y})\,,
\end{aligned} \tag{3.26}$$

where

$$\varepsilon_{mk}(x) = \begin{cases} v_{lmij} C_{ijkl}(x) , & x \in D , \\ 0 , & x \in D_c , \\ \frac{1}{2} V_{lmij} C_{ijkl}(x) , & x \in \partial D . \end{cases}$$

If C_{ijkl} is not continuous, we have to modify the result. We assume a jump in C_{ijkl} when crossing ∂S such that

$$\Delta C_{ijkl}(x) = \lim_{x \to x^-} C_{ijkl}(x) - \lim_{x \to x^+} C_{ijkl}(x) \tag{3.27}$$

where x^- and x^+ imply the minus and plus sides of ∂S. The integral equation for the domain in which a discontinuity surface in C_{ijkl} is included can be easily obtained by adding two equations, each of which is formulated for each domain of either side of ∂S, i.e.

$$\int_{\partial D} \{\mathring{U}_{mi}(x, y) t_i(y) - \mathring{W}_{mi}(x, y) u_i(y)\} \, dS(y) - \int_{\partial S} \Delta \mathring{W}_{mi}(x, y) u_i(y) \, dS(y)$$
$$+ \int_D \{\mathring{U}_{mi.j}(x, y) C_{ijkl}(y)\}_{.l} u_k(y) \, dV(y)$$
$$+ \int_D \mathring{U}_{mi}(x, y) \{\rho(y) b_i(y) + \rho(y) \omega^2 u_i(y)\} \, dV(y)$$
$$= \begin{cases} v_{lmij} C_{ijkl}(x) u_k(x) , & x \in D - \partial S \\ \frac{1}{2} v_{lmij} \{C^-_{ijkl}(x) + C^+_{ijkl}(x)\} u_k(x) , & x \in \partial S \\ \frac{1}{2} v_{lmij} C_{ijkl}(x) u_k(x) , & x \in \partial D \\ 0 , & x \in D_c , \end{cases} \tag{3.28}$$

where

$$\Delta \mathring{W}_{mi}(x, y) = n_l(y) \Delta C_{ijkl}(y) \mathring{U}_{km.j}(x, y) , \qquad y \in \partial S .$$

In the special case, when the elastic medium is homogeneous and isotropic, Eq. (3.26) reduces to the integral equation for elastostatics with body force $\rho \boldsymbol{b} + \rho \omega^2 \boldsymbol{u}$. When the body force $\boldsymbol{b}$ is disregarded, as it is often the case in elastodynamics, Eq. (3.26) shows that time-harmonic elastodynamic problems can also be solved by the use of domain-type integral equations with virtual body forces of the d'Alembert type. Synthesizing time-harmonic solutions, we have of course transient solutions.

The domain integrals, however, sometimes can be converted to boundary integrals by the technique which will be mentioned in section 4.4.4. A formulation in the time–space domain of such a type was studied by Nardini and Brebbia [88, 89].

Two-dimensional problems can be also similarly formulated

3.3.3. *Eigenfrequency problems*

Eigenfrequencies can be obtained as the real and non-negative parameters ω (circular frequencies) which assure the existence of non-trivial solutions of the homogeneous boundary integral equations derived by just deleting all the terms relevant to the prescribed boundary values. This leads to the well known condition that the determinant of the coefficient matrix of the discretized equations of the system of boundary integral equations is equal to zero. In the numerical treatment, it is hard to obtain the true roots of the determinant, since ω is implicitly contained in the fundamental solutions $\hat{U}(x, y; \omega)$ and $\hat{W}(x, y; \omega)$. However, if the frequency ω is very close to one of the eigenfrequencies, the value of the determinant becomes very close to zero. This phenomenon can easily be detected during the triangular (L − U) decomposition process by the Crout method [66]. Hence, roots of the determinant are searched using the (L − U) decomposition by direct substitution of successively increased values of ω into the fundamental solutions [57, 98].

When we use the domain-type formulation such as Eq. (3.26), we may obtain eigenfrequencies more easily, since the parameter ω is explicitly expressed.

It is noted here that

(i) for the boundary integral equation of the second kind, the integral equations for the first (second) exterior problems determine the eigenfrequencies of the second (first) interior problems, and

(ii) for the boundary integral equations of the first kind, the integral equations for the first (second) exterior problems determine the eigenfrequencies of the first (second) interior problems [57, 61, 98].

Eigenmodes are determined from the discretized (boundary) equations by substituting the eigenfrequencies [57, 98].

Free vibration problems of thin plates have been also extensively studied [57, 88, 89, 98, 126].

3.3.4. *Fictitious eigenfrequencies*

It is well known that the Fourier-transformed domain boundary integral formulation for the exterior problems in acoustics fails to yield a unique solution at certain wave-numbers which correspond to the eigenfrequencies of the interior region of the closed surface [22, 73, 112, 125]. A similar fact is also known in elastodynamics [66, 70, 106]. These special wavenumbers are called "eigen-" or "characteristic" wavenumbers and the corresponding eigenfrequencies are termed "fictitious" eigenfrequencies. When we solve transient exterior problems, we must pay special attention to the treatment of the fictitious eigenfrequencies.

In order to circumvent the fictitious eigenfrequencies, the following four types of techniques have been proposed:

(i) To use the interior representation, say Eq. (3.5) for $\boldsymbol{x} \in D_-$, that is, with null displacements in the interior, [22, 70, 120], or use it as a supplementary condition for the boundary integral equation [69, 104, 106, 112].

(ii) To utilize two integral equations, for example (a) to solve simultaneously both the first and the second kind integral equations for the same problem, (b) to solve the combined equation of them with a suitably selected coupling constant [12, 85], or (c) to use a mixed potential [45].

(iii) To formulate a boundary integral equation using modified fundamental solutions which satisfy a certain vanishing condition inside the interior domain [10, 50, 65, 125].

(iv) To use interpolation techniques and numerically approximate behaviors corresponding to the fictitious eigenfrequencies by using several values of ω other than the eigenfrequencies of the interior domain [66].

3.3.5. *Peripheral stresses* [66]

The evaluation of the peripheral stresses may be the most important problem. The boundary integral equation method is suited for this purpose.

(i) For the three-dimensional case, expressing the boundary surface ∂D by a curvilinear coordinate system $\xi^K (K = 1, 2)$ and ξ^3 in the positive normal direction $\boldsymbol{n}$ to ∂D, the peripheral stresses have the form

$$\left.\begin{aligned} &\sigma_{KL} = \lambda a_{KL} \frac{2\mu u_{K|L} a^{KL} + t_3}{\lambda + 2\mu} + \mu(u_{K|L} + u_{L|K}) \\ &\sigma_{K3} = t_K , \qquad \sigma_{33} = t_3 , \end{aligned}\right\} \tag{3.29}$$

where $u_{K|L}$ indicates the (three-dimensional) covariant derivative defined as

$$u_{K|L} = \frac{\partial u_K}{\partial \xi^L} - u_M \Gamma^M_{KL} - u_3 b_{KL} \tag{3.30}$$

and a_{KL}, b_{KL}, Γ^M_{KL} stand for the first and second fundamental forms of the surface and the Christoffel symbol calculated from a_{KL}. They are defined as

$$\left.\begin{aligned} &\mathrm{d}s^2 = \mathrm{d}x^i\, \mathrm{d}x^i = a_{KL}\, \mathrm{d}\xi^K\, \mathrm{d}\xi^L , \\ &b_{KL} = \frac{1}{\sqrt{a}}\, \varepsilon_{ijk} \frac{\partial^2 x^i}{\partial \xi^K \partial \chi^L} \frac{\partial x^j}{\partial \xi^1} \frac{\partial x^k}{\partial \xi^2} , \\ &\Gamma^M_{KL} = \tfrac{1}{2} a^{MN} \left(\frac{\partial a_{KN}}{\partial \xi^L} + \frac{\partial a_{LN}}{\partial \xi^K} - \frac{\partial a_{KL}}{\partial \xi^N} \right) , \\ &a^{KL} a_{LM} = \delta^K_M . \end{aligned}\right\} \tag{3.31}$$

for the rectangular Cartesian coordinate system x^i, where ds is a line element, ε_{ijk} means the permutation symbol and $a = \det(a_{KL})$.

(ii) For the two-dimensional case, we can write down the result in physical components as

$$\left.\begin{aligned} \sigma_{\xi\xi} &= \frac{4\mu(\lambda+\mu)}{(\lambda+2\mu)a} \frac{\partial u_\alpha}{\partial \xi} \frac{\partial x_\alpha}{\partial \xi} + \frac{\lambda n_\alpha t_\alpha}{\lambda + 2\mu} \\ \sigma_{\xi n} &= \frac{1}{\sqrt{a}} t_\alpha \frac{\partial x_\alpha}{\partial \xi}, \qquad \sigma_{nn} = n_\alpha t_\alpha , \end{aligned}\right\} \tag{3.32}$$

where ξ is the parameter which describes the boundary curve $x_\alpha(\xi)$ and

$$u = \frac{\partial x_\alpha}{\partial \xi} \frac{\partial x_\alpha}{\partial \xi} .$$

Note that the vectors on the right-hand sides of Eqs. (3.32) are referred to the Cartesian coordinate system.

3.3.6. Use of Green's functions

As a kernel function of integral equations, it is advisable to use a Green's function, if available, in place of the fundamental solution. Even if it satisfies only a certain amount of the boundary conditions of the problem, it may be helpful to reduce the dimension of the discretized equations and to improve the solutions at the same time.

Green's functions for the half-space are available in [11, 49] for the time domain and in [13, 60, 76] for the frequency domain. A semi-infinite element for Rayleigh waves has also been devised [67].

For the anti-plane problem, the Green function for a half-space is easily obtained by superposing the fundamental solution of the infinite space to its mirror image with respect to the traction free boundary.

4. Numerical procedures

4.1. Discretization of integral equations

The integral equation thus obtained is usually solved numerically by discretizing it into a system of algebraic equations with the aid of a scheme similar to that used in the finite element method.

The boundary integral is evaluated for each boundary element numerically, say by the use of the Gaussian quadrature formula. Usually quadratic isoparametric elements are used for it, although constant elements may be also applicable. For the evaluation of domain integrals, a similar technique is adopted.

In what follows, we show a discretization procedure of the integral equation in conjunction with the collocation technique [72, 128]. In an isoparametric element, the local coordinate ξ_i in the natural coordinate system is related to the global coordinate x_i in the rectangular Cartesian coordinate system by the expression

$$x_i(\xi) = N^\alpha(\xi)x_i^\alpha , \tag{4.1}$$

where x_i^α is the coordinate of the nodal point $\alpha(\alpha = 1, 2, 3)$ and N^α implies the shape function with respect to the node. The shape of the boundary element is approximated by Eq. (4.1) and functions on the element can be similarly expressed by

$$u_i(\boldsymbol{\xi}) = N^\alpha(\boldsymbol{\xi})u_i^\alpha , \qquad t_i(\boldsymbol{\xi}) = N^\alpha(\boldsymbol{\xi})t_i^\alpha . \tag{4.2}$$

The function $\boldsymbol{b}(\boldsymbol{x})$ in the domain is also similarly approximated by

$$b_i(\boldsymbol{\xi}) = M^\gamma(\boldsymbol{\xi})b_i^\gamma , \tag{4.3}$$

where the shape function M^γ may not be in general the same as the N^α.

By discretizing now the boundary into M segments with N nodes and the domain into P elements having Q nodes, and using the above-mentioned approximation, we have a system of algebraic equation. Equation (3.2), for example, becomes in discretized form

$$\begin{aligned} C_{ik}^{c}(\boldsymbol{x}^a)\hat{u}_i(\boldsymbol{x}^a; \omega) &+ \sum_{\beta=1}^{M}\sum_{\alpha=1}^{N} \hat{u}_i(\boldsymbol{x}^{k(\beta,\alpha)}; \omega) \int_{\partial D^\beta} N^\alpha(\boldsymbol{\xi})\hat{W}_{ki}^{\beta}(\boldsymbol{x}^a, \boldsymbol{y}(\boldsymbol{\xi}); \omega)J_S(\boldsymbol{\xi})\,\mathrm{d}S(\boldsymbol{\xi}) \\ &= \sum_{\beta=1}^{M}\sum_{\alpha=1}^{N} \hat{t}_i(\boldsymbol{x}^{k(\beta,\alpha)}; \omega) \int_{\partial D^\beta} N^\alpha(\boldsymbol{\xi})\hat{U}_{ki}^{\beta}(\boldsymbol{x}^a; \boldsymbol{y}(\boldsymbol{\xi}); \omega)J_S(\boldsymbol{\xi})\,\mathrm{d}S(\boldsymbol{\xi}) \\ &+ \sum_{\delta=1}^{P}\sum_{\gamma=1}^{Q} \hat{b}_i(\boldsymbol{x}^{h(\delta,\gamma)}; \omega) \int_{D^\delta} \rho M^\gamma(\xi)\hat{U}_{ki}^{\delta}(\boldsymbol{x}^a, \boldsymbol{y}(\boldsymbol{\xi}); \omega)J_V(\boldsymbol{\xi})\,\mathrm{d}V(\boldsymbol{\xi}) , \end{aligned} \tag{4.4}$$

where we denote the number k of the boundary node α of the element β in the global coordinates by $k(\beta, \alpha)$ and the number h of the node γ in the domain element δ in the global coodinates by $h(\delta, \gamma)$, and $J_S(\boldsymbol{\xi})$ and $J_V(\boldsymbol{\xi})$ imply the Jacobians with respect to the surface and the volume integrals, respectively. It is also noted that the summation convention rule is implied over the repeated index i.

This equation can be simply rewritten in matrix form for the unknown nodal values $\hat{\boldsymbol{u}}$ and $\hat{\boldsymbol{t}}$ as

$$\mathbf{W}\hat{\boldsymbol{u}} - \mathbf{U}\hat{\boldsymbol{t}} = \boldsymbol{F} , \tag{4.5}$$

where $\boldsymbol{F}$ contains values due to prescribed boundary conditions and the last integral in Eq. (4.4).

4.2. Time-marching process

Making use of the space and time interpolation (basis) functions N and $\boldsymbol{\phi}$, respectively, we can express the time–space function $f(\boldsymbol{x}, t)$ in the form [81]

$$f_i(\boldsymbol{\xi}, \tau) = N^{\alpha}(\boldsymbol{\xi})\phi_{\beta}(\tau) f_i(\boldsymbol{x}^{\alpha}, t_{\beta}) . \tag{4.6}$$

If the boundary is discretized into M elements having N nodes and the time interval is divided into K steps having L nodes, substitution of Eq. (4.6) into Eq. (3.7), for example, leads to the following equation:

$$\begin{aligned} & C^{c}_{ki}(\boldsymbol{x}^{a}) u_i(\boldsymbol{x}^{a}, t) \\ & \quad + \sum_{m=1}^{M} \sum_{\alpha=1}^{N} \int_{\partial D^m} \left\{ \sum_{k=1}^{K} \sum_{\beta=1}^{L} \left(\int_{t_{k-1}}^{t_k} W_{ki}(\boldsymbol{x}^{a}, \tau; \boldsymbol{y}(\boldsymbol{\xi}), 0) \phi_{\beta}(t-\tau)\, \mathrm{d}\tau \right) \right. \\ & \quad \left. \times\, u_i(x^{(m,\alpha)}, t_{(k,\beta)}) \right\} N^{\alpha}(\boldsymbol{\xi}) J_S(\boldsymbol{\xi})\, \mathrm{d}S(\boldsymbol{\xi}) \\ & = \sum_{m=1}^{M} \sum_{\alpha=1}^{N} \int_{\partial D^m} \left\{ \sum_{k=1}^{K} \sum_{\beta=1}^{L} \left(\int_{t_{k-1}}^{t_k} U_{ki}(\boldsymbol{x}^{a}, \tau; \boldsymbol{y}(\boldsymbol{\xi}), 0) \phi_{\beta}(t-\tau)\, \mathrm{d}\tau \right) \right. \\ & \quad \times\, t_i(\boldsymbol{x}^{(m,\alpha)}, t_{(k,\beta)}) N^{\alpha}(\boldsymbol{\xi}) J_S(\boldsymbol{\xi})\, \mathrm{d}S(\boldsymbol{\xi}) , \end{aligned} \tag{4.7}$$

where the body force is disregarded, the quiescent past is assumed for simplicity, and $\boldsymbol{x}^{(m,\alpha)}$ and $t_{(k,\beta)}$ refer to the node α of the space element m and the node β of the time element k.

As an example, if we assume a linear variation in time within each time-step according to

$$\phi_1 = \frac{t_k - t}{\Delta t_k} , \qquad \phi_2 = \frac{t - t_{k-1}}{\Delta t_k} , \qquad \Delta t_k = t_k - t_{k-1} ,$$

the time convolution integral in Eq. (4.7) in each time-step reduces to

$$\begin{aligned} & \left\{ \int_{t_{k-1}}^{t_k} W_{ki}(\boldsymbol{x}^{a}, \tau; \boldsymbol{y}(\boldsymbol{\xi}), 0) \frac{t_k - \tau}{\Delta t_k}\, \mathrm{d}\tau \right\} u_i(\boldsymbol{x}^{(m,\alpha)}, t_{(k,1)}) \\ & \quad + \left\{ \int_{t_{k-1}}^{t_k} W_{ki}(\boldsymbol{x}^{a}, \tau; \boldsymbol{y}(\boldsymbol{\xi}), 0) \frac{\tau - t_{k-1}}{\Delta t_k} \right\} u_i(\boldsymbol{x}^{(m,\alpha)}, t_{(k,2)}) \end{aligned}$$

with a similar result for terms involving U.

Since $\boldsymbol{U}$ and $\boldsymbol{W}$ involve time functions as shown in Eqs. (1.46) and (1.48), the convolution integral may be evaluated by the use of the following results:

$$\int_{t_{k-1}}^{t_k} \tau \frac{t_k - \tau}{\Delta t_k} H\left(\tau - \frac{r}{c_\alpha}\right) \mathrm{d}\tau = \begin{cases} \dfrac{1}{6\Delta t_k}(t_k^3 - 3t_k t_{k-1}^2 + 2t_{k-1}^3)\,, & t_{k-1} \geqslant \dfrac{r}{c_\alpha}\,, \\ \dfrac{1}{6\Delta t_k}\left\{t_k^3 - 3t_k\left(\dfrac{r}{c_\alpha}\right)^2 + 2\left(\dfrac{r}{c_\alpha}\right)^3\right\}, & t_k \geqslant \dfrac{r}{c_\alpha} \geqslant t_{k-1}\,, \\ 0\,, & \dfrac{r}{c_\alpha} \geqslant t_k\,, \end{cases}$$

$$\int_{t_{k-1}}^{t_k} \tau \frac{\tau - t_{k-1}}{\Delta t_k} H\left(t - \frac{r}{c_\alpha}\right) \mathrm{d}\tau$$

$$= \begin{cases} \dfrac{1}{6\Delta t_k}(2t_k^3 - 3t_k^2 t_{k-1} + t_{k-1}^3)\,, & t_{k-1} \geqslant \dfrac{r}{c_\alpha}\,, \\ \dfrac{1}{6\Delta t_k}\left\{2t_k^3 - 3t_k^2 t_{k-1} + 3t_{k-1}\left(\dfrac{r}{c_\alpha}\right)^2 - 2\left(\dfrac{r}{c_\alpha}\right)^3\right\}, & t_k \geqslant \dfrac{r}{c_\alpha} \geqslant t_{k-1}\,, \\ 0\,, & \dfrac{r}{c_\alpha} \geqslant t_k\,, \end{cases}$$

$$\int_{t_{k-1}}^{t_k} \frac{t_k - \tau}{\Delta t_k} \delta\left(\tau - \frac{r}{c_\alpha}\right) \mathrm{d}\tau = \begin{cases} \dfrac{1}{\Delta t_k}\left(t_k - \dfrac{r}{c_\alpha}\right), & t_{k-1} \leqslant \dfrac{r}{c_\alpha} \leqslant t_k\,, \\ 0\,, & \dfrac{r}{c_\alpha} \geqslant t_k,\ \dfrac{r}{c_\alpha} \leqslant t_{k-1}\,, \end{cases}$$

$$\int_{t_{k-1}}^{t_k} \frac{\tau - t_{k-1}}{\Delta t_k} \delta\left(\tau - \frac{r}{c_\alpha}\right) \mathrm{d}\tau = \begin{cases} \dfrac{1}{\Delta t_k}\left(\dfrac{r}{c_\alpha} - t_{k-1}\right), & t_{k-1} \leqslant \dfrac{r}{c_\alpha} \leqslant t_k\,, \\ 0\,, & \dfrac{r}{c_\alpha} \geqslant t_k,\ \dfrac{r}{c_\alpha} \leqslant t_{k-1}\,, \end{cases}$$

$$\int_{t_{k-1}}^{t_k} \frac{t_k - \tau}{\Delta t_k} \dot{\delta}\left(\tau - \frac{r}{c_\alpha}\right) \mathrm{d}\tau = \begin{cases} 1\,, & t_{k-1} \leqslant \dfrac{r}{c_\alpha} \leqslant t_k\,, \\ 0\,, & \dfrac{r}{c_\alpha} \geqslant t_k,\ \dfrac{r}{c_\alpha} \leqslant t_{k-1}\,, \end{cases}$$

where the relation

$$\int_a^b \delta^{(k)}(\tau - \alpha) f(\tau)\, \mathrm{d}\tau = (-1)^{(k)} f^{(k)}(\alpha)\,, \qquad a \leqslant \alpha \leqslant b$$

has been used.

It should be noted that the convolution integral has meaning only in the interval of the past time characteristic cone. It may also be said that the stability condition for time stepping is less stringent than that of conventional explicit difference marching process, since the time is implicitly contained in the integral equation.

A more comprehensive discussion on the numerical treatment of elastodynamic boundary integrals can be found in the next chapter of this book.

4.3. Combination of integral equation method and finite element method

In order to solve non-homogeneous or nonlinear problems for infinite or semi-infinite domains that are frequently encountered in practical applications, the combination of the boundary integral equation method (BIEM) and the finite element method (FEM) is most useful [56, 83]. Of course, the finite element method is used for the interior domain where non-homogeneity or nonlinearity is predominant and the boundary integral equation method is utilized for the rest of the infinite or semi-infinite domain which may well be regarded as homogeneous and linear.

A weak form (virtual work) of the elastodynamic problem in the Fourier-transformed domain can be expressed for the interior domain with an arbitrary vector $\hat{\boldsymbol{v}}(\boldsymbol{x};\omega)$ as [136]

$$\int_{D_-} \hat{\boldsymbol{\varepsilon}} : \hat{\boldsymbol{\sigma}}\, \mathrm{d}V = \int_{D_-} \rho\omega^2 \hat{\boldsymbol{v}} \cdot \hat{\boldsymbol{u}}\, \mathrm{d}V + \int_{\partial D_2} \hat{\boldsymbol{v}} \cdot \hat{\bar{\boldsymbol{t}}}\, \mathrm{d}S \tag{4.8}$$

where $\hat{\boldsymbol{\varepsilon}}$ means strain given in matrix form as

$$\hat{\boldsymbol{\varepsilon}} = \mathbf{C}\hat{\boldsymbol{v}} \tag{4.9}$$

$\mathbf{C}$ is an operator whose components are given by

$$\tfrac{1}{2}\left(\delta_{im}\frac{\partial}{\partial x_j} + \delta_{im}\frac{\partial}{\partial x_j}\right) \tag{4.10}$$

and $\bar{\boldsymbol{t}}$ is the prescribed traction on ∂D_2.

By expressing now the displacement vectors and the surface traction vectors for each element in terms of the shape function N^k and the respective nodal values $\tilde{\boldsymbol{u}}$, $\tilde{\boldsymbol{v}}$ and $\tilde{\boldsymbol{t}}$ as in Eq. (4.2),

$$\hat{\boldsymbol{u}} = \mathbf{N}\tilde{\boldsymbol{u}}\,, \qquad \hat{\boldsymbol{v}} = \mathbf{N}\tilde{\boldsymbol{v}}\,, \qquad \hat{\boldsymbol{t}} = \mathbf{N}\tilde{\boldsymbol{t}} \tag{4.11}$$

with

$$\mathbf{N} = [\mathbf{N}^1 \quad \mathbf{N}^2 \quad \mathbf{N}^3 \quad \cdots]\,, \qquad \mathbf{N}^k = \begin{bmatrix} N_1^k & 0 & 0 \\ 0 & N_2^k & 0 \\ 0 & 0 & N_3^k \end{bmatrix},$$

$$\tilde{\boldsymbol{u}}^{\mathrm{T}} = [\boldsymbol{u}^1 \quad \boldsymbol{u}^2 \quad \boldsymbol{u}^3 \quad \cdots]\,, \qquad \hat{\boldsymbol{u}}^{\mathrm{T}} = [u_1 \quad u_2 \quad u_3]$$

and substituting them into Eq. (4.8) together with the constitutive relation

$$\hat{\boldsymbol{\sigma}} = \mathbf{D}\hat{\boldsymbol{\varepsilon}} = \mathbf{D}\mathbf{C}\tilde{\boldsymbol{u}} \tag{4.12}$$

(where $\mathbf{D}$ is the elastic matrix), we have the following finite element equation

$$\sum_e \left(\int_{D^e_-} \mathbf{B}^{\mathrm{T}}\mathbf{D}\mathbf{B}\,\mathrm{d}V - \rho\omega^2 \int_{D^e_-} \mathbf{N}^{\mathrm{T}}\mathbf{N}\,\mathrm{d}V \right)\tilde{\boldsymbol{u}}^e = \sum_e \left(\int_{\partial D^e} \mathbf{N}^{\mathrm{T}}\mathbf{N}\,\mathrm{d}S \right)\tilde{\boldsymbol{t}}^e \tag{4.13}$$

where $\mathbf{B}$ is the strain matrix given by

$$\mathbf{B} = \mathbf{C}\mathbf{N} \tag{4.14}$$

and $\sum_e$ means to take sum over the relevant elements.

We simply write the system of algebraic equations (4.13) as

$$\mathbf{K}\tilde{\boldsymbol{u}} = \mathbf{H}\tilde{\boldsymbol{t}}\,. \tag{4.15}$$

Combining Eqs. (4.15) and (4.5) with the aid of compatibility and equilibrium conditions at the interface between the interior and the exterior domains, we have a system of equations. In order to fix ideas, consider an example as shown in Fig. 2, in which a part of the interior domain is embedded in a half-space exterior domain.

We denote the boundaries of the interior and the exterior domains by S_{i} and S_{e}, respectively, their common interface by S, and also use these symbols as subscripts for their related quantities. Here the boundary S_{e} is assumed temporarily finite.

We have then the finite element equation for the interior domain and the boundary integral equation for the exterior domain as

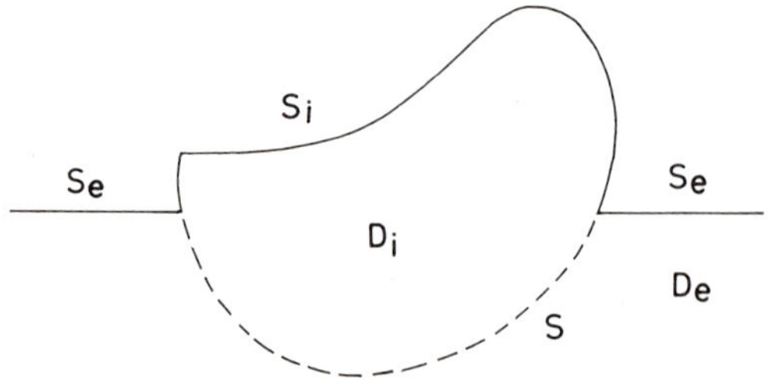

Fig. 2. A model for the BIE–FEM combined method.

$$\begin{bmatrix} \mathbf{K}_{FF} & \mathbf{K}_{FS_i} & \mathbf{K}_{FS} \\ \mathbf{K}_{FS_i} & \mathbf{K}_{S_iS_i} & \mathbf{K}_{S_iS} \\ \mathbf{K}_{FS} & \mathbf{K}_{S_iS} & \mathbf{K}_{SS} \end{bmatrix} \begin{bmatrix} \tilde{u}_F^{\mathrm{i}} \\ \tilde{u}_{S_i}^{\mathrm{i}} \\ \tilde{u}_S^{\mathrm{i}} \end{bmatrix} = \begin{bmatrix} \mathbf{0} \\ \mathbf{H}_{S_i}\tilde{t}_{S_i}^{\mathrm{i}} \\ \mathbf{H}_S\tilde{t}_S^{\mathrm{i}} \end{bmatrix} \tag{4.16}$$

$$[\mathbf{W}_{S_e} \quad \mathbf{W}_S] \begin{bmatrix} \tilde{u}_{S_e}^{\mathrm{e}} \\ \tilde{u}_S^{\mathrm{e}} \end{bmatrix} - [\mathbf{U}_{S_e} \quad \mathbf{U}_S] \begin{bmatrix} \tilde{t}_{S_e}^{\mathrm{e}} \\ \tilde{t}_S^{\mathrm{e}} \end{bmatrix} = [F_{\mathrm{I}}] \tag{4.17}$$

where F_{I} is a field due to the incident wave (cf. Eq. (3.4)) and the subscript F implies quantities related to the interior domain.

Using the continutity conditions on S (note that the positive unit normal is taken as pointing from the interior to the exterior domain)

$$\tilde{u}_S^{\mathrm{i}} = \tilde{u}_S^{\mathrm{e}} = \tilde{u}_S , \qquad \tilde{t}_S^{\mathrm{i}} + \tilde{t}_S^{\mathrm{e}} = \mathbf{0} \tag{4.18}$$

and the boundary conditions, say,

$$\tilde{t}_{S_i}^{\mathrm{i}} = \bar{t}_{S_i} , \qquad \tilde{t}_{S_e}^{\mathrm{e}} = \bar{t}_{S_e} , \tag{4.19}$$

we have the combined system of equations as follows:

$$\begin{bmatrix} \mathbf{K}_{FF} & \mathbf{K}_{FS_i} & \mathbf{K}_{FS} & 0 & 0 \\ \mathbf{K}_{FS_i} & \mathbf{K}_{S_iS_i} & \mathbf{K}_{S_iS} & 0 & 0 \\ \mathbf{K}_{FS} & \mathbf{K}_{S_iS} & \mathbf{K}_{SS} & -\mathbf{H}_S & 0 \\ \mathbf{0} & \mathbf{0} & \mathbf{W}_S & -\mathbf{U}_S & \mathbf{W}_{S_e} \end{bmatrix} \begin{bmatrix} \tilde{u}_F^{\mathrm{i}} \\ \tilde{u}_{S_i}^{\mathrm{i}} \\ \hat{u}_S \\ \tilde{t}_S^{\mathrm{e}} \\ \tilde{u}_{S_e}^{\mathrm{e}} \end{bmatrix} = \begin{bmatrix} \mathbf{0} \\ \mathbf{H}_{S_i}\bar{t}_{S_i} \\ \mathbf{0} \\ F_{\mathrm{I}} + \mathbf{U}_{S_e}\bar{t}_{S_e} \end{bmatrix} \tag{4.20}$$

As the right-hand side of the above equation is known from the incident wave field and prescribed boundary conditions, we can solve this equation for the nodal displacements $\hat{u}$ in D_{i}, on S_{i} and S_{e}, and for the displacement $\hat{u}$ and the traction $\hat{t}$ on S.

4.4. *Miscellaneous topics in numerical treatment*

4.4.1. *Numerical integration*

In numerical procedures, evaluation of integrals over each element is of basic importance. Element size must be chosen in general sufficiently small so as to express the shortest wavelength appropriately, that is, at least four or five nodal points must be included in the shortest wavelength.

The boundary integrals, whenever non-singular, are usually evaluated with the aid of Gaussian quadrature formula, or double exponential formula [119].

The double exponential formula is such as to compute the integral

$$I = \int_{-1}^{1} f(x)\,\mathrm{d}x \tag{4.21}$$

approximately by using the trapezoidal rule with step size h according to the formula

$$I_h = \frac{\pi}{2} h \sum_{n=-\infty}^{\infty} f(\tanh(\tfrac{1}{2}\pi \sinh nh)) \frac{\cosh nh}{\cosh^2(\tfrac{1}{2}\pi \sinh nh)}, \qquad (4.22)$$

where the integration variable x has been changed by

$$x = \tanh(\tfrac{1}{2}\pi \sinh t)$$

and the integral has been transformed to an infinite one. This formula can easily be extended to the two-dimensional case.

Since the evaluation of boundary integrals occupies a significant amount of computing time, it is advisable to optimize, if possible, by specifying certain maximum error bounds [72].

When the field point coincides with the node of the same element of integration, special attention must be paid. In the numerical integration of the simple layer kernel $\boldsymbol{U}$ for that case, the singular element is divided into a number of triangular subelements in such a manner as the common apex of all subelements to be located on the field point. Then each triangular subelement is mapped on a flat triangle and numerical integration is performed by introducing polar coordinates centered at the singular apex. In the integration of the double layer kernel $\boldsymbol{W}$, since the fundamental solutions consist of the sum of the static counterparts and the regular parts as will be seen in section 4.4.2, the integral can be obtained as a sum of the integral of the regular part and the static contribution, which is actually obtainable as the sum, with minus sign, of the off-diagonal submatrices of the elastostatics. This is based on the fact that the sum of the free-term $\boldsymbol{C}^{c}$ and the singular integral of elastostatics can be evaluated altoghether by applying arbitrary rigid-body displacements [72, 78, 105–107].

4.4.2. Expansion of the time-harmonic fundamental solutions

In the evaluation of the integrals containing time-harmonic fundamental solutions, it is advised to expand them and delete unfavourable terms beforehand, particularly for small argument. Some manipulations can yield the following results: From Eq.(1.40),

$$\hat{U}_{jk}(\boldsymbol{x}, \boldsymbol{y}; \omega) = A(\hat{U}_1 \delta_{jk} - \hat{U}_2 r_{,j} r_{,k}), \quad \hat{U}_1 = \mathring{U}_1 + \hat{U}_1^{\mathrm{D}}, \quad \hat{U}_2 = \mathring{U}_2 + \hat{U}_2^{\mathrm{D}}, \qquad (4.23a)$$

where

(i) for the three-dimensional case:

$$\left.\begin{aligned}
&A=\frac{1}{4\pi\mu}\,,\qquad \mathring{U}_1=\frac{\lambda+3\mu}{2(\lambda+2\mu)}\,\frac{1}{r}\,,\qquad \mathring{U}_2=-\frac{\lambda+\mu}{2(\lambda+2\mu)}\,\frac{1}{r}\,,\\
&\hat{U}_1^{\mathrm{D}}=\sum_{n=1}^{\infty}\frac{i^n k_{\mathrm{T}}^n r^{n-1}}{n!(n+2)}\left\{(n+1)+\left(\frac{k_{\mathrm{L}}}{k_{\mathrm{T}}}\right)^{n+2}\right\},\\
&\hat{U}_2^{\mathrm{D}}=\sum_{n=1}^{\infty}\frac{i^n k_{\mathrm{T}}^n r^{n-1}}{n!}\,\frac{n-1}{n+2}\left\{1-\left(\frac{k_{\mathrm{L}}}{k_{\mathrm{T}}}\right)^{n+2}\right\},
\end{aligned}\right\}\qquad(4.23\mathrm{b})$$

$$\mathring{U}_{jk}=\frac{1}{8\pi\mu(\lambda+2\mu)r}\{(\lambda+3\mu)\delta_{jk}+(\lambda+\mu)r_{,j}r_{,k}\}\,,\qquad(4.24)$$

(ii) for the two-dimensional case:

$$A=\frac{\mathrm{i}}{4\mu}\,,\qquad \mathring{U}_1=\frac{\mathrm{i}(\lambda+3\mu)}{\pi(\lambda+2\mu)}\ln r\,,\qquad \mathring{U}_2=\frac{\mathrm{i}(\lambda+\mu)}{\pi(\lambda+2\mu)}\,,$$

$$\begin{aligned}
\hat{U}_1^{\mathrm{D}}=\frac{1}{2(\lambda+2\mu)}\Bigg[&(\lambda+3\mu)\\
&+\frac{\mathrm{i}}{\pi}\left\{(\lambda+\mu)+2(\lambda+3\mu)\gamma+2(\lambda+2\mu)\ln\frac{k_{\mathrm{T}}}{2}+2\mu\ln\frac{k_{\mathrm{L}}}{2}\right\}\Bigg]\\
&+\sum_{n=0}^{\infty}\frac{(-1)^n}{(n!)^2}\left(\frac{k_{\mathrm{T}}r}{2}\right)^{2n}\Bigg[\frac{1}{2(n+1)}\left\{2n+1+\left(\frac{k_{\mathrm{L}}}{k_{\mathrm{T}}}\right)^{2n+2}\right\}A+B\\
&-\frac{1}{2(n+1)}D+\frac{1}{2(n+1)}\left\{1-\left(\frac{k_{\mathrm{L}}}{k_{\mathrm{T}}}\right)^{2n+2}\right\}E\Bigg],
\end{aligned}\qquad(4.23\mathrm{c})$$

$$\begin{aligned}
\hat{U}_2^{\mathrm{D}}=\sum_{n=1}^{\infty}\frac{(-1)^n}{(n!)^2}\left(\frac{k_{\mathrm{T}}r}{2}\right)^{2n}\Bigg[&\frac{n}{n+1}\left\{1-\left(\frac{k_{\mathrm{L}}}{k_{\mathrm{T}}}\right)^{2n+2}\right\}A+B\quad\left(\frac{k_{\mathrm{L}}}{k_{\mathrm{T}}}\right)^{2n+2}C\\
&-\frac{n}{n+1}D+\frac{1}{n+1}\left\{1-\left(\frac{k_{\mathrm{L}}}{k_{\mathrm{T}}}\right)^{2n+2}\right\}E\Bigg],
\end{aligned}$$

$$A=1+\frac{2\mathrm{i}}{\pi}\{\ln r+\gamma\}\,,\qquad B=\frac{2\mathrm{i}}{\pi}\left\{\ln\frac{k_{\mathrm{T}}}{2}-\phi(n)\right\},$$

$$C=\frac{2\mathrm{i}}{\pi}\left\{\ln\frac{k_{\mathrm{L}}}{2}-\phi(n)\right\},\qquad D=\frac{2\mathrm{i}}{\pi}\left\{\ln\frac{k_{\mathrm{T}}}{2}-\left(\frac{k_{\mathrm{L}}}{k_{\mathrm{T}}}\right)^{2n+2}\ln\frac{k_{\mathrm{L}}}{2}\right\},$$

$$E=\frac{\mathrm{i}}{\pi}\{\phi(n)+\phi(n+1)\}\,,\qquad \gamma=0.57721\ \text{(Euler constant)}\,,$$

$$\phi(n)=\sum_{m=1}^{n}\frac{1}{m}\,,$$

$$\mathring{U}_{jk}=\frac{1}{4\pi\mu(\lambda+2\mu)}\left\{(\lambda+3\mu)\ln\frac{1}{r}\,\delta_{jk}+(\lambda+\mu)r_{,j}r_{,k}\right\},\qquad(4.25)$$

(iii) for the anti-plane case:

$$\hat{U}_{33}(\boldsymbol{x}, \boldsymbol{y}; \omega) = \mathring{U}_{33} + \hat{U}^{\mathrm{D}}_{33} ,$$

$$\mathring{U}_{33} = \frac{1}{2\pi\mu} \ln(1/r) , \tag{4.23d}$$

$$\hat{U}^{\mathrm{D}}_{33} = \frac{\mathrm{i}}{4\mu} - \frac{1}{2\pi\mu}\left(\ln\frac{k_{\mathrm{T}}}{2} + \gamma\right) + \frac{\mathrm{i}}{4\mu}\sum_{n=1}^{\infty}\frac{(-1)^n}{(n!)^2}\left(\frac{k_{\mathrm{T}}r}{2}\right)^{2n}(A + B) .$$

In these expressions, $\mathring{U}_{jk}$ are fundamental solutions for elastostatics. They show that the order of singularity of elastodynamic fundamental solutions is the same as that of elastostatics.

The second fundamental solutions defined by Eq. (1.44) can also be expanded with simultaneous deletion of unfavourable terms by the use of $\hat{U}_1$, $\hat{U}_2$ given above.

For the anti-plane case, we have

$$\hat{W}_{33}(\boldsymbol{x}, \boldsymbol{y}; \omega) = \mathring{W}_{33} + \hat{W}^{\mathrm{D}}_{33} , \tag{4.26a}$$

$$\mathring{W}_{33} = \frac{1}{2\pi r}\frac{\partial r}{\partial n} ,$$

$$\hat{W}^{\mathrm{D}}_{33} = \frac{\mathrm{i}k_{\mathrm{T}}}{4}\left[\sum_{n=0}^{\infty}\frac{(-1)^n}{n!(n+1)!}\left(\frac{k_{\mathrm{T}}r}{2}\right)^{2n+1}\left\{A - E + \frac{2\mathrm{i}}{\pi}\ln\frac{k_{\mathrm{T}}}{2}\right\}\right]\frac{\partial r}{\partial n} . \tag{4.26b}$$

4.4.3. Circumvention of singularities

The existence of a geometric singularity such as a corner, an edge, etc. and of an abrupt change in imposed boundary conditions may pollute the solution of integral equations. The standard method to improve the solution is to take properties of singularities into account. Eigenfunction expansion solutions may be used at the singularity which are incorporated into the boundary integral equation [48, 129]. Singular boundary elements are also devised to evaluate boundary integrals [5, 74].

At a corner where several subdomains meet or two subdomains meet the exterior domain, the "double point" technique is also available [14, 108, 127].

Singularity problems are sometimes circumvented by introducing "discontinuous elements" or "partially discontinuous elements" [103], in which boundary functions (quantities) are expressed by using free nodes (arbitrarily chosen points) other than geometric nodes.

4.4.4. Volume integrals

Volume integrals can sometimes be converted into surface integrals by the use of Gauss's divergence theorem or the formula given by Eq. (3.24) for the constant element [25, 118].

4.4.5. Numerical inversion of the Fourier transform

The numerical Fourier transform and its inversion are conveniently carried out with the aid of the Fast Fourier Transform (FFT) algorithm [9, 21]. The numerical Laplace transform inversion has also been studied in some detail in [87] and the conclusion has been drawn that among the many proposed methods, Durbin's method [34] seems the most suitable one for dynamic problems. Ref. [87] also discusses the numerical direct Laplace transform based on Durbin's method.

5. Dynamic problems in viscoelastic media

5.1. Viscoelastic media

With the assumption of infinitesimal deformations, the constitutive relation for a linear viscoelastic medium is expressed by [15]

$$\sigma_{ij}(\boldsymbol{x}, t) = \int_{-\infty}^{t} G_{ijkl}(t-\tau)\, \frac{\mathrm{d}\varepsilon_{kl}(\boldsymbol{x}, \tau)}{\mathrm{d}\tau}\, \mathrm{d}\tau\,, \tag{5.1a}$$

where $G_{ijkl}(t)$ characterizes the mechanical behavior of the viscoelastic medium and is termed the relaxation tensor. Using the convolution integral defined over $(-\infty, t]$, we may write Eq. (5.1a) as

$$\sigma_{ij}(\boldsymbol{x}, t) = G_{ijkl}(t) * \mathrm{d}\varepsilon_{kl}(\boldsymbol{x}, t)\,. \tag{5.1b}$$

For the homogeneous isotropic medium,

$$G_{ijkl}(t) = \lambda(t)\delta_{ij}\delta_{kl} + \mu(t)(\delta_{ik}\delta_{jl} + \delta_{il}\delta_{jk}) \tag{5.2}$$

where $\lambda(t)$, $\mu(t)$ are two independent relaxation functions. Hence,

$$\sigma_{ij}(\boldsymbol{x}, t) = \lambda(t) * \mathrm{d}\varepsilon_{kk}(\boldsymbol{x}, t)\delta_{ij} + 2\mu(t) * \mathrm{d}\varepsilon_{ij}(\boldsymbol{x}, t)\,. \tag{5.3}$$

Here we assume that

$$G_{ijkl}(t) = \mathring{G}_{ijkl} + \bar{G}_{ijkl}(t) \tag{5.4}$$

and $\bar{G}_{ijkl}(t) \to 0$ as $t \to \infty$.

Specifying the strain history as being a time-harmonic function

$$\boldsymbol{\varepsilon}(\boldsymbol{x}, t) = \boldsymbol{\varepsilon}_0(\boldsymbol{x})\, \mathrm{e}^{-\mathrm{i}\omega t} \tag{5.5}$$

and introducing it into Eq. (5.1), we have, with the change of variables $t-\tau$ to η

$$\boldsymbol{\sigma}(\boldsymbol{x}, t) = \mathrm{i}\omega\left[\int_0^\infty \boldsymbol{G}(\eta)\, \mathrm{e}^{\mathrm{i}\omega\eta}\, \mathrm{d}\eta\right]\boldsymbol{\varepsilon}_0(\boldsymbol{x})\, \mathrm{e}^{-\mathrm{i}\omega t}\,, \tag{5.6}$$

Since $\boldsymbol{G}(t) = \boldsymbol{0}$ for $t < 0$, we can replace the lower limit of the integtration by $-\infty$. Hence the integral in (5.6) can be expressed as the Fourier integral

$$\hat{\boldsymbol{G}}(\omega) = \int_{-\infty}^{\infty} \boldsymbol{G}(\eta)\, \mathrm{e}^{\mathrm{i}\omega\eta}\, \mathrm{d}\eta\,. \tag{5.7}$$

Rewriting Eq. (5.6) as

$$\boldsymbol{\sigma}(\boldsymbol{x}, t) = \boldsymbol{G}^*(\omega)\boldsymbol{\varepsilon}_0(\boldsymbol{x})\, \mathrm{e}^{-\mathrm{i}\omega t} \tag{5.8}$$

we have

$$\boldsymbol{G}^*(\omega) = \mathrm{i}\omega\hat{\boldsymbol{G}}(\omega)\,, \tag{5.9}$$

which is called the complex relaxation tensor. This tensor may be also written as

$$\boldsymbol{G}^*(\omega) = \boldsymbol{G}^0(\omega) + \mathrm{i}\boldsymbol{G}^1(\omega)\,. \tag{5.10}$$

At zero frequency, it follows that for exponentially decaying type relaxation functions

$$\boldsymbol{G}^0(0) = \mathring{\boldsymbol{G}} = \bar{\boldsymbol{G}}(t)\big|_{t=\infty}\,, \qquad \boldsymbol{G}^1(0) = \boldsymbol{0}\,. \tag{5.11}$$

At infinite frequency, it can be shown with the change of variable $\omega\eta = \tau$, that

$$\boldsymbol{G}^0(\infty) = \mathring{\boldsymbol{G}} + \bar{\boldsymbol{G}}(0) = \boldsymbol{G}(t)\big|_{t=0}\,, \qquad \boldsymbol{G}^1(0) = \boldsymbol{0}\,. \tag{5.12}$$

These expressions imply that viscoelastic solids behave as elastic solids for very high and low frequencies.

From the more general point of view, the Fourier transform of Eq. (5.1) reduces to

$$\hat{\boldsymbol{\sigma}}(\boldsymbol{x}, \omega) = \mathrm{i}\omega\hat{\boldsymbol{G}}(\omega)\hat{\boldsymbol{\xi}}(\boldsymbol{x}; \omega) \tag{5.13}$$

and hence, using Eq. (5.9), we have

$$\hat{\boldsymbol{\sigma}}(\boldsymbol{x}, \omega) = \boldsymbol{G}^*(\omega)\hat{\boldsymbol{\varepsilon}}(\boldsymbol{x}; \omega)\,. \tag{5.14}$$

5.2. Dynamic problems for viscoelastic media

When the dynamic problems for viscoelastic media are considered in the Fourier-transformed domain, the same formulations as in elastodynamics can be immediately derived by simply replacing elastic tensors by the complex relaxation tensors.

The complex relaxation tensor for an isotropic homogeneous viscoelastic medium is given by

$$G^*_{ijkl}(\omega) = \lambda^*(\omega)\delta_{ij}\delta_{kl} + \mu^*(\omega)(\delta_{ik}\delta_{jl} + \delta_{il}\delta_{jk}) , \tag{5.15}$$

the Fourier-transformed equation of motion becomes, ignoring the body force,

$$\mu^*(\omega)\nabla^2\hat{\boldsymbol{u}}(\boldsymbol{x};\omega) + [\lambda^*(\omega) + \mu^*(\omega)]\boldsymbol{\nabla\nabla}\cdot\hat{\boldsymbol{u}}(\boldsymbol{x};\omega) + \rho\omega^2\hat{\boldsymbol{u}}(\boldsymbol{x};\omega) = \boldsymbol{0} , \tag{5.16}$$

and the boundary conditions are the same as those in elastodynamics. Therefore, we can utilize the same techniques as those developed for elastodynamics.

Consider the time-harmonic displacement wave

$$\boldsymbol{u} = A\boldsymbol{d}\exp[\mathrm{i}(k\boldsymbol{x}\cdot\boldsymbol{p} - \omega t)] , \tag{5.17}$$

where ω is the circular frequency, k is the wave-number and $\boldsymbol{p}$ and $\boldsymbol{d}$ are unit vectors in the directions of propagation and motion, respectively. Substituting Eq. (5.17) into Eq. (5.16), we have

$$(\lambda^* + \mu^*)k^2(\boldsymbol{p}\cdot\boldsymbol{d})\boldsymbol{p} + (\mu^*k^2 - \rho\omega^2)\boldsymbol{d} = \boldsymbol{0} . \tag{5.18}$$

This equation can be satisfied by the alternatives
(i) $\boldsymbol{d} = \boldsymbol{p}$, and hence $(\lambda^* + 2\mu^*)k^2 = \rho\omega^2$,
(ii) $\boldsymbol{p}\cdot\boldsymbol{d} = 0$, and hence $\mu^*k^2 = \rho\omega^2$.

In Eq. (5.18) either k or ω can be considered as the real-valued independent variable. In the following we take ω as the independent real variable. Hence k must be a complex number of the form

$$k^* = k^0 + \mathrm{i}k^1 , \qquad k^0, k^1 : \text{real} . \tag{5.19}$$

Case (i) implies a P-wave with a displacement $\boldsymbol{u}$ given as

$$\boldsymbol{u} = A\boldsymbol{d}\exp(-k^1_{\mathrm{L}}\boldsymbol{x}\cdot\boldsymbol{d})\exp[\mathrm{i}(k^0_{\mathrm{L}}\boldsymbol{x}\cdot\boldsymbol{d} - \omega t)] \tag{5.20}$$

and a wavenumber k^*_{L} determined from

$$k_{\mathrm{L}}^{*}=\left(\frac{\rho}{\lambda^{*}+2\mu^{*}}\right)^{1/2}\omega=k_{\mathrm{L}}^{0}+\mathrm{i}k_{\mathrm{L}}^{1}\,. \tag{5.21}$$

Now putting

$$\frac{\rho}{\lambda^{*}+2\mu^{*}}=\frac{1}{c_{\mathrm{L}}^{2}}\,\mathrm{e}^{2\mathrm{i}\alpha}\,,\qquad c_{\mathrm{L}},\alpha:\mathrm{real}\,, \tag{5.22}$$

we have

$$k_{\mathrm{L}}^{0}=\frac{\omega}{c_{\mathrm{L}}}\cos\alpha\,,\qquad k_{\mathrm{L}}^{1}=\frac{\omega}{c_{\mathrm{L}}}\sin\alpha\,. \tag{5.23}$$

Case (ii) expreses an S-wave with a displacement $\boldsymbol{u}$ given as

$$\boldsymbol{u}=A\boldsymbol{d}\exp(-k_{\mathrm{T}}^{1}\boldsymbol{x}\cdot\boldsymbol{p})\exp[\mathrm{i}(k_{\mathrm{T}}^{0}\boldsymbol{x}\cdot\boldsymbol{p}-\omega t)] \tag{5.24}$$

and a wave-number k_{T}^{*} determined from

$$k_{\mathrm{T}}^{*}=(\rho/\mu^{*})^{1/2}\omega=k_{\mathrm{T}}^{0}+\mathrm{i}k_{\mathrm{T}}^{1} \tag{5.25}$$

with

$$k_{\mathrm{T}}^{0}=\frac{\omega}{c_{\mathrm{T}}}\cos\beta\,,\qquad k_{\mathrm{T}}^{1}=\frac{\omega}{c_{\mathrm{T}}}\sin\beta\,, \tag{5.26}$$

where β is given by $\rho/\mu^{*}=\mathrm{e}^{2\mathrm{i}\beta}/c_{\mathrm{T}}^{2}$.

Furthermore, since $\boldsymbol{d}$ can be decomposed into the horizontal and vertical directions, $\boldsymbol{p}\cdot\boldsymbol{d}=0$ implies the two cases such as motion in the vertical plane (SV-wave) and in the horizontal plane (SH-wave).

Fundamental solutions can be derived from the ones for elastodynamics by simply replacing the wavenumbers k_{L} and k_{T} by k_{L}^{*} and k_{T}^{*}, respectively. Finally we have the same integral representations and integral equations as the ones for time-harmonic elastodynamics, i.e. Eqs. (2.10), (3.5) and (3.2), (3.6) and (3.26), (3.28), using the fundamental solutions for viscoelastodynamics.

For the time–space domain, we have integral representations similar to those of Eqs. (2.12)–(2.14) and integral equations similar to those of Eqs. (3.7)–(3.9) [38], in which we must replace the elastodynamic fundamental solutions by the corresponding viscoelastic ones, which may be derived from the time-harmonic viscoelastodynamic fundamental solutions by the aid of inverse Fourier transform.

6. Examples

In this section, we present some numerical examples, which may help to understand the usefulness and applicability of the integral equation method.

6.1. Time-harmonic solutions

In order to have a general idea for the applicability of the integral equation method and also to confirm the accuracy of the numerical results some simple time-harmonic example problems are solved first.

(1) A circular hole subject to a time-harmonic SH-wave

As the simplest case, we computed displacements and stresses around a circular hole of radius a in an infinite domain subject to a time-harmonic plane displacement SH wave. The results were obtained with quadratic isoparametric boundary elements and an 8-point Gaussian quadrature formula. Figure 3 shows some results for different wave-numbers and for different numbers of elements, in which analytical curves are calculated from the formula given in the series expansion form [101] by taking 21 terms into account.

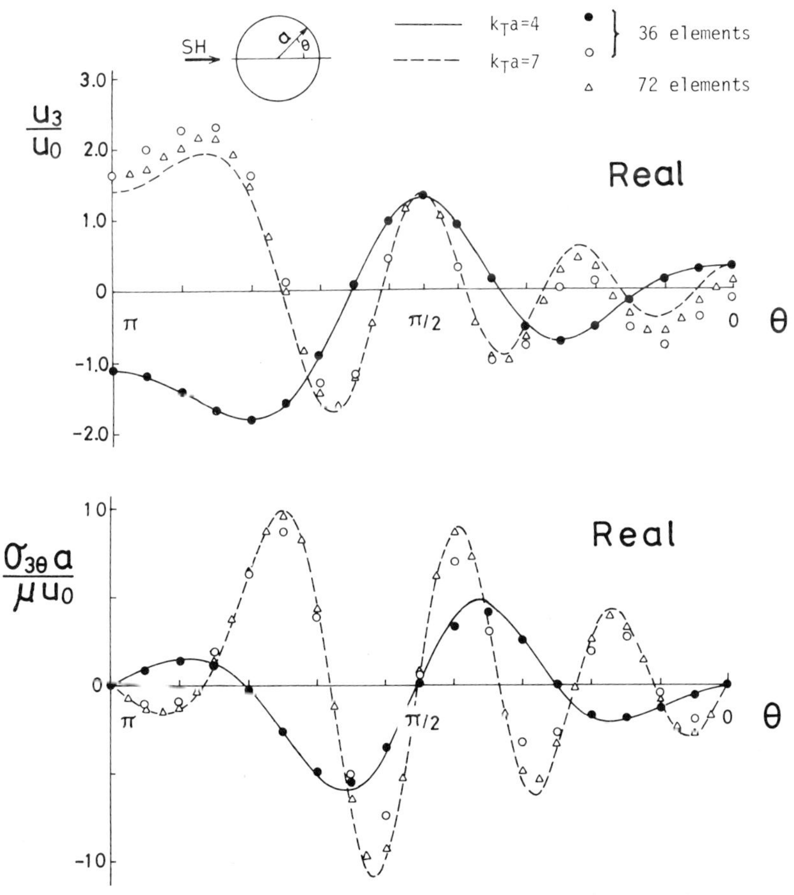

Fig. 3. Boundary displacements and hoop stresses around a hole due to a plane harmonic displacement SH-wave. Symbols: BIEM, Curves: analytical. u_0: amplitude of the incident wave.

The real part implies the solution at the moment when the peak of the incident wave arrives at the center of the hole. The results show reasonable agreement with each other. The wave-number $k_T a = 7$ corresponds to a wavelength of about 0.86 times the radius of the hole.

(2) A circular hole subject to a time-harmonic P-wave

In Fig. 4, hoop stresses obtained by the use of BIEM are compared with the analytical ones [100] when the hole is subject to a time-harmonic plane stress P-wave. The analysis was carried out with 24 quadratic isoparametric boundary elements and an 8-point Gaussian quadrature formula. The results show a good agreement with each other.

(3) A circular lined tunnel in a half-space subject to a time-harmonic SH wave

Figure 5 shows the anti-plane displacement amplitudes on the outer surface of the lining and on the ground surface due to a time-harmonic plane displacement SH wave with different incident angles. The analysis was done with 120 constant boundary elements on the basis of the half-space Green's function for the integral kernels. These results by the BIEM agree very well with the analytical ones.

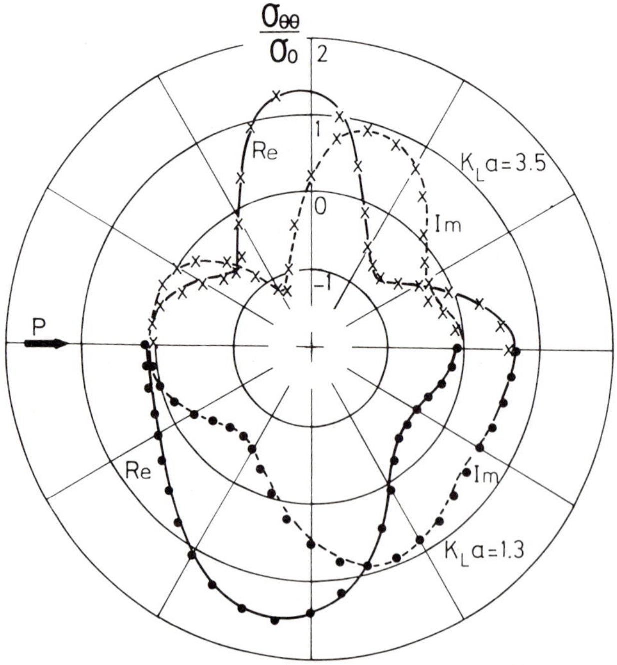

Fig. 4. Hoop stresses around a hole due to the plane harmonic stress P-wave. Symbols: BIEM, Curves: analytical. σ_0: magnitude of the incident wave.

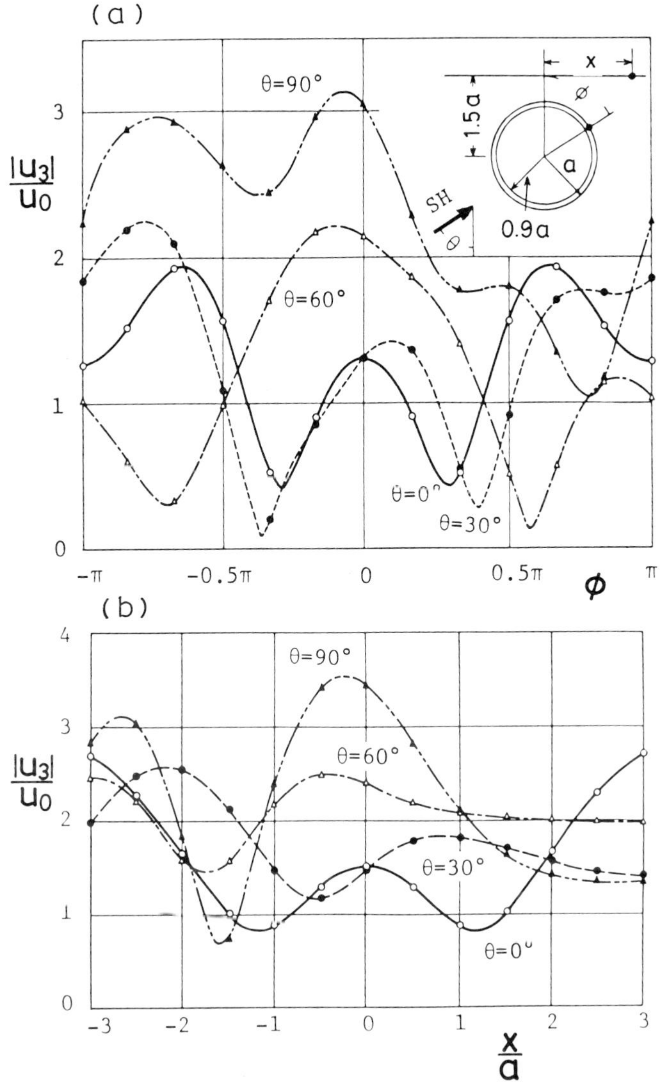

Fig. 5. Displacements on the outer surface of the linning and on the ground surface due to a plane harmonic displacement SH-wave. Symbols: BIEM, Curves: analytical. ($c_T^i/c_T^e = 9$, $k_{Te}a = 1.57$; i: lining, e: soil). u_0: amplitude of the incident wave.

(4) A spherical cavity subject to a plane time-harmonic P wave [59]

As an attempt to provide results from a three-dimensional analysis, displacements around a spherical cavity due to a plane time-harmonic P-wave were analysed. Figure 6 shows the results in comparison with the analytical ones [101]. The results show a good agreement even for the rather coarse discretization of the boundary (30 quadratic isoparametric elements and 16-point Gaussian quadratic formula) that was used.

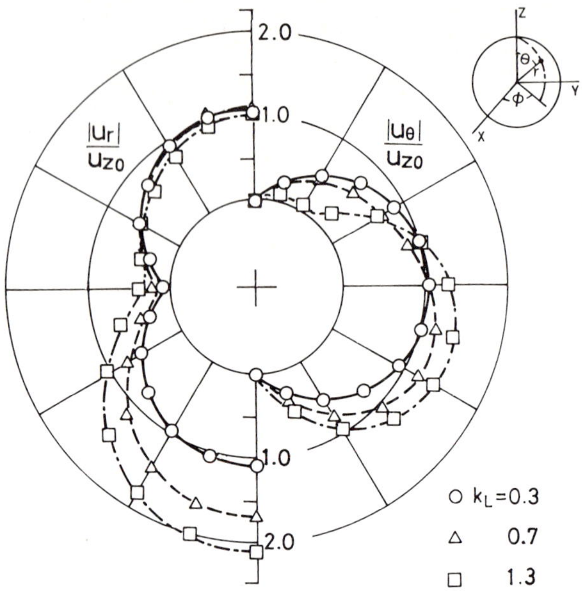

Fig. 6. Boundary displacements around a spherical cavity due to a plane harmonic displacement P-wave. Poisson's ratio $\nu = 0.25$. Symbols: BIEM, Curves: analytical. u_{z0}: amplitude of the incident wave.

6.2. *Transient responses*

6.2.1. *BIEM in the Fourier-transformed domain*

(1) A circular hole in an infinite space subject to a step SH-wave

Figure 7 shows displacement distributions on a circular hole subject to a step plane SH-wave as compared with anaytical ones calculated from a formula [101]

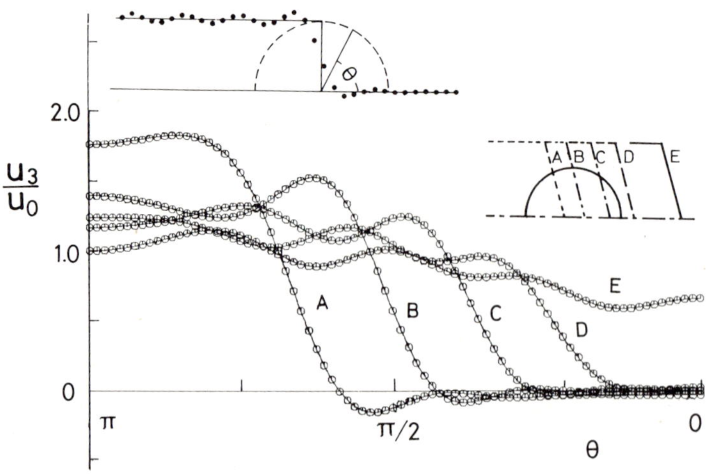

Fig. 7. Transient displacement and hoop stress distributions around a hole subject to a plane step displacement SH-wave. Symbols: BIEM, Curves: analytical. u_0: amplitude of the incident wave.

with 21 terms taken into account. In the analysis, 160 quadratic isoparametric boundary elements with an 8-point Gaussian quadrature formula were used. The shape of the incident wave is shown in the small figure, which accommodates wave-numbers up to $k_T a = 7$. The BIEM solutions show an excellent agreement with the analytical ones.

(2) A circular hole in an infinite plane subject to a step P-wave

Figure 8 shows the displacement and hoop stress histories calculated at several boundary points of a hole subject to a step P-wave in comparison with analytical

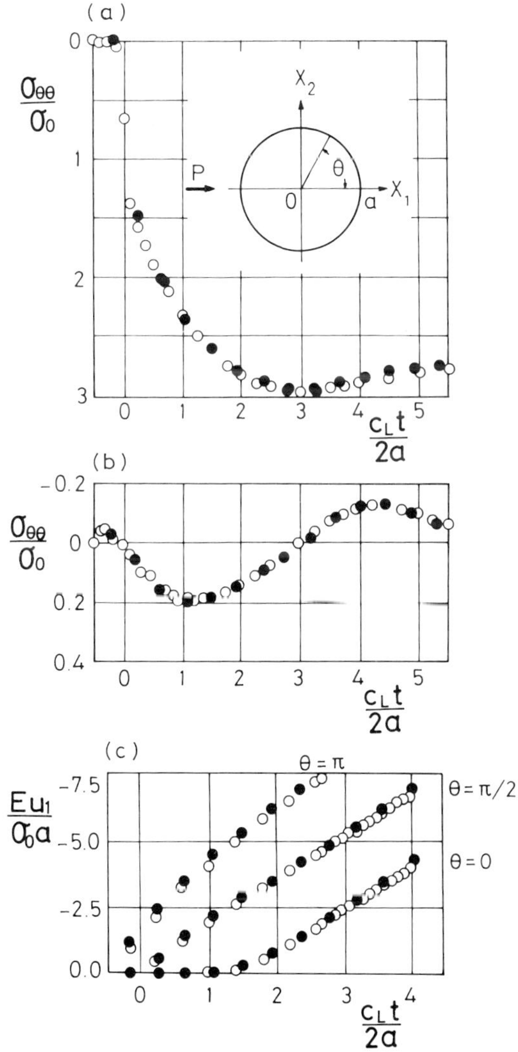

Fig. 8. Displacement and hoop stress distributions at several boundary points on the hole subject to a plane step stress P-wave. Poisson's ratio $\nu = 0.25$. (a) Hoop stress at $\theta = \pi/2$, (b) Hoop stress at $\theta = \pi$, (c) Displacement u_1. Solid circles: BIEM, Open circles: Garnet and Crouzet-Pascal [42]. σ_0: magnitude of the incident wave.

ones obtained by Garnet and Crouzet-Pascal [42]. The numerical results with 24 isoparametric quadratic elements agree well with the analytical ones for up to approximately $k_T a = 7$.

6.2.2. BIEM in time–space domain

(1) A circular hole subject to a step SH-wave [41]

A time-space domain formulation is suitable for transient analysis, specifically for the analysis of the early stage of responses to travelling waves.

Figure 9 shows histories of displacements and stresses at several points on the

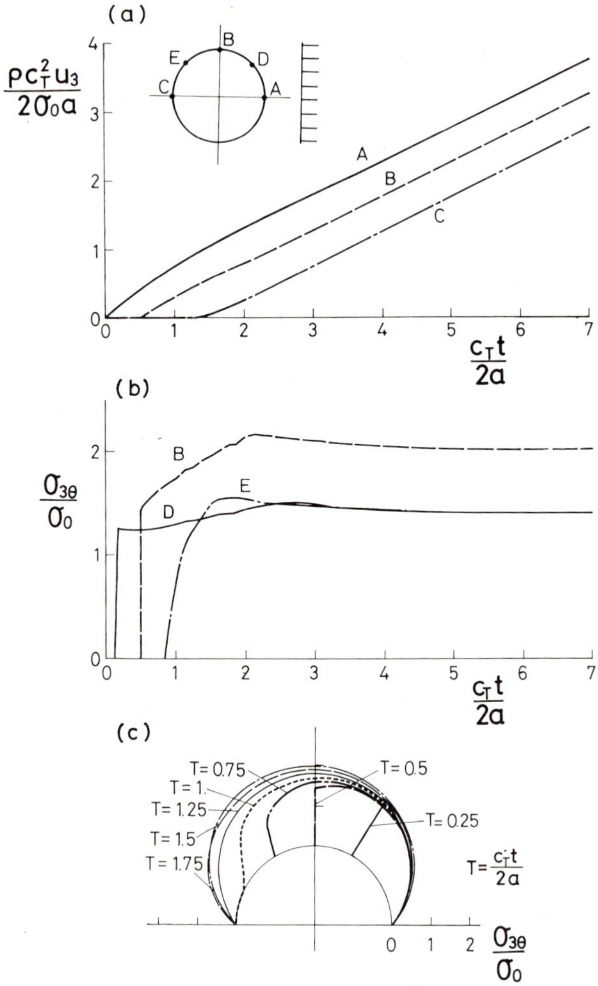

Fig. 9. Displacement and hoop stress histories at several points on the hole subject to a step stress SH-wave analysed by the time–space domain BIEM. (a) Displacements. (b) Hoop stresses. (c) Hoop stresses. σ_0: magnitude of the incident wave.

hole in an infinite space subject to a step SH-wave. The hoop stress distributions are also shown in Fig. 9. In the analysis, 16 boundary elements were used with a linear time shape function and the time-step was selected such that the wave-front crosses the hole in 8 steps. The results show good stability.

(2) A circular hole subject to a step SV-wave

Figure 10 shows stress histories at several points on a circular hole due to a step SV-wave calculated by the BIEM in time–space domain [66] as compared with those by the BIEM in the Fourier-transformed domain. These results show again quite good agreement with each other.

6.3. Eigenfrequency problems [57, 98]

Eigenfrequencies are roots of the determinant of the discretized homogeneous boundary integral equation. The roots are usually searched for by the direct substitution of presumed values.

Table 1 shows the eigenfrequencies of the Helmholtz equation in a rectangular domain with $2a \times a$, which is fixed along the long edge and traction-free on the short edge. In Table 1, m and n mean the nodal numbers on the long and short edges, respectively. In the calculation, the boundary was discretized into 54 equal constant elements and an 8-point Gaussian quadrature formula was used with a increment of wavenumber $k_T a = 0.1$.

Figure 11 shows eigenmodes of a dam and reactions from the rigid base. The numerical figures denote the eigenfrequencies. The figures in parentheses imply

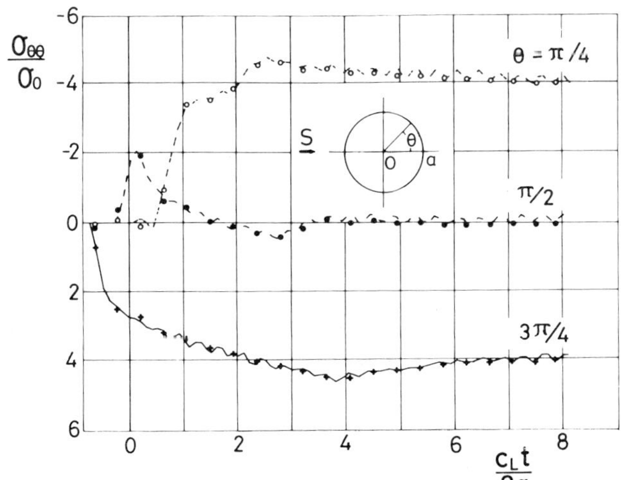

Fig. 10. Hoop stress histories at several points on a circular hole due to a step stress SV-wave. Comparison was made for the results by the time–space type BIEM (curves) and those by the Fourier-transformed type BIEM (symbols). σ_0: magnitude of the incident wave.

Table 1
Eigenfrequencies of the Helmholtz equation in a $2a \times a$ rectangular domain, which is fixed along the long edge and traction-free on the short edge. $\Delta k_T a = 0.1$.

	$n=1$		$n=2$		$n=3$		$n=4$	
m	BIEM	Anal.	BIEM	Anal.	BIEM	Anal.	BIEM	Anal.
0	6.3	6.283	12.6	12.566	18.9	18.850	25.1	25.133
1	7.0	7.025	13.0	12.953	19.1	19.110	25.3	25.328
2	8.9	8.886	14.0	14.050	19.9	19.869	25.9	25.906
3	11.3	11.327	15.7	15.708	21.1	21.074	26.8	26.842
4	14.0	14.050	17.8	17.772	22.6	22.654	28.1	28.099
5	16.9	16.918	20.1	20.116	24.5	24.537	29.6	29.638
6	19.9	19.869	22.6	22.654	26.6	26.657	31.4	31.416

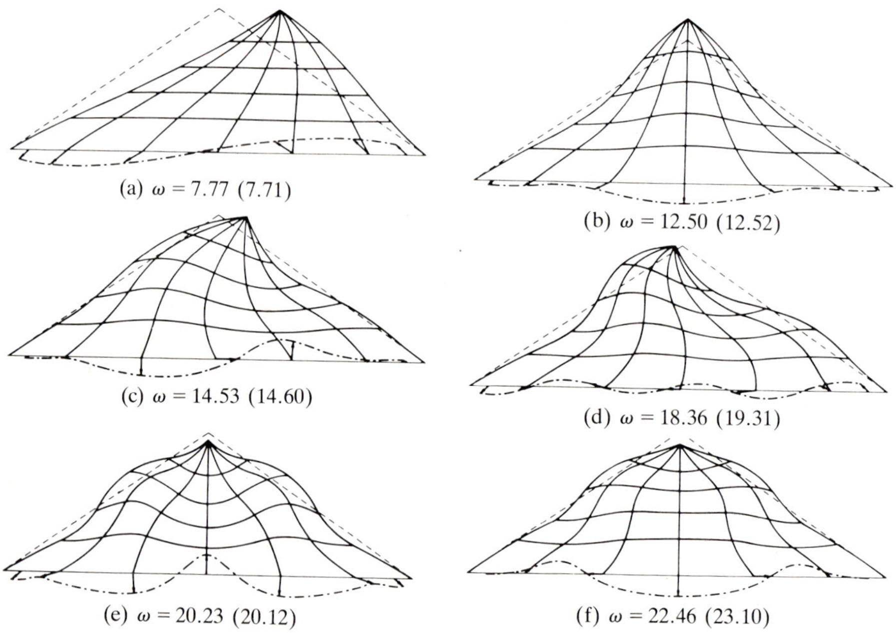

(a) $\omega = 7.77$ (7.71)
(b) $\omega = 12.50$ (12.52)
(c) $\omega = 14.53$ (14.60)
(d) $\omega = 18.36$ (19.31)
(e) $\omega = 20.23$ (20.12)
(f) $\omega = 22.46$ (23.10)

Fig. 11. Eigenmodes, reactions and eigenfrequencies of a dam structure on a rigid base. Poisson's ratio $\nu = 0.45$. Eigenfrequencies: BIEM (FEM [19]).

the values obtained by the use of the FEM [19]. The results agree well with each other.

6.4. *Miscellaneous examples*

(1) Displacement response curves of a column

Figure 12 shows response curves for the horizontal displacement at the top of a column subject to SV-waves. The critical angle of incidence is $\theta_{cr} = 35.26°$ in this

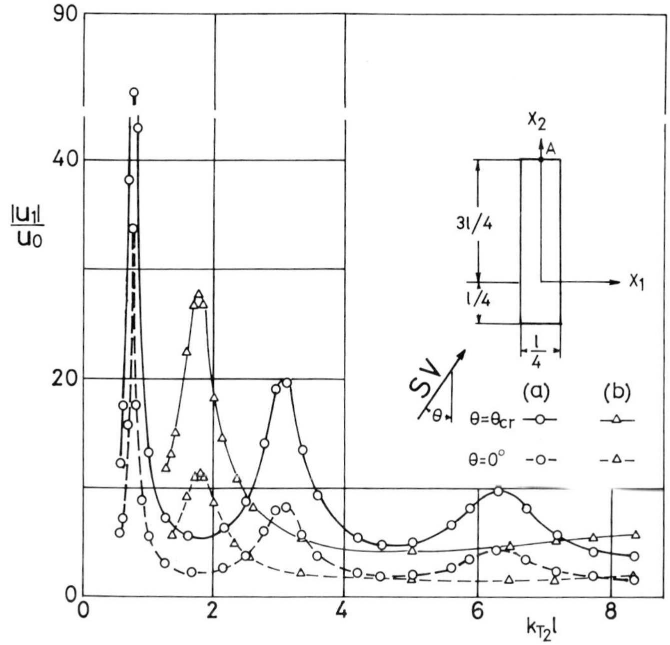

Fig. 12. Response curves of the horizontal displacement at the top of a column A due to SV-waves. (a) Young's modulus $E_1/E_2 = 1.0$, density $\rho_1/\rho_2 = 1.0$. (b) $E_1/E_2 = 10$, $\rho_1/\rho_2 = 1.0$ (1: column, 2: soil, Poisson's ratio $\nu_1 = \nu_2 = 0.25$). u_0: amplitude of the incident wave.

example. The magnification factor in the displacement becomes very large when the wave comes in at the critical angle. This means that dominant frequencies can be obtained more accurately by means of waves with the critical incident angle.

(2) Displacement distributions in a deposit-filled valley due to a time-harmonic SV-wave [61]

As a rather practical application, an analysis was made for the determination of the dynamic response of a filled valley with two layers of deposits subject to a time-harmonic plane SV-wave.

Figure 13 shows the displacement distributions. The real (imaginary) part implies that the peak (node) of the incident wave arrives at the origin of the coordinates (see the small figure). In the analysis the combined method of BIEM with FEM is applied using quadratic isoparametric elements as shown in the small figure. The boundaries on both sides of the valley are extended to more than twice the incident wavelength.

(3) Viscoelastic analsis of a deposit-filled valley

A similar problem as in (2) but with the two layers of the deposits being viscoelastic was analysed using the BIE–FE combined method. In this case, the

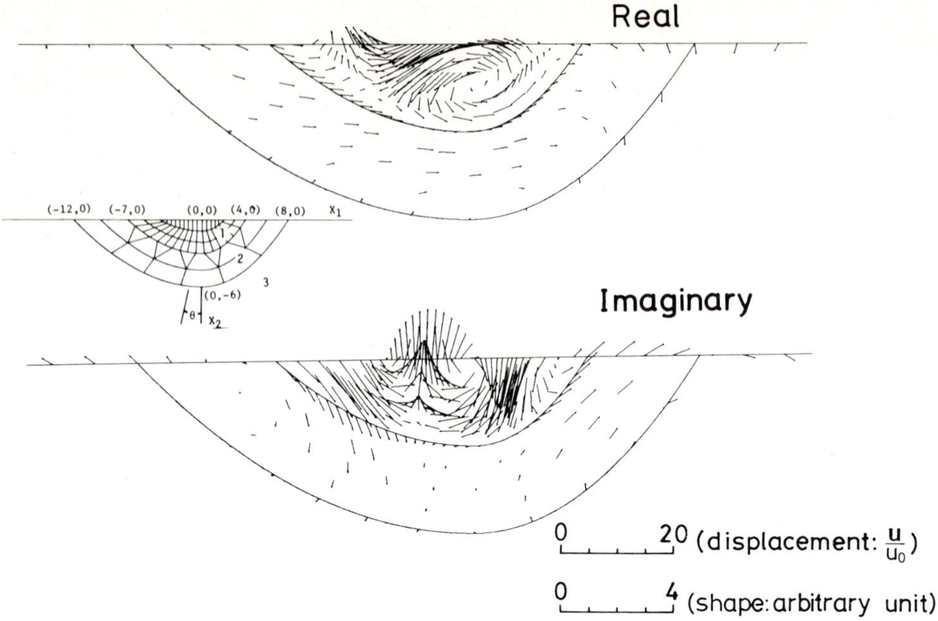

Fig. 13. Displacements in an alluvial valley subject to a plane harmonic SV-wave with wavelength $\lambda_0 = 24$ (arbitrary unit) and amplitude u_0. Young's modulus: $E_1 = E_3/25$. $E_2 = E_3/5$, $E_3 = 10$ GPa, Density: $\rho_1 = 2\,\mathrm{t/m^3}$, $\rho_2 = \rho_3 = 2.4\,\mathrm{t/m^3}$, Poisson's ratio: $\nu_1 = 0.4$, $\nu_2 = 1/3$, $\nu_3 = 1/4$. Incident angle $\theta = 20°$.

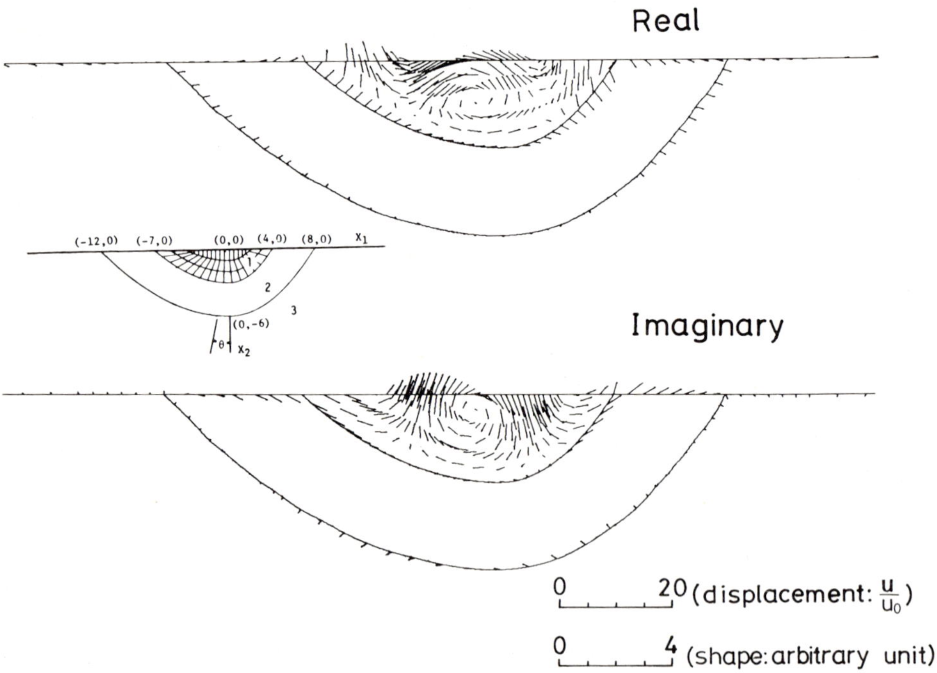

Fig. 14. Displacements in a viscoelastic alluvial valley subject to a plane harmonic SV-wave with wavelength $\lambda_0 = 24$ (arbitrary units) and amplitude u_0. Material constants are the same as those in Fig. 13 with $k_1^1/k_1^0 = 0.07$, $k_2^1/k_2^0 = 0.05$, $k_3^1/k_3^0 = 0$ in the complex wave-number $k_\alpha^* = k_\alpha^0 + \mathrm{i}k_\alpha^1$. Incident angle $\theta = 20°$.

model was modified slightly so as to use the FEM only in the central deposit domain denoted by 1 and the BIEM for the rest.

The imaginary wave-numbers used in this calculation were $k_2^1 = 0.07k_2^0$ and $k_1^1 = 0.05k_1^0$ for deposit domains 2 and 1, respectively. Figure 14 shows the displacement distributions due to the same incident wave as the one mentioned in Fig. 13.

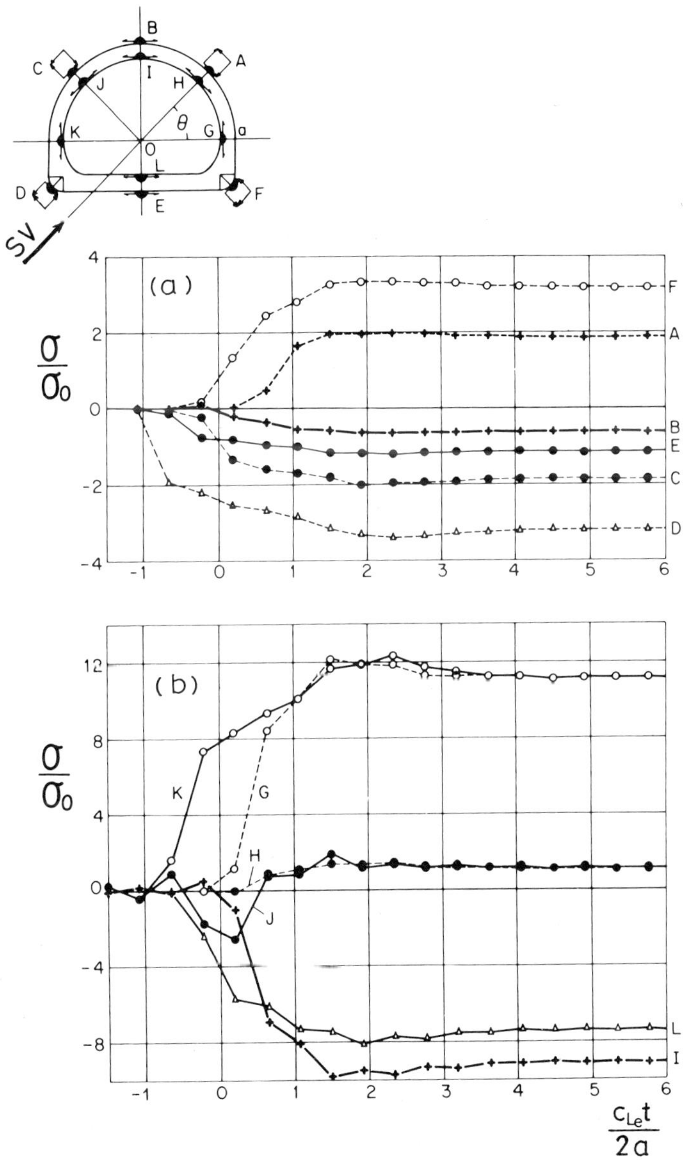

Fig. 15. Transient stresses around a lined tunnel subject to a step stress SV-wave. (a) Stresses on the surrounding medium. (b) Stresses on the inner surface of the lining. Poisson's ratio: $\nu_e = 1/4$, $\nu_i = 1/6$, Density: $\rho_i/\rho_e = 1.0$, Young's modulus: $E_i/E_e = 5$. (i: lining, e: soil).

(4) Transient stresses around a lined tunnel [65, 66]

In Fig. 15, we show transient stresses around a horseshoe-shaped lined tunnel during the passage of an incident step SV-wave. The model is depicted in the small figure. Use is made of 48 quadratic isoparametric boundary elements and an 8-point Gaussian quadrature formula.

(5) Transient responses of an earth dam [94]

Transient responses of an earth dam with a center-core due to the passage of the plane shear Ricker wavelet were analysed. The model is shown in Fig. 16(a). The shear modulus μ and Poisson's ratio ν vary linearly as shown in Fig. 16(a) with μ^0 = arbitrary constant and $\nu^0 = 0.28$. The plane Ricker wavelet

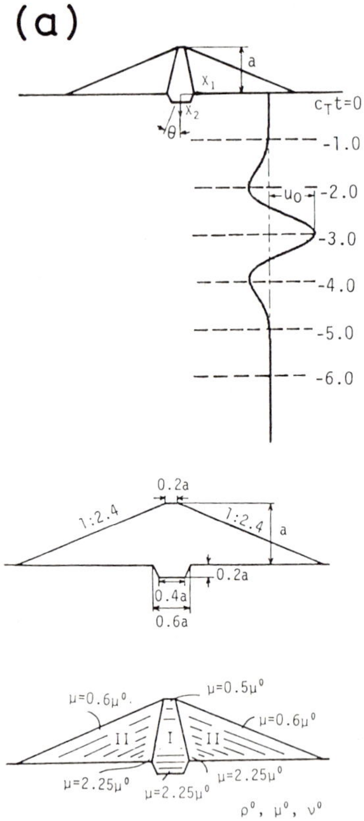

	ν	ρ
I	$1.492\nu^0$	$1.038\rho^0$
II	$1.248\nu^0$	$1.038\rho^0$

Fig. 16. Transient displacement of an earth dam subject to a shear Ricker wavelet of oblique incidence with angle $\theta = 30°$. (a) Model and the incident wave. (b) Movements of the dam. u_0: the maximum amplitude of the incident wave.

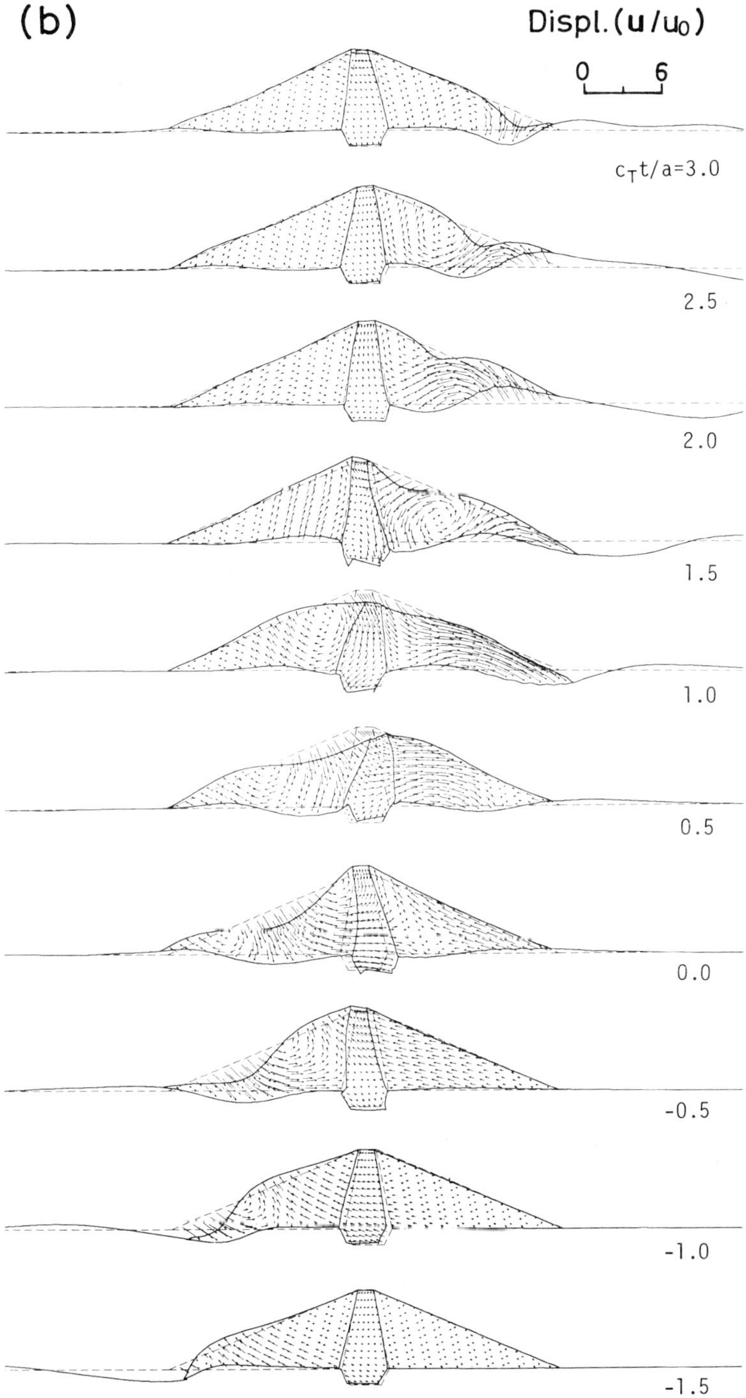

Fig. 16 (continued).

propagating in the $\boldsymbol{p}$ direction is defined as

$$\boldsymbol{u}^{\mathrm{I}}(\boldsymbol{x}, t) = u_0 \boldsymbol{d}(\zeta^2 - \tfrac{1}{2})\,\mathrm{e}^{-\zeta^2}$$

with

$$\zeta = \pi\left(t - \frac{\boldsymbol{x}\cdot\boldsymbol{p}}{c_{\mathrm{T}}}\right)\Big/ t_{\mathrm{p}}\,, \qquad \frac{t_{\mathrm{p}} c_{\mathrm{T}}^0}{a} = \frac{5\pi}{a k^0_{\mathrm{T\,max}}},$$

where $\boldsymbol{d}$ is the unit displacement vector, c_{T} is the incident wave velocity, c_{T}^0 is the wave velocity in the half-plane, t_{p} stands for the characteristic time and $k^0_{\mathrm{T\,max}}$ denotes the maximum value of the transverse wave-number used in the inverse Fourier transform which was $2\pi/a$ in this analysis.

Figure 16(b) shows the transient displacements of the center-cored earth dam subject to the transverse Ricker wavelet of oblique incidence with angle $\theta = 30°$.

For some other applications, the reader is advised to see Refs. [1, 29–33, 36, 43, 68, 84, 86, 99, 109–111, 121–124, 132–135].

7. Concluding remarks

In this chapter we discussed some fundamentals of boundary integral equation methods for elasto- and viscoelastodynamics. Integral equations for elastodynamic problems both in the time–space and the Fourier-transformed domains were formulated. The latter case was also extended to viscoelastic problems with a slight modification in the fundamental solutions. The combination technique of integral equation method with the finite element method was also briefly presented.

Numerical implementations for these formulations were considered and the efficiency and applicability of these formulations for engineering problems were demonstrated by some examples.

In conclusion, boundary integral equation methods as well as their combination with the finite element method can be very effectively applied to linear elasto- and viscoelastodynamic problems. Even for nonlinear problems, these methods may also be applicable with the aid of the equivalent linearization technique.

Recent progress in the area of integral equation methods is remarkable. Further research will find out more efficient and versatile ways to apply them to elasto- and viscoelastodynamic problems.

Acknowledgement

The author would like to express his sincere gratitude to Professor D.E. Beskos for his valuable comments and suggestions.

References

[1] R. Abascal and J. Dominguez, Dynamic Compliances of Strip Footings on Non-Homogeneous Soils, in: *Boundary Elements* VI, C.A. Brebbia, Ed. (Springer-Verlag, Berlin, 1984) pp. 7.15–7.26.

[2] J.D. Achenbach, *Wave Propagation in Elastic Solids* (North-Holland, Amsterdam, 1973).

[3] K. Aki and P.G. Richards, *Quantative Seismology – Theory and Methods* (W.H. Freeman and Co., San Francisco, 1980).

[4] E. Alarcón, J. Dominguez and F. del Cano, Dynamic Stiffness of Foundations, in: *New Developments in Boundary Element Methods*, C.A. Brebbia, Ed., (CML Publications, Southampton, 1980) pp. 264–280.

[5] C. Atkinson, Fracture Mechanics Stress Analysis, in: *Progress in Boundary Element Methods*, Vol. 2, C.A. Brebbia, Ed. (Pentech Press, London, 1983) pp. 53–100.

[6] R.P. Banaugh, Application of Integral Representations of Displacement Potentials in Elastodynamics, *Bulletin of the Seismological Society of America* **54**, 1073–1086 (1964).

[7] R.P. Banaugh and W. Goldsmith, Diffraction of Steady Elastic Waves by Surfaces of Arbitrary Shape, *Journal of Applied Mechanics* **29**, 589–597 (1963).

[8] A. Ben-Menahem and S.J. Singh, *Seismic Waves and Sources* (Springer-Verlag, Berlin, 1981).

[9] E.O. Brigham, *The Fast Fourier Transform* (Prentice-Hall, Englewood Cliffs, NJ, 1974).

[10] K. Brod, On the Uniqueness of Solution for All Wavenumbers in Acoustic Radiation, *Journal of the Acoustical Society of America* **76**, 1238–1243 (1984).

[11] P.W. Buchen, The Elastodynamic Green's Tensor for the 2-D Half-Space, *Journal of the Australian Mathematical Society* **20**, (Ser. B), 385–400 (1978).

[12] A.J. Burton and G.F. Miller, The Application of Integral Equation Methods to the Numerical Solution of Some Exterior Boundary-Value Problems, *Proceedings of the Royal Society of London*, Ser. A, **323**, 201–210 (1971).

[13] F. Chapel and C. Tsakalidis, Computation of the Green's Function of Elastodynamics for a Layered Half Space through a Hankel Transformation – Applications to Foundation Vibration and Seismology, in: *Numerical Methods in Geomechanics Nagoya*, T. Kawamoto and Y. Ichikawa, Eds. (Balkema, Rotterdam, 1985) pp. 1311–1318.

[14] M. Chaudonneret, On the Discontinuity of the Stress Vector in the Boundary Integral Equation Method for Elastic Analysis, in: *Recent Advances in Boundary Element Methods*, C.A. Brebbia, Ed. (Pentech Press, London, 1978), pp. 233–249.

[15] R.M. Christensen, *Theory of Viscoelasticity: An Introduction* (Academic Press, New York, 1971).

[16] L.L. Chu, A. Askar and A.S. Cakmak, An Approximate Method for Scattering in Elastodynamics – The Born Approximation I: The General Formulation, *Soil Dynamics and Earthquake Engineering* **1**, 59–66 (1982).

[17] L.L. Chu, A. Askar and A.S. Cakmak, An Approximate Method for Scattering in Elastodynamics – The Born Approximation II: SH-Waves in Infinite and Half-Space, *Soil Dynamics and Earthquake Engineering* **1**, 102–116 (1982).

[18] L.L. Chu, A. Askar and A.S. Cakmak, Scattering of Plane Strain P and S waves by Rigid Inclusions: An Iterative Approximation, *Soil Dynamics and Earthquake Engineering* **1**, 178–182 (1982).

[19] R.W. Clough and A.K. Chopra, Earthquake Stress Analysis in Earth Dams, *Journal of the Engineering Mechanics Division of the ASCE* **92**, (EM2), 197–211 (1966).

[20] D.M. Cole, D.D. Kosloff and J.B. Minster, A Numerical Boundary Integral Equation Method for Elastodynamics, I, *Bulletin of the Seismological Society of America* **68**, 1331–1357 (1978).

[21] J.W. Cooley and J.W. Tukey, An Algorithm for Machine Calculation of Complex Fourier Series, *Mathematics of Computation* **19**, 297–301 (1965).

[22] L.G. Copley, Integral Equation Method for Radiation from Vibrating Bodies, *Journal of the Acoustical Society of America* **41**, 807–816 (1967).

[23] T.A. Cruse and F.J. Rizzo, A Direct Formulation and Numerical Solution of the General Transient Elastodynamic Problem: I, *Journal of Mathematical Analysis and Applications* **22**, 244–259 (1968).

[24] T.A. Cruse, A Direct Formulation and Numerical Solution of the General Transient Elastodynamic Problem: II, *Journal of Mathematical Analysis and Applications* **22**, 341–355 (1968).

[25] D. Danson, Linear Isotropic Elasticity with Body Forces, in: *Progress in Boundary Element Methods*, Vol. 2, C.A. Brebbia, Ed. (Pentech Press, London, 1983) pp. 101–135.

[26] B. Davies and B. Martin, Numerical Inversion of the Laplace Transform: A Survey and Comparison of Methods, *Journal of Computational Physics* **33**, 1–32 (1979).

[27] J. Dominguez and E. Alarcón, Elastodynamics, in: *Progress in Boundary Element Methods*, Vol. 1, C.A. Brebbia, Ed. (Pentech Press, London, 1981) pp. 213–257.

[28] J.M. Doyle, Radiation Conditions in Elasticity, *Zeitschrift für Angewandte Mathematik und Physik* **16**, 527–531 (1965).

[29] M. Dravinski, Scattering of SH Waves by Subsurface Topography, *Journal of the Engineering Mechanics Division of the ASCE* **108** (EM1), 1–17 (1982).

[30] M. Dravinski, Scattering of Elastic Waves by an Alluvial Valley, *Journal of the Engineering Mechanics Division of the ASCE* **108** (EM1), 19–31 (1982).

[31] M. Dravinski, Ground Motion Amplification due to Elastic Inclusion in a Half-Space, *Earthquake Engineering and Structural Dynamics* **11**, 313–335 (1983).

[32] M. Dravinski, Amplification of P, SV and Rayleigh Waves by Two Alluvial Valleys, *Soil Dynamics and Earthquake Engineering* **2**, 66–77 (1983).

[33] M. Dravinski, Scattering of Plane Harmonic SH Wave by Dipping Layers of Arbitrary Shape, *Bulletin of the Seismological Society of America* **73**, 1303–1319 (1983).

[34] F. Durbin, Numerical Inversion of Laplace Transforms: An Efficient Improvement to Dubner and Abate's Method, *Computer Journal* **17**, 371–376 (1974).

[35] G. Eason, J. Fulton and I.N. Sneddon, The Generation of Waves in an Infinite Elastic Solid by Variable Body Forces, *Philosophical Transactions of the Royal Society*, Ser. A **248**, 575–607 (1956).

[36] R. England, F.J. Sabina and I. Herrera, Scattering of SH Waves by Surface Cavities of Arbitrary Shape Using Boundary Methods, *Physics of the Earth and Planetary Interiors* **21**, 148–157 (1980).

[37] A.C. Eringen and E.S. Suhubi, *Elastodyamics*, Vol. 2 (Academic Press, New York, 1975).

[38] S.F. Felszeghy, J.L. Sackman and W. Goldsmith, Dynamic Response of a Bounded, Three-Dimensional, Viscoelastic Body, *Journal of the Acoustical Society of America* **59**, 551–556 (1976).

[39] I. Fredholm, Sur une classe d'equations functionalles, *Acta Mathematica*, Sweden **27**, 365–390 (1903).

[40] M.B. Friedman and R.P. Shaw, Diffraction of Pulses by Cylindrical Obstacles of Arbitrary Cross-Section, *Journal of Applied Mechanics* **29**, 40–46 (1962).

[41] T. Fukui, Time Marching Analysis of Boundary Integral Equations in Two Dimensional Elastodynamics, in: *Innovative Numerical Methods in Engineering*, *Proceedings of the 4th International Symposium*, R.P. Shaw, J. Periaux, A. Chaudouet, J. Wu, C. Marino and C.A. Brebbia, Eds. (Springer-Verlag, Berlin, 1986) pp. 405–410.

[42] H. Garnet and J. Crouzet-Pascal, Transient Response of a Circular Cylinder of Arbitrary Thickness, in an Elastic Medium, to a Plane Dilatational Wave, *Journal of Applied Mechanics* **33**, 521–531 (1966).

Boundary Element Hybrid Method, in: *Numerical Methods in Geomechanics Nagoya*, T. Kawamoto and Y. Ichikawa, Eds. (Balkema, Rotterdam, 1985) pp. 1519–1524.

[44] D. Graffi, Sul Teorema di Reciprocita nella Dinamica dei Corpi Elastici, *Memoria della Academia dells Scienze*, Bologna, Ser. 10, **4**, 103–111 (1946–7).

[45] D. Greenspan and P. Werner, A Numerical Method for the Exterior Dirichlet Problem for Reduced Wave Equation, *Archive for Rational Mechanics and Analysis* **23**, 288–316 (1966).

[46] F. Hartmann, Computing the C-matrix in Non-Smooth Boundary Points, in: *New Development in Boundary Element Methods*, C.A. Brebbia, Ed., (CML Publications, Southampton, 1980) pp. 367–379.

[47] J.A. Hudson and J.R. Heritage, The Use of the Born Approximation in Seismic Scattering Problems, *Geophysical Journal of the Royal Astronomical Society* **66**, 221–240 (1981).

[48] M.A. Jaswon and G. Symm, *Integral Equation Methods in Potential Theory and Elastostatics* (Academic Press, New York, 1977).

[49] L.R. Johnson, Green's Function for Lamb's Problem, *Geophysical Journal of the Royal Astronomical Society* **37**, 99–131 (1974).

[50] D.S. Jones, Integral Equations for the Exterior Acoustic Problem, *Quarterly Journal of Mechanics and Applied Mathematics* **27**, 129–142 (1974).

[51] D.L. Karabalis and D.E. Beskos, Dynamic Response of 3-D Rigid Surface Foundation by Time Domain Boundary Element Method, *Earthquake Engineering and Structural Dynamics* **12**, 73–93 (1984).

[52] D.L. Karabalis and D.E. Beskos, Dynamic Response of 3-D Flexible Foundation by Time Domain BEM and FEM, *Soil Dynamics and Earthquake Engineering* **4**, 91–101 (1985).

[53] D.L. Karabalis and D.E. Beskos, Dynamic Response of 3-D Rigid Embedded Foundations by the Boundary Element Method, *Computer Methods in Applied Mechanics and Engineering* **56**, 91–119 (1986).

[54] D.L. Karabalis, C.C. Spyrakos and D.E. Beskos, Dynamic Response of Surface Foundations by Time Domain Boundary Element Method, in: *Dynamic Soil–Structure Interaction*, D.E. Beskos, T. Krauthammer and I. Vardoulakis, Eds. (Balkema, Rotterdam, 1984) pp. 19–24.

[55] O.D. Kellogg, *Foundations of Potential Theory* (Springer-Verlag, New York, 1929).

[56] D.E. Kelly, G.G.W. Mustoe and O.C. Zienkiewicz, Coupling Boundary Element Methods with Other Numerical Methods, in: *Development in Boundary Element Methods*, Vol. 1, P.K. Banerjee and R.P. Butterfield, Eds. (Applied Science Publishers, London, 1979) pp. 251–285.

[57] M. Kitahara, *Boundary Integral Equation Methods in Eigenvalue Problems of Elastodynamics and Thin Plates* (Elsevier, Amsterdam, 1985).

[58] M. Kitahara, M. Hamada, K. Nakagawa and Y. Muranishi, Transient Wave Fields around Elastic Inclusions in a Semi-Infinite Foundation, in: *Boundary Elements VI*, C.A. Brebbia, Ed. (Springer-Verlag, Berlin, 1984) pp. 7.3–7.14.

[59] M. Kitahara and K. Nakagawa, Boundary Integral Equation Methods in Three Dimensional Elastodynamics, in: *Boundary Element Methods in Engineering VII*, C.A. Brebbia, Ed. (Springer-Verlag, Berlin, 1985) pp. 6.27–6.36.

[60] S. Kobayashi, Some Problems of Boundary Integral Equation Method in Elastodynamics, in: *Boundary Elements*, C.A. Brebbia, T. Futagami and M. Tanaka, Eds. (Springer-Verlag, Berlin, 1983) pp. 775–784.

[61] S. Kobayashi, Fundamentals of Boundary Integral Equation Methods in Elastodynamics, in: *Topics in Boundary Element Research*, Vol. 2, C.A. Brebbia, Ed. (Springer-Verlag, Berlin, 1985) pp. 1–54.

[62] S. Kobayashi and T. Kishima, Dynamic Analysis of Non-Homogeneous Ground Movements by the Boundary Integral Equation – Finite Element Hybrid Method, in: *Numerical Methods in Geomechanics Nagoya*, T. Kawamoto and Y. Ichikawa, Eds. (Balkema, Rotterdam, 1985) pp. 135–142.

[63] S. Kobayashi and N. Nishimura, Green's Tensor for Elastic Half-Spaces, *Memoirs of the Faculty of Engineering, Kyoto University* **42**, 228–241 (1980).

[64] S. Kobayashi and N. Nishimura, Dynamic Analysis of Underground Structures by the Integral Equation Method, in: *Numerical Methods in Geomechanics Edmonton*, Z. Eisenstein, Ed. (Balkema, Rotterdam, 1982) pp. 401–409.

[65] S. Kobayashi and N. Nishimura, On the Indeterminancy of BIE Solutions for the Exterior Problems of Time-Harmonic Elastodynamics and Incompressible Elastostatics, in: *Boundary Element Methods in Engineering*, C.A. Brebbia, Ed. (Springer-Verlag, Berlin, 1982) pp. 282–296.

[66] S. Kobayashi and N. Nishimura, Transient Stress Analysis of Tunnels and Cavities of Arbitrary Shape due to Travelling Waves, in: *Developments in Boundary Element Methods*, Vol. 2, P.K. Banerjee and R.P. Shaw, Eds. (Applied Science Publishers, London, 1982) pp. 177–210.

[67] S. Kobayashi and N. Nishimura, Analysis of Dynamic Soil-Structure Interactions by Boundary Integral Equation Method, in: *Numerical Methods in Engineering, Proceedings of the 3rd International Symposium on Numerical Methods in Engineering*, Vol. 1, P. Lascaux, Ed. (Pluraris, Paris, 1983) pp. 353–362.

[68] T. Kobori and Y. Shinozaki, Applications of the Boundary Integral Equation Method to Dynamic Soil–Structure Interaction Analysis under Topographic Site Condition, in: *Boundary Elements*, C.A. Brebbia, T. Futagami and M. Tanaka, Eds. (Springer-Verlag, Berlin, 1983) pp. 731–740.
[69] G.H. Koopmann and H. Benner, Method for Computing the Sound Power of Machines Based on the Helmholtz Integral, *Journal of the Acoustical Society of America* **71**, 78–89 (1982).
[70] V.D. Kupradze, *Potential Methods in the Theory of Elasticity* (Israel Program of Scientific Translation, Jerusalem, 1965).
[71] V.D. Kupradze (Ed.), *Three-Dimensional Problems of the Mathematical Theory of Elasticity and Thermoelasticity* (North-Holland Publ. Co., Amsterdam, 1979).
[72] J.C. Lachat and J.D. Watson, Effective Numerical Treatment of Boundary Integral Equations; A Formulation for Three-Dimensional Elastostatics, *International Journal for Numerical Methods in Engineering* **10**, 991–1005 (1976).
[73] H. Lamb, *Hydrodynamics* (Dover, New York, 1954) Sect. 290.
[74] S.G. Lera, E. Paris and E. Alarcon, Treatment of Singularities in 2-D Domains Using BIEM, *Applied Mathematical Modelling* **6**, 111–118 (1982).
[75] A.E. Love, *A Treatise on the Mathematical Theory of Elasticity* (Dover, New York, 1944) Sect. 121, 169, 210.
[76] J.E. Luco and R.J. Apsel, On the Green's Functions for a Layered Half-Space, Parts I & II, *Bulletin of the Seismological Society of America* **73**, 909-929, 931–951 (1983).
[77] G.D. Manolis, A Comparative Study on Three Boundary Element Method Approaches to Problems in Elastodynamics, *International Journal for Numerical Methods in Engineering* **19**, 73–91 (1983).
[78] G.D. Manolis, Boundary Element Method for Soil-Structure Interaction, in: *Dynamic Soil–Structure Interaction*, D.E. Beskos, T. Krauthammer and I. Vardoulakis, Eds. (Balkema, Rotterdam, 1984) pp. 85–91.
[79] G.D. Manolis and D.E. Beskos, Dynamic Stress Concentration Studies by Boundary Integrals and Laplace Transform, *International Journal for Numerical Methods in Engineering* **17**, 573–599 (1981).
[80] G.D. Manolis and D.E. Beskos, Dynamic Response of Lined Tunnels by an Isoparametric Boundry Element Method, *Computer Methods in Applied Mechanics and Engineering*, **36**, 291–307 (1983).
[81] W.J. Mansur and C.A. Brebbia, Numerical Implementation of the Boundary Element Method for Two Dimensional Transient Scalar Wave Propagation Problems, *Applied Mathematical Modelling* **6**, 299–306 (1982).
[82] W.J. Mansur and C.A. Brebbia, Transient Elastodynamics using a Time-Stepping Technique, in: *Boundary Elements*, C.A. Brebbia, T. Futagami and M. Tanaka, Eds. (Springer-Verlag, Berlin, 1983) pp. 677–698.
[83] M. Margulies, Combination of the Boundary Element and Finite Element Methods, in: *Progress in Boundary Element Methods*, Vol. 1, C.A. Brebbia, Ed. (Pentech Press, London, 1981) pp. 258–288.
[84] O. Matsuoka, T. Yokoi and K. Torii, The Fundamental Solution for Periodically Oscillating Line Torsional Loads on a Circular Ring Interior of a Semi-Infinite Solid and Its Application, in: *Boundary Elements*, C.A. Brebbia, T. Futagami and M. Tanaka, Eds. (Springer-Verlag, Berlin, 1983) pp. 741–749.
[85] W.L. Meyer, W.A. Bell, B.T. Zinn and M.P. Stallybrass, Boundary Integral Solutions of Three Dimensional Acoustic Radiation Problems, *Journal of Sound and Vibration* **59**, 245–262 (1978).
[86] A. Mita and W. Takanashi, Dynamic Soil–Structure Interaction Analysis by Hybrid Method, in: *Boundary Elements*, C.A. Brebbia, T. Futagami and M. Tanaka, Eds. (Springer-Verlag, Berlin, 1983) pp. 785–794.
[87] G.V. Narayanan and D.E. Beskos, Numerical Operational Methods for Time-Dependent Linear Problems, *International Journal for Numerical Methods in Engineering* **18**, 1829–1854 (1982).
[88] D. Nardini and C.A. Brebbia, A New Approach to Free Vibration Analysis using Boundary Elements, *Applied Mathematical Modelling* **7**, 157–162 (1983).

[89] D. Nardini and C.A. Brebbia, Transient Dynamic Analysis by the Boundary Element Method, in: *Boundary Elements*, C.A. Brebbia, T. Futagami and M. Tanaka, Eds. (Springer-Verlag, Berlin, 1983) pp. 719–730.

[90] N. Nishimura and S. Kobayashi, Elastoplastic Analysis by Indirect Methods, in: *Developments in Boundary Element Methods*, Vol. 3, P.K. Banerjee and S. Mukherjee, Eds. (Elsevier Applied Science Publishers, London, 1984) pp. 59–85.

[91] Y. Niwa, T. Fukui, S. Kato and K. Fujiki, An Application of the Integral Equation Method to Two-dimensional Elastodynamics, in: *Theoretical and Applied Mechanics* **28** (Univ. of Tokyo Press, 1980) pp. 281–290.

[92] Y. Niwa and S. Hirose, Three Dimensional Analysis of Ground Motion by Integral Equation Method in Wave Number Domain, in: *Numerical Methods in Geomechanics Nagoya*, T. Kawamoto and Y. Ichikawa, Eds. (Balkema, Rotterdam, 1985) pp. 143–149.

[93] Y. Niwa, M. Kitahara and S. Hirose, Elastodynamic Problems for Inhomogeneous Bodies, in: *Boundary Elements*, C.A. Brebbia, T. Futagami and M. Tanaka, Eds. (Springer-Verlag, Berlin, 1983) pp. 751–763.

[94] Y. Niwa, M. Kitahara and S. Hirose, Transient Analysis of Inhomogeneous Structures by Integral Equation Method, in: *Boundary Elements VI*, C.A. Brebbia, Ed. (Springer-Verlag, Berlin, 1984) pp. 6.15–6.26.

[95] Y. Niwa, M. Kitahara and S. Hirose, Dynamic Analysis of Inhomogeneous Anisotropic Bodies by Integral Equation Method, in: *Innovative Numerical Methods in Engineering*, *Proceedings of the 4th International Symposium*, R.P. Shaw, J. Periaux, A. Chaudouet, J. Wu, C. Marino and C.A. Brebbia, Eds. (Springer-Verlag, Berlin, 1986) pp. 361–366.

[96] Y. Niwa, S. Kobayashi and N. Azuma, An Analysis of Transient Stresses Produced around Cavities of an Arbitrary Shape during the Passage of Travelling Waves, *Memoirs of the Faculty of Engineering*, Kyoto University **37**, 28–46 (1975).

[97] Y. Niwa, S. Kobayashi and T. Fukui, Application of Integral Equation Method to Some Geomechanical Problems, in: *Proceedings of the 2nd International Conference on Numerical Methods in Geomechanics*, C.S. Desai, Ed. (ASCE, New York, 1976) pp. 120–131.

[98] Y. Niwa, S. Kobayashi and M. Kitahara, Determination of Eigenvalues by Boundary Element Methods, in: *Developments in Boundary Element Methods*, Vol. 2, P.K. Banerjee and R.P. Shaw, Eds. (Applied Science Publishers, London, 1982) pp. 143–176.

[99] M. Ohtsu and S. Uesugi, Analysis of Elastic Wave Field in a Half-Space and Some Applications to Earthquake Engineering, in: *Boundary Elements* VI, C.A. Brebbia, Ed. (Springer-Verlag, Berlin, 1984) pp. 7.37–7.47.

[100] Y.H. Pao, Dynamic Stress Concentration in an Elastic Plate, *Journal of Applied Mechanics* **29**, 299–305 (1962).

[101] Y.H. Pao and C.C. Mow, Diffraction of Elastic Waves and Dynamic Stress Concentrations (Crane-Russak, New York, 1973).

[102] Y.H. Pao and V. Varatharajulu, Huygens' Principle, Radiation Conditions, and Integral Formulas for the Scattering of Elastic Waves, *Journal of the Acoustical Society of America* **59**, 1361–1371 (1976).

[103] C. Patterson and M.A. Sheikh, Interelement Continuity in the Boundary Element Method, in: *Topics in Boundary Element Research*, Vol. 1, C.A. Brebbia, Ed. (Springer-Verlag, Berlin, 1984) pp. 123–142.

[104] C.M. Piaszczyk, Acoustic Radiation from Vibrating Surfaces at Characteristic Frequencies, *Journal of the Acoustical Society of America* **75**, 363–375 (1984).

[105] F.J. Rizzo and D.J. Shippy, An Advanced Boundary Integral Equation Method for Three-Dimensional Thermoelasticity, *International Journal for Numerical Methods in Engineering* **11**, 1753–1768 (1977).

[106] F.J. Rizzo, D.J. Shippy and M. Rezayat, Boundary Integral Equation Analysis for a Class of Earth-Structure Interaction Problems, *Final NSF Report*, Department of Engineering Mechanics, University of Kentucky, Lexington, Kentucky (1985).

[107] F.J. Rizzo, D.J. Shippy and M. Rezayat, A Boundary Integral Equation Method for Radiation and Scattering of Elastic Waves in Three Dimensions, *International Journal for Numerical Methods in Engineering* **21**, 115–129 (1985).

[108] T.J. Rudolphi, An Implementation of the Boundary Element Method for Zoned Media with Stress Discontinuities, *International Journal for Numerical Methods in Engineering* **19**, 1–15 (1983).
[109] F.J. Sánchez-Sesma, Diffraction of Elastic Waves by Threes-Dimensional Surface Irregularities, *Bulletin of the Seismological Society of America* **73**, 1621–1636 (1983).
[110] F.J. Sánchez-Sesma and E. Rosenblueth, Ground Motions at Canyons of Arbitrary Shapes under Incident SH Waves, *Earthquake Engineering and Structural Dynamics* **7**, 441–450 (1979).
[111] T. Sato, H. Kawase and J. Yoshida, Dynamic Response Analysis of Rigid Foundations Subjected to Seismic Waves by Boundary Element Method, in: *Boundary Elements*, C.A. Brebbia, T. Futagami and M. Tanaka, Eds. (Springer-Verlag, Berlin, 1983) pp. 765–774.
[112] H.A. Schenk, Improved Integral Formulation for Acoustic Radiation Problems, *Journal of the Acoustical Society of America* **44**, 45–58 (1968).
[113] R.P. Shaw, Retarded Potential Approach to the Scattering of Elastic Pulses, by Rigid Obstacles of Arbitrary Shape, *Journal of the Acoustical Society of America* **44**, 745–748 (1967).
[114] A. Sommerfeld, *Partial Differential Equation in Physics* (Academic Press, New York, 1949).
[115] C.C. Spyrakos and D.E. Beskos, Dynamics Response of Rigid Strip Foundations by Time Domain Boundary Element Method, *International Journal for Numerical Methods in Engineering* **23**, 1547–1565 (1986).
[116] C.C. Spyrakos and D.E. Beskos, Dynamic Response of Flexible Strip Foundations by Boundary and Finite Elements, *Soil Dynamic and Earthquake Engineering* **5**, 84–96 (1986).
[117] E. Sternberg and M.E. Gurtin, On the Completeness of Certain Stress Function in the Linear Theory of Elasticity, in: *Proceedings of the 4th U.S. National Congress of Applied Mechanics* **2** (ASME, New York, 1962) pp. 793–797.
[118] M. Stippes and F.J. Rizzo, A Note on the Body Force Integral of Classical Elastostatics, *Zeitschrift für Angewandte Mathematik und Physik* **28**, 339–341 (1977).
[119] H. Takahasi and M. Mori, Double Exponential Formulas for Numerical Integration, *Publications of the Research Institute of Mathematical Sciences*, Kyoto University **9**, 721–741 (1974).
[120] P.S. Theocaris, N. Karayanopoulas and G.A. Tsamasphyros, Numerical Method for the Solution of Static and Dynamic Three-Dimensional Elasticity Problems, *Computers and Structures* **16**, 777–784 (1983).
[121] K. Toki and T. Sato, Seismic Response Analysis of Surface Layer with Irregular Boundaries, *Proceedings of the 6th World Conference on Earthquarke Engineering* (New Dehli, India, 1977) Vol. 1, pp. 409–415.
[122] K. Toki and T. Sato, Seismic Response Analyses of Ground with Irregular Profiles by the Boundary Element Method, in: *Boundary Elements*, C.A. Brebbia, T. Futagami and M. Tanaka, Eds. (Springer-Verlag, Berlin, 1983) pp. 699–708.
[123] O. Tullberg, BEMDYN – A Boundary Element Program for Two-Dimensional Elastodynamics, in: *Boundary Elements*, C.A. Brebbia, T. Futagami and M. Tanaka, Eds. (Springer-Verlag, Berlin, 1983) pp. 835–845.
[124] O. Tullberg and N.-E. Wiberg, Boundary Element Solutions for Some Problems in Soil Dynamics, in: *Numerical Methods in Geomechanics Nagoya*, T. Kawamoto and Y. Ichikawa, Eds. (Balkema, Rotterdam, 1985) pp. 157–164.
[125] F. Ursell, On the Exterior Problems of Acoustics, *Proceedings of the Cambridge Philosophical Society* **74**, 117–125 (1973).
[126] J. Vivoli and P. Filippi, Eigenfrequencies of Thin Plates and Layer Potentials, *Journal of the Acoustical Society of America* **55**, 562–567 (1974).
[127] L.J. Wardle and M. Crotty, Two-Dimensional Boundary Integral Equation Analysis for Non-Homogeneous Mining Applications, in: *Recent Advances in Boundary Element Methods*, C.A. Brebbia, Ed. (Pentech Press, London, 1978) pp. 233–249.
[128] J.O. Watson, Advanced Implementation of the Boundary Element Method for Two- and Three-Dimensional Elastostatics, in: *Developments in Boundary Element Methods*, Vol. 1, P.K. Banerjee and R.P. Butterfield, Eds. (Applied Science Publishers, London, 1979) pp. 31–63.
[129] J.O. Watson, Hermitian Cubic Boundary Elements for Plane Problems of Fracture Mechanics, *Res Mechanica* **4**, 23–42 (1982).

[130] L. Wheeler, On the Integral Representations of Stokes and Love for Elastodynamic Displacement Field, *Society for Industrial and Applied Mathematics* (*SIAM*) *Journal on Applied Mathematics* **21**, 331–338 (1971).

[131] L.T. Wheeler and E. Sternberg, Some Theorems in Classical Elastodynamics, *Archive for Rational Mechanics and Analysis* **31**, 51–90 (1968).

[132] H.L. Wong, Effect of Surface Topography on the Diffraction of P, SV and Rayleigh Waves, *Bulletin of the Seismological Society of America* **72**, 1167–1183 (1982).

[133] H.L. Wong and P.C. Jennings, Effects of Canyon Topography on Strong Ground Motion, *Bulletin of the Seismological Society of America* **65**, 1239–1257 (1975).

[134] H.L. Wong and J.E. Luco, Dynamic Response of Rigid Foundations of Arbitrary Shape, *Earthquake Engineering and Structural Dynamics* **4**, 579–587 (1976).

[135] H.L. Wong, M.D. Trifnac and B. Westermo, Effects of Surface and Subsurface Irregularities on the Amplitudes of Monochromatic Waves, *Bulletin of the Seismological Society of America* **67**, 353–368 (1977).

[136] O.C. Zienkiewicz and K. Morgan, *Finite Elements and Approximation* (John Wiley & Sons, New York, 1983).

CHAPTER 5

Advanced Elastodynamic Analysis

Prasanta K. BANERJEE, Shahid AHMAD and George D. MANOLIS

Department of Civil Engineering
State University of New York at Buffalo
Buffalo, New York 14260, U.S.A.

Boundary Element Methods in Mechanics
Edited by D.E. Beskos

Introduction

Problems arising in three-dimensional elastodynamics are challenging and difficult ones [18, 26]. Analytic techniques are applicable only to highly idealized geometric representations and to linear elastic material behavior, i.e. to cases that hardly ever arise in practice. Experiments, on the other hand, are expensive to conduct and involve elaborate apparatuses.

Work of engineering significance on the dynamic stress concentration phenomenon manifested around cavities engulfed by travelling elastic waves by analytical methods was done in the early 1960s. Attention was focused on the displacement and stress fields produced around two-dimensional circular, elliptic, and parabolic cylindrical cavities in infinitely extending elastic or viscoelastic media due to the passage of harmonic or transient waves. Much of this work is summarized in Pao and Mow [31], where it is observed that the class of problems amenable to analytical solutions is small and recourse must be made to numerical methods of solution.

There are currently two major categories of numerical techniques for dynamic analyses, namely approximate continuum and discrete (lumped parameter) models. The most widely used approximate continuum methods at present and in order of popularity are the FEM and the FDM. In principle, the FEM appears to be a very versatile technique because it can handle complex structure geometry, medium inhomogeneities and complicated material behavior in both two and three dimensions. The finite element formulation results in a system of equations that may be solved by modal analysis, Fourier transform techniques, or step-by-step integration schemes [41].

The major deficiency of the FEM is, as mentioned above, that a semi-infinite medium must be modelled by a mesh of finite size. This situation is remedied by use of transmitting boundaries [20] and of hybrid techniques [38]. In recent years considerable effort has been expended therefore in developing transmitting boundaries [20] that are placed at the ends of the mesh to allow for energy radiation. A transmitting boundary encompassing all possible cases of waves impinging at the ends of a mesh has yet to be devised. The FEM method employs solid, axisymmetric and planar elements. Use of the first type of elements results in enormous computational requirements that few researchers can afford. The FDM has been used less frequently than the FEM, primarily because of the difficulties associated with handling complicated geometries.

Discrete models have successfully competed with approximate continuum methods in certain classes of problems [19]. The key idea behind a lumped parameter approach is the determination of values for the mass, stiffness and damping coefficients that essentially represent the medium. Use of these frequency-dependent coefficients, which are known as impedance functions, effectively uncouples the structure from the medium with due consideration to the interaction phenomenon. Thus, an efficient analysis of the structure alone is possible. Exact expressions for impedance functions can be obtained for very few cases only. In most instances, analytic expressions for these functions are obtained under relaxed boundary conditions, i.e. by assuming uncoupling. These ideas have found extensive use in foundation problems [1, 39].

During the last ten to fifteen years the Boundary Element method (BEM) [4] has emerged as a strong candidate for numerically investigating dynamic problems. The reasons are that the radiation damping condition is automatically (and correctly) satisfied, and for linear problems only the surface of the problem needs to be discretized. Thus, a tremendous reduction in the size of the problem can be achieved. BEM is based on integral equation formulations that first appeared about a hundred years ago [4]. The solution of the general elastodynamic or viscoelastodynamic problem in two dimensions by this method can be accomplished via one of the following basic approaches available.

(a) Determination of the steady-state solution by the BEM followed by reconstitution of the transient response using Fourier synthesis, as done by Banaugh and Goldsmith [3], and Niwa et al. [27, 28].

(b) Solution of the problem in the Laplace transform domain by the BEM, followed by a numerical inverse transformation to obtain the response in the time domain, as was originally outlined by Doyle [15], and done by Cruse and Rizzo [12], Cruse [13], and Manolis and Beskos [23].

(c) Time-domain formulation and solution of the problem by the BEM in conjunction with step-by-step time integration schemes. This has been done by Cole et al. [10] for the anti-plane strain case, and by Niwa et al. [29] for the two-dimensional case.

Most of the above mentioned work suffers either from the lack of generality or from unacceptable level of accuracy. For example, in Ref. [10], Cole et al. find the transient dynamic formulation to be unstable, leading to a building up of errors as the time stepping progresses. In Ref. [29], the same implementation was applied to problems which do not reveal the basic flaw in their approach, i.e. the numerical treatment of the convolutions involved. These have been corrected in the present work.

A comparison of the above three approaches on the basis of their accuracy and efficiency was done by Manolis [24]. It should be noted that in the above papers some simple two-dimensional elastodynamic problems were solved such as: (1) the case of an unlined or lined circular cylindrical cavity under the passage of longitudinal or transverse waves; (2) the cases of square or horseshoe-shaped cylindrical cavities under longitudinal waves, (3) the case of wave propagation in half-planes, etc.

Three-dimensional problems of elastodynamics have not been attempted in the past, principally because of formidable computing requirements. In order to reduce the computations involved, approximate formulations of the boundary integral equations have also been developed by a number of workers. Dominguez [14] simplified the kernel functions for the special case of periodic surface loading on rectangular foundations. More complex problems of the periodic response of piles and pile groups [21, 34, 35] have been attempted by simplifying the boundary integral formulation such that only displacement kernels are involved in the formulation. Some authors [2, 16] have also introduced the potentially unstable method involving an "auxiliary boundary" so that no singular integrations are necessary. Problems involving the determination of dynamic stiffness influence coefficients for surface and embedded, rigid and flexible foundations were addressed by Beskos and his co-workers [7, 8] using a time-domain approach. Despite the fact that they employed rectangular elements with nodal collocation at the centroid of an element and the time derivatives in the formulations were neglected, the relatively simple geometries involved enabled them to obtain satisfactory solutions.

As far as the general periodic dynamic analysis is concerned, the authors and co-workers [5] have implemented the boundary integral formulations for a general multi-region three-dimensional solid. The work was first reported by Wilson et al. [40] in an annual status report in February 1984. Somewhat similar implementation has been recently described by Rizzo et al. [33], who additionally introduced an excellent technique to cater for the problems of fictitious eigenfrequency in certain exterior problems with homogeneous boundary conditions.

The work described in this chapter is a comprehensive attempt to solve the general three-dimensional dynamic problem using the direct BEM. The bulk of the effort went on the development and computer implementation of the method. In what follows, the governing equations of motion are presented and the BEM formulation is developed in both the time- and the transformed domain. Subsequently, surface discretization aspects are discussed, along with the numerical integration techniques. After some comments on programming aspects, a number of representative examples are solved to illustrate the methodology. To the best of the authors' knowledge, a comparable system for steady-state and time dependent analyses by the BEM has not yet appeared in the published literature.

1. Governing equations

The fundamental equations of elastodynamics, namely equilibrium, the constitutive law, and the kinematic equations, can be combined to give the Navier–Cauchy equations

$$(c_1^2 - c_2^2)u_{i,ij} + c_2^2 u_{j,ii} + b_j = \ddot{u}_j \tag{1.1}$$

where $u_i(\boldsymbol{x}, t)$ is the displacement vector at $\boldsymbol{x}$ and at the time t and b_j is the body force per unit mass. The propagation velocities of the pressure and shear waves in the body are given respectively as

$$c_1^2 = (\lambda + 2\mu)/\rho \qquad \text{and} \qquad c_2^2 = \mu/\rho \tag{1.2}$$

where λ and μ are the Lame constants and ρ the mass density. Commas and dots indicate space and time differentiation, respectively, and the summation convention is implied for repeated indices. Equation (1.1) must be accompanied by the appropriate boundary and initial conditions

$$t_i = \sigma_{ij} n_j$$

$$u_i(\boldsymbol{x}, t=0) = U_{i0} \qquad \text{and} \qquad \dot{u}_i(x, t=0) = \dot{U}_{i0} \tag{1.3}$$

where t_i are the tractions, σ_{ij} the stresses, n_j the outward normal at the surface, U_{i0} the initial displacement, and $\dot{U}_{i0}$ the initial velocity.

The general form of the integral equations replacing (1.1) can be derived by combining the fundamental point force solution of Eq. (1.1) (Stoke's solution) with the dynamic extension of Beti's reciprocal theorem [18, 26]

$$c_{ij}(z)u_i(z, t) = \int_S \{G_{ij}(\boldsymbol{x}, t, z) * t_i(\boldsymbol{x}, t) - F_{ij}(\boldsymbol{x}, t, z) * u_i(\boldsymbol{x}, t)\} \, \mathrm{d}S(\boldsymbol{x}) \,. \tag{1.4}$$

In (1.4), $*$ denotes time convolution, and z and $\boldsymbol{x}$ are the receiver (field) and source points, respectively. For simplicity, it is assumed that the body force and the initial displacements and velocities are zero. The value of the discontinuity (or jump) term c_{ij} is defined by the tangent plane at z. For Liapunov smooth surfaces S, $c_{ij} = 0.5\delta_{ij}$, where δ_{ij} is the Kronecker delta. Also, the fundamental solutions G_{ij} and F_{ij} are the displacements and tractions at $\boldsymbol{x}$, respectively, resulting from a unit impulse of the form $\delta(t)$ at z. They are listed in Appendix A. It is known [26] that Eq. (1.4) is valid for both regular and unbounded regions. Equation (1.4) can also be used to find interior displacements once the boundary solution is obtained by assigning $c_{ij} = \delta_{ij}$.

The process of constructing an integral equation representation can also be done in a transformed domain. The advantage is that the transformed equations of motion become elliptic, in contrast to the hyperbolic nature of Eq. (1.1). Thus, transformed domain integral equation representations do not required time convolutions, and instead are parametric in the transformed variable. To that end, the Laplace transform of a function $f(\boldsymbol{x}, t)$ with respect to t is defined as

$$L\{f(\boldsymbol{x}, t)\} = \bar{f}(\boldsymbol{x}, s) = \int_0^\infty f(\boldsymbol{x}, t)\, e^{-st}\, dt \tag{1.5}$$

where s is the transform parameter. Application of the Laplace transform to Eqs. (1.1) under zero initial conditions yields

$$(c_1^2 - c_2^2)\bar{u}_{i,ij} + c_2^2\bar{u}_{j,ii} - s^2\bar{u}_j = 0 \tag{1.6}$$

The constitutive law, as well as the boundary conditions (1.3), have the same form in the Laplace domain as they do in the time domain, i.e.

$$\bar{\sigma}_{ij} = \rho[(c_1^2 - 2c_2^2)\bar{u}_{k,k}\delta_{ij} + c_2^2(\bar{u}_{i,j} + \bar{u}_{j,i})]$$

and

$$\bar{\sigma}_{ij}n_j = \bar{t}_i \,. \tag{1.7}$$

The integral equation formulation corresponding to (1.4) can be derived in a similar manner and reads as

$$\bar{c}_{ij}(z)\bar{u}_i(z, s) = \int_S [\bar{G}_{ij}(\boldsymbol{x}, z, s)\bar{t}_i(\boldsymbol{x}, s) - \bar{F}_{ij}(\boldsymbol{x}, z, s)\bar{u}_i(\boldsymbol{x}, s)]\, dS(\boldsymbol{x})\,. \tag{1.8}$$

The transformed fundamental singular solutions are also listed in Appendix A. It should be noted that although the functions $\bar{G}$ and $\bar{F}$ become identical to their static counterparts as s tends to zero, it is important to evaluate this limit carerfully because of the presence of s in the denominator. As s increases, these functions remain well behaved.

Under zero initial conditions, the Laplace transform becomes identical to the Fourier transform if s is replaced by $-i\omega$, ω being the circular frequency. Thus, the methodology that will be subsequently developed in equally valid for the frequency domain as well. Finally, both Eqs. (1.4) and (1.8) can be differentiated in accordance with the constitutive law to give an identity for computing the interior stresses in terms of the boundary tractions and displacements. This is elaborated in Ref. [5]. For stresses at the surface, it is best to combine the constitutive law, the boundary conditions, and the directional derivatives of the displacement vector in an accurate matrix formulation [25].

2. Time-domain formulation

Equation (1.4) represents an exact formulation involving integrations over the surface as well as the time history. It is also of interest to note that it is an implicit time-domain formulation, since the displacements at a time t are being calculated by taking account of the history of surface tractions and displacements up to and including the time t. Therefore, it we do not make grossly simplified assumptions in the time variations of quantities, the stability problems one encounters in the finite difference and the finite element methods would be absent here.

In order to obtain the response of current time t_N, the time axis is discretized as

$$t_N = \sum_{n=1}^{N} n\Delta t \tag{2.1}$$

where Δt is the time step. Keeping in mind that the convolution is defined as

$$G_{ij} * t_i = \int_0^t G_{ij}(x, t; z, \tau) t_i(x, \tau)\, \mathrm{d}\tau \tag{2.2}$$

and similarly for the traction kernel, Eq. (1.4) can be rewritten as

$$\begin{aligned} c_{ij}u_i(z, t_N) - \int_{t_{N-1}}^{t_N} \int_S [G_{ij}t_i - F_{ij}u_i]\, \mathrm{d}S\, \mathrm{d}\tau &= \int_0^{t_{N-1}} \int_S [\,]\, \mathrm{d}S\, \mathrm{d}\tau \\ &= \int_{t_1=0}^{t_2} \int_S [\,]\, \mathrm{d}S\, \mathrm{d}\tau + \int_{t_2}^{t_3} \int_S [\,]\, \mathrm{d}S\, \mathrm{d}\tau + \cdots + \int_{t_{N-2}}^{t_{N-1}} \int_S [\,]\, \mathrm{d}S\, \mathrm{d}\tau \end{aligned} \tag{2.3}$$

where [ ] denotes $[G_{ij}t_i - F_{ij}u_i]$ with $t = t_N$.

Note that Eq. (2.3), like Eq. (1.4), still remains an exact formulation of the problem since no approximation has yet been introduced. The equation for the second time node (first time step) is

$$c_{ij}u_i(z, t_2) - \int_{t_1=0}^{t_2} \int_S [G_{ij}t_i - F_{ij}u_i]\, \mathrm{d}S\, \mathrm{d}\tau = 0\,. \tag{2.4}$$

Both displacements and tractions are now assumed to remain constant during a time step, i.e.

$$u_i(\boldsymbol{x}, \tau) = \sum_{n=1}^{N} u_i^n(\boldsymbol{x}) \phi_n(\tau)$$

and

$$t_i(\boldsymbol{x}, \tau) = \sum_{n=1}^{N} t_i^n(\boldsymbol{x}) \phi_n(\tau)\,. \tag{2.5}$$

In the above, $u_i^n(\boldsymbol{x})$ and $t_i^n(\boldsymbol{x})$ encompass the spatial variation of u_i and t_i, respectively, that will be discussed in a subsequent section. The time shape functions $\phi_n(\tau)$ are equal to 1 if $(n-1)\Delta t \leqslant \tau \leqslant n\Delta t$ and equal to 0 otherwise. The time integration in (2.4) is done analytically. In reference to Appendix A, the fundamental singular solutions are of the general form $a_{ij}(r)\delta(t - r/c)$, where $a_{ij}(r)$ is a function of the radial distance $r = |\boldsymbol{x} - z|$ and c is a wave velocity. In view of the space–time discretization and the properties of the impulse δ, a representative term of Eq. (2.4) becomes

$$\int_S \left\{ \int_0^{t_2} a_{ij}(r)\delta(t_2 - r/c_1 - \tau)t_i^n(\boldsymbol{x})\phi_n(\tau)\,\mathrm{d}\tau \right\} \mathrm{d}S(\boldsymbol{x})$$
$$= \int_S \{a_{ij}(r)t_i^n(\boldsymbol{x})\phi_n(t_2 - r/c_1)\}\,\mathrm{d}S(\boldsymbol{x})\,. \tag{2.6}$$

Once the surface integration is numerically performed and the resulting system of algebraic equations rearranged in view of the boundary conditions, Eq. (2.4) reads as

$$A^1X^1 = B^1Y^1 \tag{2.7}$$

In the above, A and B are coefficient matrices, X and Y are the unknown components of the boundary tractions and displacements, and the superscript pertains to the time step.

The equation for the third time node (second time step) is

$$c_{ij}u_i(z, t_3) - \int_{t_2}^{t_3} \int_S [G_{ij}t_i - F_{ij}u_i]\,\mathrm{d}S\,\mathrm{d}\tau = \int_{t_1=0}^{t_2} \int_S [G_{ij}t_i - F_{ij}u_i]\,\mathrm{d}S\,\mathrm{d}\tau \tag{2.8}$$

which can once again be written as

$$A^2X^2 = B^2Y^2 + R^1 \tag{2.9}$$

where R^1 is the effect of the dynamic history of the first time step on the current time node and is given by the right-hand side of Eq. (2.8). Thus, for the Nth time node we can write

$$A^NX^N = B^NY^N + \sum_{k=1}^{N-1} R^k \tag{2.10}$$

which can be solved for X^N. It may appear at first glance that a prodigious amount of coefficient calculations are involved. However, a closer examination will reveal that:

(a) If the time step size is constant, the A and B matrices do not change from time step to time step.

(b) For each time step, a new ΣR needs to be formed. This involves the evaluation of a new set of coefficients involving the effects of the dynamic history of the first time step on the current time node. After some time stepping intervals, this contribution to ΣR reduces to zero and from that point onwards no new coefficients need to be evaluated.

Two items remain to be mentioned. The first, the contributions of the terms (see Appendix A)

$$\int_{c_1^{-1}}^{c_2^{-1}} \delta(t-\lambda r)\lambda \, d\lambda \qquad \text{and} \qquad \dot{\delta}(t-r/c_k)\,, \qquad k=1,2 \tag{2.11}$$

to the time integrals are also evaluated analytically [25]. The $\dot{\delta}$ term gives rise to a velocity term that must be evaluated by a backward finite difference scheme. Second, the accuracy of the time marching scheme, Eq. (2.10), can be improved by averaging the displacements and tractions at every two consecutive time nodes [25].

3. Transformed-domain formulation

Since the basic governing equations for a dynamic analysis in the transformed space (either in s or ω space) are virtually identical to the corresponding equations for the static analysis, the numerical implementation developed for the static case can be used to extract solutions for the dynamic problem for one value of the Laplace transform parameter s or frequency parameter ω. It should also be noted that any internal viscous dissipation of energy (damping) can easily be accounted for by replacing the elastic moduli λ and μ by their complex counterparts λ^* and μ^* of the form

$$\begin{aligned} \mu^* &= \mu(1+2i\gamma) \\ \lambda^* &= \lambda(1+2i\gamma) \end{aligned} \tag{3.1}$$

leaving Poisson's ratio unaltered. By analogy with single degree-of-freedom systems, the damping ratio γ is equal to $\omega\eta/2\mu$, where η is the coefficient of viscosity for a Kelvin–Voight model.

Upon numerically integrating Eq. (1.8) over the surface, the final system equations can be assembled to read

$$\bar{A}\bar{X} = \bar{B}\bar{Y}\,. \tag{3.2}$$

All expressions in the above equation are dependent on the transform parameter s (or ω), and are thus complex. Equation (3.2) is solved for $\bar{X}$ and for a spectrum of values of the transform parameter. What remains to be done is to invert the solution back to the real time domain. The inverse Laplace transformation is defined as

$$f(\boldsymbol{x}, t) = \frac{1}{2\pi \mathrm{i}} \int_{\beta - \mathrm{i}\infty}^{\beta + \mathrm{i}\infty} \bar{f}(\boldsymbol{x}, s)\, \mathrm{e}^{st}\, \mathrm{d}s\,, \qquad \mathrm{i} = \sqrt{-1} \tag{3.3}$$

where $\beta(>0)$ is arbitrary, but greater than the real part of all the singularities of $\bar{f}(s)$, and s is a complex number with $\mathrm{Re}(s) > 0$.

The inverse transformation is done numerically using the inversion formula by Durbin [17]

$$f(t_j) = 2\left(\frac{\exp(\beta j \Delta t)}{T}\right)\left[-\tfrac{1}{2}\,\mathrm{Re}[\bar{f}(\beta)] + \mathrm{Re}\left\{\sum_{n=0}^{N-1} (A(n) + \mathrm{i}B(n))W^{jn}\right\}\right] \tag{3.4}$$

where

$$s_n = \beta + \mathrm{i}n(2\pi/T)\,, \qquad W = \exp(\mathrm{i}2\pi/N)\,,$$

$$A(n) = \sum_{l=0}^{L} \mathrm{Re}\{\bar{f}(\beta + \mathrm{i}(n + lN)2\pi/T)\}\,,$$

and

$$B(n) = \sum_{l=0}^{L} \mathrm{Im}\{\bar{f}(\beta + \mathrm{i}(n + lN)2\pi/T)\}$$

$$n = 0, 1, 2, 3, \ldots, N\,; \qquad l = 0, 1, 2, 3, \ldots, L\,. \tag{3.5}$$

The numerical values of $f(\boldsymbol{x}, t)$ are computed for N equidistant points $t_j = j\Delta t = jT/N$, $j = 0, 1, 2, \ldots, N-1$. It is suggested that for $L \times N$ ranging from 50 to 5000 one should select $\beta T = 5$ to 10 for good results, where T is the total time interval of interest. In the case of the Fourier transform ($s = -\mathrm{i}\omega$), the above algorithm is equivalent to a Fourier synthesis. In this work we have used $L = 1$, $N = 20$ and $\beta T = 6$, although as mentioned above these values can be adjusted to improve acuracy. The computations involved in Eq. (3.4) are performed by employing the Fast Fourier algorithm of Cooley and Tukey [11]. The algorithm employed here was used with considerable success by Manolis and Beskos [23].

The only topic remaining for discussion is the direct transform, i.e. to find the Laplace or Fourier transform of a given forcing function in time. If the forcing function is piecewise linear in time, which is an excellent approximation for functions that are densely sampled, then the following exact formula can be used:

$$\bar{f}(s) = \sum_{i=1}^{N-1} \frac{1}{\Delta t s^2} \{\Delta F(e^{-st_i} - e^{-st_{i+1}}) + s\Delta t(F_i e^{-st_i} - F_{i+1} e^{-st_{i+1}})\} \tag{3.6}$$

where F_i is the value of f at time t_i and $\Delta F = F_{i+1} - F_i$.

While the transform approach outlined above is capable of providing accurate solutions to transient elastodynamic problems, it suffers from some defects. These are:

(a) The Laplace transform solution is essentially a superposition of a series of steady-state solutions and is therefore applicable only to linear elastodynamic problems. For nonlinear problems the solution must be obtained in the real time domain.

(b) Since the Laplace/Fourier transform casts the entire problem in the complex domain, the storage and computer time requirements are considerably increased. Typically, the solution for a single value of the transform parameter will cost two to four times a single static analysis for the same boundary mesh.

(c) Any problem with non-periodic loads must be solved by taking transform of the loads, solving the transformed BEM problem for multiple values of the transform parameter and then numerically inverted to reconstruct the time-domain solution. Since the transform parameter is embedded nonlinearly in the point load solution, the BEM solution for each transform parameter value requires a complete reconstruction of the system equations, leading to extremely long computing times.

4. Surface discretization

The surface discretization is the same irrespective of whether the time-domain or the transformed-domain approach is pursued. To that purpose we use an isoparametric representation for the geometry and the field variables (boundary displacements and tractions) that is similar to those used in the FEM [41]. In the past, flat boundary elements with field variable collocation in the centroid of the element were used. This approach, however, did not yield accurate results and convergence was slow.

The Cartesian coordinates of an arbitrary point P on a surface boundary element are given in terms of the nodal coordinates X_i as

$$x_i(\mathrm{P}) = M_\alpha(\boldsymbol{\eta}) X_{i\alpha} \tag{4.1}$$

where $i = 1, 2, 3$ and $\alpha = 1, 2, \ldots, A$, with A the number of nodal points necessary to describe the element. Furthermore, M_α are the shape functions defined in the local or intrinsic coordinate system (η_1, η_2). Two basic quadratic elements are used here: (a) the six-node triangular "patch" and (b) the eight-node quadrilateral "patch". These two quadratic elements are shown in Fig. 1 and more information may be found in Appendix B. The Jacobian matrix relating the transformation from the Cartesian coordinate system (x, y, z) to the element's intrinsic coordinate system (η_1, η_2) is

$$J_{ij} = (\partial M_\alpha / \partial \eta_j) X_{i\alpha} \tag{4.2}$$

where $j = 1, 2$ and the summation convention is again implied for repeated indices such as α.

The field variables are also represented by the same shape functions, i.e.

$$u_i^n(\boldsymbol{x}) = M_\alpha(\boldsymbol{\eta}) U_{i\alpha}^n \quad \text{and} \quad t_i^n(\boldsymbol{x}) = M_\alpha(\boldsymbol{\eta}) T_{i\alpha}^n \tag{4.3}$$

where $U_{i\alpha}^n$ and $T_{i\alpha}^n$ are nodal values of the displacements and tractions, respectively, at time step n. The superscript n should be neglected in the transformed-domain solution.

Infinite elements, which are esential if problems involving half-spaces are to be solved, can be constructed by modifying the eight-node quadrilateral as shown in Fig. 2. The intrinsic coordinate along the dimension of the quadrilateral that we

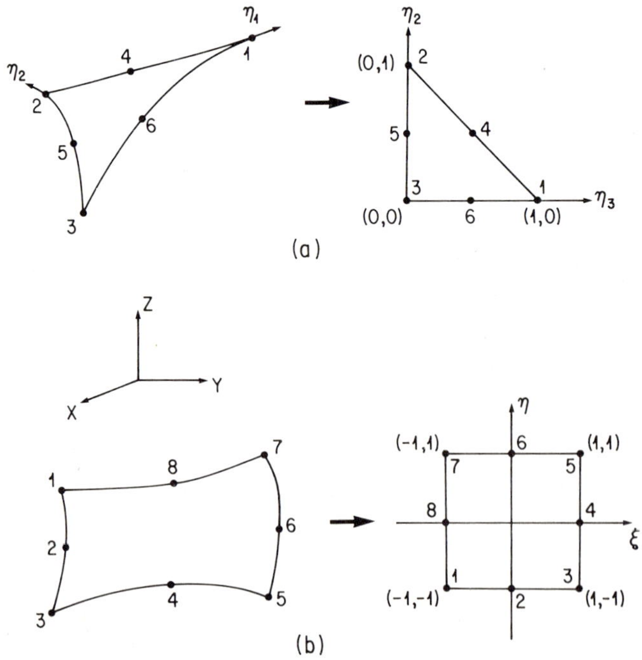

Fig. 1. (a) Triangular and (b) quadrilateral surface elements.

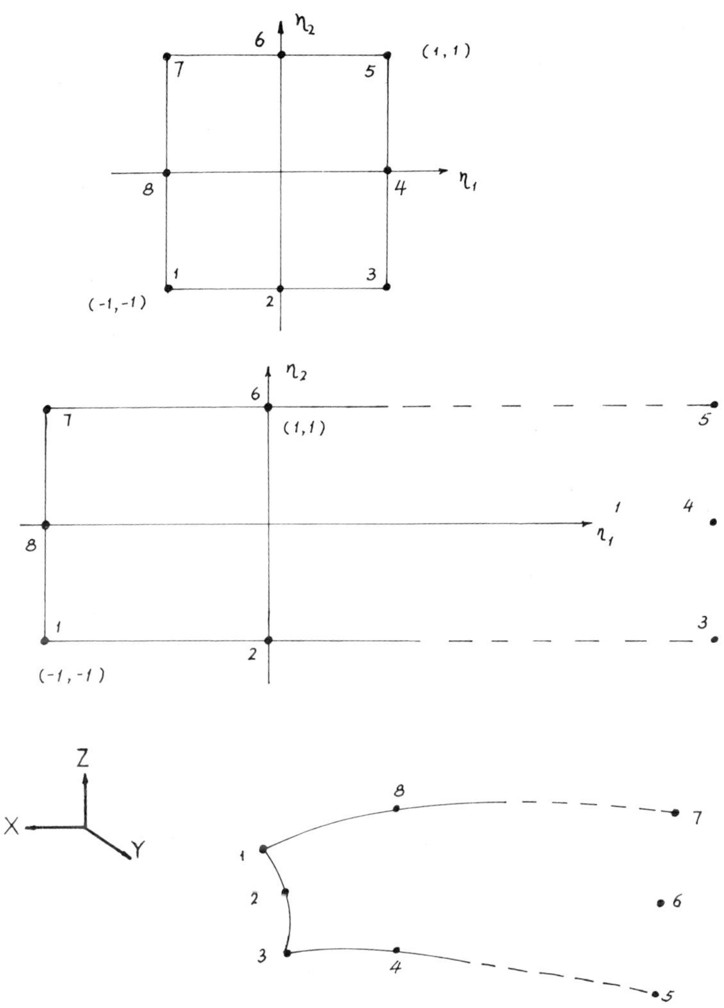

Fig. 2. Infinite surface element.

want extended to infinity (say η_1) is modified as

$$\eta_1' = (3\eta_1 + 1)/(1 - \eta_1) . \tag{4.4}$$

This way, the original interval $(-1, 1)$ is mapped into $(-1, +\infty)$. It should be noted that only the three nodes on the side of the infinite element that is adjacent to a surface element belonging to the "core" region contribute to the system equations. The original shape functions M_α for these three nodes are then modified by the ratio $d = [(x_i - z_i)(x_i - z_i)/(y_i - z_i)(y_i - z_i)]^{1/2}$ for the displacement kernel and d^2 for the traction kernel, where x_i are the Cartesian coordinates of the integration points, y_i their projection on the common side with the core, and z_i an arbitrary reference point. This type of stretching of a quadrilateral results in a Jacobian determinant equal to $4/(1-\eta_1)^2$ that must be included in the kernel integrations. The infinite element thus obtained reproduces the correct spatial decay of the fundamental singular solutions as $r \rightarrow \infty$. Finally, two other types of surface elements, the linear three-node triangle and the linear four-node quadrilateral are used in the quadratic element subdivisions that will be discussed subsequently.

5. Numerical integration

In view of the surface elements introduced in the previous section, either Eq. (1.4) or Eq. (1.8), when integrated over the surface of the problem in question, assumes the following form:

$$c_{ij}(z)u_i(z) = \sum_{q=1}^{Q} \left\{ \int_{S_q} G_{ij}(\boldsymbol{x}(\boldsymbol{\eta}), M_\alpha(\boldsymbol{\eta}) \,\mathrm{d}S(\boldsymbol{x}(\boldsymbol{\eta}))T_{i\alpha} - \int_{S_q} F_{ij}(\boldsymbol{x}(\boldsymbol{\eta}), z)M_\alpha(\boldsymbol{\eta}) \,\mathrm{d}S(\boldsymbol{x}(\boldsymbol{\eta}))U_{i\alpha} \right\}. \tag{5.1}$$

In the above equation, S_q is the surface of the qth element and Q is the total number of elements. Furthermore, dependence on either time t or the transform parameter s has been tacitly omitted since it was described extensively in the previous sections. The global system of boundary element equations at a given time step Δt or at a given value of s is obtained by the usual nodal collocation scheme, i.e. by allowing point z to coincide sequentially with all the nodal points of the boundary.

All of the surface integrals in the numerical implementations have been calculated numerically. Since this is the most time consuming portion of the analysis it is essential to optimize this effort. Essentially two types of integrals, singular and non-singular, are involved. The integrals are singular if the field point for the equations being constructed lies on the element being integrated. Otherwise, the integrals are non-singular, although numerical evaluation is still difficult if the field point and the element being integrated are close together.

In both the singular and non-singular cases Gausian integration is used. The basic technique is developed in Ref. [4] and was first applied in the three-dimensional boundary element method by Lachat and Watson [22]. In the non-singular case an approximate error estimate for the integral was developed based on the work of Stroud and Secrest [36]. This allows the determination of element subdivisions and orders of Gaussian integration which will retain a consistent level of error throughout the structure. Numerical tests have shown that the use of 3×3, 4×4, and 5×5 Gauss rules provide the best combination of accuracy and efficiency. In the present implementation the 4×4 rule is used for non-singular integration, and error is controlled through element subdivision. The element subdivision is controlled by the distance between the field point and to the point closest to the field point on the element being integrated. For the transform case, higher values of s require lower integration tolerance, leading to more element subdivision. If the field point is very close to the element being integrated, use of a uniform subdivision of the element can lead to excessive computing time. In order to improve efficiency while still retaining accuracy, a graded element subdivision is employed. Based on one-dimensional tests, it was found that the subelement divisions could be allowed to grow geometrically away from the origin of the element subdivision. Numerical tests on a complex three-dimensional problem have shown that a mesh expansion factor as high as 4.0 can be employed without significant degradation of accuracy.

In the case of singular integration, which arises when the field point is on the element being integrated, the element is first divided into triangular subelements. The integration over each subelement is carried out in a polar coordinate system with origin at the field point. This coordinate transformation produces non-singular behavior in all except one of the required integrals. Normal Gauss rules can then be employed. The integral of the traction kernel times the isoparametric shape function which is 1.0 at the source point is still singular, and cannot be numerically evaluated with reasonable efficiency and accuracy. Its calculation is carried out indirectly, as discussed below. It has been found that subdivision in the circumferential (angular) direction is required to preserve accuracy in the singular integration. A maximum included angle of 15° is used. Subdivision in the radial direction has been found necessary only with the time domain approach. This process is illustrated in Fig. 3 for a quadrilateral element.

If we assume, for illustration purposes, that the singularity arises in the $q = 1$ element and the $\alpha = 1$ node, then the corresponding strictly diagonal 3×3 block that contains the jump term c_{ij} as well as the Cauchy principal value of the traction kernel integral is evaluated as follows. First define

$$d_{ij}^{\mathrm{st}} = c_{ij} + \int_{S_1} F_{ij}^{\mathrm{st}} M_1 \, \mathrm{d}S \tag{5.2}$$

and

$$d_{ij}^{\mathrm{dyn}} = c_{ij} + \int_{S_1} F_{ij}^{\mathrm{dyn}} M_1 \, \mathrm{d}S \tag{5.3}$$

where d_{ij}^{st} and d_{ij}^{dyn} are the strictly diagonal blocks for the static and dynamic (steady-state or time-dependent) systems, respectively. It should be noted that the jump term depends only on the geometry at the singular node. Next, the d_{ij}^{st} block is evaluated using the rigid-body motion approach. Specifically, if a rigid-body

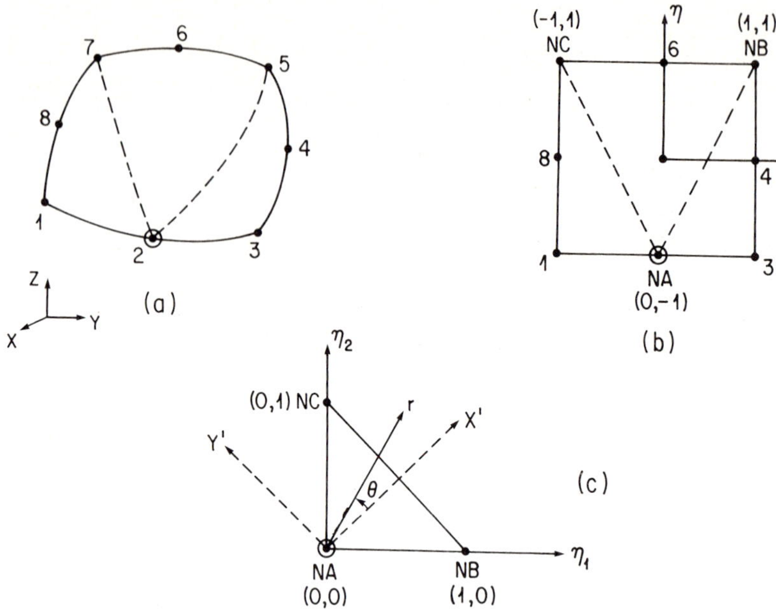

Fig. 3. Singularity integration process for a quadrilateral element.

motion $u_j = \{1\}$ is imposed on the static version of (5.1), no tractions result and

$$d_{ij}^{\mathrm{st}} = c_{ij} + \int_{S_1} F_{ij}^{\mathrm{st}} M_1 \,\mathrm{d}S = -\Big(\sum_{\alpha=2}^{A} \int_{S_1} F_{ij}^{\mathrm{st}} M_\alpha \,\mathrm{d}S + \sum_{q=2}^{Q} \sum_{\alpha=1}^{A} \int_{S_q} F_{ij}^{\mathrm{st}} M_\alpha \,\mathrm{d}S \Big) . \tag{5.4}$$

Finally, by combining (5.2) and (5.3),

$$d_{ij}^{\mathrm{dyn}} = d_{ij}^{\mathrm{st}} + \int_{S_1} (F_{ij}^{\mathrm{dyn}} - F_{ij}^{\mathrm{st}}) M_1 \,\mathrm{d}S . \tag{5.5}$$

In the above equation, the integral involving $F_{ij}^{\mathrm{dyn}} - F_{ij}^{\mathrm{st}}$ is non-singular and can easily be evaluated numerically. Equation (5.4) indicates that d_{ij}^{st} is found by summing the off-diagonal (i.e. non-singular) terms of the row pertaining to the receiver z. A somewhat similar approach is also used in Ref. [33].

There is a fundamental difference between steady-state and transient approaches when it comes to the numerical integration schemes outlined above. In the steady-state case the integrands in all of the non-singular surface integrals are infinitely differentiable and solution accuracy can, therefore, always be improved by the use of increased integration order. In the transient case, however, the point load solutions are only continuous. Physically this corresponds to the fact that the disturbance due to an impulse applied to a given time and spatial location is nonzero, at some later time, in only a finite portion of space. This means that the kernel function may be nonzero over only part of a given surface element. While the integrand is infinitely differentiable within both the zero and nonzero regions considered separately, its overall continuity over the entire element is only C_0. The use of higher-order quadrature rules is, therefore, of little use in improving accuracy. Based on these observations a revised integration strategy was adopted for the transient case. All surface elements are subdivided into a relatively large number of subelements, and relatively low (usually second or third) order quadrature rules, together with the usual distortion in mapping so that the kernel shape functions and Jacobian products remain well behaved are used over each subelement. This has led to much improved acuracy in the transient analysis.

6. Some programming aspects

Detailed programming aspects concerning the BEM can be found elsewhere [25, 40]. Only a few minor comments will be here regarding the numerical integration schemes, the assembly of the system equations, and discretization questions.

(a) *Numerical integrations*. The numerical integration schemes outlined in the previous section have been programmed and a number of cases for which analytic results exist were tried. As an example, the surface of a sphere was discretized into three quadrilateral elements per octant and the receiver point z was placed on a number of locations to reproduce various singular and non-singular cases. Functions such as $f(\boldsymbol{x}, z) =$ constant, $=r$, $=1/r$, $\boldsymbol{n}/r$, etc., were integrated over the entire surface of the sphere or portions of it. $\boldsymbol{n}$ is the normal at the surface. For this example, the maximum integration errors did not exceed 1%, and they are attributed to the most part in the inability of the surface elements to exactly conform to the curvature of the sphere. Similar results were obtained using triangular elements. In another example, the above types of numerical experiments were performed on a flat surface (half-space) discretized into triangular and quadrilaterals. In this case, the results were exact to within three or four significant digits.

(b) *Assembly of system equations*. The assembly, as well as the imposition of the boundary conditions, are quite general. Any combination of tractions and displacements may be prescribed at a node. Furthermore, the system equations are scaled so that the coefficients of A in (2.10) and (3.2) are of the same magnitude. Reduction of A is done using the standard Gauss elimination algorithm with pivoting in the transient case. In the steady-state case, a standard complex block banded solver is used. Finally, interior, exterior, and half-space type of problems may all be solved by the present programs.

(c) *Discretization aspects*. The steady-state BEM methodology is essentially of quasi-static nature. Some care, however, must be exercised at very high values of the transformed parameter. As a general rule finer meshes are required at higher frequencies. The transient methodology results in an unconditionally stable implicit time marching scheme. For better accuracy, however, it is recommended that the time step remain smaller than L/c_1, L being a typical surface dimension such as the smallest distance measured along the surface between two nodal points. This is rather unfortunate, because if one refines the boundary element mesh to get a better surface discretization, one must also decrease the time step. The time required for execution of such problems is substantial and depends heavily on the number of time or transform parameter steps used.

7. Numerical examples

A number of representative problems were chosen to test the steady-state solution with the numerical inverse transformation and the time-stepping solution. In all cases, SI units are used with meters (m) for length, kilograms (kg) for force, and seconds (s) for time.

(a) *Prismatic beam*. A slender beam fixed at one end is modelled by a total of 18 quadrilateral surface elements resulting in 56 nodes. The modulus of elasticity of the beam is $E = 1.1 \times 10^5$, Poisson's ratio $\nu = 0.0$, and the mass density $\rho = 2.0$. Two examples are solved:

(1) Catilever subjected to a harmonic end shear: The boundary conditions here are complete fixity at one end, a uniformly distributed traction $t_y = 1000\, e^{i\omega t}$ ($\omega = 314$ rad/s) at the other end, and traction-free conditions along the sides. The length of the beam $L = 10$, the width $w = 1$, and the depth $d = 3$. The mesh used is shown in Fig. 4, where it will be observed that the surface elements are arranged closer to the loaded end. This is so because the displacement function varies more sharply at the loaded end than at the fixed end. The transformed domain algorithm is used, with $s = -i\omega$. The absolute value of the displacement $\bar{u}_y$ along the length of the beam at a frequency $\Omega = \omega = 314$ rad/s is also plotted in Fig. 4, and it is in excellent agreement with the exact solution for a flexural beam which was developed from Ref. [9].

(2) Bar subjected to a transient end load: The bar is held along its sides by lubricated rollers and is supported on a frictionless base. The free end is subjected to a suddenly applied and maintained uniform compression $t_z = -1000$. The dimensions of the beam are $L = 8$ and $w = d = 2$. In view of the material properties, the characteristic time required for the compressive wave to reach the fixed end is 0.0356s. Figure 5 shows the discretization and the numerical results for the normal stress σ_{zz}. There, the results from the time-domain algorithm for two different time steps Δt are compared with the exact analytical solution for one-dimensional stress wave propagation [37]. Although the numerical results are in good agreement with the analytical solution, it is clearly very difficult to

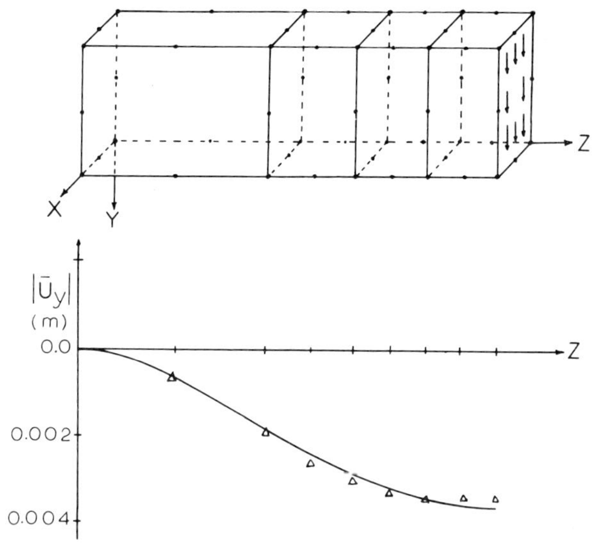

Fig. 4. Cantilever subject to harmonic end shear: analytical solution –; numerical solution △.

reproduce the sharp jump in the stress as the disturbance reaches the point initially and when the reflected stress wave returns to the same location. This difficulty has been observed elsewhere as well [6].

(b) *Spherical cavity*. A spherical cavity is embedded in an infinitely extending medium with $E = 8.993 \times 10^6$, $\nu = 0.25$, and $\rho = 2.5 \times 10^{-4}$. The radius of the cavity $R = 212$ and its surface is discretized into three triangular elements per octant for a total of 50 nodes. This is mesh 2 of Fig. 6. The characteristic times

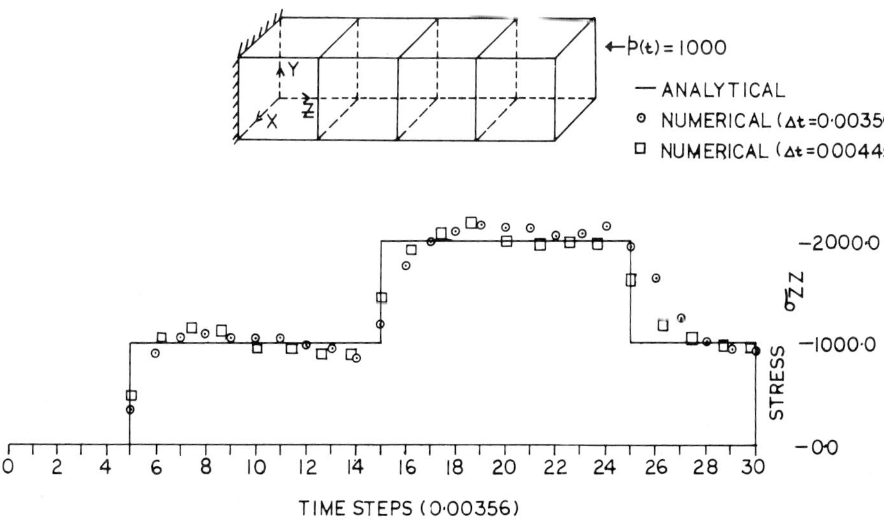

Fig. 5. Longitudinal stress at the mid-length of a column subjected to a suddenly applied end load.

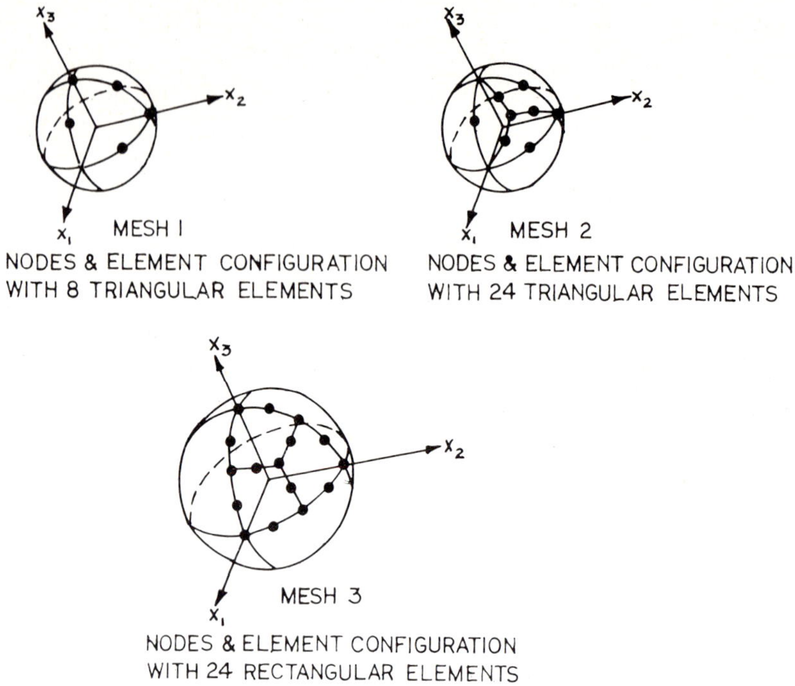

Fig. 6. Spherical cavity surface discretization using three different meshes.

required for the pressure and shear waves to travel a cavity radius are 0.00102s and 0.00177s, respectively. Again two cases are considered:

(1) Spherical cavity under sudden radial expansion: A radial pressure $p = 1000$ is suddenly applied and maintained at the cavity surface. Figure 7 shows the radial displacement obtained by both Laplace transform and time-domain algorithms.

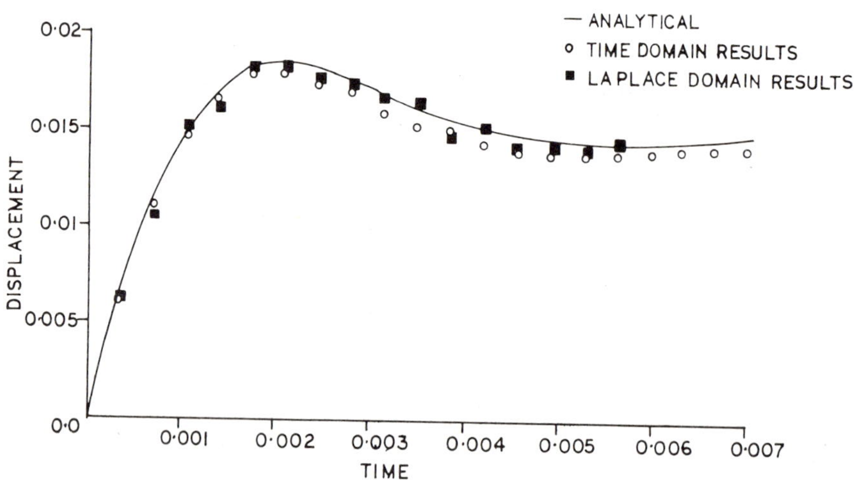

Fig. 7. Comparison of the Laplace transform and the time-domain transient results with the analytical solution for cavity expansion.

Concurrently plotted is the exact solution [37]. In general, the numerical results are in good agreement with the analytical solution. These is some oscillation in the results of the Laplace transform solution which becomes progressively worse beyond $t = 0.0056$ s. The transient time-domain solution, on the other hand, remains stable and reaches the expected static solution at larger times. The maximum error in the displacement time history did not exceed 9%. The maximum error dropped to 6% when mesh 3 of Fig. 6 with three quadrilaterals per octant (74 nodes) was used. Mesh 1 of Fig. 6 is too coarse and was not used.

(2) Spherical cavity engulfed by a pressure wave: A propagating plane pressure wave whose front is perpendicular to the *Z*-axis first impinges on the pole with coordinates (0, 0, 212). The resulting nonzero incident stresses are $\sigma_{zz}^{(i)} = -1000$, $\sigma_{xx}^{(i)} = \sigma_{yy}^{(i)} = (\nu/(1-\nu))\sigma_{zz}^{(i)}$. These incident stresses are suddenly applied and maintained at a station reached by the wave. This wave propagation type of problem is solved by superposition [26]. The three quadrilaterals per octant mess are used here in conjunction with the time-domain approach. Figure 8 plots the hoop stresses $\sigma_{\phi\phi}^{*}$ and $\sigma_{\theta\theta}^{*}$ normalized by the magnitude of the incident stress $\sigma_{zz}^{(i)}$ versus the non-dimensional time $\tau^{*} = tR/c_1$. The plots are for three locations on

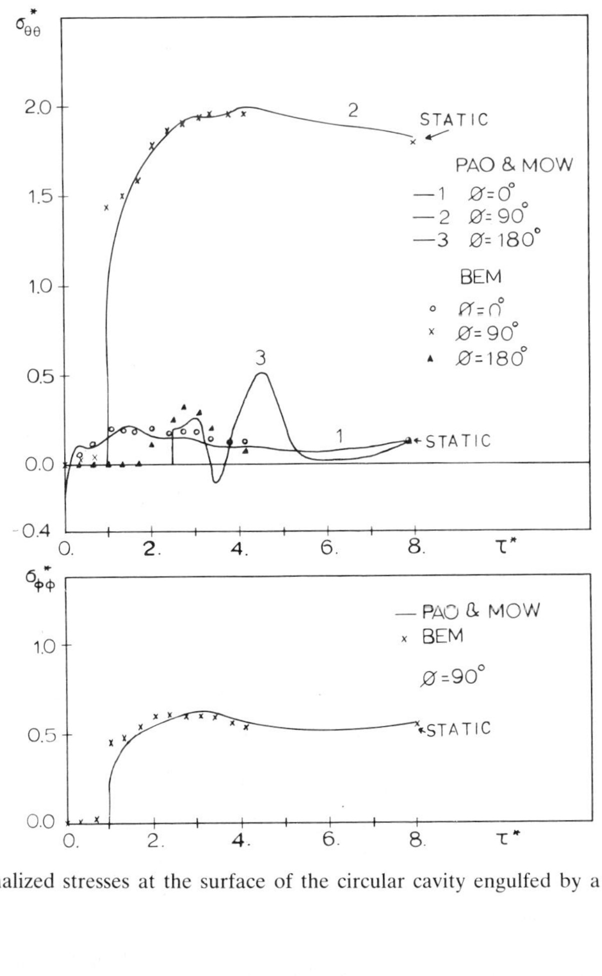

Fig. 8. Normalized stresses at the surface of the circular cavity engulfed by a pressure wave.

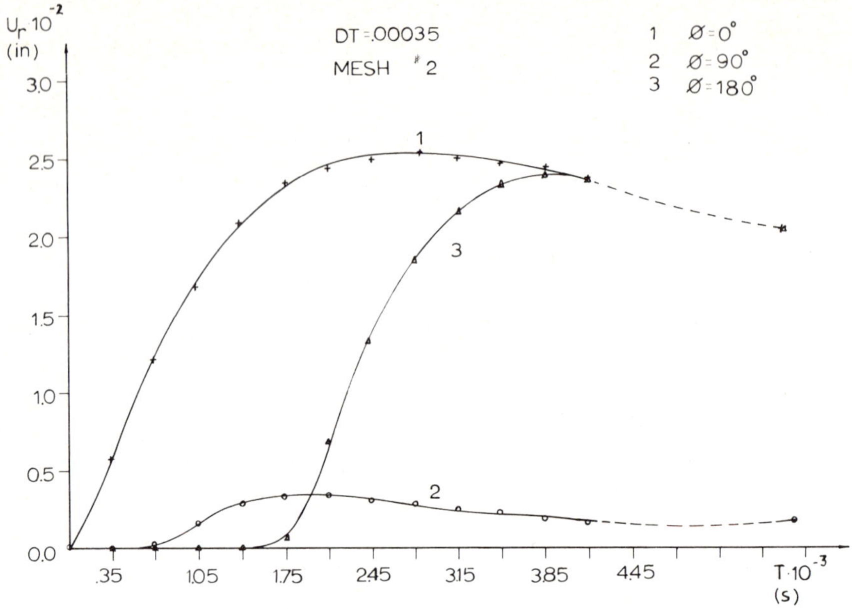

Fig. 9. Radial displacements of the surface of the circular cavity engulfed by a pressure wave.

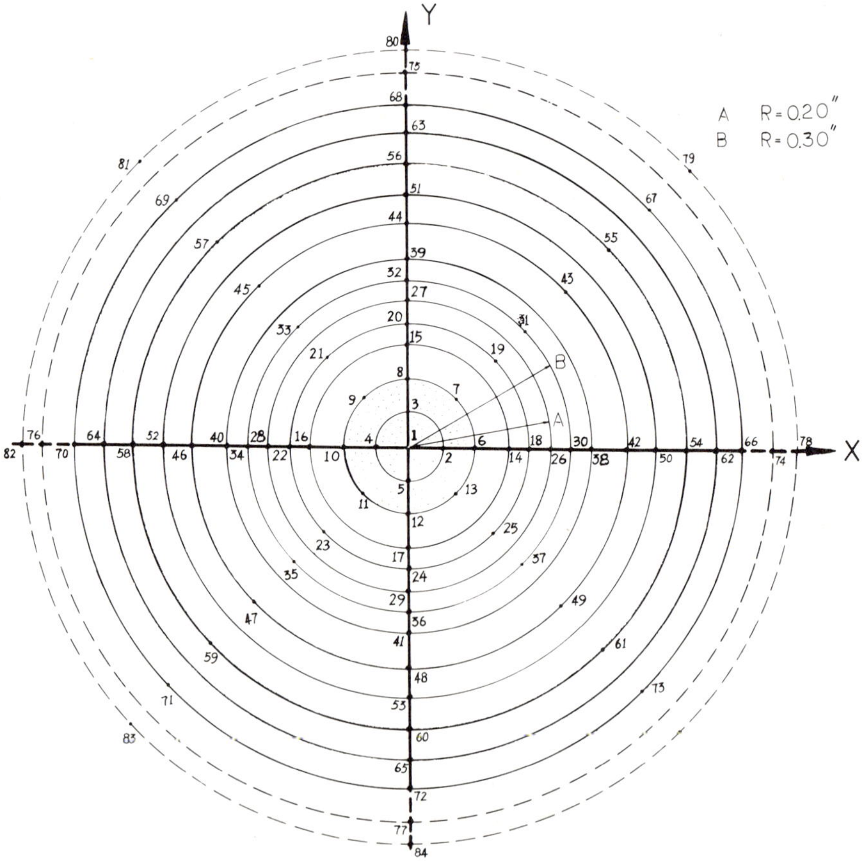

Fig. 10. Discretization pattern for the half-space.

the surface of the cavity: the two poles ($\phi = 0, \pi$) and the equator ($\phi = \pi/2$). Concurrently plotted are the analytic results [30, 31], obtained by analytical inversion of the Fourier-transformed solution. Good agreement is observed between the two solutions. Finally, Fig. 9 plots the radial displacement time history at the same three locations as before.

(c) *Half-space*. This example is Lamb's problem for an impulsive vertical point force on the surface of a semi-infinite solid [32]. The modelling difficulty encountered here is that the point load must be represented by a finite (yet small) area, and hence the complicated mesh shown in Fig. 10. There are seven concentric rings, with four elements per ring, resulting in a total of 85 nodes. Uniform vertical tractions $t_z = 1000$ are prescribed on the triangular elements of the inner ring, which has an outer radius of 0.05. The outer ring has an inner radius of 3.0 and is modelled by infinite elements. Obviously the small circular load solution behaves differently from the analytical point force solution (labelled solution A). The static solution showed that the results agree well at a radial distance of 0.2, where there is a 4.5% and a 1.0% difference in the horizontal and vertical displacements, respectively. Therefore, the results from the time-domain algorithm shown in Fig. 11 are for the normalized horizontal displacements at $r = 0.2$. When the exact solution [32] was used to calculate the superimposed effects of multiple point forces to reproduce a finite area loading (labelled as analytic solution B), good agreement with the BEM results was obtained.

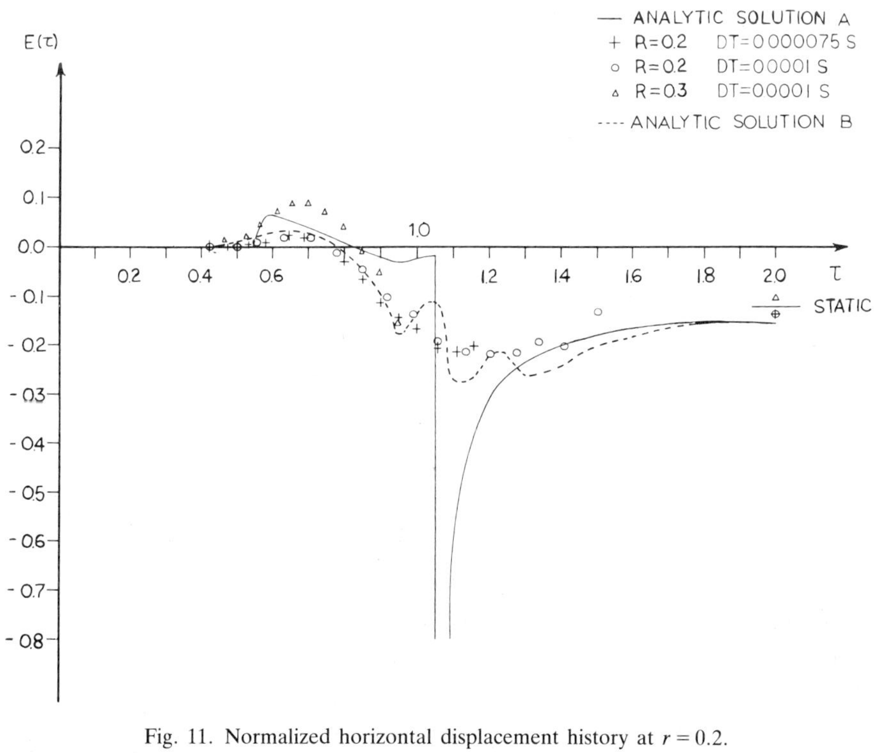

Fig. 11. Normalized horizontal displacement history at $r = 0.2$.

8. Concluding remarks

An advanced algorithm based on the boundary element method for both steady-state and time-dependent dynamic elastic analysis has been presented. The numerical implementation employed is one of the most general currently available and can be used in conjunction with substructuring to treat three-dimensional solids of complicated geometry and connectivity. The algorithm is unconditionally stable and capable of producing very accurate results. Thus, the algorithm is a viable alternative to that based on the finite element methodology. Furthermore, because of the incremental character of the time-dependent approach, the stage has been set to address problems exhibiting nonlinearities, through the use of volume integrals and incremental/iterative schemes.

Acknowledgements

The computations reported were done on the Computational Mechanics Division's HP9000 (series 500) systems. The work was funded by the NSF Grant No: CEE 82-02604 and in part by the NASA Contract No: NAS3-23697. The authors are indebted to Dr. C. Babendreier, the NSF program manager, Dr. Chris Chamis, the NASA program manager, and Dr. Edward Todd, the Pratt and Whitney program manager for their support and encouragement. We are also indebted to Dr. R.B. Wilson and Miss Nancy Miller of Pratt and Whitney for many valuable discussions.

Appendix A. Fundamental singular solutions

The tensors G_{ij} and F_{ij} in the time domain are of the form [18]

$$G_{ij}(\boldsymbol{x}, t, z, \tau = 0) = \frac{1}{4\pi\rho}\left\{\left(\frac{3r_i r_j}{r^3} - \frac{\delta_{ij}}{r}\right)\int_{c_1^{-1}}^{c_2^{-1}} \lambda\delta(t - \lambda r)\,\mathrm{d}\lambda\right.$$

$$\left. + \frac{r_i r_j}{r^3}\left(\frac{1}{c_1^2}\,\delta(t - r/c_1) - \frac{1}{c_2^2}\,\delta(t - r/c_2)\right) + \frac{\delta_{ij}}{rc_2^2}\,\delta(t - r/c_2)\right\} \tag{A.1}$$

where $r_i = x_i - z_i$, $i = 1, 2, 3$, $r = |\boldsymbol{x} - z|$, and

$$F_{ij} = H_{ijk}n_j = [\lambda G_{mk,m}\delta_{ij} + \mu(G_{ik,j} + G_{jk,i})]n_j \tag{A.2}$$

with n_j the normal vector and

$$H_{ijk}(\boldsymbol{x}, t, z, \tau = 0) = \frac{1}{4\pi} \left\{ -6c_2^2 \left[5 \frac{r_i r_j r_k}{r^5} - \frac{\delta_{ij} r_k + \delta_{ik} r_j + \delta_{jk} r_i}{r^3} \right] \right.$$

$$\times \int_{c_1^{-1}}^{c_2^{-1}} \lambda \delta(t - \lambda r) \, \mathrm{d}\lambda + 2 \left(6 \frac{r_i r_j r_k}{r^5} - \frac{\delta_{ij} r_k + \delta_{ik} r_j + \delta_{jk} r_i}{r^3} \right)$$

$$\times \left(\delta(t - r/c_2) - \frac{c_2^2}{c_1^2} \delta(t - r/c_1) \right)$$

$$+ 2 \frac{r_i r_j r_k}{r^4 c_2} \left(\dot{\delta}(t - r/c_2) - \frac{c_2^3}{c_1^3} \dot{\delta}(t - r/c_1) \right)$$

$$- \frac{r_k \delta_{ij}}{r^3} \left(1 - 2 \frac{c_2^2}{c_1^2} \right) \left(\delta(t - r/c_1) - \frac{r}{c_1} \dot{\delta}(t - r/c_1) \right)$$

$$\left. - \frac{\delta_{ik} r_j + \delta_{jk} r_i}{r^3} \left(\delta(t - r/c_2) + \frac{r}{c_2} \dot{\delta}(t - r/c_2) \right) \right\}. \tag{A.3}$$

The above fundamental singular solutions obey the causality condition, i.e.

$$G_{ij}(\boldsymbol{x}, t, z, \tau) = 0 \qquad \text{if } c_1(t - \tau) < r \tag{A.4}$$

and have the following time translation property:

$$G_{ij}(\boldsymbol{x}, t + \tau_0, z, \tau + \tau_0) = G_{ij}(\boldsymbol{x}, t, z, \tau). \tag{A.5}$$

The tensors $\bar{G}_{ij}$ and $\bar{F}_{ij}$ in a transformed domain are of the form [12]

$$\bar{G}_{ij}(\boldsymbol{x}, z, s) = \frac{1}{4\pi\mu} [\psi \delta_{ij} - \phi r_{,i} r_{,j}] \tag{A.6}$$

and

$$\bar{F}_{ij}(\boldsymbol{x}, z, s) = \frac{1}{4\pi} \left[\alpha \left(\delta_{ij} \frac{\partial r}{\partial n} + r_{,i} n_j \right) + \beta \left(n_i r_{,j} - 2 r_{,i} r_{,j} \frac{\partial r}{\partial n} \right) \right.$$

$$\left. + \gamma r_{,i} r_{,j} \frac{\partial r}{\partial n} + \delta r_{,j} r_{,i} \right] \tag{A.7}$$

where s is the Laplace transform parameter. In addition,

$$\phi = \left(\frac{3c_2^2}{s^2 r^2} + \frac{3c_2}{sr} + 1 \right) \frac{\mathrm{e}^{-sr/c_2}}{r} - \frac{c_2^2}{c_1^2} \left(\frac{3c_1^2}{s^2 r^2} + \frac{3c_1}{sr} + 1 \right) \mathrm{e}^{-sr/c_1} \tag{A.8}$$

and

$$\psi = \left(\frac{c_2^2}{s^2 r^2} + \frac{c_2}{sr} + 1 \right) \frac{\mathrm{e}^{-sr/c_2}}{r} - \frac{c_2^2}{c_1^2} \left(\frac{c_1^2}{s^2 r^2} + \frac{c_1}{sr} \right) \frac{\mathrm{e}^{-sr/c_1}}{r} \tag{A.9}$$

where e is the exponential function. Furthermore,

$$\alpha = \frac{\partial\psi}{\partial r} - \frac{\phi}{r}, \qquad \beta = -\frac{2\phi}{\partial r}, \qquad \gamma = -2\frac{\partial\phi}{\partial r},$$

and

$$\delta = \left(\frac{c_1^2}{c_2^2} - 2\right)\left(\frac{\partial\psi}{\partial r} - \frac{\partial\phi}{\partial r} - \frac{2\phi}{r}\right). \tag{A.10}$$

Appendix B. Isoparametric boundary elements

Both the six-node triangular and the eight-node rectangular surface elements were depicted in Fig. 1. It is worth noticing that these intrinsically planar elements become curved in three-dimensional space. The shape functions for the six-node triangle are:

$$M_\alpha = \begin{cases} \eta_\alpha(2\eta_\alpha - 1) & \text{if } \alpha = 1, 2, 3 \\ 4\eta_\beta\,\eta_\gamma & \text{if } \alpha = 4, 5, 6 \quad \text{with } \beta = \alpha - 3 \\ & \text{and } \gamma = \alpha - 2 \end{cases} \tag{B.1}$$

where η_1 and η_2 are two linearly independent coordinates and $\eta_3 = 1 - \eta_1 - \eta_2$.

The shape functions for the eight-node rectangle are:

$$M_\alpha = \begin{cases} 0.25(1 + \xi_0)(1 + \eta_0)(\xi_0 + \eta_0 - 1) & \text{if } \alpha = 1, 3, 5, 7 \\ 0.50(1 - \xi^2)(1 - \eta_0) & \text{if } \alpha = 2, 6 \\ 0.50(1 + \xi_0)(1 - \eta^2) & \text{if } \alpha = 4, 8 \end{cases} \tag{B.2}$$

where $\xi_0 = \xi\xi_\alpha$ and $\eta_0 = \eta\eta_\alpha$, with ξ and η being the two linearly independent coordinates and $(\xi_\alpha, \eta_\alpha)$ the coordinates of node α.

Two base vectors along the intrinsic coordinates ξ, η (or η_1, η_2) can be defined as

$$\begin{aligned} \boldsymbol{e}_1 &= \frac{\partial x}{\partial\xi}\,\mathrm{d}\xi\,\boldsymbol{i} + \frac{\partial y}{\partial\xi}\,\mathrm{d}\xi\,\boldsymbol{j} + \frac{\partial z}{\partial\xi}\,\mathrm{d}\xi\,\boldsymbol{k} \\ \boldsymbol{e}_2 &= \frac{\partial x}{\partial\eta}\,\mathrm{d}\eta\,\boldsymbol{i} + \frac{\partial y}{\partial\eta}\,\mathrm{d}\eta\,\boldsymbol{j} + \frac{\partial z}{\partial\eta}\,\mathrm{d}\eta\,\boldsymbol{k} \end{aligned} \tag{B.3}$$

where $\boldsymbol{i}$, $\boldsymbol{j}$, and $\boldsymbol{k}$ are unit vectors along the x, y, and z coordinates, respectively. Their cross product

$$\boldsymbol{e}_3 = \boldsymbol{e}_1 \times \boldsymbol{e}_2 \tag{B.4}$$

is a vector normal to the surface of the element and its magnitude is equal to the value of the determinant of the Jacobian matrix.

References

[1] R.N. Arnold, G.N. Bycroft and G.B. Warburton, Forced Vibrations of a Body on an Infinite Elastic Solid, *Journal of Applied Mechanics* **22**, 391–400 (1955)

[2] R.J. Apsel, Dynamic Green's Functions for Layered Media and Applications to Boundary-Value Problems, Doctoral Dessertation, University of California, San Diego (1979).

[3] R.P. Banaugh and W. Goldsmith, Diffraction of Steady Elastic Waves by Surface of Arbitrary Shape, *Journal of Applied Mechanics* **30**, 589–597 (1963).

[4] P.K. Banerjee and R. Butterfield, *Boundary Element Methods In Engineering Science* (McGraw-Hill, London, 1981).

[5] P.K. Banerjee and S. Ahmad, Advanced Three-dimensional Dynamic Analysis by Boundary Element Methods, in: *Advanced Topics in Boundary Element Analysis*, T.A. Cruse, A.B. Pifko and H. Armen, Eds. (ASME, New York, 1985) pp. 65–81.

[6] T. Belytschko, R.L. Chiapetta and H.D. Bartel, Efficient Large Scale Nonlinear Transient Analysis by Finite Elements, *International Journal for Numerical Methods in Engineering*, **10**, 579–596 (1976).

[7] D.E. Beskos and D.L. Karabalis, Dynamic Response of 3-D Foundations by Time Domain Boundary Element Method, *Final Report Part A*, *National Science Foundation Grant No.* CEE-8024725, University of Minnesota, Minneapolis (1984).

[8] D.E. Beskos and C.C. Spyrakos, Dynamic Response of Strip Foundations by the Time Domain BEM-FEM Method, *Final Report Part B*, *National Science Foundation Research Grant No.* CEE-8024725, University of Minnesota, Minneapolis (1984).

[9] R.W. Clough and J. Penzien, *Dynamics of Structures* (McGraw-Hill, New York, 1975).

[10] D.M. Cole, D.D. Kosloff and J.B. Minster, A Numerical Boundary Integral Equation Method for Elastodynamics I, *Bulletin of the Seismological Society of America* **68**, 1331–1357 (1978).

[11] J.W. Cooley and J.W. Tukey, An Algorithm for Machine Calculation of Complex Fourier Series, *Mathematics of Computation* **19**, 297–310 (1965).

[12] T.A. Cruse and F.J. Rizzo, A Direct Formulation and Numerical Solution of the General Transient Elastodynamic Problem I, *Journal of Mathematical Analysis and Applications* **22**, 244–259 (1968).

[13] T.A. Cruse, A Direct Formulation and Numerical Solution of the General Elastodynamic Problem II, *Journal of Mathematical Analysis and Applications* **22**, 341–355 (1968).

[14] J. Dominguez, Dynamic Stiffness of Rectangular Foundations, *Final Report*, *National Science Foundation-RANN Grant No.* ENV77-18339, Massachusetts Institute of Technology, Cambridge, Massachusetts (1978).

[15] J.M. Doyle, Integration of the Laplace Transformed Equations of Classical Elastokinetics, *Journal of Mathematical Analysis and Applications* **13**, 118–131 (1966).

[16] M. Dravinski, Amplification of P, SV and Rayleigh Waves by Two Alluvial Valleys, *International Journal of Soil Dynamics and Earthquake Engineering* **2**, 65–77 (1983).

[17] F. Durbin, Numerical Inversion of Laplace Transforms: An Efficient Improvement to Dubner and Abates' Method, *Computer Journal* **17**, 371–376 (1974).

[18] A.C. Eringen and E.S. Suhubi, *Elastodynamics*, Vols. 1 and 2 (Academic Press, New York, 1975).

[19] A.H. Hadjian, J.E. Luco and N.C. Tsai, Soil–Structure Interaction: Continuum or Finite Element?, *Nuclear Engineering and Design* **31**, 151–167 (1974).

[20] E. Kausel, J.M. Roesset and G. Waas, Dynamic Analysis of Footings on Layered Media, *Journal of the Engineering Mechanics Division of the ASCE* **101**, (EM5), 679–693 (1975).

[21] A.M. Kaynia and E. Kausel, Dynamic Behavior of Pile Groups, *Proceedings of the 2nd International Conference in Numerical Methods in Offshore Piling*, Austin, Texas (1982).

[22] J.C. Lachat and J.O. Watson, Effective Numerical Treatment of Boundary Integral Equations: A Formulation for Three-dimensional Elastostatics, *International Journal for Numerical Methods in Engineering* **10**, 991–1005 (1976).

[23] G.D. Manolis and D.E. Beskos, Dynamic Stress Concentration Studies by Boundary Integrals and Laplace Transform, *International Journal for Numerical Methods in Engineering* **17**, 573–599 (1981).

[24] G.D. Manolis A Comparative Study on Three Boundary Element Method Approaches to Problems in Elastodynamics, *International Journal for Numerical Methods in Engineering* **19**, 71–93 (1983).

[25] G.D. Manolis, S. Ahmad and P.K. Banerjee, Boundary Element Method Implementation for Three-dimensional Elastodynamics, in: *Developments in Boundary Element Methods*–4, P.K. Banerjee and J.O. Watson, Eds (Elsevier Applied Science Publishers, London, 1986) pp. 29–65.

[26] J. Miklowitz, *The Theory of Elastic Waves and Waveguides* (North Holland, Amsterdam, 1978).

[27] Y. Niwa, S. Kobayashi and N. Azuma, Analysis of Transient Stresses Produced around Cavities of Arbitrary Shape during the Passage of Travelling Waves, *Memoirs of the Faculty of Engineering, Kyoto University* **36**, 28–46 (1975).

[28] Y. Niwa, S. Kobayashi and T. Fukui, Applications of Integral Equation Method to Some Geotechnical Problems, in: *Numerical Methods in Geomechanics*, C.S. Desai, Ed. (American Society of Civil Engineers, New York, 1976) pp. 120–131.

[29] Y. Niwa, T. Fukui, S. Kato and K. Fujiki, An Application of the Integral Equation Method to Two-Dimensional Elastodynamics, *Theoretical and Applied Mechanics* **28**, 281–290 (1980).

[30] F.R. Norwood and J. Miklowitz, Diffraction of Transient Elastic Waves by a Spherical Cavity, *Journal of Applied Mechanics* **34**, 735–741 (1967).

[31] Y.H. Pao and C.C. Mow, *Diffraction of Elastic Waves and Dynamic Stress Concentration* (Crane Russak, New York, 1973).

[32] C.L. Pekeris, The Seismic Surface Pulse, *Proceedings of the National Academy of Science* **41**, 469–480 (1955).

[33] F.J. Rizzo, D.J. Shippy and M. Rezayat, Boundary Integral Equation Analysis for a Class of Earth-structure Interaction Problems, *Final Report to National Science Foundation Grant No.* CEE 80-13461, University of Kentucky, Lexington (1985).

[34] R. Sen, T.G. Davies and P.K. Banerjee, Dynamic Analysis of Piles and Pile Groups Embedded in Homogeneous Soils, *Earthquake Engineering and Structural Dynamics* **13**, 53–65 (1985).

[35] R. Sen, E. Kausel and P.K. Banerjee, Dynamic Analysis of Piles and Pile Groups Embedded in Nonhomogeneous Soils, *International Journal for Numerical and Analytical Methods in Geomechanics* **9**, 507–524 (1985).

[36] A.H. Stroud and D. Secrest, *Gaussian Quadrature Formulas* (Prentice Hall, Englewood Cliffs, N.J., 1966).

[37] S.P. Timoshenko and J.N. Goodier, *Theory of Elasticity*, 3rd Edition (McGraw-Hill, New York, 1970).

[38] T.J. Tzong, S. Gupta and J. Penzien, Two Dimensional Hybrid Modelling of Soil–Structure Interaction, *Earthquake Engineering Research Center Report* 81/11, University of California, Berkeley (1981).

[39] A.S. Veletsos andd Y.T. Wei, Lateral and Rocking Vibration of Footings, *Journal of the Soil Mechanics and Foundations Division of the ASCE* **97**, (SM9), 1227–1248 (1971).

[40] R. Wilson, M.J. Bak, S. Nakazawa and P.K. Banerjee, 3-D Inelastic Analysis Methods for Hot Section Components, *Base program, First Annual Status Report, NASA Contractor Report* 174700, Lewis, Ohio (1984).

[41] O.C. Zienkiewicz, *The Finite Element Method* (McGraw-Hill, London, 1977).

CHAPTER 6

Nonlinear Solid Mechanics

Subrata MUKHERJEE

Department of Theoretical and Applied Mechanics
Cornell University
Ithaca, New York 14853, U.S.A.

and

Abhijit CHANDRA

Department of Aerospace and Mechanical Engineering
University of Arizona
Tucson, Arizona 85721, U.S.A.

Boundary Element Methods in Mechanics
Edited by D.E. Beskos

Introduction

The subject of this chapter is a discussion of applications of the Boundary Element Method (BEM) in nonlinear solid mechanics. This includes problems that are geometrically linear (i.e. problems with small strains, rotations and displacements) but are materially nonlinear due to the presence of plastic or viscoplastic deformations in addition to elastic ones; as well as fully nonlinear problems that involve large strains and rotations as well as the above-mentioned material nonlinearities. The discussion of the fully nonlinear problems is limited, however, to those cases where the elastic strains remain small even though the rotations and non-elastic strains can be large. Such is often the case in many applications, especially when deformations of metallic solids are under consideration.

The BEM is rooted in classical integral equation formulations of boundary-value problems such as the work of famous mathematicians like Fredholm and Volterra. As has been described in the earlier chapters of this book, it is a "natural" approach for linear problems, in that many linear partial differential operators can be transformed into equivalent integral forms through the use of appropriate Green's functions. It is not surprising to observe, therefore, that the principal BEM applications in computational solid mechanics in the 1960s dealt primarily with linear problems such as torsion of elastic bars [24], bending of elastic plates [25] or classical elastostatics [46, 20]. The method was carried forward during the 1970s to include problems with material nonlinearities such as plasticity [44, 47, 31, 45, 36, 5, 4, 28, 7, 10, 43, 22, 30, 42] or viscoplasticity [27, 18, 39, 32, 38, 35, 49]. Very recently, the BEM has been applied to fully nonlinear problems including both material and geometrical nonlinearities [12, 17, 41].

This chapter reviews Boundary Element Method formulations and numerical applications first for small-strain and then for large-strain–large-rotation problems of solid mechanics. Elastic plastic and elastic–viscoplastic materials are considered in this chapter. The elastic strains are assumed to remain small in all cases but rotations and non-elastic strains can be large in the geometrically nonlinear problems. Bulk compressibility effects, which are often ignored for large-strain problems of metallic solids, are explicitly included here. The large-strain Boundary Element Method formulation is based on updated Lagrangian approach.

Numerical results are presented here for some small-strain and planar large-strain–large-rotation problems. The large-strain examples are concerned with soil mechanics as well as metal forming. Results from the boundary element analysis are compared, in most cases, with those from the finite element analysis.

1. BEM for nonlinear problems

It is instructive, at this stage, to consider the basic issue in the extension of the BEM approach from linear to nonlinear problems, namely, how does one adapt a method with Green's functions, which is a natural one for problems with linear operators, to analyse nonlinear problems. Such an adaptation is by no means obvious, nor is it clear how to do this in complete generality. It is possible in many nonlinear solid mechanics problems, however, to recast the original initial boundary-value problem in rate form, such that the spatial partial differential operator, now operating on the rates of the dependent variables, remains linear. It may then be possible to solve this equivalent rate problem, at any instant of time, by the BEM, and march forward in time. This idea can be illustrated here through the simple example of Saint Venant torsion of a prismatic solid shaft for which the simple Laplacian operator appears. It is also noted here that BEM solutions for the Laplace equation have been discussed in the second chapter of this book.

If the shaft material is linear elastic, a stress function formulation of the problem is [51]

$$\nabla^2 \Phi = -2G\alpha \tag{1.1}$$

inside the two-dimensional uniform cross-section B of the shaft, together with the boundary condition

$$\Phi = \text{constant} \tag{1.2}$$

on the boundary of the cross-section. Here Φ is the stress function with the nonzero stress components defined as

$$\sigma_{31} = \frac{\partial \Phi}{\partial x_2}, \qquad \sigma_{32} = -\frac{\partial \Phi}{\partial x_1} \tag{1.3}$$

(with the x_1 and x_2 axes in the plane of the cross-section and x_3 measured in the axial direction), ∇^2 is the two-dimensional Laplacian operator, G is the shear modulus and α is the twist per unit length of the shaft. For simplicity, only a simply connected cross-section (solid shaft) is considered in this discussion.

Using a standard approach such as Green's identity, Eq. (1.1) can be transformed into an equivalent boundary integral equation [40]

$$\pi\Phi(\mathrm{P}) = \int_{\mathrm{B}} \ln r_{\mathrm{P}q} f_1(q)\,\mathrm{d}A_q + \int_{\partial B} \left[\frac{\partial \ln r_{\mathrm{PQ}}}{\partial n_{\mathrm{Q}}} \Phi(\mathrm{Q}) - \ln r_{\mathrm{PQ}} \frac{\partial \Phi(\mathrm{Q})}{\partial n_{\mathrm{Q}}} \right] \mathrm{d}c_{\mathrm{Q}} \tag{1.4}$$

where $f_1 = -2G\alpha$, $\ln r$ is the "free-space" Green's function for the Laplacian operator in two dimensions, n is an unit outward normal to ∂B at a point on it and $\mathrm{d}A$ and $\mathrm{d}c$ are differential area and boundary elements, respectively. The explicit notation, which will be used throughout this chapter, denotes p as a source and q as a field point (with capital letters denoting points on the boundary ∂B and lower case letters denoting points inside B). Also, r is the Euclidian distance between p and q and Eq. (1.4) is valid at a point P on ∂B where it is locally smooth.

Turning now to the elastic–viscoplastic problem with small strains and rotations, it is customary to assume additive decomposition of strains in the form

$$\varepsilon_{ij} = \varepsilon_{ij}^{(\mathrm{e})} + \varepsilon_{ij}^{(\mathrm{n})} \tag{1.5}$$

where the superscripts denote elastic and non-elastic parts respectively. It is important to mention here that while the elastic strain is the usual one, the non-elastic strain is history dependent. Also, it is assumed here that the rate of non-elastic strain $\dot{\varepsilon}_{ij}^{(\mathrm{n})}$ (where the superposed dot denotes the derivative with respect to real time t) is a function of stress and perhaps certain other internal or material structure variables but not of the rates of any of these variables. This is a key assumption and will be discussed further later in this chapter.

Using a rate form of Eq. (1.5), it can be shown [40] that

$$\nabla^2\dot{\Phi} = -2G\dot{\alpha} - 2G(\dot{\varepsilon}_{31,2}^{(\mathrm{n})} - \dot{\varepsilon}_{32.1}^{(\mathrm{n})}) \tag{1.6}$$

(where, $i \equiv \partial/\partial x_i$, $i = 1, 2$) so that the only differences between Eqs. (1.1) and (1.6) are limited to the facts that the rates are used in the latter and that some extra terms appear on the right-hand side of Eq. (1.6). The Laplacian operator, which is linear, is identical in both cases. It is easy to show, therefore, that a boundary integral equation, analogous to Eq. (1.4), can be derived as

$$\pi\dot{\Phi}(\mathrm{P}) = \int_{B} \ln r_{\mathrm{P}q} f_2(q)\,\mathrm{d}A_q + \int_{\partial B} \left[\frac{\partial \ln r_{\mathrm{PQ}}}{\partial n_{\mathrm{Q}}} \dot{\Phi}(\mathrm{Q}) - \ln r_{\mathrm{PQ}} \frac{\partial \dot{\Phi}(\mathrm{Q})}{\partial n_{\mathrm{Q}}} \right] \mathrm{d}c_{\mathrm{Q}} \tag{1.7}$$

where the modified right-hand side function is defined as

$$f_2 = -2G\dot{\alpha} - 2G(\dot{\varepsilon}_{31,2}^{(\mathrm{n})} - \dot{\varepsilon}_{32,1}^{(\mathrm{n})})\,. \tag{1.8}$$

Further, by virtue of the above-mentioned constitutive assumptions, the integrand in the domain integral on the right-hand side of Eq. (1.7), at any time, is explicitly known once the stresses and internal variables (and, of course, the prescribed rate of twist per unit length) are known at this time. This sets the stage for solving the BEM equations at each discretized time together with a scheme for marching forward in time. It should be mentioned here that the situation changes somewhat if the constitutive assumptions are altered and an elastic time-independent plastic model, for example, is used instead to depict material behavior. The essential concept of adaptation of a Green's function type of approach to a certain class of nonlinear problems, however, remains as illustrated above. Indeed, this idea provides the central theme for the rest of the chapter.

The remainder of this chapter has two main parts. The first is a discussion of geometrically linear problems with material nonlinearities. Elastic–plastic and elastic–viscoplastic problems are considered here. BEM formulations for such problems are presented first. The constitutive assumptions, cited above, are elaborated upon. Finally, BEM numerical results for sample problems are presented and discussed. The last part of the chapter follows the same general pattern except that the more complicated class of problems with both material and geometrical nonlinearities is the subject under review. This time, recently obtained sample numerical results from soil mechanics as well as metal forming, are discussed.

2. Problems with material nonlinearities

2.1. Governing differential equations

The governing differential equations in rate form, for small elastic–plastic or elastic–viscoplastic deformations of a solid body, in the absence of body forces, can be written as:

equilibrium: $$\dot{\sigma}_{ji,j} = 0 \qquad \text{in } B \tag{2.1}$$

$$n_j \dot{\sigma}_{ji} = \dot{\tau}_i \qquad \text{on } \partial B \tag{2.2}$$

kinematic: $$\dot{\varepsilon}_{ij} = \dot{\varepsilon}_{ij}^{(e)} + \dot{\varepsilon}_{ij}^{(n)} = \tfrac{1}{2}(\dot{u}_{i,j} + \dot{u}_{j,i}) \tag{2.3}$$

constitutive: $$\dot{\sigma}_{ij} = \lambda \dot{\varepsilon}_{kk}^{(e)} \delta_{ij} + 2G\dot{\varepsilon}_{ij}^{(e)} \, . \tag{2.4}$$

In the above equations, u_i and τ_i are the components of the displacement and traction vectors, and ε_{ij} and σ_{ij} are the components of the strain and stress tensors, respectively. Eq. (2.3) assumes the additive decomposition of strain rates into elastic and non-elastic parts. Also, λ and G are Lamé constants and a superposed dot denotes a real time derivative for elastic–viscoplastic and pseudo-time derivative for elastic–plastic problems. Finally, δ_{ij} is the Kronecker delta. These equations are written for three-dimensional situations so that the range of indices is 1, 3.

In order to get a complete description of the problem, it is necessary to prescribe a constitutive description for the non-elastic strain rate $\dot{\varepsilon}_{ij}^{(n)}$ in terms of relevant variables. This topic will be discussed later in the section called constitutive models. The exact form of such a model is kept unspecified for the time being.

An integral formulation for the above problem can be obtained by using, as a reference problem, the infinitesimal deformations of an isotropic linear elastic solid of infinite extent, due to a point body force. Using a superscript R to denote the different fields in this reference problem, the governing equations for this case can be written as:

equilibrium: $$\sigma_{ji,j}^{(R)} = -\rho F_i^{(R)} = -\Delta(\mathrm{p}, \mathrm{q})\delta_{ij}e_j \tag{2.5}$$

$$\sigma_{ji}^{(R)} n_j = \tau_i^{(R)} \tag{2.6}$$

kinematic: $$\varepsilon_{ij}^{(R)} = \tfrac{1}{2}(u_{i,j}^{(R)} + u_{j,i}^{(R)}) \tag{2.7}$$

constitutive: $$\sigma_{ij}^{(R)} = \lambda\varepsilon_{kk}^{(R)}\delta_{ij} + 2G\varepsilon_{ij}^{(R)}\,. \tag{2.8}$$

The new symbols in the above are the body force components $F_i^{(R)}$, the Dirac delta function Δ, unit orthogonal Cartesian base vectors e_j and density ρ. The solution of the above problem, due to Lord Kelvin, can be written as

$$u_i^{(R)} = U_{ij}e_j \tag{2.9}$$

$$\tau_i^{(R)} = T_{ij}e_j \tag{2.10}$$

where U_{ij} and T_{ij} are the usual two point kernels of linear elasticity. They are given in many references (e.g. [40]) and are repeated here for completeness

$$U_{ij} = \frac{1}{16\pi(1-\nu)Gr}\,[(3-4\nu)\delta_{ij} + r_{,i}r_{,j}] \tag{2.11}$$

$$T_{ij} = -\frac{1}{8\pi(1-\nu)r^2}\left[[(1-2\nu)\delta_{ij} + 3r_{,1}r_{,j}]\,\frac{\partial r}{\partial n} + (1-2\nu)(r_{,i}n_j - r_{,j}n_i)\right] \tag{2.12}$$

A comma in the above denotes a derivative with respect to a field point and ν is Poisson's ratio.

2.2. Boundary element formulation for displacement rates

An integral formulation for the problem described by Eqs. (2.1)–(2.4) can be obtained from the indentity, adapted from the reciprocal theorem of Betti

$$\int_B \dot{\sigma}_{ij}\varepsilon_{ij}^{(R)}\,\mathrm{d}V = \int_B \sigma_{ij}^{(R)}\dot{\varepsilon}_{ij}^{(e)}\,\mathrm{d}V\,. \tag{2.13}$$

The integrands on either side of Eq. (2.13) are identical in view of Eqs. (2.4) and (2.8).

Using Eqs. (2.7), (2.1), (2.9), and (2.2), together with the divergence theorem, one can show that

$$\int_B \dot{\sigma}_{ij}\varepsilon_{ij}^{(R)}\,\mathrm{d}V = e_j\int_{\partial B} U_{ij}\dot{\tau}_i\,\mathrm{d}S\,. \tag{2.14}$$

Similarly, using Eqs. (2.3), (2.8), (2.5), (2.7), (2.6), (2.10), and (2.9), together with the divergence theorem, one gets

$$\int_B \sigma_{ij}^{(R)}\dot{\varepsilon}_{ij}^{(e)}\,\mathrm{d}V = e_j\Bigg[\int_{\partial B} T_{ij}\dot{u}_i\,\mathrm{d}S + \int_B \Delta(\mathrm{p},\mathrm{q})\delta_{ij}\dot{u}_i\,\mathrm{d}V - \int_B [\lambda U_{ij,i}\dot{\varepsilon}_{kk}^{(\mathrm{n})} + 2GU_{ij,k}\dot{\varepsilon}_{ik}^{(\mathrm{n})}]\,\mathrm{d}V\Bigg]. \tag{2.15}$$

By virtue of Eq. (2.13), the right-hand sides of Eqs. (2.14) and (2.15) are equal. Equating these quantities and using the relevant property of the Dirac delta function, one obtains the equation

$$\dot{u}_j(\mathrm{p}) = \int_{\partial B} [U_{ij}(\mathrm{p},\mathrm{Q})\dot{\tau}_i(\mathrm{Q}) - T_{ij}(\mathrm{p},\mathrm{Q})\dot{u}_i(\mathrm{Q})]\,\mathrm{d}S_{\mathrm{Q}} + \int_B [\lambda U_{ij,i}(\mathrm{p},\mathrm{q})\dot{\varepsilon}_{kk}^{(\mathrm{n})}(\mathrm{q}) + 2GU_{ij,k}(\mathrm{p},\mathrm{q})\dot{\varepsilon}_{ik}^{(\mathrm{n})}(\mathrm{q})]\,\mathrm{d}V_{\mathrm{q}}\,. \tag{2.16}$$

A boundary integral equation for the unknown components of the traction and displacement rates in terms of the prescribed ones can be obtained by taking the limit as p approaches a boundary point P. This leads to the equation [40]

$$C_{ij}(\mathrm{P})\dot{u}_i(\mathrm{P}) = \int_{\partial B} [U_{ij}(\mathrm{P},\mathrm{Q})\dot{\tau}_i(\mathrm{Q}) - T_{ij}(\mathrm{P},\mathrm{Q})\dot{u}_i(\mathrm{Q})]\,\mathrm{d}S_{\mathrm{Q}} + \int_B [\lambda U_{ij,i}(\mathrm{P},\mathrm{q})\dot{\varepsilon}_{kk}^{(\mathrm{n})}(\mathrm{q}) + 2GU_{ij,k}(\mathrm{P},\mathrm{q})\dot{\varepsilon}_{ik}^{(\mathrm{n})}(\mathrm{q})]\,\mathrm{d}V_{\mathrm{q}}\,. \tag{2.17}$$

The surface integrals in the above Eqs. (2.16) and (2.17) must be interpreted in the sense of Cauchy principal values. The coefficients C_{ij} multiplying $\dot{u}_i$ in the free term arise from the integration of $T_{ij}\dot{u}_i$. If ∂B is locally smooth at P, $C_{ij} = \frac{1}{2}\delta_{ij}$. Otherwise, C_{ij} can be evaluated in closed form for two-dimensional problems [40] and indirectly for three-dimensional problems [21, 29].

Once the non-elastic strain rate components are prescribed by a suitable constitutive model, Eq. (2.17) can be solved, at any time, for the unspecified boundary components of the displacement and traction rates, in terms of the prescribed ones. This can only be done, in general, by numerical discretization of Eq. (2.17) and solving of the resultant algebraic system. Further details of this procedure are given later in this chapter. It is important to note, however, that if an internal variable type of constitutive model is used, the non-elastic strain rate components are known at any time as functions of stress components and internal variables at that time, so that the unknowns in Eq. (2.17) only lie on the boundary ∂B. The size of coefficient matrices in resultant algebraic systems, therefore, depend only on the number of boundary nodes, rather than on the total number of nodes in the entire three-dimensional solid, as is typical in domain-type methods such as the Finite Element Method (FEM).

The next step is the use of Eq. (2.16) to obtain the velocity distribution inside the body B. As many internal points as necessary can now be sampled for this purpose. The right-hand side of Eq. (2.16) only involves known quantities at this stage.

2.3. *Internal stress rates*

Strain rate components inside the body B must now be determined by differentiation of Eq. (2.16). This can be achieved by several methods, the easiest and least accurate of which is numerical differentiation of the velocity field by interpolation of $\dot{u}_j$ over internal cells. This method, in general, leads to discontinuities in velocity gradients across boundaries of internal cells, and, therefore, can lead to significant numerical errors in the differentiated quantities.

A much more elegant approach is the analytical differentiation of Eq. (2.16) at a source point p. This gives

$$\dot{u}_{j,\bar{l}}(\mathrm{p}) = \int_{\partial B} [U_{ij,\bar{l}}(\mathrm{p}, \mathrm{Q})\dot{\tau}_i(\mathrm{Q}) - T_{ij,\bar{l}}(\mathrm{p}, \mathrm{Q})\dot{u}_i(\mathrm{Q})]\, \mathrm{d}S_Q$$
$$+ \frac{\partial}{\partial x_{\bar{l}}} \int_B [\lambda W_{iji}(\mathrm{p}, \mathrm{q})\dot{\varepsilon}^{(\mathrm{n})}_{kk}(\mathrm{q}) + 2GW_{ijk}(\mathrm{p}, \mathrm{q})\dot{\varepsilon}^{(\mathrm{n})}_{ik}(\mathrm{q})]\, \mathrm{d}V_q \quad (2.18)$$

where $W_{ijk} = U_{ij,k}$ and the letter $\bar{l}$ following a comma denotes differentiation with respect to a source point. The differentiated kernels are given in Ref. [40] and will not be repeated here.

An immediate difficulty is that the kernel W_{ijk} has a singularity of order r^{-2} and differentiation of the domain integral must be handled with great care. One possibility is to evaluate the volume integral analytically for an arbitrary source point p and then differentiate this integral at p. This typically requires interpolation of the non-elastic strain rate field within a volume cell, evaluation of the integral over the cell in which the integrand is singular, and then differentiation of this integral at a source point p. This has been carried out successfully for planar problems [39].

Another alternative [11] is to carry out a convected derivative of the domain integral in Eq. (2.17) by dividing the domain B into $B - B_\eta(\mathrm{p})$ and $B_\eta(\mathrm{p})$, where $B_\eta(\mathrm{p})$ is a sphere of radius η centered at p. Taking the limit as $\eta \to 0$ yields the appropriate free term and a principal-value integral over $B - B_\eta$. Numerical evaluation of this principal-value integral, for the case of planar and axisymmetric problems, has been discussed in a paper and in a recent book by Brebbia et al. [50, 9]. Evaluation of this integral for three-dimensional problems can be carried out in analogous fashion. Once the total strain rate at a point is determined, it is a simple matter to determine the elastic strain rate from Eq. (2.3) and then the stress rate from Hooke's Law (Eq. (2.4)).

2.4. Boundary stress rates

The boundary stress rate components are best obtained by using the method described below rather than by trying to take the limit of Eq. (2.18) as $\mathrm{p} \to \mathrm{P}$. Also, for the purpose of this discussion, P is a point on ∂B where it is locally smooth.

The calculation involves six vectors and two tensors at P. The vectors are the traction rate $\dot{\boldsymbol{\tau}}$, the tangential derivatives of the velocity $\partial \dot{\boldsymbol{u}}/\partial s_1$ and $\partial \dot{\boldsymbol{u}}/\partial s_2$ along two linearly independent directions s_1 and s_2 on ∂B at P, and the unit normal and tangent vectors $\boldsymbol{n}$, $\boldsymbol{t}^{(1)}$ and $\boldsymbol{t}^{(2)}$ at P. The tensors are the stress rate $\dot{\boldsymbol{\sigma}}$ and the velocity gradient $\partial \dot{\boldsymbol{u}}/\partial \boldsymbol{x}$. Once Eq. (2.17) is solved, the velocity gradients $\partial \dot{\boldsymbol{u}}/\partial s_1$, $\partial \dot{\boldsymbol{u}}/\partial s_2$ must be evaluated numerically by differentiation of the appropriate shape functions for $\dot{\boldsymbol{u}}$ along s_1 and s_2. Now, these six vectors $\partial \dot{\boldsymbol{u}}/\partial s_1$, $\partial \dot{\boldsymbol{u}}/\partial s_2$, $\dot{\boldsymbol{\tau}}$, $\boldsymbol{n}$, $\boldsymbol{t}^{(1)}$ and $\boldsymbol{t}^{(2)}$ (18 scalar components) are known at P. The 15 unknown tensor components $\dot{\sigma}_{ij}$ and $\dot{u}_{i,j}$ ($\dot{\sigma}_{ij}$ is symmetric) can now be obtained from the following set of four equations (15 scalar linear algebraic equations) at P.

$$\dot{\sigma}_{ij} = \lambda \dot{u}_{k,k} \delta_{ij} + G(\dot{u}_{i,j} + \dot{u}_{j,i}) - \lambda \dot{\varepsilon}_{kk}^{(n)} \delta_{ij} - 2G\dot{\varepsilon}_{ij}^{(n)} \tag{2.19}$$

$$\dot{\tau}_i = \dot{\sigma}_{ji} n_j \tag{2.20}$$

$$\frac{\partial \dot{u}_i}{\partial s_1} = \dot{u}_{i,j} t_j^{(1)}, \qquad \frac{\partial \dot{u}_i}{\partial s_2} = \dot{u}_{i,j} t_j^{(2)} \qquad \text{(2.21a,2.21b)}$$

This procedure must often be carried out since the surface tangential stress rate components at P, which, in general, cannot be obtained from the traction rates at that point, can be very important in design.

2.5. Plane strain

The equations for plane strain can be obtained from the three-dimensional ones by setting $u_3 = 0$ and using the corresponding plane strain solution in an infinite domain with $u_3^{(R)} = 0$. Also, the quantities are assumed invariant in the x_3 direction, i.e. partial derivatives of the relevant variables in the x_3 direction are assumed to vanish.

An immediate consequence of the above assumptions are the equations

$$\varepsilon_{13} = \varepsilon_{23} = \varepsilon_{33} = 0 \tag{2.22}$$

$$\varepsilon_{13}^{(R)} = \varepsilon_{23}^{(R)} = \varepsilon_{33}^{(R)} = 0\,. \tag{2.23}$$

The strain and reference strain tensors are, of course, symmetric.

It is important to realize that quantities such as $\varepsilon_{33}^{(e)}$ and $\varepsilon_{33}^{(n)}$ do not, in general, vanish separately.

An examination of the starting Eq. (2.13) reveals that by virtue of Eq. (2.23) the range of indices for the left-hand side of Eq. (2.13) is 1, 2. The right-hand side, however, retains the term $\sigma_{33}^{(R)}\dot{\varepsilon}_{33}^{(e)}$ $(\sigma_{13}^{(R)} = \sigma_{23}^{(R)} = 0)$ in addition, of course, to the terms with $i, j = 1, 2$. Following the derivation of the three-dimensional equations, the resulting plane strain equation is

$$\dot{u}_j(\mathrm{p}) = \int_{\partial B} [U_{ij}(\mathrm{p}, \mathrm{Q})\dot{\tau}_i(\mathrm{Q}) - T_{ij}(\mathrm{p}, \mathrm{Q})\dot{u}_i(\mathrm{Q})]\, \mathrm{d}S_{\mathrm{Q}}$$

$$+ \int_{\mathrm{B}} [\lambda U_{ij,i}(\mathrm{p}, \mathrm{q})\dot{\theta}^{(\mathrm{n})}(\mathrm{q}) + 2G U_{ij,k}(\mathrm{p}, \mathrm{q})\, \dot{\varepsilon}_{ik}^{(\mathrm{n})}(\mathrm{q})]\, \mathrm{d}V_{\mathrm{q}} \tag{2.24}$$

where $\dot{\theta}^{(\mathrm{n})} = \dot{\varepsilon}_{11}^{(\mathrm{n})} + \dot{\varepsilon}_{22}^{(\mathrm{n})} + \dot{\varepsilon}_{33}^{(\mathrm{n})}$ and the range of indices in Eq. (2.24) is 1, 2.

The plane strain kernels in the above equations are the usual ones and are given by

$$U_{ij} = -\frac{1}{8\pi(1-\nu)G}\,[(3-4\nu)\ln r\delta_{ij} - r_{,i}r_{,j}] \tag{2.25}$$

$$T_{ij} = -\frac{1}{4\pi(1-\nu)r}\left[[(1-2\nu)\delta_{ij} + 2r_{,i}r_{,j}]\frac{\partial r}{\partial n} + (1-2\nu)(r_{,i}n_j - r_{,j}n_i\right] \tag{2.26}$$

where r is now the distance between two points p and q in the x_1x_2 plane.

Taking the limit as p→P leads one to the boundary integral equation

$$C_{ij}(\mathrm{P})\dot{u}_i(\mathrm{P}) = \int_{\partial B} [U_{ij}(\mathrm{P},\mathrm{Q})\dot{\tau}_i(\mathrm{Q}) - T_{ij}(\mathrm{P},\mathrm{Q})\dot{u}_i(\mathrm{Q})]\,\mathrm{d}S_{\mathrm{Q}}$$

$$\int_{B} \lambda U_{ij,i}(\mathrm{P},\mathrm{q})\dot{\theta}^{(\mathrm{n})}(\mathrm{q}) + 2GU_{ij,k}(\mathrm{P},\mathrm{q})\dot{\varepsilon}^{(\mathrm{n})}_{ik}(\mathrm{q})]\,\mathrm{d}V_{\mathrm{q}} \qquad (2.27)$$

where the coefficients C_{ij} can now be written in closed form as

$$C_{11} = \frac{\beta}{2\pi} + \frac{\cos 2\gamma \sin\beta}{4\pi(1-\nu)}$$

$$C_{12} = C_{21} = \frac{\sin 2\gamma \sin\beta}{4\pi(1-\nu)} \qquad (2.28)$$

$$C_{22} = \frac{\beta}{2\pi} - \frac{\cos 2\gamma \sin\beta}{4\pi(1-\nu)}$$

In the above, P lies on a boundary corner of included angle β and γ is the angle between the bisector of β and the x_1 axis, measured positive in the anticlockwise direction [40].

The internal and boundary stress rates can now be obtained from the displacement rate in a manner which is entirely analogous to the three-dimensional situation. Determination of the boundary stress rate at a point in this case, of course, only requires the use of a single tangential direction s along ∂B.

2.6. Plane stress

The plane stress situation considered here is

$$\sigma_{13} = \sigma_{23} = \sigma_{33} = 0 \qquad (2.29)$$

with the corresponding reference solution

$$\sigma^{(\mathrm{R})}_{13} = \sigma^{(\mathrm{R})}_{23} = \sigma^{(\mathrm{R})}_{33} = 0\,. \qquad (2.30)$$

In this case, by virtue of Eqs (2.29) and (2.30), the range of indices on both sides of Eqs. (2.13) becomes 1, 2. Equation (2.8), however, contains $\varepsilon^{(\mathrm{R})}_{\mathrm{kk}} = \varepsilon^{(\mathrm{R})}_{11} + \varepsilon^{(\mathrm{R})}_{22} + \varepsilon^{(\mathrm{R})}_{33}$. It is advantageous to eliminate $\varepsilon^{(\mathrm{R})}_{33}$ in favour of $\varepsilon^{(\mathrm{R})}_{11} + \varepsilon^{(\mathrm{R})}_{22}$. Thus, from Eq. (2.8) with $i = j = 3$, one gets

$$\varepsilon^{(\mathrm{R})}_{33} = -\frac{\nu}{1-\nu}(\varepsilon^{(\mathrm{R})}_{11} + \varepsilon^{(\mathrm{R})}_{22}) \qquad (2.31)$$

so that

$$\sigma_{11}^{(R)} = \bar{\lambda}(\varepsilon_{11}^{(R)} + \varepsilon_{22}^{(R)}) + 2G\varepsilon_{11}^{(R)} \tag{2.32}$$

and

$$\sigma_{22}^{(R)} = \bar{\lambda}(\varepsilon_{11}^{(R)} + \varepsilon_{22}^{(R)}) + 2G\varepsilon_{22}^{(R)}$$

where $\bar{\lambda} = 2G\bar{\nu}/(1-2\bar{\nu})$ with $\bar{\nu} = \nu/(1+\nu)$.

Following the derivation of the three-dimensional equations and using the above, the plane stress equation for $\dot{u}_j$ becomes

$$\dot{u}_j(\mathrm{p}) = \int_{\partial B} [U_{ij}(\mathrm{p}, \mathrm{Q})\dot{\tau}_i(\mathrm{Q}) - T_{ij}(\mathrm{p}, \mathrm{Q})\dot{u}_i(\mathrm{Q})]\, \mathrm{d}S_{\mathrm{Q}}$$

$$+ \int_{B} [\bar{\lambda} U_{ij,i}(\mathrm{p}, \mathrm{q})\dot{\varepsilon}_{kk}^{(\mathrm{n})}(\mathrm{q}) + 2GU_{ij,k}(\mathrm{p}, \mathrm{q})\dot{\varepsilon}_{ik}^{(\mathrm{n})}(\mathrm{q})]\, \mathrm{d}V_{\mathrm{q}} \tag{2.33}$$

where the range of indices is 1, 2, as for plane strain.

The plane stress kernels U_{ij} and T_{ij} have the same forms as those for plane strain from Eqs (2.2)–(2.26) with ν replaced by $\bar{\nu}$. The boundary integral equation corresponding to (2.33) leads to the tensor **C**, whose components C_{ij} are given by Eq. (2.28) with ν replaced by $\bar{\nu}$. Finally, the stress rate is obtained in a manner entirely analogous to the plane strain case.

2.7. Constitutive models

As mentioned previously, it has been assumed that the strain rates can be additively decomposed into elastic and non-elastic parts (Eq. (2.3)) and that Hooke's law relates the rate of stress to that of the elastic strain (Eq. (2.4)). The non-elastic strain rate components must now be determined from an appropriate constitutive model. This can be a conventional rate independent plasticity model, in which case $\dot{\varepsilon}_{ij}^{(\mathrm{n})}$ is the rate of plastic strain with respect to an appropriate loading variable (pseudo-time). It can also be a combined creep–plasticity model with internal or state variables, in which case $\dot{\varepsilon}_{ij}^{(\mathrm{n})}$ is the time rate of change of the non-elastic strain. This non-elastic strain, for the latter type of model, includes both irreversible as well as anelastic strains. The reader is referred to Mukherjee's book [40] for a further discussion of such constitutive models. A brief discussion of these models, together with some examples, follow.

The mathematical structure of many of such constitutive models can be written in the form

$$\dot{\varepsilon}_{ij}^{(\mathrm{n})} = f_{ij}(\sigma_{ij}, \dot{\sigma}_{ij}, q_{ij}^{(\mathrm{k})}) \tag{2.34}$$

and

$$\dot{q}_{ij}^{(k)} = g_{ij}^{(k)}(\sigma_{ij}, \dot{\sigma}_{ij}, q_{ij}^{(l)})\,, \tag{2.35}$$

where $q_{ij}^{(k)}$ are the state variables.

2.7.1. *Time-independent plasticity*

The plastic strain rate is generally a function of stress as well as stress rate. The quantities $q_{ij}^{(k)}$ might denote, for example, the location of the center of the yield surface for a kinematic hardening model. A suitable equation for the yield surface must be added to Eqs (2.34) and (2.35) with the condition that $\dot{\varepsilon}_{ij}^{(n)} = 0$ if the stress state lies inside the yield surface. An an example, kinematic hardening with a von Mises yield criterion can be considered here. One may write [3]:

Yield surface:

$$\tfrac{1}{2}\hat{\sigma}'_{ij}\hat{\sigma}'_{ij} - \tfrac{1}{3}\bar{\sigma}_0^2 = 0 \tag{2.36}$$

where the effective stress $\sigma'_{ij} = \sigma_{ij} - \alpha_{ij}$ and its deviator $\hat{\sigma}'_{jj} = \sigma'_{ij} - \frac{1}{3}\sigma'_{kk}\delta_{ij}$. Further, the stress invariant is given as

$$\bar{\sigma}_0 = +(\tfrac{3}{2}\sigma'_{ij}\sigma'_{ij})^{1/2}$$

and α_{ij} locates the center of the yield surface.

Flow rule:

$$\dot{\varepsilon}_{ij}^{(n)} = \frac{3\hat{\sigma}'_{ij}\hat{\sigma}'_{kl}}{2c\bar{\sigma}_0^2}\,\dot{\sigma}_{kl} \tag{2.37}$$

where $c = 2\bar{H}/3$ in terms of the uniaxial hardening modulus $\bar{H}$.

Evolution of α_{ij}:

$$\dot{\alpha}_{ij} = \frac{3\hat{\sigma}'_{ij}\hat{\sigma}'_{kl}}{2\bar{\sigma}_0^2}\,\dot{\sigma}_{kl}\,. \tag{2.38}$$

It is easy to see that Eqs. (2.37) and (2.38) fit into the general format of Eqs. (2.34) and (2.35) with $k = 1$ and $q_{ij}^{(1)} = \alpha_{ij}$.

2.7.2. *Viscoplasticity with state variables*

In this case Eq. (2.34) states that the history dependence of the non-elastic strain rate at any time can be represented by the current values of the stress and suitably chosen state variables q_{ij}. The stress rate $\dot{\sigma}_{ij}$ is absent from Eqs (2.34) and (2.35). The non-elastic strain rate contains both slowly recoverable as well as irrecoverable components and thus includes both plastic and creep effects. Permanent as well as recoverable strains are considered to be time dependent. There is no explicit yield surface and $\dot{\varepsilon}_{ij}^{(\mathrm{n})}$ is present at all stages of the deformation process. Its value, however, must be very small at low values of stress.

The state variables $q_{ij}^{(k)}$ (usually k equals 1 or 2) must be chosen carefully. They can be scalars or tensors and their choices vary in different models. They can physically represent quantities like average dislocation density, back stress, plastic work, etc. They are akin to quantities like yield stress and location of the centre of the yield surface in classical plasticity. The important assumption here is that the current values of the state variables completely characterize the present deformation state of the material.

The model due to Hart [23] has been used in some of the numerical applications presented later in this chapter. This model has two state variables, a scalar and a tensor. A brief review of this model is presented below.

The model in uniaxial tension is shown in Fig. 1. Material response in dilatation is assumed to be elastic. According to this model, the non-elastic strain rate is decomposed into two (time-dependent) components

$$\dot{\varepsilon}_{ij}^{(\mathrm{n})} = \dot{\varepsilon}_{ij}^{(\mathrm{a})} + \dot{\varepsilon}_{ij}^{(\mathrm{p})} \tag{2.39}$$

where $\dot{\varepsilon}_{ij}^{(\mathrm{a})}$ is the anelastic strain which reflects the magnitude and direction of prior deformation history and $\varepsilon_{ij}^{(\mathrm{p})}$ is the completely irrecoverable and path-dependent permanent strain. The two state variables in the model are the anelastic strain and a scalar H, called hardness, which is similar to an isotropic strain hardening parameter or curent yield stress.

The deviatoric component of the stress, $\hat{\sigma}_{ij}$, is decomposed into two auxillary tensors

$$\hat{\sigma}_{ij} = \hat{\sigma}_{ij}^{(\mathrm{a})} + \hat{\sigma}_{ij}^{(\mathrm{f})} . \tag{2.40}$$

The flow rules for the strains or strain rates are of the form

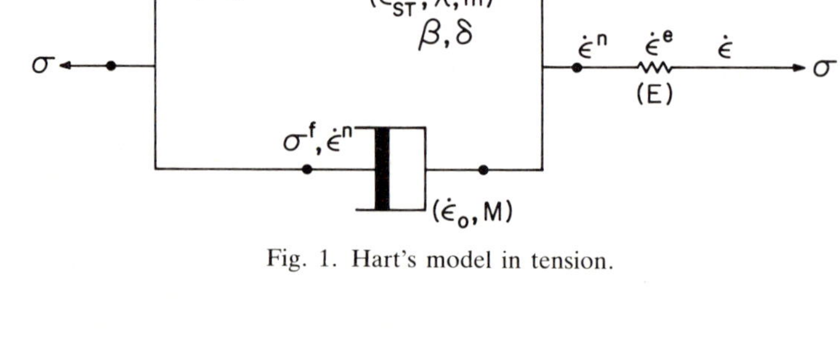

Fig. 1. Hart's model in tension.

$$\varepsilon_{ij}^{(a)} = \frac{3\varepsilon^{(a)}}{2\sigma^{(a)}} \hat{\sigma}_{ij}^{(a)} \tag{2.41}$$

$$\dot{\varepsilon}_{ij}^{(a)} = \frac{3\varepsilon^{(a)}}{2\sigma^{(a)}} \hat{\sigma}_{ij}^{(a)} \tag{2.42}$$

$$\dot{\varepsilon}_{ij}^{(n)} = \frac{3\dot{\varepsilon}^{(n)}}{2\sigma^{(f)}} \hat{\sigma}_{ij}^{(f)} \tag{2.43}$$

where $\dot{\varepsilon}^{(n)}$, $\dot{\varepsilon}^{(p)}$, $\varepsilon^{(a)}$, σ, $\sigma^{(a)}$, and $\sigma^{(f)}$ are scalar invariants of the corresponding tensors defined as

$$\dot{\varepsilon}^{(n)} = (\tfrac{2}{3}\dot{\varepsilon}_{ij}^{(n)}\dot{\varepsilon}_{ij}^{(n)})^{1/2}, \qquad \dot{\varepsilon}^{(p)} = (\tfrac{2}{3}\dot{\varepsilon}_{ij}^{(p)}\dot{\varepsilon}_{ij}^{(p)})^{1/2}, \qquad \varepsilon^{(a)} = (\tfrac{2}{3}\varepsilon_{ij}^{(a)}\varepsilon_{ij}^{(a)})^{1/2}$$

$$\sigma = (\tfrac{3}{2}\hat{\sigma}_{ij}\hat{\sigma}_{ij})^{1/2}, \qquad \sigma^{(a)} = (\tfrac{3}{2}\hat{\sigma}_{ij}^{(a)}\hat{\sigma}_{ij}^{(a)})^{1/2}, \qquad \sigma^{(f)} = (\tfrac{3}{2}\hat{\sigma}_{ij}^{(f)}\hat{\sigma}_{ij}^{(f)})^{1/2}. \tag{2.44}$$

The scalar invariants are related to each other through the uniaxial equations (Fig. 1)

$$\sigma^{(a)} = \mathcal{M}\varepsilon^{(a)}, \qquad \dot{\varepsilon}^{(n)} = \varepsilon_0(\sigma^{(f)}/\sigma_0)^M \tag{2.45,2.46}$$

$$\dot{\varepsilon}^{(p)} = E_0[\ln(H/\sigma^{(a)})]^{-1/\lambda} \tag{2.47}$$

$$E_0 = E_{0s}(H/H_s)^m \tag{2.48}$$

$$\dot{H} = \dot{\varepsilon}^{(p)} H\Gamma(H, \sigma^{(a)}) \tag{2.49}$$

$$\Gamma(H, \sigma^{(a)}) = (\beta/H)^{\delta}(\sigma^{(a)}/H)^{\beta/H} \tag{2.50}$$

Equation (2.45) represents a linear anelastic element, (2.46) a nonlinear dashpot, (2.47) and (2.48) a "plastic" element and, finally, (2.49) and (2.50) describe strain hardening. The flow parameters are $\mathcal{M}$, M, m, λ, ε_0 (at a reference stress level σ_0) and E_{0s} (at a reference hardness level H_s); and β and δ are strain hardening parameters. The above equations are written for a uniform temperature throughout the body. These material parameters, for a given material or alloy, must be determined from relaxation and tension tests [40, 23].

2.8. Numerical implementation

Boundary-value problems with irregular geometry must, in general, be solved by numerical methods. Some remarks on the numerical implementation of the integral equations are made in this section. The three-dimensional equations are considered here. Of course, similar procedures would apply to the two-dimensional equations.

The first step is the discretization of the three-dimensional body into surface elements and internal cells. As mentioned before, the internal discretization is necessary for the evaluation of volume integrals with known integrands. A discretized version of the boundary integral equation (2.17) for the displacement rate can be written as

$$\begin{aligned} C_{ij}(\mathrm{P}_M)\dot{u}_i(\mathrm{P}_M) = & \sum_{N_s} \int_{\Delta s_N} U_{ij}(\mathrm{P}_M, \mathrm{Q})\dot{\tau}_i(\mathrm{Q})\,\mathrm{d}S_\mathrm{Q} \\ & - \sum_{N_s} \int_{\Delta s_N} T_{ij}(\mathrm{P}_M, \mathrm{Q})\dot{u}_i(\mathrm{Q})\,\mathrm{d}S_\mathrm{Q} \\ & + \sum_{n_i} \int_{\Delta V_n} 2GW_{ijk}(\mathrm{P}_M, \mathrm{Q})\dot{\varepsilon}_{ik}^{(\mathrm{n})}(\mathrm{q})\,\mathrm{d}V_\mathrm{q} \\ & + \sum_{n_i} \int_{\Delta V_n} \lambda W_{iji}(\mathrm{P}_M, \mathrm{Q})\dot{\varepsilon}_{kk}^{(\mathrm{n})}(\mathrm{q})\,\mathrm{d}V_\mathrm{q} \end{aligned} \tag{2.51}$$

where the surface of the body ∂B has been divided into N_s boundary elements and the interior into n_i internal cells and $\dot{U}_i(\mathrm{P}_M)$ are the components of the displacement rates at a point P which coincides with node M.

Suitable shape functions must now be chosen for the variations of traction and displacement rates on the surface elements ΔS_N and the variation of the non-elastic strain rates over an internal cell ΔV_n. Special care must be taken for integration over an element in which a kernel is singular, i.e. over an element which contains the source point P_M. If the elements have simple geometry and the shape functions have simple form, it might be possible to carry out integrals of kernels analytically. Cruse [20] for example, has solved three-dimensional elasticity problems with flat triangular surface elements and piecewise uniform tractions and displacements on these elements. Each surface element is denoted by its centroidal point so that $C_{ij} = \frac{1}{2}\delta_{ij}$. In this special case, it is possible to evaluate the integrals

$$\Delta U_{ij} = \int_{\Delta s_N} U_{ij}(\mathrm{P}_M, \mathrm{Q})\,\mathrm{d}S_\mathrm{Q}$$

$$\Delta T_{ij} = \int_{\Delta s_N} T_{ij}(\mathrm{P}_M, \mathrm{Q})\,\mathrm{d}S_\mathrm{Q}$$

exactly, knowing the size, orientation and location of the surface element ΔS_N and the point P_M.

Analytical integration of this type is not possible in general. Integration of these kernels over elements which are far from the source point can generally be carried out conveniently and very accurately by Gaussian integration. Integration in singular and near-singular situations, however, is another matter and special methods such as use of rigid-body displacements [40] are useful in such cases.

Another comment pertains to the numerical modelling of possible jumps in normals or prescribed tractions across boundaries of surface elements. A simple way around this difficulty is to place source points inside rather than on the boundaries of surface elements. This is easily done in the Boundary Element Method since the source points are really sampling points (or boundary collocation points). These points, for example, need not lie on the vertices of a triangular surface element over which traction and displacement rates are assumed to vary in a linear fashion. An alternative is to specify, if necessary, a "zero area" element between coincident source points and assign different values of physical quantities on these points. This latter approach has been used for the solution of planar and axisymmetric viscoplastic problems which are described, for example, in Ref. [40].

Accurate evaluation of the domain integral in Eq. (2.51) presents a significant challenge. First, it should be noted that if a viscoplastic model with state variables is employed, the integrands in the domain integrals are known at any time and it is then a matter of accurate evaluation of known integrals. Use of a plasticity-type model, on the other hand, introduces the unknown velocity field inside the domain integrals. In this case, only the boundary velocities can be treated as primary unknowns and an iterative procedure at each time step can be used. Alternatively, the velocities inside can also be treated as primary unknowns. The size of coefficient matrices in the latter case would then reflect the total number of boundary as well as internal nodes.

Even for viscoplasticity problems with known domain integrands, the accurate evaluation of the domain integrals is by no means trivial, since the kernel W_{ijk} becomes singular when the source point lies inside an internal cell over which the integral must be evaluated. The non singular portions of the integrals, of course, can be accurately obtained by numerical quadrature. One way to determine the singular portions of the integrals accurately is to use simple shape functions for $\dot{\varepsilon}_{ij}^{(n)}$ over internal cells and evaluate the integrals analytically. Combined analytical/numerical methods are more versatile in that one is then not limited to internal cells of simple shape or to simple shape functions for the non-elastic strain rates. Such methods are discussed in books by Mukherjee [40] or Brebbia et al. [9].

Equation (2.51) is eventually reduced to an algebraic system of the type

$$[A]\{\dot{u}\} + [B]\{\dot{\tau}\} = \{\dot{b}\} \tag{2.52}$$

where the coefficient matrices $[A]$ and $[B]$ contain integrals of the kernels and the shape functions and the vector $\{\dot{b}\}$ contains the non-elastic strain rates and the kernel W_{ijk}. The vector $\{\dot{b}\}$, for viscoplasticity problems, is known at any time through the constitutive equations, and the size of the matrices $[A]$ and $[B]$

depend only on the boundary discretization. A simple switching of the columns of $[A]$ and $[B]$ is carried out in order to determine the unknown components of the traction and displacement rates in terms of the prescribed ones. The displacement rate field throughout the body can be obtained from Eq. (2.16) discretized in an analogous manner.

The stress rates on the boundary and inside the body must now be determined from the velocities. The boundary stress rates can easily be determined by using the algorithm set forth earlier in this section. Determination of internal stress rates requires discretization of Eq. (2.18) over boundary elements and internal cells as before. There are no more unknowns at this stage but kernels with very strong singularities must be treated carefully. One possible approach, for three-dimensional problems, is the use of convected derivatives as suggested by Bui [11]. Numerical realization of this approach is discussed by Brebbia et al. [9]. Another possibility is the use of the strain rate gradient method as discussed by Mukherjee [40]. A third approach is analytical integration over internal cells and then differentiation at a source point. Mukherjee and his group have successfully used the third approach for planar and the second approach for axisymmetric problems [40]. Brebbia and his group, on the other hand, have used the Bui approach successfully for planar problems [9].

2.9. *Solution strategy*

The solution strategy for three-dimensional viscoplastic boundary value problems is best summarized in Fig. 2. The initial values of the non-elastic strain components are taken to be zero and the initial values of the state variables, if any, are prescribed. The initial values of the stress and strain components are obtained from the solution for an identical body with boundary conditions same as those prescribed for the original non-elastic problem at zero time. This elastic solution is obtained by the Boundary Element Method by solving the appropriate integral equations.

The rates of the non-elastic strains at zero time are next obtained from the constitutive equations (2.34). These rates are used in Eq. (2.52) which is solved for the unspecified components of the boundary traction and displacement rates at $t = 0$. The displacement rates throughout the body are next obtained from a discretized version of Eq. (2.16). Displacement rate gradients are then determined from a differentiated version of the displacement rate equation (Eq. (2.18)), or by direct numerical differentiation of the displacement rate field. The stress rate components throughout the body follow from the components of the displacement rate gradients through Hooke's law. Finally, the rates of the state variables, if any, are determined from Eq. (2.35).

Since the rate equations often give rise to a stiff system of differential equations, it is very important, for these problems, to select an efficient time integration strategy with a suitable algorithm for time step control. Many such strategies–both implicit and explicit–exist in the literature and some of them have

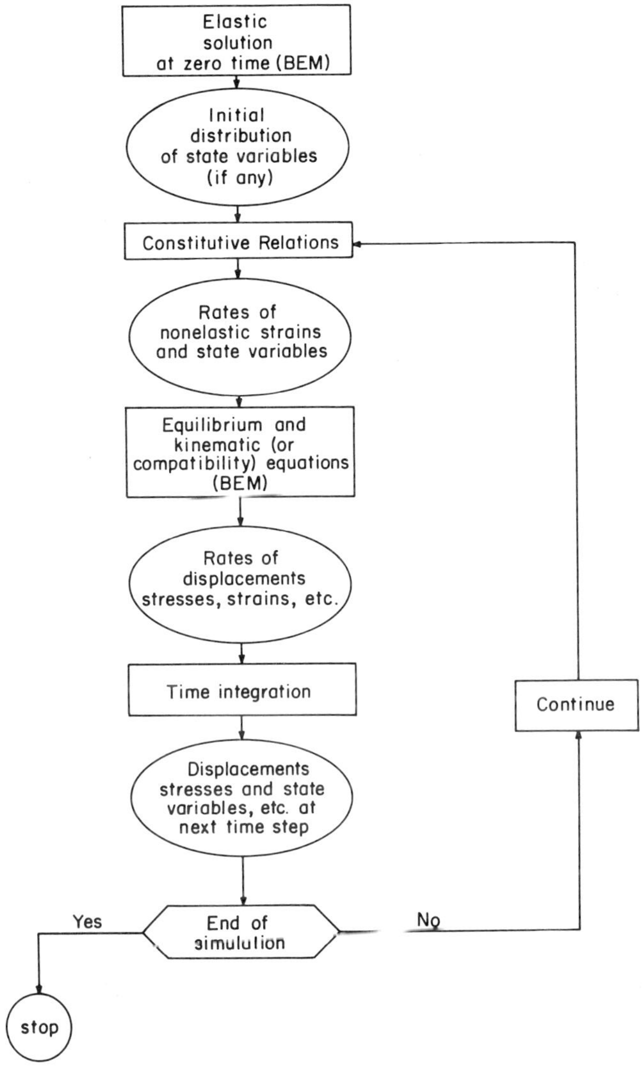

Fig. 2. Flow chart for the solution of small-strain elastic–viscoplastic problems (from [14]).

been reviewed in Ref. [40]. By and large, implicit methods are unconditionally stable but are more difficult to implement than explicit ones. An example of recent research using the implicit scheme, for Hart's constitutive model, can be found in Ref. [19].

2.10. Sample numerical results

This section is concluded by presenting a few numerical results for planar elastic–viscoplastic problems with material behavior governed by the constitutive model of Hart [23]. These results are taken from Mukherjee's book [40].

Figure 3 shows a comparison of BEM and FEM results for the internal-pressure = internal-displacement response of a ring in plane stress subjected to internal pressure increasing at a constant rate. The material parameters chosen for this similation are representative of 304 stainless steel at 200°C and are available in Ref. [40]. Only a quarter of the annular plate is modelled due to symmetry. The BEM calculations use a piecewise linear approximation of boundary variables on straight boundary elements and piecewise uniform non-elastic strain rates on polygonal internal cells. The finite element solutions are obtained using linear strain triangles and area integrals for pseudo loads in the FEM formulation are evaluated by Gaussian quadrature. The BEM mesh has 30 boundary nodes and 20 internal cells while the FEM calculation has 48 elements

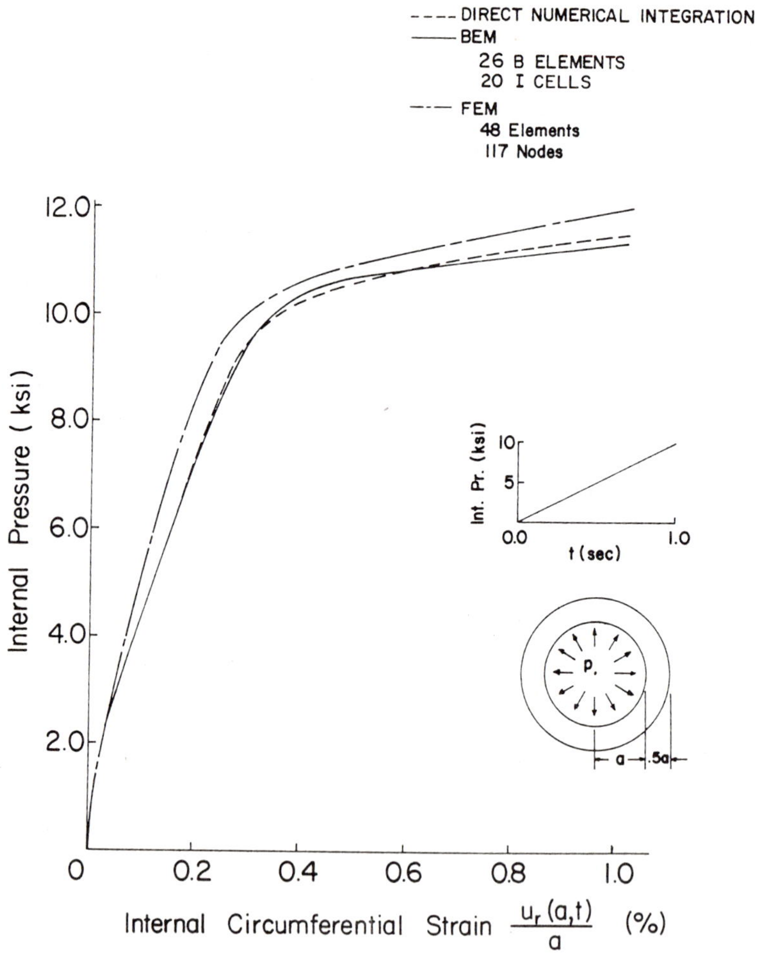

Fig. 3. Comparison of direct, BEM and FEM solutions for a circular plate with a concentric circular cut-out under internal pressure increasing at a constant rate. $\dot{p} = 10$ ksi/s Annealed 304 SS at 200°C (from [14]).

and 117 nodes. The meshes for the two cases are given in Ref. [40]. Details of the FEM formulation, used here, are given in Refs. [33, 34]. The direct numerical solution procedures is discussed in Ref. [37].

Another example is that of a square plate with a circular hole, in plane stress, subjected to a constant remote uniaxial tensile stress. A quarter of the plate is modelled again. This time the BEM mesh has 37 boundary nodes and 20 internal cells while the FEM mesh has 102 nodes and 41 elements.

The redistribution of σ_{22}, with time, along the line BC ($x_2 = 0$) appears in Fig. 4. A direct solution cannot be obtained for this problem. The results obtained from the two methods are in fair agreement. The zero-time elastic stress distributions in Fig. 4 are within 3% of the correct solution for the BEM and within 6% of the correct solution for the FEM, in a region near the cut-out. The solutions are more accurate away from the cut-out. The presence of stress concentration near the cut-out causes substantial inelastic deformation in this region. This causes rapid unloading of stress close to the cut-out while the stress distribution remains essentially unaltered beyond about twice the cut-out radius. A vertical equilibrium check on the two long-time solutions reveals that the FEM solution overpredicts the remote stress resultant by about 2% while the BEM

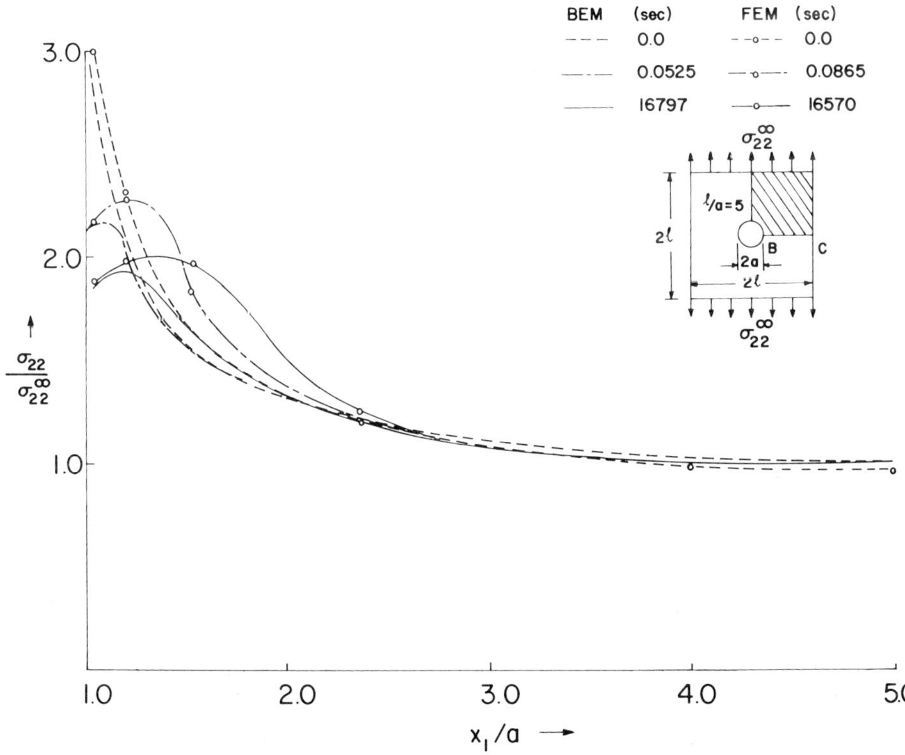

Fig. 4. Comparison of BEM and FEM solutions for stress redistribution along BC in an annealed 304SS square plate, with a circular cut-out, under uniaxial tension at 400°C. $\sigma_{22}^{\infty} = 10$ ksi (from [14]).

solution underpredicts this quantity by about 3%. This is considered quite good considering that the BEM solution requires 367 and the FEM 420 time steps to reach the final simulated time. It is interesting to compare the results in Fig. 4 with results for the same geometry when finite strains and rotations are imposed on the plate. These results are presented later in this chapter.

These problems were run on an IBM 370/168 computer at Cornell University. The CPU times for the two cases in Figs. 3 and 4 were 63 and 91 s for the BEM and 127 and 205 s respectively, for the FEM calculations.

The idea in this subsection has been to give the reader a flavour for BEM numerical results for elastic–viscoplastic problems. The interested reader is referred to, for example, recent books by Mukherjee [40], Brebbia et al. [9] or Banerjee and Butterfield [2], for more detailed discussion on the subject.

3. Problems with material and geometrical nonlinearities

This section of this chapter is devoted to a discussion of solutions of large-strain–large-rotation problems, in the presence of plasticity or viscoplasticity, by the Boundary Element Method. It is assumed throughout that the elastic strains are infinitesimal while the non-elastic strains, as well as the rotations, can be arbitrary. The formulation presented here is based on an updated Lagrangian approach in which the configuration of a body of time t is used as reference for the deformation between t and $t + \Delta t$. This necessitates updating of the geometry during computer simulation of deformation but provides great simplifications in the BEM formulation. Numerical results for applications in soil mechanics and metal forming are presented at the end of the section.

3.1. Kinematics

A three-dimensional body is considered in this section. Referring to a set of spatially fixed rectangular Cartesian coordinates, a material particle in the body in a reference configuration is assumed to have coordinates X_i. The same particle has coordinates x_i in the current configuration. Unless otherwise indicated, the range of indices in this section is 1, 2, 3.

The displacement vector u_i is defined by the equation

$$x_i = X_i + u_i \,. \tag{3.1}$$

The velocity of this material point during deformation is denoted by v_i. The deformation rate d_{ij} is the symmetric part of the velocity gradient.

$$d_{ij} = \tfrac{1}{2}\left[\frac{\partial v_i}{\partial x_j} + \frac{\partial v_j}{\partial x_i}\right] \tag{3.2}$$

whereas the spin ω_{ij} is the antisymmetric part

$$\omega_{ij} = \tfrac{1}{2}\left[\frac{\partial v_i}{\partial x_j} - \frac{\partial v_j}{\partial x_i}\right]. \tag{3.3}$$

3.2. Constitutive assumptions

The boundary element formulation presented next is valid for a wide range of elastic–viscoplastic constitutive models. While some details about these models are given later in this section, the key assumptions used in the BEM formulation are presented here.

The first assumption is that the deformation rate gradient tensor can be linearly decomposed into an elastic and a non-elastic part

$$d_{ij} = d_{ij}^{(e)} + d_{ij}^{(n)}. \tag{3.4}$$

The next step relating $d_{ij}^{(e)}$ to the stress rate is very important. A hypoelastic material is one in which the components of a (proper) stress rate are homogeneous linear functions of the components of a (proper) rate of deformation. One form of such a law, under conditions of material isotropy (of the virgin specimen) can be written as

$$\overset{*}{\sigma}' = \lambda d_{kk}^{(e)}\delta_{ij} + 2Gd_{ij}^{(e)} \tag{3.5}$$

where $\overset{*}{\sigma}'_{ij}$ is the Jaumann rate of the effective Cauchy stress.

Several important points must be noted here. The first is that for many applications the effective Cuachy stress is simply the Cauchy stress, while for other cases, notably soil mechanics, it might be given by

$$\sigma_{ij} = \sigma'_{ij} - P\delta_{ij} \tag{3.6}$$

where P is the pore pressure. Thus, allowance is being made here for this difference between total and effective stress. The second point is that Eq. (3.5) is one of the simplest hypoelastic laws imaginable and considerable controversy exists in the mechanics community regarding the proper form for such a law one should use as well as the appropriate choice of the rate of effective stress. A spirited debate on these issues ensued recently at an ASME meeting and the interested reader is referred to a recent ASME publication [54] containing papers presented there. The chosen stress rate must, of cource, be objective with respect to rigid-body rotations, but, unfortunately, many such rates exist. The general feeling among the researchers currently is that, unless very large shearing strains are involved, Eq. (3.5) is probably sufficient, and further experimentation and analysis, both at the micro and macro levels, is necessary before the verdict with regard to the best constitutive assumption is out. Equation (3.5) is adopted in this research.

The Jaumann (sometimes called the Zaremba–Jaumann–Noll rigid-body rate, or the corotational rate) [26] of a second-rank tensor is related to its material rate by the formula

$$\overset{*}{\mathbf{T}} = \dot{\mathbf{T}} + \mathbf{T} \cdot \boldsymbol{\omega} - \boldsymbol{\omega} \cdot \mathbf{T} \tag{3.7}$$

or, in component form,

$$\overset{*}{t}_{ij} = \dot{t}_{ij} + t_{ik}\omega_{kj} - \omega_{ik}t_{kj} \tag{3.8}$$

where the material derivative is defined as

$$\dot{t}_{ij} = \frac{\partial t_{ij}}{\partial t} + \frac{\partial t_{ij}}{\partial x_k} v_k \tag{3.9}$$

It is clear that unlike in the case of small deformations discussed in the previous section, the choice of tensor rates is no longer unambiguous and must be given very careful attention. Physically, the Jaumann rate is the rate of change of a tensor as evidenced by an observer which translates with a material point as well as rotates, in some sense, with the differential material element around that point.

The above constitutive assumptions are sufficient for the purpose of developing a BEM formulation for this class of problems. Further assumptions regarding the specifications of $d_{ij}^{(\mathrm{n})}$ are, of course, necessary for the solution of the complete problem, but this discussion is being postponed till later in this chapter.

3.3. BEM formulation for velocities

The key to the BEM formulation is an appropriate form of Betti's reciprocal theorem. Analogous to Eq. (2.13), the applicable version for the large-strain problems is

$$\int_B \overset{*}{\sigma}{}'_{ij} \varepsilon_{ij}^{(\mathrm{R})} \, \mathrm{d}V = \int_B \sigma_{ij}^{(\mathrm{R})} d_{ij}^{(\mathrm{e})} \, \mathrm{d}V \tag{3.10}$$

where the reference solution is exactly the same as the one used for small deformations (Eqs (2.5)–(2.10)). Again, the integrands on either side of the above equation are identical by virtue of Eqs (2.8) and (3.5). Also, it should be noted that the domain of integration of Eq. (3.10) is the current volume.

Equation (3.10) must next be written in terms of the known reference region B^0. Use of the updated Lagrangian formulation renders this procedure particularly simple. A consequence of this assumption is that during each small time step, since the current configuration at the start of the step is also the reference configuration, the deformation gradient

$$\mathbf{F} = \partial x / \partial X \tag{3.11}$$

is approximately equal to the identity tensor $\mathbf{I}$. The situation for the rates, of course, is different.

The next step requires the determination of the Jaumann rate of the Kirchhoff stress and that of the same rate of the Cauchy stress. Starting from the equation

$$J\boldsymbol{\sigma} = \boldsymbol{\tau} \tag{3.12}$$

where $J = \det(\mathbf{F})$, differentiating Eq. (2.11) and noting that [52]

$$\dot{J} = J\,\mathrm{tr}(\mathbf{D}) \tag{3.13}$$

(where det denotes determinant and tr denotes trace) it is easy to show that

$$\dot{\boldsymbol{\tau}} = J(\mathrm{tr}(\mathbf{D})\boldsymbol{\sigma} + \dot{\boldsymbol{\sigma}})\,. \tag{3.14}$$

In the updated Lagrangian approach, $J \approx 1$ and $\boldsymbol{\sigma} \approx \boldsymbol{\tau}$. Thus, using (3.13) and the definition of the Jaumann rate (Eq. (3.7)), one easily obtains

$$\overset{*}{\boldsymbol{\sigma}} = \overset{*}{\boldsymbol{\tau}} - \mathrm{tr}(\mathbf{D})\boldsymbol{\sigma} \tag{3.15}$$

or, in component form

$$\overset{*}{\sigma}_{ij} = \overset{*}{\tau}_{ij} - v_{k,k}\sigma_{ij}\,. \tag{3.16}$$

It should also be noted that

$$d_{kk} = -\dot{\rho}/\rho \tag{3.17}$$

where ρ is the current density at a material point. Thus, if the deformation is incompressible, this term can be ignored and $\overset{*}{\sigma}_{ij}$ set equal to $\overset{*}{\tau}_{ij}$. This is usually done for metal deformation with infinitesimal elastic and finite non-elastic strains, since the non-elastic portion of metal deformation is often volume-preserving. Such, however, is not the case for soils and Eq. (3.16) must be used for compressible materials.

In order to proceed, a relationship between the rate of pore pressure P and rate of volume change must be assumed at this stage. It is usual to assume [6]

$$\dot{P} = -k_{\mathrm{b}}d_{kk} \tag{3.18}$$

where k_{b} is the bulk solid–fluid compressibility coefficient. It should be again mentioned here that for applications like metal forming the pore pressure P is zero.

The idea here is to replace $\overset{*}{\sigma}'_{ij}$ in Eq. (3.10) by $\overset{*}{\tau}_{ij}$. To this end, use of Eqs. (3.6), (3.15) and (3.18) gives

$$\overset{*}{\sigma}'_{ij} = \overset{*}{\tau}_{ij} - v_{k,k}\sigma_{ij} - k_{\mathrm{b}} v_{k,k}\delta_{ij} \,. \tag{3.19}$$

It is more convenient to recast the above equation in the form

$$\overset{*}{\sigma}'_{ij} = \overset{*}{\tau}_{ij} + C_{ijkl}v_{k,l} - M^{0}_{ijkl}v_{k,l} \tag{3.20}$$

where $C_{ijkl} = -\sigma_{ij}\delta_{kl}$ and $M^{0}_{ijkl} = k_{\mathrm{b}}\delta_{ij}\delta_{kl}$. Finally

$$\overset{*}{\sigma}'_{ij} = \overset{*}{\tau}_{ij} + \bar{C}_{ijkl}v_{k,l} \tag{3.21}$$

where

$$\bar{C}_{ijkl} = C_{ijkl} - M^{0}_{ijkl} \,. \tag{3.22}$$

The rate of equilibrium equation in terms of the material rate of the Lagrange (non-symmetric, nominal) stress, in the absence of body forces, is given as

$$\dot{s}_{ji,j} = 0 \,. \tag{3.23}$$

Strictly speaking, the derivative of the rate of Lagrange stress should be taken with respect to reference coordinates X_j, but here, in the updated Lagrangian frame, these derivatives are approximately the same as those with respect to x_j. Also, in the updated Lagrangian frame,

$$d_{ij} = \dot{E}_{ij} \tag{3.24}$$

where E_{ij} is the Green–Saint Venant strain.

The next step requires the relationship between the Jaumann rate of the Kirchhoff stress and the material rate of the Lagrange stress. Starting from

$$\boldsymbol{\tau} = \mathbf{F} \cdot \mathbf{S} \tag{3.25}$$

one differentiates this equation with respect to time and uses Eq. (3.7) for the definition of the Jaumann rate to get

$$\overset{*}{\boldsymbol{\tau}} - \boldsymbol{\tau} \cdot \boldsymbol{\omega} + \boldsymbol{\omega} \cdot \boldsymbol{\tau} = \dot{\mathbf{F}} \cdot \boldsymbol{\tau} \cdot \mathbf{F}^{-1} + \mathbf{F} \cdot \dot{\mathbf{S}} \tag{3.26}$$

so that

$$\dot{\mathbf{S}} = \mathbf{F}^{-1} \cdot (\overset{*}{\boldsymbol{\tau}} - \boldsymbol{\tau} \cdot \boldsymbol{\omega} + \boldsymbol{\omega} \cdot \boldsymbol{\tau}) - \mathbf{F}^{-1} \cdot \dot{\mathbf{F}} \cdot \boldsymbol{\tau} \cdot \mathbf{F}^{-1} \,. \tag{3.27}$$

This complicated expression simplifies immensely in the updated Lagrangian frame. One gets

$$\dot{\mathbf{S}} = \overset{*}{\boldsymbol{\tau}} - \boldsymbol{\tau}\cdot\boldsymbol{\omega} + \boldsymbol{\omega}\cdot\boldsymbol{\tau} - \dot{\mathbf{F}}\cdot\boldsymbol{\tau}\,. \tag{3.28}$$

Now,

$$\mathbf{L} = \frac{\partial \boldsymbol{v}}{\partial \boldsymbol{x}} = \dot{\mathbf{F}}\cdot\mathbf{F}^{-1} = \mathbf{D} + \boldsymbol{\omega}$$

and here this simply becomes

$$\dot{\mathbf{F}} = \mathbf{D} + \boldsymbol{\omega}\,. \tag{3.29}$$

Substituting Eqs (3.29) into (3.28) and, once again, identifying $\boldsymbol{T}$ with $\boldsymbol{\sigma}$ in the updated Lagrangian frame, the final equation is

$$\dot{\mathbf{S}} = \overset{*}{\boldsymbol{\tau}} - \boldsymbol{\sigma}\cdot\boldsymbol{\omega} - \mathbf{D}\cdot\boldsymbol{\sigma}\,. \tag{3.30}$$

In component form,

$$\dot{s}_{ji} = \overset{*}{\tau}_{ji} - \sigma_{jk}\omega_{ki} - d_{jk}\sigma_{ki}\,. \tag{3.31}$$

Again, it is convenient to write this equation in the form

$$\overset{*}{\tau}_{ji} = \dot{s}_{ji} + G_{jikl}v_{k,l} \tag{3.32}$$

where

$$2G_{ijkl} = \sigma_{ik}\delta_{jl} + \sigma_{jl}\delta_{ik} + \sigma_{jk}\delta_{li} - \sigma_{li}\delta_{jk}\,.$$

The tensor G_{ijkl} is, therefore, a function only of the components of the Cauchy stress σ_{ij}. Using the convention

$$\{v_{k,l}\}^{\mathrm{T}} = [v_{1,1}v_{2,1}v_{3,1}v_{1,2}v_{2,2}v_{3,2}v_{1,3}v_{2,3}v_{3,3}]$$

the matrix corresponding to G_{ijkl} is given in Table 1.

The form of Eq. (3.10) in the reference configuration B^0 is now

$$\int_{B_0} [\dot{s}_{ji} + G_{jikl}v_{k,l} + \bar{C}_{jikl}v_{k,l}]\varepsilon_{ij}^{(\mathrm{R})}\,\mathrm{d}V = \int_{B^0} \sigma_{ij}^{(\mathrm{R})}\dot{E}_{ij}^{(\mathrm{e})}\,\mathrm{d}V\,. \tag{3.33}$$

The rest of the derivation parallels that of the small deformation situation. Using Eqs (2.7), (3.23), (2.9) and the divergence theorem, the left-hand side of Eq. (3.33) becomes

Table 1.
The matrix $[G]$ for three-dimensional problems

$$[G]=\left[\begin{array}{ccc|ccc|ccc}
\sigma_{11} & & & & \text{(symmetric)} & & & & \\
\sigma_{12} & -\frac{1}{2}(\sigma_{11}-\sigma_{22}) & & & & & & & \\
\sigma_{13} & \frac{1}{2}\sigma_{23} & -\frac{1}{2}(\sigma_{11}-\sigma_{33}) & & & & & & \\
0 & \frac{1}{2}(\sigma_{11}+\sigma_{22}) & \frac{1}{2}\sigma_{23} & -\frac{1}{2}(\sigma_{22}-\sigma_{11}) & & & & & \\
0 & 0 & 0 & \sigma_{12} & \sigma_{22} & & & & \\
0 & \frac{1}{2}\sigma_{13} & -\frac{1}{2}\sigma_{12} & \frac{1}{2}\sigma_{13} & \sigma_{23} & -\frac{1}{2}(\sigma_{22}-\sigma_{33}) & & & \\
0 & \frac{1}{2}\sigma_{23} & \frac{1}{2}(\sigma_{11}+\sigma_{33}) & -\frac{1}{2}\sigma_{23} & 0 & \frac{1}{2}\sigma_{12} & -\frac{1}{2}(\sigma_{33}-\sigma_{11}) & & \\
0 & -\frac{1}{2}\sigma_{13} & \frac{1}{2}\sigma_{12} & \frac{1}{2}\sigma_{13} & 0 & \frac{1}{2}(\sigma_{22}+\sigma_{33}) & \frac{1}{2}\sigma_{12} & -\frac{1}{2}(\sigma_{33}-\sigma_{22}) & \\
0 & 0 & 0 & 0 & 0 & 0 & \sigma_{13} & \sigma_{23} & \sigma_{33}
\end{array}\right]$$

$$\text{L.H.S.} = e_j \left[\int_{\partial B^0} U_{ij} \dot{s}_{ki} n_k^0 \, \mathrm{d}S^0 + \int_{B^0} U_{ij,m} G_{mikl} v_{k,l} \, \mathrm{d}V^0 \right.$$

$$\left. + \int_{B^0} U_{ij,m} \bar{C}_{mikl} v_{k,l} \, \mathrm{d}V^0 \right]. \tag{3.34}$$

Similarly using Eqs (3.24), (3.4), (3.2), (2.8), (2.5), (2.7), (2.6), (2.10) and (2.9) and the divergence theorem, the right-hand side of Eq. (3.33) becomes

$$\text{R.H.S.} = e_j \left[\int_{\partial B^0} T_{ij} v_i \, \mathrm{d}S^0 + \int_{B^0} \Delta(\mathrm{p}, \mathrm{q}) \delta_{ij} v_i \, \mathrm{d}V^0 \right.$$

$$\left. - \int_{B^0} [\lambda U_{ij,i} d_{kk}^{(\mathrm{n})} + 2G U_{ij,k} d_{ik}^{(\mathrm{n})}] \, \mathrm{d}V^0 \right]. \tag{3.35}$$

It is interesting to compare Eqs. (3.34) and (3.35) with Eqs. (2.14) and (2.15).

Finally, equating the expressions in Eqs. (3.34) and (3.35) and using the relevant property of the Dirac delta function, one obtains the equation

$$v_j(\mathrm{p}) = \int_{\partial B^0} [U_{ij}(\mathrm{p}, \mathrm{Q}) \dot{\tau}_i(\mathrm{Q}) - T_{ij}(\mathrm{p}, \mathrm{Q}) v_i(\mathrm{Q})] \, \mathrm{d}S_{\mathrm{Q}}^0$$

$$+ \int_{B^0} [\lambda U_{ij,i}(\mathrm{p}, \mathrm{q}) d_{kk}^{(\mathrm{n})}(\mathrm{q}) + 2G U_{ij,k}(\mathrm{p}, \mathrm{q}) d_{ik}^{(\mathrm{n})}(\mathrm{q})] \, \mathrm{d}V_q^0$$

$$+ \int_{B^0} U_{ij,m}(\mathrm{p}, \mathrm{q}) G_{mikl}(\mathrm{q}) v_{k,l}(\mathrm{q}) \, \mathrm{d}V_{\mathrm{q}}^0$$

$$+ \int_{B^0} U_{ij,m}(\mathrm{p}, \mathrm{q}) \bar{C}_{mikl}(\mathrm{q}) v_{k,l}(\mathrm{q}) \, \mathrm{d}V_{\mathrm{q}}^0 \tag{3.36}$$

where $\dot{\tau}_i = n_j^0 \dot{s}_{ji}$.

Comparing Eq. (3.36) with the corresponding Eq. (2.16) for small strains, it is seen that the first two integrals on the right-hand side of the above equation are analogous to the small-strain formulation. The traction rate term $\dot{\tau}_i$ requires special care and is discussed a little later. The third term is the so called "geometric correction" term which arises due to finite deformations and rotations inside the body. The last integral is a consequence of bulk compressibility and consideration of the pore pressure in soils. If one is interested in incompressible metal deformation, for example, there is no pore pressure and the tensor $\bar{\mathbf{C}}$ in Eq. (3.36) reduces to $\mathbf{C}$ (eq. (3.22)). If incompressible metal deformation is being studied, this last integral vanishes altogether, and so does the term containing $\mathrm{d}_{kk}^{(\mathrm{n})}$ in the second integral [41].

Using Eq. (3.31), the traction rate $\dot{\tau}_i$ in Eq. (3.36) can be written as

$$\dot{\tau}_i = n_j \dot{s}_{ji} = \dot{t}_i - n_j \sigma_{jk} \omega_{ki} - n_j d_{jk} \sigma_{ki} = \dot{t}_i - n_j G_{jikl} v_{k,l} \tag{3.37}$$

where

$$\dot{t}_i = n_j \overset{*}{\tau}_{ji} = n_j \overset{*}{\sigma}_{ji} - n_j C_{jikl} v_{k,l} \,. \tag{3.38}$$

The presence of the **G** tensor in the boundary traction expression is sometimes referred to as "load correction" and is a consequence, during deformation, of the change in the area of a surface element and the rotation of the unit normal at a point on it. The rate $\dot{t}_i$ can be interpreted as a component of the rate of the prescribed follower force, per unit deformed surface area, on the deforming boundary.

It can be seen that, unlike elastic–viscoplastic problems with small strains and rotations, the unknown velocity gradient now occurs in the surface as well as some of the domain integrals in Eq. (3.36). Thus, iterations now become necessary at each time step. This has been carried out in order to obtain the numerical results presented later in this chapter.

As before, the limit p→P is taken so that Eq. (3.36) becomes

$$\begin{aligned} C_{ij}(\mathrm{P})v_i(\mathrm{P}) &= \int_{\partial B^0} [U_{ij}(\mathrm{P},\mathrm{Q})\dot{\tau}_i(\mathrm{Q}) - T_{ij}(\mathrm{P},\mathrm{Q})v_i(\mathrm{Q})]\,\mathrm{d}S^0_{\mathrm{Q}} \\ &+ \int_{B^0} [\lambda U_{ij,i}(\mathrm{P},\mathrm{q})d^{(\mathrm{n})}_{kk}(\mathrm{q}) + 2GU_{ij,k}(\mathrm{P},\mathrm{q})d^{(\mathrm{n})}_{ik}(\mathrm{q})]\,\mathrm{d}U^0_{\mathrm{q}} \\ &+ \int_{B^0} U_{ij,m}(\mathrm{P},\mathrm{q})G_{mikl}(\mathrm{q})v_{k,l}(\mathrm{q})\,\mathrm{d}V^0_{\mathrm{q}} \\ &+ \int_{B^0} U_{ij,m}(\mathrm{P},\mathrm{q})\bar{C}_{mikl}(\mathrm{q})v_{k,l}(\mathrm{q})\,\mathrm{d}V^0_{\mathrm{q}} \end{aligned} \tag{3.39}$$

The coefficient C_{ij} is the same as in the case of small-strain–small-rotation problems.

3.4. Internal stress rates

As mentioned before, analytical differentiation of Eq. (3.36) at a source point is the best way to obtain velocity rates at internal points. To this end, a differentiated version of Eq. (3.36) is

$$v_{j,\bar{l}}(\mathrm{p}) = \int_{\partial B^0} [U_{ij,\bar{l}}(\mathrm{p},\mathrm{Q})\dot{\tau}_i(\mathrm{Q}) - T_{ij,\bar{l}}(\mathrm{p},\mathrm{Q})v_i(\mathrm{Q})]\,\mathrm{d}S_Q^0$$

$$+ \frac{\partial}{\partial x_{\bar{l}}} \int_{B^0} [\lambda U_{ij,i}(\mathrm{p},\mathrm{q})d_{kk}^{(\mathrm{n})}(\mathrm{q}) + 2GU_{ij,k}(\mathrm{p},\mathrm{q})\,\mathrm{d}_{ik}^{(\mathrm{n})}(\mathrm{q})]\,\mathrm{d}V_\mathrm{q}^0$$

$$+ \frac{\partial}{\partial x_{\bar{l}}} \int_{B^0} U_{ij,m}(\mathrm{p},\mathrm{q})G_{mikn}(\mathrm{q})v_{k,n}(\mathrm{q})\,\mathrm{d}V_\mathrm{q}^0$$

$$+ \frac{\partial}{\partial x_{\bar{l}}} \int_{B^0} U_{ij,m}(\mathrm{p},\mathrm{q})\bar{C}_{mikn}(\mathrm{q})v_{k,n}(\mathrm{q})\,\mathrm{d}V_q^0 . \tag{3.40}$$

As before, one must be careful with differentiation of kernels like $U_{ij,m}$. The extra integrals in the large-strain formulation, however, present no added difficulties relative to the small-strain problem. Thus, all the earlier discussion regarding numerical evaluation of the domain integrals applies here also. So far, numerical results from these equations have only been obtained for planar problems. In such cases, it has been possible to evaluate each of the volume integrals analytically (for a specific simple choice of shape functions for the unknowns in these domain integrals) and then differentiate these integrals at an arbitrary source point p.

Once the velocity gradients and then d_{ij} have been determined, the nonelastic deformation rate $d_{ij}^{(\mathrm{n})}$ is subtracted from it to get the clastic deformation rate $d_{ij}^{(\mathrm{e})}$. The components of thc Jaumann rates of the effective stress are now obtained from the hypoelasticity relationship Eq. (3.5) and then the Jaumann rates of the stress components are obtained from a rate form of Eq. (3.6)

$$\overset{*}{\sigma}_{ij} = \overset{*}{\sigma}{}'_{ij} + k_\mathrm{b} d_{kk}\delta_{ij} \tag{3.41}$$

where Eq. (3.18) has been used. It is then a simple matter to obtain the material rates of the Cauchy stress from Eq. (3.8) written for σ_{ij}.

3.5. *Plane strain*

As before, the plane strain equations can be obtained from the three-dimensional ones with $v_3 = 0$, $\partial/\partial x_3 = 0$ and using the corresponding solution with $u_3^{(\mathrm{R})} = 0$. This means that the left-hand side of Eq. (3.10) is now summed only for $i, j = 1, 2$. The right-hand side of Eq. (3.10) still retains the term $\sigma_{33}^{(\mathrm{R})} d_{33}^{(\mathrm{e})}$. Following the derivation of the three-dimensional equations, the plane strain equation is

Table 2
The matrix $[G]$ for plane strain or plane stress

$$[G]=\begin{bmatrix} \sigma_{11} & & \text{(symmetric)} & \\ \sigma_{12} & -\frac{1}{2}(\sigma_{11}-\sigma_{22}) & & \\ 0 & \frac{1}{2}(\sigma_{11}+\sigma_{22}) & -\frac{1}{2}(\sigma_{22}-\sigma_{11}) & \\ 0 & 0 & \sigma_{12} & \sigma_{22} \end{bmatrix}$$

$$\begin{aligned} v_j(\mathrm{p}) = & \int_{\partial B^0} [U_{ij}(\mathrm{p},\mathrm{Q})\dot{\tau}_i(\mathrm{Q}) - T_{ij}(\mathrm{p},\mathrm{Q})v_i(\mathrm{Q})]\,\mathrm{d}S^0_{\mathrm{Q}} \\ & + \int_{B^0} [\lambda U_{ij,i}(\mathrm{p},\mathrm{q})\eta^{(\mathrm{n})}(\mathrm{q}) + 2GU_{ij,k}(\mathrm{p},\mathrm{q})d^{(\mathrm{n})}_{ik}(\mathrm{q})]\,\mathrm{d}V^0_{\mathrm{q}} \\ & + \int_{B^0} U_{ij,m}(\mathrm{p},\mathrm{q})G_{mikl}(\mathrm{q})v_{k,l}(\mathrm{q})\,\mathrm{d}V^0_{\mathrm{q}} \\ & + \int_{B^0} [U_{ij,m}(\mathrm{p},\mathrm{q})\bar{C}_{mikl}(\mathrm{q})v_{k,l}(\mathrm{q})\,\mathrm{d}V^0_{\mathrm{q}} \end{aligned} \tag{3.42}$$

where now $\eta^{(\mathrm{n})} = d^{(\mathrm{n})}_{11} + d^{(\mathrm{n})}_{22} + d^{(\mathrm{n})}_{33}$ and the range of indices in Eq. (3.42) is 1, 2. The plane strain kernels U_{ij} and T_{ij} must be used. Using the convention

$$\{v_{k,l}\} = [v_{1,1}v_{2,1}v_{1,2}v_{2,2}]$$

the matrix corresponding to G_{ijkl}, for plane strain, is given in Table 2. The components of $\bar{C}_{ijkl}$ can be easily obtained from Eq. (3.22).

3.6. *Plane stress*

The plane stress situation considered here is that for a compressible metal, so that $P = 0$ and $\sigma_{ij} = \sigma'_{ij}$. Although generally it is sufficient to consider $d^{(\mathrm{n})}_{kk} = 0$ for metal deformation (and hence $d_{kk} \simeq 0$), compressibility effects can be important in some situations, e.g. if the effect of distributed voids inside the metallic body is important.

For plane stress $\sigma_{13} = \sigma_{23} = \sigma_{33} = 0$. It is also usual to assume that for plane stress $v_{1,3} = v_{2,3} = v_{3,1} = v_{3,2} = 0$ so that $\omega_{13} = \omega_{23} = 0$. This means that from the definition of the Jaumann rates (Eq. (3.8)), $\overset{*}{\sigma}_{13} = \overset{*}{\sigma}_{23} = \overset{*}{\sigma}_{33} = 0$. Thus, the left-hand side of Eq. (3.10) (with $\overset{*}{\sigma}'_{ij} = \overset{*}{\sigma}_{ij}$) needs only to be summed for i, $j = 1, 2$. The right-hand side, of course, must also be summed for i, $j = 1, 2$, and here $\varepsilon^{(\mathrm{R})}_{33}$ can be eliminated in favour of $\varepsilon^{(\mathrm{R})}_{11} + \varepsilon^{(\mathrm{R})}_{22}$ as for the small-deformation case. Finally, the resulting plane stress equation is

$$
\begin{aligned}
v_j(\mathrm{p}) = & \int_{\partial B^0} [U_{ij}(\mathrm{p},\mathrm{Q})\dot{\tau}_i(\mathrm{Q}) - T_{ij}(\mathrm{p},\mathrm{Q})v_i(\mathrm{Q})]\,\mathrm{d}S^0_{\mathrm{Q}} \\
& + \int_{B^0} [\bar{\lambda}U_{ij,i}(\mathrm{p},\mathrm{q})d^{(\mathrm{n})}_{kk}(\mathrm{q}) + 2GU_{ij,k}(\mathrm{p},\mathrm{q})d^{(\mathrm{n})}_{ik}(\mathrm{q})]\,\mathrm{d}V^0_{\mathrm{q}} \\
& + \int_{B^0} U_{ij,m}(\mathrm{p},\mathrm{q})G_{mikl}(\mathrm{q})v_{k,l}(\mathrm{q})\,\mathrm{d}V^0_{\mathrm{q}} \\
& - \int_{B^0} U_{ij,m}(\mathrm{p},\mathrm{q})\sigma_{mi}(\mathrm{q})\eta(\mathrm{q})\,\mathrm{d}V^0_{\mathrm{q}}
\end{aligned} \tag{3.43}
$$

where the range of indices is $i, j = 1, 2$. The plane stress kernels U_{ij}, T_{ij} must be used in the above. The plane stress version of G_{mikl}, with the range of indices $1, 2$, is the same as that for plane strain. However,

$$
\eta = d_{11} + d_{22} + d_{33} \tag{3.44}
$$

and it is convenient to eliminate d_{33} in terms of d_{11} and d_{22}. Using Eq. (3.5) (with $\sigma'_{ij} = \sigma_{ij}$) for $i = j = 3$ and $\overset{*}{\sigma}_{33} = 0$, one gets

$$
d_{33} = -\frac{\nu}{1-\nu}(d_{11} + d_{22}) + \frac{\nu}{1-\nu}(d^{(\mathrm{n})}_{11} + d^{(\mathrm{n})}_{22}) + d^{(\mathrm{n})}_{33}\,. \tag{3.45}
$$

Thus the presence of $d^{(\mathrm{n})}_{33}$ in Eq. (3.43) cannot be avoided. For the incompressible plane stress problem, however, the last integral in Eq. (3.43) vanishes and the equation is expressible only in terms of components in the 1 and 2 directions.

3.7. *Constitutive models*

As mentioned before, additive decomposition of d_{ij} (Eq. (3.4)) and the hypoelastic law (Eq. (3.5)) have been assumed previously. The non-elastic velocity gradient tensor $d^{(\mathrm{n})}_{ij}$ must be determined from an appropriate plastic or viscoplastic constitutive law.

A general form of such constitutive models can be written as

$$
d^{(\mathrm{n})}_{ij} = f_{ij}(\sigma'_{ij}, \overset{*}{\sigma}'_{ij}, q^{(k)}_{ij}) \tag{3.46}
$$

$$
\overset{*}{q}{}^{(k)}_{ij} = g^{(k)}_{ij}(\sigma'_{ij}, \overset{*}{\sigma}'_{ij}, q^{(l)}_{ij})\,. \tag{3.47}
$$

3.7.1. *A plastic constitutive model for soils*

A typical constitutive model for soils, which includes compressible plasticity, is the one due to Banerjee and Stipho [6], here generalized for large deformations. This model uses an yield surface Y and a plastic potential Q where

$$Y(p', p_0, \bar{\sigma}_0) \qquad \text{and} \qquad Q(p', p_0, \bar{\sigma}_0)$$

with $p' = \sigma'_{kk}/3$ and $\bar{\sigma}_0$ the second invariant of the effective deviatoric stress

$$\bar{\sigma}_0 = +(\tfrac{3}{2}\hat{\sigma}'_{ij}\hat{\sigma}'_{ij})^{1/2}$$

with $\hat{\sigma}'_{ij}$ the deviator of σ'_{ij}.

For an associated flow rule $Y = Q$ and for a non-associated rule $Y \neq Q$. The plastic flow rule is

$$d_{ij}^{(n)} = \dot{\varphi} a_{ij} \tag{3.48}$$

with

$$\dot{\varphi} = + \frac{b_{ij}\overset{*}{\sigma}'_{ij}}{\gamma a_{kk}}$$

and

$$a_{ij} = \frac{\partial Q}{\partial \sigma'_{ij}}, \qquad b_{ij} = \frac{\partial Y}{\partial \sigma'_{ij}}.$$

The parameter γ has the form

$$\gamma = \frac{\partial Y}{\partial p_0}\left[\frac{p_0(1 + e_0)}{(\alpha - \beta)}\right] \tag{3.49}$$

where p_0 and e_0 are the compressive pressure and void ratio. Also, α is the slope of the compressive line and β is the slope of the rebound curve.

Finally, the total stress is given by Eq. (3.6). The law for the evolution of the pore pressure is Eq. (3.18) which can be written as

$$\dot{P} = -k_b[d_{kk}^{(e)} + d_{kk}^{(n)}]$$

or

$$\dot{P} = -k_b\left(-\frac{\dot{p}'}{\lambda + \frac{2}{3}G} + \frac{b_{ij}\overset{*}{\sigma}'_{ij}}{\gamma}\right). \tag{3.50}$$

A form of the yield function suggested for an associated cam clay model [6] is

$$Y = Q = \mp(\bar{\sigma}_0/M_f p') + \ln(2.7182p'/p_0) - 1 \tag{3.51}$$

with $Y < 0$ defining the elastic regime and $Y = 0$ signifying yielding. The parameter M_f is the slope of the failure line.

3.7.2. A viscoplastic constitutive model for metals

Viscoplasticity models with state variables for large deformation are obtained by extending the ones for small deformation. Thus, again there is no explicit yield surface and the stress rate $\overset{*}{\sigma}'_{ij}$ is absent from Eqs. (3.46) and (3.47).

An extended version [12, 41] of Hart's model [23], for incompressible non-elastic deformation of metals, has been employed to obtain most of the numerical results that are presented later in this section. The differences between the large-strain constitutive model and the small-strain one presented earlier in Eqs. (2.39)–(2.50) is that $\dot{\varepsilon}_{ij}^{(p)}$ and $\dot{\varepsilon}_{ij}^{(n)}$ are now replaced by $d_{ij}^{(p)}$ and $d_{ij}^{(n)}$, respectively. One key point is that Eq. (2.39) is replaced by the equation

$$\overset{*}{\varepsilon}_{ij}^{(a)} = d_{ij}^{(n)} - d_{ij}^{(p)} . \tag{3.52}$$

Since $\overset{*}{\varepsilon}_{ij}^{(a)}$ is a proper state variable, its Jaumann rate is used in the above equations. The rest of the equations have the same form as Eqs. (2.40)–(2.50). Thus, given the Cauchy stress and the stress variables (hardness and anelastic strains) at any time, it delivers, among other things, $d_{ij}^{(n)}$ and the state variable rates $\dot{H}$ and $\overset{*}{\varepsilon}_{ij}^{(a)}$ at that time.

3.8. Numerical Implementation

Most of the earlier discussion regarding numerical implementation of the small deformation BEM equations apply here to the case of large deformations also. A discretized version of Eq. (3.39) for the velocity is now

$$\begin{aligned} C_{ij}(P_M)v_i(P_M) = {} & \sum_{N_S} \int_{\Delta S_N} [U_{ij}(P_M, Q)\dot{\tau}_i(Q) - T_{ij}(P_M, Q)v_i(Q)] \, dS_Q^0 \\ & + \sum_{n_i} \int_{\Delta A_n} [\lambda U_{ij,i}(P_M, q) d_{kk}^{(n)}(q) + 2G U_{ij,k}(P_M, q)] d_{ik}^{(n)}(q)] \, dV_q^0 \\ & + \sum_{n_i} \int_{\Delta A_n} U_{ij,m}(P_M, q) G_{mikl}(q) v_{k,l}(q) \, dV_Q^0 \\ & + \sum_{n_i} \int_{\Delta A_n} U_{ij,m}(P_M, q) \bar{C}_{mikl}(q) v_{k,l}(q) \, dV_q^0 \end{aligned} \tag{3.53}$$

where the boundary of the body in the reference configuration ∂B^0 is divided into N_s boundary segments and the interior into n_i internal cells and $v_i(P_M)$ are the components of velocities at a point P which coincides with node M.

Suitable shape functions must now be chosen for the variation of tractions and velocities on the surface element Δs_N and the variation of the non-elastic strain rates and velocities over an internal cell ΔA_n.

Following essentially the same procedure as for small-strain problems, one obtains the algebraic system

$$[A]\{v\} + [B]\{\dot{\tau}\} = \{b\} \tag{3.54}$$

where the coefficient matrices $[A]$ and $[B]$ contain integrals of the kernels and the shape functions; and the vector $\{b\}$ contains the contributions of various quantities from the three domain integrals.

Equations (3.36) and (3.40) for the velocity field and the velocity gradients at an internal point are discretized in similar fashion.

3.9. Solution strategy

The solution strategy for these problems is best explained by referring to Fig. 5. The discussion here is about compressible elastic–viscoplastic deformation of metals using a state variable type of constitutive model. Soil mechanics problems with compressible plasticity constitutive models can be solved with a slight variation of the solution strategy described below [13].

The presence of velocity gradients in the boundary traction rates (Eq. (3.37)) and in some of the domain integrals (Eq. (3.39)) requires interations within each time step. As mentioned before, the unknown velocity gradient components occur in a domain integral for elastic–plastic problems, even for the case of small strain. Here, this situation occurs whether elastic–plastic or elastic–viscoplastic constitutive models are used.

The solution strategy for these problems can be described as

(a) The elasticity problem is first solved to obtain initial displacements. For this step, Eq. (3.54) is solved with $\{b\} = 0$ for the unknown components of $\boldsymbol{u}$ and $\boldsymbol{\tau}$ in terms of the prescribed ones.

(b) The initial values of the displacement gradients are obtained from a truncated version of Eq. (3.40) with $d_{ik}^{(\mathrm{n})} = 0$ and $v_{k,n} = 0$ and $\boldsymbol{v}$ and $\dot{\boldsymbol{\tau}}$ replaced by $\boldsymbol{u}$ and $\boldsymbol{\tau}$ in the rest of the equation. The initial stresses are determined from strains through Hooke's Law. Also, initial rotations are determined from the displacement gradients.

(c) The tensor $d_{ik}^{(\mathrm{n})}$ at zero time is obtained from the constitutive equations.

(d) The first approximation to $v_j(\mathrm{P})$ at $t = 0$ is calculated from Eq. (3.39) with $v_{k,l}$ set to zero. It must be noted that the velocity gradients occur in two of the domain integrals as well as in the expression for $\dot{\tau}_i$.

(e) The first approximation to $v_{j,\bar{l}}(\mathrm{p})$ at $t = 0$ is obtained from Eq. (3.40) with all the velocity gradients in the integrals on the right-hand side of Eq. (3.40) set to zero. The $v_{j,\bar{l}}(\mathrm{p})$ values are extrapolated to get $v_{j,\bar{l}}(\mathrm{P})$.

(f) The velocity gradients are inserted into Eq. (3.39) and the full equation is solved to determine a second approximation to $v_j(\mathrm{P})$. This $v_j(\mathrm{P})$ and the first approximation of velocity gradients are then used to obtain a second approximation for $v_{j,\bar{l}}(\mathrm{p})$.

(g) Step (f) is repeated until the required convergence is achieved and $v_j(\mathrm{P})$ is determined at zero time.

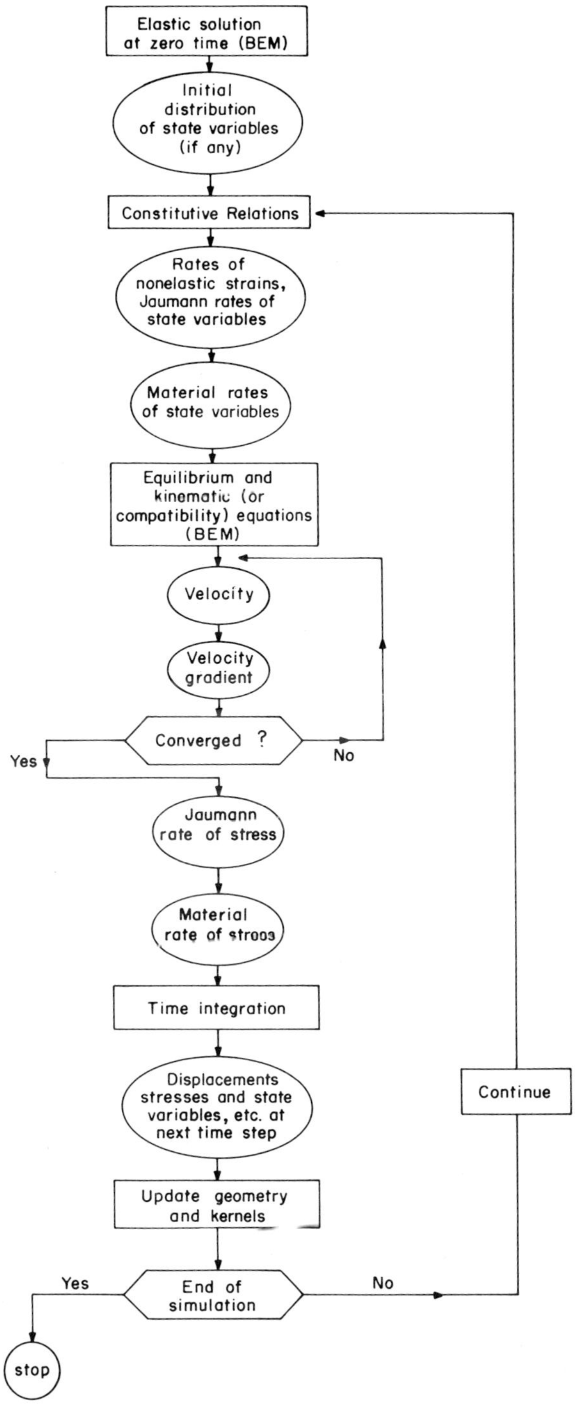

Fig. 5. Flow chart for the solution of large-strain elastic–viscoplastic problems (from [12]).

(h) The quantities $v_j(\mathrm{p})$, $v_{j,\bar{l}}(\mathrm{p})$ and then $\overset{*}{\sigma}_{ij}$ are calculated at zero time. The material rate of the Cauchy stress is obtained from the Jaumann rate using Eq. (3.8).

(i) The relevant quantities are calculated at time Δt from the values and rates at $t = 0$. The steps (c) through (h) are repeated to obtain the rates at $t = \Delta t$ and so on. The time histories of the various quantities are thus obtained by marching forward in time and suitable updating of the geometry and the kernels.

Ideally, the geometry and the kernels should be updated at each time step. In practice, updating is done as often as is necessary during the integration process. It should be noted that the time marching process must be carried out with the material derivatives of the tensors in order to obtain their integrated values in the global coordinate frame (original vector basis $\boldsymbol{e}_1$, $\boldsymbol{e}_2$) as functions of $\boldsymbol{x}$ and time.

3.10. Sample numerical results for soil mechanics

Recently obtained numerical results for a plane strain simple shear problem are presented here. The results are taken from Ref. [13].

The BEM program uses straight boundary elements and polygonal internal cells. The velocity and traction rate $\boldsymbol{v}$ and $\dot{\boldsymbol{\tau}}$ are taken to be piecewise linear on the boundary elements and $d_{ij}^{(\mathrm{n})}$ and $v_{k,n}$ in Eq. (2.34) are taken to be piecewise constant in the internal cells. The values of the boundary variables are assigned at nodes which lie at the intersections of boundary segments. Possible discontinuities in tractions are taken care of by placing a zero length element between nodes and assigning different values of traction at each of these nodes.

All integrations of kernels are carried out analytically. The last two integrals in Eq. (2.33) are evaluated by first performing the integrations for an arbitrary source point p_m and then differentiating the integrals at p_m.

The constitutive model of Banerjee and Stipho [6] is used to describe the behavior of the soil. The material parameters used here are

$$E = 3p_0(1 + e_0)(1 - 2\nu)/\beta , \qquad \nu = 0.3$$

where E is the Young's modulus.

Also, for cam clay (from [6])

$$\alpha = 0.14 , \qquad \beta = 0.05 , \qquad M_\mathrm{f} = 0.85 ,$$

$$p_0 = +53 \text{ psi} , \qquad e_0 = 0.9513 , \qquad k_\mathrm{b} = 5E .$$

The problem considered here is that of simple shear in plane strain. The initially square slab of side L_0 is deformed into a sequence of parallelograms. This involves large strains as well as large rotations. Hence, both the load correction and the geometric correction terms become important. Figure 6 shows the

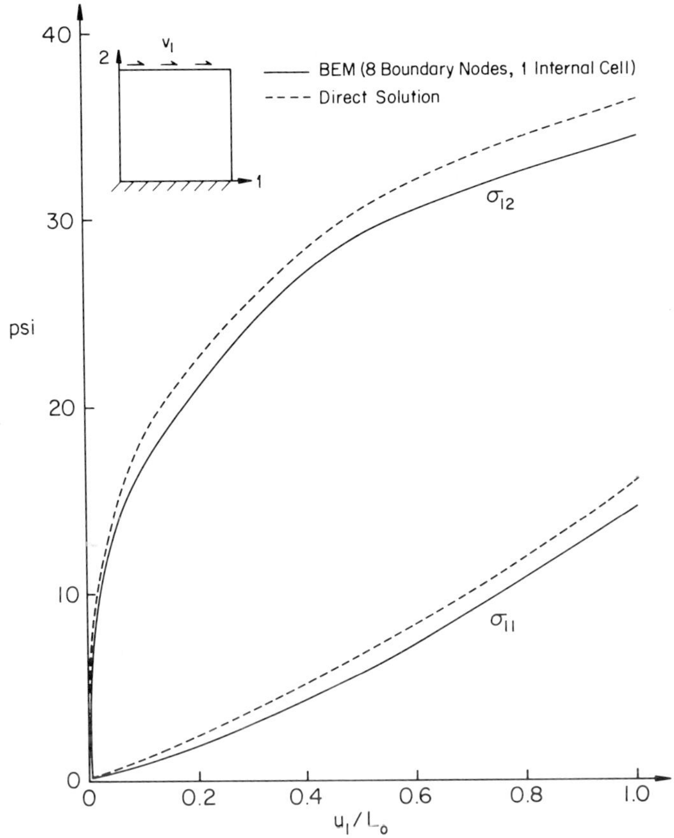

Fig. 6. Stress–strain behavior of soil in simple shear (plane strain) from [32]).

shearing stress (σ_{12}) and the nominal stress σ_{11} up to a shearing strain of unity. For this simple problem, a direct solution can be obtained and at $u_1/L_0 = 1.0$, σ_{12} obtained from the BEM differs by less than 5% from that obtained from the direct solution. The normal stresses from the BEM agree well with the direct solution and at $u_1/L_0 = 1.0$, the σ_{11} from the BEM differs by less than 5% from that obtained from the direct scheme.

3.11. *Sample numerical results for metal forming*

The numerical results presented in this subsection are taken from Refs. [41] and [14]. The generalized Hart model is used to describe incompressible elastic–viscoplastic deformation of metallic solids. The planar BEM program uses the same shape functions, etc., as described in the previous subsections. In fact, it is the same general-purpose computer program with a different constitutive model.

Results for large deformation of a slab with a hole in plane strain are shown in Figs 7–9. The material properties used here for Hart's constitutive model are

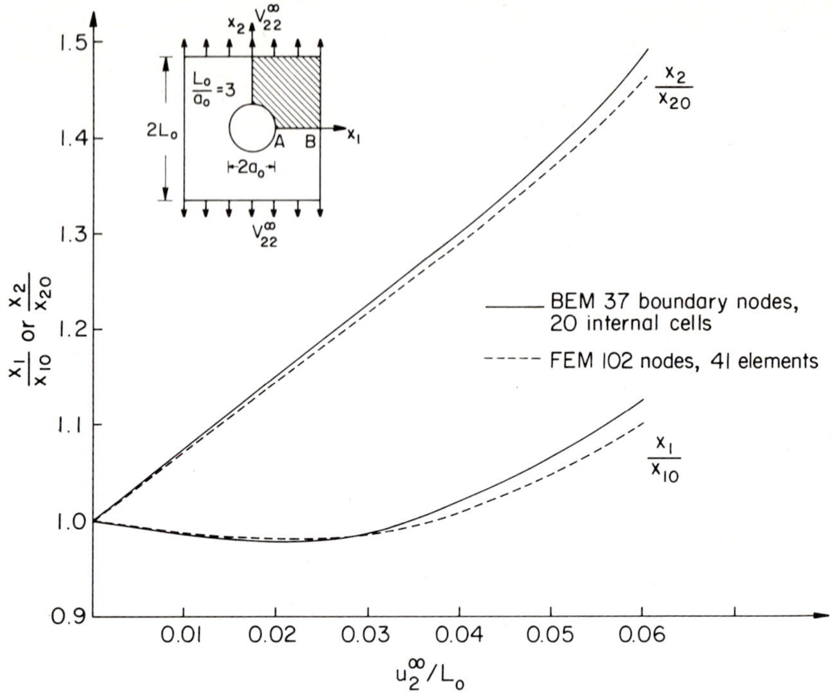

Fig. 7. Comparison of BEM and FEM solutions for hole growth in a square block in plane strain extension at constant velocity. $v_2^\infty = 10^{-3}$ in/s. 304 SS at 400°C (from [12]).

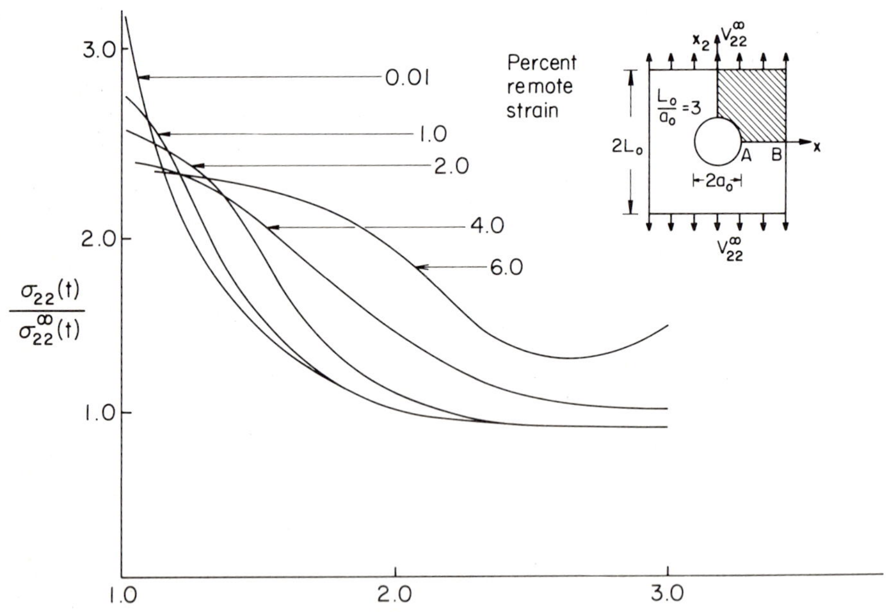

Fig. 8. Stress redistribution along AB at various stages of plane strain hole growth in a square block in plane strain extension at constant velocity. $v_2^\infty = 10^{-3}$ in/s. 304 SS at 400°C (from [12]).

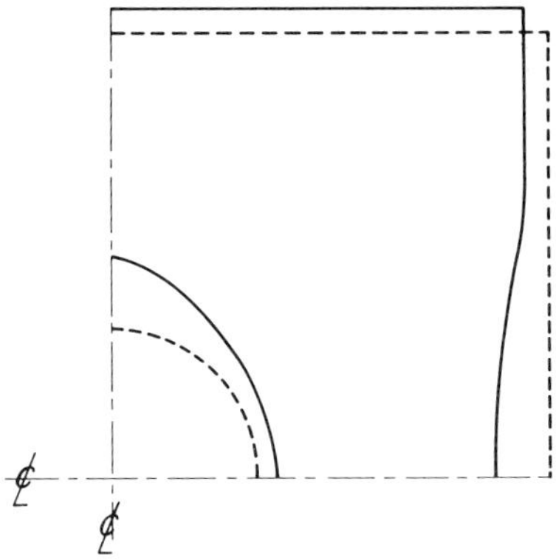

Fig. 9. Deformation of a block with a circular hole for the problem described in Figs. 7 and 8. $\varepsilon_{22}^{\infty} = 0.06$ (from [12]).

representative of 304 stainless steel at 400°C. These are available in Chapter 5 of Mukherjee's book [40]. Only a quarter of the slab is modelled because of symmetry. The BEM mesh, with 37 boundary nodes and 20 internal cells and the FEM mesh, with 102 nodes and 41 elements, are available in Ref. [41]. The FEM program is described in Ref. [15]. It uses a piecewise quadratic description of velocities over triangular finite elements.

Displacements and stresses for the problem are shown in Figs. 7 and 8. The locations of the ends of the horizontal and vertical diameters of the hole, points $(a_0, 0)$ and $(0, a_0)$, as functions of u_2^{∞}/L_0 (with $u_2^{\infty} = v_2^{\infty}t$) are shown in Fig. 7. The horizontal diameter of the hole is first seen to decrease due to Poisson effect and then increase with deformation. The redistribution of stress along the line AB in Fig. 8 shows the expected drop in stress concentration from approximately 3.16 to 2.35. The stress becomes more evenly distributed with deformation. In contrast to the small-strain problem discussed earlier (Fig. 4), a special feature of the large deformation problem is that the stress concentration at a remote strain of 6% increases near the free edge of the slab. This is a consequence of "caving in" of the free edge of the slab as shown in Fig. 9. A vertical equilibrium check for the stress distribution at a remote strain of 6% in Fig. 8 shows that the BEM result underpredicts the remote stress resultant by about 6%. This is considered quite good in view of the fact that the slab has undergone a substantial amount of deformation at this stage.

Researchers at US Steel [1, 8] have recently carried out some FEM analysis and large-deformation experiments on a slab with a hole. The calculations described in this section are able to predict the important qualitative features of the experiments in very satisfactory fashion. Bourcier et al. [8], for example, measured an initial decrease of the horizontal diameter of the hole in a slab, followed by an increase of this diameter with deformation. They also observed a caving in of the free edge of the slab. The magnitudes of the calculated displacements are quite close to the measured ones. Detailed quantitative comparisons of theory

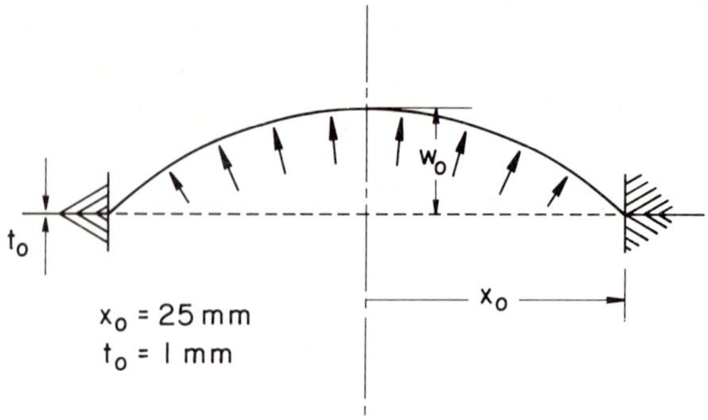

Fig. 10. Schematic of the plane strain bulge test (from [33]).

and experiment have not been carried out yet since the Hart model parameters for the materials used by Bourcier et al. are not known at present.

The CPU times on an IBM 370/178 computer are 297 s and 502 s, respectively, for the BEM and FEM calculations.

The next set of numerical results are taken from Ref. [14]. This time the material parameters for Hart's model are representative of commercially pure aluminum at 24°C. The material parameters for this case are available in Ref. [16].

Results for the problem of plane strain bulging is shown in Figs. 10–13. As shown in Fig. 10, the sheet is clamped at two ends and the normal pressure on the

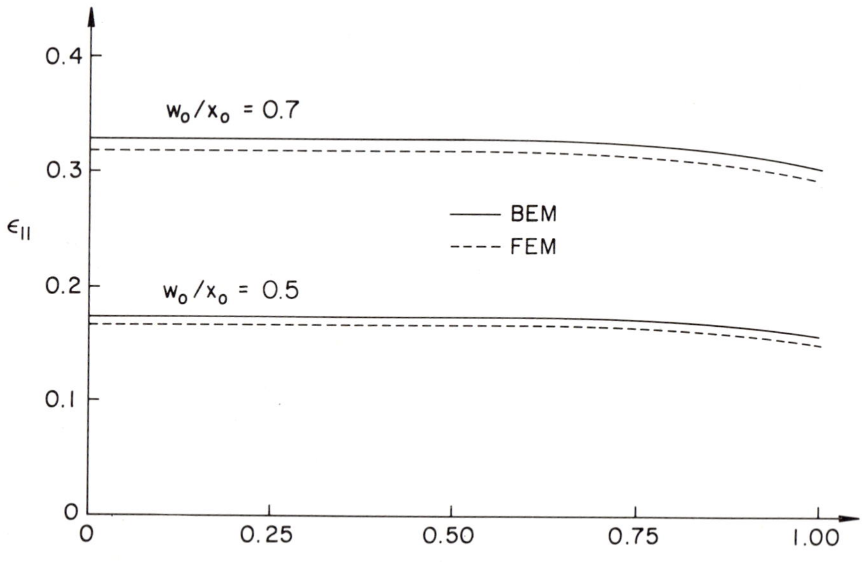

Fig. 11. Longitudinal strain distribution for the bulge test at different bulge heights. BEM and FEM solutions. Commercially pure aluminum at 24°C (from [33]).

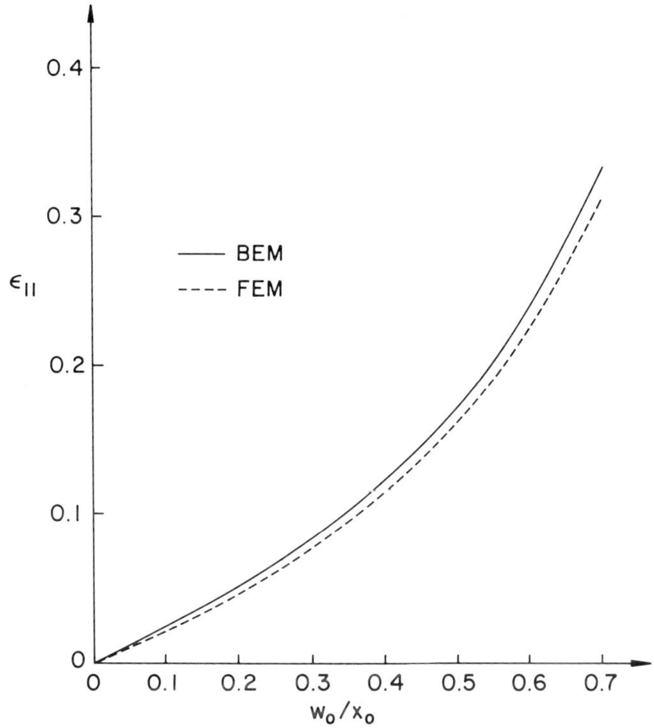

Fig. 12. Variation of central longitudinal strain for the bulge test as a function of bulge height. BEM and FEM solutions. Commercially pure aluminum at 24°C (from [33]).

sheet is increased at the rate of 0.5 MPa/s (72.5 psi/s). The specified tractions are follower loads and change with deformation of the sheet.

The longitudinal strain distributions along the sheet at different stages of deformation are shown in Fig. 11. For the case of plane strain bulging, the strain distribution remains almost uniform throughout the sheet. At $w_0/x_0 = 0.7$, ε_{11} at $x_1 = 0$ is 0.325. Hence, unlike axisymmetric bulging [53], the material will not tend to thin at the center. Thus, higher dome heights may be obtained from the same blank size under plane strain conditions. The firm line denotes the results obtained from the BEM analysis and the dotted line shows the results obtained from a finite element analysis of the same problem. As observed in Fig. 11, the BEM and the FEM predictions are quite close.

Figure 12 shows the increase in longitudinal strain at the center ($x = 0$) with bulge height. It is observed that the slope of the ε_{11} versus w_0/x_0 curve increases as bulge height is increased. Thus, as the deformation proceeds, the strain at the center increases at a progressively faster rate and the structure becomes more and more prone to failure. The firm line representing the BEM analysis and the dotted line representing the FEM analysis agree quite well.

The rise in bulge height with increase in pressure is shown in Fig. 13. It is seen that for normalized pressure in excess of 7.0, the bulge height is quite sensitive to

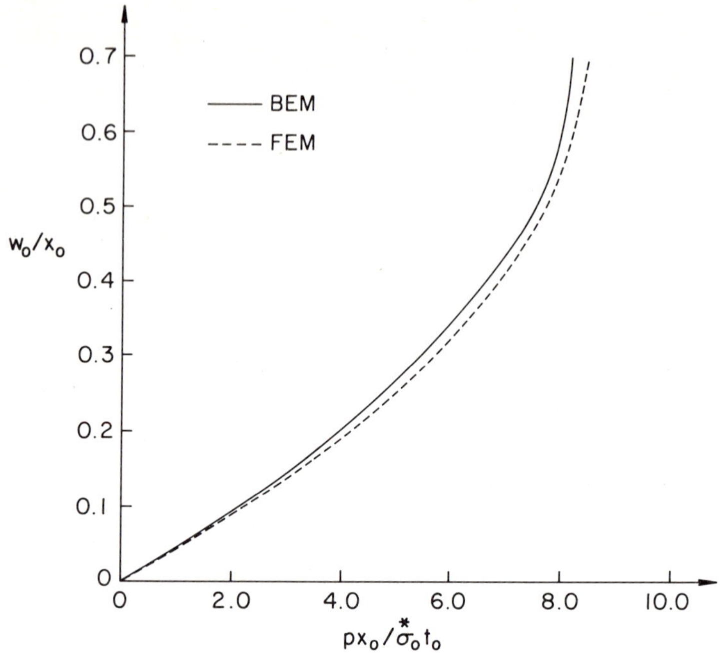

Fig. 13. Variation of bulge height with pressure from the BEM and FEM. Commercially pure aluminum at 24°C (from [33]).

small changes in pressure. This sets the stage for potential instability. Once again, it is observed that the BEM and FEM results are in good agreement.

The secondary variables such as strains and stresses are obtained pointwise in the BEM analysis through analytic differentiation of the relevant primary variables. These quantities (specially $\varepsilon^{(a)}$, σ, and H) play a very important role in this class of problems and their accuracy is critical to the stability of the time integration scheme. The higher obtainable accuracy for these secondary variables permits bigger time steps in the BEM analysis than those allowable in the FEM analysis. This results in improved computing times for the BEM analysis. To obtain a normalized bulge height of $w_0/x_0 = 0.7$, the BEM program (with 28 boundary nodes and 10 internal cells) needs 205 s of CPU time on an IBM 370/168 whereas the FEM program (with 105 nodes and 40 elements) requires 484 s.

Acknowledgements

Most of the research discussed in this chapter has been supported by contracts from the US Department of Energy and grants from the US National Science Foundation (CME-7918658 and CEE-8206344) to Cornell University.

References

[1] E.T. Appleby, et al., Stretching of Notched Steel Sheets: A Theoretical–Experimental Comparison, *Proceedings of International Conference on Numerical Methods in Industrial Forming Processes*, Swansea, UK (1982).

[2] P.K. Banerjee and R. Butterfield, *Boundary Element Methods in Engineering Science* (McGraw-Hill, London, 1981).

[3] P.K. Banerjee, D.N. Cathie and T.G. Davies, Two and Three Dimensional Problems in Elasto–Plasticity, in: *Developments in Boundary Element Methods–1*, P.K. Banerjee and R. Butterfield, Eds. (Elsevier Applied Science Publishers, London, 1979) pp. 65–95.

[4] P.K. Banerjee and D.N. Cathie, A Direct Formulation and Numerical Implementation of the Boundary Element Method for Two-Dimensional Problems of Elasto–Plasticity, *International Journal of Mechanical Sciences* **22**, 233–245 (1980).

[5] P.K. Banerjee and G.C.W. Mustoe, The Boundary Element Method for Two-Dimensional Problems of Elasto–Plasticity, in: *Recent Advances in Boundary Element Methods*, C.A. Brebbia, Ed. (Pentech Press, London, 1978) pp. 283–300.

[6] P.K. Banerjee and A.S. Stipho, Associated and Non-Associated Constitutive Relations for Undrained Behavior of Isotropic Clays, *International Journal for Numerical and Analytical Methods in Geomechanics* **2**, 35–66 (1978).

[7] F.G. Benitez, E. Alarcón, C.A. Brebbia and J. Telles, Tridimensional Plasticity using BIEM, *Applied Mathematical Modelling* **5**, 442–447 (1981).

[8] R. Bourcier, et al., Deformation and Failure at Isolated Holes in Plane Strain Tension, *Proceedings of the 9th National Congress of Theoretical and Applied Mechanics*, Cornell University, Ithaca, NY (1982).

[9] C.A. Brebbia, J.C.F. Telles and L.C. Wrobel, *Boundary Element Techniques: Theory and Applications in Engineering* (Springer-Verlag, Berlin, 1984).

[10] M. Brunet, Numerical Analysis of Cyclic Plasticity Using the Boundary Integral Equation Method, in: *Boundary Element Methods*, C.A. Brebbia, Ed. (Springer-Verlag, Berlin, 1981) pp. 337–349.

[11] H.D. Bui, Some Remarks About the Formulation of Three-Dimensional Thermoelastoplastic Problems by Integral Equations, *International Journal of Solids and Structures* **14**, 935–939 (1978).

[12] A. Chandra and S. Mukherjee, Applications of the Boundary Element Method to Large Strain Large Deformation Problems of Viscoplasticity, *Journal of Strain Analysis* **18**, 261–270 (1983).

[13] A. Chandra and S. Mukherjee, *A Boundary Element Formulation for Large Strain Problems of Compressible Plasticity*, *Engineering Analysis* **3**, 77–78 (1986).

[14] A. Chandra and S. Mukherjee, A Boundary Element Formulation for Sheet Metal Forming, *Applied Mathematical Modelling* **9**, 175–182 (1985).

[15] A. Chandra and S. Mukherjee, A Finite Element Analysis of Metal-Forming Problems with an Elastic–Viscoplastic Material Model, *International Journal for Numerical Methods in Engineering* **20**, 1613–1628 (1984).

[16] A. Chandra, *A Generalized Finite Element Analysis of Sheet Metal Forming with an Elastic-Viscoplastic Material Model*, General Motors Research Laboratories Publication GMR-4642 (1983).

[17] A. Chandra and S. Mukherjee, Boundary Element Formulations for Large Strain–Large Deformation Problems of Viscoplasticity, *International Journal of Solids and Structures* **20**, 41–53 (1984).

[18] M. Chaudonneret, Structure Computation in Viscoplasticity. Application to Two-Dimensional Calculation of Stress Concentration, *Addendum to Proceedings for the Second International Symposium on Innovative Numerical Analysis in Applied Engineering Science*, Montreal, Canada, R.P. Shaw et al., Eds., University of Virginia Press, Charlottesville, Virginia, (1980); see also, Methode des Equations Integrales Appliquées á la Resolution de Probléme de Viscoplasticité, *Journal de Mécanique Appliqué* **1**, 113–132 (1977).

[19] D. Cordts and F.G. Kollmann, An Implicit Time Integration Scheme for Inelastic Constitutive Equations with Internal State Variables, *International Journal for Numerical Methods in Engineering* **23**, 533–554 (1986).

[20] T.A. Cruse, Numerical Solutions in Three Dimensional Elastostatics, *International Journal of Solids and Structures* **5**, 1259–1274 (1969).

[21] T.A. Cruse, An Improved Boundary-Integral Equation Method for Three-Dimensional Elastic Stress Analysis, *Computers and Structures* **4**, 741–754 (1974).

[22] M. Doblare, J.C.F. Telles, C.A. Brebbia and E. Alarcón, Boundary Element Formulation for Elastoplastic Analysis of Axisymmetric Bodies, *Applied Mathematical Modelling* **6**, 130–135 (1982).

[23] E.W. Hart, Constitutive Relations for the Non-Elastic Deformation of metals, *ASME Journal of Engineering Materials and Technology* **98**, 193–202 (1976).

[24] M.A. Jaswon and A.R. Ponter, An Integral Equation Solution of the Torsion Problem, *Proceedings of the Royal Society of London*, Series A, **273**, 237–246 (1963).

[25] M.A. Jaswon and M. Maiti, An Integral Equation Formulation of Plate Bending Problems, *Journal of Engineering Mathematics* **2**, 83–93 (1968).

[26] G. Jaumann, Geschlossenes System Physikalischer und Chemischer Differentialgesetze Sitzber, *Akademische Wissenschaft Wien* (IIa) **120**, 385–530 (1911).

[27] V. Kumar and S. Mukherjee, A Boundary-Integral Equation Formulation for Time-Dependent Inelastic Deformation in Metals, *International Journal of Mechanical Sciences* **19**, 713–724 (1977).

[28] S. Kobayashi and N. Nishimura, Elastoplastic Analysis by the Integral Equation Method, *Memoirs of the Faculty of Engineering, Kyoto University* **42**, 324–334 (1980).

[29] J.C. Lachat, A Further Development of the Boundary-Integral Technique for Elastostatics, Ph.D. Thesis, University of Southampton, England (1975).

[30] G. Maier and G. Novati, Elastic–Plastic Boundary Element Analysis as a Linear Complementary Problem, *Applied Mathematical Modelling* **7**, 74–82 (1983).

[31] A. Mendelson, Boundary Integral Methods in Elasticity and Plasticity, NASA Technical Note TND-7418, Washington, D.C. (1973).

[32] M. Morjaria and S. Mukherjee, Inelastic Analysis of Transverse Deflection of Plates by the Boundary Element Method, *Journal of Applied Mechanics ASME* **47**, 291–296 (1980).

[33] M. Morjaria, V. Sarihan and S. Mukherjee, Comparison of Boundary Element and Finite Element Methods in Two-Dimensional Inelastic Analysis, *Res Mechanica* **1**, 3–20 (1980).

[34] M. Morjaria and S. Mukherjee, Finite Element Analysis of Time-Dependent Inelastic Deformation in the Presence of Transient Thermal Stresses, *International Journal for Numerical Methods in Engineering* **17**, 909–921 (1981).

[35] S. Mukherjee, Time-Dependent Inelastic Deformation of Metals by Boundary Element Methods, in: *Developments in Boundary Element Methods*–2, P.K. Banerjee and R.P. Shaw, Eds. (Applied Science Publishers, London, 1982) pp. 111–142.

[36] S. Mukherjee, Corrected Boundary Integral Equations in Planar Thermoelastoplasticity, *International Journal of Solids and Structures* **13**, 331–335 (1977).

[37] S. Mukherjee, Thermoviscoplastic Response of Cylindrical Structures Using a State Variable Theory, in: *Mechanical Behavior of Materials–Proceedings of ICM* 3, Cambridge, England, K.J. Miller and R.F. Smith, Eds. (Pergamon Press, Oxford, 1979) pp. 233–242.

[38] S. Mukherjee and M. Morjaria, Comparison of Boundary Element and Finite Element Methods in the Inelastic Torsion of Prismatic Shafts, *International Journal for Numerical Methods in Engineering* **17**, 1576–1588 (1981).

[39] S. Mukherjee and V. Kumar, Numerical Analysis of Time-Dependent Inelastic Deformation in Metallic Media Using the Boundary-Integral Equation Method, *ASME Journal of Applied Mechanics* **45**, 785–790 (1978).

[40] S. Mukherjee, *Boundary Element Methods in Creep and Fracture* (Elsevier Applied Science Publishers, London, 1982).

[41] S. Mukherjee and A. Chandra, Boundary Element Formulations for Large Strain–Large Deformation Problems of Plasticity and Viscoplasticity, in: *Developments in Boundary Element Methods–3*, P.K. Banerjee and S. Mukherjee, Eds. (Elsevier Applied Science Publishers, London, 1984) pp. 27–58.

[42] N. Nishimura and S. Kobayashi, Elastoplastic Analysis by Indirect Methods, in: *Developments in Boundary Element Methods–3*, P.K. Banerjee and S. Mukherjee, Eds. (Elsevier Applied Science Publishers, London, 1984) pp. 59–81.

[43] L.M. Oliveira Faria, C.A. Mota Soares, M.F. Seabra Pereira and C.A. Brebbia, Boundary Elements in 2D Plasticity Using Quadratic Shape Functions, *Applied Mathematical Modelling* **5**, 371–375 (1981).

[44] A.R.S. Ponter, On Plastic Torsion, *International Journal of Mechanical Sciences* **8**, 227–235 (1966).

[45] P. Riccardella, *An Implementation of the Boundary Integral Technique for Plane Problems of Elasticity and Elasto–Plasticity*, Ph.D. Thesis, Carnegie Mellon University, Pittsburgh, PA (1973).

[46] F.J. Rizzo, An Integral Equation Approach to Boundary Value Problems of Classical Elastostatics, *Quarterly of Applied Mathematics* **25**, 83–95 (1967).

[47] J.L. Swedlow and T.A. Cruse, Formulation of Boundary Integral Equations for Three-Dimensional Elasto–Plastic Flow, *International Journal of Solids and Structures* **7**, 1673–1681 (1971).

[48] M. Tanaka and K. Tanaka, On a New Boundary Element Solution Scheme for Elastoplasticity, *Ingenieur Archiv* **50**, 289–295 (1981).

[49] J.C.F. Telles and C.A. Brebbia, Viscoplasticity and Creep Using Boundary Elements, in: *Progress in Boundary Element Methods*, Vol. 2, C.A. Brebbia, Ed. (Pentech Press, London, 1983) pp. 200–215.

[50] J.C.F. Telles and C.A. Brebbia, The Boundary Element Method in Plasticity, in: *New Developments in Boundary Element Methods*, C.A. Brebbia, Ed. (CML Publications, Southampton, 1980) pp. 295–317.

[51] S.P. Timoshenko and J.N. Goodier, *Theory of Elasticity*, 3rd edition (McGraw–Hill, New York, 1970).

[52] C. Truesdell and W. Noll, The Nonlinear Field Theories of Mechanics, *Handbuch der Physik*, S. Flugge, Ed., Vol. III/3 (Springer Verlag, Berlin, 1965).

[53] N.M. Wang and M.L. Wenner, Effects of Strain-Hardening Representation in Sheet Metal Forming Calculations of 2036-T4 Aluminum. Formability of Metallic Materials–2000AD, *American Society for Testing and Materials. Special Technical Publication*–753, 84–104 (1982).

[54] K.J. Willam, Ed., *Constitutive Equations–Macro and Computational Aspects* (The American Society of Mechanical Engineers, New York, 1984).

CHAPTER 7

Fracture Mechanics

Thomas A. CRUSE

Department of Engineering Mechanics
Southwest Research Institute
San Antonio, Texas 78284, U.S.A.

Boundary Element Methods in Mechanics
Edited by D.E. Beskos

Introduction

The use of the direct potential formulation of the Boundary Integral Equation (BIE) method provided the earliest, general-purpose numerical solution of the three-dimensional elastic fracture mechanics problem [17]. That this was possible derives from the accuracy and efficiency of the BIE formulation for three-dimensional problems. The numerical solution of fracture mechanics problems requires an accurate resolution of the singular stress/strain field near the crack tip.

Finite element methods by their nature must approximate this field singularity by spatial interpolations which in 1971 did not include a three-dimensional (3D) singular modeling capability. The 3D BIE formulation does not interpolate over the interior (singular) stress/strain field. Rather, as well illustrated by the other authors in this volume, only the boundary quantities are interpolated. The interior behavior is recovered by a quadrature of the boundary solution which has been found to be very effective for both fracture mechanics and stress concentration analyses.

However, these attributes carry their own limitations. Cruse [24] first detailed the mathematical degeneration in the BIE formulation that occurs for the direct, numerical solution of two co-planar surfaces for crack problems. A number of numerical or formulation strategies evolved to avoid this fundamental limitation, firmly rooted in the basic mathematics needed for well-posed problems (i.e. singly defined surfaces). Cruse [18] outlined the crudest of these as the notch method (model the actual crack as a narrow notch), and the symmetric-half solution. In the latter, the plane of the crack for symmetric crack problems (mode I problems) is modeled in its entirety. Thus, the non-singular displacements of the crack surface and the singular crack plane tractions are both interpolated. While this interpolation of crack plane tractions spoils the true boundary formulation, the interior stresses are modeled only on a single plane, not throughout the volume.

Other, more sophisticated approaches to the fracture mechanics problem have now become widely used. These approaches include the use of a special Green's function for two-dimensional (2D) problems [42–46, 58, 62], crack tip interpolation functions [2, 76, 79], displacement discontinuity potentials [13–15, 36], traction integral equation [7, 20, 32, 53, 54, 71, 77], energy release rate [22, 67, 68], multidomain boundary elements [8, 40, 60], and quadratic isoparametric boundary elements with quarter-point interpolations [23, 33, 38, 41, 67, 68, 75]. Earlier overviews of various BIE approaches to the fracture mechanics have been published by Cruse [25] and Atkinson [1].

The BIE formulation of problems in solid mechanics relies on the use of fundamental solutions to the governing equations, generally point load solutions. The superposition of the fundamental solutions through the reciprocal work relationship results in a boundary integral equation which serves as a constraint equation between physical boundary data, referred to as the direct formulation by Cruse [16]. Other integral equation formulations of the fracture mechanics problem are also used. In two-dimensional elasticity, complex variable techniques result in integral equations with Cauchy singularities. Theocaris [71] has made extensive use of this formulation approach. Others use integral transform methods [27] to obtain dual integral equations to match the applied boundary data. Generally, transform methods are limited to special geometries in two and three dimensions.

The present chapter seeks to describe the essentials of the BIE approach to solving fracture mechanics problems. In general, the methods all utilize a form of the BIE. The numerical solution of boundary data models is generally referred to as boundary element methodology (BEM). The essential point to be made is that the BEM techniques for the elastic fracture mechanics and stress concentration problems are well developed and are becoming more commonly used in engineering analysis and design applications.

The following sections review the fundamentals of linear elastic fracture mechanics with an overview of stress intensity factors, two-dimensional, and three-dimensional BEM solutions for fracture mechanics problems. The section on two-dimensional modeling includes an overview of non-BEM stress intensity factor solutions, the very effective Green's function approach, and the displacement discontinuity method used for geomechanics problems and three-dimensional fracture mechanics modeling. The second section closes with two brief discussions on inelastic fracture mechanics modeling and multi-region modeling. The third section discusses three-dimensional fracture mechanics. The emphasis is on direct modeling of symmetric crack problems and on the formulation of the displacement discontinuity approach for surface and buried cracks.

1. Fracture mechanics fundamentals

The earliest effort to characterize the strength of brittle materials in a manner that recognized the size of the defect or crack was that of Griffith [30]. In his analysis, the basic energy balance for a virtual crack length extension was first derived in terms of the elastic stored energy per unit thickness, U

$$G = -\partial U / \partial a \tag{1.1}$$

where a is the half crack length, ∂a is the crack length change, and G is the strain energy release rate. Equating the release in stored energy to the increase in crack surface area energy (material property), for the Inglis [34] plane strain problem of a central slit in tension, results in the basic relation

$$G = (1 - \nu^2)\pi\sigma^2 a/E \tag{1.2}$$

which relation applies up to the point of crack growth instability. In (1.2), σ is the applied far-field stress and E, ν are the usual material constants.

Irwin [35] considered the state of stress ahead of the crack under load and recognized that the Griffith energy release rate could be described in terms of a parameter which characterized the state of stress or strain in the region around the crack tip. This parameter K, the stress intensity factor, is related to the energy release rate by the following:

$$G = \begin{cases} K^2/E \text{ (plane stress)} \\ (1 - \nu^2)K^2/E \text{ (plane strain)} \end{cases} . \tag{1.3}$$

The key feature recognized by Irwin which became the foundation of linear elastic fracture mechanics (LEFM) is the fact that, for limited plasticity, the stress intensity factor (SIF) characterizes the state of material loading over a volume of sufficient size that the fracture strength of many engineering materials can be given in terms of the critical (maximum) stress intensity factor, K_{IC}.

Williams [78] treated the general class of problems associated with a vertex in the material under load for various types of boundary conditions on the edges of the vertex as an eigenvalue problem. He deduced that the transcendental equation

$$\sin z = \pm[(\sin \alpha)/\alpha]z\,; \qquad z = \lambda\alpha \tag{1.4}$$

gives the power λ of the inverse singularity of stress or strain at the vertex tip for the case of stress free edges near the tip. From (1.4), the stress singularity at the tip of the vertex is given by $r^{\lambda-1}$; α is the total included material angle at the vertex. The crack tip stress solution for the case of $\alpha = 2\pi$ for remote loading is given by

$$\sigma = a_0 r^{-1/2} + a_1 + a_2 r^{+1/2} + \cdots . \tag{1.5}$$

The leading term in (1.5) is seen to be singular at the crack tip, and thus for some region near the crack tip dominates the near-field stress and strain field. The coefficient of this leading term in Williams' solution is proportional to the SIF defined by Irwin. Using the current SIF definition, the leading term for the two-dimensional stress and displacement field near the crack tip (crack along

x-axis) is given for plane strain by

$$\left.\begin{aligned}\sigma_x &= [K_{\mathrm{I}}/(2\pi r)^{1/2}] \cos \tfrac{1}{2}\theta(1 - \sin \tfrac{1}{2}\theta \sin \tfrac{3}{2}\theta)\\ \sigma_y &= [K_{\mathrm{I}}/(2\pi r)^{1/2}] \cos \tfrac{1}{2}\theta(1 + \sin \tfrac{1}{2}\theta \sin \tfrac{3}{2}\theta)\\ \tau_{xy} &= [K_{\mathrm{I}}/(2\pi r)^{1/2}] \sin \tfrac{1}{2}\theta \cos \tfrac{1}{2}\theta \cos \tfrac{3}{2}\theta\end{aligned}\right\} \tag{1.6}$$

$$\left.\begin{aligned}u &= (K_{\mathrm{I}}/G)(r/2\pi)^{1/2} \cos \tfrac{1}{2}\theta(1 - 2\nu + \sin^2 \tfrac{1}{2}\theta)\\ v &= (K_{\mathrm{I}}/G)(r/2\pi)^{1/2} \sin \tfrac{1}{2}\theta(2 - 2\nu - \cos^2 \tfrac{1}{2}\theta)\end{aligned}\right\} \tag{1.7}$$

$$\left.\begin{aligned}\sigma_x &= -[K_{\mathrm{II}}/(2\pi r)^{1/2}] \sin \tfrac{1}{2}\theta(2 + \cos \tfrac{1}{2}\theta \cos \tfrac{3}{2}\theta)\\ \sigma_y &- [K_{\mathrm{II}}/(2\pi r)^{1/2}] \cos \tfrac{1}{2}\theta \sin \tfrac{1}{2}\theta \cos \tfrac{3}{2}\theta\\ \tau_{xy} &= [K_{\mathrm{II}}/(2\pi r)^{1/2}] \cos \tfrac{1}{2}\theta(1 - \sin \tfrac{1}{2}\theta \sin \tfrac{3}{2}\theta)\end{aligned}\right\} \tag{1.8}$$

$$\left.\begin{aligned}u &= (K_{\mathrm{II}}/G)(r/2\pi)^{1/2} \sin \tfrac{1}{2}\theta(2 - 2\nu + \cos^2 \tfrac{1}{2}\theta)\\ v &= (K_{\mathrm{II}}/G)(r/2\pi)^{1/2} \cos \tfrac{1}{2}\theta(1 - 2\nu + \sin^2 \tfrac{1}{2}\theta)\end{aligned}\right\} \tag{1.9}$$

$$\left.\begin{aligned}\tau_{xz} &= -[K_{\mathrm{III}}/(2\pi r)^{1/2}] \sin \tfrac{1}{2}\theta\\ \tau_{yz} &= [K_{\mathrm{III}}/(2\pi r)^{1/2}] \cos \tfrac{1}{2}\theta\end{aligned}\right\} \tag{1.10}$$

$$w = (K_{\mathrm{III}}/G)(2r/\pi)^{1/2} \sin \tfrac{1}{2}\theta\,. \tag{1.11}$$

In (1.6)–(1.11), G is the shear modulus; θ is the angle from the plane of crack prolongation given in Fig. 1. The subscripts on the SIF terms in these equations refer to the three kinematically independent modes of crack tip motion as shown in Fig. 2.

The above results apply to the case of isotropic material behavior. While the essential singular behavior of the crack tip stresses and strains is retained in the rectilinearly anisotropic case, the leading term results in (1.6)–(1.11) are somewhat more complicated. Let β_{ij} describe the full (2D) material compliance matrix (up to 21 constants in the general 3D case). The governing differential equations for planar problems with elastic mid-plane symmetry are integrable along the complex directions given by the roots of the following characteristic equation (Sih and Liebowitz [59])

$$\beta_{11}\mu^4 - 2\beta_{16}\mu^3 + (2\beta_{12} + \beta_{66})\mu^2 - 2\beta_{26}\mu + \beta_{22} = 0\,. \tag{1.12}$$

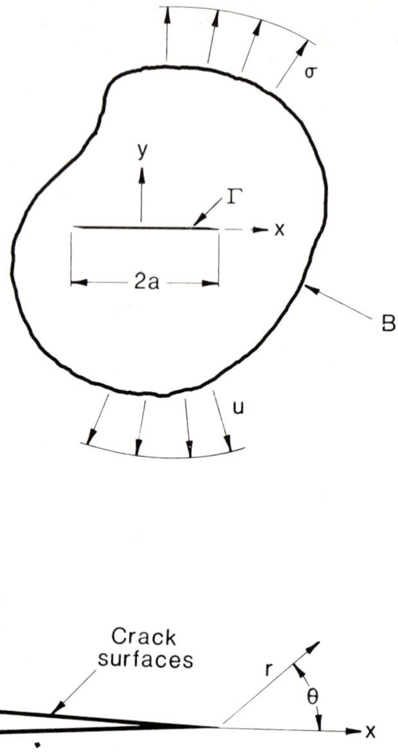

Fig. 1. Crack geometry definition.

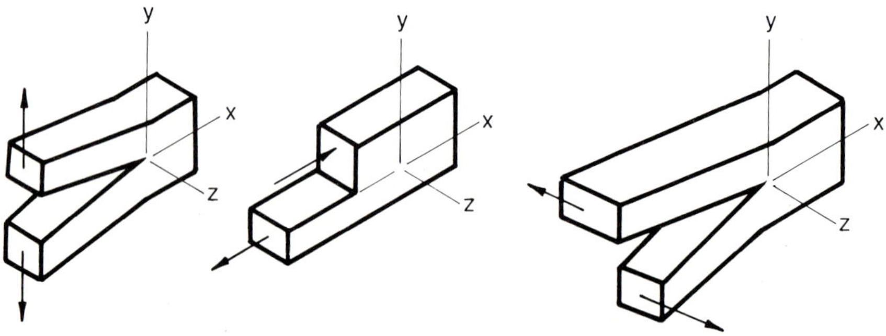

Fig. 2. Three modes of crack tip motion.

The roots of (1.12) given in complex conjugates μ_k, $\bar{\mu}_k$ are dimensionless quantities. The energy release rate is given for mode I, II loading as

$$G = -\frac{K_{\mathrm{I}}^2}{2\pi}\,\beta_{22}\,\mathrm{Im}\left(\frac{(\mu_1+\mu_2)}{\mu_1\mu_2}\right) + \frac{K_{\mathrm{II}}^2}{2\pi}\,\beta_{11}\,\mathrm{Im}(\mu_1+\mu_2)\,. \tag{1.13}$$

The crack tip stresses are given by

$$\left.\begin{aligned}
\sigma_x &= \frac{K_{\mathrm{I}}}{(2\pi r)^{1/2}} \operatorname{Re}\left[\frac{\mu_1\mu_2}{\mu_1-\mu_2}\left(\frac{\mu_2}{d_2}-\frac{\mu_1}{d_1}\right)\right] + \frac{K_{\mathrm{II}}}{(2\pi r)^{1/2}} \operatorname{Re}\left[\frac{1}{\mu_1-\mu_2}\left(\frac{\mu_2^2}{d_2}-\frac{\mu_1^2}{d_1}\right)\right] \\
\tau_{xy} &= \frac{K_{\mathrm{I}}}{(2\pi r)^{1/2}} \operatorname{Re}\left[\frac{\mu_1\mu_2}{\mu_1-\mu_2}\left(\frac{1}{d_1}-\frac{1}{d_2}\right)\right] + \frac{K_{\mathrm{II}}}{(2\pi r)^{1/2}} \operatorname{Re}\left[\frac{1}{\mu_1-\mu_2}\left(\frac{\mu_1}{d_1}-\frac{\mu_2}{d_2}\right)\right] \\
\sigma_y &= \frac{K_{\mathrm{I}}}{(2\pi r)^{1/2}} \operatorname{Re}\left[\frac{1}{\mu_1-\mu_2}\left(\frac{\mu_1}{d_2}-\frac{\mu_2}{d_1}\right)\right] + \frac{K_{\mathrm{II}}}{(2\pi r)^{1/2}} \operatorname{Re}\left[\frac{1}{\mu_1-\mu_2}\left(\frac{1}{d_2}-\frac{1}{d_1}\right)\right]
\end{aligned}\right\} \tag{1.14}$$

where

$$d_j = (\cos\theta + \mu_j \sin\theta)^{1/2} . \tag{1.15}$$

The SIF terms in (1.6)–(1.11) for two-dimensional results have been widely published in a number of handbooks. Other results, for cases not included in the handbooks, are developed by numerical stress analysis of the cracked geometry and loading conditions. The use of BIE methods for such problems is well developed and will be discussed in detail in the next section.

Three-dimensional SIF results are not as widely available as for 2D problems. The earliest, classical result is by Green and Sneddon [28] for the case of the buried, elliptical plan-form crack. The solution, in terms of the quantities defined in Fig. 3, is given by

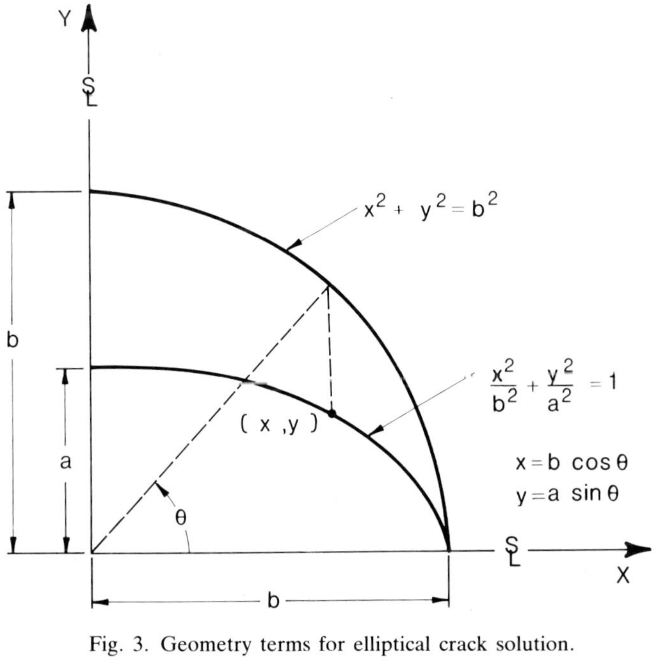

Fig. 3. Geometry terms for elliptical crack solution.

$$K_\mathrm{I} = \frac{\sigma\sqrt{\pi a}}{E(k)}\{\sin^2\theta + a^2/b^2\cos^2\theta\}^{1/4} \qquad (a < b)$$

with

$$E(k) = \int_0^{\pi/2} (1 - k^2 \sin^2\theta)^{1/2}\, d\theta = \sqrt{Q} \tag{1.16}$$

and

$$k^2 = 1 - a^2/b^2 .$$

The elliptic integral in (1.16) is the square root of the shape factor (Q) for elliptical cracks and is often used as a normalizing parameter for surface crack results. An extensive set of finite element numerical results for surface cracks in plates under tension and bending was curve fit by Newman and Raju [49]. Their results are reproduced below

$$\left.\begin{aligned}
&K_\mathrm{I} = (S_\mathrm{t} + HS_\mathrm{b})(\pi a/Q)^{1/2}F(a/t, a/c, c/b, \phi)\\
&Q = 1 + 1.464(a/c)^{1.65} \qquad (a/c \leq 1)\\
&F = [M_1 + M_2(a/t)^2 + M_3(a/t)^4]f_\phi g f_w\\
&M_1 = 1.13 - 0.09(a/c)\\
&M_2 = -0.54 + \frac{0.89}{0.2 + (a/c)}\\
&M_3 = 0.5 - \frac{1.0}{0.65 + (a/c)} + 14\left(1.0 - \frac{a}{c}\right)^{24}\\
&g = 1 + [0.1 + 0.35(a/t)^2](1 - \sin\phi)^2\\
&f_\phi = [(a/c)^2\cos^2\phi + \sin^2\phi)]^{1/4}\\
&f_w = [\sec\{(\pi c/2b)(a/t)^{1/2}\}]^{1/2}\\
&H = H_1 + (H_2 - H_1)\sin^p\phi\\
&p = 0.2 + a/c + 0.6a/t\\
&H_1 = 1 - 0.34a/t - 0.11(a/c)(a/t)\\
&H_2 = 1 + G_1(a/t) + G_2(a/t)^2\\
&G_1 = -1.22 - 0.12(a/c)\\
&G_2 = 0.55 - 1.05(a/c)^{0.75} + 0.47(a/c)^{1.5} .
\end{aligned}\right\} \tag{1.17}$$

In (1.17), t is the section thickness, a the crack depth, c the half-crack surface length, ϕ the elliptic angle (θ in Fig. 3), b the plate half width, Q the shape factor (1.16), $H = E$, or $E/(1-\nu^2)$, and S_t, S_b the remote applied tensile and bending stresses. These results generally agree, for shallow cracks in tension, with the earlier BIE results of Cruse and Meyers [22], but are more comprehensive and are of great engineering utility.

Extensive collections of stress intensity factors have been published in handbook form by Tada [66] and Cartwright and Rooke [11]. Most of these solutions are for two-dimensional problems, while Newman and Raju [49] should be used for surface cracks. Very powerful methods for constructing K-solutions from the uncracked stress fields have become popular, including the weight function method as summarized by Besuner [6] and the influence function method, also called the Green's function method by Cartwright and Rooke [10]. In the latter approach, K-solutions for point loads on the two-dimensional crack surface are superimposed to match the uncracked stresses. The weight function method uses an integral dependent on the crack displacement solution for a reference problem, operating on the uncracked stress field. Besuner [6] has been a developer of this methodology along with Cruse [21] and uses the BIE method to calculate the weight function.

2. Two-dimensional fracture mechanics modeling

2.1. *Overview*

There are various numerical solution methods for the 2D LEFM problem including finite elements, collocation, and boundary integral equations (BIE). Of these, finite element methods are the most widely used due to the wide availability of general purpose computer programs. The greatest problem in finite element modeling is the need to capture the singularity in the crack tip behavior. A substantial number of specialized interpolation methods have been developed for the finite element method including special crack tip elements [9], [74], quadratic, isoparametric elements with "quarter-point" mapping by Barsoum [5], and a variety of hybrid (mixed variational principle) formulations [73].

While each of these approaches has resulted in considerable improvement over non-singular modeling, they suffer from the following common limitations. The size of the elastic singularity field is not known, a priori, for a particular problem, and the singular term is not a complete representation of the field solution of Eq. (1.5). The extent of the singular fields is set by the user in the selection of the modeling grid, and the model is rarely tested by trying different mesh sizes or transition region sizes for the singular region.

The easiest engineering approach is to use the quadratic isoparametric element with quarter-point mapping, and as fine a grid as possible. The use of transition regions for the singularity improves the results, but at the expense of greater modeling complexity. As will be seen in the next section, the quarter-point mapping approach still serves as the basis of both finite element and BIE models for 3D cracks.

Collocation is the earliest development of 2D solution methods of very high accuracy. The approach is generally to take the Williams' [78] solution in Eq. (1.5) for the eigenvalue problem with a large number of terms. The coefficients in (1.5) are determined by a least-squares fitting of the eigenfunction expansions for stresses with the applied loading along the physical, non-cracked boundary. The collocation results of Gross and Srawley [31], Newman [47], and Barone and Robinson [4] demonstrate the efficacy of this approach. However, the method is somewhat sensitive to the technique of boundary data matching (often, over-determined systems are needed), and collocation is not an engineering design tool of the generality of finite element and boundary element methods. Nikooyeh and Robinson [50] have merged collocation and BIE methods to generate improved numerical solutions for geometries with severe notches. Stern and Kim [65] have applied collocation to the problem of the cracked plate in flexure. Another collocation method has been developed based on higher-order singularities than those of Williams [78]. Quinlan has been a long-term proponent of the edge function method and has recently published an approach to the three-dimensional fracture mechanics problem [55]. The edge function approach in [55] uses advanced analytical solution functions for the loaded elliptical crack. The functions are superimposed to satisfy the boundary data in a least-squares sense.

A number of different approaches for boundary element models have been developed. These include non-singular, extrapolation modeling by Cruse and Van Buren [17] and Cruse [18, 19], crack tip singular interpolation functions of Xanthis et al. [79], Watson [76], Atkinson et al. [2], and Van der Weeën [75]; quarter-point modeling by Martinez and Dominguez [41], and the special Green's function approach of Snyder and Cruse [62] and Cruse [24], Rudolphi and Ashbaugh [58], Marjoria and Mukherjee [42, 45, 46]. The remainder of this section will focus on the limitations of the standard BIE formulation for fracture mechanics problems, followed by the development of the special Green's function formulation for the straight, stress-free crack.

2.2. *Special Green's function formulation*

The usual 2D BIE, written using the notation of Cruse [24] for the geometry in Fig. 4 is given by

$$C_{ij}u_j(\mathrm{P}) + \int_{S+\Gamma^+ +\Gamma^-} T_{ij}(\mathrm{P},\mathrm{Q})u_j(\mathrm{Q})\,\mathrm{d}S = \int_{S+\Gamma^+ +\Gamma^-} U_{ij}(\mathrm{P},\mathrm{Q})t_j(\mathrm{Q})\,\mathrm{d}S \tag{2.1}$$

where $\mathrm{P}\in S$, and $\mathrm{Q}\in S+\Gamma$. The crack surface Γ is represented as a narrow slit in Fig. 4 with upper and lower surfaces slightly separated. The fundamental problem

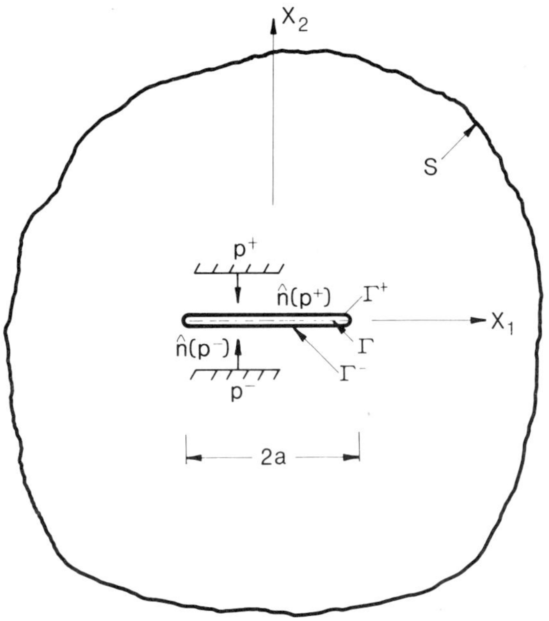

Fig. 4. Flat crack modeling terms definition.

in using (2.1) for fracture mechanics applications is seen as $\Gamma^+, \Gamma^- \to \Gamma$. The usually posed problems are associated with $t_j(Q^+) = -t_j(Q^-)$; for this case the integral of the tractions in (12) can be seen to become zero, as $U_{ij}(P, Q^+) = U_{ij}(P, Q^-)$.

As discussed in more detail by Cruse [24], the usual BIE (Eq. (2.1)) degenerates for the flat crack and is not a valid basis for numerical modeling. This degeneracy is fundamentally linked to the ill-posed nature of problems with two coplanar surfaces.

The models in Fig. 5 illustrate the approaches used to circumvent the above state degeneracy. For the symmetric center-cracked example in Fig. 5(a), the degeneracy occurs if the model in Fig. 5(b) is attempted. Figure 5(c) models the crack as an open notch, resulting in a loss of crack tip modeling accuracy. Figure 5(d) is the approach used almost exclusively for a direct implementation of (2.1) and is clearly limited to symmetric geometries.

Geometrically defined Green's functions have been widely used in potential theory (Greenberg [29]) as a means for analytical solutions for limited classes of geometries. Such applications are essentially limited to geometrically/topologically simple two-dimensional shapes. The broad use of complex variable mapping functions for two-dimensional potential theory and elasticity solutions is linked in a fundamental, mathematical sense to the Green's function methodology.

The essential idea behind the use of Green's function is to modify the kernel functions in the potential representation or the BIE (Eq. (2.1)) so as to

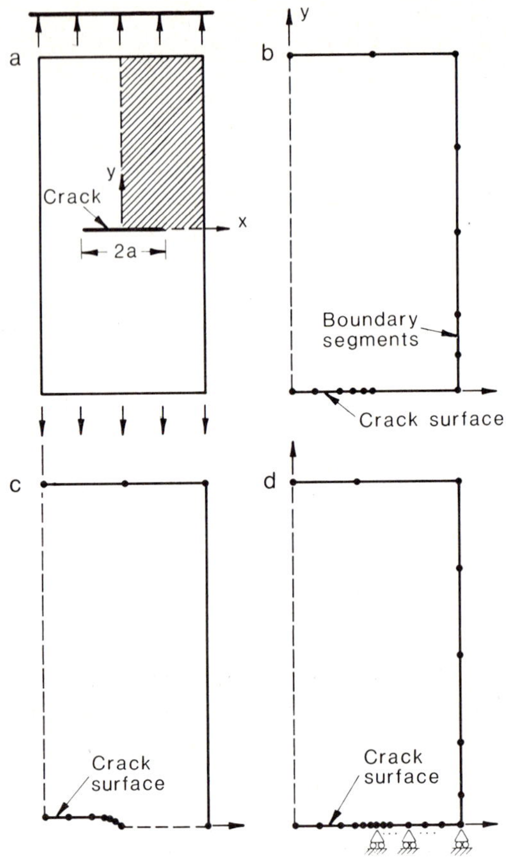

Fig. 5. Flat crack modeling approaches.

identically reduce the unknown boundary data terms to a null condition. While such approaches are fundamentally linked to simple geometric shapes, the essential benefit can be obtained for (2.1) in the case of fracture mechanics problems.

Consider the case of an anisotropic, elastic body containing a single, traction-free crack of length $2a$. Let Γ represent the crack surface and the regular surface, as in Fig. 1.

The basic BIE formulation is sufficiently well published that it need not be repeated herein. The essential element in the formulation is the use of the traction and displacement solutions for the Kelvin problem of an infinite body internally loaded by discrete forces at $p(x)$. The work of Snyder and Cruse [62] modified this by including an additional condition on the infinite region: that it contain a straight, traction-free crack of length $2a$ at an arbitrary location and orientation. Thus, for the physical problem of modeling a finite body with an internal or edge crack, the BIE may be written, again using the notation of Cruse [24].

$$C_{ji}u_i(\mathrm{P}) + \int_S T^*_{ji}(\mathrm{P},\mathrm{Q})u_i(\mathrm{Q})\,\mathrm{d}S + \int_\Gamma T^*_{ji}(\mathrm{P},\mathrm{Q})u_i(\mathrm{Q})\,\mathrm{d}S$$

$$= \int_S U^*_{ji}(\mathrm{P},\mathrm{Q})t_i(\mathrm{Q})\,\mathrm{d}S + \int_\Gamma U^*_{ji}(\mathrm{P},\mathrm{Q})t_i(\mathrm{Q})\,\mathrm{d}S\,. \tag{2.2}$$

The kernels U^*_{ij} and T^*_{ij} represent the corresponding Kelvin solution tensors for displacement and traction, respectively.

The formulation of Snyder and Cruse [62] was for anisotropic materials and determined that the displacement field corresponding to the point load in an infinite cracked plate is given by

$$U_{ji}(\mathrm{P},\mathrm{Q}) = 2\,\mathrm{Re}\Big\{\sum_{k=1}^{2}\Big[P_{ik}A_{jk}\log(z_k - c_k) + \frac{P_{ik}}{2(\mu_k - \mu_l)}$$

$$\times\{(\mu_l - \mu_k)A_{jk}J(z_k, c_k) + (\mu_l - \bar{\mu}_l)\bar{A}_{jl}J(z_k, \bar{c}_l)$$

$$+ (\mu_l - \bar{\mu}_k)\bar{A}_{jk}J(z_k, \bar{c}_k)\}\Big]\Big\}\,; \qquad l = 3 - k\,. \tag{2.3}$$

Equation (2.3) is written in the form of complex variables c_k, z_k corresponding to the physical points p(x) and q(x). These variables map the physical coordinates according to $z_k = x + \mu_k y$ where μ_k denotes the corresponding root to the fourth-order material characteristic direction written in terms of the material compliances (β_{ij}) as given in Eq. (1.12).

The constants P_{ik} and A_{jk} in Eq. (2.3) are material constants given by Snyder and Cruse [62]. The logarithm in Eq. (2.3) represents the point load solution for the uncracked body. All of the remaining terms serve to represent the presence of the traction-free crack. The principal function in these equations is $J(z, c)$ given, following Cruse [24], by the following

$$J(z,c) = \pi\log\left(2\,\frac{(z^2 - a^2)^{1/2}(c^2 - a^2)^{1/2} + cz - a^2}{[z + (z^2 - a^2)^{1/2}][c + (c^2 - a^2)^{1/2}]}\right). \tag{2.4}$$

The traction kernel in Eq. (2.2) contains derivatives of the $J(z, c)$ function given as

$$\frac{\partial J}{\partial z} = \pi\left(\frac{(z^2 - a^2)^{1/2} - z - (c^2 - a^2)^{1/2} + c}{(z - c)(z^2 - a^2)^{1/2}}\right). \tag{2.5}$$

The internal strain identity for the BIE formulation is derived from the Somigliana identity for the interior displacements, an equation of the same form as (2.2) except with $C_{ji} = \delta_{ji}$. Following the usual differentiation of $u_j(\mathrm{p})$ in (1) the interior strain is given by

$$\varepsilon_{ij}(\mathrm{p}) = \int_S S^*_{kij}(\mathrm{p}, \mathrm{Q}) t_k(\mathrm{Q})\, \mathrm{d}s + \int_S D^*_{kij}(\mathrm{p}, \mathrm{Q}) u_k(\mathrm{Q})\, \mathrm{d}s\,. \tag{2.6}$$

In Eq. (2.6), the kernel functions are merely the symmetric displacement gradient derivatives of the U^*_{ij} and T^*_{ij} kernels with respect to the field point $\mathrm{p}(x)$. These kernels contain derivatives of the crack terms corresponding to

$$\partial J/\partial c = \pi\left(\frac{(c^2 - a^2)^{1/2} - (z^2 - a^2)^{1/2} + z - c}{(c - z)(c^2 - a^2)^{1/2}}\right) \tag{2.7}$$

$$\partial^2 J/\partial z \partial c = \pi\left(\frac{(c^2 - a^2)^{1/2}(z^2 - a^2)^{1/2} - cz + a}{(z - c)^2(z^2 - a^2)^{1/2}(c^2 - a^2)^{1/2}}\right). \tag{2.8}$$

Of particular concern is the behavior of the expressions in Eqs. (2.7)–(2.8) as c approaches the crack tips $(+/-a)$. It has been found that in Eqs. (2.7)–(2.8) both functions explicitly contain the square-root of crack tip distance ($\rho = |c - a|$). Thus, Snyder and Cruse [62] found it possible to explicitly solve for the crack tip stress intensity factors K_{I}, K_{II} which are proportional to the limit of the product of strain at $\mathrm{p}(x)$ times $(c^2 - a^2)^{1/2}$, as c approaches the crack tips. The form of this result is given from Snyder and Cruse as

$$K_{\mathrm{I,II}} = -\int_S R_i^{\mathrm{I,I}}(\mathrm{Q}) u_i(\mathrm{Q})\, \mathrm{d}S + \int_S L_i^{\mathrm{I,II}}(\mathrm{Q}) t_i(\mathrm{Q})\, \mathrm{d}S\,. \tag{2.9}$$

The benefits deriving from the use of this Green's function approach are elimination of the need to model the crack, and the direct computation of K_{I}, K_{II} from boundary data using (2.9). This latter result was also developed, using a different analytical approach by Stern et al. [64]. Soni and Stern [63] extended their work to the orthotropic material.

Other authors have used special Green's functions as means for developing BIE or related formulations for the half-space [69], elliptical slits [42–45], and circular cut-outs [46]. These latter two papers concern inelastic response analysis. Rudolphi and Ashbaugh [58] solve the two-dimensional isotropic problem using much the same approach as Snyder and Cruse [62], but without the direct addition of the two solutions in the kernel functions used in (2.2). Rather, they use the Green's function solution for the loaded crack to establish additional boundary loading solutions associated with the crack and superimpose these with the regular BIE. The solution of both systems of equations is, while numerically implemented, roughly equivalent to (2.2).

The enhanced accuracy of the Green's function approach is often worth the greater complexity associated with coding the augmented kernel functions. While these approaches are strongly similar to collocation methods, they benefit from the region of the basic BIE formulation. That is, the stability, conditioning and model insensitivity of the BIE method over the collocation method are retained for the cases of special Green's function formulations.

As an example of the general modeling capability of the special Green's function approach, consider the problem of a cracked balance flange for the rotating wheel shown in Fig. 6. The illustrated flange has been machined to various notch conditions as shown in Fig. 7. Three levels of strain concentration severity were modeled using the standard BIE formulation (2.1) with linear interpolation of the boundary data.

The three BIE meshes are illustrated in Fig. 7. A section of the balance flange is modeled as a plane stress region subject to a uniform, applied strain imposed by tractions on the two ends and the lower surface. This uniform strain model simulates the interaction of the flange and wheel by approximating the stiffness of the flange as zero, relative to that of the wheel.

Strain concentrations are obtained from the boundary solution and are plotted in Fig. 8. In order to compute the stress intensity factors for cracks due to these high strains, it was first necessary to analyze the uncracked geometry. The fracture mechanics analyses were performed by specifying the boundary node where the crack is to be placed, as well as the size and orientation of the crack.

The BIE boundary meshes for the fracture mechanics analyses are identical to those in Fig. 7, for the uncracked case. This, of course, derives from the use of the special Green's function which adds the full field solution for the crack to the kernel functions. The results in Fig. 9 show that the cracks reach a peak value of stress intensity factor before decreasing slightly as the compliance increase for the longer crack depths reduces the crack tip driving force. The additional solutions for different crack lengths are obtained by the change of a single field (crack length) in the input data file.

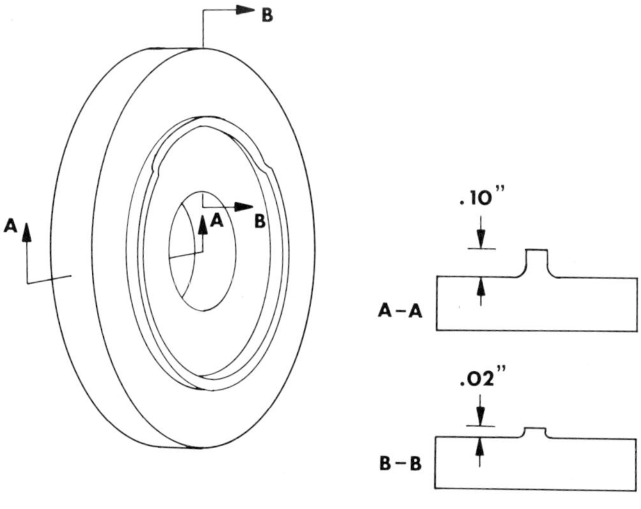

Fig. 6. Rotating wheel with balance flange.

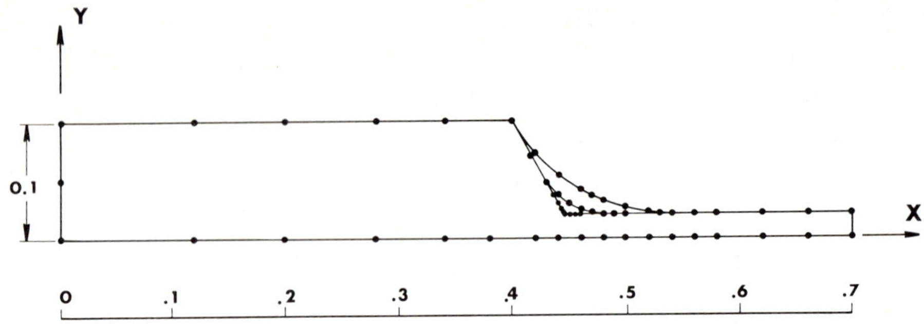

Fig. 7. BIE meshes for three notch geometries.

Figure 10 compares the meshes used for the fracture mechanics analysis of an edge notched specimen with an edge crack from the notch. Again, the BIE model for the crack is the same boundary mesh as for the uncracked geometry. The finite element map is based on eight-noded isoparametric elements, with the quarter-point mapping used for the crack tip elements.

The two solution meshes were drawn up by experienced users to obtain comparable results for K_I. Table 1 shows that the comparisons indeed were close to each other. The computer time shows some advantage to the BIE computer code BIE/CRX, where CRX phonetically connotes the cracks solution capability

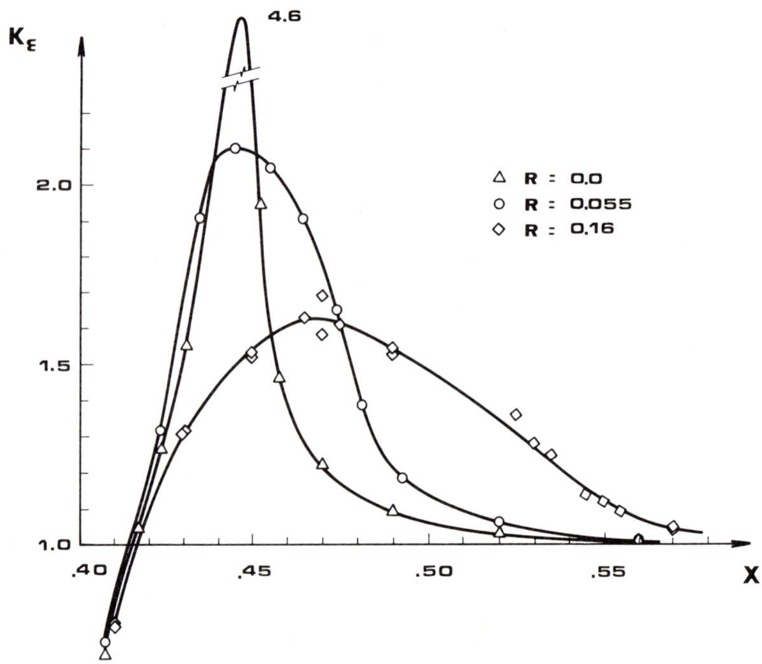

Fig. 8. Notch strain concentration factor results.

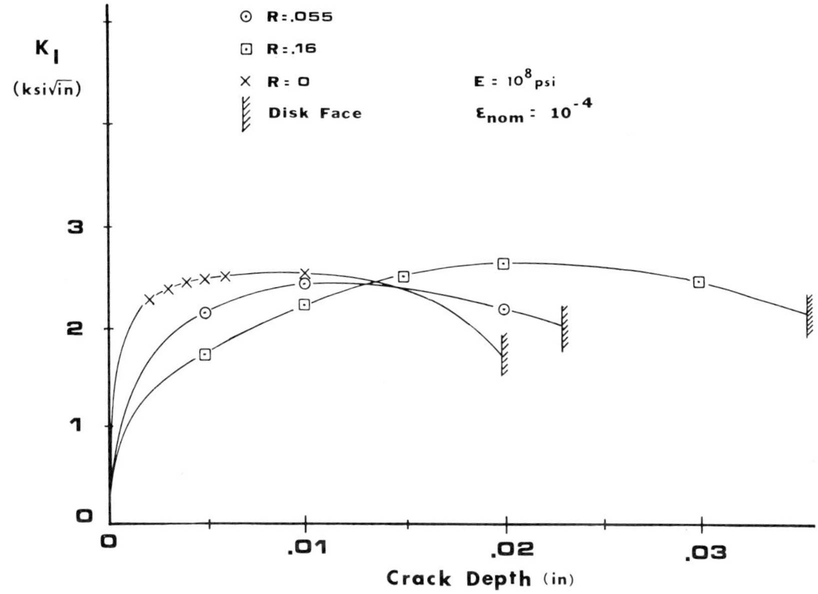

Fig. 9. Stress intensity factors for notch cracks.

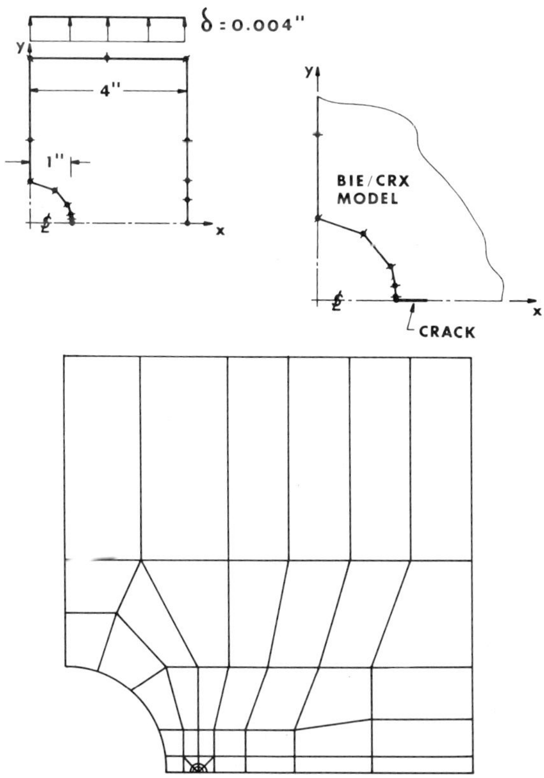

Fig. 10. Comparison of BEM/FEM fracture mechanics model.

Table 1
BIE versus FEM analysis

	BIE	FEM
K_{I} (ksi$\sqrt{\text{in.}}$)	23.2	24.0
Solution time (s)	23.0	41.0

of the special Green's function formulation. The real advantage in BIE/CRX for engineering purposes is in the reduced engineering time necessary to model the geometry. The relative advantage for the first mesh was about 4:1 in favor of the BIE mesh. Additional crack length models require essentially no additional engineering time for BIE/CRX, while each crack length requires a new finite element mesh.

2.3. Displacement discontinuity method

As discussed in the previous section, the basic BIE formulation for elasticity problems has an inherent mathematical degeneracy for the case of crack problems. The current section will amplify upon this limitation, formulate a higher-order (traction) integral equation without the limitation, and discuss a general numerical implementation of the displacement discontinuity method of Crouch [14]. This approach will be further discussed for three-dimensional fracture mechanics problems at a later point in this chapter.

An extensive discussion of the displacement discontinuity method is given by Crouch and Starfield [15]. Their work focuses on the geomechanics problems associated with modeling fractured rock. Further application of Crouch's formulation to soils was given by Cornet [13], and to crack problems by Ishijima et al. [36]. In all cases, the displacement discontinuities are taken as piecewise constant, resulting in the simplest numerical implementation.

The geometry of concern is again that shown in Fig. 4, where Γ may now be referred to as the surface of displacement discontinuity. In general, Γ may be a segment of a much larger, curved line of displacement discontinuity. The basic BIE for this geometry was given by (2.1). The stress quadrature relationship for the usual BIE formulation at an arbitrary *interior* point is then given, following Cruse, [24] as

$$\sigma_{ij}(\mathrm{p}) = \int_S D_{kij}(\mathrm{p},\mathrm{Q})t_k(\mathrm{Q})\,\mathrm{d}S - \int_S S_{kij}(\mathrm{p},\mathrm{Q})u_k(\mathrm{Q})\,\mathrm{d}S + \int_\Gamma D_{kij}(\mathrm{p},\mathrm{Q})\Sigma t_k(\mathrm{Q})\,\mathrm{d}S - \int_\Gamma S_{kij}(\mathrm{p},\mathrm{Q})\Delta u_k(\mathrm{Q})\,\mathrm{d}S\,. \tag{2.10}$$

In (2.10) the terms on Γ are written by the convention that $\Delta u_j = u_j^+ - u_j^-$, $\Sigma t_j = t_j^+ + t_j^-$.

To formulate the basic relations of the displacement discontinuity method let the outer surface S be remote and unloaded, and let $\Sigma t_j \equiv 0$. Then, (2.10) reduces to the form

$$\sigma^0_{ij}(\mathrm{p}) = -\int_\Gamma S_{kij}(\mathrm{p}, \mathrm{Q}) \Delta u_k(\mathrm{Q})\, \mathrm{d}S\,. \tag{2.11}$$

The kernel $S_{kij}(\mathrm{p}, \mathrm{Q})$ in (2.11) has a fundamental singularity for $\mathrm{p} \to \mathrm{P} \in \Gamma$ which precludes, in its present form, the use of (2.11) as a boundary integral equation. However, Cruse [20], Guidera and Lardner [32], Bui [7], Weaver [77], Balas and Sladek [3], Theocaris [71], and Putot [53, 54] determined that by a proper integration by parts, the fundamental singularity in (2.10) can be reduced by one order such that, in three dimensions (x_1, x_2 in the plane of the crack; x_3 normal to the crack)

$$\begin{aligned}\sigma^0_{13}(\mathrm{p}^+) &= [\sigma^0_{13}(\mathrm{p}^+) + \sigma^0_{13}(\mathrm{p}^-)]/2 \\ &= \frac{\mu}{4\pi(1-\nu)} \left[\int_\Gamma \left((\Delta u_{1,1} + \Delta u_{2,2}) \frac{r_{,1}}{r^2} + (1-\nu)(\Delta u_{1,2} - \Delta u_{2,1}) \frac{r_{,2}}{r^2} \right) \mathrm{d}S \right]\end{aligned} \tag{2.12}$$

$$\sigma^0_{23}(\mathrm{p}^+) = \frac{\mu}{4\pi(1-\nu)} \left[\int_\Gamma \left((\Delta u_{1,1} + \Delta u_{2,2}) \frac{r_{,2}}{r^2} + (1-\nu)(\Delta u_{2,1} - \Delta u_{1,2}) \frac{r_{,1}}{r^2} \right) \mathrm{d}S \right] \tag{2.13}$$

$$\sigma^0_{33}(\mathrm{p}^+) = \frac{\mu}{4\pi(1-\nu)} \left[\int_\Gamma \left(\Delta u_{3,1} \frac{r_{,1}}{r^2} + \Delta u_{3,2} \frac{r_{,2}}{r^2} \right) \mathrm{d}S \right]. \tag{2.14}$$

The referenced articles provide differing forms for (2.12)–(2.14). These differences can be traced to the non-unique integration-by-parts process.

For the case of two-dimensional dislocation fields, (2.12)–(2.14) can be further integrated to remove dependency on the x_2 (thickness) dimension. Taking the stresses to be independent of x_2, and performing the x_2 integration on the σ^0_{13} and σ^0_{33} terms in (22), we obtain (for $x_3 = 0$)

$$\sigma^0_{i3}(x_1, x_3 = 0) = \frac{\mu}{2\pi} \frac{1}{1-\nu} \int_{-a}^{a} \frac{1}{x_1 - \xi_1} \Delta u_{i,1}\, \mathrm{d}\xi_1\,. \tag{2.15}$$

The result in (2.15) may be interpreted for any interior point or for the point taken to the crack surface Γ. In this latter case, (2.15) becomes an integral equation for the dislocation distribution for an arbitrary state of crack surface applied stress. This form is a fundamental result also obtained using transform methods and has the well known solution for the pressure loaded crack ($\sigma_{13} = 0$) with crack opening displacement v given by

$$\frac{\mu}{\pi}\frac{1}{1-\nu}\frac{\partial v}{\partial x_1} = \frac{1}{\pi^2}\left(\frac{a+x_1}{a-x_2}\right)\int_{-a}^{a}\left(\frac{a-\xi_1}{a+\xi_1}\right)^{1/2}\frac{\sigma_{33}(\xi_1)\,\mathrm{d}\xi_1}{x_1-\xi_1} - \frac{\mathrm{i}C_1}{(a^2-x_1^2)^{1/2}}\,, \tag{2.16}$$

where C_1 is a constant of integration, taken to eliminate a free term from the integration. For the case of uniform crack pressure p, (2.16) reduces to the solution given by

$$\mu \Delta v_3 = 2(1-\nu)pa(1-(x/a)^2)^{1/2}\,. \tag{2.17}$$

Displacement discontinuity models are obtained, following Crouch and Starfield [15], by first reintegrating (2.15) by parts with the result (for $x_3 \neq 0$)

$$\sigma^0_{33}(x_1, x_3) = \frac{\mu}{2\pi}\frac{1}{1-\nu}\int_{-a}^{a}\Delta u_3(\xi_1)\frac{1}{(x_1-\xi_1)^2+x_3^2}\,\mathrm{d}\xi_1\,. \tag{2.18}$$

Letting $\Delta u_3(\xi_1)$ be constant over the surface of discontinuity, we obtain the basic result

$$\sigma^0_{33}(x_1, 0) = -\frac{a\mu}{\pi(1-\nu)}D_3\frac{1}{x_1^2-a^2} \tag{2.19}$$

where $D_3 = \Delta u_3$, a constant. Crouch and Starfield [15] give the comparable results for the shear loaded discontinuity.

In the displacement discontinuity method, the boundary element model is obtained by the superposition of such "fundamental solutions" as (2.19) over the entire crack surface. For the general problem, including the curved crack surface, the superposition process must include in the traction conditions the applied stress as well as the correct stresses at the local surface due to the displacement discontinuities for all of the other surface patches.

The application of this model has been shown by Crouch and Starfield [15] to generate the essential crack opening behavior for the reference problem of a center cracked panel. However, it does not generate accurate solutions near the crack tip. The approach taken is to introduce a special crack tip solution for (2.6) by letting $\Delta u_3 = D_3(\xi_1/a)^{1/2}$. This results in a significant, but not sufficient, increase in accuracy. A more general extension of the displacement discontinuity method for 3D crack problems will be presented in the next section of this chapter.

The multi-region modeling approach was used by Smith and Mason [61] for modeling curved cracks. Traction singularity elements were used for numerical accuracy. Their work was extended [62] to the prediction of crack growth direction by computing the next term in the stress expansion near the crack tip (see Eq. (1.5)), and including this non-singular stress in the criterion. Luchi and Poggialini [40] used multi-region modeling for the three-dimensional problem.

2.4. *Inelastic fracture mechanics modeling*

The inelastic fracture mechanics problem using BEM methods is still largely under development and is thus considered to be beyond the scope of this review. Most of the contributions have been by Mukherjee and his co-workers [42–46] with more recent work by Cruse and Polch [26]. Mukherjee [43] and Telles [70] published excellent reviews of the general inelastic solution methods problem.

Morjaria and Mukherjee used the special Green's function modeling approach for the various ellipse as an approximation to the crack [42, 45] and the circular cut-out [46]. This work is summarized by Mukherjee and Banthia [44]. The formulations are based on viscoplastic, state variable models of material behavior. Inelastic and time-dependent response ahead of the "crack" is computed, and in many cases compared with the finite element results. Their BEM formulations are based on solutions of the biharmonic elasticity equations, rather than the Navier equations.

Cruse and Polch [26] have formulated the classical elastoplasticity problem with BIE, using the special Green's function approach described in section 2.2. While these results are relatively fresh, some new insights into the elastoplastic crack tip response have been generated. In particular, their formulation leads to a modification to Eq. (2.9) that includes a volumetric integral of inelastic strains. This term shows great promise in modeling problems of elastic cracks in residual (inelastic) strains of notch plasticity or welding, as well as for thermal gradient strains.

2.5. *Multi-region and singularity modeling approaches*

A brief comment on the multi-region method of Blandford et al. [8] is in order. In order to avoid the flat crack formulation degeneracy described above, these authors have modeled the crack body as sets of regions, linked at the material interface ahead and behind the crack. Thus, the two crack surfaces are formally single surfaces, but of two different regions. No degeneracy occurs in this way, but physical surfaces linking the modeled regions are required. Blandford et al. [8] use singular and singular-transition interpolation functions to model the displacements and tractions along these interface surfaces. The results show good accuracy and are independent of crack or loading symmetry requirements.

The accurate application of the multi-region modeling approach requires special numerical modeling of the crack tip region. Behind the crack tip the crack opening displacements are proportional to the square root of distance from the crack front; ahead of the crack tip the stresses are singular, being inversely proportional to the square-root of the crack tip distance.

The treatment of these problems was first developed for the three-dimensional problem by Cruse and Wilson [23] and is discussed in detail in section 3.2. The approach has been successfully developed by Cheng and Liggett [12] for the special equations for linear porous-elastic response of soils. Martinez and Dominguez [41] review the quarter-point modeling approach for two-dimensional cracks and include a discussion of traction singularity modeling. These were earlier used by Xanthis et al. [79] and Atkinson et al. [2] for the mode III fracture mechanics problem. Watson [76] treats the two-dimensional fracture mechanics problem uniquely through the use of cubic Hermetian interpolation functions.

3. Three-dimensional fracture mechanics modeling

3.1. Overview

The three-dimensional fracture mechanics formulation for boundary element methods is hampered by the lack (at least in closed form) of a suitable Green's function for 3D cracks. Thus, the numerical strategies are limted to direct numerical modeling (symmetric cracks), and to displacement discontinuity modeling. Fortunately, for engineering applications, 3D surface cracks are most nearly always subject only to mode I (symmetric) loading. Thus, the literature is replete with various, effective strategies for mode I modeling.

The two principal competing fracture mechanics modeling strategies are finite elements with singularity interpolation elements, and the finite element alternating method. The most comprehensive set of numerical results using the singularity element strategy for various 3D crack problems is due to Newman and Raju [48, 49] and Raju and Newman [56].

The finite element alternating method of Nishioka and Atluri [52] has recently been established as an extension of the earlier alternating method models of Thresher and Smith [72]. The method relies on the existence of the "closed-form" solution of the buried elliptical crack, loaded by surface stresses of some degree of artibitrariness. The recent developent by Nishioka and Atluri [52] of very general (infinite series form) solutions for the loaded ellipse has greatly enhanced the power of this method.

The following subsection will review the various strategies for direct computation of fracture mechanics results for surface cracks including extrapolation, crack extension/energy release rate, "quarter-point" interpolation, influence function, and polynomial loading function methods. The last subsection will briefly comment on past and current work to establish a general-purpose displacement discontinuity approach to the 3D crack problem.

3.2. Direct numerical modeling

The results of Cruse and Meyers [22] were the first comprehensive numerical study of surface cracks with the BIE formulation. This study used the piecewise linear interpolation of boundary data of Cruse [19] which has no special capability for modeling cracks. Table 2 reproduces the estimated stress intensity factor results from this study for various aspect ratios of buried ellipse geometries. The crack tip stress intensity factor is estimated (1.7, 1.9, 1.11).

A substantial improvement in modeling accuracy for the same models has been obtained by use of the crack extension (strain energy release rate) procedure, as outlined by Cruse and Meyers [22] and demonstrated by Tan and Fenner [67, 68]. In this approach, the elastic strain energy is computed for a given loading and crack size from the boundary work calculation. A second value of elastic strain energy is calculated for the case of a small increase in crack size (e.g., 0.1%). The difference in work done is equal to the negative value of strain energy release rate, as given in (1.1).

In fact, it is possible to estimate the variation in K_I around the crack front using the relationships developed by Cruse and Meyers [22]. Table 2 presents the improved numerical results obtained from the linear BIE code for the crack extension methodology. The first column shows the level of accuracy achieved, based on crack opening displacements, and the second column shows the values obtained using the energy release rate procedure.

Cruse and Wilson [23] demonstrated a significant enhancement in the accuracy of crack-front stress intensity factor calculations through the use of quadratic isoparamctrc shape functions, together with the "quarter-point" geometry interpolation developed first for finite elements by Barsoum [5].

The use of quadratic isoparametric functions for interpolating BIE boundary

Table 2
Comparison of method for opening values of K for 2:1 buried ellipse

θ^{a} (1)	$K^{cod}/K^{exact\ b,c}_{(2)}$ (2)	$K_G/K^{exact\ b,d}_{(4)}$ (4)
0	0.916	0.951
15	0.923	0.958
30	0.923	0.959
45	0.924	0.960
60	0.925	0.961
75	0.929	0.965
90	0.937	0.974

[a]elliptical angle from ellipse major axis
[b]K^{exact} = exact value of K from Eq. (1.16)
[c]K^{cod} = values of K using crack opening displacements
[d]K^{G} = values of K using strain energy release rate procedure

data was introduced by Lachat and Watson [39] and improved by Rizzo and Shippy [57]. These higher-order models, based on effective numerical integration of the singular kernel functions, moved 3D BIE analysis from a research tool to a design tool.

An example of a design application of the 3D BIE stress concentration analysis is given for the idealized compound stress concentration problem shown in Fig. 11. The piecewise linear interpolation model (BINTEQ) is shown in Fig. 12(a) and the piecewise quadratic interpolation model (BASQUE) is shown in Fig. 12(b). The statistics for these nearly comparable models are given in Table 3. The models differ in the number of symmetry planes used. Table 3 clearly illustrates the enhanced efficiency of the quadratic model, while producing an equivalent answer.

To illustrate the "quarter-point" interpolation procedure consider the isoparametric representation for boundary displacements

$$u_j(x_i) = M^\alpha(\xi_i)u_j^\alpha \,. \tag{3.1}$$

In (3.1) the $M^\alpha(\xi_i)$ are the shape functions, while u_j^α are the nodal values of the boundary displacements.

The nodal coordinates are given similarly as

$$x_j = M^\alpha(\xi_i)x_j^\alpha \,. \tag{3.2}$$

Now let the rectangular element adjacent to the crack front have a length (from the crack front) equal to L, and select the quadratic shape functions, $M^\alpha(\xi_i)$, $\alpha = 1, \ldots, 8$. Further, without loss of generality, let the x-axis lie along the crack front and let X_2 be in the plane of the crack and normal to the crack front line. Letting the two mid-side nodes for the sides of the element which are perpendicular to the crack front actually be located at $L/4$, the quarter-point location, the

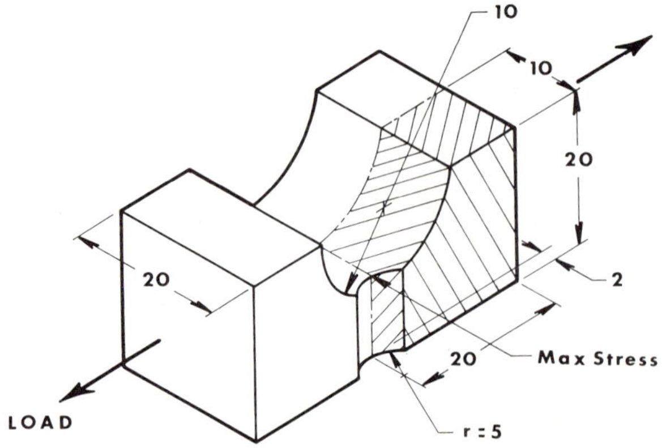

Fig. 11. Compound stress K_T problem due to intersecting slots.

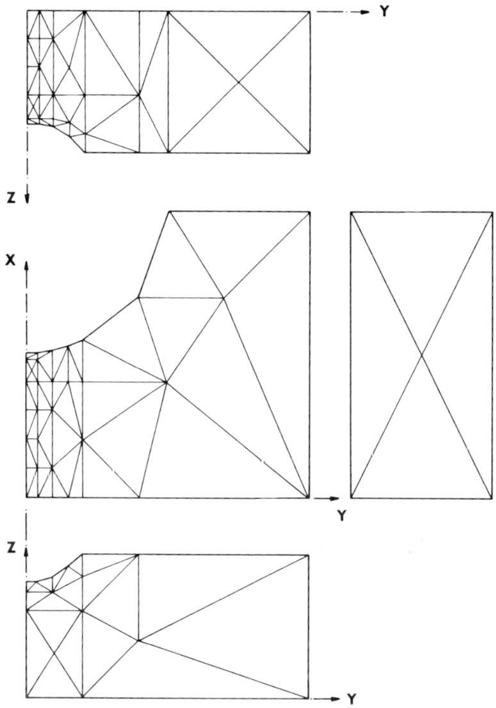

(a)

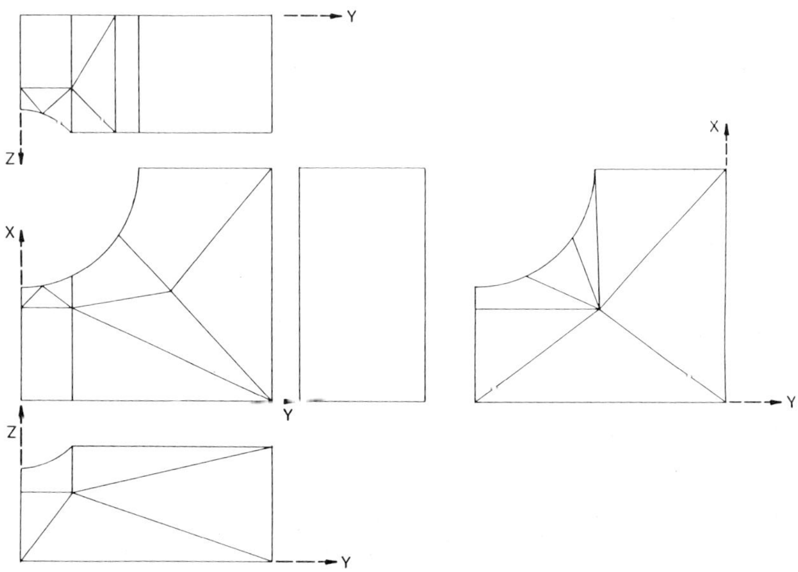

(b)

Fig. 12. Boundary mesh models: (a) piecewise linear interpolation model; (b) piecewise quadratic model.

Table 3
Basque superior to BINTEQ in 3D notch problem

	BINTEQ	BASQUE
Number of elements	115	31
Number of nodes	70	80
Planes of symmetry	2	1
Degrees of freedom	840	480
CPU seconds	4790	785
K_T	7.03	7.31

geometry mapping becomes

$$x_2 = L(\eta - 1)^2/4 \tag{3.3}$$

where η is the second intrinsic coordinate function. Rewriting (3.3), we obtain

$$\eta = 1 + 2(x_2/L)^{1/2} . \tag{3.4}$$

Since the crack surfaces displacements are written from (3.2) in terms of linear and quadratic terms in η, the geometry mapping from (3.4) produces a representation in the form

$$u_j \sim A_1 + A_2(x_2/L)^{1/2} + A_3(x_2/L) + \cdots . \tag{3.5}$$

The representation in (3.5) closely represents the near-crack-tip displacement field, resulting in enhanced modeling accuracy. Figure 13(a) shows a finite element model for a tension specimen with a center cracked geometry with finite thickness as modeled by Newman and Raju [48]. Figure 13(b) shows a crude BIE model of the same geometry, where the "quarter-point" interpolation is used for the crack front elements. The comparison of crack opening displacements in Fig. 14 shows excellent agreement between the two methods. Extensive and very successful use of this modeling approach has been demonstrated by Tan and Fenner [67, 68] and by Heliot et al. [33].

In finite element modeling, the use of the "quarter-point" procedure further results through differentiation in a $(x_2/L)^{-1/2}$ singularity in the near-crack-tip stresses. There is no such differentiation process in BIE modeling. Thus, Cruse and Wilson [23] added a further term to introduce the stress singularity for the traction elements ahead of the crack front. The singularity was achieved by multiplication of the traction shape function by

$$\Phi(x_2) = x_2^{-1/2}[1 - (x_2/L)(1 - L^{1/2})] . \tag{3.6}$$

Since (3.6) is equal to one at the edge of the element at a distance L from the crack front, (3.6) does not require any other changes to the shape functions or

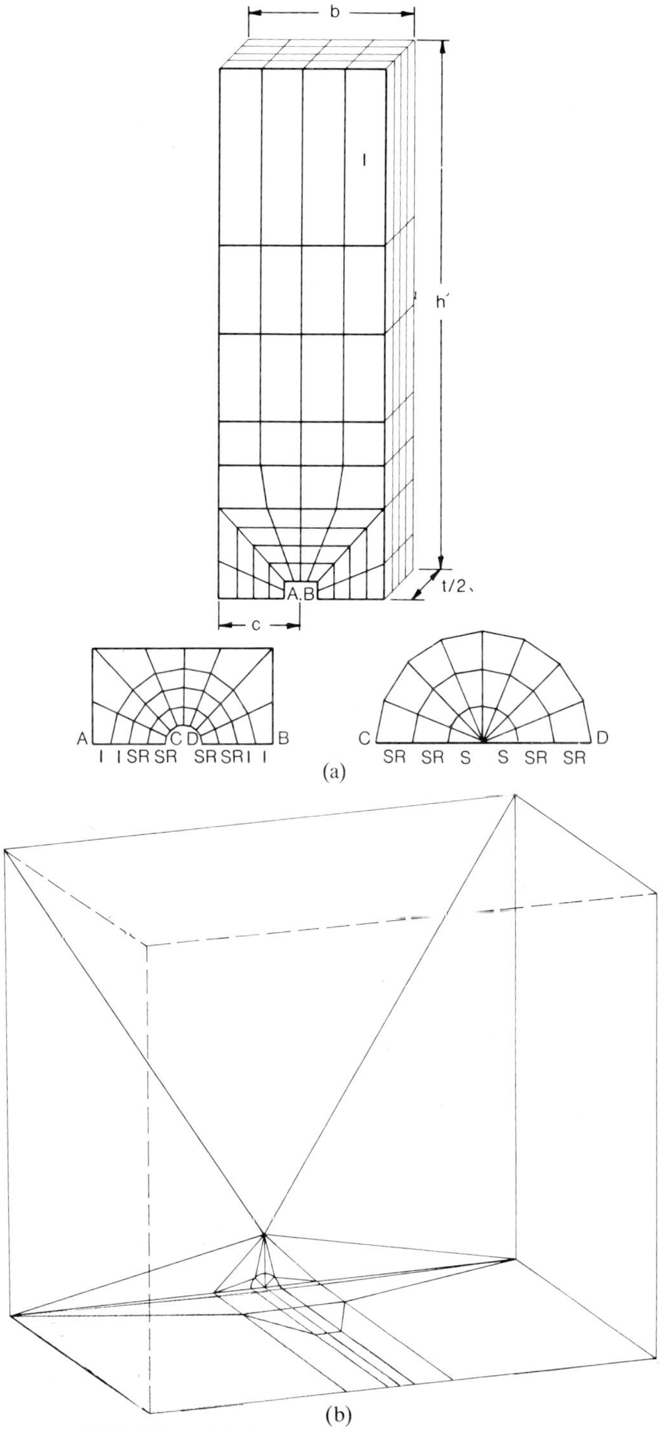

Fig. 13. Comparison of BIE/FE models for center-cracked specimen: (a) Finite Element Model (Ref. [48]); (b) BASQUE Model.

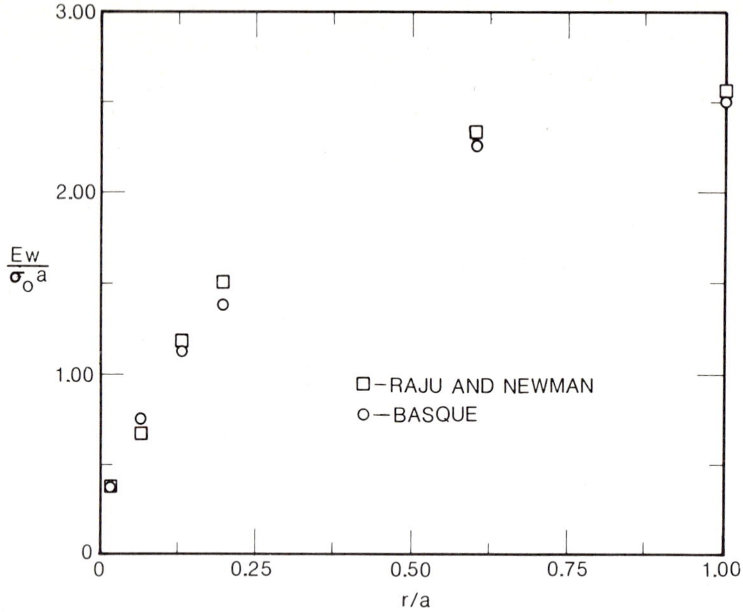

Fig. 14. Crack opening displacement results.

numerical integration. The numerical results by Cruse and Wilson [23] show a further improvement in modeling accuracy.

3.3. Displacement discontinuity method

Guidera and Lardner [32] first established the formulation of the traction integral equation for displacement discontinuities given in (2.12)–(2.14). Cruse [20] had the additional line integral terms necessary to model surface cracks where $\Delta u_j \neq 0$ at the perimeter of the crack. Numerical implementations by Weaver [77] for a rectangular crack and Bui [7] for the circular crack met with some success.

Weaver [77] used (x, y) polynomial series to represent the crack opening displacements and included the crack tip singularities. By using polynomials, Weaver was satisfying the requirement that the displacement gradients in the integral equations be continuous. Further, his approach allowed a direct computation of the stress intensity factor in terms of the polynomial coefficients.

Weaver's and Bui's models were limited to the buried crack in an infinite body. For this case, the traction integral equations are given by

$$\sigma_{33} = \frac{E}{8\pi(1-\nu^2)} \int_{\langle A \rangle} \frac{r_{,\alpha} \Delta u_{3,\alpha}}{r^2} \, \mathrm{d}A \tag{3.7}$$

$$\sigma_{\alpha 3} = \frac{E(1-2\nu)}{16\pi(1-\nu^2)} \int_{\langle A \rangle} \left(\delta_{\beta\alpha} r_{,\gamma} - \delta_{\alpha\gamma} r_{,\beta} + \frac{3 r_{,\alpha} r_{,\beta} r_{,\gamma}}{1-2\nu} \right) \frac{\Delta u_{\beta,\gamma}}{r^2} \, \mathrm{d}A \tag{3.8}$$

where $\alpha, \beta = 1, 2$ represent the in-plane coordinates and x_3 is the direction normal to the crack. The area integrals are to be taken in the sense of the Cauchy principal value, denoted $\langle A \rangle$ in (3.7), (3.8). Cruse [20] includes the terms which augment (3.7), (3.8) for surface cracks in finite bodies. Differences in the form of (3.8) between the various authors result from the non-unique integration-by-parts process and all are equivalent.

Bui [7] used essentially the same terms as Weaver, but approached the formulation in terms of single- and double-layer potentials. Bui analyzes the buried circular crack subject to uniform normal pressure. In his numerical modeling, Bui used piecewise linear variation approximations to the density of the potential, following Cruse [19]. He did not satisfy continuity of the displacement gradients and did not use special crack tip interpolations. Nevertheless, he was able to come within 20% of the exact solution. Balas and Sladek [3] analyzed the same equations as Weaver [77] but also did not satisfy continuity requirements on the crack surface strains.

Putot [53, 54] undertook to expand on and improve the results of Bui. He superimposed potentials to model the intersection of the crack and a free surface. Further, he satisfied displacement gradient continuity in a least-squares sense. His stress intensity factor results were within a few per cent of other solutions. No crack tip singularity modeling was used.

4. Conclusions

Considerable progress has been made in the exploitation of BIE and related BEM methods for fracture mechanics problems since the first numerical solution by Cruse and Van Buren [17]. The most extensive use of BIE methods for engineering design is in 3D analysis of surface cracks for various structural problems (e.g. Heliot et al. [33]). Two-dimensional BIE models have had less impact on design/failure analyses due to the widely available handbook results and the ready availability of finite element codes.

The future emphasis should be placed on developing very general-purpose codes for various shaped cracks, with the least need for user manipulation for reliable results. The two most promising approaches appear to be multi-region modeling (2D) and the traction BIE or displacement discontinuity methods (2D, 3D).

Acknowledgements

The partial support of the Southwest Research Institute Internal Research Panel and the Air Force Office of Scientific Research is gratefully acknowledged. Portions of the current report reflect work being done under AFOSR contract F49620-84-C-0042 with Major David Glasgow, program manager. This manuscript is submitted for publication with the understanding that the United States Government is authorized to reproduce and distribute reprints for governmental purposes.

The author also wishes to thank Ms. Tanya Jackson, Ms. Chris Joyce, and Mrs. Cathy Dean for their excellent assistance in the preparation of this manuscript.

References

[1] C. Atkinson, Fracture Mechanics Stress Analysis, in: *Progress in Boundary Element Methods*, Vol. 2, C.A. Brebbia, Ed. (Pentech Press, London, 1983) pp. 53–100.

[2] C. Atkinson, L.S. Xanthis and M.J.M. Bernal, Boundary Integral Equation Crack-Tip Analysis and Applications to Elastic Media With Spatially Varying Elastic Properties, *Computer Methods in Applied Mechanics and Engineering* **29**, 35–49 (1981).

[3] J. Balas and J. Sladek, Method of Boundary Integral Equations for Analysis of Three Dimensional Crack Problems, in: *Boundary Element Methods*, C.A. Brebbia, Ed. (Springer-Verlag, Berlin, 1981) pp. 183–205.

[4] M.R. Barone and A.R. Robinson, Determination of Elastic Stresses at Notches and Corners by Integral Equations, *International Journal of Solids and Structures* **8**, 1319–1338 (1972).

[5] R.S. Barsoum, On the Use of Isoparametric Finite Elements in Linear Fracture Mechanics, *International Journal for Numerical Methods in Engineering* **10**, 25–37 (1976).

[6] P.M. Besuner, The Application of the Boundary-Integral Equation Method to the Solution of Engineering Stress Analysis and Fracture Mechanics Problems, *Nuclear Engineering and Design* **43**, 161–173 (1977).

[7] H.D. Bui, An Integral Equations Method for Solving the Problem of a Plane Crack of Arbitrary Shape, *Journal of the Mechanics and Physics of Solids* **25**, 29–39 (1977).

[8] G.E. Blandford, A.R. Ingraffea and J.A. Liggett, Two-Dimensional Stress Intensity Factor Computations Using the Boundary Element Method, *International Journal for Numerical Methods in Engineering* **17**, 387–404 (1981).

[9] E. Byskov, The Calculation of Stress Intensity Factors Using the Finite Element Method with Cracked Elements, *International Journal of Fracture Mechanics* **6**, 159–167 (1970).

[10] D.J. Cartwright and D.P. Rooke, Green's Functions, in: *Fracture Mechanics: Current Status, Future Prospects*, R.A. Smith, Ed. (Pergamon Press, Oxford, 1979) pp. 91–123.

[11] D.P. Rooke and D.J. Cartwright, *Compendium of Stress Intensity Factors* (Her Majesty's Stationery Office, London, 1976).

[12] A.H.D. Cheng and J.A. Liggett, Boundary Integral Equation Method for Linear Porous-Elasticity With Applications to Fracture Propagation, *International Journal for Numerical Methods in Engineering* **20**, 279–296 (1984).

[13] F.H. Cornet, Comparative Analysis by the Displacement-Discontinuity Method of Two Energy Criteria of Fracture, *Journal of Applied Mechanics* **46**, 349–355 (1979).

[14] S.L. Crouch, Solution of Plane Elasticity Problems by the Displacement Discontinuity Method, *International Journal for Numerical Methods in Engineering* **10**, 301–343 (1976).

[15] S.L. Crouch and A.M. Starfield, *Boundary Element Methods in Solid Mechanics* (George Allen and Unwin, London, 1983).

[16] T.A. Cruse, Numerical Solutions in Three-Dimensional Elastostatics, *International Journal of Solids and Structures* **5**, 1259–1274 (1969).

[17] T.A. Cruse and W. Van Buren, Three-Dimensional Elastic Stress Analysis of a Fracture Specimen with an Edge Crack, *International Journal of Fracture Mechanics* **7**, 1–15 (1971).

[18] T.A. Cruse, Numerical Evaluation of Elastic Stress Intensity Factors by the Boundary-Integral Equation Method, in: *The Surface Crack: Physical Problems and Computational Solutions*, J.L. Swedlow, Ed. (American Society of Mechanical Engineers, New York, 1972) pp. 153–170.

[19] T.A. Cruse, An Improved Boundary-Integral Equation Method for Three Dimensional Elastic Stress Analysis, *Computer and Structures* **4**, 741–754 (1974).

[20] T.A. Cruse, Boundary-Integral Equation Method for Three-Dimensional Elastic Fracture Mechanics Analysis, AFOSR-TR-75-0813, Accession No. ADA01160, 13–20 (1975).
[21] T.A. Cruse and P.M. Besuner, Residual Life Prediction for Surface Cracks in Complex Structural Details, *Journal of Aircraft* **12**, 369–375 (1975).
[22] T.A. Cruse and G.J. Meyers, Three-Dimensional Fracture Mechanics Analysis, *Journal of the Structural Division, ASCE* **103**, 309–320 (1977).
[23] T.A. Cruse and R.B. Wilson, Advanced Applications of Boundary-Integral Equation Methods, *Nuclear Engineering and Design* **46**, 223–234 (1978).
[24] T.A. Cruse, Two Dimensional BIE Fracture Mechanics Analysis, *Applied Mathematical Modeling* **2**, 287–293 (1978).
[25] T.A. Cruse, Two- and Three-Dimensional Problems of Fracture Mechanics, in: *Developments in Boundary Element Methods – 1*, P.K. Banerjee and R. Butterfield, Eds. (Applied Science, London, 1979) pp. 97–119.
[26] T.A. Cruse and E.Z. Polch, Elastoplastic BIE Analysis of Cracked Plates and Related Problems, *Part 1: Formulation, Part 2: Numerical Results, International Journal for Numerical Analysis in Engineering* **23**, 429–437, 439–452 (1986).
[27] F. Erdogan, G.D. Gupta and T.S. Cook, Numerical Solution of Integral Equations, in: *Mechanics of Fracture, Methods of Analysis and Solutions of Crack Problems*, G.C. Sih, Ed. (Noordhoff, Leyden, The Netherlands, 1973) pp. 368–425.
[28] A.E. Green and I.N. Sneddon, The Stress Distribution in the Neighborhood of a Flat Elliptical Crack in an Elastic Solid, *Proceedings of the Cambridge Philosophical Society* **46**, 159–163 (1950).
[29] M.D. Greenberg, *Application of Green's Functions in Science and Engineering* (Prentice-Hall, Englewood Cliffs, NJ, 1971).
[30] A.A. Griffith, The Phenomena of Rupture and Flow in Solids, *Philosophical Transactions of the Royal Society* **A221**, 163–198 (1920).
[31] B. Gross and J. Srawley, Stress Intensity Factors for Three Point Bend Specimens by Boundary Collocation, *NASA Technical Note* D-3092 (1965).
[32] J.T. Guidera and R.W. Lardner, Penny-shaped Cracks, *Journal of Elasticity* **5**, 59–73 (1975).
[33] J. Heliot, R.C. Labbens and A. Pellissier-Tanon, Semi-Elliptical Cracks in a Cylinder Subjected to Stress Gradients, in: *Fracture Mechanics*, C.W. Smith, Ed. (American Society for Testing and Materials, STP 677, 1979) pp. 341–364.
[34] C.E. Inglis, Stresses in a Plate Due to the Presence of Cracks and Sharp Corners, *Proceedings of the Institute of Naval Architects* **60**, 219–230 (1913).
[35] G.R. Irwin, Analysis of Stresses and Strains Near the End of a Crack Traversing a Plate, *Journal of Applied Mechanics* **24**, 361–364 (1957).
[36] Y. Ishijima, K. Sato and S. Konoshita, Application of the Displacement Discontinuity Method to Some Crack Problems, *Theoretical and Applied Mechanics* **28**, 167–186 (1980).
[37] K. Kishitani, T. Hirai and K. Murakami, J-Integral Calculations With Boundary Elements, in: *Boundary Elements*, C.A. Brebbia, T. Futagami and M. Tanaka, Eds. (Springer-Verlag, Berlin, 1983) pp. 481–493.
[38] S. Kobayashi and N. Nishimura, Stress Analysis Around a Surface Crack by the Integral Equation Method, *Transactions of the Japanese Society of Civil Engineers* **11**, 74–75 (1979).
[39] J.C. Lachat and J.O. Watson, Effective Numerical Treatment of Boundary Integral Equations: A Formulation for Three-Dimensional Elastostatics, *International Journal for Numerical Methods in Engineering* **10**, 991–1005 (1976).
[40] M.L. Luchi and A. Poggialini, Computation of 3-Dimensional Stress Intensity Factors Using Special Boundary Elements, in: *Boundary Elements*, C.A. Brebbia, T. Futagami and M. Tanaka, Eds. (Springer-Verlag, Berlin, 1983) pp. 461–470.
[41] J. Martinez and J. Dominguez, On the Use of Quarter-Point Boundary Elements for Stress Intensity Factor Computations, *International Journal for Numerical Methods in Engineering* **20**, 1941–1950 (1984).
[42] M. Morjaria and S. Mukherjee, Numerical Analysis of Planar, Time-Dependent Inelastic Deformation of Plates with Cracks by the Boundary Element Method, *International Journal of Solids and Structures* **17**, 127–143 (1981).
[43] S. Mukherjee, *Boundary Element Methods in Creep and Fracture* (Applied Science, London, 1982).

[44] S. Mukherjee and V. Banthia, Non-Linear Problems of Fracture Mechanics, in: *Developments in Boundary Element Methods* – 3, P.K. Banerjee and S. Mukherjee, Eds. (Elsevier Applied Science Publishers, London, 1984) pp. 87–114.

[45] S. Mukherjee and M. Morjaria, Boundary Element Analysis of Time-Dependent Inelastic Deformation of Cracked Plates Loaded in Anti-Plane Shear, *International Journal of Solids and Structures* **17**, 753–763 (1981).

[46] S. Mukherjee and M. Morjaria, A Boundary Element Formulation for Planar Time-Dependent Inelastic Deformation of Plates with Cutouts, *International Journal of Solids and Structures* **17**, 115–126 (1981).

[47] J.C. Newman, An Improved Methods of Collocation for the Stress Analysis of Cracked Plates with Various Shaped Boundaries, *NASA Technical Note* D-6376 (1971).

[48] J.C. Newman and I.S. Raju, Analysis of Surface Crack in Finite Plates Under Tension or Bending Loads, NASA TP-1578 (1979).

[49] J.C. Newman and I.S. Raju, An Empirical Stress Intensity Factor Equation for the Surface Crack, *Engineering Fracture Mechanics* **15**, 185–192 (1981).

[50] H. Nikooyeh and A.R. Robinson, Approximate Determination of Stresses and Displacements Near a Rounded Notch, *International Journal of Solids and Structures* **17**, 669–682 (1981).

[51] N. Nishimura and S. Kobayashi, Elastoplastic Analysis by Indirect Methods, in: *Developments in Boundary Element Methods – 3*, P.K. Banerjee and S. Mukherjee, Eds. (Elsevier Applied Science, London, 1984) pp. 59–85.

[52] T. Nishioka and S.N. Atluri, Analysis of Surface Flaw in Pressure Vessels by a New 3-Dimensional Alternating Method, *Journal of Pressure Vessel Technology* **104**, 199–307 (1982).

[53] C.J. Putot, An Integral Equation Method for Resolution, in Opening Mode, of the Problem of Plane Cracks at Free Surface, *Proceedings of the Fifth International Conference on Fracture* (*ICFS*) Vol 1, Cannes, France, 141–150 (1981).

[54] C.J. Putot, Une Nouvelle Methode d'Equations Integrales pour Certains Problems de Fissures Planes, Ph.D. Thesis, University Pierre et Marie Curie, Paris VI (1980).

[55] P.M. Quinlan, J.J. Grannel, S.N. Atluri and J.E. Fitzgerald, The Edge Function Method for Three-Dimensional Stress Analysis, Including Embedded Elliptical Cracks and Surface Flaws, in: *Boundary Element Methods in Engineering*, C.A. Brebbia, Ed. (Springer-Verlag, Berlin, 1982) pp. 457–471.

[56] I.S. Raju and J.C. Newman, Stress Intensity Factors for a Wide Range of Semi-elliptical Surface Cracks in Finite-Thickness Plates, *Engineering Fracture Mechanics* **11**, 817–829 (1979).

[57] F.J. Rizzo and D.J. Shippy, An Advanced Boundary Integral Equation Method for Three Dimensional Thermoelasticity, *International Journal for Numerical Methods in Engineering* **11**, 1753–1768 (1977).

[58] T.J. Rudolphi and N.E. Ashbaugh, An Integral-Equation Solution for a Bounded Elastic Body Containing a Crack: Mode I Deformation, *International Journal of Fracture* **14**, 527–541 (1978).

[59] G.C. Sih and H. Liebowitz, *Fracture*: *An Advanced Treatise* Vol. 2 (Academic Press, New York, 1968).

[60] R.N.L. Smith and J.C. Mason, A Boundary Element Method for Curved Crack, Problems in Two Dimensions, in: *Boundary Element Methods in Engineering*, C.A. Brebbia, Ed. (Springer-Verlag, Berlin, 1982) pp. 472–484.

[61] R.N.L. Smith and J.C. Mason, Prediction of Crack Growth Direction Using the Boundary Element Method, in: *Boundary Elements*, C.A. Brebbia, T. Futagami and M. Tanaka, Eds. (Springer-Verlag, Berlin, 1983) pp. 495–503.

[62] M.D. Snyder and T.A. Cruse, Boundary-Integral Equation Analysis of Cracked Anisotropic Plates, *International Journal of Fracture* **11**, 315–328 (1975).

[63] M.L. Soni and M. Stern, On the Computation of Stress Intensity Factors in Fiber Composite Media Using a Contour Integral Method, *International Journal of Fracture* **12**, 331–334 (1976).

[64] M. Stern, E.B. Becker and R.S. Dunham, A Contour Integral Computation of Mixed Mode Stress Intensity Factors, *International Journal of Fracture* **12**, 359–368 (1976).

[65] M. Stern and J.W. Kim, The Calculation of Stress Intensity Factors in Plate Flexure, in: *Innovative Numerical Analysis for the Engineering Sciences*, R.P. Shaw et al., Eds. (University Press of Virginia, Charlottesville, 1980) pp. 777–786.

[66] H. Tada, *The Stress Analysis of Cracks Handbook*, Del Research Corporation, Hellertown, PA (1973).

[67] C.L. Tan and R.T. Fenner, Elastic Fracture Mechanics Analysis by the Boundary-Integral Equation Method, *Proceedings of the Royal Society of London, A, Mathematical and Physcial Sciences* **369**, 243–260 (1979).

[68] C.L. Tan and R.T. Fenner, Stress Intensity Factors for Semi-Elliptical Surface Cracks in Pressurized Cylinders Using the Boundary Integral Equation Method, *International Journal of Fracture* **16**, 233–245 (1980).

[69] J.C.F. Telles and C.A. Brebbia, Boundary Element Solution for Half-Plane Problems, *International Journal of Solids and Structures* **17**, 1149–1158 (1981).

[70] J.C.F. Telles, *The Boundary Element Method Applied to Inelastic Problems* (Springer-Verlag, Berlin, 1983).

[71] P.S. Theocaris, Numerical Solution of Singular Integral Equations: Parts I and II, *Journal of the Engineering Mechanics Division, ASCE* **107** (EM5), 733–752 and 753–771 (1981).

[72] R.W. Thresher and F.W. Smith, Stress Intensity Factors for a Surface Crack in a Finite Solid, *Journal of Applied Mechanics* **39**, 195–200 (1972).

[73] P. Tong, T.H.H. Pian and S.J. Lasry, A Hybrid-Element Approach to Crack Problems in Plane Elasticity, *International Journal for Numerical Methods in Engineering* **7**, 297–308 (1973).

[74] D.M. Tracey, Finite Elements for Determination of Crack Tip Elastic Stress Intensity Factors, *Engineering Fracture Mechanics* **3**, 255–265 (1971).

[75] F. Van der Weeën, Mixed Mode Fracture Analysis of Rectilinear Anisotropic Plates Using Singular Boundary Elements, *Computers and Structures* **17**, 469–474 (1983).

[76] J.O. Watson, Hermitian Cubic Boundary Elements for Plane Problems of Fracture Mechanics, *Research Mechanica* **4**, 23–42 (1982).

[77] J. Weaver, Three Dimensional Crack Analysis, *International Journal of Solids and Structures* **13**, 321–330 (1977).

[78] M.L. Williams, Stress Singularities Resulting from Various Boundary Conditions in Angular Corners of Plates in Extension, *Journal of Applied Mechanics* **19**, 526–528 (1952).

[79] L.S. Xanthis, M.J.M. Bernal and C. Atkinson, The Treatment of Singularities in the Calculation of Stress Intensity Factors Using the Boundary Integral Equation Method, *Computer Methods in Applied Mechanics and Engineering* **26**, 285–304 (1981).

CHAPTER 8

Fluid Mechanics

James A. LIGGETT

School of Civil and Environmental Engineering
Cornell University
Ithaca, New York 14853, U.S.A.

Boundary Element Methods in Mechanics
Edited by D.E. Beskos

Introduction

The list of complications in solving fluid mechanics problems is long: large, three-dimensional problems; nonlinearities; turbulence; instabilities; time dependence; free surfaces; complex and multiple boundary conditions. At one time it appeared that most of these difficulties might be overcome by finite difference techniques, but that hope fell far short of its goal. Finite element methods form a more flexible system which can be adapted to solve many fluid mechanics problems. Indeed, the progress has been remarkable; however, it still falls short of the goals.

It is tempting to say that the Boundary Element Method is another step toward the problem-solving goal. In fact, it probably is. Certainly it aids in the economical solution of many problems and makes many solutions practicable which would otherwise be outside of the available resources. In principle, however, the Boundary Element Method falls far short of solving even those problems that can be treated by finite elements or finite differences. The limitation of the method is that it does not solve the really nasty, nonlinear problems and those that are complicated by turbulence. It has made some inroads on such problems and in the future there is hope that the Boundary Element Method, or a derivative of it, may become much more powerful.

At this date the primary advantage of the Boundary Element Method with respect to fluid mechanics problems is not its raw power in complicated problems, and neither is it the often mentioned efficiency, although that is certainly a point in its favor. The primary advantage, in my opinion, is its ability to create good, usuable software. The saving in engineering time over finite elements or finite differences is outstanding, even when those methods are equipped with grid generators and other preprocessing conveniences. As computers become less expensive and more powerful, the efficiency advantages of the BEM will become of less importance.

The Boundary Element Method is not just another step in the progression of numerical techniques exemplified by finite differences and finite elements. Instead, it is one of the initial steps in the emerging techniques in which analysis is combined with numerics in order to take advantage of both in the solution of problems. While finite elements can be a very sophisticated method, it begins the problem solution at an elementary stage and uses a brute force technique. The BEM, on the other hand, uses elementary solutions to begin and resorts to numerics at a later stage. More importantly, it provides a vehicle in which the analyst can use much of what he knows about the problem to aid in the efficiency and integrity of the solution. Examples are the inclusion of analytical representations of singularities and far-field behavior. That sort of technique obviously has a long way to go before it can be applied to very general classes of problems, but (if I can make a prediction) the combination of analysis with numerical methods is the future of scientific and engineering calculation.

At the present the BEM carries some of the disadvantages of both analysis and numerical methods. The analysis disadvantage is that it is limited in the type of problem it can solve. That can be compared to the Finite Element Method, which originally was thought to be limited to those problems that can be expressed in a variational statement. Like the Finite Element Method, it is now obvious that the BEM is much more general than was originally thought, but unlike the Finite Element Method there has not been nearly as much progress in making the BEM general.

As pointed out elsewhere [40] the BEM predates the Finite Element Method and its recent popularity represents a rebirth. The origin of the method depends somewhat on definition, but it is generally conceded that the first calculation was done by Trefftz [67]. That calculation was a masterpiece in that it solved a problem which would be somewhat of a challenge to most of us today. Trefftz solved the first fluid mechanics problem by the BEM (the numerical value of the contraction coefficient of a round jet issuing from an infinite tank), the first axisymmetric problem in which the Green's function was expressed by elliptic integrals, and the first nonlinear free surface problem. In many respects it is remarkable that there has been so little additional progress in the 68 years since Trefftz's paper was published, especially considering the vast advances in computing resources.

It is obviously impossible to cover in one chapter the fluid mechanics applications of the BEM. In what follows there are some glaring omissions which reflect the writer's preference in what to include in a limited space. Primary among the omissions are the applications made by the aircraft industry in aerodynamic calculations. The panel methods were pioneered in the late 1950s and early 1960s [60, 18]. These techniques are amply reviewed by Hess [19]. Neither have I included the integral equation method developed by J.C. Wu and his colleagues. Those techniques stem from the early 1970s and have also been amply reviewed [75, 76].

Further, this text takes the point of view that the reader is not intimately familiar with boundary element methods; therefore, some of the methods of application that appear herein would not ordinarily appear in a review of the subject. Unfortunately, there is insufficient space to adequately cover the subject and the techniques. What does appear is, again, written from my experience. For example, very little consideration is given to indirect boundary element calculations even though they form a valuable part of the method for many applications, especially the aerodynamic applications.

The fundamentals of the Boundary Element Method are discussed in the next section. These include the basic equations and methods of discretization for numerical calculation, along with the treatment of singularities, infinite fields and inhomogeneous domains. The third section briefly reviews some of the literature on potential flow. The porous media applications, one of the rapidly growing areas that employ the method, are explained in the fourth section. The various methods used in that section constitute many of the techniques used in boundary element calculations. The advantage of the reduction of the computational dimension of problems is most apparent in free surface calculations, which are treated in the section on porous media and the following section on water waves. Finally, some of the applications of viscous flow are treated in section 6.

1. Fundamentals

1.1. Integral equations, elements, and nodes

A general explanation of the Boundary Element Method usually begins with the example of the Laplace equation. That is the most logical beginning point for fluid mechanics applications because it is the simplest practical application and because the greatest use of the BEM is still in potential flow problems. The integral equation that is equivalent to the Laplace equation in two dimensions is

$$2\pi\Phi(\mathrm{p}) = \int_{\partial D} \left(\Phi(\mathrm{Q}) \frac{\partial}{\partial n} \ln r_{\mathrm{pQ}} - \ln r_{\mathrm{pQ}} \frac{\partial \Phi(\mathrm{Q})}{\partial n} \right) \mathrm{d}s \qquad (1.1)$$

in which ∂D is the boundary of the domain D (Fig. 1a), n is the unit outward normal and r_{pQ} is the distance between point p (at which the Φ of the left-hand side is evaluated) and the boundary point. Equation (1.1) states that if the variable Φ and its normal derivative, $\partial\Phi/\partial n$, are both known everywhere on the boundary, then the solution at any interior point consists of a boundary integration. One of these quantities, Φ or $\partial\Phi/\partial n$, or a relationship between them, is given as a part of the boundary condition in any well-posed problem. Moving point p to P, a point on the boundary of the domain D (Fig. 1b), Eq. (1.1) is used

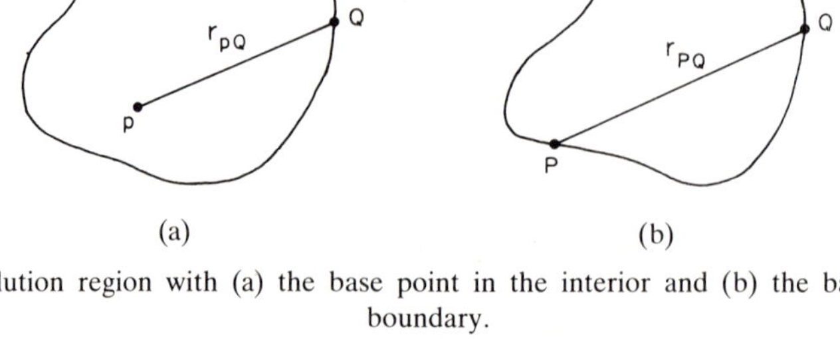

Fig. 1. The solution region with (a) the base point in the interior and (b) the base point on the boundary.

to find the "missing" boundary data, the Φ or $\partial\Phi/\partial n$ which is not given by the boundary conditions. In doing so the boundary of the region is discretised into elements, which may be straight lines or polynomials or whatever the analyst desires. The known boundary conditions are approximated in some manner (as constants on each element, linearly, quadratically, or otherwise) along the elements. For the Laplace equation the only approximation is in the boundary condition and the ultimate representation of the solution on the boundary; in the interior the governing equation is solved "exactly" (i.e. the boundary-value problem using the approximate boundary conditions is exactly satisfied). The same statement can be made for any linear equation in which an explicit Green's function is known. Expansion of the method to linear and nonlinear problems for which the Green's function cannot be written explicitly requires some approximation of the governing differential equation. That approximation is, however, quite different than is made by finite element or finite different methods.

Consider a scheme wherein the boundary of the domain is divided into a number of segments (Fig. 2). On any one segment the boundary conditions are of the same type: Dirichlet (Φ given), Neumann ($\partial\Phi/\partial n$ given), mixed (a relationship between and $\partial\Phi/\partial n$), or free surface. The segments are, in turn, divided into elements. In Fig. 3 there are several types of elements: a "constant" element containing one node in which the boundary condition and the solution are approximated as constants for the entire length of the element, one- and two-noded linear elements in which the boundary condition and the solution are interpolated linearly along the element, quadratic three- or four-noded elements, and cubic spline elements which (isoparametrically) represent cubically the geometry, the boundary condition, and the solution and with continuous derivatives between elements. Obviously, there are an infinite variety of elements. One of the primary strengths of the method lies in the variety of elements and the fact that the analyst can choose an element which best represents the known or expected behavior of the boundary conditions and of the solution.

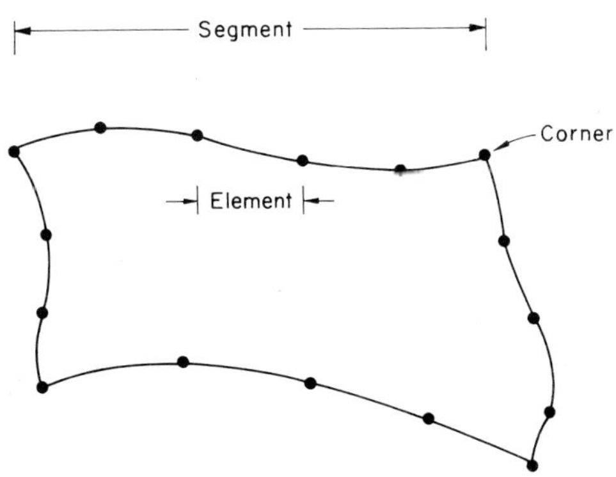

Fig. 2. Division of the solution region into segments and elements.

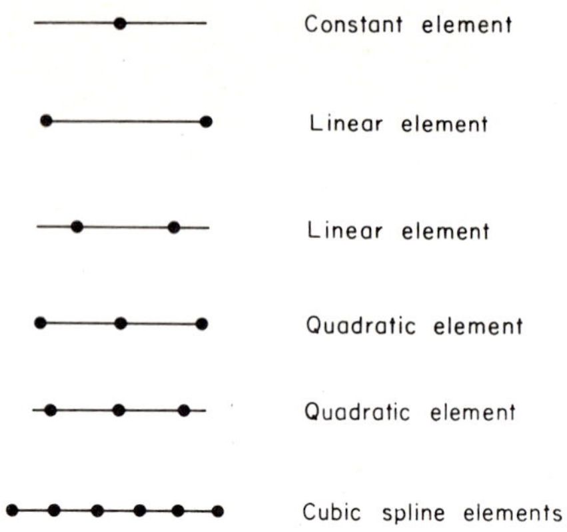

Fig. 3. Element types.

The purpose is, of course, to devise a method which will convert Eq. (1.1) into a series of linear algebraic equations. Consider the linear element. The unknowns of Eq. (1.1) are interpolated between nodes as

$$\Phi = \frac{(\Phi_{j+1} - \Phi_j)\xi + \xi_{j+1}\Phi_j - \xi_j\Phi_{j+1}}{\xi_{j+1} - \xi_j} \qquad \xi_j \leqslant \xi \leqslant \xi_{j+1} \tag{1.2}$$

$$\frac{\partial \Phi}{\partial n} = \frac{\left[\dfrac{\partial \Phi_{j+1}}{\partial n} - \dfrac{\partial \Phi_j}{\partial n}\right]\xi + \xi_{j+1}\dfrac{\partial \Phi_j}{\partial n} - \xi_j\dfrac{\partial \Phi_{j+1}}{\partial n}}{\xi_{j+1} - \xi_j} \qquad \xi_j \leqslant \xi \leqslant \xi_{j+1} \tag{1.3}$$

in which ξ represents distance along the element (Fig. 4) and the subscript indicates that the quantity is to be evaluated at the node. Using the free-space Green's function, $\ln r$, for the two-dimensional Laplace equation, Eq. (1.1) becomes

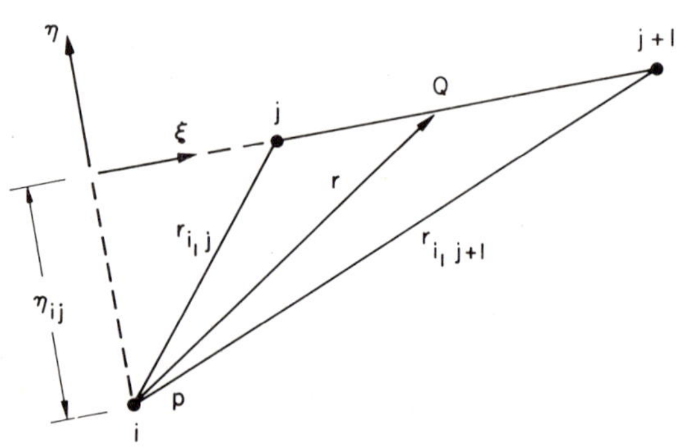

Fig. 4. The local coordinates system for an element.

$$\alpha\Phi_i = \sum_{j=1}^{N} \frac{1}{\xi_{j+1} - \xi_j} \left[(\Phi_{j+1} - \Phi_j) \int_j^{j+1} \xi \frac{\partial}{\partial n} \ln r_{ij} \, d\xi + (\xi_{j+1}\Phi_j - \xi_j\Phi_{j+1}) \right.$$

$$\times \int_j^{j+1} \xi \ln r_{ij} \, d\xi - \left[\frac{\partial \Phi_{j+1}}{\partial n} - \frac{\partial \Phi_j}{\partial n} \right] \int_j^{j+1} \xi \ln r_{ij} \, d\xi$$

$$\left. - \left(\xi_{j+1} \frac{\partial \Phi_j}{\partial n} - \xi_j \frac{\partial \Phi_{j+1}}{\partial n} \right) \int_j^{j+1} \ln r_{ij} \, d\xi \right] \tag{1.4}$$

in which r_{ij} is the distance between the "base point" (or "source point"), p or P, and the "field point", Q, and $\alpha = 2\pi$ if p (point i) is in the interior or $\alpha =$ the boundary angle at P (point i) when the base point is on the boundary. With the simple kernel of the two-dimensional Laplace equation the integral can be expressed in closed form

$$\int_j^{j+1} \xi \frac{\partial}{\partial n} \ln r_{ij} \, d\xi = \tfrac{1}{2}\eta_{ij} \ln \frac{\eta_{ij}^2 + \xi_{j+1}^2}{\eta_{ij}^2 + \xi_j^2} \tag{1.5a}$$

$$\int_j^{j+1} \frac{\partial}{\partial n} \ln r_{ij} \, d\xi = \tan^{-1} \frac{\xi_{j+1}}{\eta_{ij}} - \tan^{-1} \frac{\xi_j}{\eta_{ij}} \tag{1.5b}$$

$$\int_j^{j+1} \xi \ln r_{ij} \, d\xi = \tfrac{1}{4}\{(\eta_{ij}^2 + \xi_{j+1}^2)[\ln(\eta_{ij}^2 + \xi_{j+1}^2) - 1]$$

$$- (\eta_{ij}^2 + \xi_j^2)[\ln(\eta_{ij}^2 + \xi_j^2) - 1]\} \tag{1.5c}$$

$$\int_j^{j+1} \ln r_{ij} \, d\xi = \tfrac{1}{2}\left\{ \xi_{j+1} \ln(\eta_{ij}^2 + \xi_{j+1}^2) - \xi_j \ln(\eta_{ij}^2 + \xi_j^2) - 2(\xi_{j+1} - \xi_j) \right.$$

$$\left. + 2\eta_{ij}\left[\tan^{-1} \frac{\xi_{j+1}}{\eta_{ij}} - \tan^{-1} \frac{\xi_j}{\eta_{ij}} \right] \right\} \tag{1.5d}$$

where r_{ij}, η_{ij}, and ξ_j are defined in Fig. 4. With other, more complex, kernels or with more complex geometric representation of the domain or with more complex interpolation of the boundary values, the integration would have to be done numerically. With linear geometric representation and polynomial interpolation and with $\ln r$ or $1/r$ (the three-dimensional free-space Green's function for the Laplace equation) kernels the integral can be expressed in closed form and there is usually considerable advantage in doing so.

For any given geometry the integrals become numerical values which can be assembled into a matrix. The position of the elements of the matrix depends on the location of the base point and the location of the target element of the field point. The result is the matrix equation

$$[R]\{\Phi\} = [L]\{\partial\Phi/\partial n\} . \tag{1.6}$$

Since the boundary conditions supply additional information on the values of the Φ and $\partial\Phi/\partial n$, Eqs. (1.6) can be assembled into

$$[C]\{u\} = \{r\} \tag{1.7}$$

in which $\{u\}$ represents the unknown parts of the $\{\Phi\}$ and $\{\partial\Phi/\partial n\}$ vectors, $\{r\}$ is composed of the appropriate parts of $[L]$ and $[R]$ multiplied into the known parts of the $\{\Phi\}$ and $\{\partial\Phi/\partial n\}$ vectors, and $[C]$ is constituted by the remainder of $[L]$ and $[R]$. The solution of Eq. (1.7) furnishes the unknowns on the boundary.

In most BEM problems the normal derivatives, $\partial\Phi/\partial n$, are discontinuous at various points on the boundary. We call such points "corners". They are often, but not always, located at physical corners where the boundary changes direction. In order to define the discontinuity, the normal derivative is split into two parts: $\partial\Phi^-/\partial n$, which applies on the element immediately before the node in the direction of integration, and $\partial\Phi^+/\partial n$, which applies on the element just after the node. Equation (1.6) is then divided into two parts

$$[R]\{\Phi\} = [L^-]\{\partial\phi^-/\partial n\} + [L^+]\{\partial\Phi^+/\partial n\} \tag{1.8}$$

in which the $[L^-]$ and $[L^+]$ matrices are the appropriate parts of the $[L]$ matrix of (1.6); that is, L_{ij}^- represents the integration along the element from $j-1$ to j and L_{ij}^+ is the contribution of the element from j to $j+1$ [42].

In the assembly of the BEM equations, the nodes with discontinuous derivatives must be identified. The actual process is to provide the boundary data in segments, which are defined as parts of the boundary between corners. Each segment must have the same boundary condition (Neumann, Dirichlet, mixed, or free surface) for its entire length. The types of segments are identified by number in the top of Fig. 5. There are only three shown since the Dirichlet type is taken as a special case of the mixed type.

After the segment types have been identified, each node is assigned a node type. This latter operation can be done automatically in the program without intervention of the analyst. The node types are shown on Fig. 5. The first ten types are sufficient for any fluid mechanics problem that has been encountered thus far (although it has been found convenient, but not necessary, to identify an eleventh type for flow through porous media with applies on a seepage surface).

All of the boundary combinations are represented in the node types of Fig. 5. The only possible missing type is where Dirichlet conditions are imposed on both

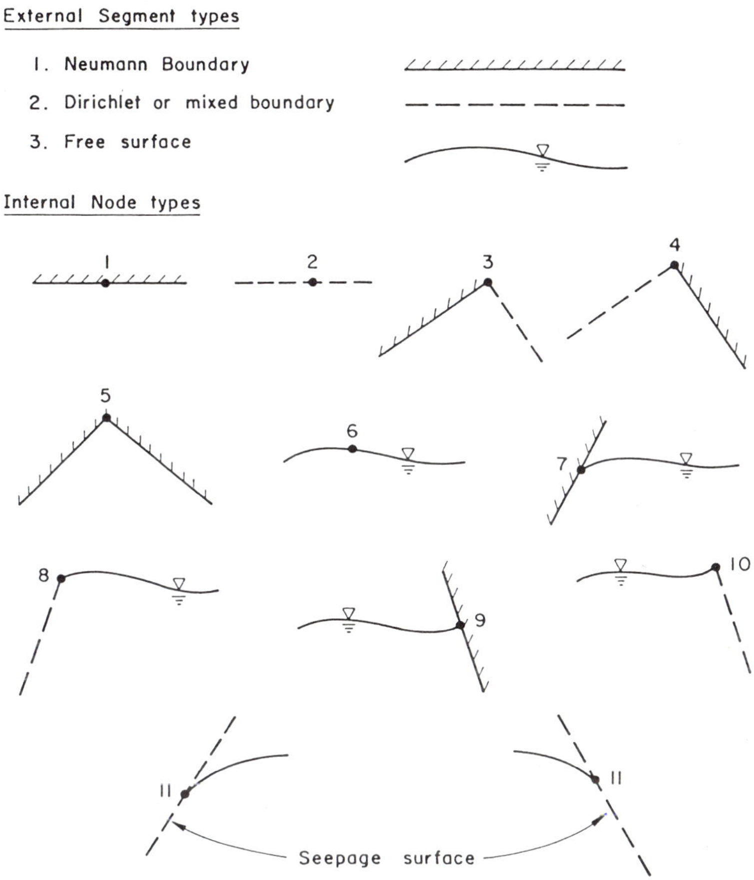

Fig. 5. Segment types and internally generated node types.

sides of the node and where the potential is discontinuous at the node. If such a condition appears in the middle of a straight line segment, it represents a singularity and should be treated as such. If it appears at a change in direction on the boundary, it is also a singularity and probably overspecifies the boundary condition at the point. Where the boundary changes direction, the derivative in any direction can be determined by differentiating the Dirichlet condition. Thus, the potential and all derivatives would be known at the point, over specifying the boundary condition. Fortunately, such a situation does not occur in physical problems.

In many, perhaps the majority, of fluid mechanics problems the boundary solution is all that the analyst needs; the solution consists of finding the fluid interaction with the boundary. If the interior solution is of interest, then Eq. (1.1), discretised in the same manner but with the source point at the position of the desired solution, supplies the answer. Another strength of the BEM is that the derivatives of the unknown can be found in the interior of the domain without numerical differentiation and its subsequent loss of accuracy. This property is especially important in potential flow problems where the velocity is often the relevant unknown. Differentiating Eq. (1.1) with respect to the coordinate directions gives (see Ref. [41] for details)

$$2\pi \frac{\partial}{\partial x} \Phi(\mathrm{p}) = \int_{\partial D} \Big[\frac{\Phi(\mathrm{Q}) \sin \mu}{r^2} + \frac{2\Phi(\mathrm{Q})\eta(\xi \cos \mu - \eta \sin \mu)}{r^4} + \frac{\xi \cos \mu - \eta \sin \mu}{r^2} \frac{\partial}{\partial n} \Phi(\mathrm{Q}) \Big] \mathrm{d}s \tag{1.9}$$

$$2\pi \frac{\partial}{\partial y} \Phi(\mathrm{p}) = \int_{\partial D} \Big[-\frac{\Phi(\mathrm{Q}) \cos \mu}{r^2} + \frac{2\Phi(\mathrm{Q})\eta(\eta \cos \mu + \xi \sin \mu)}{r^4} + \frac{\eta \cos \mu + \xi \sin \mu}{r^2} \frac{\partial}{\partial n} \Phi(\mathrm{Q}) \Big] \mathrm{d}s\,. \tag{1.10}$$

In obtaining Eqs. (1.9) and (1.10) the differentiation was with respect to the position of the base point only; the variables which refer to the target point remain undifferentiated. The latter two equations can be discretised in the same manner as Eq. (1.1) and their solution is a direct boundary integration with (now) known values $\{\Phi\}$ and $\{\partial\Phi/\partial n\}$ along the boundary.

2.2. Singularities and infinite fields

The flexibility of specifying boundary elements provides a system to handle accurately both singularities and boundary conditions at infinity in a finite computational domain. Consider the corner singularity. For potential flow around a corner (Fig. 6)

$$\Phi \sim R^{\pi/\theta} \tag{1.11}$$

in which R is the distance from the corner and θ is defined in Fig. 6. The velocity, being the derivative of the potential, is zero at the corner for $\theta < \pi$ and infinity

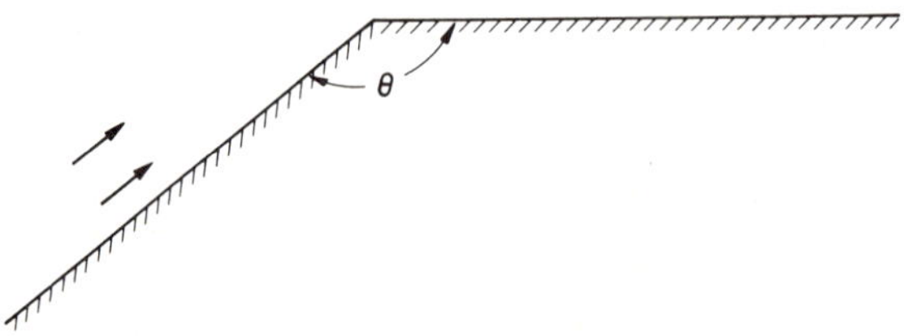

Fig. 6. Flow around the corner singularity.

for $\theta > \pi$. Obviously, such behavior, especially for $\theta > \pi$, is not well modeled by the regular polynomial interpolation function.

Rewriting Eq. (1.4)

$$\alpha\Phi_i = \sum_{j=1}^{N} \left[\Phi_j \int_j^{j+1} N_1 \frac{\partial}{\partial n} \ln r_{ij}\, d\xi + \Phi_{j+1} \int_j^{j+1} N_2 \frac{\partial}{\partial n} \ln r_{ij}\, d\xi \right.$$

$$\left. + \frac{\partial \Phi_j}{\partial n} \int_j^{j+1} N_1' \ln r_{ij}\, d\xi + \frac{\partial \Phi_{j+1}}{\partial n} \int_j^{j+1} N_2' \ln r_{ij}\, d\xi \right]. \tag{1.12}$$

The interpolation functions, N_1 and N_2, may be linear, in which case

$$N_1 = N_1' = (1 - \xi/L) \qquad N_2 = N_2' = \xi/L \tag{1.13}$$

where L is the length of the element. The interpolations can be quite general, however. In the case of flow around a corner (now using "corner" to designate the geometrical change in direction and not as in the last section) we choose the following:

$$N_1 = (1 - \xi/L)^{\pi/\theta} \qquad N_2 = 1 - (1 - \xi/L)^{\pi/\theta} \tag{1.14a}$$

when approaching the corner in the direction of integration and

$$N_1 = 1 - (\xi/L)^{\pi/\theta} \qquad N_2 = (\xi/L)^{\pi/\theta} \tag{1.14b}$$

when leaving the corner in the direction of integration. Since the normal derivative is zero along the wall, it is immaterial what is used for N_1' and N_2'. The combination of (1.14a) and (1.14b) force the potential to behave in the proper exponential fashion with distance from the corner. The worst possible singularity represented by a corner is when $\theta = 2\pi$, the case that is common in flow through porous media in the form of a sheetpile cutoff wall. In that case the velocity tends to infinity as $r^{-1/2}$.

Not only can the elements which include the sudden change in direction represent the behavior, but the next elements can also reflect the solution near the corner. A "transition" element of length L approaching the corner would have the interpolation functions

$$N_1 = [1 - (\xi + L)/L]^{\pi/\theta} \qquad N_2 = 1 - [1 - (\xi + L)/L]^{\pi/\theta}. \tag{1.15}$$

The same case can be used to illustrate how the solution can be extended to infinity. Approaching the first node from the left only one interpolation function is needed. The second of (1.15) will serve that purpose if the first L is replaced by the distance from the node to the corner, d,

$$N_2 = 1 - [1 - (\xi + d)/L]^{\pi/\theta} . \tag{1.16}$$

The integration along the infinite element is carried out along the wall from $-\infty$ to the location of the first node only in the second term of (1.12).

The above problem is not very practical in the sense that it would not be solved by a numerical method unless it had other complicating features. When viewing the problem far from the complicating features, however, it can be reduced to the basic problem of flow around a corner. There are a great many such examples. In potential flow around an object the flow in the far field is independent of the shape of the object. An example showing that flow under a dam in porous media can be replaced by a series of vortices is given by [41].

The idea of an infinite element is further illustrated in problems dealing with water waves in a later section. In that case an approximate (linear) solution is used to represent the far-field behavior of a nonlinear problem.

2.3. Zones

In dividing the solution region into zones the problem is treated in parts instead of a whole. There are several reasons for zoning in the Boundary Element Method. The most important occurs in porous media problems where the hydraulic conductivity of the medium varies from place to place. It is easier to solve such problems if the physical properties are constants. In a porous medium it is sufficient to divide the solution region into zones in which the hydraulic conductivity is taken as constant. Usually the properties of the medium are not known in sufficient detail to justify a more detailed treatment.

In other fluid mechanics problems the analyst may wish to divide the problem into zones in order that the separate zones have a better aspect ratio than the problem as a whole. Or different equations may apply to different parts of the problem. Zoning also has the advantage, at least in one method, that the coefficient matrix of the simultaneous equations becomes block banded, and thus one can take advantage of the efficiency of banded solvers. The last attribute comes at the price of a larger system of equations, however. Finally, internal solutions are obtained on the first pass along zonal lines, which may be of limited advantage in some problems.

Consider a solution region such as is shown in Fig. 7. The internal conditions along the interzonal dividing line are obvious: the potential is the same at any node on that line whether viewed from zone 1 or from zone 2, and the flux across the dividing line calculated from zone 1 must be the same as the flux calculated from zone 2. In equation form

$$\Phi^1 = \Phi^2 \quad \text{and} \quad K_1 \frac{\partial \Phi^1}{\partial n} = K_2 \frac{\partial \Phi^2}{\partial n} \quad \text{on } \Gamma_D \tag{1.17}$$

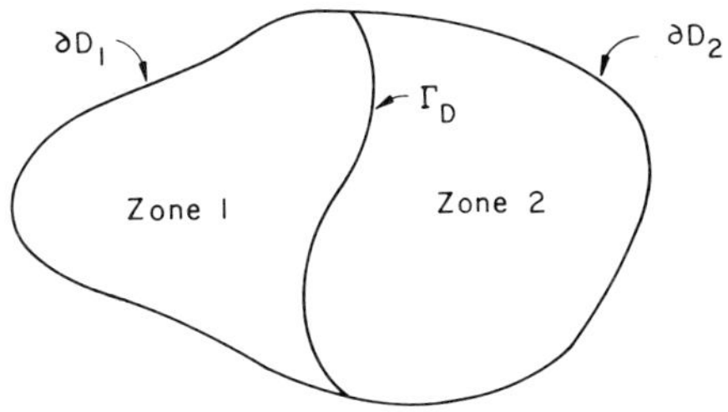

Fig. 7. A solution region divided into zones.

in which the superscripts represent the zone numbers. The K_1 and K_2 would represent the hydraulic conductivities in porous media and could simply be unity in a potential flow problem. In the second of Eqs. (1.17) the normal vector, n, is defined in the same direction for both adjacent zones.

Both the Φ and the $\partial\Phi/\partial n$ are unknown at the interzonal points. Thus, Eqs. (1.17) are sufficient, when added to the normal BEM equations for nodes on the external and internal boundaries, to calculate both the potential and the normal derivative. The total equations are assembled into a matrix which has the dimensions of

$$D = \sum N_e^i + 2N_z \tag{1.18}$$

in which N_e^i is the number of exterior nodes in zone i (some zones may not have exterior nodes; that is, they may be interior to the solution domain) and N_z is the total number of interior nodes on all the interzonal lines.

Using a slightly different method, modeled after Haitjema and Strack [16, 17], the number of equations can be reduced to [71]

$$D = \sum N_e^i + N_z \tag{1.19}$$

which could represent a substantial savings in those problems that contain many interior nodes. In order to illustrate the method, we use the equation

$$\nabla \cdot K\nabla\Phi = 0 \tag{1.20}$$

with K constant within a zone but different in the various zones.

Equation (1.20) is written as

$$\nabla^2\Phi = -\frac{1}{K}\nabla K \cdot \nabla\Phi = -\frac{1}{K}\left(\frac{\partial K}{\partial s}\frac{\partial\Phi}{\partial s} + \frac{\partial K}{\partial n}\frac{\partial\Phi}{\partial n}\right) \tag{1.21}$$

which represents the Laplace equation within the zones but is the Poisson equation on the interzonal boundaries. Applying Green's second identity in the usual way

$$\alpha\Phi = \int_{\partial D} \left(\Phi \frac{\partial G}{\partial n} - G \frac{\partial \Phi}{\partial n}\right) \mathrm{d}s = \int_{D} \frac{G}{K} \nabla K \cdot \nabla \Phi \, \mathrm{d}V \tag{1.22}$$

where α is the boundary angle, G is the Green's function ($\ln r$), and ∂D is the total exterior boundary. The volume integral is zero everywhere except along the interzonal boundaries, Γ_D. In order to evaluate the volume integral, the interzonal boundary is broadened into a transition region of width δ in which K changes continuously from K_1 to K_2. Then

$$I = \int_{V} \frac{G}{K} \nabla K \cdot \nabla \Phi \, \mathrm{d}V = \lim_{\delta \to 0} \int_{\Gamma_D} \mathrm{d}s \int_{0}^{\delta} \frac{G}{K} \frac{\partial K}{\partial n} \frac{\partial \Phi}{\partial n} \, \mathrm{d}n \tag{1.23}$$

in which V represents the transition region. From the second of Eqs. (1.17) the flux is constant across the transition area and (1.23) can be written

$$I = \lim_{\delta \to 0} \int_{\Gamma_D} gK \frac{\partial \Phi}{\partial n} \, \mathrm{d}s \int_{0}^{\delta} \frac{1}{K^2} \frac{\partial K}{\partial n} \, \mathrm{d}n \, . \tag{1.24}$$

The last integral is easily integrated. Since the flux term can be written for either zone 1 or zone 2, it is arbitrarily written for zone 1, giving

$$I = \int_{\Gamma_D} G\left(1 - \frac{K_1}{K_2}\right)\left(\frac{\partial \Phi^1}{\partial n}\right) \mathrm{d}s \, . \tag{1.25}$$

Thus, the integration can be carried out over the internal and the external boundaries

$$\alpha\Phi = \int_{\partial D} \left(\Phi \frac{\partial G}{\partial n} - G \frac{\partial \Phi}{\partial n}\right) \mathrm{d}s - \int_{\Gamma_D} G\left(1 - \frac{K_1}{K_2}\right) \frac{\partial \Phi^1}{\partial n} \, \mathrm{d}s \, . \tag{1.26}$$

If $K_1 = K_2$, the second integral disappears so that the normal, homogeneous situation applies. Although the volume integral has been eliminated in favor of line integrals, the equation involves unknowns of both Φ and $\partial \Phi / \partial n$ on the external and internal boundaries. We now write $\alpha\Phi = \alpha^1 \Phi^1 + \alpha^2 \Phi^2$ where $\alpha^1 + \alpha^2 = 2\pi$ and $\Phi^1 = \Phi^2$ and differentiate (1.26) with respect to n to give

$$\left(\alpha^1 + \alpha^2 \frac{K_1}{K_2}\right) \frac{\partial \Phi^1(\mathrm{P})}{\partial n_\mathrm{P}} = \int_{\partial D} \left[\Phi(\mathrm{Q}) \frac{\partial^2 G(\mathrm{P},\mathrm{Q})}{\partial n_\mathrm{P} \partial n_\mathrm{Q}} - \frac{\partial \Phi(\mathrm{Q})}{\partial n_\mathrm{Q}} \frac{\partial G(\mathrm{P},\mathrm{Q})}{\partial n_\mathrm{P}}\right] \mathrm{d}s$$

$$- \int_{\Gamma_D} \left(1 - \frac{K_1}{K_2}\right) \frac{\partial \Phi^1(\mathrm{Q})}{\partial n_\mathrm{Q}} \frac{\partial G(\mathrm{P},\mathrm{Q})}{\partial n_\mathrm{P}} \, \mathrm{d}s \, , \tag{1.27}$$

in which the derivatives and the variables are now clearly identified as belonging to the field point, Q, or to the source point P. The unknown $\Phi(P)$ does not appear in (1.27); the only unknowns are the derivative $\partial\Phi/\partial n$ on the internal boundary, $\Phi(Q)$ on the external boundary, and $\partial\Phi(Q)/\partial n$ on the external boundary. Thus, the number of internal unknowns has been reduced to one per node. Unfortunately, the resulting coefficient matrix is fully populated. The discretization of this method is described in Wei and Liggett [71].

Dong [11] reports a variation of the foregoing method which is particularly simple. Writing the basic boundary integral (1.22) for either zone and multiplying the equation by K

$$K_i\alpha_i\Phi^i = \int_{\partial D_i} K_i\left(\Phi^i \frac{\partial G}{\partial n} - G\frac{\partial \Phi^i}{\partial n}\right) ds \tag{1.28}$$

in which i can take on the number (1 or 2) for either zone and ∂D_i represents of the entire boundary of the zone which may consist of a portion of the external and the internal boundary. Subtracting the equation for $i = 2$ from the equation for $i = 1$ yields

$$\begin{aligned} K_1\alpha_1\Phi^1 + K_2\alpha_2\Phi^2 = &\int_{\partial D} K\left(\Phi \frac{\partial G}{\partial n} - G\frac{\partial \Phi}{\partial n}\right) ds \\ &+ \int_{\Gamma_D} K_1\left(\Phi^1 \frac{\partial G}{\partial n_1} - G\frac{\partial \Phi^1}{\partial n_1}\right) ds \\ &+ \int_{\Gamma_D} K_2\left(\Phi^2 \frac{\partial G}{\partial n_2} - G\frac{\partial \Phi^1}{\partial n_2}\right) ds \end{aligned} \tag{1.29}$$

in which the normal directions are always out of the respective zone. The value of K in the first integral must be K_1 or K_2 depending on which part of the boundary the integration is taking place. Using the second of (1.17) and realizing that n_2 points in the negative direction of n_1

$$\begin{aligned} K_1\alpha_1\Phi^1 + K_2\alpha_2\Phi^2 = &\int_{\partial D} K\left(\Phi \frac{\partial G}{\partial n} - G\frac{\partial \Phi}{\partial n}\right) ds \\ &+ \int_{\Gamma_D} \Phi(K_1 - K_2)\frac{\partial G}{\partial n} ds\,. \end{aligned} \tag{1.30}$$

Again, the integral along the interzonal boundary disappears if the K are the same in each zone. The base point is to take on all the nodes of the exterior and internal boundaries, providing $N_e + N_z$ equations in the $N_e + N_z$ values of Φ on the external and interzonal boundaries and the N_e values of $\partial\Phi/\partial n$ on the external boundary. The N_e boundary conditions applied to the external boundary close the system. Again, the coefficient matrix, although smaller than that for two unknowns per interzonal node, is fully populated.

2. Potential flow

The Boundary Element Method is often illustrated with potential flow problems. Indeed, many people have the mistaken impression that it can only be applied to potential problems. The method was first used for potential problems and developed to its highest state in their solution. In this section such problems are treated only briefly since their formulation and solution appears in many, many sources and of course in the second chapter of this book. Indeed, the previous section in this chapter was written assuming that the reader has some familiarity with the use of the BEM to solve potential problems.

The free surface flow solved by Trefftz [67] was, of course, a potential flow. Prager [52] was also one of the early users of boundary methods. The "surface theory" or "surface source" method of Lotz [46] continued such developments. Later users developed axisymmetric solutions [32, 54, 68–70]. The use of integral equation methods up to the middle 1950s probably exceeded the use of finite elements.

True three-dimensional problems had to await the development of the digital computer into a practical tool for large-scale calculation. Smith and Pierce [60] may have been the first to use boundary methods on a digital computer although the field had become active at about that time in the aircraft industry and in the National Advisory Committee on Aeronautics (NACA, the predecessor to NASA). Later three-dimensional problems were solved by Hess and Smith [18], Davenport [8], Jaswon [27], Symm [62], and Chaplin [5]. The "panel methods" of the pioneers are very much in use today [19, 21]. The early uses of boundary methods considered only incompressible, irrotational flow. Extensions to compressible flow are presented by Carey [4] and Carey and Kim [3]. Zuosheng [77] reports a successful calculation for the entire Mach number range, including transonic flow, using perturbation theory. Rotational but frictionless flow has been solved by several authors [23]. Flows outside the boundary layer constituted natural problems for boundary methods. Such an axisymmetric flow in a diffuser was solved by White and Kline [74]. The use of boundary and domain methods by Wu [75, 76] has been quite extensive in solving problems with both potential and viscous regions.

One of the most valuable uses of the BEM has been in the solution of potential free surface problems. The problem formulations are easy (there are, of course, some difficult problems) because the problem is one of potential flow. The movement of the free surface constitutes the difficult part. That change in the configuration of the solution region requires, for nonlinear problems, the reformulation of the numerical grid. Such reformulation is much easier using boundary methods than using domain methods. Moreover, the inherent efficiency of a boundary method is often important due to the density of the grid and the small size required for the time step because of the stability and accuracy requirements. Much of this chapter is devoted to free surface problems with applications to water waves and flow in porous media. There are a great many free surface problems that do not fall into the category of potential flow. These, too, can often be solved advantageously by the Boundary Element Method.

3. Porous media flow

A large part of porous media flow falls into the classification of potential flow and boundary methods began to be used to solve such problems for that reason. The technique proved so effective that it has been extended to many porous media problems that are not potential flows.

3.1. Equations of porous media flow

The calculation of flow through a porous matrix usually assumes the dimensions of the particular problem are large compared with the dimensions of the individual grains of the medium. The problem then ignores the microscopic processes and treats the macroscopic process by Darcy's law

$$\boldsymbol{v} = -\mathbf{K}\cdot\nabla\Phi \tag{3.1}$$

in which $\boldsymbol{v}$ is the macroscopic velocity vector, $\mathbf{K}$ is the tensor of hydraulic conductivity, and Φ is the head (elevation plus pressure). Applying conservation of mass, $\nabla\cdot\boldsymbol{v} = 0$, gives

$$(\nabla\cdot\mathbf{K})\cdot\nabla\Phi + \mathbf{K}:\nabla\nabla\Phi = 0\,. \tag{3.2}$$

If the flow is isotropic, K can be represented by a scalar and (3.2) becomes

$$\nabla\cdot(K\nabla\Phi) = 0\,. \tag{3.3}$$

If, in addition, the hydraulic conductivity is independent of position, (3.3) becomes the Laplace equation and the flow is treated by the usual potential flow methods. The latter is often the case, or at least assumed. Unfortunately, the data used to determine K are almost always poor and the distribution of K with position is known only in a general sense. In such cases the hydraulic conductivity is assumed constant within large areas or zones. Fortunately, such an assumption is very often valid (at least to the accuracy to which K can be determined experimentally) because the geological history of the soil can be similarly divided into periods or episodes. The case of anisotropy can be treated as an isotropic flow in a solution region of distorted coordinates.

Usually groundwater aquifers are thin compared to their horizontal extent or the horizontal distances between features. In such cases a simplification called the Dupuit–Forchheimer approximation is used to suppress the vertical dimension. That approximation assumes that the vertical velocity is small compared with the horizontal velocity and thus the head (Φ) is constant in the z-direction. Equation (3.3) then becomes

$$\nabla \cdot (T \nabla \Phi) = 0 \tag{3.4}$$

in which $T = Kb$ (where b is the thickness of the aquifer) is called the "transmissibility". Although K is often taken as a constant, it is sometimes a poor approximation to take T as a constant because of variations in the aquifer thickness. In the case of unconfined flow the aquifer thickness is governed by the elevation of the lower confining layer (the "basement") and a free surface, in which case $T = K\Phi$ and (3.4) is written

$$\nabla \cdot (K \Phi \nabla \Phi) = 0 \,. \tag{3.5}$$

The governing differential equation is then nonlinear, but if K is constant then

$$\nabla^2 \Phi^2 = 0 \tag{3.6}$$

so that the equation is Laplace's in Φ^2. Unfortunately, the boundary conditions are expressed in terms of Φ so that a nonlinearity remains.

Another complication results from "leaky" aquifers. In case water passes through the top or bottom boundary of the aquifer, (3.3) becomes

$$\nabla \cdot (K \nabla \Phi) - k\Phi = 0 \tag{3.7}$$

in which k is a constant. A similar term is also added to (3.6) which means that it is no longer linear in Φ^2:

$$K \nabla^2 \Phi^2 - k\Phi = 0 \,. \tag{3.8}$$

Often the compressibility of the medium is important in which case the governing equation becomes the diffusion equation

$$\nabla \cdot (K \nabla \Phi) = S \frac{\partial \Phi}{\partial t} \tag{3.9}$$

in which S is the "storativity" of the medium.

In summary, there are about five equations that arise in porous media flow that are solved by the BEM. These include the Laplace equation, (3.3) in which zones are often used, (3.6) which is the Laplace equation in Φ^2 but with boundary conditions expressed in Φ, (3.7) which for constant K becomes a modified Helmholtz equation, and the diffusion equation (3.9). In dealing with the transport of material (pollution) through porous media, the advective terms must be added to (3.9). The BEM has generally not been considered suitable for the solution to that equation, but effective BEM solutions should emerge in the near future. In addition many porous media problems have a free surface which must be explicitly treated in detail. The set of equations listed above does not include those for unsaturated flow as no BEM solution has yet emerged in that highly nonlinear case.

3.2. *Free surface flow*

In many cases where we need to know local detail about the flow, the Dupuit–Forchheimer approximation is not valid. In the case that the aquifer is unconfined, the free surface must be treated as such. Two boundary conditions are applied to a free surface (see Fig. 8), i.e.

$$\Phi = \eta \tag{3.10}$$

$$\frac{\partial \eta}{\partial t} = -(1 + |\nabla_2 \eta|^2)^{1/2}(\nabla \cdot n) \tag{3.11}$$

in which η is the elevation of the free surface above an arbitrary datum, and ∇_2 is the horizontal gradient. (Also, t in (3.11) is really time divided by the "effective porosity" of the aquifer.) The first condition (3.10) expresses the fact that the pressure is zero on the free surface. The second condition results from the kinematic consequence that a particle on the free surface must remain on the free surface.

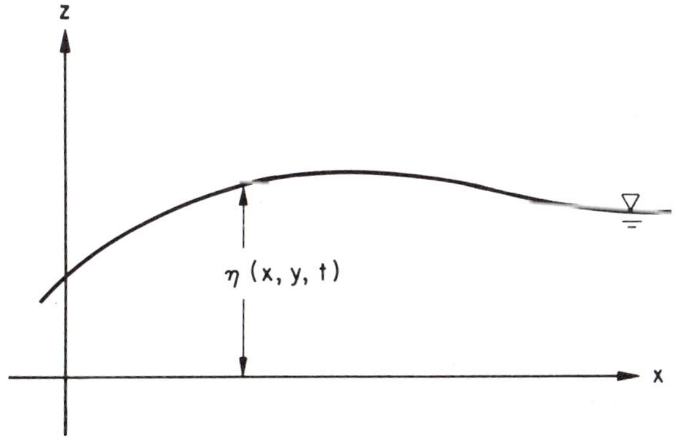

Fig. 8. The coordinate system for free surface flow.

Since free surface flow is very much affected by gravity, the coordinate system is taken where the z-axis is vertical. Using β as the angle between the free surface and the xy plane (horizontal), the two free surface conditions can be combined as

$$\left[\frac{\partial \Phi}{\partial t}\right]_{x,y} = -\frac{1}{\cos \beta}\frac{\partial \Phi}{\partial n} \qquad \text{on } z = \eta \tag{3.12}$$

where the subscripts indicate that the horizontal coordinates are to be held constant when taking the derivative but the derivative is expressed for Φ remaining on the free surface, $z = \eta$. Using time steps k ($t = k\Delta t$), (3.12) is discretised as

$$\Phi^{k+1} = \Phi^k - \frac{\Delta t}{\cos \beta}\left[\theta \frac{\partial \Phi^{k+1}}{\partial n} + (1-\theta)\frac{\partial \Phi^k}{\partial n}\right] \tag{3.13}$$

in which θ is a weighting factor that positions the normal derivative between the two time levels and is used to optimize the stability and accuracy of the calculation [41]. Equation (3.13) provides a relationship between Φ and $\partial\Phi/\partial n$ and is assembled together with the integral equations. Due to the angle β, (3.13) is nonlinear. In most calculations β is taken at time step k instead of weighting it between the two time steps. If the angle changes rapidly, an iterative solution will provide sufficient accuracy.

The free surface problem is nonlinear in another way that is often not recognized. The boundary element equations are written at the unknown time level, $k + 1$. Yet the integrals are performed on the known geometry at time level, k. This inconsistency can also be rectified by iteration, but it is usually of sufficient accuracy to take small time steps, which are necessary in any method that approximates the derivatives.

As with other sorts of free surface calculations, there are stability and accuracy considerations when using the BEM. The analysis of the stability criteria starts with the linearized problem where the free surface boundary condition is expressed as

$$\frac{\partial \Phi}{\partial t} = -\frac{\partial \Phi}{\partial z} \qquad \text{on } z = 0\,. \tag{3.14}$$

Both the nonlinearity of the kinematic condition and the nonlinearity caused by applying the condition to the unknown free surface have been eliminated in (3.14). In the usual manner a solution of the Laplace equation is given in a Fourier series which satisfies (3.14) and a lower boundary condition that $\partial\Phi/\partial z = 0$ on $z = -h$. A typical term is

$$\Phi = A \sin kx \frac{\cosh k(z+h)}{\cosh kh}\, \mathrm{e}^{-\sigma t} \tag{3.15}$$

where h is the distance between the equilibrium free surface ($z = 0$) and the bottom boundary, k is a wave-number, and σ is a (real) damping factor which must satisfy the relationship

$$\sigma = k \tanh kh \,. \tag{3.16}$$

Using (3.16) in (3.13) produces

$$\exp(-\sigma_N \Delta t) = 1 - \Delta t\, k \tanh kx[\theta \exp(-\sigma_N \Delta t) + (1 - \theta)] \tag{3.17}$$

in which σ_N is the numerical damping factor to be compared with the actual damping given in (3.16). Defining the difference in the numerical and actual damping as ε produces the plot on Fig. 9 [41] where the parameter is defined as $q = k\Delta t \tanh kh$. Figure 9 indicates that θ should be between 0.5 and 0.6 and that Δt should be small in order that q is not too large for any wavelength. The stability diagram is shown as Fig. 10. The analysis can be empirically verified by calculation. More details of a similar analysis for water wave problems is given in section 4.

3.3. *Axisymmetric flow*

In dealing with well problems the geometry of the problem often is axially symmetric. Although such problems can be treated three-dimensionally, it is

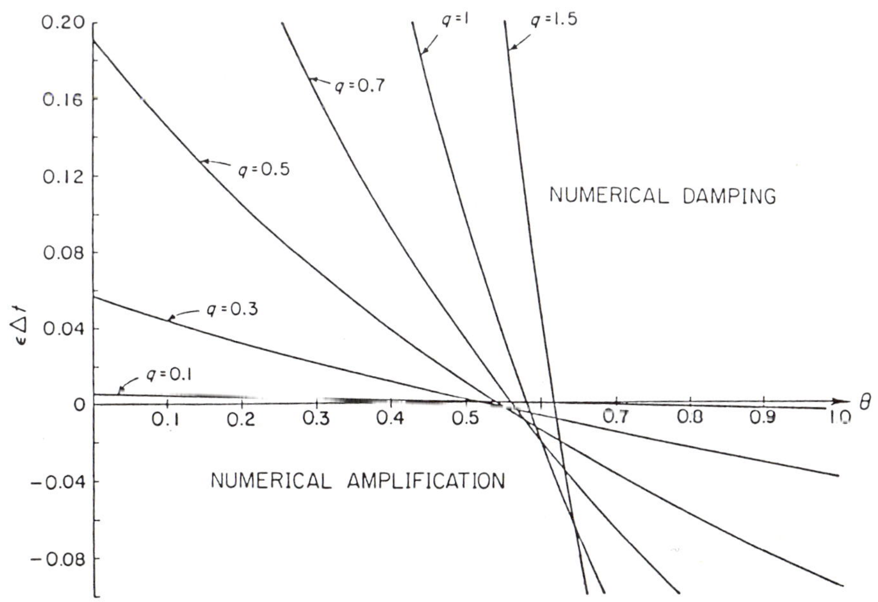

Fig. 9. Amplification and damping of free surface flow in porous media.

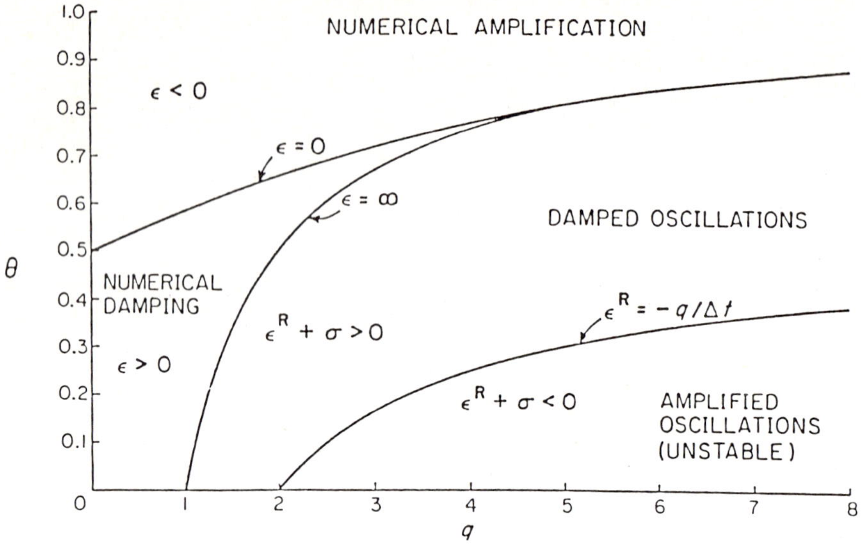

Fig. 10. The stability diagram for free surface flow in porous media.

much more efficient to take advantage of the symmetry. The counterpart of (1.1) in three dimensions is

$$-\lambda\Phi = \int_{\Omega} \left(\Phi \frac{\partial}{\partial n} \frac{1}{r} - \frac{1}{r} \frac{\partial \Phi}{\partial n} \right) \mathrm{d}A \tag{3.18}$$

in which $\lambda = 4\pi$ for interior points, $\lambda = 2\pi$ on smooth boundaries and is the solid angle of the boundary at corners. The Green's function has been specified as $1/r$ for the Laplace equation. Taking advantage of the axial symmetry in the θ-direction, (3.18) can be written

$$-\lambda\Phi = \int_{\Gamma} \left[\Phi \int_0^{2\pi} \frac{\partial}{\partial n} \frac{1}{r} \,\mathrm{d}\theta - \frac{\partial \Phi}{\partial n} \int_0^{2\pi} \frac{1}{r} \,\mathrm{d}\theta \right] r \,\mathrm{d}\xi \tag{3.19}$$

where Γ is the contour in the zr plane and $\mathrm{d}\xi$ is an increment of length along Γ (Fig. 11). The contour Γ may or may not be closed as shown in the figure. In the case of well problems, the well casing becomes a part of Γ and it is a closed contour entirely on one side of the z-axis. In other situations Γ may intersect the z-axis and thus remain open; the indicated integration does not include any part of the z-axis.

From (3.19) an "axisymmetric Green's function" can be defined

$$g(r, z; r_i, z_i) = \int_0^{2\pi} \frac{1}{r} \,\mathrm{d}\theta \,. \tag{3.20}$$

(We note parenthetically that there is nothing new about the axisymmetric

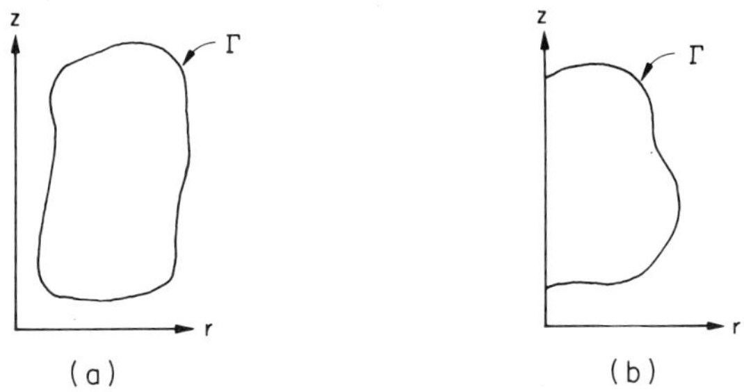

Fig. 11. Solution regions for axisymmetric flow with (a) a region with a hole along the axis of rotation and (b) a region without the hole.

Green's function; it was known to Trefftz [67] in 1917.) The function can be written as an elliptic integral

$$g = \int_0^{2\pi} \frac{d\theta}{(a - b\cos\theta)^{1/2}} = \frac{4}{(a+b)^{1/2}} K(m) \tag{3.21}$$

where

$$a = r_i^2 + r^2 + z_d^2 \qquad b = 2rr_i \qquad z_d = z - z_i \tag{3.22}$$

$$K(m) = \int_0^{2\pi} \frac{d\theta}{(1 - m\sin^2\theta)^{1/2}} \qquad 0 \leqslant m < 1\,. \tag{3.23}$$

The modulus, m, of the elliptic integral and its complimentary modulus, m_1, arc

$$m = \frac{4rr_i}{(r+r_i)^2 + z_d^2} \qquad m = \frac{(r-r_i)^2 + z_d^2}{(r+r_i)^2 + z_d^2}\,. \tag{3.24}$$

The elliptic integral can most easily and efficiently be evaluated using a rational approximation (see, e.g., Ref. [1])

$$K(m) = [a_0 + a_1 m_1 + \cdots + a_4 m_1^4] + [b_0 + b_1 m_1 + \cdots + b_4 m_1^4]$$
$$\times \ln(1/m_1) + \varepsilon(m) \qquad |\varepsilon(m)| \leqslant 2 \times 10^{-8}\,. \tag{3.25}$$

The values of the coefficients are

$$a_0 = 1.38629436112 \qquad b_0 = 0.50000000000$$

$$a_1 = 0.09666344259 \qquad b_1 = 0.12498593597$$

$$a_2 = 0.03590092383 \qquad b_2 = 0.06880248576$$

$$a_3 = 0.03742563713 \qquad b_3 = 0.03328355346$$

$$a_4 = 0.01451196212 \qquad b_3 = 0.0044178012\,.$$

The use of the elliptic integral requires that the boundary integrals be evaluated using numerical quadrature and caution should be taken because the function has a logarithmic singularity when $r \to 0$. Specially weighted quadrature [1] can be used to evaluate such integrals accurately.

In the term $\partial g/\partial n$ an elliptic integral of the second kind, $E(m)$, appears

$$\begin{aligned}\frac{\partial g}{\partial n} &= \frac{\partial g}{\partial r} n_r + \frac{\partial g}{\partial z} n_z \\ &= \frac{2n_r}{r(a+b)^{1/2}} [E(m) - K(m)] - 4 \frac{[n_r(r-r) + n_z z_d]}{(a-b)(a+b)^{1/2}} E(m)\end{aligned} \tag{3.26}$$

in which n_r and n_z are direction cosines and $E(m)$ is a complete elliptic integral of the second kind. $E(m)$ is evaluated using the rational approximation [1]

$$\begin{aligned}E(m) = [1 + c_1 m_1 + \cdots + c_4 m_1^4] + [d_1 m + \cdots + d_4 m^4] \\ \times \ln(1/m_1) + \varepsilon(m) \qquad |\varepsilon(m)| < 2 \times 10^{-8} .\end{aligned} \tag{3.27}$$

The coefficients are

$$\begin{aligned} c_1 &= 0.44325141463 & d_1 &= 0.24998368310 \\ c_2 &= 0.06260601220 & d_2 &= 0.02900180037 \\ c_3 &= 0.04757483546 & d_3 &= 0.04069697526 \\ c_4 &= 0.01746596451 & d_4 &= 0.00426449639 . \end{aligned}$$

Well problems in confined aquifers are solved directly by the methods cited above. In unconfined aquifers the free surface must be moved in much the same way as it is for two-dimensional flow [36, 37].

3.4. *Leaky aquifers – the modified Helmholtz equation*

With K constant, Eq. (3.7) becomes

$$\nabla^2 \Phi - \lambda \Phi = 0 \tag{3.28}$$

which has the Green's function in two dimensions

$$G = -K_0(\sqrt{s} r) \tag{3.29}$$

where K_0 is a modified Bessel function of the second kind and of order zero. (The integral equation is the boundary integral part of (3.33) below.) Accurate polynomial approximations are available to evaluate this function and its derivative, $K_1(\sqrt{s}r)$. As the counterpart for the Laplace equation K_0 has a logarithmic singularity as $r \to 0$; thus, some care must be exercised in the boundary integration when the source point is on the target element. Generally, Gaussian quadrature has been used for the nonsingular integration. When the integration is singular, one method of integration is to use special Gaussian quadrature with logarithmic weights. Alternately, the logarithm of r can be added and subtracted from the modified Bessel function. The term where the log is added becomes regular and can be integrated by normal quadrature methods, i.e.

$$K_0(\mathrm{p}) = [K_0(\mathrm{p}) + \ln r] - \ln r \tag{3.30}$$

where the term in brackets is not singular as $r \to 0$. The other term contains only the log and can be integrated analytically for polynomial interpolation functions.

Problems using the modified Helmholtz equation can be zoned in much the same way as is described for the Laplace equation in section 1. Also, it is frequently necessary to join zones of the Laplace equation to zones of leaky aquifers.

3.5. Unsteady flow – the diffusion equation

Taking K as constant and absorbing K and S into t, (3.9) becomes

$$\nabla^2 \Phi = \frac{\partial \Phi}{\partial t} . \tag{3.31}$$

3.5.1. The Laplace transform method

One method to solve the diffusion problem has been to remove the time dependence from (3.31) by means of the Laplace transform [53],

$$\tilde{\Phi}(x, y, s) = \int_0^\infty \Phi \, \mathrm{e}^{-st} \, \mathrm{d}t . \tag{3.32}$$

Applying (3.32) to (3.31) yields

$$\nabla^2 \tilde{\Phi} - s\tilde{\Phi} + \Phi_0 = 0 \tag{3.33}$$

where Φ_0 is the initial value of Φ. The homogeneous part of the equation is a modified Helmholtz equation. The resulting integral is

$$2\pi\tilde{\Phi} = \int_\Gamma \left[\tilde{\Phi}\sqrt{s}K_1(\sqrt{s}r) \frac{\partial r}{\partial n} + K_0(\sqrt{s}r) \frac{\partial \tilde{\Phi}}{\partial n} \right] \mathrm{d}s + \int_\Omega K_0(\sqrt{s}r)\Phi_0 \, \mathrm{d}A \tag{3.34}$$

in which the domain integral reflects the influence of the initial conditions on the solution. Although the integration must be done numerically, there is usually no problem in solving (3.34). Usually from eight to ten values of s are taken. These values are selected from the plot of $\tilde{\Phi}/\Phi_s$ where Φ_s is the steady-state value of Φ. The values of s are selected in logarithmic progression (e.g. 1, 10, 100, . . .) along the steep part of the curve.

The difficulty with using the Laplace transform method comes in the inversion of the transform to recover the physical variables. Rizzo and Shippy [53] used the method of Schapery [56] which assumes that the solution progresses smoothly from an initial condition to steady state. If another behavior of the solution is known, the inversion can be tailored to that function. If, however, the behavior is unknown or if the solution does not progress smoothly from one state to another, the inversion is likely to give bad results. The lack of generality of the method has prevented it from coming into wide use. Narayanan and Beskos [51] have examined eight methods of numerical inversion of the Laplace transform for applicability, accuracy and computational efficiency using vibration problems as the application basis. They differentiate between methods that use little computer time but are of low accuracy and those that are of greater accuracy but use more computer time. The method of Schapery is in the former category where Narayanan and Beskos recommend the method of Durbin [12] in the latter category.

3.5.2. *The Green's function method*

Equation (3.31) possesses the Green's function

$$G = \frac{H(t-\tau)}{(t-\tau)^{d/2}} \exp\left(-\frac{r^2}{4(t-\tau)}\right) \tag{3.35}$$

where d is the dimensionality of the problem ($d=2$ in two dimensions, $d=3$ in three dimensions) and τ is the variable of integration in the two-dimensional equation

$$\lambda\Phi = \int_{t_0}^{t}\int_{\Gamma} \left(G\frac{\partial \Phi}{\partial n} - \Phi\frac{\partial G}{\partial n}\right) \mathrm{d}s\,\mathrm{d}\tau + \int_{\Omega} \Phi(r, t_0) G\,\mathrm{d}A \tag{3.36}$$

where $\lambda = 4\pi$ in the interior, $\lambda = 2\pi$ on a smooth boundary, and λ is twice the boundary angle at a boundary corner. The domain integral reflects the influence of the initial conditions on the solution. The solution of (3.36) using boundary elements was given by Shaw [57].

In general the time integral in (3.36) must be done numerically, leading to the necessity of employing a time stepping technique. There are two primary ways to do the time stepping, both with disadvantages. First, the problem can be solved using (3.36) up to some short time Δt, then the solution at this time used to advance another step Δt. The difficulty with this method is that the domain integral must be done at each time step. The method then loses the efficiency of a boundary method and, in fact, is no more efficient than the Finite Element Method [65].

The second method of solution is to solve (3.36) for a target time, say t_t, that is far from the initial conditions. In order to perform accurately the time integration, intermediate solutions must be established on the boundary at frequent intervals between t_0 and t_t. Thus, (3.36) has to be solved for short time intervals, Δt. Assuming that the boundary solution is known from solving (3.36) up to time Δt, the solution cannot be advanced directly to $2\Delta t$ because the intermediate target time ($2\Delta t$) appears not only in the limits but also in the integrand, which means that the integral is different at time Δt for a target time of $2\Delta t$ than for a target time of Δt. The solution at time Δt can be used, however, to improve the accuracy of the time integration beginning at t_0 with target time $2\Delta t$. In order to reach the time of $2\Delta t$, the section between t_0 and Δt must be integrated twice, once to find the solution at time Δt and once to find the solution at time $2\Delta t$. In finding the solution at time $t_t = n\Delta t$ the boundary integration has to be done $n\Delta t + (n-1)\Delta t + (n-2)\Delta t + \cdots + \Delta t$ times. This method avoids the domain integration (except at the initial time which can also be avoided if the initial condition is $\Phi = \text{constant}$), but it loses efficiency due to the repeated boundary integrations at the early time steps. The result is that, like the straight time stepping method, it has no advantage in efficiency over the Finite Element Method.

Fortunately, in either method the domain integrations can be done automatically with a technique that is transparent to the user [65]. The discretization is simpler than with the Finite Element Method since the internal elements do not have the same connectivity requirements. In doing the automatic domain integration the user sees the technique as a boundary method and it retains all of the advantages of a boundary method except for the efficiency.

The best programs use a combination of the two methods described above. If the first method (where the domain integration is done frequently) is used, the domain integration must be done accurately since it (that is, the initial condition) is the primary factor in determining the solution a short time step away. If a long interval occurs between the domain integration and the target time, the solution is not so dependent on the accuracy of the domain integration.

3.6. Nonlinear flow – the Dupuit–Forchheimer approximation

In almost all regional groundwater calculations the reduction of the problem to two horizontal dimensions is an excellent approximation. The groundwater aquifers are thin compared with their horizontal dimensions and the vertical flow can be neglected compared with the horizontal flow. This approximation, the Dupuit–Forchheimer approximation, leads to (3.6), which is simply the Laplace equation in the square of the potential and can be solved by standard methods. A problem arises in the frequent case of a leaky aquifer which is governed by (3.8). In that case the equation contains both linear and nonlinear terms. There are two general ways to solve the problem by the BEM [63]. The first is to write (3.8) as

$$\nabla^2\Phi^2 - k\frac{\Phi^2}{\Phi_a} = 0 \tag{3.37}$$

where Φ_a is an average value of Φ in some sense (often the average of Φ at the boundary nodes) and is treated as a constant. Equation (3.37) then becomes a modified Helmholtz equation in Φ^2. This approximation can be made more accurate by zoning so that the average more nearly represents the local value of the potential. Unless the leak dominates the solution, it is usually of sufficient accuracy and is reasonably efficient.

The second method is to put the "nonlinear" term on the right side of the equation

$$\nabla^2\Phi^2 = k\Phi = f(\Phi^2)\,. \tag{3.38}$$

Then (3.38) is solved as a Poisson equation

$$2\pi\Phi^2 = \int_\Gamma \left(\frac{\Phi^2}{r}\frac{\partial r}{\partial n} - \ln r\,\frac{\partial \Phi^2}{\partial n}\right)\mathrm{d}s - \int_\Omega f \ln r\,\mathrm{d}A\,. \tag{3.39}$$

Using (3.39) the solution is iterative. The function f is first estimated and then Φ^2 is found, leading to better estimate of f, etc., until the solution is considered sufficiently accurate. This method does have the disadvantage of a domain integration, but such integrations can be done automatically. Usually, the solution is dominated by the left-hand side of the equation.

3.7. Stream aquifer interaction

An application of the above techniques is the problem of the mutual response of a stream and an aquifer which are interconnected (Fig. 12 in which one half of the problem is shown and symmetry is assumed for the other half). The aquifer may be "connected" or "unconnected". In the first instance the phreatic surface intersects the streambed; in the second instance the stream recharges the aquifer through a section that is probably unsaturated. If the phreatic surface connects the streambed, the flow may be from the stream into the groundwater or it may be out of the aquifer into the stream. In the latter case there appears a seepage surface on the sides of the stream in which the intersection of the free surface and the channel is an unknown point. All of these different conditions may appear in a single episode. The problem is a moving-boundary problem, but one with the

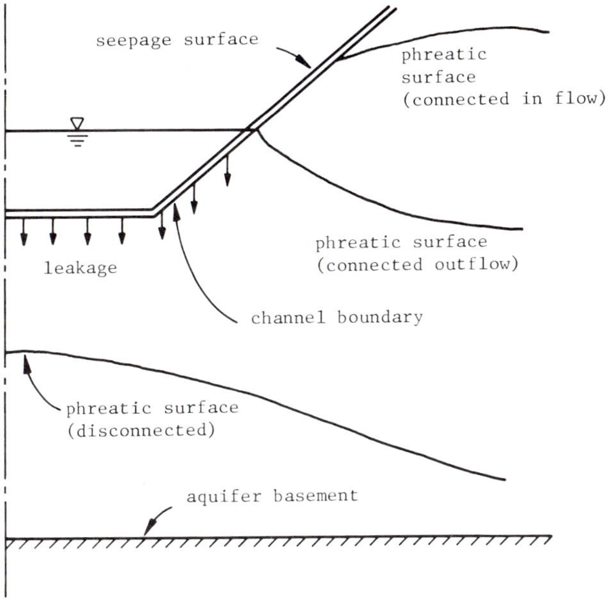

Fig. 12. Definition sketch for stream–aquifer interaction showing both the connected and unconnected states.

additional complications of multiple boundary conditions which change from one type to another over the section of boundary near the stream.

The outflow of the channel is governed by the permeability of a "blanket" that lines the channel. The blanket is less permeable than the underlying porous media. The boundary condition is

$$\frac{\partial \Phi}{\partial n} = K_b(p_s - p) \tag{3.40}$$

where p_s is the pressure in the stream, p is the pressure in the porous media adjacent to the stream, and K_b is a constant dependent on the permeability and thickness of the blanket.

As the free surface contacts or leaves the channel, the surface grid must be regenerated and the boundary conditions reset. Since the free surface positions and normal derivatives of the previous time step are important, the regeneration process must not forget those features. The grid regeneration, while retaining the features and accurate geometrical description, is the challenging part of the computer code. The actual code is somewhat self-checking in that the conservation of mass is constantly monitored. Mass conservation is difficult to achieve during the connection and disconnection processes but can be done with sufficiently small time steps and care in the solution.

More detailed descriptions of these problems are found in Dillon and Liggett [9], Dillon [10], and Liggett [42].

3.8. Miscellaneous

There are many other problems in porous media which can be solved by the BEM. Multiphase flows consist of two or more fluids of different densities. The problem becomes a free surface case where the interface between the two fluids is free to move [41]. Applications include intrusion of seawater into coastal aquifers and flow in oil–water systems.

The seawater intrusion problem is solved more thoroughly by Taigbenu and Liggett [63]. The Dupuit–Forchheimer approximation is used together with the Ghyben–Herzberg approximation, which states that the motion of the fluid in the heavier, seawater region can be neglected. These approximations lead to diffusion-type equations. A "special potential" [61] can be defined so that the two flow regions, seawater and freshwater, can be treated as one.

The BEM has not been applied to unsaturated flow in porous media due to the nonlinearity of the problem. It has, however, been coupled to the Finite Element Method for the solution of such problems [14, 41]. The BEM is used in regions of saturated flow and the Finite Element Method is used in regions of unsaturated flow. In free surface problems which require many time steps, a large part of the coefficient matrix for the simultaneous equations can be decomposed at the beginning of the calculation so that subsequent time steps require only the solution of a tightly banded coefficient matrix.

Three-dimensional BEM solutions in porous media are not yet common, perhaps due to the length of the program. Details of such calculation in porous media are found in Lennon et al. [38].

Perhaps the primary challenge of the method occurs in calculating problems in groundwater basins in which there are a variety of different conditions including confined and unconfined aquifers, wells, streams, leaky aquitards, recharge, anisotropy, and inhomogeneities. The computer program for such cases is necessarily large and complex. The BEM appears to hold a considerable advantage over domain methods for reasons of efficiency, and the ease of data input. Such a program for steady-state calculations was coded by Lafe [30] (see [31, 41]). That program, however, is not as efficient in terms of computer time and storage as is possible and it does not consider the important unsteady problem.

4. Water waves

One of the most elegant and advantageous uses of the BEM is in the calculation of waves on a free surface. The problem in its primitive form is governed by the Laplace equation since viscous damping is truly negligible for the vast majority of cases. The problem is challenging in that, since there is no damping, any numerical instability will destroy the solution. Waves have natural instabilities, especially in the three-dimensional form, which from the point of view of numerical calculation can be almost indistinguishable from numerical instabilities.

There are several types of equations which govern the linearized version of the wave problem.

4.1. *The nonlinear free surface problem*

In the usual manner the velocity vector is defined as the gradient of a potential

$$\boldsymbol{v} = -\nabla\Phi \tag{4.1}$$

leading to the Laplace equation

$$\nabla^2\Phi = 0\,. \tag{4.2}$$

The free surface boundary conditions consist of the kinematic condition

$$\frac{\partial\eta}{\partial t} = \frac{\partial\Phi}{\partial z} - \frac{\partial\Phi}{\partial x}\frac{\partial\eta}{\partial x} - \frac{\partial\Phi}{\partial y}\frac{\partial\eta}{\partial y} \quad \text{on } z = \eta \tag{4.3}$$

where η is the vertical distance from an arbitrary datum (usually the equilibrium free surface) to the free surface, and the coordinate system is aligned with z vertical. The dynamic free surface condition is a form of the Bernoulli equation which states that the pressure is zero

$$\frac{\partial\Phi}{\partial t} = B - g\eta - \frac{1}{2}\left[\left(\frac{\partial\Phi}{\partial x}\right)^2 + \left(\frac{\partial\Phi}{\partial y}\right)^2 + \left(\frac{\partial\Phi}{\partial z}\right)^2\right] \quad \text{on } z = \eta \tag{4.4}$$

in which B is the Bernoulli constant and the pressure has been set to zero.

As stated in this manner the problem is highly nonlinear. Both the kinematic and dynamic boundary conditions arc nonlincar, but thc primary difficulty is that these conditions must be applied to the unknown free surface.

4.2. *Other wave equations*

The difficulties with the nonlinear free surface problem have caused the derivation of a number of approximate equations for special cases. Linearizing (4.3) and (4.4) and applying the boundary conditions to the equilibrium surface gives

$$\frac{\partial\eta}{\partial t} = \frac{\partial\Phi}{\partial z} \quad \text{on } z = 0 \tag{4.5}$$

$$\frac{\partial\Phi}{\partial t} = -g\eta \quad \text{on } z = 0 \tag{4.6}$$

which can be combined to yield

$$\frac{\partial^2 \Phi}{\partial t^2} = -g \frac{\partial \Phi}{\partial z} \qquad \text{on } z = 0 . \tag{4.7}$$

If ω is the frequency of a wave component, then the harmonic time dependence can be removed

$$\Phi = \phi \exp(-\mathrm{i}\omega t) \tag{4.8}$$

where i is the the unit imaginary number. The frequency, ω, must satisfy the dispersion equation,

$$\omega^2 = gk \tanh kh \tag{4.9}$$

when (4.8) is used in (4.7). The wave-number is $k = 2\pi/L$ where L is the wavelength, g is the acceleration of gravity, and h is the equilibrium water depth. The reduced potential, ϕ, satisfies the Laplace equation.

Two important two-dimensional approximations further simplify the problem. In problems in which there is a rigid obstacle (wall, breakwater, etc.) parallel to the y-axis, let

$$\phi(z, y, z) = \phi_{\mathrm{v}}(x, z) \exp(\mathrm{i}k_y y) . \tag{4.10}$$

Using (4.10) in the three-dimensional Laplace equation produces a modified Helmholtz equation (see section 3.4 for solution methods)

$$\frac{\partial^2 \phi_{\mathrm{v}}}{\partial x^2} + \frac{\partial^2 \phi_{\mathrm{v}}}{\partial z^2} - k_y^2 \phi = 0 . \tag{4.11}$$

If the water depth is everywhere constant, then the incident waves approaching any region may be described as

$$\phi_{\mathrm{i}} = -\frac{\mathrm{i}gA}{\omega} \frac{\cosh k(z+h)}{\cosh kh} \exp(\mathrm{i}kx) \tag{4.12}$$

in which A is a constant related to the amplitude of the wave. If the incoming wave interacts with obstacles that are vertical and penetrate the fluid from the bottom through the free surface (the horizontal plane problem), then a reduced potential can be defined as ϕ_{h} in the equation

$$\phi = \phi_{\mathrm{i}} - \frac{\mathrm{i}gA \cosh k(z+h)}{\omega \cosh kh} \phi_{\mathrm{h}} . \tag{4.13}$$

Using (4.13) in the Laplace equation produces the Helmholtz equation

$$\frac{\partial^2 \phi_{\mathrm{h}}}{\partial x^2} + \frac{\partial^2 \phi_{\mathrm{h}}}{\partial y^2} + k^2 \phi = 0 \tag{4.14}$$

which has a Green's function

$$G = H_0^1(kr) \qquad r = [(x - x_i)^2 + (y - y_i)^2]^{1/2} \tag{4.15}$$

where H_0^1 is a Hankel function of the first kind and of order zero. Like the modified Bessel function used in the modified Helmholtz equation, the Hankel function has a logarithmic singularity as the argument passes to zero. The integration should be handled in a similar manner to that explained in section 3.4 and the Hankel function can be divided according to Eq. (3.30) except with the signs on the ln terms reversed.

Water wave problems described by the above equations have a number of applications. One of the first such papers appears to be that of Shaw [59] which treated such problems as wave scattering about an island. Other applications include the harbor resonance problem (which was actually computed by Lee [33] and Lee and Raichlen [34] before boundary element methods became popular) and the extensive applications of sloshing within a container (e.g. the series by Nakayama and Washizu [47–50]. The equations of this section all represent linear problems and have applications far wider than that of water waves. Some of the primary advances in the field have had acoustic problems as a motivation. Shaw [58] gives a general review of the Boundary Integral Method applied to wave problems. His paper includes an extensive bibliography. Due to the extensive coverage of these types of problems in other papers and to the highly developed nature of linear wave theory, the present treatment will proceed to the subject of nonlinear waves.

There is, however, one other application that will be mentioned briefly. That is the important problem of fluid–structure interaction, which has been treated from a linear point of view from both the fluid side and structural side. It is unfortunate that more has not been done with nonlinear fluid–structure interaction. From the point of view of calculating hydrodynamic force, the greatest forces and the most critical conditions from the standpoint of safety or design come about when the wave forces exceed those which can be generated by waves small enough to be approximated by linear theory. Also, the structural deformations that apply to the failure conditions exceed the limits of linearity. The linear theory does, however, lead to useful conclusions, especially in identifying resonance conditions.

Eatock Taylor [13] provides a review of the problem of calculating hydrodynamic loads, including those on fixed and floating objects. He uses not only boundary elements but also coupled finite elements and boundary elements, and he speculates that future developments will make further use of the coupling of the two methods. Many such problems can be treated by boundary element methods alone. Liu and Cheng [43] examine the two-dimensional response of a dam and reservoir subject to earthquake loads. Obviously, there is much more to be done in the area of fluid–solid interaction.

4.3. Nonlinear, two-dimensional waves

Defining β as the angle the free surface makes with the horizontal plane, the derivatives of Φ are

$$\frac{\partial \Phi}{\partial x} = -\frac{\partial \Phi}{\partial n} \sin \beta + \frac{\partial \Phi}{\partial s} \cos \beta \tag{4.16a}$$

$$\frac{\partial \Phi}{\partial z} = \frac{\partial \Phi}{\partial n} \cos \beta + \frac{\partial \Phi}{\partial s} \sin \beta \tag{4.16b}$$

where n is normal to the free surface and s is tangent to the free surface. Using (4.16) in the kinematic condition (4.3) gives

$$\frac{\partial \eta}{\partial t} = -\frac{1}{\cos \beta} \frac{\partial \Phi}{\partial n} . \tag{4.17}$$

The same quantities substituted into the dynamic boundary condition yield

$$\left(\frac{\partial \Phi}{\partial t}\right)_x = B - \frac{1}{2}\left[\left(\frac{\partial \Phi}{\partial s}\right)^2 - \left(\frac{\partial \Phi}{\partial n}\right)^2 - 2\frac{\partial \Phi}{\partial s}\frac{\partial \Phi}{\partial n} \tan \beta\right] - g\eta \tag{4.18}$$

where the subscript on the time derivative indicates that it is to be taken with x held constant but allowing z to vary with the free surface, or

$$\left(\frac{\partial \Phi}{\partial t}\right)_x = \frac{\partial \Phi}{\partial t} + \frac{\partial \Phi}{\partial z}\frac{\partial \eta}{\partial t} . \tag{4.19}$$

Equations (4.17) and (4.18) need to be discretised in a finite difference scheme in terms of the unknowns at the advanced time step, Φ^{k+1} and $(\partial\Phi/\partial n)^{k+1}$. The BEM equations should be written at the advanced time step. From (1.6)

$$[L]^{k+1}\{\Phi\}^{k+1} = [R]^{k+1}\{\partial\Phi/\partial n\}^{k+1} . \tag{4.20}$$

But the geometry is unknown at time $k+1$ so that (1.6) is actually written

$$[L]^{k}\{\Phi\}^{k+1} = [R]^{k}\{\partial\Phi/\partial n\}^{k+1} . \tag{4.21}$$

As in the case of the free surface in porous media, this difficulty can be resolved by iteration. At the same time the difference equations from the free surface conditions are nonlinear and thus create a set of nonlinear simultaneous equations, giving a second reason to iterate. In spite of these two types of nonlinearities, many problems can be solved directly. The free surface conditions are discretised using finite differences in time. The equations are then linearized in terms of the unknowns at the advanced time step and (4.21) is used directly without iteration. Discretization schemes are given in Liu and Liggett [44, 44a], Lennon et al. [39], Nakayama and Washizu [47–50] and Faltinsen [15].

It is often convenient to move the nodal points along nonvertical lines instead of vertically. That can be done by replacing β in the above equations by $\beta + \gamma$ where γ is the angle of rotation of the line along which the nodes are to move [41]. If $\gamma = -\beta$, the nodes can be moved normal to the free surface, which is convenient for highly curved surfaces such as in the case of breaking waves. Longuet-Higgens and Cokelet [45] and Cokelet [7] were able to compute the shape of the free surface well into the breaking phase. They used a complex variable formulation and a transformation which transforms the periodic free surface into a closed contour. In the complex plane the integral equation becomes Cauchy's equation

$$\beta(z_0, t) = \frac{1}{2\pi \mathrm{i}} \int_C \frac{\beta(z, t)}{z_c - z_0} \, \mathrm{d}z_c \tag{4.22}$$

in which $\beta = \Phi + \mathrm{i}\Psi$ is the complex potential and $z_c = x + \mathrm{i}y$ (see also Hunt and Isaacs [22]). The key to the calculation appears not to be the complex formulation but the method of moving the free surface along its own normal.

4.4. Stability and accuracy of the free surface problem

The stability of the method probably depends on the exact finite difference discretization, but the analysis of stability must come from a linearized scheme. Using the linearized boundary conditions, Salmon et al. [55] and Liu and Liggett [44a] expressed the change in the free surface and the potential over a time step as

$$\eta^{k+1} - \eta^k + \frac{\Delta t}{2}\left(\frac{\partial \Phi^k}{\partial n} + \frac{\partial \Phi^{k+1}}{\partial n}\right) + O(\Delta t)^3 \tag{4.23}$$

$$\Phi^{k+1} = \Phi^k - g\eta^k \Delta t - \frac{g\Delta t^2}{2}\left(a \frac{\partial \Phi^{k+1}}{\partial n} + b \frac{\partial \Phi^k}{\partial n}\right) + O(\Delta t)^3 \tag{4.24}$$

in which a and b are constants that can be adjusted to provide the maximum stability and accuracy to the scheme. Although (4.24) does not contain η^{k+1}, it is completely general; a term containing η^{k+1} could be replaced with its value from (4.23) leading to the same equation with different definitions of a and b.

In order to proceed with the analysis, (4.23) and (4.24) are combined to form a single equation with terms written at three time levels

$$\begin{aligned} \Phi^{k+1} &= 2\Phi^k - \Phi^{k-1} \\ &\quad - \frac{g\Delta t^2}{2}\left((1-b)\frac{\partial \Phi^{k-1}}{\partial n} + (1-a+b)\frac{\partial \Phi^k}{\partial n} + a\frac{\partial \Phi^{k+1}}{\partial n}\right) \\ &\quad + O(\Delta t)^3 . \end{aligned} \tag{4.25}$$

The linear equation is solved by

$$\Phi = A\exp(-\mathrm{i}\hat{\sigma}t)\sin Kx\cosh K(h+z) \tag{4.26}$$

in which A is the amplitude of the potential, $\hat{\sigma}$, is the frequency, h is the depth from the solid bottom to the undisturbed free surface, and K is the wave-number. The right-hand side of (4.26) could be considered as one term of a Fourier series. An equation for the numerical value of the frequency is found by substituting (4.26) into (4.25)

$$(1+ap)\exp(-2\mathrm{i}\hat{\sigma}\Delta t) - [2-(1-a+b)]\exp(-\mathrm{i}\tilde{\sigma}\Delta t) + 1 + (1-b)p = 0 \tag{4.27}$$

where $2p = g(\Delta t)^2 K\tanh Kh$. After solving for $\exp(-\mathrm{i}\hat{\sigma}t)$

$$\exp(-\mathrm{i}\hat{\sigma}t) = \{2-(1-a+b)p - \mathrm{i}[8p - p^2(1-6a+2b+2ab+a^2+b^2)]^{1/2}\}/[2(1+ap)]\,. \tag{4.28}$$

In solving for the exponential, the negative sign was taken of a square root and the imaginary quantity was written so that it is positive for small p, which is of the order of Δt^2.

The frequency of $\hat{\sigma} = \sigma_r + \mathrm{i}\sigma_i$. The amplification factor is

$$\exp(-\mathrm{i}\sigma_i\Delta t) = \frac{[1+p(1+a-b)+ap^2(1-b)]^{1/2}}{1+ap} \tag{4.29}$$

which is unity, meaning that there is no numerical amplification or damping, if

$$a+b=1\,. \tag{4.30}$$

Although the analyst may occasionally want a slight numerical damping, the sum of a and b should be close to, but not exceeding, unity. The numerical frequency distortion can be investigated by examining the real part of the frequency

$$\tan\sigma_r\Delta t = \frac{[8p - p^2(1-6a+2b+2ab+a^2+b^2)]^{1/2}}{2-p(1-a+b)}\,. \tag{4.31}$$

The numerical frequency should be equal to the analytical frequency which is

$$\sigma = (gK\tanh Kh)^{1/2}\,. \tag{4.32}$$

Using (4.30) in (4.31) gives the numerical frequency as a function of the parameter a. Figure 13 shows a plot of the frequency error. A value of $a = 0.17$ indicates that the numerical frequency is nearly the same as the true frequency.

In order minimize the accumulation of error, Nakayama and Washizu [50] have used an idea from the finite difference marker and cell (MAC) method of Hirt and Harlow [20] (see Welch et al. [73]). They define D such that

$$\frac{\partial D}{\partial t} = \left(\frac{\partial \Phi}{\partial t}\right)_x + \frac{1}{2}\left[\left(\frac{\partial \Phi}{\partial s}\right)^2 - \left(\frac{\partial \Phi}{\partial n}\right)^2 - 2\,\frac{\partial \Phi}{\partial s}\,\frac{\partial \Phi}{\partial n}\tan\beta\right] + g\eta \tag{4.33}$$

which is the free surface condition (4.18) with the Bernoulli constant set equal to zero. The quantity D should be identically equal to zero so that if the boundary condition is satisfied, then $\partial D/\partial t = 0$. Actually, D is discretised with the derivative for D approximated as

$$\frac{\partial D}{\partial t} = \frac{D^{k+1} - D^k}{\Delta t}\,. \tag{4.34}$$

Although D^k should be zero, some numerical error in the calculation for time

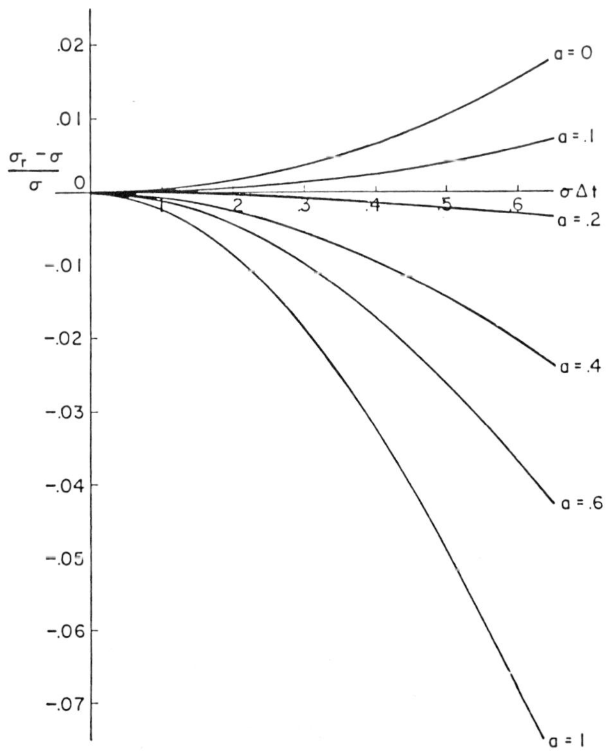

Fig. 13. Frequency error for water wave calculations.

level k has given it a small value. This error is used to correct the calculation by setting D^{k+1} equal to zero and writing (4.34) as

$$\frac{\partial D}{\partial t} = \frac{-D^k}{\Delta t} . \tag{4.35}$$

The value of D^k is calculated at each free surface nodal point from

$$D^k = D^{k-1} + \frac{\partial D}{\partial t} \Delta t \tag{4.36}$$

where $\partial D / \partial t$ is obtained from (4.35).

4.5. Far-field boundary conditions

The region of the numerical calculation must be finite even though the wave problem takes place in an infinite ocean. For hydrodynamic problems, perhaps more than other types of problems, the boundary conditions at the boundaries of the numerical region can contaminate the solution to an unacceptable degree. The radiation boundary condition expresses the condition that the boundaries admit incoming waves unchanged and permit outgoing waves to exit the computational region without reflection. In many engineering calculations the placement of the boundaries sufficiently far from the region of interest is accurate enough. In fact, the common method of placing the computational boundaries is to recalculate the solution with the boundaries further and further from the region of interest until there is no change in the solution. That method will not, in general, work for water wave problems because the problem is completely undamped. No matter how far away the computational boundary is, any unwanted reflections will eventually travel back to the region of interest. The problem is further complicated by economics. Wave problems usually require on the order of a dozen nodal points (using linear elements) per wavelength to ensure sufficient accuracy (more if the waves are particularly steep). Placing the computational boundaries far away can be expensive.

In a three-dimensional problem reflections from the computational boundaries are not so serious in the sense that the reflected energy is scattered over an area and becomes weaker as the reciprocal of distance from the boundary. On the other hand it pays to terminate the solution region as close as possible to the region of interest in view of the expense of large regions, in which case the unwanted reflections are particularly close.

The general problem of a nonlinear radiation boundary condition is unsolved. One approach is to linearize the problem outside of the computational region and use the linearized solution to extend the nonlinear solution. That approximate approach is used in the next section for the case of the steady solution of flow around a submerged object.

The problem of a radiation boundary condition is not confined to numerical calculation. In physical experiments a beach or energy-absorbing material is placed around the region of experimental interest to absorb the energy of outgoing waves and prevent reflection. Even if the waves are absorbed by beaches, the solution can still be contaminated by reflections from the wave maker.

In the case of linear, monochromatic waves it is easy to write a radiation condition. In that case the potential can be written

$$\Phi = f(x - ct) \tag{4.37}$$

in which c is the wave speed given by

$$c = (gL \tanh 2\pi h/L)^{1/2} \tag{4.38}$$

where L is the wavelength and h is the equilibrium water depth. The radiation condition is

$$\frac{\partial \Phi}{\partial t} = -c\,\frac{\partial \Phi}{\partial x} \tag{4.39}$$

which can serve as an adequate boundary condition. Obviously, (4.39) is a problem in the more usual case where there are waves of many lengths and, thus, no unique value for c. For shallow water waves the wave speed is independent of length and is

$$c = (gh)^{1/2} \tag{4.40}$$

and (4.39) also applies.

The beach idea can also apply to numerical calculations, including the nonlinear calculations. Like the experiment, the damping is not applied to the wave maker and so reflections from that source may be a problem. LeMehaute [35] suggested adding an irrotational damping term to the linearized combined free surface boundary conditions. In the present, nonlinear context that would make the dynamic boundary condition as follows [2]:

$$\frac{\partial \Phi}{\partial t} = B - g\eta - \frac{1}{2}\left[\left(\frac{\partial \Phi}{\partial x}\right)^2 + \left(\frac{\partial \Phi}{\partial y}\right)^2 + \left(\frac{\partial \Phi}{\partial z}\right)^2\right] - \lambda(x, y)\Phi \qquad \text{on } z = \eta \tag{4.41}$$

in which λ is a damping factor and can be a function of position. In physical experiments LeMehaute describes the best wave absorbers as having "progressive damping" which increases slowly over some length rather than suddenly imposing a region of maximum damping which would reflect energy. The same procedure can be followed in the numerical calculations by choosing $\lambda = 0$ in the region of interest and then increasing it (say, linearly) outside of the region of interest until the wave energy is absorbed. Betts and Mohamad [2] used a value of λ which increased from zero to 0.07 in a linear manner. They were able to reduce the reflection coefficient to less than 10%. Although this technique can be used with any computational method, it is especially simple in the case of the Boundary Element Method.

4.6. *Steady free surface flow*

The steady-state problem can be approached from the unsteady problem by the calculation through time under steady boundary conditions until there is no change in the solution. That calculation can be expensive, however, and there should be a more direct method. The direct method, unfortunately, is full of pitfalls. First, the nonlinear problem does not necessarily, and often not actually, have a unique solution; the solution depends on the history. Second, the steady-state solution for a given set of boundary conditions does not necessarily exist, at least with a smooth solution. Either of these conditions would probably be discovered in the course of the unsteady computation but could cause a more subtle difficulty in the direct solution of the steady problem. On the question of uniqueness, the solution would converge to the correct steady-state form if the initial conditions are correct. As for the existence, the solution algorithm fails if the surface becomes discontinuous, that is if the waves break at some point in the evolutionary development. Nevertheless, there are many instances where the direct steady-state solution is valuable.

The problem is most easily formulated in terms of a stream function

$$\frac{\partial \Psi}{\partial x} = -v_z \qquad \text{and} \qquad \frac{\partial \Psi}{\partial z} = v_x \,. \tag{4.42}$$

The boundary conditions are then

$$\Psi = 0 \qquad \text{on } z = -h \text{ (the solid bottom)} \tag{4.43}$$

$$\frac{1}{2}\left(\frac{\partial \Psi}{\partial n}\right)^2 + g\eta = \frac{U^2}{2} \qquad \text{on } \eta = z \text{ (the free surface)} \tag{4.44a}$$

$$\Psi = Q \qquad \text{on } \eta = z \text{ (the free surface)} \tag{4.44b}$$

in which U is the free stream velocity and $Q = Uh$ is the quantity of flow per unit width per unit time between the bottom and the free surface.

Consider the problem of flow about a stationary, submerged object as shown in

Fig. 14. On the object the boundary condition is

$$\Psi = \Psi_c \qquad \text{on the object} . \tag{4.45}$$

Also the circulation about the object must be prescribed. It is arbitrarily set to zero

$$\int_{\text{obj}} \frac{\partial \Phi}{\partial n} \, ds = 0 . \tag{4.46}$$

The problem is made more general by the introduction of dimensionless variables

$$\begin{aligned} x^* &= x/h \qquad & z^* &= z/h \qquad & n^* &= n/h \\ t^* &= tU/h \qquad & \Psi^* &= \Psi/Uh \qquad & \eta^* &= \eta/h \end{aligned} \tag{4.47}$$

in which h is the undisturbed water depth. Also a perturbation stream function is defined as

$$\psi^* = \Psi^* - z^* . \tag{4.48}$$

The difficulty with the far-field boundary conditions which was discussed in the last section also applies to the steady problem, perhaps with even more importance. In unsteady problems the lateral boundaries can sometimes be taken far enough away from the region of interest so that a wave does not have time to travel to the boundary and back to the region of interest within the time of interest for the computation. In the steady-state problem even this expensive

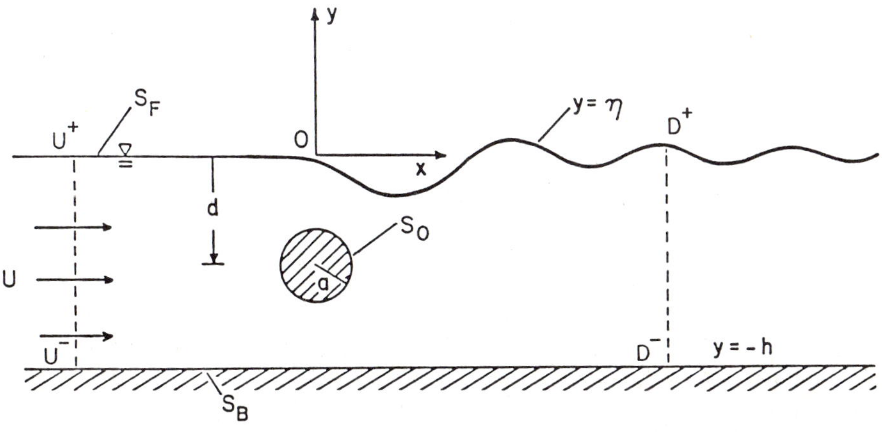

Fig. 14. Free surface flow around a submerged object.

remedy is not available. In the present example the solution is continued to infinity by linearizing the boundary condition outside of the region computed by boundary elements. The linearized free surface conditions become

$$\eta^* = -F^2 \frac{\partial \psi^*}{\partial x^*} \qquad \text{on } z^* = 0 \tag{4.49}$$

$$\frac{\partial \eta^*}{\partial x^*} = -\frac{\partial \psi^*}{\partial x^*} \qquad \text{on } z^* = 0 \tag{4.50}$$

in which $F = U/(gh)^{1/2}$ is the Froude number of the free stream flow.

The stream function downstream of the computational boundary is written

$$\psi^{*+} = 1 + C^+(1 + z^*) + (a \cos kx^* + B \sin kx^*) \frac{\sinh k_0(z^* + 1)}{\sinh k_0}$$

$$+ \sum_{m=1}^{N^+} C_m^+ \exp(-k_m x^*) \sin k_m(z^* + 1) \tag{4.51}$$

where C_0^+, A, B, and C_m^+ are unknown constants. The wave-number, k_0, is given by

$$F^2 k_0 \tanh k_0 \tag{4.52}$$

and

$$F^2 k_i \tan k_i \,. \tag{4.53}$$

On the upstream side of the computation region

$$\psi^{*-} = 1 + C_0^-(1 + z^*) + \sum_{m=1}^{N^-} C_m^- \exp(k_m x^*) \sin k_m(z^* + 1) \,. \tag{4.54}$$

Equations (4.51) and (4.54) constitute the linearized solution in the far-field regions with unknown coefficients. The determination of the constants C_0^+, C_0^-, k_0, k_i, A and B become part of the solution. The BEM is formulated so that these constants become unknowns in addition to the velocities and free surface position (the $\partial\psi^*/\partial n^*$ and η^*) around the solution region.

The boundary of Fig. 14 is discretised using linear boundary elements. In order to connect the BEM solution to the far-field solution, the following additional conditions are imposed:

$$\left(\frac{\partial \Psi^{*-}}{\partial x^*}\right)_{\mathrm{U}_-,\mathrm{U}_+} = 0 \tag{4.55}$$

$$\left(\frac{\partial \Psi^{*+}}{\partial x^*}\right)_{\mathrm{D}_-,\mathrm{D}_+} = 0 \tag{4.56}$$

$$\frac{\partial}{\partial x^*}\left(\frac{\partial \Psi^{*-}}{\partial z^*}\right)_{U_-,U_+} = 0 \tag{4.57}$$

$$\frac{\partial}{\partial x^*}\left(\frac{\partial \Psi^{*+}}{\partial z^*}\right)_{D_-,D_+} = 0 \tag{4.58}$$

in which U_+, U_-, D_+, D_- represent the points on the upper (U) and lower (D) corners of the upstream (+) and downstream (−) ends of the computational region (Fig. 14). The number of terms in each series is chosen in conjunction with the number of nodes and the ends of the region in order that there are the same number of equations as unknowns. The steps in the solution algorithm are [29]:

(1) The free surface position is estimated from an initial guess or from the linearized solution for the entire region. Care must be taken if using the linearized solution since it can give quite wrong results. For example, it is not uncommon for the linearized solution to show that the free surface passes through the object. Such an obviously incorrect solution is possible since the linearized boundary conditions are applied at the equilibrium free surface instead of on the actual free surface.

(2) The BEM equations are solved using the estimated free surface as the upper boundary. The conditions on the limits of the computational boundaries are as cited above.

(3) The Bernoulli constant

$$B_i = \frac{F^2}{2}\left(\frac{\partial \Phi_i^*}{\partial n^*}\right)^2 + z_i^* \tag{4.59}$$

is computed for all of the free surface nodes. The B_i are different for each of the free surface nodes unless the solution has converged to its proper value. The quantity $\Delta B_i = F^2/2 - B_i$ is the error in the dynamic condition due to the inaccurate estimate of the free surface position.

(4) In order to adjust the free surface toward the actual solution, a perturbation matrix is computed. Each of the free surface nodes are perturbed upward an amount δz^* and the entire BEM solution is redone for the new location of the node. Each solution gives a row in the matrix $\delta B_i/\delta z_j^*$ which represent a change of B at the point i due to the nodal location change at the point j. This same method was used by Cheng et al. [6] to calculate the steady-state flow over a spillway.

(5) The perturbation matrix is used to form the set of linear algebraic equations

$$[\partial B_i/\partial z_j^*]\{\Delta z_j^*\} = \{\Delta B_i\} \tag{4.60}$$

in which Δz_j^* is the adjustment to be made at node j in order to better satisfy the Bernoulli condition.

(6) The free surface is adjusted by applying the increment of elevation computed by (4.60) to the equation

$$z_j^{*k+1} = z_j^{*k} + \Delta z_j^* \tag{4.61}$$

in which the superscript indicates the iteration number. The new position of the free surface is used as a new estimate until the Bernoulli constants as calculated by (4.59) are equal to all the nodes to within a predetermined accuracy.

Step (4) is the time-consuming part of the algorithm since an entire BEM solution must be generated multiple times, once for each perturbation of a free surface node. The generation of the perturbation matrix is sometimes skipped in the iteration process since the matrix usually changes little from one iteration to the next and can be calculated only every three or four iterations.

For some problems the convergence of the scheme is poor or the accuracy is not all that is desired. The convergence can be enhanced if the Bernoulli equation is satisfied at points between the BEM nodes. A number of "pseudo-nodes" are created between the BEM nodes and the perturbation matrix is calculated at these points even though only the actual BEM nodes are perturbed. The result is a rectangular (non-square) perturbation matrix which leads to more equations than unknowns in the system (4.60). These equations are then solved in the least-squares sense as

$$\{\Delta z_j^*\} = \left[\left(\frac{\delta B_i}{\delta z_j^*}\right)^{\mathrm{T}}\left(\frac{\delta B_i}{\delta z_j^*}\right)\right]^{-1}\left[\frac{\delta B_i}{\delta z_j^*}\right]\{\Delta B_i\}\,. \tag{4.62}$$

Also convergence can be enhanced for this type of problem by using higher-order elements. In calculating flows over spillways, Cheng et al. [6] used cubic spline elements, which have the advantage of providing continuous derivatives on the free surface.

5. Viscous flow

The BEM has been used to compute flows outside of the boundary layer while coupled to the boundary layer flows [74–76]. In this section, however, we consider flows at low Reynolds numbers. Defining the vorticity vector as $\boldsymbol{\omega} = \nabla \times \boldsymbol{v}$, the Navier–Stokes equations reduce to the equation which describe the diffusion of vorticity

$$\frac{\mathrm{D}\boldsymbol{\omega}}{\mathrm{D}t} = \boldsymbol{\omega} \cdot \nabla \boldsymbol{v} + \nu \nabla^2 \boldsymbol{\omega} \tag{5.1}$$

in which the capital D represent the substantial derivative, $\boldsymbol{v}$ is the velocity vector, and ν is the kinematic viscosity. The first term on the right represents a gain in vorticity due to vortex stretching (only applicable to three-dimensional flow) and the last term is the vorticity production from viscous action.

The steady-state equations of motion are

$$\boldsymbol{v} \cdot \nabla \boldsymbol{v} = -\nabla(p/\rho) + \nu\nabla^2 \boldsymbol{v} \tag{5.2}$$

$$\nabla \cdot \boldsymbol{v} = 0 \tag{5.3}$$

$$\boldsymbol{v} \cdot \nabla \boldsymbol{\omega} = \boldsymbol{\omega} \cdot \nabla \boldsymbol{v} + \nu\nabla^2 \boldsymbol{\omega} \tag{5.4}$$

where the gravity term has been absorbed by the pressure. In the case that the Reynolds number is small, the inertial terms can be neglected. If, in addition, we restrict the flow to two dimensions, the set of equations becomes (5.3) plus

$$\nabla(p/\rho) = \nu\nabla^2 \boldsymbol{v} \tag{5.5}$$

$$\nabla^2 \omega = 0\,. \tag{5.6}$$

The common method of solution is to introduce a stream function

$$v_x = \frac{\partial \Psi}{\partial y} \qquad v_y = -\frac{\partial \Psi}{\partial x} \tag{5.7}$$

which automatically satisfies (5.3). Taking the curl of (5.5) produces

$$\nabla^4 \Psi = 0 \tag{5.8}$$

and from the definition of the stream function

$$\nabla^2 \Psi = -\omega \tag{5.9}$$

where ω is now written as a scalar since the vector $\boldsymbol{\omega}$ can have only one component, the one out of the plane of the two-dimensional flow.

The free-space Green's function for (5.8) in two dimensions is

$$G = r^2(\ln r - 1)/4\,. \tag{5.10}$$

Applying Green's theorem yields the boundary integral equation

$$\alpha\Psi = \int_\Gamma \left[\Psi \frac{\partial}{\partial n}(\ln r) - \ln r \frac{\partial \Psi}{\partial n} - \omega \frac{\partial}{\partial n}\left(\frac{r^2}{4}(\ln r - 1)\right) + \frac{r^2}{4}(\ln r - 1)\frac{\partial \omega}{\partial n}\right] \mathrm{d}s \tag{5.11}$$

in which α is 2π in the interior or the boundary angle on the boundary (π on a smooth boundary). There are four boundary unknowns represented in (5.11): Ψ, $\partial\Psi/\partial n$, ω, and $\partial\omega/\partial n$. The boundary conditions of a well-posed problem should give two of the unknowns. The system is closed by considering the boundary integral equation equivalent to (5.6)

$$\alpha\omega = \int_{\Gamma} \left(\omega \frac{\partial}{\partial n} \ln r - \ln r \frac{\partial \omega}{\partial n} \right) \mathrm{d}s \, . \tag{5.12}$$

These equations will solve a variety of viscous flows in which the inertial terms can be neglected to a good approximation.

The BEM with its variety of elements and ability to handle singularities is particularly suited to this type of calculation. With the biharmonic equation the accuracy may deteriorate when using finite elements or finite differences due to the necessity of approximating the fourth-order derivatives. The boundary integral avoids that difficulty. Ingham and Kelmanson [24, 26] have illustrated these points by solving a problem in a bifurcating channel where the vorticity becomes infinite at a point.

Of greater interest Kelmanson [29] uses similar equations to solve viscous free surface problems [26]. There are three boundary conditions on the free surface: (1) The free surface is a streamline; (2) the shear stress condition that $\sigma_{tn} = 0$ (in which n is the normal to the surface and t is the tangent); and (3) the condition that the normal stress on the surface is balanced by the pressure and the surface tension. Kelmanson shows that the latter condition is given by

$$T \frac{\partial \kappa}{\partial t} = \frac{\partial \omega}{\partial n} - \kappa\omega + 2 \frac{\partial^3 \Psi}{\partial n \partial^2 s} \qquad \text{on } y = \eta \tag{5.13}$$

in which T is the surface tension and κ is the curvature. The second boundary condition can be expressed as

$$\omega = -2\kappa \frac{\partial \Psi}{\partial n} \qquad \text{on } y = \eta \, . \tag{5.14}$$

Solid boundaries carry the condition that the boundary is a streamline plus the no-slip criterion, $\partial\Psi/\partial n = 0$.

Kelmanson [28] used an iterative technique to locate the free surface. The first two conditions are satisfied in the solution of the boundary integrals and an iteration scheme is used to satisfy the pressure condition.

The possibilities of boundary element solutions for many such problems are obvious. Lubrication is a prime example. Ingham and Kelmanson [24–26] provide an example where they computed the flow in a concentric bearing while imposing the periodicity conditions on pressure and vorticity

$$\int_C \frac{\partial p}{\partial s}\, ds = 0 \tag{5.15}$$

$$\int_C \frac{\partial \omega}{\partial n}\, ds = 0 \tag{5.16}$$

where C is the closed curve around the bearing. In this simple case the BEM results agreed with analysis. Apparently the BEM has not been applied to complex bearing geometries although the flows within actual bearing geometries are not well known. For axisymmetric Stoke's flow Tozeren [66] has developed axisymmetric Green's functions. He has computed the flow around objects in tubes with the goal of studying flow in capillaries with white and red blood cells.

References

[1] M. Abramowitz and I.A. Stegun, *Handbook of Mathematical Functions* (Dover, New York, 1970).

[2] P.L. Betts and T.T. Mohamad, A Boundary Approach to Water Waves, in: *Finite Element Analysis*, T. Kawai, Ed. (University of Tokyo Press, Tokyo, 1982) pp. 923–929.

[3] G.F. Carey and S.W. Kim, Extension of Boundary Element Method to Lifting Subcritical Flows, *19th Annual Meeting of the Society of Engineering Science*, University of Missouri-Rolla (1982).

[4] G.F. Carey, Extension of Boundary Element Method to Lifting Compressible Aerodynamics, in: *Finite Element Flow Analysis*, T. Kawai, Ed. (University of Tokyo Press, Tokyo, 1982) pp. 939–943.

[5] H.R. Chaplin, A Method for Numerical Calculation of Slip Stream Contraction of a Shrounded Impulse Disk in the Static Case with Application to Other Axisymmetric Potential Flow Problems, *David Taylor Model Basin Report No.* 1857 (1964).

[6] A.H-D. Cheng, J.A. Liggett and P.L-F. Liu, Boundary Calculations of Sluice and Spillway Flows, *Journal of the Hydraulics Division, ASCE* **107**, HY10, 1163–1178 (1981).

[7] E.D. Cokelet, Breaking Waves – the Plunging Jet and Interior Flow-Field, *Proceedings of the Symposium on Mechanics of Wave-Induced Forces on Cylinders*, Bristol (1978).

[8] F.J. Davenport, Singularity Solutions to General Potential Flow Airfoil Problem, D-7207, Boeing Airplane Company, Seattle, Washington (1963).

[9] P.J. Dillon and J.A. Liggett, An Ephemeral Stream-Aquifer Interaction Model, *Water Resources Research* **19**, 621–626 (1983).

[10] P.J. Dillon, Ephemeral Stream-Aquifer Interaction, Ph.D. Thesis, University of Adelaide, Australia (1984).

[11] Z.J. Dong, Location of Free Surface for Zoned Problems – A New Economical Boundary Element Method, in: *Boundary Elements*, C.A. Brebbia, T. Futagami and M. Tanaka, Eds (Springer-Verlag, Berlin, 1983) pp. 85–94.

[12] F. Durbin, Numerical Inversion of Laplace Transform: an Efficient Improvement to Dubner and Abate's Method, *Computer Journal* **17**, 371–376 (1974).

[13] R. Eatock Taylor, Analysis of Hydrodynamic Loads by Boundary Element Methods, in: *Developments in Boundary Element Methods – 2*, P.K. Banerjee and R.P. Shaw, Eds. (Applied Science Publishers, Ltd., London, 1982) pp. 211–243.

[14] H.M. Erdemir, A Combination of the Boundary Integral Equation Method and the Finite Element Method for the Analysis of 2-Dimensional Groundwater Problems, M.S. Thesis, Cornell University, Ithaca, New York (1981).

[15] O.M. Faltinsen, A Numerical Nonlinear Method of Sloshing in Tanks with Two-Dimensional Flow, *Journal of Ship Research* **23**, 193–202 (1978).

[16] H.M. Haitjema and O.D.E. Strack, Modeling Double Aquifer Flow Using a Comprehensive Potential and Distributed Singularities, 1. Solutions for Homogeneous Permeability, *Water Resources Research* **17**, 1535–1549 (1981).

[17] H.M. Haitjema and O.D.E. Strack, Modeling Double Aquifer Flow Using a Comprehensive Potential and Distributed Singularities, 2. Solutions for Inhomogeneous Permeabilities, *Water Resources Research* **17**, 1551–1560 (1981).

[18] J.L. Hess and A.M.O. Smith, Calculation of Nonlifting Potential Flow About Arbitrary Three-Dimensional Bodies, ES 40622, Douglas Aircraft Corp., Long Beach, California (1962); also in *Journal of Ship Research* **8**, 22–44 (1964).

[19] J.L. Hess, Review of Integral-Equation Techniques for Solving Potential-Flow Problems with Complicated Boundaries, in: *Innovative Numerical Analysis for the Applied Engineering Sciences*, R. Shaw et al., Eds. (University Press of Virginia, Charlottesville, 1980) pp. 131–143.

[20] C.W. Hirt and F.W. Harlow, A General Corrective Procedure for the Numerical Solution of Initial-Valued Problems, *Journal of Computational Physics* **2**, 114–119 (1967).

[21] B. Hunt, The Mathematical Basis and Numerical Principles of the Boundary Integral Equation Method for Incompressible Potential Flow over 3-D Aerodynamic Configurations, in: *Numerical Methods in Applied Fluid Dynamics*, B. Hunt, Ed. (Academic Press, London, 1980) pp. 49–135.

[22] B. Hunt and L.T. Isaacs, Integral Equation Formulation for Ground-Water Flow, *Journal of the Hydraulics Division, ASCE* **107**, HY10, 1197–1209 (1981).

[23] T. Inamuro, T. Adachi and H. Sakata, A Numerical Analysis of Unsteady Separated Flow by Discrete Vortex Model Using Boundary Element Method, in: *Finite Element Flow Analysis*, T. Kawai, Ed. (University of Tokyo Press, Tokyo, 1982) pp. 931–938.

[24] D.B. Ingham and M.A. Kelmanson, Modified Integral Equation Solution of Steady Viscous Flow in a Bifurcating Channel, in: *Boundary Element Methods in Engineering*, C.A. Brebbia, Ed. (Springer-Verlag, Berlin, 1982) pp. 218–229.

[25] D.B. Ingham and M.A. Kelmanson, A Boundary Integral Equation Method for the Study of Slow Flow Within Bearing Geometries, in: *Boundary Elements*, C.A. Brebbia, T. Futagami and M. Tanaka, Eds. (Springer-Verlag, Berlin, 1983) pp. 311–321.

[26] D.B. Ingham and M.A. Kelmanson, *Boundary Integral Equation Analysis of Singular, Potential and Biharmonic Problems* (Springer-Verlag, Berlin, 1984).

[27] M.A. Jaswon, Integral Equation Methods in Potential Theory: I, *Proceedings of the Royal Society* **A275**, 23–32 (1963).

[28] M.A. Kelmanson, Boundary Integral Equation Solution of Viscous Flows with Free Surfaces, *Journal of Engineering Mathematics* **17**, 329–343 (1983).

[29] S.K. Kim, J.A. Liggett and P.L-F. Liu, Steady Free Surface Flow About a Submerged Object, in: *Boundary Elements VI*, C.A. Brebbia, Ed. (Springer-Verlag, Berlin, 1984) pp. 4.9–4.16.

[30] O.E. Lafe, Boundary Integral Solutions to Nearly Horizontal Flows in Multiply Zoned Aquifers, Ph.D. Thesis, Cornell University, Ithaca, New York (1981).

[31] O.E. Lafe, J.A. Liggett and P.L-F. Liu, BIEM Solutions to Combinations of Leaky, Layered, Confined, Unconfined Nonisotropic Aquifers, *Water Resources Research* **17**, 1431–1444 (1981).

[32] L. Landweber, The Axially Symmetric Potential Flow About Elongated Bodies of Revolution, *David Taylor Model Basin Report No.* 761 (1951).

[33] J-J. Lee, Wave-Induced Oscillation in Harbors of Arbitrary Geometry, *Journal of Fluid Mechanics* **45**, 375–394 (1971).

[34] J-J. Lee and F. Raichlen, Oscillations in Harbor with Connection Basins, *Journal of the Waterways, Harbors, and Coastal Engineering Division, ASCE* **98**, 311–332 (1972).

[35] B. LeMehaute, Progressive Wave Absorber, *Journal of Hydraulic Research* **10**, 153–169 (1972).

[36] G.P. Lennon, P.L-F. Liu and J.A. Liggett, Boundary Integral Equation Solution to Axisymmetric Potential Flows, 1, Basic Formulation, *Water Resources Research* **15**, 1102–1106 (1979).

[37] G.P. Lennon, P.L-F. Liu and J.A. Liggett, Boundary Integral Equation Solution to Axisymmetric Potential Flows, 2, Recharge and Well Problems in Porous Media, *Water Resources Research* **15**, 1107–1115 (1979).

[38] G.P. Lennon, P.L-F. Liu and J.A. Liggett, Boundary Integral Solutions to Three-Dimensional, Unconfined Darcy Flow, *Water Resources Research* **16**, 651–658 (1980).
[39] G.P. Lennon, P.L-F. Liu and J.A. Liggett, Boundary Integral Solutions to Water Wave Problems, *Journal of the Hydraulics Division, ASCE* **108**, HY8, 921–931 (1982).
[40] J.A. Liggett and P.L-F. Liu, Applications of Boundary Element Methods to Fluid Mechanics, in: *Topics in Boundary Element Research*, Vol. 1, C.A. Brebbia, Ed. (Springer-Verlag, Berlin, 1984) pp. 78–96.
[41] J.A. Liggett and P.L-F. Liu, *The Boundary Integral Equation Method for Porous Media Flow* (George Allen and Unwin, London, 1983).
[42] J.A. Liggett, Multiple Boundary Conditions, Free Surface Flow, and the Boundary Element Method, *Communications in Applied Numerical Methods* **1**, 105–112 (1985).
[43] P.L-F. Liu and A.H-D. Cheng, Boundary Solutions for Fluid-Structure Interaction, *Journal of Hydraulic Engineering, ASCE* **110**, 51–64 (1984).
[44] P.L-F. Liu and J.A. Liggett, Applications of Boundary Element Methods to Problems of Water Waves, in: *Developments in Boundary Element Methods – 2*, P.K. Banerjee and R.P. Shaw, Eds. (Elsevier Applied Science Publishers, Ltd., London, 1982) pp. 37–67.
[44a] P.L.-F. Liu and J.A. Liggett, Boundary Element Formulations and Solutions for some Non-linear Water Wave Problems, in: *Developments in Boundary Element Methods* – 3, P.K. Banerjee and S. Mukherjee, Eds. (Elsevier Applied Science Publishers, Ltd., London, 1982) pp. 172–190.
[45] M.S. Longuet-Higgens and E.D. Cokelet, The Deformation of Steep Surface Waves on Water. I. A Numerical Method of Computation, *Proceedings of the Royal Society*, **A350**, 1–25 (1976).
[46] I. Lotz, Calculation of Potential Flow Past Airship Bodies in Yaw, NACA TM 675 (1932); also in *Ingenieur Archive* **2**, 507–527 (1931).
[47] T. Nakayama and K. Washizu, Nonlinear Analysis of Liquid Motion in a Container Subjected to Force Pitching Oscillation, *International Journal for Numerical Methods in Engineering* **15**, 1207–1220 (1980).
[48] T. Nakayama and K. Washizu, The Boundary Element Method Applied to the Analysis of Two-Dimensional Nonlinear Sloshing Problems, *International Journal for Numerical Methods in Engineering* **17**, 1631–1646 (1981).
[49] T. Nakayama and K. Washizu, Boundary Element Analysis of Nonlinear Water Wave Problems, *International Journal for Numerical Methods in Engineering* **19**, 953–970 (1983).
[50] T. Nakayama and K. Washizu, Boundary Element Analysis of Non-Linear Sloshing Problems, in: *Developments in Boundary Element Analysis – 3*, P.K. Banerjee and S. Mukherjee, Eds. (Elsevier Applied Science Publishers, London, 1984) pp. 191–211.
[51] G.V. Narayanan and D.E. Beskos, Numerical Operational Methods for Time-Dependent Linear Problems, *International Journal for Numerical Methods in Engineering* **18**, 1828–1854 (1982).
[52] W. Prager, Die Druckverteilung an Körpern in Ebener Potentialströmung, *Physikalische Zeitschrift* **29**, 865–869 (1928).
[53] F.J. Rizzo and D.J. Shippy, A Method of Solution for Certain Problems of Transient Heat Conduction, *Journal of the American Institute for Aeronautics and Astronautics* **8**, 2004–2009 (1970).
[54] M.A. Sadowsky and E. Sternberg, Elliptic Integral Representation of Axially Symmetric Flows, *Quarterly of Applied Mathematics* **8**, 113–126 (1950).
[55] J.R. Salmon, P.L-F. Liu and J.A. Liggett, Integral Equation Method for Linear Water Waves, *Journal of the Hydraulics Division, ASCE* **106**, HY12, 1995–2010 (1980).
[56] R.A. Shapery, Approximate Methods of Transform Inversion for Viscoelastic Stress Analysis, in: *Proceedings of the Fourth US National Congress of Applied Mechanics*, R.M. Rosenberg, Ed. (ASME, New York, 1962) pp. 1075–1085.
[57] R.P. Shaw, An Integral Equation Approach to Diffusion, *International Journal of Heat and Mass Transfer* **17**, 693–699 (1974).
[58] R.P. Shaw, Boundary Integral Equation Methods Applied to Wave Problems, in: *Developments in Boundary Element Methods – 1*, P.K. Banerjee and R. Butterfield, Eds. (Applied Science Publishers, Ltd., London, 1979) pp. 121–153.
[59] R.P. Shaw, Boundary Integral Equation Methods Applied to Water Waves, in: *Boundary Integral*

Equation Method: *Computational Applications in Applied Mechanics*, AMD-11, T.A. Cruse and F. Rizzo, Eds. (ASME, New York, 1975) pp. 7–14.

[60] A.M.O. Smith and J. Pierce, Exact Solution of the Neumann Problem. Calculation of Plane and Axially Symmetric Flows About or Within Arbitrary Boundaries, *Douglas Aircraft Company Report No.* 26988 (1958); for a summary see in: *Proceedings of the Third US National Congress of Applied Mechanics*, R.M. Haythornthwaite, Ed. (ASME, New York, 1958) pp. 807–815.

[61] O.D.E. Strack, A Single-Potential Solution for Regional Interface Problems in Coastal Aquifers, *Water Resources Research* **12**, 1165–1174 (1976).

[62] G.T. Symm, Integral Equation Methods in Potential Theory: II, *Proceedings of the Royal Society* **A275**, 33–46 (1963).

[63] A.E. Taigbenu, J.A. Liggett and A.H-D. Cheng, Boundary Integral Solution to Seawater Intrusion into Coastal Aquifers, *Water Resources Research* **20**, 1150–1158 (1984).

[64] A.E. Taigbenu and J.A. Liggett, Boundary Element Solutions of the Diffusion Equation, in: *Finite Elements and Flow Problems*, G.F. Carey and J.T. Oden, Eds. (University of Texas, Austin, 1984) pp. 137–141.

[65] A.E. Taigbenu and J.A. Liggett, Boundary Element Calculations of the Diffusion Equation, *Journal of Engineering Mechanics*, *ASCE* **111**, 311–328 (1985).

[66] H. Tozeren, Boundary Integral Equation Method for Some Stokes Problems, *International Journal for Numerical Methods in Fluids* **4**, 159–170 (1984).

[67] E. Trefftz, Über die Kontraktion Kreisförmiger Flüssigkeitsstrahlen, *Zeitschrift für Mathematik und Physik* **64**, 34–61 (1917).

[68] F. Vandrey, A Direct Iteration Method for the Calculation of Velocity Distribution of Bodies of Revolution and Symmetrical Profiles, *Admiralty Research Laboratory Report No.* R1/G.HY/12/2 (1951).

[69] F. Vandrey, On the Calculation of the Transverse Potential Flow Past a Body of Revolution with the Aid of the Method of Mrs. Flügge-Lotz, *Astia*, AD-40089 (1951).

[70] A. Van Tuyl, On the Axially Symmetric Flow Around a New Family of Half Bodies, *Quarterly of Applied Mathematics* **7**, 399–409 (1949).

[71] L-Y. Wei and J.A. Liggett, Zoned Boundary Elements – An Economical Calculation, in: *Proceedings of the International Conference on Finite Element Methods* (Science Press, Beijing, China, 1982) pp. 816–820.

[72] A. Weinstein, On Axially Symmetric Flows, *Quarterly of Applied Mathematics* **5**, 429–444 (1948).

[73] J.E. Welch, F.H. Harlow, J.P. Shannon and B.J. Daly, The MAC Method, Los Alamos Scientific Laboratory, Los Alamos, New Mexico (1966).

[74] J.W. White and S.J. Kline, A Calculation Method for Incompressible Axisymmetric Flows, Including Unseparated, Fully Separated, and Free Surface Flows, Report MD-35, US Air Force Office of Scientific Research, Mechanics Division, Contract AF-F44620-74-C-0016 (1975); also, Thermo-Sciences Division, Department of Mechanical Engineering, Stanford University (1975).

[75] J.C. Wu, Problems of General Viscous Flow, in: *Developments in Boundary Element Methods – 2*, P.K. Banerjee and R.P. Shaw, Eds. (Applied Science Publishers, London, 1982) pp. 69–109.

[76] J.C. Wu, Y.M. Rizk and N.L. Sankar, Problems of Time-Dependent Navier–Stokes Flow, in: *Developments in Boundary Element Methods – 3*, P.K. Banerjee and S. Mukherjee, Eds. (Elsevier Applied Science Publishers, London, 1984) pp. 137–169.

[77] Y. Zuosheng, A Special Boundary Element Technique in Transonic Flow, in: *Boundary Elements*, C.A. Brebbia, T. Futagami and M. Tanaka, Eds. (Springer-Verlag, Berlin, 1983) pp. 293–300.

CHAPTER 9

Acoustics

Richard P. SHAW

Department of Civil Engineering
State University of New York at Buffalo
Buffalo, New York 14260, U.S.A.

Boundary Element Methods in Mechanics
Edited by D.E. Beskos

Introduction

Acoustics is such a broad field that it would seem appropriate in any discussion to give some general description of those problems that will be discussed. The series *Physical Acoustics* edited by Mason and Thurston [28] will provide some concept of the range involved. In the present context, acoustics will be thought of as the study of wave motion in a slightly compressible fluid. For the application of boundary integral equation methods, emphasis will be placed on linearized problems in homogeneous media. While some extension can be made to non-linear and/or inhomogeneous problems, the very foundation of the boundary integral equation approach is linear, i.e. superposition of fundamental solutions, and requires media for which those fundamental solutions are known, e.g. Brebbia et al. [5], Banerjee and Butterfield [4].

Even with this reduction in scope, there are still a wide range of problems to be considered. Some classification can be made on the basis of the domain of the medium, the type of time dependence, the geometry and boundary conditions, etc.; this will be useful for any further discussion. The first classification is that of domain in the sense of interior or finite domains and exterior or infinite domains. Clearly there exist problems such as those for semi-infinite domains which do not fall neatly into either category, but this classification can usually be made. Typical of the interior problem is that of architectural acoustics where studies are made of the propagation of sound waves within a closed chamber, while exterior problems would include applications such as "sonar" in an ocean environment so large that its actual finite boundaries will not influence the problem solution. Another categorization would be on the basis of time dependence. While the concept of a time-dependent problem includes the special case of a time harmonic dependence, this particular dependence has come to be treated as a class in its own right. Thus, acoustic problems are generally classified as transient, i.e. general time dependence, or time harmonic, i.e. all variables are proportional to $\exp(-i\omega t)$ where ω is the frequency of the time dependence. Clearly, one form is related to the other by either the Duhamel integral, (time harmonic solutions obtained from the transient "pulse" solution) or by the Fourier transform, (transient problems transformed to resemble the corresponding time harmonic case). A further division can be made in the interior time harmonic case between forced problems and "free" or eigenvalue problems where acoustic resonances are considered. Next, for the case of forced motion, especially for exterior problems, the source of the forcing mechanism must be considered. Here the distinction is made on whether the impetus is provided by the finite boundary through some inhomogeneous boundary condition, i.e. radiation problems, or by some mechanism of waves generated at some distance, frequently infinite, and interacting with an otherwise passive finite body, i.e. scattering problems. Distinctions based on boundary conditions are relatively straightforward when Dirichlet, Neumann or Robin conditions are treated and only a single acoustic medium is involved, but some special consideration must be given to those problems involving coupled media wherein the boundaries serve to couple several domains, as for example in fluid–structure interaction problems. The shape of the boundary is also of significance in that the classical approach of separation of variables will only work for a limited class of geometries; this is of particular significance for the boundary integral equation approach which was designed to relieve this restriction and apply to all geometries.

All of the above classes of problems have been studied by a number of methods, including boundary integral equations, e.g. see review articles by Shaw [39, 43]. In fact, it is not surprising that acoustic problems were among the very first to be treated by the boundary integral equation approach. As will be seen in the next section, the concept of fundamental solutions and their superposition has had a long history in this field.

1. General formulation and Green's function methods

The governing equation for linearized wave motion in a slightly compressible fluid is the classical wave equation, e.g. Lamb [26]

$$\nabla^2\phi(r, t) = (1/c^2)\partial^2\phi/\partial t^2 + Q(r, t) \tag{1.1}$$

where ϕ is a velocity potential defined by the velocity vector, $\boldsymbol{v}$, equal to +grad ϕ and the pressure, p, equal to $-\rho \, \mathrm{d}\phi/\mathrm{d}t$. The sound speed, c, is determined from physical considerations but for linearized wave theory in slightly compressible fluids it is defined by c^2 equal to $\mathrm{d}p/\mathrm{d}\rho$. This allows for the possibility of fairly substantial presures within the context of a linearized theory for water since this particular fluid is only slightly compressible; linearization actually requires only small changes in density or condensation and fluid velocity. Acoustics in air would require small pressure changes to keep within a linearlized theory.

The corresponding time harmonic problem requires that all variables have a time dependence proportional to $\exp(-\mathrm{i}\omega t)$, leading to the Helmholtz equation,

$$\nabla^2\phi(r) + k^2\phi(r) = Q(r) \tag{1.2}$$

where k is the wave-number equal to ω/c. Note that the choice of sign in the exponential, i.e. $-\mathrm{i}\omega t$, is arbitrary but once chosen must be maintained throughout. Many authors, including Lamb, prefer to chose $+\mathrm{i}\omega t$ instead and care must be exercised when reading books and papers in this field. A similar problem of sign convention exists with the definition of the velocity potential, i.e. some authors, again including Lamb, prefer to use the negative of the convention used here.

These equation are well known in classical physics and much effort has gone into their solution for a wide variety of geometries, boundary conditions, forcing functions, etc., e.g. Morse and Feshbach [31]. Solutions based on separation of variables were the mainstay of these efforts, but two alternatives were developed for those problems not well suited to separation of variables. The first was the theory of geometrical acoustics which was basically a high-frequency approximation to the wave equation. This approach did allow for the variation of material properties throughout the medium and is still one of the most powerful approaches for such problems, e.g. Friedlander [18].

The second approach was that of Green's function solutions, which became physically interpreted as point source and doublet distributions. In fact, one of the classical principles of acoustics, Huygens' principle, may be interpreted in terms of such distributions, e.g. Baker and Copson [2]. These formulations, Kirchhoff's Integral Equation for the transient case and Helmholtz's Integral Equation for the time harmonic case, were developed in the last century. However, their application lay primarily in the solution of problems for which "the" Green's function was known or to the development of theoretical interpretation of physical phenomena, e.g. asymptotic solutions. It has only been in the last 25 years, essentially since the development of the digital computer for general usage, that these equations have been employed for direct numerical solutions of acoustics problems through the boundary integral equation and boundary element methods.

2. Integral equation formulations and Green's functions

The Kirchhoff and Helmholtz integral equations mentioned in the previous section may be obtained from the original wave equation by the use of Green's theorem and a fundamental or point source solution called a Green's function. Consider the time harmonic case, i.e. the Helmholtz equation, as an illustration since this case is the simpler of the two. The classical Helmholtz equation, Eq. (1.2), for a general wave problem is supplemented by the Helmholtz equation for a point source forcing function, i.e.

$$\nabla^2 G(\boldsymbol{r}, \boldsymbol{r}') + k^2 G(\boldsymbol{r}, \boldsymbol{r}') = -(4\pi)\delta(\boldsymbol{r} - \boldsymbol{r}') \tag{2.1}$$

where $(\pi)\delta(\boldsymbol{r} - \boldsymbol{r}')$ is the Dirac delta function corresponding to a point source at $\boldsymbol{r} = \boldsymbol{r}'$. Derivatives are taken with respect to $\boldsymbol{r}$. Note that some authors do not place the factor of 4π into the definition of a point source, but leave it to appear as part of G. This factor is usually included as shown to correspond to a unit "flux" from a small sphere surrounding the point $\boldsymbol{r}'$. The choice of sign is made for the same reason, i.e. to have an "outflux" from the source.

Green's third theorem provides a relationship between a volume integral and a surface integral over the surface bounding the original volume, i.e.

$$\iiint_V [f(\boldsymbol{r})\nabla^2 g(\boldsymbol{r}) - g(\boldsymbol{r})\nabla^2 f(\boldsymbol{r})]\,\mathrm{d}V(\boldsymbol{r})$$

$$= \oiint_S [f(\boldsymbol{r})\,\mathrm{grad}\{g(\boldsymbol{r})\} - g(\boldsymbol{r})\,\mathrm{grad}\{f(\boldsymbol{r})\}\,\mathrm{d}\boldsymbol{S}(\boldsymbol{r}) \tag{2.2}$$

where the vector surface element is simply the scalar $\mathrm{d}S$ mustiplied by the outward (from the volume V) unit normal vector. Identifying $f(\boldsymbol{r})$ with $\phi(\boldsymbol{r})$ and $g(\boldsymbol{r})$ with $G(\boldsymbol{r}, \boldsymbol{r}')$ and carrying out the integration in terms of the coordinate $\boldsymbol{r}$ provides the following result:

$$\iiint_V -4\pi\phi(\boldsymbol{r})\delta(\boldsymbol{r}-\boldsymbol{r}')\,\mathrm{d}V(\boldsymbol{r}) - \iiint G(\boldsymbol{r}, \boldsymbol{r}')Q(\boldsymbol{r})\,\mathrm{d}V(\boldsymbol{r})$$

$$= \oiint_S \left(\phi(\boldsymbol{r})\,\frac{\partial G(\boldsymbol{r}, \boldsymbol{r}')}{\partial n} - G(\boldsymbol{r}, \boldsymbol{r}')\,\frac{\partial \phi(\boldsymbol{r})}{\partial n}\right)\mathrm{d}S(\boldsymbol{r})\,. \tag{2.3}$$

The value of the volume integral of the delta function depends on the location of the point $\boldsymbol{r}'$; the corresponding value of ϕ is taken at $\boldsymbol{r}'$ since the integral is nonzero only at the point $\boldsymbol{r} = \boldsymbol{r}'$ and may thus be taken outside of the volume integration. For points, $\boldsymbol{r}'$, which lie wholly within the volume $V(\boldsymbol{r})$, the integral of the delta function is 4π while for points, $\boldsymbol{r}'$, which lie outside of V the integral is zero. The question of what happens when a point, $\boldsymbol{r}'$, lies on the surface $S(\boldsymbol{r})$ and is thus neither "inside" nor "outside" of V is the heart of the boundary integral equation method. The usual and probably most physically appealing approach to these integrals is to consider the point $\boldsymbol{r}'$ to be excluded from the volume of integration by a small sphere if $\boldsymbol{r}'$ indeed lies within V. Green's theorem is then considered to have been applied to this modified volume and an integration over the modified surface, i.e. the original plus the surface of the small sphere, leads directly to the result of $4\pi\phi(\boldsymbol{r}')$. Clearly, points lying outside of V never have the delta function different from zero within V. For a point $\boldsymbol{r}'$ on S, the exclusion can only involve a partial sphere and the corresponding surface integral yields only a fraction of the 4π factor found for the complete sphere. For a point $\boldsymbol{r}'$ on a smooth surface, the exclusion is made with a hemisphere and the resulting factor is clearly 2π. Points which lie on edges or corners will lead to other fractions of the 4π factor. The mathematical property of removing a finite contribution from a single point under an integration is called taking the "principal value" of the integral; the remaining integrals in Eq. (2.3) are then referred to as principal-value integrals.

Thus the integral equation equivalent to the original Helmholtz differential equation is

$$\alpha\phi(\boldsymbol{r}') = -\frac{1}{4\pi}\iiint_V G(\boldsymbol{r},\boldsymbol{r}')Q(\boldsymbol{r})\,\mathrm{d}V(\boldsymbol{r}) + \oiint_S \left(G(\boldsymbol{r},\boldsymbol{r}')\frac{\partial\phi(\boldsymbol{r})}{\mathrm{d}n} - \phi(\boldsymbol{r})\frac{\partial G(\boldsymbol{r},\boldsymbol{r}')}{\mathrm{d}n}\right)\mathrm{d}S(\boldsymbol{r}) \tag{2.4}$$

where α is 0 for $\boldsymbol{r}'$ outside of $V(\boldsymbol{r})$, 1/2 for $\boldsymbol{r}'$ on a smooth $S(\boldsymbol{r})$ and 1 for $\boldsymbol{r}'$ inside of $V(\boldsymbol{r})$. The classical use of these integral equations was for problems where "the" Green's function was known for the given geometry and boundary condition. For example, if the original problem were of the Dirichlet form, i.e. values of ϕ were given on the boundary surface S as $\phi(\boldsymbol{r}) = f(\boldsymbol{r})$ for $\boldsymbol{r} \in S$, "the" Green's function would be that which satisfied the Helmholtz equation with a point source and the boundary condition $G(\boldsymbol{r},\boldsymbol{r}') = 0$ for $\boldsymbol{r} \in S$. Then for any point $\boldsymbol{r}'$ in V, $\phi(\boldsymbol{r}')$ would be given by the quadrature

$$\phi(\boldsymbol{r}') = -\frac{1}{4\pi}\iiint_V G(\boldsymbol{r},\boldsymbol{r}')Q(\boldsymbol{r})\,\mathrm{d}V(\boldsymbol{r}) - \oiint_S f(\boldsymbol{r})\frac{\partial G(\boldsymbol{r},\boldsymbol{r}')}{\partial n}\,\mathrm{d}S(\boldsymbol{r}) \tag{2.5}$$

where all terms on the right-hand side of this equation are known. As long as $\boldsymbol{r}'$ is not on $S(\boldsymbol{r})$, these are ordinary rather than principal-value integrals. Unfortunately, the number of known Green's functions for specific geometries and boundary conditions is quite small and corresponds essentially to those problems which could have been solved by separation of variables plus a few solutions which can be constructed by image methods. This is still an active area of study however and new Green's functions are being developed, especially for inhomogeneous media. It must be pointed out, however, that the Green's function approach may well be superior to the separation of variables approach in some cases, e.g. the acoustic piston in an infinite flange to be discussed below, even when both are applicable. For the purposes of the present discussion, the most important Green's function is that for free space, i.e. a solution for a point source with no specified boundary conditions other than outgoing waves at infinity. This will be referred to as the "fundamental solution" and is well known to be

$$G_\infty(\boldsymbol{r},\boldsymbol{r}') = \exp(+\mathrm{i}kR)R \tag{2.6}$$

where R is the scalar distance between the integration point $\boldsymbol{r}$ and the observation point, i.e. solution point, $\boldsymbol{r}'$. Current terminology would refer to $\boldsymbol{r}$ as the "source" point and to $\boldsymbol{r}'$ as the "field" point even though $\boldsymbol{r}'$ was originally the location of the point source which produced the fundamental solution. This terminology is in keeping with a physical interpretation of these equations to be discussed in the next section. Frequently the integration variable, $\boldsymbol{r}$ here, is written as $\boldsymbol{r}'$ and the field point variable, $\boldsymbol{r}'$ here, is written as $\boldsymbol{r}$. The distinction between various notations is important, especially when the indirect method, to be discussed in the next section, is introduced. The "standard" Helmholtz integral equation then would be Eq. (2.4) with the fundamental solution, G, used for G, i.e.

$$\alpha\phi(\boldsymbol{r}') = -\frac{1}{4\pi}\iiint_V [\exp(\mathrm{i}kR)/R]Q(\boldsymbol{r})\,\mathrm{d}V(\boldsymbol{r}) + \oiint_S \left([\exp(\mathrm{i}kR)/R]\frac{\partial\phi(\boldsymbol{r})}{\mathrm{d}n} - \phi(\boldsymbol{r})\frac{\partial[\exp(\mathrm{i}kR)/R]}{\partial n}\right)\mathrm{d}S(\boldsymbol{r})\,. \tag{2.7}$$

The corresponding transient or Kirchhoff integral equation may be obtained in a number of ways. The simplest, although not the most general, is to return the time harmonic term to the Helmholtz equation and consider it to represent the Fourier transform of the original transient wave equation. Then the Helmholtz integral equation with the $\exp(-\mathrm{i}\omega t)$ dependence explicitly included and with the free-space fundamental solution, is

$$\alpha\phi(\boldsymbol{r}')\exp(-\mathrm{i}\omega t) = \iiint_V Q(\boldsymbol{r})(\exp[-\mathrm{i}\phi(t-R/c)]R)\,\mathrm{d}V(\boldsymbol{r}) + \frac{1}{4\pi}\oiint_S \left[\frac{1}{R}\frac{\partial\phi(\boldsymbol{r})}{\partial n} - \phi(\boldsymbol{r})\left(\frac{\mathrm{i}\omega}{cR} - \frac{1}{R^2}\right)\frac{\partial R}{\mathrm{d}n}\right] \times \exp[-\mathrm{i}\omega(t-R/c)]\,\mathrm{d}S(\boldsymbol{r}) \tag{2.8}$$

recognizing that $k = \omega/c$. The inverse transform may then be taken formally using the concept of "retarded time", i.e. $T = t - R/c$, as the appropriate time variable under the integral. This retarded time has physical meaning in that it rpresents the value of time at which a wave was initiated at a point $\boldsymbol{r}$ such that, travelling with wave speed c, it reaches the observation point $\boldsymbol{r}'$ at the present time t. This yields the Kirchhoff time-retarded integral equation,

$$\alpha\phi(\boldsymbol{r}') = \iiint_V \frac{Q(\boldsymbol{r},T)}{R}\,\mathrm{d}V(\boldsymbol{r}) + \frac{1}{4\pi}\oiint_S \left[\frac{1}{R}\frac{\partial\phi(\boldsymbol{r},T)}{\partial n} + \left(\frac{1}{cR}\frac{\partial\phi(\boldsymbol{r},T)}{\partial T} + \frac{\phi(\boldsymbol{r},T)}{R^2}\right)\frac{\partial R}{\mathrm{d}n}\right]\mathrm{d}S(\boldsymbol{r})\,. \tag{2.9}$$

Alternative derivations which remove restrictions implicit in a Fourier transform approach are given by Morse and Feshbach [31] using singular solutions or by Friedman and Shaw [19] using a four-dimensional Green's theorem.

There are of course two-dimensional analogs to these three-dimensional formulations. The Helmholtz equation reduces readily to the Weber equation in two dimensions:

$$\alpha\phi(\boldsymbol{r}') = \frac{i}{4} \iint_A H_0^{(1)}(kR) Q(\boldsymbol{r})\, \mathrm{d}A(\boldsymbol{r}) + \frac{i}{4} \oint_B \left(H_0^{(1)}(kR) \frac{\partial \phi(\boldsymbol{r})}{\partial n} - \phi(\boldsymbol{r}) \frac{\partial H_0^{(1)}(kR)}{\partial n} \right) \mathrm{d}s(\boldsymbol{r}) \tag{2.10}$$

where $A(\boldsymbol{r})$ is the two-dimensional area bounded by the curve $B(\boldsymbol{r})$ with arclength element $\mathrm{d}s$ and $H_0^{(1)}$ is the Hankel function of the first kind of order zero.

The corresponding two-dimensional transient formulation does not carry over as smoothly. The fundamental solution for a point source in a two-dimensional medium corresponds to a line source in three dimensions. The corresponding solution then is expressed as an integral over time, integral still leaving a triple integral to carry out. This phenomenon holds for the wave equation in all even dimension spaces, e.g. Hadamard [21]. For this reason, most two-dimensional transient integral equation formulations are either carried out as three spatial dimensional problems, i.e. the area is extended into a cylinder with solutions identical at each cross-section for the same value of time (realize that the time retardation plays a role here), e.g. Friedman and Shaw [19] or alternatively a Fourier transform may be taken leading to the Weber equation but still requiring the inversion to be made.

Now that these formulations have been developed, the next section will discuss a physical interpretation and some alternative formulations. Following sections give some applications both for the exact solution method, i.e. "the" Green's function, and for approximate solution methods.

3. Physical interpretation

There is a physical interpretation of these integral equations which is of both historical and practical interest. The Helmholtz (time harmonic) form will be used as an example, but the same interpretations may be made for the Kirchhoff (transient) from. The terms $\exp(\mathrm{i}kR)/R$ and $\mathrm{d}[\exp(\mathrm{i}kR)/R]/R]/\partial n$ in Eq. (2.7) have a physical interpretation which is a mathematical counterpart to Huygen's Principle, e.g. Baker and Copson [2]. These may be interpreted as a point source and a doublet of strength $\partial\phi/\partial n$ and ϕ respectively. Then the surface integral in Eq. (2.7) may be thought of as a continuous distribution of sources and doublets, each of which is producing a "wavelet", i.e. a portion of the total wave, of strength proportional to $\partial\phi/\partial n$ and ϕ respectively. These sources and doublets produce discontinuities in $\partial\phi/\partial n$ and ϕ respectively as a field point passes from one side of the surface S to the other. One approach to these equations then is to identify the "strengths" $\partial\phi/\partial n$ and ϕ with the actual value of the variables $\partial\phi/\partial n$ and ϕ in the fluid as the field point approaches the surface S and, with the given discontinuities in these variables, pass to identically zero values of these variables on the "other", i.e. non-fluid, side of S. The advantage (at least in visualization) is that the actual physical surface S with its appropriate boundary conditions may be replaced by a "saltus" problem, i.e. by a geometrically identical surface of sources and doublets in a fluid which extends everywhere, inside and outside of the surface S but where the solution inside (outside) of S is identically zero for an exterior (interior) problem.

One of the main advantages to this point of view is that, once the physical concept of sources and doublets and an infinite fluid extending everywhere has been introduced, alternative formulations in terms of other combinations, e.g. sources alone or doublets alone or even some other combination, become easier to understand, e.g. Chertock [12]. These alternative approaches have become known as "indirect" methods in that a ficticious source and/or doublet strength is sought from which the final potential solution may be found. A usual derivation of this approach is contained in Lamb [26]. Consider a surface S separation an "exterior" from an "interior" domain defined by V_e and V_i respectively. Take the exterior time harmonic case with no inhomogeneous term, i.e. $Q=0$, for simplicity. Then Eq. (2.7) written for a field point $\boldsymbol{r}'$ within V_e yields an expression for the potential $\phi(\boldsymbol{r}')$:

$$\alpha\phi(\boldsymbol{r}') = \frac{1}{4\pi} \oiint_S \left([\exp(ikR)/R] \frac{\partial\phi(\boldsymbol{r})}{\partial n} - \phi(\boldsymbol{r}) \frac{\partial[\exp(ikR)/R]}{\sigma n} \right) dS(\boldsymbol{r}) \tag{3.1}$$

where α is 1 since the field point A lies within the volume of integration, V_e, but not on the surface S. Next consider as a different, but complementary, problem, the same surface but for an interior fluid. The solution for this new potential, ψ, at the same field point will give

$$\alpha\psi(\boldsymbol{r}') = \frac{1}{4\pi} \oiint_S \left([\exp(ikR)/R] \frac{\partial\psi(\boldsymbol{r})}{\partial n^1} - \psi(\boldsymbol{r}) \frac{\partial[\exp(ikR)/R]}{\partial n^1} \right) dS(\boldsymbol{r}) \tag{3.2}$$

where α is now zero since the field point A lies outside this "new" volume of integration. Note that the normal direction n^1 in Eq. (3.2) is opposite to the normal direction n in Eq. (3.1) since the normal in all cases must point away from the domain of interest; note that Lamb [26] uses an inward-pointing normal rather than the more usual outward-pointing normal, another sign convention difficulty which may lead to errors. These equations may be added, using $-n$ for n^1 in one term, to give

$$\phi(\boldsymbol{r}') = \frac{1}{4\pi} \oiint_S [\exp(ikR)/R] \left(\frac{\partial\phi(\boldsymbol{r})}{\partial n} + \frac{\partial\psi(\boldsymbol{r})}{\partial n^1} \right)$$

$$- [\phi(\boldsymbol{r}) - \psi(\boldsymbol{r})] \frac{\partial[\exp(ikR)/R]}{\partial n} \Bigg) dS(\boldsymbol{r}) . \tag{3.3}$$

The interior solution, ψ, is available to modify Eq. (3.3) to alternative forms. For example, if ψ on S is chosen to be equal to ϕ, the potential itself will be continuous across S but its normal derivative will be discontinuous leaving an integral equation in terms of surface sources

$$\phi(\boldsymbol{r}') = \frac{1}{4\pi} \oiint_S [\exp(\mathrm{i}kR)/R]\mu(\boldsymbol{r})\,\mathrm{d}S(\boldsymbol{r})\,. \tag{3.4}$$

The source strength, $\mu = \partial\phi/\partial n + \partial\psi/\partial n^1$, is then a new "indirect" unknown from which ϕ may be determined. Similarly, if $\partial\phi/\partial n$ were chosen equal to $\partial\psi/\partial n = -\partial\psi/\partial n^1$ on S, the formulation would yield an expression in terms of surface doublets:

$$\phi(\boldsymbol{r}') = \frac{1}{4\pi} \oiint_S -\sigma(\boldsymbol{r}) \frac{\partial[\exp(\mathrm{i}kR)/R]}{\partial n}\,\mathrm{d}S(\boldsymbol{r}) \tag{3.5}$$

where $\sigma = \phi - \psi$ on S.

Equations on these source and doublet strengths may be obtained. For example, if the field point in the original domain described by Eq. (3.1) were allowed to approach the surface S, and similarly Eq. (3.2) for ψ where formed for a point exterior to the original volume V_e, i.e. applied to an interior point in V_i, were to have that point approach the surface S, these equations would yield

$$\tfrac{1}{2}\phi(\boldsymbol{r}'(S)) = \frac{1}{4\pi} \oiint_S \left([\exp(\mathrm{i}kR)/R]\frac{\partial\phi(\boldsymbol{r})}{\partial n} - \phi(\boldsymbol{r})\,\frac{\partial[\exp(\mathrm{i}kR)/R]}{\partial n} \right) \mathrm{d}S(\boldsymbol{r}) \tag{3.6}$$

and

$$(\tfrac{1}{2}\psi(\boldsymbol{r}'(S)) = \frac{1}{4\pi} \oiint_S \left(-[\exp(\mathrm{i}kR)/R]\,\frac{\partial\psi(\boldsymbol{r})}{\partial n} + \psi(\boldsymbol{r})\,\frac{\partial[\exp(\mathrm{i}kR)/R]}{\partial n} \right) \mathrm{d}S(\boldsymbol{r}) \tag{3.7}$$

using $-n$ as the "outward" normal in Eq. (3.7). For the case of a continuous ϕ, i.e. Eq. (3.4), these would add to

$$\phi(\boldsymbol{r}'(S)) = \frac{1}{4\pi} \oiint_S [\exp(\mathrm{i}kR)/R]\nu(\boldsymbol{r})\,\mathrm{d}S(\boldsymbol{r}) \tag{3.8}$$

on S as well as in V_e. This represents an integral equation of the first kind on ν for a Dirichlet problem where $\phi(\boldsymbol{r}(S))$ is given. An integral equation of the second kind could have been formed by taking the derivative of Eqs. (3.1) and (3.2) with respect to their appropriate normals, in the $\boldsymbol{r}'$ coordinate system, before taking the limit as the respective field points approached the surface S. Actually, an infinite number of indirect integral equation forms can be produced but different choices of the relationship between ϕ and ψ, e.g. Chertock [12]. Only three basic forms are really unique, i.e. the simple source, the doublet and the standard Helmholtz integral equation with a null field outside of the original domain. The alternatives may be useful in some circumstances, but the vast majority of the work in this field has dealt with these three forms. For the purposes of numerical solution, integral equations of the second kind are generally much preferred to those of the first kind, indeed those of the first kind may not have useful solutions, e.g. Fox [17]. (The author has found the standard Helmholtz form most appropriate for his own work.)

One other advantage of considering the exterior scattering problem as a "saltus" problem is described by Friedman and Shaw [19] in that it allows the use of the total potential rather than the scattered potential as a direct variable. The basic integral equation, either Helmholtz or Kirchhoff, for an exterior scattering problem must be described in terms of the scattered potential in that this is the potential which satisfies the Sommerfeld radiation condition of outgoing waves at infinity. Thus the problem must be phrased in terms of $\phi(\text{scattered}) = \chi$ which is $\phi(\text{total}) - \phi_\omega(\text{incident wave potential})$. The boundary conditions must then be rewritten in terms of χ; frequently, they would be homogeneous in terms of the total wave potential and this advantage would be lost when a scattered potential is used. However, as a saltus problem, with "fluid" everywhere, the incident wave is considered to pass through the obstacle with appropriate discontinuities in the wave potential caused by the surface of sources and doublets replacing the actual scattering surface. Since the incident wave is continuous across this surface, the discontinuities in the scattered potential are equivalent to those for the total potential. Then Eq. (3.1) written for the scattered potential would be

$$\begin{aligned}\chi(\boldsymbol{r}') &= \frac{1}{4\pi} \oiint_S \left([\exp(\mathrm{i}kR)/R] \frac{\partial \chi(r)}{\partial n} - \chi(\boldsymbol{r}) \frac{\partial [\exp(\mathrm{i}kR)/R]}{\partial n} \right) \mathrm{d}S(\boldsymbol{r}) \\ &= \frac{1}{4\pi} \oiint_S \left([\exp(\mathrm{i}kR)/R] \frac{\partial \phi(\boldsymbol{r})}{\partial n} - \phi(\boldsymbol{r}) \frac{\partial [\exp(\mathrm{i}kR/R]}{\partial n} \right) \mathrm{d}S(\boldsymbol{r})\end{aligned} \tag{3.9}$$

which could be written in terms of ϕ as

$$\phi(\boldsymbol{r}') = \phi_\omega(\boldsymbol{r}') + \frac{1}{4\pi} \oiint_S \left([\exp(\mathrm{i}kR)/R] \frac{\partial \phi(\boldsymbol{r})}{\partial n} - \phi(\boldsymbol{r}) \frac{\partial [\exp(\mathrm{i}kR)/R]}{\partial n} \right) \mathrm{d}S(\boldsymbol{r}) . \tag{3.10}$$

Now when the limit is taken as $\boldsymbol{r}'$ approaches the surface S, the singular integral will contribute a value of $\frac{1}{2}\phi(\boldsymbol{r}')$ leaving as a governing equation

$$\tfrac{1}{2}\phi(\boldsymbol{r}'(S)) = \phi_\omega(\boldsymbol{r}'(S)) + \frac{1}{4\pi}\left(\oiint_S ([\exp(\mathrm{i}kR)/R]\,\frac{\partial\phi(\boldsymbol{r})}{\partial n} - \phi(\boldsymbol{r})\,\frac{\partial[\exp(\mathrm{i}kR)/R]}{\partial n}\right)\mathrm{d}S(\boldsymbol{r})\,. \tag{3.11}$$

This form allows the use of boundary conditions on the total potential directly. As mentioned above this can be advantageous when homogeneous boundary conditions are specified.

Finally, the original form of the integral equations may be modified before taking the limit as the field point approaches the boundary surface. Consider a Dirichlet problem where the potential itself is specified on the boundary. The usual form of the boundary integral equation would then lead to an integral equation of the first kind. These are very difficult to solve numerically, e.g. Fox [17], although there has been some success with the transient case, which lead to a Volterra integral equation as opposed to the Fredholm equation found in the time harmonic case. It is possible, however, to again manipulate the integral equation before the field point is allowed to approach the boundary. Again consider Eq. (3.1) for a field point in V_e,

$$\phi(\boldsymbol{r}') = \frac{1}{4\pi}\oiint_S \left([\exp(\mathrm{i}kR)/R]\,\frac{\partial\phi(\boldsymbol{r})}{\partial n} - \phi(\boldsymbol{r})\,\frac{\partial[\exp(\mathrm{i}kR)/R]}{\partial n}\right)\mathrm{d}S(\boldsymbol{r})\,. \tag{3.12}$$

This equation may be differentiated with respect to the boundary normal direction, n', just as described above for the indirect formulations, to give

$$\frac{\partial\phi(\boldsymbol{r}')}{\partial n'} = \frac{1}{4\pi}\oiint_S \left(\frac{\partial[\exp(\mathrm{i}kR)/R]}{\partial n'}\,\frac{\partial\phi(r)}{\partial n} - \phi(\boldsymbol{r})\,\frac{\partial^2[\exp(\mathrm{i}kR)/R]}{\partial n\partial n'}\right)\mathrm{d}S(\boldsymbol{r})\,. \tag{3.13}$$

Now as the field point approaches the boundary, the same type of singular contribution is found, but this time it provides a $\frac{1}{2}\partial\phi(\boldsymbol{r}')/\partial n'$ term, as described by Shaw and Friedman [37]. This was especially useful for scattering by a free, i.e. $\phi = 0$, boundary obstacle.

Now that some discussion has been presented concerning the foundation of this approach, it would be appropriate to consider some examples, as in the next two sections.

4. Examples of exact integral equation solutions

This section will discuss three examples of exact solutions to time harmonic acoustic problems using integral equation approaches; analogous transient problems are possible, but usually more difficult to solve. The first, the acoustic piston problem, is an illustration of an acoustic problem which is best solved by the integral equation approach. The second discusses eigenvalues for a two-dimensional circular domain using a known Green's function. Finally, a circular radiator with a prescribed surface velocity in an infinite region is discussed.

The classical problem of a moving finite piston in an otherwise infinite rigid flat surface has been extensively studied and is frequently used as an illustration of the exact Green's function approach, e.g. Harris [22]. Consider Eq. (2.4) applied to a field point at A in the fluid region, $z > 0$, defining the half-space above the infinite flat surface. The Green's function for this problem is formed by images such that $\mathrm{d}G(\boldsymbol{r}, \boldsymbol{r}')\mathrm{d}n$ is zero on $z = 0$. Then α is 1 and the potential at A is given by

$$\phi(\boldsymbol{r}'(A)) = \frac{1}{4\pi} \oiint_S \left(\frac{\exp(\mathrm{i}kR)}{R} + \frac{\exp(\mathrm{i}kR')}{R'} \right) V(\boldsymbol{r}) \, \mathrm{d}S(\boldsymbol{r}) \,, \tag{4.1}$$

where Q is taken as zero, i.e. no distributed sources within the fluid. The distances R and R' are the scalar distances from the integration point, $\boldsymbol{r}$, on $z = \theta$ to the field point, $\boldsymbol{r}'$, at (x', y', z') and the image field point at $(x', y', -z')$. The velocity on $z = 0$ is specified as V over the piston area, S, and zero elsewhere. This approach is valid for arbitrary piston shapes and illustrates the power of the exact Green's function approach. The well-known separation of variables solution for a circular piston is less straightforward and does not apply to arbitrary geometries. A "boundary element" solution, which is really just a numerical quadrature using surface elements and specified shape functions for the known velocity on the elements is discussed by Shaw and Wang [44]. This problem contains many of the same numerical difficulties as found in the Boundary Element Method such as a singular integrand, Gaussian integrations, etc., but with a "known" solution, i.e. the pressure is found directly, not from a matrix inversion resulting from a system of simultaneous equations.

A second "exact" solution is that for eigenvalues for a circular geometry, Tai and Shaw [45], which includes in its derivation a caution concerning the proper Green's functions to be used. Consider a two-dimensional circular region of radius a with a Neumann boundary condition, $\mathrm{d}\phi/\mathrm{d}n = 0$ on the circular boundary, S. Then the Weber integral equation becomes

$$\tfrac{1}{2}\phi(\theta) = \frac{\mathrm{i}ka}{4} \int_0^{2\pi} \phi(\theta + 2\beta) H_0^{(1)}(2ka \sin \beta) \sin \beta \, \mathrm{d}\beta \tag{4.2}$$

where $\beta = (\theta - \theta')/2$ and θ' denotes the basic integration variable or "source" point. $\phi(\theta)$ may be expanded into a Fourier series

$$\phi(\theta) = \sum_{m=0}^{\infty} [A_m \sin(m\theta) + B_m \cos(m\theta)] \tag{4.3}$$

with

$$A_m = \mathrm{i}ka \int_0^{2\pi} A_m \cos(2m\beta) H_0^{(1)}(2ka \sin\beta) \sin\beta \, \mathrm{d}\beta \tag{4.4}$$

$$B_m = \mathrm{i}ka \int_0^{2\pi} B_m \cos(2m\beta) H_0^{(1)}(2ka \sin\beta) \sin\beta \, \mathrm{d}\beta \, . \tag{4.5}$$

Either A_m or VB_m could vanish but not both in order to have a non-trivial solution. Since these equations are identical, only one need be considered. In real and imaginary parts, this gives

$$1 = -ka \int_0^{2\pi} \cos(2m\beta) Y_0(2ka \sin\beta) \, \mathrm{d}\beta = 1 + \pi ka Y_m(ka) J_m'(ka) \tag{4.6a}$$

$$0 = ka \int_0^{2\pi} \cos(2m\beta) J_0(2ka \sin\beta) \, \mathrm{d}\beta = +\pi ka J_m(ka) J_m'(ka) \, . \tag{4.6b}$$

The only common root is $J_m'(ka) = 0$ as expected. However, if only the imaginary part of the fundamental Greeen's function was used, as has been suggested by some authors, e.g. DeMey [15], Hutchinson [23], an additional set of spurious roots is introduced.

Finally, the two-dimensional radiation problem from a circle with a prescribed surface velocity given in Shaw [43] may be discussed. This problem shows that the integral equation for this particular exterior Neumann problem fails to have a solution are the eigenvalues of the corresponding interior Dirichlet problem; this question is discussed in general by Ursell [46].

Take the axisymmetric case for simplicity. Eq (2.10) with constant velocity, v_0, and constant potential, ϕ_0, on the circular boundary curve, B, gives

$$\alpha\phi(r') = -\frac{\mathrm{i}av}{4} \int_0^{2\pi} H_0^{(1)}(kR) \, \mathrm{d}\theta - \frac{\mathrm{i}a\phi}{4} \int_0^{2\pi} \frac{\mathrm{d}H_0^{(1)}(kR)}{\partial n} \, \mathrm{d}\theta \, . \tag{4.7}$$

As long as $\boldsymbol{r}$ and $\boldsymbol{r}'$ are distinct, section 7.2 of Morse and Feshbach [31] provides an appropriate expansion for this problem of the form

$$H_0^{(1)}(kR) = \sum_{m=0}^{\infty} \varepsilon_m \cos(m(\theta' - \theta)) J_m(kr') H_m(kr) \, ; \qquad r' < r \tag{4.8}$$

where the r coordinate of the field point is at a^- infinitessimally less than the r-value for the integration variable. Take θ' equal to zero for convenience since the potential is independent of the actual value for θ'. The coefficient ε_m is 1 if $m = 0$ and 2 otherwise. Then Eq. (4.8) reduces to

$$0 = -(\mathrm{i}av_0/4)J_0(ka^-)H_0^{(1)}(ka) - (\mathrm{i}a\phi_\sigma/4)kJ_0(ka^-)H_{0'}^{(1)}(ka) \tag{4.9}$$

or

$$\phi_0 = -(v_0/k)[H_0^{(1)}(ka)/H_{0'}^{(1)}(ka)] \tag{4.10}$$

i.e. the classical separation of variables solution unless $J_0(ka)$ is zero in which case the integral equation fails, i.e. unless k is an eigenvalue of the corresponding interior Dirichlet problem.

These examples have been given to provide some confidence in the integral equation formulation as well as to point out some pitfalls that exist in this approach. The generality of the integral equation approach, however, must be kept in mind when considering alternatives. It is in the approximate or numerical solution of these boundary integral equations that the method shows its real value as will be seen in the next section.

5. Approximate solution methods for boundary integral equations

There are several approximate methods by which the boundary integral equation formulations of Eqs. (2.7) and (2.9) may be solved. The most common is that of the "Boundary Element Method" whereby the boundary integrals in these equations are approximated by finite summations through approximations of the unknown variables by elementary shape functions over individual boundary elements which together form an approximation to the original boundary surface. This approach is so popular that alternative solution methods are frequently overlooked and the Boundary Integral Equation Method is "reduced" to the Boundary Element Method. It is clear, however, that a number of solution methods for integral equations exist and the BEM approach is just one. In fairness, it must be said that it has been the most successful one.

Before discussion the BEM approach, consider some of the alternatives. Chertock [11] describes an iterative approach to the solution of the governing integral equation in terms of an extension of the classical Neumann series. Kleinman [24] and Ar and Kleinman [1] give low-frequency expansion solutions to radiation problems. High-frequency approximations have been the basis of the general theory of geometric acoustics, e.g. Friedlander [18]. This approach is of course not based on boundary integral equation methods per se but is far more general, i.e. can be applied to problems with inhomogeneous material properties, etc. However, it is possible to carry out high-frequency expansions within these integral equations and this is an area of possible further development. Global solution techniques based on Galerkin methods or least-squares methods have been used, e.g. Sharma [36], but do not seem to enjoy the popularity one might expect.

The actual details of the Boundary Element Method as a numerical technique are given in a number of texts for general problems, e.g. Brebbia et al. [5], Banerjee and Butterfield [4], etc., and a number of specific acoustic wave review articles, e.g. Shaw [23, 43]. The essence of this approach for time harmonic problems is very similar to that for potential problems, e.g. the type of singularity is the same, the discretization of the surface would be very similar, the shape functions used for the unknown variables would be much the same, etc. One important distinction, however, would be the accuracy requirements, i.e. the characteristic size of the surface elements would have to be small compared with the wavelengths involved. This may cause some difficulty when very short wavelengths are involved, but these in turn lend themselves to high-frequency approximation methods. Another distinction for wave problems is the possibility of eigenvalues, i.e. resonances. These problems are well understood and some general-purpose BEM programs are available for radiation and scattering problems in infinite domains, e.g. Schenck's CHIEF program [32], etc. A bibliography of BEM codes has recently been compiled by Mackerle and Andersson [27] which contains some acoustic (and elastic) wave scattering codes. Chen [9] describes some of the US Navy research codes developed for acoustic radiation and scattering as well as acoustic–structure interaction.

The first numerical solutions using boundary element methods appear to have been carried out in the early 1960s by Friedman and Shaw [19] for the transient scattering case and Banaugh and Goldsmith [3] and Chen and Schweikert [8] for the time harmonic case for scattering and radiation problems respectively. Some typical results are shown in Figs. 1 and 2 for transient scattering, [19], and time harmonic scattering, [3], respectively. These were followed by a number of papers continuing this effort, e.g. Shaw [38], Mitzner [30], Farn and Huang [16] and Shaw and English [40], Chertock [10], Brundit [6], Greenspan and Werner [20], Copley [14] and Kleinman and Roach [25]. These all essentially replaced the boundary surface by discrete surface elements on which the potential was assumed to be of a specific form, e.g. constant, linear, etc. The field points were then taken at appropriate element locations, e.g. the midpoint of the surface element, the nodes of the element, etc. giving rise to linear algebraic equations where the coefficients of these equations were simply integrals of the integral kernels over the individual surface elements. Some care has to be taken with the "self-influence coefficients" since their integrands are singular, but this is now a fairly standard procedure, e.g. [5, 4]. By chosing one field point for each unknown potential nodal value, the resulting system of linear algebraic equations was complete and dense. Problems developed with the time harmonic exterior prob-

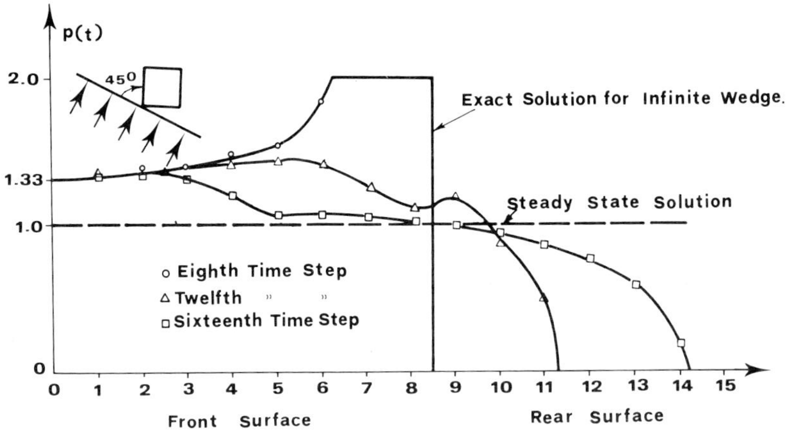

Fig. 1. Transient scattering from a rigid square cylinder.

lem due to the difficulty at interior eigenvalues described above and several authors proposed methods to circumvent this difficulty, e.g. Schenck [32] with an added interior solution point and a least-squares solution to the resulting overdetermined system, with the improvement in relative error as shown in Fig. 3, Burton and Miller [7] by adding the original Helmholtz equation to its normal derivative multiplied by a complex coefficient and Meyer, Bell, Zinn and Stallybrass [29] who used a pointwise variational principle to avoid the strong singularities of Burton and Miller's approach. A collection of various direct approaches is given by Schuster et al. [33–35].

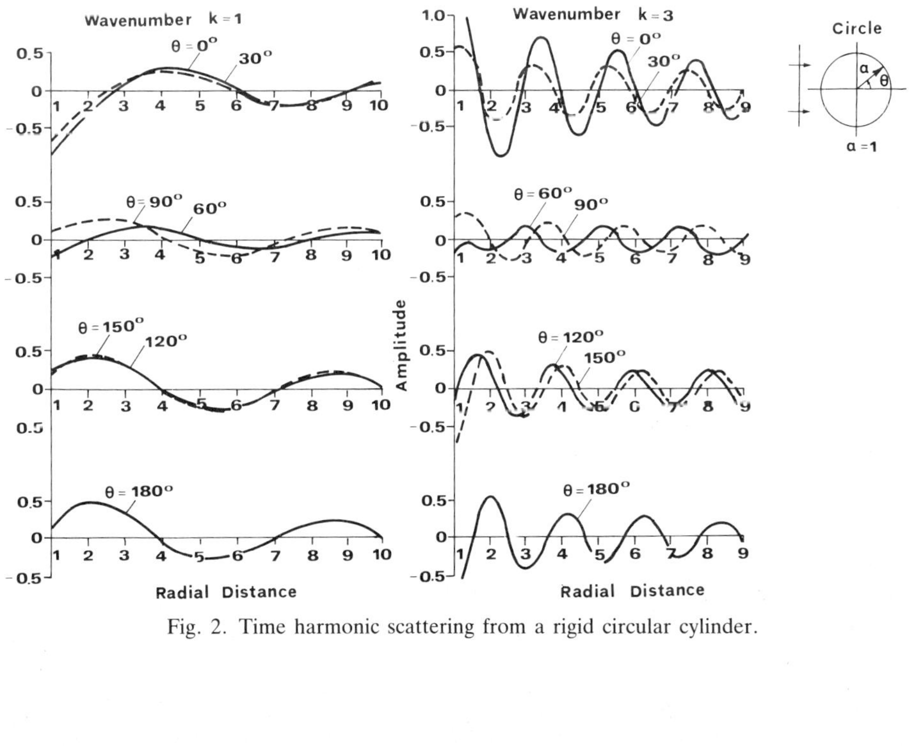

Fig. 2. Time harmonic scattering from a rigid circular cylinder.

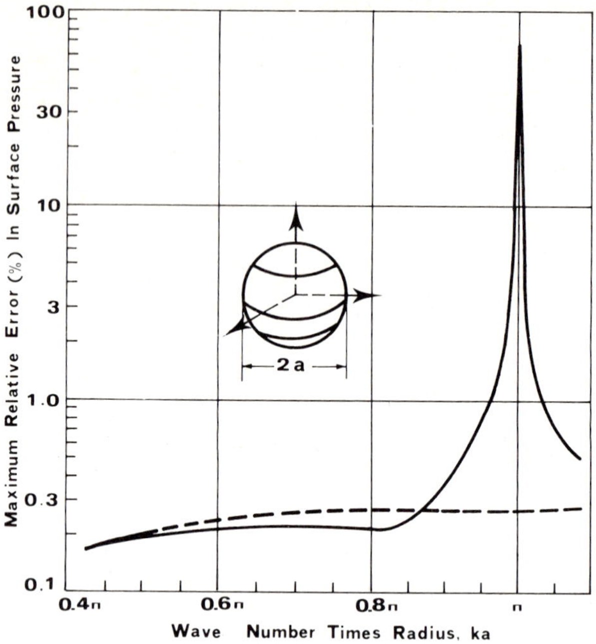

Fig. 3. Improvement in relative error near an interior eigenvalue for exterior scattering by a sphere.

The transient problem leads to successive linear algebraic equations rather than a simultaneous system due to the time-retarded effect. Such systems are easier to solve, but do suffer from some potential error accumulations. It is of interest, however, that a Volterra integral equation of the first kind, which arises in the direct solution of a transient Dirichlet problem, does allow a direct numerical solution by boundary elements, e.g. Shaw and English [40], as shown in Fig. 4 for

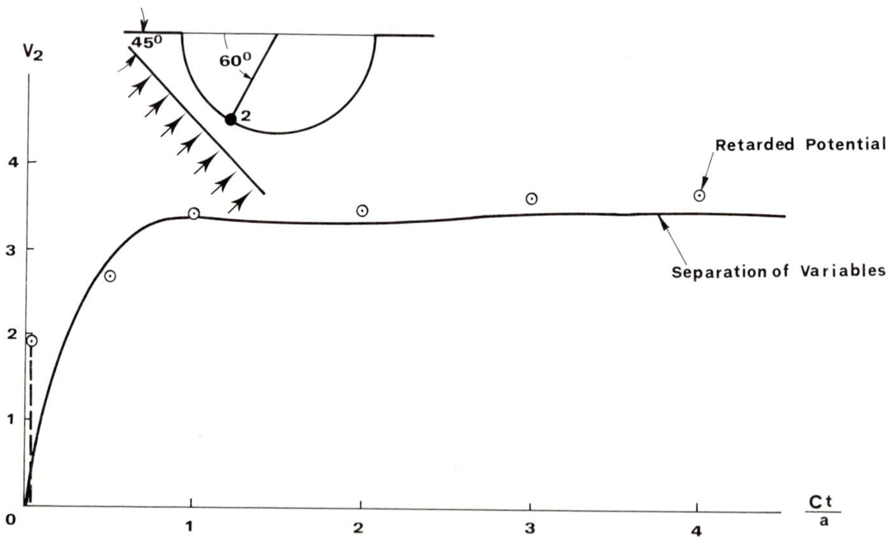

Fig. 4. Velocity history in transient scattering from a half-submerged sphere.

transient scattering by a sphere, whereas the numerical boundary element found in the equivalent time harmonic problem would be expected to be unstable, e.g. Fox [17].

While there is clearly much more work to be done in the development of computer BEM codes for general obstacle shapes, boundary conditions, forcing functions, etc., there is also still room for a deeper understanding of the foundations of this method and its application to related acoustic problems such as those involving inhomogeneous media, acoustic–structure interaction, etc.; these concepts will close this chapter.

6. Conclusion

The most obvious conclusion to be reached concerning boundary integral equation methods is that they work for a wide range of acoustics problems and have been a mainstay of numerical and approximate solution for such problems. Much of this effort, especially as related to underwater sound, has been carried out by various research centers of the US Navy and any survey must take into account the many papers and reports generated by these centers. Other applications are of course widespread, e.g. architectural acoustics, acoustic–structure interaction, etc.; these may not have been given appropriate consideration here since much of the author's research efforts have been related to hydroacoustics problems. Nevertheless, some points should be raised concerning the future development of this field. Probably the most important in the authors's opinion is the use of "non-element" solution methods for boundary integral equations. Such methods include asymptotic expansions for both low- and high-frequency, global solution methods such as those based on Galerkin methods or least-squares methods, e.g. Sharma [36] and other methods. Alternative "generalized Born" series approaches, e.g. Schuster [34, 35] should be examined as well. Hybrid methods, e.g. Schuster [34], should be expanded. General-purpose boundary element codes have been developed for arbitrary shaped radiators and scatterers, e.g. Schenck [32]; these must be expanded and improved as more and more complex problems are studied. An interesting mathematical treatment of this topic is given by Colton and Kress [13]. Finally, a word of warning concerning the general approach. The boundary integral equation method is essentially linear, i.e. it depends on superposition, and primarily homogeneous, i.e. it depends on known fundamental solutions. When nonlinear and/or inhomogeneous problems are to be solved, it may be best to consider coupled approaches, e.g. finite element–boundary element or even finite difference–boundary element approaches, e.g. Shaw [41, 42], to take advantage of those aspects of these methods that are best suited for a particular part of the problem. Those aspects for the boundary integral equation are basically the reduction in dimensionality and the ability to deal with infinite domains using a finite range, i.e. the boundary, on which the unknowns must be found. Both of these aspects are important in the field of acoustics and it may well be expected that the boundary integral equation method will continue to play a vital role in this field.

References

[1] E. Ar and R.E. Kleinmann, The Exterior Neumann Problem for the Three Dimensional Helmholtz Equation, *Archive for Rational Mechanics and Analysis* **23**, 218–236 (1966).

[2] B.B. Baker and E.T. Copson, *The Mathematical Theory of Huygens' Principle* (Clarendon Press, Oxford, UK, 1939).

[3] R.P. Banaugh and W. Goldsmith, Diffraction of Steady Acoustic Waves by Surfaces of Arbitrary Shape, *Journal of the Acoustical Society of America* **35**, 1590–1601 (1963).

[4] P.K. Banerjee and R. Butterfield, *Boundary Element Methods in Engineering Science* (McGraw-Hill Publishing Co., London, 1981).

[5] C.A. Brebbia, J.C.F. Telles and L.C. Wrobel, *Boundary Element Techniques* (Springer-Verlag, Berlin, 1984).

[6] G.B. Brundit, A Solution to the Problem of Scalar Scattering from a Smooth Bounded Obstacle Using Integral Equations, *Quarterly Journal of Mechanics and Applied Mathematics* **18**, 473–489 (1965).

[7] A.J. Burton and G.F. Miller, The Application of Integral Equation Methods to the Numerical Solutions of Some Exterior Boundary Value Problems, *Proceedings of the Royal Society of London* **A323**, 201–210 (1971).

[8] L.H. Chen and D.G. Schweikert, Sound Radiation from an Arbitrary Body, *Journal of the Acoustical Society of America* **35**, 1626–1632 (1963).

[9] L.H. Chen, Acoustic Emissions from Submerged Structures, in: *Developments in Boundary Element Methods*–2, P.K. Banerjee and R.P. Shaw, Eds. (Applied Science Pub., London, 1982) pp. 245–281.

[10] G. Chertock, Sound Radiation from Vibrating Surfaces, *Journal of the Acoustical Society of America* **36**, 1305–1313 (1964).

[11] G. Chertock, Convergence of Iterative Solutions to Integral Equations for Sound Radiation, *Quarterly of Applied Mathematics* **26**, 268–271 (1968).

[12] G. Chertock, Integral Equation Methods in Sound Radiation and Scattering from Arbitrary Surfaces, NSRDC Rep. 3538, Washington DC (1971).

[13] D. Colton and R. Kress, *Integral Equation Methods in Scattering Theory* (J. Wiley & Sons, New York, 1983).

[14] L.G. Copley, Fundamental Results Concerning Integral Representations in Acoustic Radiation, *Journal of the Acoustical Society of America* **44**, 28–32 (1968).

[15] G. DeMey, Calculation of Eigenvalues of the Helmholtz Equation by an Integral Equation, *International Journal for Numerical Methods in Engineering* **11**, 59–66 (1976).

[16] C.L.S. Farn and H. Huang, Transient Acoustic Fields Generated by a Body of Arbitrary Shape, *Journal of the Acoustical Society of America* **43**, 253–257 (1968).

[17] L. Fox, *Numerical Solution of Ordinary and Partial Differential Equations* (Pergamon Press, London, 1962).

[18] F.G. Friedlander, *Sound Pulses* (Cambridge University Press, Cambridge, UK, 1958).

[19] M.B. Friedman and R.P. Shaw, Diffraction of a Plane Shock Wave by an Arbitrary Rigid Obstacle, *Journal of Applied Mechanics* **29**, 40–46 (1962).

[20] D. Greenspan and P. Werner, A Numerical Method for the Exterior Dirichlet Problem for the Reduced Wave Equation, *Archive for Rational Mechanics and Analysis* **23**, 288–316 (1966).

[21] J. Hadamard, *Lectures on Cauchy's Problem in Linear Partial Differential Equations* (Dover Publications, New York, 1952; reprint of 1923 edition).

[22] G.R. Harris, Review of Transient Field theory for a Baffled Planar Piston, *Journal of the Acoustical Society of America* **70**, 10–20 (1981).

[23] J.R. Hutchinson, Determination of Membrane Vibrational Characteristics by the Boundary Integral Equation Method, in: *Recent Advances in Boundary Element Methods*, C.A. Brebbia, Ed. (Pentech Press, London, 1978) pp. 301–316.

[24] R.E. Kleinmann, The Dirichlet Problem for the Helmholtz Equation, *Archive for Rational Mechanics and Analysis* **18**, 205–229 (1965).

[25] R.E. Kleinmann and G.F. Roach, Boundary Integral Equations for the Three Dimensional Helmholtz Equation, *SIAM Review* **16**, 214–236 (1974).

[26] H. Lamb, *Hydrodynamics* (Dover Publications, New York, 1945; reprint of 1932 edition).

[27] J. Mackerle and T. Andersson, Boundary Element Software in Engineering, *Advances in Engineering Software* **6**, 66–102 (1983).

[28] W.P. Mason and R.N. Thurston, Eds. *Physical Acoustics: Principles and Methods*, Vols 1–17 (Academic Press, New york, 1964–1984).

[29] W.A. Meyer, W.L. Bell, B.T. Zinn and M.P. Stallybrass, Boundary Integral Solutions of Three Dimensional Acoustic Radiation Problems, *Journal of Sound and Vibration* **65**, 631–638 (1978).

[30] K.M. Mitzner, Numerical Solution for Transient Scattering from a Hard Surface Shape; Retarded Potential Technique, *Journal of the Acoustical Society of America* **42**, 391–397 (1967).

[31] P. Morse and H. Feshbach, *Methods of Theoretical Physics*, Vols 1 and 2, (MacGraw-Hill Publishing Co., New York, 1953).

[32] H.A. Schenck, Improved Integral Equation Formulation for Acoustic Radiation Problems, *Journal of the Acoustical Society of America* **44**, 41–58 (1968).

[33] G.T. Schuster and L.C. Smith, A Comparison Among Four Direct Boundary Integral Methods, *Journal of the Acoustical Society of America* **77**, 850–864 (1985).

[34] G.T. Schuster, A Hybrid BIE + Born Series Modelling Scheme: Generalized Born Series, *Journal of the Acoustical Society of America* **77**, 865–879 (1985).

[35] G.T. Schuster, Solution of the Acoustic Transmission Problem by a Perturbed Born Series, *Journal of the Acoustical Society of America* **77**, 880–886 (1985).

[36] D.L. Sharma, Scattering of Steady Elastic Waves by Surfaces of Arbitrary Shape, *Bulletin of the Seismological Society of America* **57**, 795–812 (1967).

[37] R.P. Shaw and M.B. Friedman, Diffraction of a Plane Shock Wave by a Free Cylindrical Obstacle at a Free Surface, *Proceedings of the 4th US National Congress of Applied Mechanics*, R.M. Rosenberg, Ed. (ASME, New York, 1962) pp. 371–379.

[38] R.P. Shaw, Diffraction of Acoustic Pulses by Obstacles of Arbitrary Shape with a Robin Boundary Condition, *Journal of the Acoustical Society of America* **41**, 855–859 (1967).

[39] R.P. Shaw, An Integral Equation Approach to Acoustic Radiation and Scattering in: *Topics in Ocean Engineering* II, C. Bretschneider, Ed. (Gulf Publishing Co., Houston, Texas) pp. 143–163.

[40] R.P. Shaw and J.A. English, Transient Acoustic Scattering by a Free Sphere, *Journal of Sound and Vibration* **20**, 321–331 (1972).

[41] R.P. Shaw, Time Harmonic Scattering by Obstacles Surrounded by an Inhomogeneous Medium, *Journal of the Acoustical Society of America* **56**, 1354–1360 (1974).

[42] R.P. Shaw, Coupling Boundary Element Methods to Other Numerical Techniques, in: *Recent Advances in Boundary Element Methods*, C.A. Brebbia, Ed. (Pentech Press, London, 1978) pp. 137–147.

[43] R.P. Shaw, Boundary Integral Equation Methods Applied to Wave Problems, in: *Developments in Boundary Element Methods*–1, P.K. Banerjee and R. Butterfield, Eds. (Applied Science Publishers, London, 1980) pp. 121–153.

[44] R.P. Shaw and H-C. Wang, Acoustic Radiation from a Piston of Arbitrary Shape in an Infinite Rigid Baffle, *Microsoftware for Engineers*, in press.

[45] G.R.C. Tai and R.P. Shaw, Helmhotz Equation Eigenvalues and Eigenmodes for Arbitrary Domains, *Journal of the Acoustical Society of America* **56**, 796-804 (1974).

[46] F. Ursell, On the Exterior Problems of Acoustics, *Proceedings of the Royal Society of London* **A275**, 33–46 (1973).

CHAPTER 10

Heat Conduction, Thermoelasticity and Consolidation

Sam SHARP and Steven L. CROUCH

Department of Civil and Mineral Engineering
University of Minnesota
Minneapolis, Minnesota 55455, U.S.A.

Boundary Element Methods in Mechanics
Edited by D.E. Beskos

Introduction

In this work we are concerned with boundary integral methods for problems that occur in the fields of heat flow and thermoelasticity and of fluid flow in porous media, also known as consolidation. While these phenomena are physically different, their respective theories are mathematically analogous, and can be represented generally by the set of equations

$$\mu u_{i,jj} + (\lambda + \mu) u_{j,ji} - \rho \frac{\partial}{\partial t} u_i - A\theta_{,i} + \rho F_i = 0$$
$$B\theta_{,jj} - C \frac{\partial \theta}{\partial t} - D \frac{\partial}{\partial t} u_{j,j} + S = 0 \qquad i, j = 1, 2, 3 \tag{1}$$

where u_i are displacements and θ is a scalar field which represents either the temperature (thermoelasticity) or pressure (consolidation). A, B, C and D are constants, while ρ is the mass density of the medium, F_i and S are forcing functions and μ and λ are Lamé elasticity constants defined by

$$\mu = G \tag{2a}$$

$$\lambda = 2G\nu/(1 - 2\nu) \qquad i, j = 1, 2, 3 \tag{2b}$$

where G and ν are the shear modulus and Poisson's ratio respectively. Equation (1) constitutes a linear system of coupled partial differential equations which in general is difficult to solve. It emerges, however, that for certain important classes of problems, the system may be simplified considerably if we make some assumptions regarding its behavior with respect to time.

For example, in the case of thermoelasticity under the usual conditions of heat exchange, the rate of temperature change in the body is low and the corresponding deformations proceed slowly. Consequently the accelerations are negligible and the inertia term may be dropped. Further, since the deformations are assumed to be small, the heat produced by them will be negligible and the coupling term containing the strain rate $\dot{u}_{i,i}$ may also be ignored. The equations can thus be written as

$$\mu u_{i,jj} + (\lambda + \mu) u_{j,ji} - (3\lambda + 2\mu)\omega\theta_{,i} + \rho F_i = 0 \tag{3a}$$

$$K\theta_{,jj} - \rho c \frac{\partial \theta}{\partial t} + S = 0 \tag{3b}$$

where ω, K and c are the thermal expansion coefficient, conductivity and specific heat of the conducting medium respectively. It is apparent that no mechanical terms now appear in the heat equation. The temperature can therefore be found independently and used as input to the displacement equations. Furthermore, in the absence of inertia and strain rate terms, the displacements are time-dependent only by virtue of this input. We have then an essentially static mechanical system with time entering as a parameter. The body thus deforms slowly through a continuous sequence of equilibrium positions, each relating to a particular temperature profile. Such motion is called quasi-static.

Solution of the heat flow problem (i.e. solving the diffusion equation (3b)) is the first step towards finding the thermoelastic response of the material. As we shall see, the major obstacles arising in the construction of a boundary integral method for quasi-static thermoelasticity problems are those concerned with this prior solution. Once the details of this task have been attended to, the incorporation of the mechanical effects can be accomplished in a relatively straightforward manner.

In the case of consolidation, matters are not so simple. Although the inertia term is usually dropped, the equations cannot be decoupled. (One-dimensional consolidation is a notable exception.) This is because deformations of the solid matrix significantly affect the behavior of the pressure, i.e. the strain rate term dominates that of the pressure absorption. Hence the diffusion equation cannot be solved independently for the pressure, and alternative approaches must be followed.

We shall confine our attention to two-dimensional linear problems of plane strain. However, the approach is applicable to three dimensions [61, 45], as well as axisymmetric problems [19]. Nonlinearities of limited types can also be accommodated by the approach [45, 19]. However, these will not be discussed here.

Consider the partial differential equation

$$L(\phi) = -f \tag{4}$$

where L is a differential operator, and f a forcing function. A classical method [84, 43] for solving such equations is to represent the solution as a trial function in the form

$$\phi' = \sum_{i=0}^{n} a_i \phi_i \tag{5}$$

where the ϕ_i are basis or trial functions, and the a_i are the coefficients which must be chosen in a way that allows ϕ' to satisfy the constraints of the problem in some best sense. The various alternatives available are distinguished by the choice of functions ϕ_i and the criterion by which the a_i are determined. Finite element techniques, for example, may be viewed as being based upon a trial-function approach. Here the approximating expression is simplified by prescribing it sectionally across the body of interest, which must therefore be divided into a set of discrete partitions or elements. The performance criterion corresponds to a process of distributing the error according to some principle, and takes the form of a body integral, which, upon discretization, reduces to a system of algebraic equations in a_i.

Boundary methods may be seen as special cases of a trial-function technique. Their distinctive characteristic concerns the nature of the functions ϕ_i. Specifically, these must be chosen so that

$$L(\phi_i) = 0 \qquad i = 1, \dots, n \tag{6}$$

$$L(\phi_0) = -f \,.$$

If L is linear, these conditions allow the trial solution to satisfy the differential equation exactly, so that the error introduced is limited to that incurred in attempting to fulfill the auxiliary conditions. The coefficients a_i are then determined by distributing the error over the boundary according to some principle.

The error principles are usually classified as being of the variational or residual type and can be expressed by equating a boundary integral to zero. For example, in the case of the diffusion equation, we could minimize a functional based on a variational principle [41] or define the residual-based criterion of the form

$$\int_{\partial V} F(R) W \, \mathrm{d}s = 0 \tag{7}$$

where R is the boundary residue $B(\phi) - B(\phi')$, F is some functional of R, W is a weighting function, the operator B describes the auxiliary conditions and ∂V is the surface bounding the function space. Choices F and W correspond to various approximation principles, such as those of Galerkin and least-squares [54].

Modern boundary integral methods can be accommodated in this general framework and correspond to particular forms of R, F and W [16]. However, they are distinguished from the general class of techniques by the presence of a singular solution. As a result it would be more precise to refer to them as "singular" or "fundamental" solution methods and consider them merely as members of the general boundary integral family. In any event it is these methods that we shall be concerned with here.

Early applications of boundary integral methods are found in elliptical equations pertaining to problems of potential [44, 80] and elastostatics [64, 32, 4, 28, 14]. While refinements and generalizations have been added [65, 49, 67], the technique has also been used to solve other types of equations, arising from theories of elastodynamics [30, 31], elastoplasticity [46, 8, 9, 82], viscoelasticity [68] and diffusion [66, 22, 6, 18, 87, 35, 36, 81, 88, 10, 40, 69, 78]. The formal application of the method to these equations can be conducted in complete analogy with that for the elliptical case, with time and space dimension given similar treatment. In the case of the parabolic heat equation such "pure" formulations containing only boundary integrals are possible but, as we shall see, the unboundedness of the time axis can strain computer resources if the solution is required for sufficiently large values of the time parameter.

Alternative approaches have been attempted. These have involved the transformation of the differential equation into an elliptical one in which time appears as a parameter. For example, Rizzo and Shippy [66] applied the Laplace transform to the diffusion equation and solved the resulting elliptical problem in the transformed space. Curran et al. [36], and Roures and Alarcón [69] employed a finite difference approximation to the time derivative in the heat equation, again producing an elliptical problem which can be solved repeatedly at each time step.

Other writers [22, 18, 87, 88, 10, 40] have preferred to retain time as an explicit variable in the analysis, but have tried to avoid economic penalties incurred in carrying the entire history of the system by developing time marching algorithms that essentially slice the time axis into sections, and solve a new problem in each. We shall investigate this option later, and touch on some of the interesting and unexpected problems that arise in this type of approach.

The retention of the time domain has the merit of preserving the physical understanding of the system that the methods offer. In fact one of the most appealing aspects of boundary integral techniques is the multiple avenues they offer for approaching an understanding of them. We have already seen how an approximation theoretical approach can be used to derive these methods. The physical interpretation can be made by considering the functions ϕ_i as disturbances in the field. In elastostatic problems, they correspond to point forces whereas in the case of transient heat flow they represent the effects of distributed heat sources emitting (or absorbing) heat with time-dependent intensities. This affords a ready-made method for choosing the trial functions ϕ_i and permits a practical and intuitive grasp of the technique without resort to mathematical rigor. On the other hand, the methods can be shown to represent an approximation of an exact boundary integral expression for the solution to the problem. This gives a sound theoretical foundation to the technique, establishing its authenticity on mathematical grounds.

The methods display several other attractions. As stated earlier, conditions (6) imply that the performance integral need only range over the domains of the boundary functions rather than over that of the solution itself, as is the case with finite element methods. This offers an elegant and efficient alternative for solving partial differential equations. By exploiting the properties of the basis functions, it avoids the task of regulating the behavior of the approximation over the entire body and thus effectively reduces the dimension of the problem by one. The algorithm is also modular and can be broken down into components that can be altered or replaced without having to revise the entire process. In this way different equation solvers and boundary approximation techniques can be introduced easily.

In this work we shall show how systems of equations of the form (1) can be solved by means of boundary integral methods. We shall start with the simplest case, treating the decoupled system which allows us to solve the diffusion equation independently. This is convenient since many of the major numerical hurdles reveal themselves in this solution procedure and so can be dealt with before we address the accompanying mechanical problem. For this reason, a major part of this chapter is devoted to the solution of the diffusion equation. We shall introduce the method by deriving the underlying boundary integral equations that provide its theoretical foundation. We shall also develop it from purely physical considerations in order to promote an intuitive grasp of the technique. We shall show how the numerical techniques can be viewed both as constructive processes, piecing together elementary solutions to form more general ones, and also as simplifying processes, allowing exact but complicated analytical expressions to be decomposed into more easily handled parts. In either case, we should retain the notion of trial functions as a reminder of the approximation theoretical roots of these methods. We shall then demonstrate how the thermal stresses and displacements are calculated. We shall offer two alternative methods showing how these can be incorporated into a general framework for such solutions. Finally, we shall show how fully coupled systems such as those arising from the theories of consolidation may be solved by boundary integral methods, and will outline several different approaches that have been attempted.

1. Boundary integral methods for the diffusion equation

1.1. Mathematical foundations

1.1.1. The problem domain

We seek to solve the heat flow problem

$$\theta_{,jj} - \frac{1}{\alpha}\frac{\partial \theta}{\partial t} = -\frac{f}{\alpha} \qquad \boldsymbol{x} \in \Omega,\ t > 0 \tag{1.1}$$

subject to the general boundary conditions

$$a\theta(s, t) + h\,\frac{\partial \theta(s, t)}{\partial v} = g(s, t) \qquad a, h \geq 0,\ t > 0,\ s \in \partial\Omega \tag{1.2}$$

and the initial condition $\theta(\boldsymbol{x}, 0) = \theta_0(\boldsymbol{x})$, $\boldsymbol{x} \in \Omega$, where $\alpha = K/\rho c$ is the diffusivity, $f = S/\rho c$, g is a prescribed function and v is the direction of the outward normal.

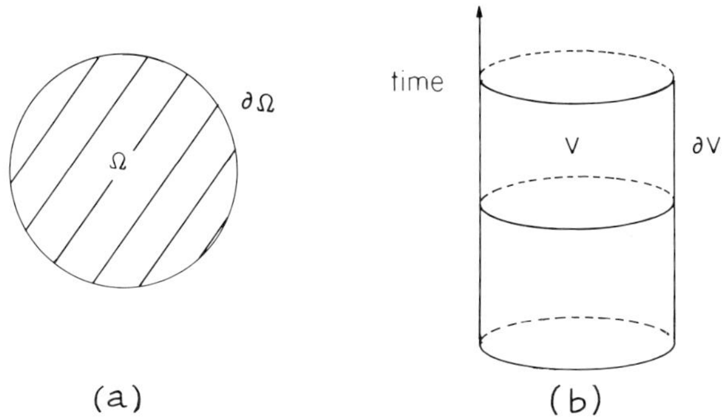

Fig. 1. Problem domains for the diffusion equation: (a) physical domain; (b) mathematical domain.

Unlike the case for time-independent problems, it is necessary to make a conceptual distinction between the physical and mathematical domains of the formulation. The physical domain, shown in Fig. 1(a), concerns the geometrical properties of the problem, and, in our case, is two-dimensional. The mathematical domain, however, represents the set of points over which the time-dependent temperature function is defined, and thus lies in a three-dimensional space–time continuum. Provided that the contour $\partial\Omega$ enclosing the region Ω in Fig. 1(a) is stationary, this domain may be viewed as a semi-infinite cylinder V bounded by the surface ∂V based upon the shape Ω, with its axis parallel to the positive time direction. This is shown in Fig. 1(b). The cylinder surface ∂V therefore consists of a lateral part $(0, \infty) \times \partial\Omega$, and a "floor" Ω at time $t = 0$. This surface features in the boundary integral expressions to be derived for temperature and, later, thermal stresses and displacements. Its inclusion of the two-dimensional spatial surface Ω should not be viewed as a departure from fundamental boundary integral methodology, but rather as a necessary consequence of the three-dimensional character of the problem. For convenience, we shall limit the height of the cylinder V to T, a time beyond which we are no longer interested in monitoring the system. The surface ∂V thus also includes a "roof" Ω at time $t = T$. The choice of T is arbitrary, and will not affect the results of the analysis.

1.1.2. The direct formulation

A formal derivation of the boundary integral method for the diffusion equation begins with the identities

$$\eta_{,i}\phi_{,i} = (\phi\eta_{,i})_{,i} - \phi\eta_{,ii} \tag{1.3a}$$

$$0 = \frac{1}{2\alpha}\frac{\partial(\eta\phi)}{\partial t} - \frac{1}{2\alpha}\left(\frac{\partial\eta}{\partial t}\phi + \frac{\partial\phi}{\partial t}\eta\right) \tag{1.3b}$$

$$\eta_{,i}\phi_{,i} = (\eta\phi_{,i})_{,i} - \eta\phi_{,ii} \tag{1.4a}$$

$$0 = -\frac{1}{2\alpha}\frac{\partial(\phi\eta)}{\partial t} + \frac{1}{2\alpha}\left(\frac{\partial\eta}{\partial t}\phi + \frac{\partial\phi}{\partial t}\eta\right) \tag{1.4b}$$

where η and ϕ are arbitrary functions for which the required derivatives are continuous. If we now add Eqs. (1.3) together, integrate over the volume of the cylinder and use the divergence theorem on the first terms on the right-hand side of each, we obtain

$$\int_V (\eta_{,i}\phi_{,i})\,\mathrm{d}V = \int_0^t \int_{\partial\Omega} \left(\phi\frac{\partial\eta}{\partial v}\right)\mathrm{d}s\,\mathrm{d}t - \int_0^T\int_\Omega [\phi\eta_{,ii}]\,\mathrm{d}\Omega\,\mathrm{d}t$$
$$+ \frac{1}{2\alpha}\int_\Omega [\phi\eta]_{t=T}\,\mathrm{d}\Omega - \frac{1}{2\alpha}\int_\Omega [\phi\eta]_{t=0}\,\mathrm{d}\Omega - \frac{1}{2\alpha}\int_0^T\int_\Omega\left(\frac{\partial\eta}{\partial t}\phi + \frac{\partial\phi}{\partial t}\eta\right)\mathrm{d}\Omega\,\mathrm{d}t\,.$$

Note that the time integrals involve convolutions. If a similar expression is written relating to Eq. (1.4) and is subtracted from the one above, the result may easily be manipulated to the form

$$\alpha\int_0^T\int_\Omega [\eta L(\phi) - \phi L^*(\eta)]\,\mathrm{d}\Omega\,\mathrm{d}t = \alpha\int_0^T\int_\Omega\left(\eta\frac{\partial\phi}{\partial v} - \phi\frac{\partial\eta}{\partial v}\right)\mathrm{d}s\,\mathrm{d}t$$
$$-\int_\Omega [\phi\eta]_{t=T}\,\mathrm{d}\Omega + \int_\Omega [\phi\eta]_{t=0}\,\mathrm{d}\Omega \tag{1.5}$$

where

$$L = {}_{,jj} - \frac{1}{\alpha}\frac{\partial}{\partial t}$$

and

$$L^* = {}_{,jj} + \frac{1}{\alpha}\frac{\partial}{\partial t}$$

are linear differential operators. L^* is known as the formal adjoint of L.

We now identify the function ϕ with θ, the desired temperature distribution. Since the heat flow problem may be stated as

$$\begin{aligned} &L(\theta) = -f/\alpha && \text{on } \Omega && t>0\\ &\theta = \theta_0 && \text{on } \Omega && t=0\\ &\theta = g && \text{on } \partial\Omega && t>0 \end{aligned}$$

(where we have assumed a Dirichlet boundary condition for the purposes of exposition) (1.5) becomes

$$\alpha \int_0^T \int_\Omega [-\theta L^*(\eta)] \, \mathrm{d}\Omega \, \mathrm{d}t = \int_0^T \int_\Omega [\eta f] \, \mathrm{d}\Omega \, \mathrm{d}t$$

$$+ \alpha \int_0^T \int_{\partial\Omega} \left(\eta \frac{\partial \theta}{\partial v} - \theta \frac{\partial \eta}{\partial v} \right) \mathrm{d}s \, \mathrm{d}t - \int_\Omega [\phi\eta]_{t=T} \, \mathrm{d}\Omega + \int_\Omega [\phi\eta]_{t=0} \, \mathrm{d}\Omega$$

which offers an explicit formula for θ at any point $\boldsymbol{\xi} \in \Omega$ and time τ if we can find a function $\eta(\boldsymbol{x}, t; \boldsymbol{\xi}, \tau)$ such that

$$\alpha L^*(\eta) = -\delta(x_1 - \xi_1)\delta(x_2 - \xi_2)\delta(\tau - t) \qquad \boldsymbol{x} \in \Omega, 0 < t < \tau \tag{1.6a}$$

$$\eta = 0 \qquad \boldsymbol{x} \in \Omega, t > \tau \tag{1.6b}$$

$$\eta = 0 \qquad \boldsymbol{x} \in \partial\Omega, t > 0 \tag{1.6c}$$

where δ is the Dirac delta function, and where $\boldsymbol{x} = (x_1, x_2)$ and $\boldsymbol{\xi} = (\xi_1, \xi_2)$. For then

$$\alpha \int_0^T \int_\Omega \theta(\boldsymbol{x}, t) L^*(\eta(\boldsymbol{x}, t; \boldsymbol{\xi}, \tau)) \, \mathrm{d}\boldsymbol{x} \, \mathrm{d}t = -\theta(\boldsymbol{\xi}, \tau) \qquad \boldsymbol{\xi} \in \Omega \tag{1.7}$$

so that

$$\theta(\boldsymbol{\xi}, \tau) = \alpha \int_0^\tau \int_{\partial\Omega} -\frac{\partial \eta(\boldsymbol{x}, t; \boldsymbol{\xi}, \tau)}{\partial v} g(\boldsymbol{x}, t) \, \mathrm{d}\boldsymbol{x} \, \mathrm{d}t \tag{1.8}$$

$$+ \int_\Omega \eta(\boldsymbol{x}, 0; \boldsymbol{\xi}, \tau)\theta_0(\boldsymbol{x}) \, \mathrm{d}\boldsymbol{x} + \int_0^\tau \int_\Omega \eta(\boldsymbol{x}, t; \boldsymbol{\xi}, \tau) f(\boldsymbol{x}, t) \, \mathrm{d}\boldsymbol{x} \, \mathrm{d}t$$

and it can be shown that $\lim_{\boldsymbol{\xi} \to \partial\Omega} \theta(\boldsymbol{\xi}, \tau) = g(\boldsymbol{\xi}, \tau)$ and $\lim_{\tau \to 0} \theta(\boldsymbol{\xi}, \tau) = \theta_0(\boldsymbol{\xi})$.

This is the standard Green's function approach [79] for solving the diffusion equation. Aside from any quadrature formulae that may be required to evaluate the integrals, (1.8) represents an analytical solution to the problem. The use of the formula is, however, clearly contingent upon finding η, known as the Green's function. This can only be done in the case of a narrow range of problems. In order to understand why this is so, we note that constraints (1.6) imply that η may be written as the sum of a singular function G^*, which is a permanent feature of the formula, and a non-singular function γ, which must be fashioned to fit the conditions of each particular problem. Thus

$$\eta(\boldsymbol{x}, t; \boldsymbol{\xi}, \tau) = G^*(\boldsymbol{x}, t; \boldsymbol{\xi}, \tau) + \gamma(\boldsymbol{x}, t; \boldsymbol{\xi}, \tau)$$

with

$$\alpha L^*(G) = -\delta(x_1 - \xi_1)\delta(x_2 - \xi_2)\delta(\tau - t) \tag{1.9}$$

and

$$\begin{aligned} &L^*(\gamma) = 0 && \text{on } \Omega,\ 0 < t < \tau \\ &\gamma = -G^* && \text{on } \partial\Omega,\ t > 0 \\ &\gamma = 0 && \text{on } \Omega,\ t = 0\,. \end{aligned}$$

The formula (1.8) is therefore only suitable for problems with geometries sufficiently simple to allow γ to be found analytically.

The boundary integral technique attempts to circumvent this limitation by relaxing (1.6c), thus broadening the scope of the method. In doing so the penalty of no longer automatically satisfying the boundary condition is incurred. For, by setting

$$\gamma = 0 \qquad \boldsymbol{x} \in \Omega,\ t > 0\,, \tag{1.10}$$

and interchanging the pairs $(\boldsymbol{x}, t)$ and $(\boldsymbol{\xi}, \tau)$, the formula becomes

$$\begin{aligned} \theta(\boldsymbol{x}, t) = \alpha \int_0^t \int_{\partial\Omega} \left(G^*(s, \tau; \boldsymbol{x}, t) \frac{\partial\theta(s, \tau)}{\partial v} - \frac{\partial G^*(s, \tau; \boldsymbol{x}, t)}{\partial v} g(s, \tau) \right) \mathrm{d}s\, \mathrm{d}\tau \\ + \int_{\Omega} G^*(\boldsymbol{\xi}, 0; \boldsymbol{x}, t)\theta_0(\boldsymbol{\xi})\, \mathrm{d}\boldsymbol{\xi} + \int_0^t \int_{\Omega} G^*(\boldsymbol{\xi}, \tau; \boldsymbol{x}, t) f(\boldsymbol{\xi}, \tau)\, \mathrm{d}\boldsymbol{\xi}\, \mathrm{d}\tau \end{aligned} \tag{1.11}$$

which now contain the unknown boundary function $\partial\theta/\partial v$. However, this can be determined so that (1.11) is made to take on the values of g when $\boldsymbol{x}$ lies on $\partial\Omega$. Thus we require that

$$\begin{aligned} g(x_s, t) = \alpha \int_0^t \int_{\partial\Omega} \left(G^*(s, \tau; \boldsymbol{x}_s, t) \frac{\partial\theta(s, \tau)}{\partial v} - \frac{\partial G^*(s, \tau; \boldsymbol{x}_s, t)}{\partial v} g(s, \tau) \right) \mathrm{d}s\, \mathrm{d}\tau \\ + \int_{\Omega} G^*(\boldsymbol{\xi}, 0; \boldsymbol{x}_s, t)\theta_0(\boldsymbol{\xi})\, \mathrm{d}\boldsymbol{\xi} + \int_0^t \int_{\Omega} G^*(\boldsymbol{\xi}, \tau; \boldsymbol{x}_s, t) f(\boldsymbol{\xi}, \tau)\, \mathrm{d}\boldsymbol{\xi}\, \mathrm{d}\tau \end{aligned} \tag{1.12}$$

which is a linear integral equation to be solved for $\partial\theta/\partial v$. Once this has been done, (1.11) provides an explicit formula for θ in terms of boundary, initial and field source data, without any restriction placed on the shape of Ω. Analogous derivations could be performed for the cases of boundary conditions involving the flux, a mix of fluxes and temperatures, or a linear combination of these.

From a theoretical standpoint then, the problem has been solved. By means of a variation of the Green's function method, the point of departure from which is represented by definition (1.10), we have obtained an expression for a function satisfying all of the problem constraints. In practice, however, it is not possible to solve the boundary integral equation (1.12) analytically and a numerical procedure must be used. It is this procedure that constitutes the Boundary Element Method, which, when viewed in this context, is simply a numerical method for solving a linear integral equation. In particular, for reasons that will become clear later, it is known as the Direct Boundary Element Method.

1.1.3. The fundamental solution

The existence of the function G^* satisfying (1.6a) and (1.6b) is clearly critical to the task of solving the heat flow problem, whether by means of the Green's function method or the boundary integral equation. However, it is more convenient to consider the integral expressions in terms of its adjoint function G which satisfies

$$\alpha L(G(\boldsymbol{x}, t; \boldsymbol{\xi}, \tau)) = -\delta(x_1 - \xi_1)\delta(x_2 - \xi_2)\delta(t - \tau) \qquad \boldsymbol{x}, \boldsymbol{\xi} \in \Omega, 0 < \tau \leq t \tag{1.13a}$$

$$G = 0 \qquad \tau > t\,. \tag{1.13b}$$

It can be shown [21] that

$$G = \frac{1}{4\pi\alpha(t-\tau)} \exp\left(\frac{-r^2}{4\alpha(t-\tau)}\right) \tag{1.14}$$

where $r = |\boldsymbol{x} - \boldsymbol{\xi}|$.

From (1.13a) it is clear that G describes the temperature distribution in an infinite plane due to a heat pulse at time τ at the point $\boldsymbol{\xi}$. G is known as a fundamental solution, and can be shown to exhibit the following properties:

$$\lim_{r\to\infty} G = 0 \tag{1.15a}$$

$$\lim_{t\to\infty} G = 0 \tag{1.15b}$$

$$\lim_{t\to\tau} G = \delta(x_1 - \xi_1)\delta(x_2 - \xi_2) \tag{1.15c}$$

$$G(\boldsymbol{x}, t; \boldsymbol{\xi}, \tau) = G^*(\boldsymbol{\xi}, \tau; \boldsymbol{x}, t)\,. \tag{1.15d}$$

Equation (1.15c) indicates that G may also be interpreted as describing the temperature in an infinite plane due to an initial concentration of temperature at point $\boldsymbol{\xi}$. It is this property that allows the initial conditions to be satisfied by the integral expressions. The equality (1.15d) allows the introduction of G into the boundary integral solution (1.11) and (1.12) which become respectively

$$\theta(\boldsymbol{x}, t) = \alpha \int_0^t \int_{\partial\Omega} \left(G(\boldsymbol{x}, t; s, \tau) \frac{\partial\theta(s, \tau)}{\partial v} - \frac{\partial G(\boldsymbol{x}, t; s, \tau)}{\partial v} \theta(s, \tau) \right) \mathrm{d}s \, \mathrm{d}\tau + \int_{\Omega} G(\boldsymbol{x}, t; \boldsymbol{\xi}, 0)\theta_0(\boldsymbol{\xi}) \, \mathrm{d}\boldsymbol{\xi} + \int_0^t \int_{\Omega} G(\boldsymbol{x}, t; \boldsymbol{\xi}, \tau) f(\boldsymbol{\xi}, \tau) \, \mathrm{d}\boldsymbol{\xi} \, \mathrm{d}\tau \tag{1.16}$$

$$g(\boldsymbol{x}_s, t) = \alpha \int_0^t \int_{\partial\Omega} \left(G(\boldsymbol{x}_s, t; s, \tau) \frac{\partial\theta(s, \tau)}{\partial v} - \frac{\partial G(\boldsymbol{x}_s, t; s, \tau)}{\partial v} \theta(s, \tau) \right) \mathrm{d}s \, \mathrm{d}\tau + \int_{\Omega} G(\boldsymbol{x}_s, t; \boldsymbol{\xi}, 0)\theta_0(\boldsymbol{\xi}) \, \mathrm{d}\boldsymbol{\xi} + \int_0^t \int_{\Omega} G(\boldsymbol{x}_s, t; \boldsymbol{\xi}, \tau) f(\boldsymbol{\xi}, \tau) \, \mathrm{d}\boldsymbol{\xi} \, \mathrm{d}\tau \tag{1.17}$$

where we have restored the general boundary function $\theta(s, \tau)$.

1.1.4. The indirect method

It is possible to arrive at boundary integral expressions for the temperature distribution which are somewhat simpler than the pair (1.16) and (1.17). In order to derive these it is necessary to form solution (1.16) for an imaginary problem posed in the complementary region Ω_c shown in Fig. 2. This region is the one that forms an infinite plane when united with Ω. Following Banerjee et al. [10], and denoting by θ_c the temperature in Ω_c we have

$$\theta_c(\boldsymbol{x}, t) = \alpha \int_0^t \int_{\partial\Omega} \left(-G(\boldsymbol{x}, t; s, \tau) \frac{\partial\theta_c(s, \tau)}{\partial v} + \frac{\partial G(\boldsymbol{x}, t; s, \tau)}{\partial v} \theta_c(s, \tau) \right) \mathrm{d}s \, \mathrm{d}\tau + \int_{\Omega_c} G(\boldsymbol{x}, t; \boldsymbol{\xi}, 0)\theta_c(\boldsymbol{x}, 0) \, \mathrm{d}\boldsymbol{\xi} \qquad \boldsymbol{x} \in \Omega_c, \ t > 0 \tag{1.18}$$

where we have assumed no heat sources to be present in Ω_c. Note that the change of sign in the lateral surface integral is due to the expression of the derivatives

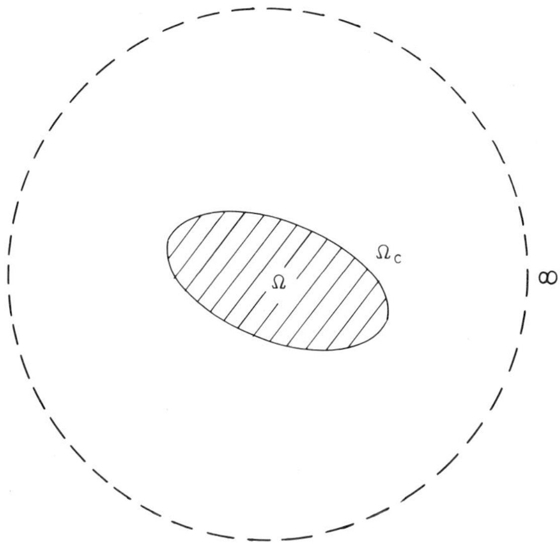

Fig. 2. The region Ω and its complement Ω_c.

with respect to the system of coordinates local to the problem on Ω. We shall specify the boundary conditions for this fictitious problem momentarily. However, we first examine the case where θ_c is evaluated at a point outside of the domain, i.e. at a point in Ω. We recall from (1.6a) that the free term in the left-hand side of (1.8) arose because the point of interest lay within the region of integration in (1.7). If the point lies in the external region, this integral is zero. Thus if in (1.18) we let $\boldsymbol{x} \in \Omega$ (i.e. external to Ω_c) we have

$$0 = \alpha \int_0^t \int_{\partial\Omega} \left(-G(\boldsymbol{x}, t; s, \tau) \frac{\partial \theta_c(s, \tau)}{\partial \upsilon} + \frac{\partial G(\boldsymbol{x}, t; s, \tau)}{\partial \upsilon} \theta_c(s, \tau) \right) \mathrm{d}s \, \mathrm{d}\tau + \int_{\Omega_c} G(\boldsymbol{x}, t; \boldsymbol{\xi}, 0) \theta_c(\boldsymbol{\xi}, 0) \, \mathrm{d}\boldsymbol{\xi} \; . \tag{1.19}$$

Summing (1.16) and (1.19) we have

$$\theta(\boldsymbol{x}, t) = \alpha \int_0^t \int_{\partial\Omega} \left(G(\boldsymbol{x}, t; s, \tau) Q(s, \tau) - \frac{\partial G(\boldsymbol{x}, t; s, \tau)}{\partial \upsilon} \mu(s, \tau) \right) \mathrm{d}s \, \mathrm{d}\tau + \int_0^t \int_{\Omega} G(\boldsymbol{x}, t; \boldsymbol{\xi}, \tau) f(\boldsymbol{\xi}, \tau) \, \mathrm{d}\boldsymbol{\xi} \, \mathrm{d}\tau + \int_{\Omega} G(\boldsymbol{x}, t; \boldsymbol{\xi}, 0) \theta_0(\boldsymbol{\xi}) \, \mathrm{d}\boldsymbol{\xi} + \int_{\Omega_c} G(\boldsymbol{x}, t; \boldsymbol{\xi}, 0) \theta_c(\boldsymbol{\xi}, 0) \, \mathrm{d}\boldsymbol{\xi} \tag{1.20}$$

where

$$Q(s, \tau) = \left(\frac{\partial \theta(s, \tau)}{\partial \upsilon} - \frac{\partial \theta_c(s, \tau)}{\partial \upsilon} \right)$$

and

$$\mu(s, \tau) = [\theta(s, \tau) - \theta_c(s, \tau)] .$$

We now choose auxiliary conditions for the fictitious problem which simplify expression (1.20) as much as possible. It is clearly most convenient to set

$$\theta_c(\boldsymbol{x}, 0) = 0 \qquad \boldsymbol{x} \in \Omega_c \tag{1.21}$$

thus confining attention to the region of interest. Further, if we enforce the temperature condition

$$\theta_c(s, \tau) = \theta(s, \tau) \qquad \tau > 0, s \in \partial\Omega$$

then (1.20) becomes

$$\theta(\boldsymbol{x}, t) = \alpha \int_0^t \int_{\partial\Omega} G(\boldsymbol{x}, t; s, \tau) Q(s, \tau) \, \mathrm{d}s \, \mathrm{d}\tau + \int_{\Omega} G(\boldsymbol{x}, t; \boldsymbol{\xi}, 0) \theta_0(\boldsymbol{\xi}) \, \mathrm{d}\boldsymbol{\xi}$$
$$+ \int_0^t \int_{\Omega} G(\boldsymbol{x}, t; \boldsymbol{\xi}, \tau) f(\boldsymbol{\xi}, \tau) \, \mathrm{d}\boldsymbol{\xi} \, \mathrm{d}\tau \tag{1.22}$$

where $Q(s, \tau)$ is an unknown boundary function. This can be determined from the integral equation that results when $\boldsymbol{x}$ is allowed to approach a general boundary point $\boldsymbol{x}_s$. Thus

$$g(\boldsymbol{x}_s, t) = \alpha \int_0^t \int_{\partial\Omega} G(\boldsymbol{x}_s, t; s, \tau) Q(s, \tau) \, \mathrm{d}s \, \mathrm{d}\tau + \int_{\Omega} G(\boldsymbol{x}_s, t; \boldsymbol{x}, 0) \theta_0(\boldsymbol{\xi}) \, \mathrm{d}\boldsymbol{\xi}$$
$$+ \int_0^t \int_{\Omega} G(\boldsymbol{x}_s, t; \boldsymbol{\xi}, \tau) f(\boldsymbol{\xi}, \tau) \, \mathrm{d}\boldsymbol{\xi} \, \mathrm{d}\tau . \tag{1.23}$$

We have therefore eliminated from (1.16) and (1.17) a term involving an integral over the boundary $\partial\Omega$ and time. However, unlike (1.17), whose solution directly yields the outward normal derivative, Eq. (1.23) determines a function not immediately related to the physical problem. Therefore if the quantity $\partial\theta/\partial\upsilon$ is required, it must be calculated indirectly by first finding Q, differentiating expression (1.22) appropriately, and then allowing the point x to approach the boundary. This explains the use of the terms "direct" and "indirect".

It is equally valid to use the flux continuity condition

$$\frac{\partial\theta(s,\tau)}{\partial\nu}=\frac{\partial\theta_c(s,\tau)}{\partial\nu}\qquad \tau>0,\ s\in\partial\Omega$$

in (1.20) yielding the equations

$$\theta(\boldsymbol{x},t)=-\alpha\int_0^t\int_{\partial\Omega}\frac{\partial G(\boldsymbol{x},t;s,\tau)}{\partial\nu}\mu(s,\tau)\,\mathrm{d}s\,\mathrm{d}\tau+\int_{\Omega}G(\boldsymbol{x},t;\boldsymbol{\xi},0)\theta_0(\boldsymbol{\xi})\,\mathrm{d}\boldsymbol{\xi}$$
$$+\int_0^t\int_{\Omega}G(\boldsymbol{x},t;\boldsymbol{\xi},\tau)f(\boldsymbol{\xi},\tau)\,\mathrm{d}\boldsymbol{\xi}\,\mathrm{d}\tau \tag{1.24}$$

and

$$g(\boldsymbol{x}_s,t)=-\alpha\int_0^t\int_{\partial\Omega}\frac{\partial G(\boldsymbol{x}_s,t;s,\tau)}{\partial\nu}\mu(s,\tau)\,\mathrm{d}s\,\mathrm{d}\tau$$
$$+\int_{\Omega}G(\boldsymbol{x}_s,t;\boldsymbol{\xi},0)\theta_0(\boldsymbol{\xi})\,\mathrm{d}\boldsymbol{\xi}+\int_0^t\int_{\Omega}G(\boldsymbol{x}_s,t;\boldsymbol{\xi},\tau)f(\boldsymbol{\xi},\tau)\,\mathrm{d}\boldsymbol{\xi}\,\mathrm{d}\tau \tag{1.25}$$

where $\mu(s,\tau)$ is unknown. The pairs (1.22), (1.23) and (1.24), (1.25) therefore represent alternative indirect methods which are known as single and double-layer potential methods respectively.

1.2. The physical approach

We have already stated that boundary integral methods admit a physical interpretation. This offers the advantage of providing an intuitive feel for the algorithm, free from any mathematical complexities that might obscure the workings of the technique. We now explore this avenue by motivating the formation of the trial solution on physical grounds. For reasons of brevity we shall confine our attention to the case of the single-layer method.

We begin by considering an infinitely long line source in a three-dimensional, homogeneous, isotropic continuum. If the source intersects a representative plane normal to it at a point $\overset{1}{\boldsymbol{p}}$ as shown in Fig. 3, and at time t_0 emits a pulse of strength $\overset{1}{q}_0$ J/m throughout its length, then the resultant temperature at an arbitrary point $\boldsymbol{x}$ on the plane at some later time t is given by

$$\theta(\boldsymbol{x},t)=G(\boldsymbol{x},t;\overset{1}{\boldsymbol{p}},t_0)\overset{1}{q}_0\alpha/K \tag{1.26}$$

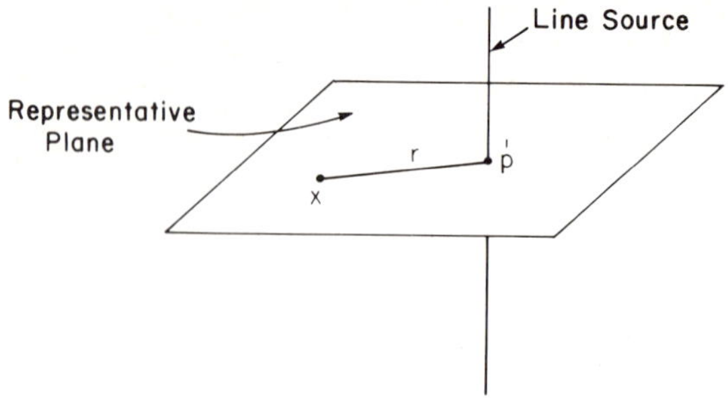

Fig. 3. An infinite line source.

where

$$G(\boldsymbol{x}, t; \overset{1}{\boldsymbol{p}}, t_0) = \frac{1}{4\pi\alpha(t - t_0)} \exp\left(\frac{-r^2}{4\alpha(t - t_0)}\right).$$

$r = |\boldsymbol{x} - \overset{1}{\boldsymbol{p}}|$, K (J/m-s-K) is the thermal conductivity coefficient and α (m^2/s) is the diffusivity coefficient. The function G was introduced previously as the fundamental solution to the diffusion equation. Except at the point $\overset{1}{\boldsymbol{p}}$ at time t_0, G is a continuous function of t and of $\boldsymbol{x}$, has continuous derivatives of all orders, and satisfies the heat equation. Now consider N such sources, with the ith one intersecting the plane at point $\overset{i}{\boldsymbol{p}}$ and acting at time τ_m, $m = 0, \ldots, M$, with intensities $\overset{i}{q}_m$ J/m. Because of the linearity of the governing equation, solutions of the form (1.26) can be superimposed yielding

$$\theta(\boldsymbol{x}, t) = \frac{\alpha}{K} \sum_{i=1}^{N} \sum_{m=0}^{M} G(\boldsymbol{x}, t; \overset{i}{\boldsymbol{p}}, \tau_m)\, \overset{i}{q}_m . \tag{1.27}$$

Equation (1.27) clearly has the form of a trial solution. Furthermore, for regions of the plane not containing the points $\overset{i}{\boldsymbol{p}}$ $i = 1, \ldots, N$, the fundamental solutions satisfy condition (6) (see Introduction) at all times (where $f = 0$) and so qualify as a set of basis functions for a boundary integral method. This suggests that (1.27) could form the foundation of a numerical procedure. We pursue this idea with reference to the diagrams that follow.

Figure 4(a) illustrates the physical body Ω, bounded by the contour $\partial\Omega$. In Fig. 4(b), this shape is embedded in the representative infinite plane which is assumed to have the same thermal properties. Thus $\partial\Omega'$ is a replica of $\partial\Omega$ surrounding the region Ω' which is then the analog of Ω. Suppose now that N line pulses intersect the plane normally at points $\overset{i}{\boldsymbol{p}}$ in the region exterior to $\partial\Omega'$. Then (1.27) obeys the differential equation at all points within $\partial\Omega'$ at all times. We further suppose that the auxiliary conditions for the problem have the form

$$\theta = g \qquad \text{on } \partial\Omega,\ t > 0$$

$$\theta = 0 \qquad \text{on } \Omega,\ t = 0$$

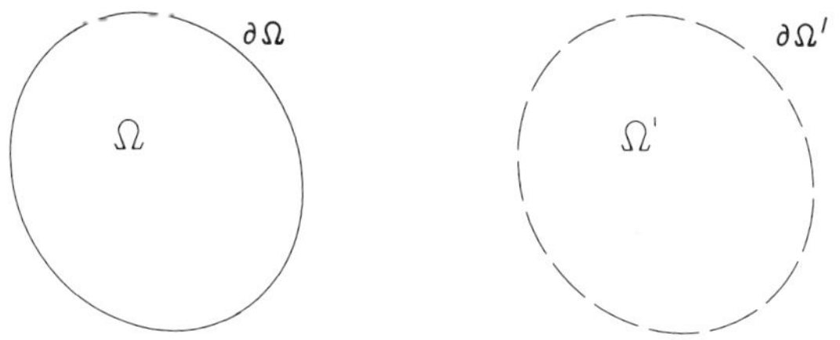

Fig. 4. Physical problem; infinite plane (plan view).

and that $f=0$. Now if we could assign values to the strengths $\overset{i}{q}_m$ in such a way that the expression (1.27) for θ takes on the value prescribed by the function g on $\partial\Omega$ for all $t>0$, this would provide a solution to the problem were it formulated on Ω' rather than on Ω. However, a temperature distribution satisfying the same initial and boundary conditions on thermally and geometrically equivalent bodies is unique. Therefore, the same temperature prevails at two corresponding points within $\partial\Omega$ and $\partial\Omega'$, regardless of the different conditions that exist beyond these boundaries. The problem for the physical case would thus also be solved.

We seek, therefore, a systematic way of determining the pulse strengths. Clearly, since these are finite in number for a given t, the boundary condition can only be satisfied at a finite set of representative points over $\partial\Omega$ and the interval $(0, t]$. In particular, if $\boldsymbol{x}$ is identified with N different boundary points $\overset{i}{s} \in \partial\Omega'$, and t with M time points, we obtain a system of linear algebraic equations

$$g(\overset{i}{s}, t_n) = \frac{\alpha}{K} \sum_{j=1}^{N} \sum_{m=0}^{N} G(\overset{i}{s}, \tau_n; \overset{j}{\boldsymbol{p}}, \tau_m)\, \overset{j}{q}_m \qquad i=1, \ldots, N;\ n=1, \ldots, M \tag{1.28}$$

which can be solved for the unknown strengths. The temperature solution is then recovered by substituting these values into (1.27). This numerical scheme is clearly primitive, and cannot be expected to produce accurate results. However, it does serve to illustrate the basic features of indirect boundary integral methods. These are:

(i) The combination of fundamental solutions to form an expression which solves general problems.

(ii) The use of fictitious devices, in this case pulsating sources, to generate elementary solutions. They are fictitious in the sense that they are not part of the physical problem. Nevertheless, the procedure is based on sound physical principles. In reality, boundary conditions involving temperature are regulated by some external means. In this case we simply assume that they are caused by a series of imaginary heat disturbances, and use our knowledge of their effects to find the strength distribution that yields this behavior.

(iii) The use of an infinite medium which facilitates the task of combining the constituent solutions. In effect we have created a simplified mathematical environment in which to solve the problem without altering its fundamental characteristics.

Having established the rationale for the construction of such solutions, we now attempt to improve their accuracy. In order to do so, we consider more general types of sources, particularly those from which heat is released continuously over the surface of a plane. This can be viewed as limiting case of assembling line sources close to each other in such a way as to form a curve in the representative plane. Also we can assume that heat is released continuously over successive time intervals rather than in quantum amounts. If the discharge rate from the jth such source γ_j is $\overset{j}{Q}_m(s, \tau)$ (J/m^2-s)] at time τ in the mth interval, (1.28) becomes

$$\theta(\boldsymbol{x}, t) = \frac{\alpha}{K} \sum_{j=1}^{N} \sum_{m=0}^{M} \int_{\gamma_j} \int_{\tau_m}^{\tau_{m+1}} G(\boldsymbol{x}, t; s, \tau) \overset{j}{Q}_m(s, \tau)\, \mathrm{d}s\, \mathrm{d}\tau\,. \tag{1.29}$$

Now if we write the unknown source intensity as

$$\overset{j}{Q}_m(s) = K \sum_r \overset{j}{b}_{mr} \overset{j}{P}_{mr}(s) \qquad j = 1, \ldots, N;\ m = 0, \ldots, M$$

where the $\overset{j}{P}_{mr}$ form a family of known approximating functions, and the $\overset{j}{b}_{mr}$ are coefficients, (1.29) may be written as

$$\theta(\boldsymbol{x}, t) = \alpha \sum_{j=1}^{N} \sum_{m=0}^{M} \sum_r \overset{j}{b}_{mr} \overset{j}{C}_{mr}(\boldsymbol{x}, t) \tag{1.30}$$

where

$$\overset{j}{C}_{mr} = \int_{\gamma_j} \int_{\tau_m}^{\tau_{m=1}} G(\boldsymbol{x}, t; s, \tau) \overset{j}{P}_{mr}(s, \tau)\, \mathrm{d}s\, \mathrm{d}\tau\,. \tag{1.31}$$

If $\boldsymbol{x}$ and t are made to take on a sufficient number of values over $\partial\Omega'$ and $(0, t]$ respectively, and the result is equated with the prescribed values given by the boundary conditions, a system of equations is formed which can be solved for the $\overset{j}{b}_{mr}$. When these have been found, Eq. (1.30) represents an explicit formula that satisfies the governing equation exactly, and the boundary conditions approximately.

Finally we need to attend to the specific placement of the curved sources with respect to the region Ω'. From a purely numerical point of view, it can be argued that they should be placed close to the body Ω', although clearly they cannot lie within its boundary if the governing equation is to be satisfied throughout the domain. Thus, let us investigate the situation when they are connected end to end to form a closed contour coincident with the boundary shape $\partial\Omega'$ as shown in Fig. 5. Suppose that, at least in theory, the piecewise approximation Q of the source

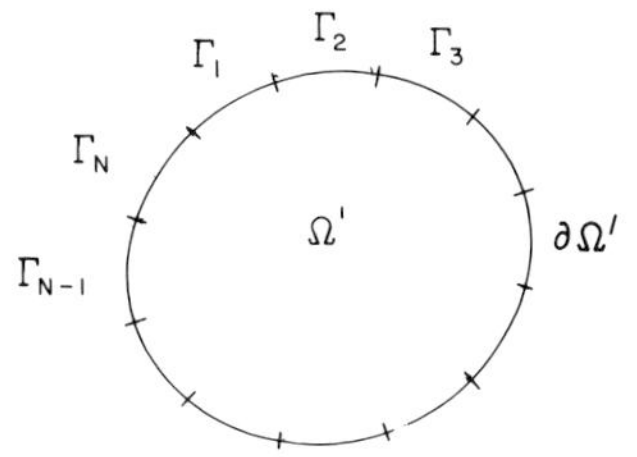

Fig. 5. Sources coincident with the boundary.

strength over each curve and time interval could be made to ensure continuity of all orders across the interfaces. Then Eq. (1.30) would no longer require individual definitions of the source strength over each time and space segment, and could be written simply as

$$\theta(x, t) = \alpha \int_{\partial\Omega'} \int_0^t G(x, t; s, \tau) Q(s, \tau) \, \mathrm{d}s \, \mathrm{d}\tau \tag{1.32}$$

where $Q(s, \tau)$ represents the source strength over $\partial\Omega'$ and $(0, t]$. This is precisely the integral expression (1.22) (for the case where $\theta_0 = 0$ and $f = 0$), which we have derived here from purely physical considerations. The positioning of the sources along the bounding surface $\partial\Omega$ is therefore consistent with the analytical derivation of the boundary integral equation. Further, it allows external body problems to be solved without alteration to the algorithm, with the general point of interest x merely being made to lie in the region exterior to $\partial\Omega$.

An interpretation analogous to the one given here can be made for the double-layer method, using dipoles rather than single sources to generate solutions. A physical view of the direct method can be obtained by combining single- and double-layer sources operating at each boundary segment. Also, by means of a combination of simple solutions, an approximation for an arbitrary initial temperature distribution can be obtained which in the limit can be shown to yield the expression containing $\theta_0(\xi)$ in the boundary integral expressions we have derived. This is discussed more fully by Sharp [74].

2. Numerical procedure

2.1. Introduction

It is now apparent that the boundary integral approximation for the temperature distribution lies in a spectrum defined by the extremes of an exact but intractable formula given for example by (1.22), and a simple but crude corresponding heat-pulse approximation (1.27). The technique may be approached from either of these starting points. On the one hand, it may be seen as a simplifying process which dismantles an integral structure into components that are easy to handle. This view then treats the method purely as a technique for solving an analytical equation numerically. From the opposite extreme, the method may be regarded as a constructive procedure that assembles a series of heat solutions in an infinite medium, and combines them to produce an approximate formula for the temperature. This is consistent with a trial solution approach in which the choice of basis functions ϕ_i in Eq. (5) is motivated by the knowledge of an exact form of the solution. In constructing the numerical procedure, it is therefore possible to adopt either of these views. In this section we shall opt for the former.

The boundary element technique for the diffusion equation can be developed in a manner strictly analogous to that used for elliptical problems. Implicit in this approach is the formal treatment of the temporal axis as a third dimension so that the integral nature of time is preserved. For the present we shall limit our attention to the single-layer indirect version of the boundary element method since the associated integral equation requires less computational effort for its solution. However, the numerical methodology is independent of this choice and can readily be carried over to the direct case.

As we have stated, the domain space for a non-steady heat flow problem is a space–time cylinder illustrated in Fig. 1. The integral expression for temperature corresponding to the indirect method was given in (1.22). The term

$$P(\boldsymbol{x}, t) = \int_0^t \int_\Omega G(\boldsymbol{x}, t; \boldsymbol{\xi}, \tau) f(\boldsymbol{\xi}, \tau)\, \mathrm{d}\boldsymbol{\xi}\, \mathrm{d}\tau$$

represents the particular solution to the non-homogeneous equation and is usually evaluated separately. If f satisfies Laplace's equation, a boundary integral formula [73] may be used. Otherwise the integration may be performed numerically, with special care given to the region containing the singularity. We shall not attend to the details of this computation.

If the point $\boldsymbol{x}$ is allowed to approach $\partial\Omega$, we obtain the integral equation

$$g(\boldsymbol{x}_s, t) - P(\boldsymbol{x}_s, t) - \int_\Omega G(\boldsymbol{x}_s, t; \boldsymbol{\xi}, 0)\theta_0(\boldsymbol{\xi})\, \mathrm{d}\boldsymbol{\xi} = \int_0^t \int_{\partial\Omega} G(\boldsymbol{x}_s, t, s, \tau) Q(s, \tau)\, \mathrm{d}s\, \mathrm{d}\tau \tag{2.1}$$

which must be solved for Q. In what follows, we prepare the way for a numerical solution of this equation.

2.2. Boundary discretization

We saw previously that, in order to obtain a discrete system, it is necessary to express the boundary functions as a finite sum of approximating functions. Thus we are required to represent the functions $Q(s, \tau)$ and $\theta_0(\xi)$ as

$$Q(s, \tau) = \sum_i \alpha_i U_i(s, \tau) \tag{2.2a}$$

$$\theta_0(\xi) = \sum_i \beta_i Y_i(\xi) \tag{2.2b}$$

where α_i and β_i are coefficients and where U_i and Y_i are approximating functions. To simplify the forms of these expressions, it is useful to prescribe the approximation in a piecewise manner over the boundary, in analogy with the finite element approach for representing functions over a domain. It follows that the first step toward the construction of a numerical algorithm involves the subdivision of the boundary into sections upon which these functions are to be defined. Thus, we must discretize the space–time surface ∂V. Since this consists of a lateral part $(0, t] \times \partial\Omega$ and the horizontal base Ω at $t = 0$, the discretization can be achieved by subdividing these constituent sets.

The contour $\partial\Omega$ may be represented by means of segments $\partial\Omega_j$, $j = 1, \ldots, N$, joined end to end as shown in Fig. 6(a), while the time axis is divided into intervals, which, for convenience, will be assumed to be of equal size Δt, as illustrated in Fig. 6(b), where we define $\tau_m = m\Delta t$, $m = 0, 1, 2, \ldots$ The resultant discretization of ∂V is shown in Fig. 7(a). The elementary planar subsurfaces are known as boundary elements. It is for this reason that the discretized form of boundary integral methods outlined in the preceding section are commonly known as boundary element methods.

A typical such element e_{mj}, shown in Fig. 7(b), comprises points on the jth segment $\partial\Omega_j$ and the mth time interval $(\tau_m, \tau_{m+1}]$. If these elements are small, it will be possible to define relatively simple functions U_i on each in order to obtain an adequate approximation of Q. The common choice for such functions is a polynomial of some limited order. This will be discussed presently.

The plane Ω, $t = 0$ of the domain will be discretized by means of a series of interlocking triangles, as shown in Fig. 8. These will be used to facilitate the

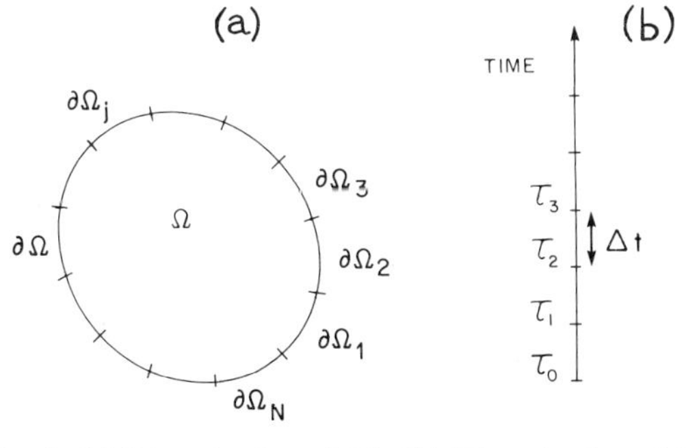

Fig. 6. (a) Discretization of $\partial\Omega$; (b) Discretization of time.

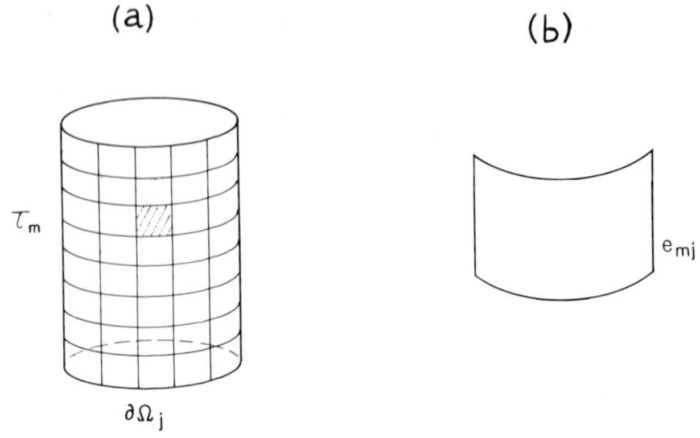

Fig. 7. (a) Discretized cylinder; (b) a boundary element.

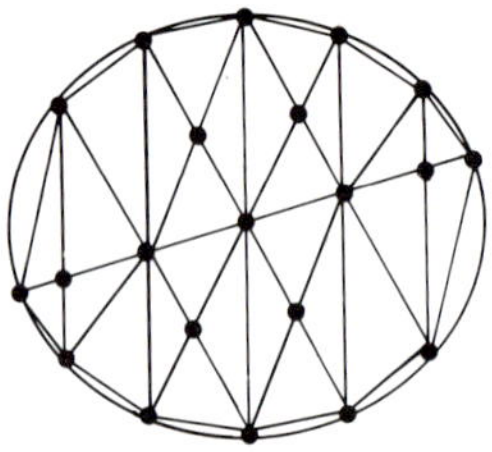

Fig. 8. Discretization of the body at $t = 0$.

evaluation of the integration term involving the known initial condition in Eq. (1.22).

Although such a configuration might be reminiscent of finite element procedures, it is to be reemphasized that Fig. 8 represents a boundary of the three-dimensional space–time domain.

In summary, the use of the boundary elements simplifies the task of handling boundary functions appearing under integral signs in (1.22). In particular, it allows

(i) a relatively simple piecewise approximation to be made for the unknown boundary function, and

(ii) a similar expression to be written for the known initial condition (and known boundary function for the case of the direct method).

2.3. *Shape functions*

We have decided that we need to approximate the function Q sectionally over each of the boundary elements. It is possible to write Q as

$$Q(s, \tau) = \sum_m \sum_j \sum_d \overset{j}{Q}{}_m^{(d)} L_{mj}^{(d)}(s, \tau) \tag{2.3}$$

where the coefficients $\overset{j}{Q}{}_m^{(d)}$ are the actual values taken on by (2.3) at different points or "nodes" d on e_{mj}, $j = 1, \ldots, N$, $m = 0, 1, 2$. The functions $L_{mj}^{(d)}(s, \tau)$ act to interpolate between these nodal values on each element and are known as shape functions. They must be defined so that

$$
\begin{aligned}
L_{mj}^{(d)}(s, \tau) &= 1 \qquad (s, \tau) \to \text{node } d \in e_{mj} \\
&= 0 \qquad (s, \tau) \to \text{node } d' \in e_{mj},\ d' \neq d \\
&= 0 \qquad (s, \tau) \not\in e_{mj}
\end{aligned}
$$

The representation (2.3) is useful since

(i) it gives meaning to the coefficients in the context of the problem;

(ii) when a boundary function is known – as in Eq. (1.16) for the direct method – the coefficients can be found immediately, and do not require the inversion of a linear system [89] as would otherwise be the case.

The approximation (2.3) is customarily represented in the form of a truncated Taylor series so that the $L_{mj}^{(d)}$ are polynomials. It is convenient to assume that these may be written in a product form

$$L_{mj}^{(d)} = B_j^{(r)}(s) T_m^{(k)}(\tau)$$

so that

$$Q(s, \tau) = \sum_r \sum_k \sum_m \sum_j B_j^{(r)}(s) T_m^{(k)}(\tau) \overset{j}{Q}_m(r, k) \tag{2.4}$$

where r labels nodes along the segment, k along the time interval, and the pairing (r, k) corresponds to d. This allows the approximations with respect to time and space to be treated independently.

As an example, consider the case where Q is assumed to behave linearly with respect to time and quadratically with respect to the boundary segment. This is shown in Fig. 9.

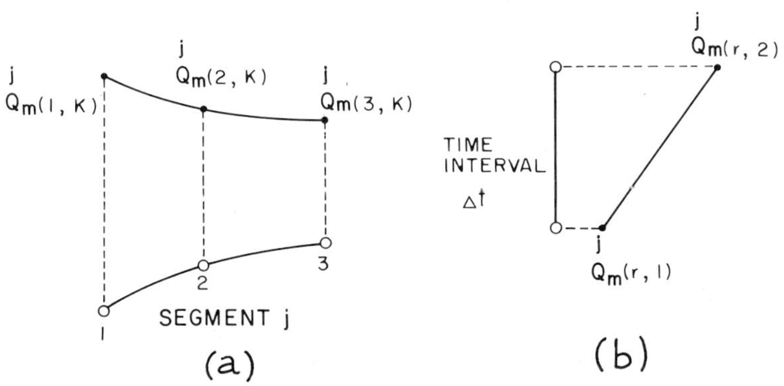

Fig. 9. (a) Quadratic spatial variation; (b) Linear temporal variation.

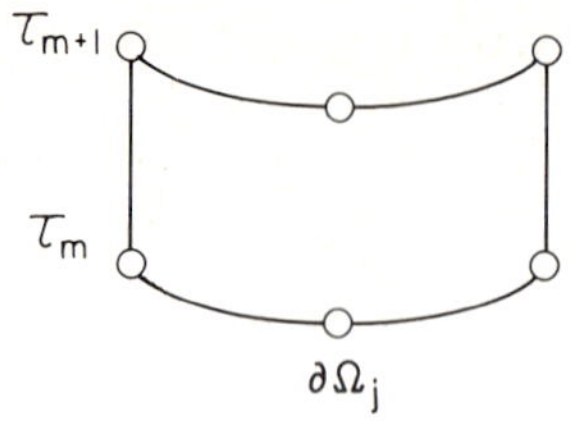

Fig. 10. Nodal arrangement for quadratic space and linear time variations.

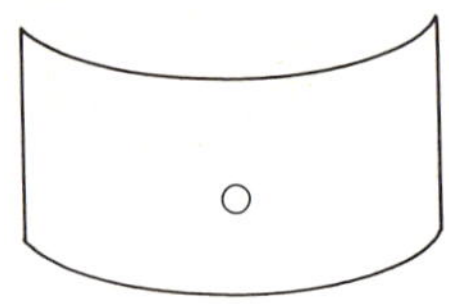

Fig. 11. Nodal arrangement for constant space and time behavior.

Clearly two nodal values are needed to determine the linear behavior while three are required to characterize the quadratic spatial variation. The resulting nodal arangement on the boundary element is shown in Fig. 10.

The location of the nodes on the element edges facilitates the application of inter-element continuity conditions for Q and certain orders of its derivatives should this be desired. Such conditions reduce the number of independent coefficients in the approximation.

Clearly a variety of combinations is possible, the simplest being constant behavior over space and time. The associated boundary element requires just one node as shown in Fig. 11, for which

$$\begin{aligned} B_j^{(1)}(s) &= 1 \qquad s \in \partial\Omega_j \\ &= 0 \qquad s \notin \partial\Omega_j \\ T_m^{(1)} &= 1 \qquad \tau \in (\tau_m, \tau_{m+1}] \\ &= 0 \qquad \tau \notin (\tau_m, \tau_{m+1}] \end{aligned} \tag{2.5}$$

so that

$$Q(s, \tau) = \sum_m \sum_j B_j(s) T_m(\tau) \overset{j}{Q}_m \,. \tag{2.6}$$

2.4. The discrete system

In the preceding discussion, the foundations were laid for the construction of a discrete analog of the integral equation (2.1). We now progress with this task by showing how the linear system of equations is assembled and solved. For simplicity we shall assume a piecewise constant behavior for the boundary function Q given by (2.5) and (2.6). Upon substituting (2.6) into (2.1), and evaluating the result at $t = \tau_n$, i.e. at the end of the nth time interval, we have

$$g(\boldsymbol{x}_s, \tau_n) - P(\boldsymbol{x}_s, \tau_n) - \int_\Omega G(\boldsymbol{x}_s, \tau_n; \boldsymbol{\xi}, 0)\theta_0(\boldsymbol{\xi})\,\mathrm{d}\boldsymbol{\xi}$$

$$= \sum_{m=0}^{n-1} \sum_{j=1}^{N} \left[\int_{\tau_m}^{\tau_{m+1}} \int_{\partial\Omega_j} G(\boldsymbol{x}_s, \tau_n; s, \tau) \right] \overset{j}{Q}_m \,.$$

If $\boldsymbol{x}_s$ is identified successively with a representative point $\overset{i}{s}$ of segment i for all n, $n = 1, 2, 3 \ldots$, we may write

$$\overset{i}{g}_n - \overset{i}{P}_n - \overset{i}{H}_n = \sum_{m=0}^{n-1} \sum_{j=1}^{N} \overset{ij}{C}_{n-m} \overset{j}{Q}_m \tag{2.7}$$

where

$$\overset{i}{g}_n = g(\overset{i}{s}, \tau_n)$$

$$\overset{i}{P}_n = P(\overset{i}{s}, \tau_n)$$

$$\overset{ij}{C}_{n-m} = \int_{\tau_m}^{\tau_{m+1}} \int_{\partial\Omega_j} G(\overset{i}{s}, \tau_n; s, \tau)\, \mathrm{d}s\, \mathrm{d}\tau \tag{2.8}$$

and where

$$\overset{i}{H}_n = \int_{\Omega} G(\overset{i}{s}, \tau_n; \xi, 0)\theta_0(\xi)\, \mathrm{d}\xi = \sum_c \overset{ci}{h}(\overset{i}{s}, \tau_n)\theta_0(\overset{c}{\xi})$$

represents a quadrature scheme for integrating over an area, with $\theta_0(\overset{c}{\xi})$ denoting the value of θ_0 at the representative point c. Equation (2.7) may be written in vector form as

$$\boldsymbol{g}_n - \boldsymbol{P}_n - \boldsymbol{H}_n = \sum_{m-0}^{m-1} \boldsymbol{C}_{n-m} \boldsymbol{Q}_m \tag{2.9}$$

where

$$\boldsymbol{g}_n = [\overset{1}{g}_n, \overset{2}{g}_n, \ldots \overset{N}{g}_n]$$

$$\boldsymbol{P}_n = [\overset{1}{P}_n, \overset{2}{P}_n, \ldots \overset{N}{P}_n]$$

$$\boldsymbol{H}_n = [\overset{1}{H}_n, \overset{2}{H}_n, \ldots \overset{N}{H}_n]$$

$$\boldsymbol{Q}_n = [\overset{1}{Q}_n, \overset{2}{Q}_n, \ldots \overset{N}{Q}_n]$$

are $N \times 1$ vectors, and $\boldsymbol{C}_{n-m}$ is an $N \times N$ matrix $\overset{ij}{C}_{n-m'}$, $i, j = 1, \ldots, N$. Now writing (2.9) for each n we have

$$\begin{bmatrix} \boldsymbol{g}_1 \\ \boldsymbol{g}_2 \\ \boldsymbol{g}_3 \\ \cdot \\ \cdot \\ \boldsymbol{g}_n \\ \cdot \\ \cdot \end{bmatrix} = \begin{bmatrix} \boldsymbol{C}_1 & \boldsymbol{0} & \boldsymbol{0} & \boldsymbol{0} \ldots\ldots & \boldsymbol{0} \\ \boldsymbol{C}_2 & \boldsymbol{C}_1 & \boldsymbol{0} & \boldsymbol{0} \ldots\ldots & \boldsymbol{0} \\ \boldsymbol{C}_3 & \boldsymbol{C}_2 & \boldsymbol{C}_1 & \boldsymbol{0} \ldots\ldots & \boldsymbol{0} \\ \cdot & \cdot & \cdot & \cdot & \cdot \\ \cdot & \cdot & \cdot & \cdot & \cdot \\ \boldsymbol{C}_n & \boldsymbol{C}_{n-1} & \boldsymbol{C}_{n-2} & \ldots \boldsymbol{C}_1 & \boldsymbol{0} \\ \cdot & \cdot & \cdot & \cdot & \cdot \\ \cdot & \cdot & \cdot & \ldots\cdot & \cdot \end{bmatrix} \begin{bmatrix} \boldsymbol{Q}_0 \\ \boldsymbol{Q}_1 \\ \boldsymbol{Q}_2 \\ \cdot \\ \cdot \\ \boldsymbol{Q}_{n-1} \\ \cdot \\ \cdot \end{bmatrix} + \begin{bmatrix} \boldsymbol{H}_1 \\ \boldsymbol{H}_2 \\ \boldsymbol{H}_3 \\ \cdot \\ \cdot \\ \boldsymbol{H}_n \\ \cdot \\ \cdot \end{bmatrix} + \begin{bmatrix} \boldsymbol{P}_1 \\ \boldsymbol{P}_2 \\ \boldsymbol{P}_3 \\ \cdot \\ \cdot \\ \boldsymbol{P}_n \\ \cdot \\ \cdot \end{bmatrix}$$

Since theoretically n is unbounded, we have a growing system of equations. Note, however, that the matrix has a lower triangular form, as a consequence of condition (1.13b). This enables the unknown subvectors Q_m to be found by forward substitution. Thus, for $n = 1$ we have

$$\boldsymbol{C}_1\boldsymbol{Q}_0 = \boldsymbol{g}_1 - \boldsymbol{H}_1 - \boldsymbol{P}_1 .$$

Each vector on the right-hand side is known so that we may solve the system of equations for the vector $\boldsymbol{Q}_0$. When this has been found, the system

$$\boldsymbol{C}_1\boldsymbol{Q}_1 = \boldsymbol{g}_2 - \boldsymbol{H}_2 - \boldsymbol{P}_2 - \boldsymbol{C}_2\boldsymbol{Q}_0$$

is assembled, and since again the entire right-hand side is known, it may be solved for $\boldsymbol{Q}_1$. This process can be continued for successive values of n so that in general we have

$$\boldsymbol{C}_1\boldsymbol{Q}_{n-1} = \boldsymbol{g}_n - \boldsymbol{H}_n - \boldsymbol{P}_n - \sum_{m=0}^{n-2} \boldsymbol{C}_{n-m}\boldsymbol{Q}_m . \tag{2.10}$$

Note that in this series of linear algebraic systems, the coefficient matrix is unchanged, a factor which may be exploited by various solution procedures to increase speed and efficiency.

When $\boldsymbol{Q}_n$ has been found it becomes possible to determine the temperature at any point $\boldsymbol{x} \in \Omega$ at $t = \tau_n$. This is done by introducing the approximation for Q into the boundary integral expression (1.22) for θ. Thus

$$\theta(\boldsymbol{x}, n\Delta t) = \sum_{m=0}^{n-1} \sum_{j=1}^{N} \overset{j}{C}_{n-m}(\boldsymbol{x})\overset{j}{Q}_m + \sum_{c} \overset{c}{h}(\boldsymbol{x}, n\Delta t)\theta_0(\overset{c}{\xi}) + P(\boldsymbol{x}, n\Delta t)$$

where

$$\overset{j}{C}_{n-m} = \int_{\tau_m}^{\tau_{m+1}} \int_{\partial\Omega_j} G(\boldsymbol{x}, n\Delta; s, \tau)\, \mathrm{d}s\, \mathrm{d}\tau .$$

For purposes of exposition we choose the simplest approximation for Q, allowing the equations to be written down fairly simply in vector notation. Although no substantial differences enter into the solution algorithm if the general expression (2.3) is used, the terms $\overset{j}{C}_{n-m}$ above (and similarly (2.8)) must be altered. The evaluation of these terms, known as influence coefficients, will be dealt with presently.

2.5. General boundary conditions

So far we have limited our attention to Dirichlet problems, i.e. those in which the boundary condition consists of a temperature specification only. We now examine the way in which the procedure handles problems with all or part of the boundary data containing terms in the outward normal derivative $\partial\theta/\partial\upsilon$. We begin by noting that the derivative of θ may be obtained directly from Eq. (1.22) and is given by

$$\frac{\partial\theta}{\partial\upsilon}(\boldsymbol{x}, t) = \alpha \int_0^t \int_{\partial\Omega} \frac{\partial G(\boldsymbol{x}, t; s, \tau)}{\partial\upsilon} Q(s, \tau)\, \mathrm{d}s\, \mathrm{d}\tau$$

$$+ \int_{\Omega} \frac{\partial G(\boldsymbol{x}, t; \xi, 0)}{\partial\upsilon} \theta_0(\boldsymbol{\xi})\, \mathrm{d}\boldsymbol{\xi} + \int_0^t \int_{\Omega} \frac{\partial G(\boldsymbol{x}, t; \boldsymbol{\xi}, \tau)}{\partial\upsilon} f(\boldsymbol{\xi}, \tau)\, \mathrm{d}\boldsymbol{\xi}\, \mathrm{d}\tau \tag{2.11}$$

where

$$\frac{\partial G}{\partial\upsilon} = G_{x,i}\upsilon_i = \frac{1}{8\alpha^2\pi(t-\tau)^2} \exp\left(\frac{-|\boldsymbol{x}-s|^2}{4\alpha(t-\tau)}\right) |\boldsymbol{x}-s| \cos(\boldsymbol{x}-s, \upsilon)$$

and where $G_{x,i}$ denotes the gradient vector with respect to the coordinates $\boldsymbol{x}$, and υ is the unit outward normal vector at $\boldsymbol{x}_s \in \partial\Omega$. The first integral on the right-hand side of (2.11) is to be understood in the sense of its principal value. It can be shown that this expression for $\partial\theta/\partial\upsilon$ has a discontinuity across the surface $\partial\Omega$ of magnitude equal to the local value of the function Q. If the boundary condition is of the form

$$\partial\theta/\partial\upsilon = g(\boldsymbol{x}_s, \tau) \qquad \boldsymbol{x}_s \in \partial\Omega,\ t>0\,, \tag{2.12}$$

then equating the expressions (2.11) and (2.12) for the normal derivative, and discretizing as before, we obtain the approximate form

$$\boldsymbol{g}_n = \sum_{m=0}^{n-1} \mathbf{D}_{n-m}\boldsymbol{Q}_m + \boldsymbol{J}_n + \boldsymbol{K}_n$$

where $\mathbf{D}_{n-m}$ represents the influence coefficients, and $\boldsymbol{J}_n$ and $\boldsymbol{K}_n$ respectively contain the normal outward derivatives of the components of $\boldsymbol{H}_n$ and $\boldsymbol{P}_n$. This system is solved as described previously. Problems in which both temperatures and fluxes enter the boundary conditions can be handled easily by combining matrices $\mathbf{C}_m$ and $\mathbf{D}_m$ appropriately.

2.6. Double-layer potentials

The boundary integral equation (1.24) for the double-layer potential involves the kernel [51]

$$\frac{\partial G}{\partial \upsilon} = G_{s,i}\upsilon_i = \frac{-1}{8\alpha^2\pi(t-\tau)^2} \exp\left(\frac{-|\boldsymbol{x}-\boldsymbol{s}|^2}{4\alpha(t-\tau)}\right) |\boldsymbol{x}-\boldsymbol{s}| \cos(\boldsymbol{x}-\boldsymbol{s}, \upsilon) \tag{2.13}$$

with $G_{s,i}$ denoting the gradient vector with respect to the coordinates of $\boldsymbol{s}$, and where μ is the unknown boundary function which may be interpreted as the strength of the double-layer source. This function differs only in sign from that related to the normal derivative discussed above. Thus it too exhibits a discontinuity across $\partial\Omega$, the size of which is equal to the local value of μ. The fluxes generated across $\partial\Omega$ are continuous. Discretization can be performed as before with only the sign differing.

It has been observed [74] that the single-layer boundary element method provides a more accurate solution of Neumann type problems than it does for the Dirichlet type, particularly for problems which display a discontinuity between initial and boundary conditions. It emerges, however, that if a double-layer source is employed, then the reverse is true. This is due to the behavior of the kernel function when $(t-\tau)$ is small. The spatial derivative of G (whether used to solve a Neumann problem with the single-layer method, or a Dirichlet problem with the double-layer method) has a zero at the time origin which is one order weaker than that for G. This allows the singularity of the boundary function for such problems to be one order weaker at $t=0$, thus improving the quality of the approximation. We therefore have two versions of the method, each capable of providing an approximation for the required temperature distribution. A strategy for selecting the appropriate source type would be to use a double layer for Dirichlet problems and a single layer for all others.

2.7. Influence coefficients

2.7.1. Boundary segments

The most general case with which we need deal involves terms of the form

$$\overset{j}{E}{}^{(d)}_{n-m} = \int_{\tau_m}^{\tau_{m+1}} \int_{\partial\Omega_j} A(\boldsymbol{x}, \tau_n; \boldsymbol{s}, \tau) L^{(d)}_{mj}(\boldsymbol{s}, \tau)\, \mathrm{d}s\, \mathrm{d}\tau$$

where $\overset{j}{E}{}^{(d)}_{n-m}$ are the coefficients $\overset{j}{C}{}^{(d)}_{n-m}$ or $\overset{j}{D}{}^{(d)}_{n-m}$,

$$L^{(d)}_{mj} = B^{(r)}_j(\boldsymbol{s}) T^{(k)}_m(\tau)$$

are shape functions, $\partial\Omega_j$ is a segment of the arbitrarily shaped contour $\partial\Omega$, $\tau_n = n\Delta t$ are where $A(x, t; s, \tau)$ is the generic form for the fundamental solutions we have required so far, i.e.

$$G = \frac{1}{4\pi\alpha(t-\tau)} \exp\left(\frac{-|x-s|^2}{4\alpha(t-\tau)}\right)$$

$$G_{x,i}v_i = \frac{1}{8\alpha^2\pi(t-\tau)^2} \exp\left(\frac{-|x-s|^2}{4\alpha(t-\tau)}\right) |x-s| \cos(x-s, v)$$

$$G_{s,i}v_i = \frac{-1}{8\alpha^2\pi(t-\tau)^2} \exp\left(\frac{-|x-s|^2}{4\alpha(t-\tau)}\right) |x-s| \cos(x-s, v) .$$

Since the last two expressions differ in sign only, we need deal only with the first two.

Although higher-order dependences have been used [15, 19], it is commonly assumed that the function L is piecewise constant with respect to time. The influence coefficients then become

$$\overset{j}{E}{}^{(r)}_{n-m}(x, t) = \int_{\Omega_j} B_j^{(r)}(s) \left[\int_{\tau_m}^{\tau_{m+1}} A(x, t; s, \tau)\, d\tau \right] ds$$

where the time integrals of the kernels above become

$$\frac{1}{4\pi\alpha} [E_i(a_{m+1}) - E_i(a_m)]$$

and

$$\frac{-1}{2\pi\alpha} \cos(x-s, v)[\exp(a_{m+1}) - \exp(a_m)]$$

respectively, where $a_m = -|x-s|/4\alpha(t-\tau)$, and

$$E_i(-y) = \gamma + \ln(y^2) + \sum_{k=1}^{\infty} \frac{(-y)^k}{(kk!)} \tag{2.14}$$

is the exponential integral, with $\gamma = 0.57721$ (Euler's constant).

The function $B_j^{(r)}(s)$ is often assumed to be constant or to vary linearly or quadratically along element j. Clearly for an arbitrary shaped segment $\partial\Omega_j$ the evaluation of the spatial integral must be performed numerically. For this purpose it is often useful if the geometric characteristics of $\partial\Omega_j$ are described by the same set of shape functions $B_j^{(r)}$ used to chart the spatial variation of the unknown boundary function Q. Thus, if the location of the segment $\partial\Omega_j$ is given parametrically by the coordinates $x_j(s)$, $y_j(s)$, then

$$x_j(s) = \sum_r B_j^{(r)}(s) x_j^{(r)}$$

$$y_j(s) = \sum_r B_j^{(r)}(s) y_j^{(r)}$$

where $x_j^{(r)}$ and $y_j^{(r)}$ are the coordinates of nodal points on the segment $\partial\Omega_j$. This is known as an isoparametric representation and is used frequently in finite element analyses. The nodes are often located at the extremes of the elements. Particular care must be taken when dealing with the elements containing the node at which the function E_{n-m}^{j} is being calculated (i.e. the diagonal term) since it becomes singular there. The case for which $A(x, t; s, \tau) = G$ involves a logarithmic singularity which does not contribute to the integral. If a quadrature scheme is used, it should automatically account for this singularity [16]. In the case of the other kernel functions, the diagonal term can only be integrated in the sense of the Cauchy principal value. It can be shown [19] that here the contribution of the singularity is $c = (\pi + e)/2\pi$ where e is the difference in the orientations of the outward normal vectors on either side of the point on the boundary. Thus if the boundary is smooth, $c = 0.5$. In general c is the ratio of the internal angle subtended by the tangents on either side of the boundary point to 2π. Thus coefficients at discontinuous boundary tangents (i.e. corner points) can be calculated quite easily. However, the discontinuity of the normal fluxes at such points presents a problem. This can be dealt with by preventing nodal values from coinciding with corner points, by employing two coincident or closely spaced nodes at corners, by employing special elements which allow the boundary functions to exhibit the appropriate singular behavior, or by interpolating the values at neighboring nodes. This problem arises in all boundary element implementations and has been discussed widely in the literature [49, 17, 1, 88, 48, 57, 50].

The task of evaluating the influence coefficients may be simplified considerably if we allow the boundary segments $\partial\Omega_j$ to be represented by straight lines. The resulting configuration, shown in Fig. 12, yields an acceptable approximation to the boundary provided that a sufficient number of segments is employed. The boundary elements forming the space–time cylinder surface then become rectangles so that the element illustrated in Fig. 10 now appears as shown in Fig. 13. If

Fig. 12. Straight line boundary segments.

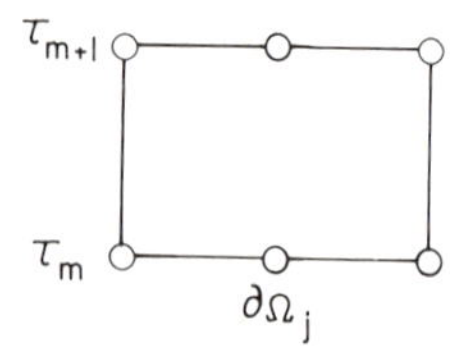

Fig. 13. Straight edged boundary elements.

the function L_{mj} is approximated sectionally by means of a polynomial, the integrals can be calculated analytically. The exponential integral (2.14) can be integrated term by term provided that the argument of E_i is sufficiently small for convergence of the series to be fast [19]. Fortunately, this covers the case of the self-influencing (diagonal) term, or at least the neighborhood of the singularity. Also, in this case the principal value of the normal derivative term vanishes by virtue of the orthogonality of the directions of the outward normal and the path of integration.

2.7.2. Body segments

We have seen that body integrals are required to account for the initial and forcing terms in the integral expressions for the solution. We shall see shortly that such integrals take on an added importance in a revised boundary element algorithm for solving the diffusion equation. These integrals are most commonly found by discretizing the domain by means of interlocking triangles and using appropriate quadrature weights and points [42, 27].

2.8. An appraisal of the method

2.8.1. Accuracy

We have already alluded to the error caused by a discontinuity at the boundary at time $t = 0$. This problem is not confined to boundary element techniques and there does not appear to be any way of avoiding it entirely. The error can be reduced significantly by reducing the time step. The solution approximation does recover from the initial aberration, and with time appears to approach the exact result in verifiable cases [74].

The accuracy of the numerical solution at points near the boundary may also be sensitive to the effects of the boundary function approximation. This is particularly noticeable for the case of the piecewise constant representation, and these results should be treated with caution.

2.8.2. Stability

The method as formulated in this section appears to be unconditionally stable.

2.8.3. Efficiency

Certain drawbacks are apparent in this regard, involving demands made on both computer storage and time. The major disadvantage concerns the requirement that a new set of influence coefficients be calculated for each time step. This is necessitated by the dependence of the solution at all times upon the entire behavior history of the boundary function, dating back to $t = 0$. Since these computations constitute the most time-consuming aspect of the procedure, runs become expensive if the process is allowed to evolve for even moderate lengths of time. This is especially the case when the number of field point calculations is high, as when contour plots are required. Further, an increasing number of terms must be accounted for as t increases. As is apparent from Eq. (2.10) these appear as matrix–vector products whose pairings are unique to each time step, and so cannot be stored in product form for later use. Thus, for the piecewise constant representation (2.6) at step n, nN^2 products are required to form the system of equations, and nMN to calculate field temperatures, where M and N are the number of field points and boundary nodes respectively. The cumulative total after n steps is $(n(n+1)N(M+N))/2$, which clearly grows quadratically in n and N.

The need to form an increasing number of matrix–vector products also implies that a growing number of matrices and vectors must be available. The storage capacity of even large computers quickly becomes inadequate if n and N are large. Furthermore, because of the convoluted arrangement of the matrix–vector pairings, one set must be accessed in the order opposite to that in which it was written, requiring the use of random access storage devices which may involve some computing overhead.

The drawbacks mentioned above are not related to any of the numerical aspects of the method we have described, but rather may be attributed to the integral treatment of time implicit in the equations upon which the procedure is based. In the following section we shall address these difficulties by attempting to rewrite the formulae in a way that allows the algorithm to operate iteratively. This will be in keeping with common practice in the modeling of transient phenomena.

2.9. *Revised numerical procedure*

2.9.1. *Introduction*

We have seen how the rate of the number of calculations required by the solution procedure increases as the heat flow process evolves. This problem is due to the treatment of time as an integral dimension like one of space. This may best be understood in geometrical terms with reference to the space–time cylinder V shown in Fig. 1. Clearly, as time advances, the area of the surface ∂V of this domain increases, requiring a growing number of elements to cover it, and a correspondingly greater effort to approximate the functions defined on it.

We now examine an alternative scheme [22, 18, 87, 10, 88, 40] that avoids this problem. It does so essentially by allowing one computational procedure to be repeated successively as the method steps through time. Each cycle corresponds to a time interval, and generates all of the information required for the next to be run, so that the earlier history of the process need not be retained. This stepwise routine is constructed as follows. We solve the problem over the first time interval as before, using the initial and boundary conditions stipulated as part of the problem data. The temperature is then generated at a series of points which form a grid over the physical domain as shown in Fig. 8. This distribution is then made to serve as an initial condition for a new problem, formulated over the second interval only, using as boundary conditions the values on $\partial\Omega$ specified in the original problem for this step. We therefore restart the process, and repeat the solution procedure with a new set of auxiliary conditions. When the second subformulation has been solved, the temperature distribution is again computed over the grid, providing both a solution after two steps for the original problem, and an initial condition for the third iteration. The procedure is continued in this fashion.

The relationship between this approach and the method discussed in previous sections is also best seen in geometrical terms. Consider a problem cylinder V divided into one-step disks as shown in Fig. 14. Instead of obtaining a solution to a single problem defined over this entire space–time domain as before, we now successively solve a subproblem in each such disk, and apply a continuity condition across the interfaces. The history effects are therefore not accounted for directly in the solution to the original formulation, but are communicated through the "initial" conditions.

Since each subproblem is defined over time intervals of equal length, they differ only in their auxiliary conditions. Thus the same boundary element method routine, with influence coefficients extending over a single step, is capable of solving each one. It follows that the coefficients need be computed only once, and can be stored and used repeatedly for the successive time intervals. In spite of the large number of these required as a result of the obligatory field point temperature calculations, this algorithm proves more economical than that implemented for the method described earlier in this section (integral time method).

We should remark that the procedure we have described here is, in a sense, a corruption of the formal boundary element method, since it betrays the "boundary only" philosphy associated with such methods. In particular, it demands that the solution be calculated at a set of points distributed thoughout the domain. This sacrifice is necessary if we are to avoid calculating a set of influence coefficients at each step. However, it should be recalled that the body temperatures are computed explicitly, and the system unknowns remain restricted to the values of functions defined on the boundary. Thus the dimension of the linear

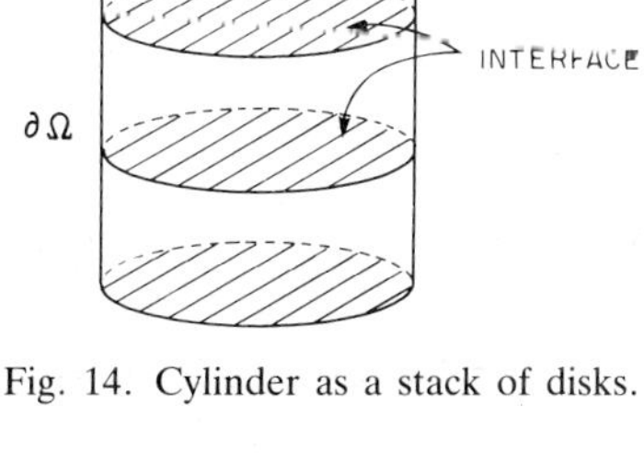

Fig. 14. Cylinder as a stack of disks.

system of equations for the method remains an order of magnitude smaller than that for a volume method such as finite elements, and the consequent advantages of this are preserved.

From the preceding discussion it appears that by inspection of integral formulae (1.16) and (1.22) respectively we could write down an exact recurrent boundary integral formula for the temperature at step t as

$$\theta(\boldsymbol{x}, t) = \alpha \int_0^{\Delta t} \int_{\partial\Omega} \left(G(\boldsymbol{x}, \Delta t; s, \tau) \frac{\partial \theta(s, \tau + t - \Delta t)}{\partial v} - \frac{\partial G(\boldsymbol{x}, \Delta t; s, \tau)}{\partial v} \theta(s, \tau + t - \Delta t) \right) \mathrm{d}s \, \mathrm{d}\tau + \int_\Omega G(\boldsymbol{x}, \Delta t; \boldsymbol{\xi}, 0) \theta(\boldsymbol{\xi}, t - \Delta t) \, \mathrm{d}\boldsymbol{\xi} + \int_0^{\Delta t} \int_\Omega G(\boldsymbol{x}, \Delta t; \boldsymbol{\xi}, \tau) f(\boldsymbol{\xi}, \tau + t - \Delta t) \, \mathrm{d}\boldsymbol{\xi} \, \mathrm{d}\tau \tag{2.15}$$

for the case of the direct method and

$$\theta(\boldsymbol{x}, t) = \alpha \int_0^{\Delta t} \int_{\partial\Omega} [G(\boldsymbol{x}, \Delta t; s, \tau) Q(s, \tau + t - \Delta t)] \, \mathrm{d}s \, \mathrm{d}\tau + \int_\Omega G(\boldsymbol{x}, \Delta t; \boldsymbol{\xi}, 0) \theta(\boldsymbol{\xi}, t - \Delta t) \, \mathrm{d}\boldsymbol{\xi} + \int_0^{\Delta t} \int_\Omega G(\boldsymbol{x}, \Delta t; \boldsymbol{\xi}, \tau) f(\boldsymbol{\xi}, \tau + t - \Delta t) \, \mathrm{d}\boldsymbol{\xi} \, \mathrm{d}\tau \tag{2.16}$$

for the indirect method. However, a more careful examination reveals that while Eq. (2.15) is correct, (2.16) requires an additional term. In order to confirm these assertions, it will be useful to define an integral operator $J\{a, b, \boldsymbol{y}, T, u, r, W\}$ to be

$$\alpha \int_a^b \int_{\partial\Omega} \left[G(\boldsymbol{y}, T; s, r + u) \frac{\partial \theta(s, r + u)}{\partial v} - \frac{\partial G(\boldsymbol{y}, T; s, r + u)}{\partial v} \theta(s, r + u) \right] \mathrm{d}s \, \mathrm{d}u + \int_\Omega G(\boldsymbol{y}, T; \boldsymbol{\xi}, 0) W(\boldsymbol{\xi}) \, \mathrm{d}\boldsymbol{\xi} + \int_a^b \int_\Omega G(\boldsymbol{y}, T; \boldsymbol{\xi}, r + u) f(\boldsymbol{\xi}, r + u) \, \mathrm{d}\boldsymbol{\xi} \, \mathrm{d}u \, .$$

Then clearly from (1.16) and the discussion in section 1.1.4

$$\theta(\boldsymbol{x}, t) = J\{0, t, \boldsymbol{x}, t, \tau, 0, \theta_0\} \qquad \boldsymbol{x} \in \Omega$$
$$= 0 \qquad \boldsymbol{x} \notin \Omega \,. \tag{2.17}$$

The definition of J implies that the expression for the temperature θ can be decomposed as

$$\theta(\boldsymbol{x}, t) = J\{0, t - \Delta t, \boldsymbol{x}, t, \tau, 0, \theta_0\} + J\{t - \Delta t, t, \boldsymbol{x}, t, \tau, 0, 0\} \qquad \boldsymbol{x} \in \Omega \,. \tag{2.18}$$

Now it is easy to show that the identity

$$G(\boldsymbol{x}, t_2; \boldsymbol{\xi}, t_0) = \int_{\Omega \cup \Omega_c} G(\boldsymbol{x}, t_2; \boldsymbol{w}, t_1) G(\boldsymbol{w}, t_1; \boldsymbol{\xi}, t_0) \, d\boldsymbol{w} \tag{2.19}$$

holds for any $t_0 \leqslant t_1 \leqslant t_2$, where Ω_c is the region complementary to Ω. If we use this identity in the first term on the right-hand side of (2.18), setting $t_2 = t$, $t_1 = t - \Delta t$ and $t_0 = 0$, and changing the order of integration, that term becomes

$$\int_{\Omega \cup \Omega_c} G(\boldsymbol{x}, t; \boldsymbol{w}, t - \Delta t) J\{0, t - \Delta t, \boldsymbol{w}, t - \Delta t, \tau, 0, \theta_0\} \, d\boldsymbol{w}$$
$$= \int_{\Omega} G(\boldsymbol{x}, t; \boldsymbol{w}, t - \Delta t) \theta(\boldsymbol{w}, t - \Delta t) \, d\boldsymbol{w} \qquad \text{by (2.17)}$$
$$= \int_{\Omega} G(\boldsymbol{x}, \Delta t; \boldsymbol{w}, 0) \theta(\boldsymbol{w}, t - \Delta t) \, d\boldsymbol{w} \qquad \text{by a property of } G \,.$$

By means of a simple transformation the second term on the right-hand of (2.18) may be written as

$$J\{0, \Delta t, \boldsymbol{x}, t, \tau, t - \Delta t, 0\} \,.$$

Thus we have

$$\theta(\boldsymbol{x}, t) = J\{0, \Delta t, \boldsymbol{x}, t, \tau, t - \Delta t, 0\} + \int_{\Omega} G(\boldsymbol{x}, \Delta t; \boldsymbol{w}, 0) \theta(\boldsymbol{w}, t - \Delta t) \, d\boldsymbol{w}$$
$$= J\{0, \Delta t, \boldsymbol{x}, t, \tau, t - \Delta t, \theta_{t - \Delta t}\} \qquad \text{by definition} \,.$$

This confirms the recurrent relation (2.15) which allows us to calculate the temperature using only information pertaining to the previous time step.

Recall from section 2.1 that indirect methods were derived by considering an imaginary diffusion flow problem in the complementary region. A direct method solution expression for this was added to one for the real problem, and the result, together with a convenient choice of boundary conditions, yielded the indirect formula. If we repeat the process, this time using the recurrent relation (2.15) above, we find that the composite expression contains an additional term

$$\int_{\Omega_c} G(\boldsymbol{x}, \Delta t; \boldsymbol{\xi}, 0)\theta(\boldsymbol{\xi}, \tau + t - \Delta t)\, \mathrm{d}\boldsymbol{\xi}\ .$$

This "initial" condition in Ω_c represents the potential for the fictitious diffusion process at time $t - \Delta t$ and cannot be arbitrarily set to zero in analogy with Eq. (1.21), since this would disrupt the continuity of the process and invalidate the solution. Thus the integration over the complementary domain is essential. This is an unexpected finding [71] which constitutes a serious computational drawback to the stepwise indirect method, particularly for finite-body problems where the additional calculations must then be made over an infinite region. For this reason we shall confine our attention to the task of constructing a numerical solution from the direct formula (2.15).

2.9.2. The numerical procedure

The required discretization process can be carried out as described previously, except that now only one-step influence coefficients are involved, and an additional boundary integral enters the equation. Thus if we write

$$\theta(s, \tau + (n-1)\Delta t) = \sum_{j=1}^{N} \sum_{d} \overset{j}{\theta}{}_n^{(d)} L_j^{(d)}(s, \tau)$$

and

$$\frac{\partial\theta(s, \tau + (n-1)\Delta t)}{\partial v} = \sum_{j=1}^{N} \sum_{d} \overset{j}{q}{}_n^{(d)} L_j^{(d)}(s, \tau)$$

in analogy with Eq. (2.3), where $L_j^{(d)}$ is the shape function for node d on boundary segment j, then Eq. (2.15) becomes

$$\theta(\boldsymbol{x}, n\Delta t) = \sum_{j=1}^{N} \sum_{d} [\overset{j}{C}{}_1^{(d)}(\boldsymbol{x})\overset{j}{q}{}^{(d)} - \overset{j}{D}{}_1^{(d)}(\boldsymbol{x})\overset{j}{\theta}{}_n^{(d)}] + \sum \overset{k}{w} G(\boldsymbol{x}, \Delta t; \overset{k}{\boldsymbol{\xi}}, 0)\theta(\overset{k}{\boldsymbol{\xi}}, (n-1)\Delta t) + P_n(\boldsymbol{x}) \tag{2.20}$$

where

$$\overset{j}{C}{}_1^{(d)}(\boldsymbol{x}) = \int_0^{\Delta t} \int_{\partial\Omega_j} G(\boldsymbol{x}, \Delta t; s, \tau) L_j^{(d)}(s, \tau)\, \mathrm{d}s\, \mathrm{d}\tau$$

$$\overset{j}{D}{}_1^{(d)}(\boldsymbol{x}) = \int_0^{\Delta t} \int_{\partial\Omega_j} \frac{\partial G(\boldsymbol{x}, \Delta t; s, \tau)}{\partial v} L_j^{(d)}(s, \tau)\, \mathrm{d}s\, \mathrm{d}\tau$$

$$P_n(\boldsymbol{x}) = \int_0^{\Delta t} \int_{\Omega} G(\boldsymbol{x}, \Delta t; \boldsymbol{\xi}, \tau) f(\boldsymbol{\xi}, \tau + (n-1)\Delta t)\, \mathrm{d}\boldsymbol{\xi}\, \mathrm{d}\tau$$

$\overset{k}{w}$ are Gaussian weights

$\overset{k}{\boldsymbol{\xi}}$ are Gaussian coordinate points

$\theta(\overset{k}{\boldsymbol{\xi}}, (n-1)\Delta t)$ are temperatures at $\overset{k}{\boldsymbol{\xi}}$ at the previous step

$\overset{j}{q}{}_n^{(d)} = \partial\theta(s, n\Delta t)/\partial v$ on element j, node d

and

$\overset{j}{\theta}{}_n^{(d)} = \theta(s, n\Delta t)$ on element j, node d.

The unknown boundary quantities can be found by allowing $\boldsymbol{x}$ to approach the boundary at a sufficient number of points, or by imposing continuity conditions at element interfaces, and solving the resulting system of equations. When this has been done, expression (2.20) provides the temperature at $\boldsymbol{x}$ at time Δt. This is the basic time marching process for the diffusion equation. We shall again opt for the simplest implementation, characterized by the piecewise constant representation of the boundary functions corresponding to Fig. 11 for which

$$L^{j(d)}(s, \tau) = 1 \qquad s \in \Omega_j$$
$$= 0 \qquad \text{otherwise}$$

with $d = 1$.

In order to generate a sufficient number of equations for the solution of the unknown values we shall need to make evaluations on one point $\overset{j}{s} \in \partial\Omega_j$ on element i, $i = 1, \ldots, N$. The resulting system can be written in vector form as

$$\overset{\mathrm{b}}{C}\overset{\mathrm{b}}{q}_n - [I + \overset{\mathrm{b}}{D}]\overset{\mathrm{b}}{\theta}_n + \overset{\mathrm{bf}}{F}\overset{\mathrm{b}}{\theta}_{n-1} + \overset{\mathrm{b}}{P}_n = 0 \tag{2.21}$$

where

$$\overset{\mathrm{b}}{\theta}_n = [\theta(\overset{1}{s}, n\Delta t), \theta(\overset{2}{s}, n\Delta t), \ldots \theta(\overset{N}{s}, n\Delta t)]$$

$$\overset{\mathrm{b}}{q}_n = \left(\frac{\partial\theta}{\partial v}(\overset{1}{s}, n\Delta t), \frac{\partial\theta}{\partial v}(\overset{2}{s}, n\Delta t), \ldots \frac{\partial\theta}{\partial v}(\overset{N}{s}, \Delta t) \right)$$

$$\overset{\mathrm{f}}{\theta}_{n-1} = [\theta(\overset{1}{\boldsymbol{\xi}}, (n-1)\Delta t), \theta(\overset{2}{\boldsymbol{\xi}}, (n-1)\Delta t), \ldots \theta(\overset{M}{\boldsymbol{\xi}}, (n-1)\Delta t)] ,$$

$$\overset{\mathrm{b}}{P}_n = [P_n(\overset{1}{s}), P_n(\overset{2}{s}), \ldots P_n(\overset{N}{s})] ,$$

$$\overset{\mathrm{b}}{C}^{(ij)} = \int_0^{\Delta t} \int_{\partial\Omega_j} G(\overset{i}{s}, \Delta t; s, \tau)\, \mathrm{d}s\, \mathrm{d}\tau ,$$

$$\overset{\mathrm{b}}{D}^{(ij)} = \int_0^{\Delta t} \int_{\partial\Omega_j} \frac{\partial G}{\partial \upsilon} (\overset{i}{s}, \Delta t; s, \tau)\, \mathrm{d}s\, \mathrm{d}\tau ,$$

$$\overset{\mathrm{b}}{F}{}_1^{ik} = \overset{k}{w} G(\overset{i}{s}, \Delta t; \overset{k}{\boldsymbol{\xi}}, 0) ,$$

where f (field) and b (boundary) denote the domain of functions and the functional space of operators.

Unlike the case for the integral time method, here all coefficients are independent of the time step number n and so need only be calculated once and stored for repeated use. The equations may be solved for the N unknown components from the vectors $\overset{\mathrm{b}}{\theta}_n$ and $\overset{\mathrm{b}}{\boldsymbol{q}}_n$ in accordance with the conditions for a well-posed heat flow problem. Thus we have a solution

$$\theta(\boldsymbol{x}, n\Delta t) = \sum_j \overset{j}{C}(\boldsymbol{x})\overset{j}{q}_n - \overset{j}{D}(\boldsymbol{x})\overset{j}{\theta}_n + \sum_k \overset{k}{w} G(\boldsymbol{x}, \Delta t; \overset{k}{\boldsymbol{\xi}}, 0)\theta(\overset{k}{\boldsymbol{\xi}}, (n-1)\Delta t) + P_n(\boldsymbol{x})$$

or in vector form

$$\overset{\mathrm{f}}{\boldsymbol{\theta}}_n = \overset{\mathrm{f\,b}}{\boldsymbol{C}\boldsymbol{q}}_n - \overset{\mathrm{f\,b}}{\boldsymbol{D}\boldsymbol{\theta}}_n + \overset{\mathrm{f\,f}}{\boldsymbol{F}\boldsymbol{\theta}}_{n-1} + \overset{\mathrm{f}}{\boldsymbol{P}}_n . \tag{2.22}$$

In particular, the values $\boldsymbol{\theta}(\overset{k}{\boldsymbol{\xi}}, (n-1)\Delta t)$ must be calculated in preparation for the succeeding step. Results for a test problem will be shown in section 3.

2.9.3. Stability

This iterative process provides an explicit formula for the temperature in terms of its values at the previous step, and of boundary and forcing function data. It has been found [36, 40, 74], however, that if the time step Δt is chosen to be too small, the accuracy of the approximation actually deteriorates. Moreover, the associated error tends to grow with time, indicating a condition of instability. It would appear then that for problems where small time behavior is important, this method should be abandoned in favor of the one discussed earlier, since that appears to be unconditionally stable.

An analysis of the behavior of the error arising from the stepwise algorithm [74] shows that a sufficient condition for stability is

$$\|\overset{\mathrm{f}}{\mathbf{M}}\| < 1 \,, \tag{2.23}$$

where

$$\overset{\mathrm{f}}{\mathbf{M}} = \left[\overset{\mathrm{f}}{\mathbf{F}} - \left(\overset{\mathrm{f}}{\mathbf{C}} + \frac{a}{h} \overset{\mathrm{f}}{\mathbf{D}} \right) \overset{\mathrm{b}}{\mathbf{B}} \overset{\mathrm{b}}{\mathbf{F}} \right] \qquad \text{for } h > 0$$

or

$$\overset{\mathrm{f}}{\mathbf{M}} = [\overset{\mathrm{f}}{\mathbf{F}} - \overset{\mathrm{f}}{\mathbf{D}} \overset{\mathrm{b}}{\mathbf{K}} \overset{\mathrm{b}}{\mathbf{F}}] \,. \qquad \text{for } h = 0$$

with

$$\overset{\mathrm{b}}{\mathbf{B}} = \left[\overset{\mathrm{b}}{\mathbf{C}} + (\overset{\mathrm{b}}{\mathbf{I}} + \overset{\mathrm{b}}{\mathbf{D}}) \frac{a}{h} \right]^{-1} , \qquad \overset{\mathrm{b}}{\mathbf{K}} + [\overset{\mathrm{b}}{\mathbf{I}} + \overset{\mathrm{b}}{\mathbf{D}}]^{-1} , \qquad \overset{\mathrm{b}}{\mathbf{I}} \text{ is the identity matrix}$$

and a and h are constants appearing in the general form of the boundary conditions given at the start of section 2.1.

The problem arises in calculating the volume integral term containing the temperature at the previous step (or the "initial" condition if the process is viewed as a series of subproblems through time). Specifically, the integrand contains the function G which is evaluated in the neighborhood of its singularity when Δt is reduced excessively, yielding erroneous results when a numerical quadrature scheme is used. If interpolation functions are used so that the integration is performed analytically, the same problem can occur, this time caused by the occurrence of negative coefficients in the matrix $\mathbf{F}$. However, in this case it has been shown [72] that, at least for one-dimensional formulations, this can be avoided by accompanying the reduction of the time step with one of $\Delta\xi$, the size of the body elements. A sufficient condition of the form

$$\Delta\xi < [4\alpha(\Delta t)]^{1/2} Z$$

for fulfilling (2.23) above – and therefore a sufficient condition for stability – is derived, where Z is a parameter reflecting the nature of the shape functions. If this result can be generalized to the multi-dimensional case, then, unlike (2.23), a stability condition that can be checked prior to calculation of the influence conditions would be available. This remains a subject for further research.

3. Thermoelasticity

3.1. The problem

We stated in section 1 that under normal conditions of heat flow, the mechanical effect upon temperature change is negligible, allowing us to decouple the heat equation in system (1), and yielding the set

$$\theta_{,jj}(\boldsymbol{x}, t) - \frac{1}{\alpha}\frac{\partial \theta}{\partial t}(\boldsymbol{x}, t) = -\frac{1}{\alpha} f(\boldsymbol{x}, t) \tag{3.1}$$

$$\mu u_{i,jj}(\boldsymbol{x}, t) + (\mu + \lambda)u_{j,ij}(\boldsymbol{x}, t) = (3\lambda + 2\mu)\omega\theta_{,i}(\boldsymbol{x}, t) - \rho F_i(\boldsymbol{x}, t) \qquad \boldsymbol{x} \in \Omega,\ t > 0 \tag{3.2}$$

where ω is the coefficient of thermal expansion, F_i the distributed body force, and where the temperature gradient $\theta_{,i}$ can be interpreted as another forcing term. Thus, independent of mechanical characteristics of the problem, the temperature distribution can be found from (3.1), whether by using the boundary element methods discussed here or by other means. We shall confine our attention to this class of problems. Boundary integral formulations for more general problems are dealt with by Sladek et al. [77, 78], Nowacki [55] and Predeleanu [62] who outlines a numerical procedure for dynamic thermoviscoelasticity problems.

Once the temperature is known, Eq. (3.2) can be solved provided they are augmented by the conditions

$$u_i(s, t) = U_i(s, t) \qquad s \in \partial\Omega,\ t > 0 \qquad \text{or} \qquad t_i(s, t) = T_i(s, t)$$

or a compatible combination of these, where t_i is the traction vector and where U_i and T_i are prescribed functions.

The solution of (3.2) for the function u_i may be written as

$$u_i = u_i^{(\theta)} + u_i^{(F)} + u_i^{(*)}$$

where $u_i^{(F)}$ and $u_i^{(\theta)}$ are the particular solutions corresponding to the terms involving F and θ respectively in Eq. (3.2), and $u_i^{(*)}$ is the homogeneous solution. The latter may be dealt with by any numerical procedure (see later) while $u_i^{(F)}$ can always be written as a volume integral, or under certain conditions [67] as a boundary integral. We focus here on determining the thermal effects upon displacements and stresses, i.e. on calculating the displacement and stress terms $u_i^{(\theta)}$ and $\sigma_{ij}^{(\theta)}$ respectively.

In accordance with the development of linear quasi-static thermoelasticity theory [56], we begin by writing

$$u_i^{(\theta)} = \phi_{,i} \tag{3.3}$$

where ϕ is known as the thermoelastic displacement potential. Substituting this into Eq. (3.2) (with $F = 0$) we obtain[1]

$$\phi_{,ii} = m\theta \tag{3.4}$$

where

$$m = \frac{(3\lambda + 2\mu)}{(\lambda + 2\mu)}\,\omega\,.$$

[1]The neglected function of integration is irrelevant in the analysis.

Since θ is known we may solve (3.4) by means of the singular fundamental function

$$\eta(x, \xi) = (1/4\pi) \ln|x - \xi|^2 .$$

Thus

$$\phi(x, t) = m \int_{\Omega} \eta(x, \xi)\theta(\xi, t) \, d\xi . \tag{3.5}$$

Now from (3.3) and (3.4), and the strain–displacement relation

$$e_{ij} = \tfrac{1}{2}(u_{i,j} + u_{j,i})$$

we have $e_{ij} = \phi_{,ij}$ and $e_{ii} = m\theta$. Since the constitutive relation for quasi-static thermoelasticity is

$$\sigma_{ij} = 2\mu e_{ij} + [\lambda e_{kk} - (3\lambda + 2\mu)\omega\theta]\delta_{ij}$$

or

$$\sigma_{ij} = 2\mu[\phi_{,ij} - m\theta\delta_{ij}] \tag{3.6}$$

the functions $u_i^{(\theta)}$ and $\sigma_{ij}^{(\theta)}$ can be found directly by differentiating the solution (3.5) of Eq. (3.4) since from (3.3) and (3.6) we could write

$$u_i^{(\theta)}(x, t) = m\left[\int_{\Omega} \eta(x, \xi)\theta(\xi, t) \, d\xi\right]_{,i} \tag{3.7}$$

and

$$\sigma_{ij}^{(\theta)}(x, t) = 2\mu m\left[\int_{\Omega} \eta(x, \xi)\theta(\xi, t) \, d\xi\right]_{,ij} - 2\mu m\theta\delta_{ij} . \tag{3.8}$$

It is important to note that the irrotational form of the displacement in (3.3) does not imply loss of generality since the balance of the temperature effects is incorporated into the homogenous solution $u_i^{(*)}$. Thus expressions (3.7) and (3.8) are equivalent to those treating the temperature gradient as a distributed body force [5, 7] and provide an apparently efficient way of determining the thermo-mechanical response to the temperature distribution since it exploits work that has already been done in determining that distribution. However there are drawbacks to this scheme. The integrands are singular and therefore require particularly careful treatment during the numerical integration process. Also the calculation of the temperature and then the mechanical effects in series necessarily compounds the error already present in θ, aside from introducing further uncertainty in determining the stresses and displacements.

3.2. A boundary integral approach

In order to avoid these problems, it is possible to calculate the thermally induced values by means of a boundáry integral. In keeping with the spirit of trying to interpret the method in physical terms, we begin by examining the thermoelastic response of an infinite elastic medium to an instantaneous heat pulse.

If we set $\theta(\boldsymbol{x}, t) = \bar{\theta}(\boldsymbol{x}, t'; \boldsymbol{w}, \tau) = G(\boldsymbol{x}, t'; \boldsymbol{w}, \tau)$ in Eq. (3.4) we find that the thermoelastic displacement potential at time t' due to a heat pulse that occurs at time $\tau < t'$ is

$$\phi(\boldsymbol{x}, t'; \boldsymbol{w}, \tau) = m[\alpha E(\boldsymbol{x}, t'; \boldsymbol{w}, \tau) + \eta(\boldsymbol{x}, \boldsymbol{w})] = \bar{\phi}(\boldsymbol{x}, t'; \boldsymbol{w}, \tau)$$

where

$$E(\boldsymbol{x}, t'; \boldsymbol{w}, \tau) = \int_{\tau}^{t'} G(\boldsymbol{x}, t'; \boldsymbol{w}, \tau)\, \mathrm{d}t' = -\frac{1}{4\alpha\pi} E_i\left(\frac{-|\boldsymbol{x} - \boldsymbol{w}|^2}{(4\alpha(t' - \tau))}\right)$$

and $E_i(-y)$ is the exponential integral which was defined in Eq. (2.14). It follows that at $\boldsymbol{x} = \boldsymbol{w}$, ϕ is continuous and differentiable to the second order at least.

From (3.3) and (3.6) we have that

$$\bar{u}_i(\boldsymbol{x}, t'; \boldsymbol{w}, \tau) = (m/2\pi) X_i h(\boldsymbol{x}, t'; \boldsymbol{w}, \tau)$$

and

$$\bar{\sigma}_{ij}(\boldsymbol{x}, t'; \boldsymbol{w}, \tau) = (\mu m/\pi r)[h(\boldsymbol{x}, t'; \boldsymbol{w}, \tau)(\delta_{ij} - 2X_i X_j)] + 2\mu m(X_i X_j - \delta_{ij})\theta$$

where $\boldsymbol{x} = (x_1, x_2)$, $\boldsymbol{w} = (w_1, w_2)$, $r = |\boldsymbol{x} - \boldsymbol{w}|$, $X_i = (x_i - w_i)/r$ and

$$h(\boldsymbol{x}, t'; \boldsymbol{w}, \tau) = \frac{1}{r}\left[1 - \exp\left(\frac{-r^2}{[4\alpha(t' - \tau)]}\right)\right] \qquad t' > \tau .$$

These expressions represent the mechanical effects of an instantaneous heat pulse occurring at a point $\boldsymbol{w}$ at time τ on a point $\boldsymbol{x}$ at time $t' > \tau$ in an infinite medium, and are analogous to the fundamental heat solution G discussed in section 1. Consequently, they can be used to generate trial solutions of the thermoelastic problem in precisely the same manner as was the function θ. By extension, they can also be generalized to comprise boundary integral formulae corresponding to Eqs. (1.16), (1.22) and (1.24).

Mathematically this can be derived by using the identity

$$\theta(\xi, t) = \int_0^t \frac{\partial\theta(\xi, t')}{\partial t}\, \mathrm{d}t' + \theta(\xi, 0)$$

in (3.5), yielding

$$\phi(x, t) = m \int_{\Omega} \eta(x, \xi)\left[\int_0^t \frac{\partial\theta(\xi, t')}{\partial t}\, dt' + \theta(\xi, 0)\right] d\xi \, .$$

If we now use the heat equation (3.1), we get

$$\phi(x, t) = m \int_{\Omega} \eta(x, \xi)\left[\int_0^t [\alpha\theta_{,jj}(\xi, t') + f(\xi, t')]\, dt' + \theta(\xi, 0)\right] d\xi \, .$$

Changing the order of integration, using Green's second theorem in the first term, and recognizing that

$$\eta_{,ii}(x, \xi) = \delta(x_1 - \xi_1)\delta(x_2 - \xi_2)$$

we obtain

$$\begin{aligned}\phi(x, t) &= m\alpha \int_0^t \int_{\partial\Omega} \left(\eta(x; s)\frac{\partial\theta(s, t')}{\partial v} - \frac{\partial\eta(x; s)}{\partial v}\theta(s, t')\right) ds\, dt' \\ &\quad + m \int_0^t \int_{\Omega} \eta(x; \xi) f(\xi, t')\, d\xi\, dt' + m \int_{\Omega} \eta(x; \xi)\theta(\xi, 0)\, d\xi \\ &\quad + m\alpha \int_0^t \theta(x, t')\, dt' \, . \end{aligned} \tag{3.9}$$

In order to avoid the use of field temperature values, we simply use expressions (1.16), (1.22) or (1.24) in place of $\theta(x, t)$ in the final term. For example, if we use the direct expression (1.16) we have that

$$\begin{aligned}\phi(x, t) &= \alpha \int_0^t \int_{\partial\Omega} \left(\bar{\phi}(x, t; s, \tau)\frac{\partial\theta(s, \tau)}{\partial v} - \frac{\partial\bar{\phi}(x, t; s, \tau)}{\partial v}\theta(s, \tau)\right) ds\, d\tau \\ &\quad + \int_{\Omega} [\bar{\phi}(x, t; \xi, 0)\theta(\xi, 0)]\, d\xi + \int_0^t \int_{\Omega} [\bar{\phi}(x, t; \xi, \tau)] f(\xi, \tau)\, d\xi\, d\tau \end{aligned} \tag{3.10}$$

and that

$$u_i^{(\theta)}(x, t) = \alpha \int_0^t \int_{\partial\Omega} \left(\bar{u}(x, t; s, \tau) \frac{\partial\theta(s, \tau)}{\partial v} - \frac{\partial \bar{u}_i(x, t; s, \tau)}{\partial v} \theta(s, \tau) \right) \mathrm{d}s \, \mathrm{d}\tau$$

$$+ \int_{\Omega} [\bar{u}_i(x, t; \xi, 0)\theta(\xi, 0)] \, \mathrm{d}\xi + \int_0^t \int_{\Omega} [\bar{u}_i(x, t; \xi, \tau) f(\xi, \tau)] \, \mathrm{d}\xi \, \mathrm{d}\tau \tag{3.11}$$

and

$$\sigma_{ij}^{(\theta)}(x, t) = \alpha \int_0^t \int_{\partial\Omega} \left(\bar{\sigma}_{ij}(x, t; s, \tau) \frac{\partial\theta(s, \tau)}{\partial v} - \frac{\partial \bar{\sigma}_{ij}(x, t; s, \tau)}{\partial v} \theta(s, \tau) \right) \mathrm{d}s \, \mathrm{d}\tau$$

$$+ \int_{\Omega} [\bar{\sigma}_{ij}(x, t; \xi, 0)\theta(\xi, 0)] \, \mathrm{d}\xi + \int_0^t \int_{\Omega} [\bar{\sigma}_{ij}(x, t; \xi, \tau) f(\xi, \tau)] \, \mathrm{d}\xi \, \mathrm{d}\tau \tag{3.12}$$

where the kernels for these expressions represent the thermoelastic effects of a heat pulse in an infinite medium, as derived previously.

Notice that the temperature, stresses and displacements are now all calculated by means of analogous integral expressions. A similar set of boundary integral formulae can also be written for the indirect method in analogy with expression (1.22) or (1.24). In addition, the recurrence formulae which correspond to (2.15) are also available. Finally, we note that formulae (3.7) and (3.8) represent limiting cases of these formulae since they can be arrived at by reducing the time step Δt to zero. Thus we have a general framework which incorporates several alternatives for calculating thermal stresses and displacements.

3.3. *The numerical procedure*

From the above it follows that the numerical procedure corresponding to the one used to calculate the temperature distribution can be used to find the thermomechanical effects [75]. Clearly, the computations for temperature, stress and displacement can be done in parallel. This arrangement does require the retrieval of thermal boundary, initial and source data and the computation of stresses and displacements involving boundary integral expressions in direct analogy with the thermal problem. This implies that the amount of computation is greater than that required in (3.7) and (3.8). It follows, however, that the expressions for displacements and stresses are of the same order of approximation as that inherent in θ. Also, the kernels in the integrands resulting from the boundary integral method are continuous, allowing computations to be made in a straightforward manner.

We shall opt here for the direct stepwise algorithm. Thus, for step M, we can write generic recurrence relations

$$R(\boldsymbol{x}, M\Delta t) = \alpha \int_0^{\Delta t} \int_{\partial\Omega} \left(\bar{R}(\boldsymbol{x}, \Delta t; s, \tau) \frac{\partial\theta(s, (M-1)\Delta t + \tau)}{\partial v} - \frac{\partial R(\boldsymbol{x}, \Delta t; s, \tau)}{\partial v} \theta(s, (M-1)\Delta t + \tau) \right) \mathrm{d}s \, \mathrm{d}\tau$$
$$+ \int_{\Omega} [\bar{R}(\boldsymbol{x}, \Delta t; \boldsymbol{\xi}, 0)\theta(\boldsymbol{\xi}, (M-1)\Delta t)] \, \mathrm{d}\boldsymbol{\xi}$$
$$+ \int_0^{\Delta t} \int_{\Omega} [\bar{R}(\boldsymbol{x}, \Delta t; \boldsymbol{\xi}, \tau) f(\boldsymbol{\xi}, (M-1)\Delta t + \tau)] \, \mathrm{d}\boldsymbol{\xi} \, \mathrm{d}\tau \qquad (3.13)$$

where $\bar{R}$ and R represent $\bar{\theta}$, $\bar{u}_i$, $\bar{\sigma}_{ij}$ and θ, $u_i^{(\theta)}$ and $\sigma_{ij}^{(\theta)}$ respectively. The surface integrals can be evaluated by using the boundary discretization discussed in section 2. If we assume as before that the boundary functions $\theta(s, \tau)$ and $\partial\theta(s, \tau)/\partial v$ can be approximated by constants $\overset{j}{\theta}_M$ and $\overset{j}{q}_M$ respectively over each segment $\partial\Omega_j$ and over each time step M, Eq. (3.13) may be written as

$$R(\boldsymbol{x}, M\Delta t) = \sum_j [\overset{j}{C}(\boldsymbol{x})\overset{j}{q}_M - \overset{j}{D}(\boldsymbol{x})\overset{j}{\theta}_M]$$
$$+ \sum_k [\overset{k}{w} R(\boldsymbol{x}, \Delta t; \overset{k}{\boldsymbol{\xi}}, 0)\theta(\overset{k}{\boldsymbol{\xi}}, (M-1)\Delta t)] + P_M(\boldsymbol{x})$$

where

$$\overset{j}{C}(\boldsymbol{x}) = \alpha \int_0^{\Delta t} \int_{\partial\Omega_j} \bar{R}(\boldsymbol{x}, \Delta t; s, \tau) \, \mathrm{d}s \, \mathrm{d}\tau$$

$$\overset{j}{D}(\boldsymbol{x}) = \alpha \int_0^{\Delta t} \int_{\partial\Omega_j} \frac{\partial R(\boldsymbol{x}, \Delta t; s, \tau)}{\partial v} \, \mathrm{d}s \, \mathrm{d}\tau$$

$\overset{k}{\boldsymbol{\xi}}$ are Gaussian points in the field

$\overset{k}{w}$ are Gaussian weights associated with field points $\overset{k}{\boldsymbol{\xi}}$

and

$$P_M(\boldsymbol{x}) = \int_0^{\Delta t} \int_{\Omega} \bar{R}(\boldsymbol{x}, \Delta t; \boldsymbol{\xi}, \tau) f(\boldsymbol{\xi}, (M-1)\Delta t + \tau) \, \mathrm{d}\boldsymbol{\xi} \, \mathrm{d}\tau$$

is the generic term resulting from the body heat source.

As we saw in section 2.9, the unknown boundary functions can be determined

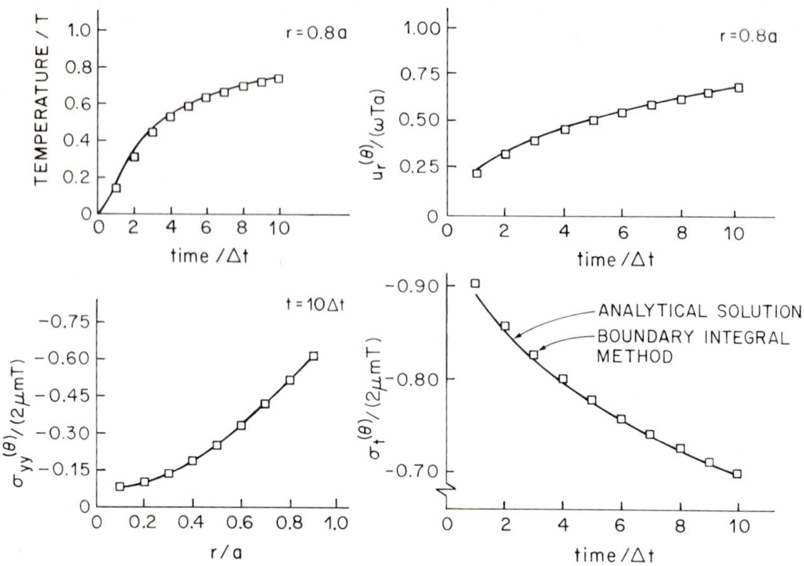

Fig. 15. Finite disk problem: results for temperature and induced stresses and displacements.

by solving, at each time step, the set of linear algebraic equations resulting from the case where

$$\bar{R} = \bar{\theta} \qquad \text{and} \qquad \boldsymbol{x} = \boldsymbol{s}\,.$$

Thereafter these values may be used in the remaining expressions enabling the thermally induced mechanical effects $u_i^{(\theta)}$ and $\sigma_{ij}^{(\theta)}$ to be calculated.

The procedure was tested on the problem of a finite cylindrical disk, free of body sources and forces, initially void of temperature, whose surface is maintained traction free at a constant temperature T over time. The approximation for the temperature distribution was calculated using the stepwise method described in section 2.9. The simple nature of this problem allows the analytical calculation [56] of the temperature θ and the components $u_i^{(\theta)}$, $\sigma_{ij}^{(\theta)}$ so that a direct comparison with the numerical solution can be made. Samples of the results are shown in Fig. 15. It should be noted that the tangential stresses $\sigma_t^{(\theta)}$ at the boundary were calculated from a numerical differentiation of the displacements obtained from (3.13) since they are found to be more accurate than when computed directly [33]. This has become standard practice in boundary element computations.

3.4. The homogeneous solution

It must be recalled that the values $u_i^{(\theta)}$ and $\sigma_{ij}^{(\theta)}$ calculated at each step will not in general agree with the prescribed mechanical boundary conditions, irrespective of the method used. The discrepancy at the boundary must be removed by superimposing a solution fulfilling the homogeneous equations

$$\mu u_{i,jj}^{(*)}(\boldsymbol{x}) + (\lambda + \mu) u_{j,ij}^{(*)}(\boldsymbol{x}) = 0$$

subject to the boundary conditions

$$u_i(s, M\Delta t) = U_i(s, M\Delta t) - u_i^{(\theta)}(s, M\Delta t)$$

or

$$t_i(s, M\Delta t) = T_i(s, M\Delta t) - t_i^{(\theta)}(s, M\Delta t)$$

or an appropriate combination of these. This solution may be found by any numerical means, the most straightforward being a boundary element method. Results obtained using an indirect (fictitious stress) [29] method and the equations

$$u_i(\boldsymbol{x}, M\Delta t) = u_i^{(\theta)}(\boldsymbol{x}, M\Delta t) + u_i^{(*)}(\boldsymbol{x}, M\Delta t)$$
$$\sigma_{ij}(\boldsymbol{x}, M\Delta t) = \sigma_{ij}^{(\theta)}(\boldsymbol{x}, M\Delta t) + \sigma_{ij}^{(*)}(\boldsymbol{x}, M\Delta t)$$

for each time step M are given in Fig. 16.

In summary, the numerical details involved in implementing boundary element methods for quasi-static thermoelasticity problems are already accounted for in solving the associated heat flow problem. The only requirements then are to replace the temperature kernels with the appropriate fundamental solutions for stresses and displacements, giving an analogous set of integral expressions whose

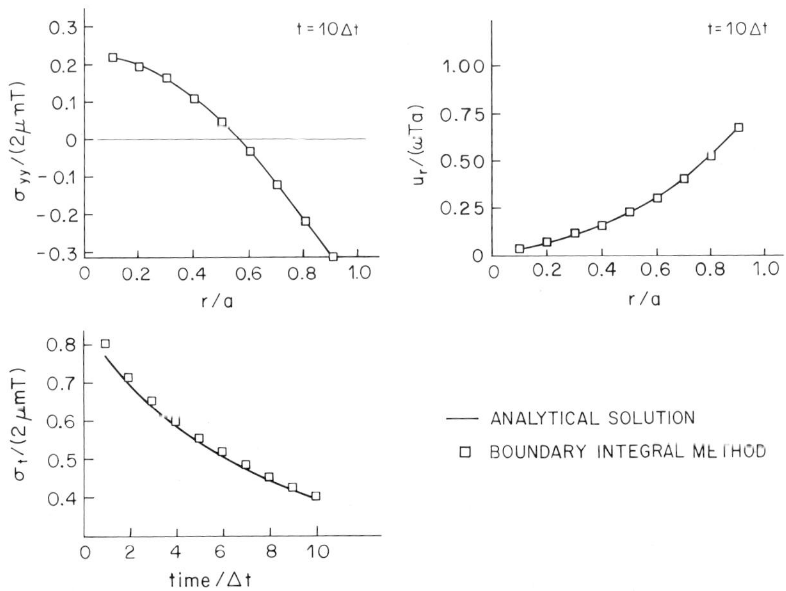

Fig. 16. Finite disk problem: final results (i.e. including homogeneous solution).

form is characterized by (3.10). Thus boundary element methods for thermoelasticity derive quite neatly from those for heat flow. The unifying framework allows us to apply the appropriate method to problems with particular properties. For example, if we are interested in the very short-term mechanical response to a sudden thermal load on the body, we would use an indirect double layer method that integrates time from the origin at each step (described in section 2.6). This is because that method uses a kernel which best handles singularities at time zero, is not too costly when the analysis is confined small times and produces results which appear to remain stable despite the use of a very small time step size.

4. Consolidation

4.1. Introduction

The deformation of saturated soils is described by the theory of consolidation. This behavior is characterized by the squeezing out of fluid from the pores between the loaded soil granules before the full settlement takes place, so that the mechanism is fundamentally a time-dependent one.

Terzaghi [83] first recognized the effect of the pore pressure of the fluid on the deformations of the saturated soil, and formulated a model for the one-dimensional case where the flow of fluid and deformation are uni-directional. Other analysts, most prominently Biot [11, 12], arrived at a set of governing partial differential equations describing the system under assumptions of small strains in an elastic medium. The coupling of these equations arises from:

(1) The heuristic definition of effective stress

$$\sigma'_{ij} = \sigma_{ij} + \delta_{ij} p \tag{4.1}$$

where σ_{ij} are the total measured stresses and p is the pore water pressure. Equation (4.1) reflects the distinction made between the stresses borne by the soil skeleton from those borne by the fluid and therefore introduces a pressure term into the mechanical equations.

(2) The relation between the volumetric strain rate and the rate of flow of the fluid. This introduces an effect of the mechanical deformation upon the diffusion process.

The system can be written as

$$\mu u_{i,jj} + (\lambda + \mu) e_{,i} - p_{,i} + \rho_s F_i = 0 \tag{4.2a}$$

$$K/(\rho_f g) p_{,jj} - n\beta \frac{\partial p}{\partial t} - \frac{\partial}{\partial t} e = 0 \tag{4.2b}$$

which is a special case of the equations in (1), where K is the permeability coefficient of the skeleton with respect to water, ρ_f and ρ_s are respectively the densities of the fluid and soil, β is the compressibility of the fluid, n is the soil porosity, g is the gravitational constant and $e = u_{j,j}$. Unlike the case for thermoelasticity, the strain rate term dominates the time-dependent behavior in the pressure equation, normally precluding the decoupling of the equations.

To examine the conditions under which the equations can be decoupled, it is instructive [86] to take the divergence of the displacement equation. This yields the relation

$$(\lambda + 2\mu)e_{,jj} = p_{,jj} .$$

Integrating we obtain

$$(\lambda + 2\mu)e = p + z$$

where z is a function of integration satisfying $z_{,jj} = 0$. Clearly we require that $z = 0$ if the equations are to be uncoupled. It has been shown [37] that in such cases the displacements are irrotational. On the other hand if we assume that the displacement vector is irrotational, i.e.

$$u_i = \phi_{,i} \tag{4.3}$$

then if we ignore body forces and substitute this into the elasticity equation in (4.2a) we obtain

$$(\lambda + 2\mu)\phi_{,ijj} = p_{,i}$$

or by integration

$$(\lambda + 2\mu)\phi_{,jj} = p + g(t) \tag{4.4}$$

where $g(t)$ is independent of spatial coordinates. Since from (4.3) we have that $\phi_{,jj} = e$, (4.4) and (4.2a) yield an equation in p and $g(t)$ only. This class of problems has been examined in detail by Cleary and Rice [26]. In the case of infinite space problems or those in one dimension, it can be shown that $g(t)$ vanishes.

4.2. Boundary element solutions

The most straightforward boundary element method for solving the coupled system of equations can be implemented by using the integral expressions for both the diffusion and elastic problems simultaneously. Thus we would employ the boundary integral expressions

$$p(\boldsymbol{x}, t) = \alpha \int_0^t \int_{\partial\Omega} \left(G(\boldsymbol{x}, t; s, \tau) \frac{\partial p(s, \tau)}{\partial \upsilon} - \frac{\partial G(\boldsymbol{x}, t; s, \tau)}{\partial \upsilon} p(s, \tau) \right) \mathrm{d}s \, \mathrm{d}\tau$$

$$+ \int_{\Omega} G(\boldsymbol{x}, t; \boldsymbol{\xi}, 0) p_0(\boldsymbol{\xi}) \, \mathrm{d}\boldsymbol{\xi}$$

$$- \frac{1}{n\beta} \int_0^t \int_{\Omega} G(\boldsymbol{x}, t; \boldsymbol{\xi}, \tau) \frac{\partial}{\partial t} e(\boldsymbol{\xi}, \tau) \, \mathrm{d}\boldsymbol{\xi} \, \mathrm{d}\tau \tag{4.5}$$

and

$$u_j(\boldsymbol{x}, t) = \int_{\partial\Omega} [T^*_{ij}(\boldsymbol{x}, s) u_i(s, t) - U^*_{ij}(\boldsymbol{x}, s) t_i(s, t)] \, \mathrm{d}\boldsymbol{\xi}$$

$$- \int_{\Omega} U^*_{ij,i}(\boldsymbol{x}, \boldsymbol{\xi}) p(\boldsymbol{\xi}, t) \, \mathrm{d}\boldsymbol{\xi} + \int_{\Omega} U^*_{ij}(\boldsymbol{x}, \boldsymbol{\xi}) \rho_s(\xi) F_i(\boldsymbol{\xi}, t) \, \mathrm{d}\boldsymbol{\xi} \tag{4.6}$$

where here

$$\alpha = K/(\rho g \beta n)$$

is the material constant. Equation (4.6) is the boundary integral expression [5, 7, 47, 2] for the elliptical equations describing the behavior of linear elastic small strain systems in two dimensions, where U^*_{ij} and T^*_{ij} are the fundamental solutions respectively for the displacement and stress for plane strain problems. Similar work for the case of axisymmetric problems is described in [3]. Notice that the coupling terms have been treated essentially as forcing functions[2]. Equation (4.5) can be discretized in the same manner described previously, using the stepwise method to march through time, and a series of boundary and volume elements that can also serve (4.6). The resulting systems of equations can be solved iteratively. The flow equation can be solved first using the initial value of p and an assumed initial value for $\partial e/\partial t$. The solution can then be generated at nodal points throughout the domain and used to solved for the displacements and stresses. The latter, in turn, can be employed to determine the new value of e, from which an approximate value for $\partial e/\partial t$ can be calculated, allowing the next value of p to be found, and so on.

There are, however, some awkward features of this scheme. First, certain compromises must be made to get the solution procedure started. The difficulty arises in obtaining an initial value for the p. One method [5, 7] calls for the assumption that the solid is incompressible at time $t = 0$. Using a Poisson's ratio of

[2] The appearance of $p(\xi, t)$ rather than its gradient in the volume integral in (4.6) is due to the use of the divergence theorem and the effective stress.

0.5, one can solve (4.6) for displacements and effective and total stresses, and use relation (4.1) to produce an initial value for p. Further assuming that $\partial e/\partial t = 0$, the diffusion equation can be solved to produce the pressure distribution after one time step. This can then be used to solve (4.6) (using soil skeleton moduli) which provides a new value for e. A finite difference formula for the time derivative then allows this scheme to be performed iteratively through time.

An alternative [47] suggests assuming that both $\partial e/\partial t$ and $\partial p/\partial t$ are zero at $t = 0$, leaving a steady-state potential problem to be solved for the initial value p. Another method [47, 2] tentatively sets the initial pressure p equal to zero and solves the elastic equation. The internal effective stresses are then calculated using a boundary integral expression involving the boundary displacements and tractions. Together with (4.1) this yields a total stress tensor σ_{ij} from which a corrected value $p = (\sigma_{11} + \sigma_{22})/2$ is computed. This process is continued iteratively until p converges, the value obtained serving as the initial one in the pressure equation.

Another source of difficulty with such algorithms involves the approximation of the time derivative. During calculations at the kth time step, i.e. at time $t = k\Delta t$, a finite difference approximation of the form

$$e((k-1)\Delta t) - e((k-2)\Delta t)$$

is used [5, 7, 47, 2] rather than the backward formula

$$e((k)\Delta t) - e((k-1)\Delta t)$$

as might be expected. This is done to avoid solving implicitly for e at points in the body at time t, i.e. to confine unknown functions to the boundary. However, this practice contributes further to the error and exacerbates the task of initiating the algorithm.

4.3. *Alternative procedures*

The procedure outlined above results from generating two boundary integral equations, one derived from (1.3) and (1.4) for the pressure, and the other from the mechanical reciprocal identity

$$\sigma_{ij}\bar{e}_{ij} = \bar{\sigma}_{ij}e_{ij}$$

where (σ_{ij}, e_{ij}) and $(\bar{\sigma}_{ij}, \bar{e}_{ij})$ are stress and strain tensors from two different loading programs. An alternative method results when a single energy relation [23, 24, 25] is used, involving both the fluid and the solid, i.e.

$$\sigma_{ij}\bar{e}_{ij} + np\bar{\varepsilon} = \bar{\sigma}_{ij}e_{ij} + n\bar{p}\varepsilon$$

where $\varepsilon = Y_{i,i}$ is the volumetric strain, with Y_i the displacement of the fluid [12]. When this relation is integrated over time and space, using the constitutive, kinetic and equilibrium equations and the divergence theorem [23, 24, 25] one obtains and single integral expression of the form

$$\int_0^t \int_{\partial\Omega} [T_i(t-\tau)\bar{u}_i(\tau) - \bar{T}_i(t-\tau)u_i(\tau) + np(t-\tau)(\bar{Y}_v(\tau) - \bar{u}_v(\tau))$$

$$- n\bar{p}(t-\tau)(Y_v(\tau) - u_v(\tau))] \, \mathrm{d}s \, \mathrm{d}\tau + \int_0^t \int_{\Omega} [\{\rho_1 \bar{u}_i(t-\tau) - \bar{\rho}_1 u_i(t-\tau)$$

$$+ \rho_2 \bar{Y}_i(t-\tau) - \bar{\rho}_2 Y_i(t-\tau)\} F_i(\tau)] \, \mathrm{d}\xi \, \mathrm{d}\tau - \bar{u}_i(\tau)$$

$$+ \int_0^t \int_{\Omega} ([\dot{Y}_i(t-\tau) - \dot{u}_i(t-\tau)][\bar{Y}_i(\tau) - \bar{u}_i(\tau)]$$

$$- [(\dot{\bar{Y}}_i(t-\tau) - \dot{\bar{u}}_i(t-\tau)][Y_i(\tau) - u_i(\tau)]) \, \mathrm{d}\xi \, \mathrm{d}\tau = 0 \, , \tag{4.7}$$

where T_i is the total traction at the boundary, $Y_v = Y_i v_i$, $u_v = u_i v_i$, v_i is the unit vector in the outward normal direction, F_i is the body force and ρ_i are the fractional densities, i.e.

$$\rho_1 = (1-n)\rho_s \qquad \rho_2 = n\rho_f$$

and where the dots denote differentiation with respect to time. Integral equations are obtained for displacements and pressure from (4.7) by replacing the bar variables with fundamental solutions obtained by solving successively the governing equations using force pulses in the usual manner [23–25].

In particular, Cheng and Ligget [23] have developed this scheme in the Laplace transform space for the anisotropic case. They solve the following example problem using this approach. Consider an infinitely deep layer of saturated clay material. The xy plane coincides with the surface while the z-axis is oriented vertically downward. At time $t = 0$ a strip load, infinitely long in the y-direction and extending uniformly from $x = -1$ to $x = +1$ with weight P per unit area is applied on the surface of the clay. This is shown in Fig. 17.

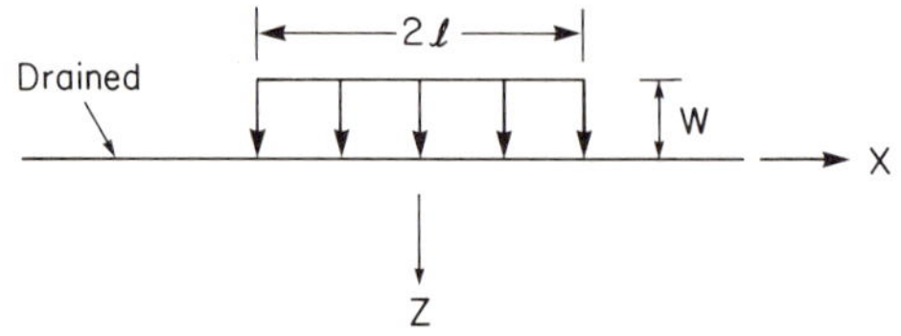

Fig. 17. Strip load on semi-infinite clay layers.

With reference to Fig. 18 the following boundary conditions hold: No shear stresses exist on the surface boundary AC while a normal load P is applied constantly after time $t = 0$ on the strip BC. There are no normal displacements or shear stresses along the line of symmetry CD. Also since this boundary is undrained, there are no relative displacements between the soil and the fluid along it. The pore pressure is zero on the horizontal surface. The solution domain was taken to be a semicircle of radius $20l$, and an approximate boundary condition based on an elasticity solution was imposed on the far-field boundary. The boundary was discretized by means of 21 quadratic elements and the Boundary Element Method was applied for each of 11 Laplace transform parameters. A sample result, that of the pore pressure at a point below the center of the strip and represented as a function of the logarithm of dimensionless time, is reproduced here from [23] as Fig. 19. The use of the Laplace transform in this method necessitates numerical inversion in order to restore the time variable. The result can be seen to agree well with the analytical solution, and shows the initial increase in pore pressure, known as the Cryer effect [34].

Vardoulakis and Harnpattanapanich [85] approached the consolidation problem by decoupling the system by means of the displacement functions E and S [52] which satisfy the equations

$$\frac{\partial}{\partial t} E = cE_{,iijj}$$

and

$$S_{,jj} = 0$$

where c is the consolidation coefficient. By applying successively Laplace and double Fourier transforms, these become ordinary second-order linear diferential equations in the transformed space. The solution to these must be trans-

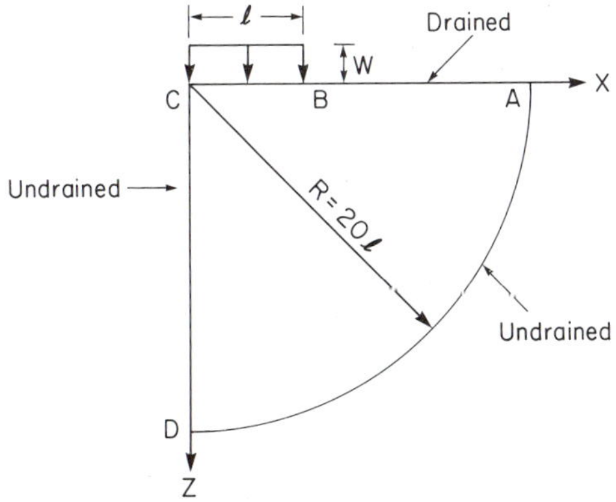

Fig. 18. The solution domain.

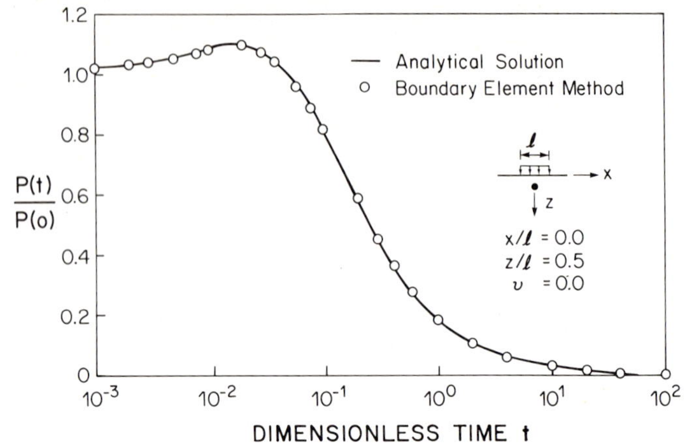

Fig. 19. Pore pressure history at a point below loading area.

formed back to the physical domain. By considering force points in the medium, singular solutions can be obtained which can be used to generate boundary integral methods for more complex problems. Although this work does not contain boundary element solution procedures (these are currently being pursued), its main thrust is the treatment of non-homogeneous half-spaces which can be used to model successive soil layers. The method clearly also requires a numerical transform inversion procedure. Further, there is no rigorous method of determining suitable ranges of values for the transform parameters to be used in the inversion. These must be found empirically.

It appears then that the question of numerical Laplace transform inversion is important in boundary element methods solving transient problems, particularly those arising in the theory of consolidation. Several analysts [66, 23, 85] have employed the algorithm due to Schapery [70] which assumes that the solution can be written as a summation of linear and exponential terms. This can therefore only be used with confidence when some knowledge of the form of the solution is available. In fact it has been shown by Narayanan and Beskos [53] that like all methods making use of real values of the transform parameter, this technique is only reliable when the relevant functions of time vary slowly. It is further shown that among several methods studied, that of Durbin [39] is best. Durbin's method treats the transform parameter as a complex variable and provides an accurate and general algorithm that can handle very complicated functions of time. Thus, it appears that boundary element procedures using the Laplace transform can be made as accurate as those operating in the time domain provided that Durbin's algorithm is used. The method does, however, require increased computational effort in its implementation.

We have described some of the approaches taken in boundary element solutions of consolidation problems. Further discussions of singular solutions and reciprocal theorems can be found in references in [63] and [13] and [38]. In addition fundamental solutions which account for dynamic effects are dealt with in Refs. [59, 60, 20], although to our knowledge no boundary element solutions for such problems have yet been implemented.

5. Concluding remarks

A discussion of boundary element methods almost inevitably invites a comparison with other established techniques, e.g. finite elements. For certain types of problems, such as those in linear isotropic elastostatics and steady-state potential theory, boundary element methods clearly offer an advantage since they perform significantly fewer calculations in producing a solution. For other formulations, i.e. those exhibiting nonlinearities or gross inhomogeneities, these methods are at a distinct disadvantage and in many cases currently cannot handle the problems at all.

It appears to us that in the cases of linear diffusion and thermoelasticity the relative advantages of these methods are less clear. From a purely algorithmic viewpoint they appear to be similar since each[3] involves a recurrence relation in time. It is apparent that boundary element methods do not offer the radically cheaper and simpler computational alternative for transient problems that they do in linear steady-state cases. Although the size of the system of equations to be solved at each step depends upon the number of nodes at the boundary only, and so is smaller, a volume integration is required to take account of conditions at the previous step.

These methods do hold an advantage in dealing with infinite-body problems since they are not required to impose an artificial boundary, nor boundary conditions, in order to approximate conditions at infinity since the solution obtained automatically ensures the correct behavior there. The magnitude of the temperature at the outer edge of the body grid can be examined at each step, the effect of ignoring the heat beyond it monitored, and the approximation improved dynamically if necessary. A similar process cannot be introduced into finite element methods without disturbing the system of equations. On the other hand, unlike finite element methods, the boundary element coefficient matrix depends explicitly upon the step size and so cannot admit dynamic modifications to account for time step refinements.

In the case of consolidation, it appears at this stage that boundary methods are less well developed. Of particular concern for problems solved in the time domain are the rather uncertain situation at time zero and the approximations that must be introduced to achieve artificial decoupling of the governing equations. Methods operating in transformed spaces must rely on an additional numerical approximation (the transform inversion) and also remove the possibility of

[3] Here we refer to the stepwise approach.

exploiting the physical intuition that boundary element procedures offer the analyst working in the space–time domain.

Boundary element methods are relatively new, having become popular during the past fifteen or twenty years. Nonetheless, they have enjoyed great success in many of the areas of application. This has mainly been due to the lower computational demands they impose, which is a consequence of their use of functions which allow the governing equations to be satisfied exactly. These methods clearly contain a strong analytical component and therein lies the key to their promise. Their potential as general tools for solving boundary-value problems is open ended; they remain constantly poised to exploit new analytic solutions which might broaden their range and power. This potential motivates the continued search for such solutions, and justifies further efforts to refine and improve the numerical aspects of the technique.

Acknowledgement

The authors would like to thank Professor I. Vardoulakis for his invaluable assistance during several discussions on consolidation.

References

[1] E. Alarcon, A. Martin and F. Paris, Boundary Elements in Potential and Elasticity Theory, *Computers and Structures* **10**, 351–362 (1979).

[2] G. Aramaki, T. Kuroki and K. Onishi, Consolidation Analysis by Boundary Element Method, in: *Boundary Element Methods in Engineering*, C.A. Brebbia, Ed. (Springer-Verlag, Berlin, 1982) pp. 363–376.

[3] G. Aramaki, T. Kuroki, K. Yasuhara and K. Onishi, Boundary Element Method in Axisymmetric Linear Consolidation, in: *Boundary Elements*, C.A. Brebbia, T. Futagami and M. Tanaka, Eds. (Springer-Verlag, Berlin, 1983) pp. 665–674.

[4] P.K. Banerjee, Integral Equation Methods for Analysis of Piecewise Non-Homogeneous Three Dimensional Elastic Solids of Arbitrary Shape, *International Journal of Mechanical Sciences* **18**, 293–303 (1976).

[5] P.K. Banerjee and R. Butterfield, *Boundary Element Methods in Engineering Science* (McGraw-Hill, New York, 1981).

[6] P.K. Banerjee and R. Butterfield, Boundary Element Methods in Geomechanics, in: *Finite Elements in Geomechanics*, G. Gudehus, Ed. (Wiley, London, 1976) pp. 529–570.

[7] P.K. Banerjee and R. Butterfield, Transient Flow Through Porous Elastic Media, in: *Developments in Boundary Element Methods* – 2, P.K. Banerjee and R.P. Shaw, Eds. (Applied Science Publishers, London, 1982), pp. 19–35.

[8] P.K. Banerjee and D.N. Cathie, A Direct Formulation and Numerical Implementation of the Boundary Element Method for Two-dimensional Problems of Elasto-plasticity, *International Journal of Mechanical Sciences* **22**, 233–245 (1979).

[9] P.K. Banerjee, D.N. Cathie and T.C. Davies, Two and Three Dimensional Problems of Elastoplasticity, in: *Developments in Boundary Element Methods* – 1, P.K. Banerjee and R. Butterfield, Eds. (Applied Science Publishers, London, 1979) pp. 65–95.

[10] P.K. Banerjee, R. Butterfield and G.R. Tomlin, Boundary Element Methods for Two Dimensional Problems of Transient Groundwater Flow, *International Journal for Numerical and Analytical Methods in Geomechanics* **7**, 269–272 (1981).

[11] M.A. Biot, General Theory of Three-Dimensional Consolidation, *Journal of Applied Physics* **12**, 155–164 (1941).
[12] M.A. Biot, Theory of Elasticity and Consolidation for a Porous Anisotropic Solid, *Journal of Applied Physics* **26**, 182–185 (1955).
[13] R.M. Bowen, Green's Functions for Consolidation Problems, *Letters in Applied and Engineering Sciences* **19**, 455–466 (1981).
[14] B.G.H. Brady and J.W. Bray, The Boundary Element Method for Determining Stresses and Displacements Around Long Openings in a Triaxial Stress Field, *International Journal for Rock Mechanics and Mining Sciences and Geomechanics Abstracts* **15**, 21–28 (1978).
[15] C.A. Brebbia, Ed., *Progress in Boundary Element Methods*, Vol. 1 (Pentech Press, London, 1981).
[16] C.A. Brebbia, *The Boundary Element Method for Engineers* (Pentech Press, London, 1978).
[17] C.A. Brebbia and J. Dominguez, Boundary Element Methods for Potential Problems, *Applied Mathematical Modelling* **7**, 372–378 (1977).
[18] C.A. Brebbia and L.C. Wrobel, Steady and Unsteady Potential Problems using the Boundary Element Method, in: *Recent Advances in Numerical Methods in Fluids*, C. Taylor and K. Morgan, Eds. (Pineridge Press, Swansea, 1979) pp. 1–25.
[19] C.A. Brebbia, J.C.F. Telles and L.C. Wrobel, *Boundary Element Techniques* (Springer-Verlag, Berlin, 1984).
[20] R. Burridge and C.A. Vargas, The Fundamental Solution in Dynamic Poroelasticity, *Geophysical Journal of the Royal Astronomical Society* **58**, 61–90 (1979).
[21] H.S. Carslaw and J.C. Jaeger, *Conduction of Heat in Solids*, 2nd Edition (Clarendon Press, Oxford, 1959).
[22] Y.P. Chang, C.S. Kang and D.J. Chen, The Use of Fundamental Green's Functions for the Solution of Problems of Heat Conduction in Anisotropic Media, *International Journal of Heat and Mass Transfer* **16**, 1905–1918 (1973).
[23] A. Cheng and J.A. Liggett, Boundary Integral Equation Method for Linear Porous-Elasticity with Applications to Soil Consolidation, *International Journal for Numerical Methods in Engineering* **20**, 255–278 (1984).
[24] M.P. Cleary, Fundamental Solutions for a Fluid-Saturated Porous Solid, *International Journal of Solids and Structures* **13**, 785–806 (1977).
[25] M.P. Cleary, *Fundamental Solutions for Fluid-Saturated Porous Media and Application to Local Rupture Phenomena*, Ph.D. Thesis, Brown University, Providence, Rhode Island (1976).
[26] M.P. Cleary and J.R. Rice, Some Basic Stress Diffusion Solutions for Fluid-Saturated Elastic Porous Media with Compressible Constituents, *Reviews of Geophysics and Space Physics* **14**, 227–241 (1976).
[27] G.R. Cowper, Gaussian Quadrature Formulas for Triangles, *International Journal for Numerical Methods in Engineering* **7**, 405–408 (1973).
[28] S.L. Crouch, Solution of Plane Elasticity Problems by Displacement Discontinuity Method, *International Journal for Numerical Methods in Engineering* **10**, 301–343 (1976).
[29] S.L. Crouch and A.M. Starfield, *Boundary Element Methods in Solid Mechanics* (George Allen and Unwin, London, 1983).
[30] T.A. Cruse and F.J. Rizzo, A Direct Formulation and Numerical Solution of the General Transient Elastodynamic Problem, I, *Journal of Mathematical Analysis and Applications* **22**, 244–259 (1968).
[31] T.A. Cruse, A Direct Formulation and Numerical Solution of the General Transient Elastodynamic Problem, II, *Journal of Mathematical Analysis and Applications* **22**, 341–355 (1968).
[32] T.A. Cruse, Numerical Solutions in Three Dimensional Elastostatics, *International Journal of Solids and Structures* **5**, 1259–1274 (1969).
[33] T.A. Cruse and W. van Buren, Three Dimensional Elastic Stress Analysis of a Fracture Specimen with an Edge Crack, *International Journal of Fracture Mechanics* **17**, 1–15 (1971).
[34] C.W. Cryer, A Comparison of the Three-Dimensional Consolidation Theories of Biot and Terzaghi, *Quarterly Journal of Mechanics and Applied Mathematics* **16**, 401–412 (1963).
[35] D.A.S. Curran, M. Cross and B.A. Lewis, A Preliminary Analysis of Boundary Element Methods Applied to Parabolic Partial Differential Equations, in: *New Developments in Boundary Element Methods*, C.A. Brebbia, Ed. (C.M.L. Publications, Southampton, 1980) pp. 179–190.

[36] D.A.S. Curran, M. Cross and B.A. Lewis, Solution of Parabolic Diferential Equations by the Boundary Element Method using Discretization in Time, *Applied Mathematical Modelling* **4**, 398–401 (1980).
[37] G. de Josselin DeJong, Consolodatie in Drie Dimensies, *Laboratorium voor Grondmechanica – Mededelingen* **7**, 57–73 (1963).
[38] W. Derski, Theorem on Reciprocity of Displacements in The Theory of Consolidation, *Bulletin de l'Academie Polonaise des Sciences, Serie des Sciences Techniques* **13**, 29–36 (1965).
[39] F. Durbin, Numerical Inversion of Laplace Transform: An Efficient Improvement to Dubner and Abate's Method, *Computer Journal* **17**, 371–376 (1974).
[40] J.L.M. Fernandes and H.L.G. Pina, Unsteady Heat Conduction Using the Boundary Element Method, in: *Boundary Element Methods in Engineering*, C.A. Brebbia, Ed. (Springer-Verlag, Berlin, 1982) pp. 156–170.
[41] M.E. Gurtin, Variational Principles for Linear Initial Value Problems, *Quarterly of Applied Mathematics* **22**, 252–256 (1964).
[42] P.C. Hammer, O.J. Marlowe and A.H. Stroud, Numerical Integration over Simplexes and Cones, *Mathematics of Computation* **10**, 130–137 (1956).
[43] I. Herrera, Trefftz Method, in: *Topics in Boundary Element Research*, Vol. 1, C.A. Brebbia, Ed. (Springer-Verlag, Berlin, 1984) pp. 225–253.
[44] M.A. Jaswon, Integral Equation Methods in Potential Theory I, *Proceedings of the Royal Society* **A275**, 23–32 (1963).
[45] M. Koizumi, M. Utamura and K. Kotani, Application of BEM to Unsteady 3-Dimensional Heat Conduction Problems with Non-Linear Boundary Conditions, in: *Boundary Elements*, C.A. Brebbia, T. Futagami and M. Tanaka, Eds. (Springer-Verlag, Berlin, 1983) pp. 153–162.
[46] V. Kumar and S. Mukherjee, A Boundary Integral Equation Formulation for Time Dependent Inelastic Deformation in Metals, *International Journal of Mechanical Sciences* **19**, 713–724 (1977).
[47] T. Kuroki, T. Ito and K. Onishi, Boundary Element Method in Biot's Linear Consolidation, *Applied Mathematical Modelling* **6**, 105–110 (1982).
[48] T. Kuroki, Y. Ohura, K. Obata and K. Onishi, Boundary Element Solutions of Heat Equation with Singularities and Nonlinearities, in: *Theoretical and Applied Mechanics*, Vol. 31 (University of Tokyo Press, Tokyo, 1982) pp. 39–48.
[49] J.C. Lachat and J.O. Watson, Effective Numerical Treatment of Boundary Integral Equations: A Formulation for Three Dimensional Elastostatics, *International Journal for Numerical Methods in Engineering* **10**, 991–1005 (1976).
[50] J.A. Ligget and P.L.-F. Liu, *The Boundary Integral Equation Method for Porous Media Flow* (George Allen and Unwin, London, 1983)
[51] W.D. MacMillan, *The Theory of Potential* (McGraw-Hill, New York, 1930).
[52] J. McNamee and R.E. Gibson, Displacement Functions and Linear Transforms Applied to Diffusion Through Porous Elastic Media, *Quarterly Journal of Mechanics and Applied Mathematics* **13**, 98–111 (1960).
[53] G.V. Narayanan and D.E. Beskos, Numerical Operational Methods for Time-Dependent Linear Problems, *International Journal for Numerical Methods in Engineering* **18**, 1829–1854 (1982).
[54] D.H. Norrie and G. de Vries, *The Finite Element Method* (Academic Press, New York, 1973).
[55] W. Nowacki, *Dynamic Problems of Thermoelasticity* (Noordhoff International Publishing, Leyden, The Netherlands, 1975).
[56] W. Nowacki, *Thermoelasticity* (Addison-Wesley, London, 1962).
[57] K. Onishi and T. Kuroki, Boundary Element Method in Singular and Nonlinear Heat Transfer, in: *Boundary Element Methods in Engineering*, C.A. Brebbia, Ed. (Springer-Verlag, Berlin, 1982) pp. 141–155.
[58] F. Paris, A. Martin and E. Alarcon, Potential Theory, in: *Progress in Boundary Element Methods*, Vol. 1, C.A. Brebbia, Ed. (John Wiley and Sons, London, 1981) pp. 45–83.
[59] S. Paul, On the Displacements Produced in a Porous Elastic Half-Space by an Impulsive Line Load (Non-dissipative Case), *Pure and Applied Geophysics* **114**, 605–614 (1976).
[60] S. Paul, On the Disturbance Produced in a Semi-Infinite Poroelastic Medium by a Surface Load, *Pure and Applied Geophysics* **114**, 615–627 (1976).

[61] H.L.G. Pina and J.L.M. Fernandez, Three Dimensional Transient Heat Conduction by the Boundary Element Method, in: *Boundary Elements*, C.A. Brebbia, T. Futagami and M. Tanaka, Eds. (Springer-Verlag, Berlin, 1983) pp. 135–142.
[62] M. Predeleanu, On a Boundary Solution Approach for the Dynamic Problem of Thermoviscoelasticity Theory, in: *Numerical Methods in Heat Transfer*, R.W. Lewis, K. Morgan and O.C. Zienkiewicz, Eds. (John Wiley and Sons, Chichester, 1981) pp. 135–150.
[63] M. Predeleanu, Boundary Integral Method for Porous Media, in: *Boundary Element Methods*, C.A. Brebbia, Ed. (Springer-Verlag, Berlin, 1981) pp. 325–334.
[64] F.J. Rizzo, An Integral Equation Approach to Boundary Value Problems of Classical Elastostatics, *Quarterly of Applied Mathematics* **25**, 83–95 (1967).
[65] F.J. Rizzo and D.J. Shippy, A Formulation and Solution Procedure for the General Non-Homogeneous Elastic Inclusion Problem, *International Journal of Solids and Structures* **4**, 1161–1179 (1968).
[66] F.J. Rizzo and D.J. Shippy, A Method of Solution for Certain Problems of Transient Heat Conduction, *AIAA Journal* **8**, 2004–2009 (1970).
[67] F.J. Rizzo and D.J. Shippy, An Advanced Boundary Integral Equation Method for Three Dimensional Thermoelasticity, *International Journal for Numerical Methods in Engineering* **11**, 1753–1768 (1977).
[68] F.J. Rizzo and D.J. Shippy, An Application of the Correspondence Principle of Linear Viscoelasticity Theory, *SIAM Journal on Applied Mathematics* **21**, 321–330 (1972).
[69] V. Roures and E. Alarcon, Transient Heat Conduction Problems Using B.I.E.M., *Computers and Structures* **16**, 717–730 (1983).
[70] R.A. Schapery, Approximate Methods for Transform Inversion for Viscoelastic Stress Analysis, in: *Proceedings for the 4th US National Congress of Applied Mechanics*, R.M. Rosenberg, Ed. (ASME, New York, 1962) pp. 1075–1085.
[71] S. Sharp, A Note on Indirect Boundary Element Methods for the Diffusion Equation, *International Journal for Numerical and Analytical Methods in Geomechanics* **7**, 269–272 (1983).
[72] S. Sharp, A Stability Analysis for Boundary Element Methods for the Diffusion Equation, *Applied Mathematical Modelling* **10**, 41–48 (1986).
[73] S. Sharp, A Condition for Simplifying the Forcing Term in Boundary Element Solutions of the Diffusion Equation, *Communications in Applied Numerical Methods* **1**, 67–70 (1985).
[74] S. Sharp, *Boundary Element Methods in Quasistatic Thermoelasticity with Applications in Rock Mechanics*, Ph.D. Thesis, University of Minnesota, Minneapolis, Minnesota (1982).
[75] S. Sharp and S.L. Crouch, Boundary Integral Methods for Thermoelasticity Problems, *Journal of Applied Mechanics* **53**, 298–302 (1986).
[76] R.P. Shaw, An Integral Equation Approach to Diffusion, *International Journal of Heat and Mass Transfer* **17**, 693–699 (1974).
[77] V. Sladek and J. Sladek, Boundary Integral Equation Method in Thermoelasticity Part I: General Analysis, *Applied Mathematical Modeling* **7**, 241–253 (1983).
[78] J. Sladek and V. Sladek, Boundary Integral Equation Methods in Thermoelasticity: Part II; Crack Analysis, *Applied Mathematical Modelling* **8**, 27–36 (1984).
[79] I. Stakgold, *Boundary Value Problems of Mathematical Physics*, Vol. II (MacMillan, New York, 1968).
[80] G.T. Symm, Integral Equation Methods in Potential Theory: II, *Proceedings of the Royal Society* **A275**, 33–46 (1963).
[81] K. Tanaka and M. Tanaka, Time Space Boundary Elements for Transient Heat Conduction Problems, *Applied Mathematical Modelling* **4**, pp. 331–334 (1980).
[82] J.C.F. Telles and C.A. Brebbia, The Boundary Element Method in Plasticity, *Applied Mathematical Modelling* **5**, 275–281 (1981).
[83] K. Terzaghi, Die Berechnung des Durchlassigkeitsziffer des Tones aus dem Verlauf der Hydrodynamischen Spannungserscheinungen, *Sitzungen der Akademie der Wissenschaften, Wien*, IIa, **132**, 125–138 (1923).
[84] E. Trefftz, Ein Gegenstück zum Ritzschen Verfahren, in: *Proceedings of the 2nd International Congress of Applied Mechanics*, E. Meissner, Ed. (Orell Füssli Verlag, Zürich, 1927) pp. 131–137.

[85] I. Vardoulakis and T. Harnpattanapanich, Numerical Laplace–Fourier Transform Inversion Technique for Layered-Soil Consolidation Problems: I. Fundamental Solutions and Validation, *International Journal for Numerical and Analytical Methods in Geomechanics* **10**, 347–365 (1986).
[86] A. Veruijt, Elastic Storage of Aquifers, in: *Flow Through Porous Media*, R.J.M. de Wiest, Ed. (Academic Press, New York, 1969) pp. 331–376.
[87] L.C. Wrobel and C.A. Brebbia, The Boundary Element Method for Steady State and Transient Heat Conduction, in: *Numerical Methods in Thermal Problems*, R.W. Lewis and K. Morgan, Eds. (Pineridge Press, Swansea, UK, 1979) pp. 58–73.
[88] L.C. Wrobel and C.A. Brebbia, Boundary Elements in Thermal Problems, in: *Numerical Methods in Heat Transfer*, R.W. Lewis, K. Morgan and O.C. Zienkiewicz, Eds. (John Wiley and Sons, Chichester, 1981) pp. 91–113.
[89] O.C. Zienkiewicz, *The Finite Element Method in Engineering Science*, 2nd Edition (McGraw-Hill, London, 1971).

CHAPTER 11

Dynamic Soil–Structure Interaction

Dimitris L. KARABALIS

Department of Civil Engineering
University of South Carolina
Columbia, South Carolina 29208, U.S.A.

and

Dimitri E. BESKOS

Department of Civil Engineering
University of Patras
Patras 26110, Greece

Boundary Element Methods in Mechanics
Edited by D.E. Beskos

Introduction

In the last twenty years one of the most important advances in the development of efficient procedures enabling a better understanding and evaluation of the dynamic behavior of structure–soil systems is undoubtedly the appearance and evolution of the Boundary Element Method (BEM). In this chapter the application of the BEM to the dynamic analysis of foundations and piles is addressed. Both the frequency and the time domain formulations of BEM are reviewed on the basis of the available literature and an attempt is made to present the subject under consideration in a systematic, tutorial manner. However, a detailed reference to the extensive body of literature dealing with the problem of soil–structure interaction (SSI) from the BEM point of view, is beyond the scope of this article. Therefore, only the major points of the BEM formulation of the SSI problem are discussed along with some selected examples that serve the dual purpose of (i) illustrating the applicability and the effectiveness of the method, and (ii) comparing the two BEM methodologies, i.e. the frequency and the time domain. Furthermore, possible directions for future research are also identified.

The popularity that the BEM enjoys lately among the researchers and practitioners in the general SSI area comes as a result of some advantages that this method presents over its main competitors, the Finite Difference Method (FDM) and, in particular, the Finite Element Method (FEM). Since in common foundation problems one is interested in finding the displacements and the stresses at the contact surface between the soil and the foundation only, use of the BEM, along with the appropriate fundamental solutions, requires a discretization of this contact surface only. As a result, the dimensions of the problem are reduced by one, a minimum amount of surface discretization is used, true three-dimensional (3D) analysis becomes feasible, and most importantly, the radiation condition is effortlessly taken into account since it is automatically contained in the fundamental solutions. These are distinct advantages over the FEM and FDM which require a discretization of not only the boundary but of the interior of the domain of interest as well. Moreover, in order to represent the semi-infinite soil medium with a finite size model both the FEM and the FDM introduce aritificial wave-reflecting boundaries and, thus, resort must be made to either extensive volume discretization of the soil medium underneath the foundation, or to special devices, e.g. non-reflecting boundaries, infinite elements, etc., in an effort to account for the radiation condition. None of these improvements succeeds in completely eliminating the problem, and in fact they complicate the application of the FEM and the FDM and make them rather uneconomical. However, in comparing the BEM with the FEM, along these general lines, one should also mention some of the advantages that contribute to the FEM's popularity, such as the straightforward handling of soil inhomogeneities, the well demonstrated potential for nonlinear analysis with regard to the soil and/or the superstructure, and of course the availability of a wide range of software for various applications.

In contrast to the FEM and FDM which usually evaluate the response of the soil–structure system in one direct stage, the BEM is generally used in conjunction with a two-stage solution also known as the substructure approach. In the first stage of the substructure approach, the superstructure, the foundation (on piles or not), and the soil medium are analyzed independently. In the second stage, the various substructures are combined by enforcing the conditions of compatibility and equilibrium at their common interfaces and thus the response of the complete system can be obtained. The main advantages of the substructure approach are that: (i) it provides the analyst with the choice of using the most suitable method of analysis for each substructure, and (ii) design modifications of the system under consideration can usually be handled at the substructure level of analysis. When the BEM is used for SSI problems its role is almost always defined in simulating the infinite part of the soil medium (exterior domain) while some other analytical or numerical method is usually used for the representation of the remaining substructures of finite size (interior domain), i.e. foundation, superstructure, backfill, etc.

The following section of this article concentrates, in the context of a substructure formulation, on the theoretical formulation and the numerical implementation of the BEM in simulating a 3D linear elastic, homogeneous and isotropic half-space. Reference will also be made to the formulation of the analogous 2D SSI problem. The representation of the foundation substructure is undertaken in the second section where both surface and embedded foundations, rigid or flexible, are studied. The discussion of the coupling of the two substructures, i.e. the soil and the foundation, takes place in the third section in which the two possible cases of dynamic excitation, i.e. external loads and seismic waves, are examined. Numerical results for selected foundation problems in both the time and the frequency domains are presented in the fourth section. Reference to some topics of special interest is made in the fifth section, where the subjects of viscoelastic, anisotropic and layered half-spaces are discussed, along with those of piles and vibration isolation. At the closure of the article some general conclusions are drawn.

1. Soil substructure – BEM representation

This section deals with the integral representation of the elastodynamic problem for the half-space and its numerical treatment. A brief description of the necessary analytical formulation of the problem is given on the basis of the book by Eringen and Suhubi [26] for the time domain, and of the work of Cruse and Rizzo [15] for the frequency domain. The emphasis, however, is on the numerical treatment of the established integral representations, with guidelines being the works of the authors [46–52] for the time domain, and Cruse [16] and Manolis and Beskos [69] for the frequency domain. The standard indicial notation has been adopted where summation is assumed over repeated indices, commas indicate spatial differentiation, dots indicate differentiation with respect to time, Latin subscripts take the values 1 to 3, and Greek subscripts take the values 1 and 2 unless otherwise indicated.

1.1. Field equations

The governing equations of a homogeneous, isotropic, and linear elastic solid occupying a regular region of volume R and surface B can be written, in terms of the displacements, as

$$(c_1^2 - c_2^2)u_{i,ij} + c_2^2 u_{j,ii} + f_j = \ddot{u}_j\,, \tag{1.1}$$

where c_1, c_2 are the dilatational and sheer wave velocities, respectively, f is the body force vector per unit mass, and

$$\boldsymbol{u} = \boldsymbol{u}(\boldsymbol{x}, t)\,, \tag{1.2}$$

with $\boldsymbol{x}$ standing for the position vector and t for time. The following conditions should be satisfied along the boundary B:

$$\begin{aligned} t_{(n)i} &= t_{ij}n_j = p_i(\boldsymbol{x}, t)\,, \qquad && \boldsymbol{x} \in B_t \\ u_i &= q_i(\boldsymbol{x}, t)\,, && \boldsymbol{x} \in B_u\,, \end{aligned} \tag{1.3}$$

where $\boldsymbol{n}$ is the unit normal vector on a differential element of the surface B, $t_{(n)i}$ and t_{ij} are the traction vector and tensor, respectively, and $B_t \cup B_u = B$. For the purposes of this article, zero initial conditons will be assumed as it is usually the case in foundation problems, i.e.

$$u_i(\boldsymbol{x}, 0^+) = \dot{u}_i(\boldsymbol{x}, 0^+) = 0\,, \qquad \boldsymbol{x} \in R \cup B\,. \tag{1.4}$$

Since the soil medium is assumed to be linear elastic, the following constitutive equation holds:

$$t_{ij} = \rho(c_1^2 - 2c_2^2)u_{m,m}\delta_{ij} + \rho c_2^2(u_{i,j} + u_{j,i})\,, \tag{1.5}$$

where ρ is the mass density, and δ_{ij} is the Kronecker delta.

Equations (1.1)–(1.5) represent the 3D time domain elastodynamic problem as it is usually formulated in the SSI area. However, the above problem is most frequently treated in its frequency domain form where the time variable is eliminated and a static-like problem emerges. Both the Fourier and the Laplace transformations can serve in this capacity.

The Fourier transform of a function $f(\boldsymbol{x}, t)$ with respect to time t is defined as

$$\bar{f}(\boldsymbol{x}, w) = \mathcal{F}\{f(\boldsymbol{x}, t)\} = \int_{-\infty}^{\infty} f(\boldsymbol{x}, t)\,\mathrm{e}^{-\mathrm{i}wt}\,\mathrm{d}t \tag{1.6}$$

where w is the Fourier transform parameter, i.e. the frequency variable. A Fourier transformation of the governing equations of motion (1.1) results in

$$(c_1^2 - c_2^2)\bar{u}_{i,ij} + c_2^2\bar{u}_{j,ii} + \bar{f}_j = -w^2\bar{u}_j\,, \tag{1.7}$$

and similarly the boundary conditions, Eq. (1.3), and the constitutive relations, Eq. (1.5), become, respectively,

$$\begin{aligned} \bar{t}_{(n)i}(\boldsymbol{x}, w) &= \bar{t}_{ij}n_j = \bar{p}_i(\boldsymbol{x}, w)\,, \qquad && \boldsymbol{x} \in B_t \\ \bar{u}_i(\boldsymbol{x}, w) &= \bar{q}_i(\boldsymbol{x}, w)\,, && \boldsymbol{x} \in B_u\,, \end{aligned} \tag{1.8}$$

$$\bar{t}_{ij} = \rho(c_1^2 - 2c_2^2)\bar{u}_{m,m}\delta_{ij} + \rho c_2^2(\bar{u}_{i,j} + \bar{u}_{j,i})\,. \tag{1.9}$$

The Laplace transform of a function $f(\boldsymbol{x}, t)$ with respect to time t is defined as

$$\bar{f}(\boldsymbol{x}, s) - \mathcal{L}\{f(\boldsymbol{x}, t)\} = \int_{0}^{\infty} f(\boldsymbol{x}, t)\,\mathrm{e}^{-st}\,\mathrm{d}t \tag{1.10}$$

where s is the Laplace transform parameter. Under the quiescent initial conditions defined previously, the Laplace transform of Eq. (1.1) will be identical to Eq. (1.7) if the Laplace parameter s is substituted instead of $-\mathrm{i}w$ in Eq. (1.7). Therefore, the integral representation of the solution of the above stated elastodynamic problem in the Fourier transform domain is identical in form to that in the Laplace transform domain. However, there are differences between these two techniques in their inverse transform procedures, i.e. from the transform domain back to the time domain.

1.2. Integral representations – time domain

Towards an integral representation of the solution of Eq. (1.1), one should first specify the fundamental solutions that will be utilized. As it will be shown later there are more than one choices of fundamental solutions that could be used. In this formulation the fundamental singular solution of Eq. (1.1) due to a concentrated body force of time dependent magnitude acting in an infinite solid region is employed. Such a force can be represented by

$$\rho f(\boldsymbol{x}, t) = f(t)\delta(\boldsymbol{x} - \boldsymbol{\xi})\boldsymbol{e}\,, \tag{1.11}$$

where $\boldsymbol{x}$ and $\boldsymbol{\xi}$ are points in the infinite medium, $\delta(z)$ is the Dirac delta function, $\boldsymbol{e}$ is the direction of application of the above force, and $f(t)$ is its time magnitude. Substitution of Eq. (1.11) as a body force in Eq. (1.1) yields the response of an infinite solid medium in the form

$$u_i = u_{ij}(\boldsymbol{x}, t; \boldsymbol{\xi} \mid f)e_j \tag{1.12}$$

where the second-order displacement tensor u_{ij}, usually called the Stokes displacement tensor or the fundamental singular solution of the elastodynamic equations, is given, for any receiver point $\boldsymbol{x}$ except the singular source point $\boldsymbol{\xi}$, by

$$u_{ij}(\boldsymbol{x}, t; \boldsymbol{\xi} \mid f) = \frac{1}{4\pi\rho}\left\{\left(\frac{3r_ir_j}{r^3} - \frac{\delta_{ij}}{r}\right)\int_{c_1^{-1}}^{c_2^{-1}} \lambda f(t - \lambda r)\,\mathrm{d}\lambda + \frac{r_ir_j}{r^3}\left[\frac{1}{c_1^2} f\left(t - \frac{r}{c_1}\right) - \frac{1}{c_2^2} f\left(t - \frac{r}{c_2}\right)\right] + \frac{\delta_{ij}}{rc_2^2} f\left(t - \frac{r}{c_2}\right)\right\}, \tag{1.13}$$

where

$$r_i = x_i - \xi_i\,, \qquad r^2 = (x_i - \xi_i)(x_i - \xi_i)\,.$$

The physical meaning of the Stokes displacement tensor consists of expressing the displacement component in the i-direction at the point $\boldsymbol{x}$ and time t due to a concentrated force of magnitude $f(t)$ acting at the point $\boldsymbol{\xi}$ in the j-direction. The corresponding Stokes stress tensor can be obtained by direct substitution of Eq. (1.12) in the constitutive Eq. (1.5), and is expressed, following the above notation, as

$$\begin{aligned} t_{ijk} &= t_{ijk}(\boldsymbol{x}, t; \boldsymbol{\xi} \mid f) \\ &= \frac{1}{4\pi}\left\{-6c_2^2\left(5\frac{r_ir_jr_k}{r^5} - \frac{\delta_{ij}r_k + \delta_{ik}r_j + \delta_{jk}r_i}{r^3}\right)\int_{c_1^{-1}}^{c_2^{-1}} \lambda f(t - \lambda r)\,\mathrm{d}\lambda \right. \\ &\quad + 2\left(6\frac{r_ir_jr_k}{r^5} - \frac{\delta_{ij}r_k + \delta_{ik}r_j + \delta_{jk}r_i}{r^3}\right)\left[f\left(t - \frac{r}{c_2}\right) - \frac{c_2^2}{c_1^2} f\left(t - \frac{r}{c_1}\right)\right] \end{aligned}$$

$$+2\frac{r_i r_j r_k}{r^4 c_2}\left[\dot{f}\left(t-\frac{r}{c_2}\right)-\frac{c_2^3}{c_1^3}\dot{f}\left(t-\frac{r}{c_1}\right)\right]$$

$$-\frac{r_k\delta_{ij}}{r^3}\left(1-2\frac{c_2^2}{c_1^2}\right)\left[f\left(t-\frac{r}{c_1}\right)+\frac{r}{c_1}\dot{f}\left(t-\frac{r}{c_1}\right)\right]$$

$$-\frac{\delta_{ik}r_j+\delta_{jk}r_i}{r^3}\left[f\left(t-\frac{r}{c_2}\right)+\frac{r}{c_2}\dot{f}\left(t-\frac{r}{c_2}\right)\right]\Bigg\}. \quad (1.14)$$

The pair of fundamental singular solutions $[u_{ij}, t_{ijk}]$ possesses the properties of causality, translation, and symmetry (only u_{ij}), and is called the Stokes state of quiescent past [26].

Utilizing the Stokes state of quiescent past as one of the two distinct elastodynamic states required in Betti's reciprocal theorem, the other one being the actual state, and under the assumption of zero body forces, one can derive a solution for the problem posed by Eqs. (1.1), (1.3) and (1.4) in the form of an integral representation, also known as Love's integral identity, which reads

$$\beta(\boldsymbol{\xi})u_k(\boldsymbol{\xi}, t) = \int_B \{u_{ik}[\boldsymbol{x}, t; \boldsymbol{\xi} \mid t_{(n)i}(\boldsymbol{x}, t)]$$

$$- t_{(n)ik}[\boldsymbol{x}, t; \boldsymbol{\xi} \mid u_i(\boldsymbol{x}, t)]\}\, \mathrm{d}B(\boldsymbol{x}) , \quad (1.15)$$

where

$$t_{(n)ik} = t_{ijk}n_j \quad (1.16)$$

is the traction tensor of the Stokes state, and expressions such as $u_{ik}[\boldsymbol{x}, t; \boldsymbol{\xi} \mid t_{(n)i}(\boldsymbol{x}, t)]$ mean that $f(t)$ in Eq. (1.12) has been replaced by $t_{(n)i}(\boldsymbol{x}, t)$. The function $\beta(\boldsymbol{\xi})$ takes the values: (i) 1 for points $\boldsymbol{\xi}$ inside the region R; (ii) 0 for points $\boldsymbol{\xi}$ outside the region R; and (iii) it is a function of the geometry of the boundary B in the vicinity of $\boldsymbol{\xi}$ in the case that ξ is on the boundary B [10, 39, 101]. In the special, but most frequently met, case of a smooth boundary B, i.e. with a unique tangent at $\boldsymbol{\xi}$, the function $\beta(\boldsymbol{\xi})$ takes the value 0.5. Since the major part of this work deals with simple plane boundaries, e.g. the surface of a half-space, for the remainder of this work the function $\beta(\boldsymbol{\xi})$ will be defined, unless otherwise indicated, as follows:

$$\beta(\boldsymbol{\xi}) = \begin{cases} 1 & \text{if } \boldsymbol{\xi} \in R \\ 0.5 & \text{if } \boldsymbol{\xi} \in B \\ 0 & \text{if } \boldsymbol{\xi} \notin R \cup B\,. \end{cases} \quad (1.17)$$

For receiver points $\boldsymbol{\xi}$ on the boundary B Eq. (1.15) relates boundary displacements and tractions and it constitutes the time domain boundary integral equation (BIE) that will be used in this formulation.

It is obvious from the above formulation that when Eq. (1.15) is used to represent a semi-infinite solid, i.e. the soil medium, one should consider as boundary B the entire free surface of the half-space. This is so because of the specific choice of fundamental singular solutions used in Eq. (1.15) which correspond to an infinite region and thus do not take into account the particular boundary conditions of the problem under consideration. There are, however, other choices of fundamental solutions which correspond to a semi-infinite solid medium. The most general of these solutions have been obtained in analytical form by Johnson [45]. His fundamental solutions of Eq. (1.1) are due to a body force expressed as

$$f(\boldsymbol{x}, t) = \delta(t)\delta(\boldsymbol{x} - \boldsymbol{\xi})\boldsymbol{e} , \tag{1.18}$$

and boundary conditions that correspond to a semi-infinite solid medium, i.e. a traction-free surface. Use of these fundamental solutions in foundation analysis problems reduces the size of the boundary B that needs to be considered to only the contact surface between the soil and the foundation, and thus Eq. (1.15) becomes

$$\beta(\boldsymbol{\xi})u_k(\boldsymbol{\xi}, t) = \int_B U_{ik}(\boldsymbol{x}, t; \boldsymbol{\xi}) * t_{(\boldsymbol{n})i}(\boldsymbol{x}, t)\, \mathrm{d}B(\boldsymbol{x}) \tag{1.19}$$

where U_{ik} is the fundamental displacement tensor (Green's function) for a semi-infinite solid medium, and the symbol $*$ is used to denote the convolution operation.

Utilization of half-space fundamental solutions in foundation problems, minimizes the extent of the domain of integration and appears, in principle, computationally advantageous over infinite space solutions. However, half-space fundamental solutions are not as popular because of their complexity, and instead, infinite space fundamental solutions are used in almost all cases. It will be shown later that the use of infinite space fundamental solutions along with a truncated part of the infinite boundary of the half-space, which is limited to a small area beneath and around the foundation, does not affect substantially the accuracy of the solution of the problem under consideration. Moreover, for surface foundations and under specific assumptions with regard to the soil–foundation contact conditions, only the contact surface between the soil and the foundation needs to be considered in the numerical solution of the soil–foundation interaction problem.

1.3. Integral representations – frequency domain

As in the time domain case, before proceeding with the integral representation solution of Eq. (1.7) one should first specify the fundamental solutions that will be used. Among a number of possible selections, the infinite space fundamental solutions due to a concentrated body force will be the most obvious choice again because of their simplicity. In this case the concentrated body force can be expressed as

$$\rho \bar{f}_i(\boldsymbol{x}, w) = \bar{f}(w)\delta(\boldsymbol{x} - \boldsymbol{\xi})e_i \, , \tag{1.20}$$

and upon substitution of which in the governing Eq. (1.7) one can obtain a frequency domain solution for the displacements in infinite solid region in the form

$$\bar{u}_i(\boldsymbol{x}) = \bar{u}_{ij}(\boldsymbol{x}, \boldsymbol{\xi}, w)e_j \, , \tag{1.21}$$

where the fundamental displacement tensor $\bar{u}_{ij}$ can be expressed as [15]

$$\bar{u}_{ij} = \frac{\bar{f}(w)}{\alpha \pi \rho c_2^2} [\psi \delta_{ij} - \chi r_{,i} r_{,j}] \, , \tag{1.22}$$

with $\alpha = 4$, and the functions ψ and χ given by

$$\phi = \frac{e^{iwr/c_2}}{r} - \left(\frac{c_2^2}{w^2 r^2} + \frac{c_2}{iwr} \right) \frac{e^{iwr/c_2}}{r} + \frac{c_2^2}{c_1^2} \left(\frac{c_1^2}{w^2 r^2} + \frac{c_1}{iwr} \right) \frac{e^{iwr/c_1}}{r} \tag{1.23}$$

$$\chi = -\left(\frac{3c_2^2}{w^2 r^2} + \frac{3c_2}{iwr} - 1 \right) \frac{e^{iwr/c_2}}{r} + \left(\frac{c_2^2}{c_1^2} \right) \left(\frac{3c_1^2}{w^2 r^2} + \frac{3c_1}{iwr} - 1 \right) \frac{e^{iwr/c_1}}{r} \, . \tag{1.24}$$

Direct insertion of Eq. (1.21) into the constitutive Eq. (1.9) yields the cognate traction tensor $\bar{t}_{ij}(\boldsymbol{x}, \boldsymbol{\xi}, w)$ in the form

$$\bar{t}_{ij} = \frac{\bar{f}(w)}{\alpha \pi} \left[\left(\frac{d\psi}{dr} - \frac{1}{r} \chi \right) \left(\delta_{ij} \frac{\partial r}{\partial n} + r_{,j} n_i \right) - \frac{2}{r} \chi \left(n_j r_{,i} - 2 r_{,i} r_{,j} \frac{\partial r}{\partial n} \right) \right.$$
$$\left. - 2 \frac{d\chi}{dr} r_{,i} r_{,j} \frac{\partial r}{\partial n} + \left(\frac{c_1^2}{c_2^2} - 2 \right) \left(\frac{d\psi}{dr} - \frac{d\chi}{dr} - \frac{\alpha}{2r} \chi \right) r_{,i} n_j \right] . \tag{1.25}$$

On the basis of Betti's reciprocal theorem in the frequency transform domain, where one of the two required dynamic states is the $[\bar{u}_{ij}, \bar{t}_{ij}]$, the other one being the actual state, one can obtain a solution to the problem posed by Eqs. (1.7)–(1.9) in the form of the integral identity

$$\beta(\boldsymbol{\xi})\bar{u}_i(\boldsymbol{\xi}) = \int_B [\bar{u}'_{ij}(\boldsymbol{x}, \boldsymbol{\xi}, w)\bar{t}_{(n)j}(\boldsymbol{x}) - \bar{t}'_{ij}(\boldsymbol{x}, \boldsymbol{\xi}, w)\bar{u}_j(\boldsymbol{x}) \, dB(\boldsymbol{x}) \, , \tag{1.26}$$

where the assumption of zero body forces has been made, the function $\beta(\boldsymbol{\xi})$ is defined as in the time domain case, e.g. for a smooth boundary B, $\beta(\boldsymbol{\xi})$ is given by Eq. (1.17), and

$$\bar{u}'_{ij} = \bar{u}_{ij}/\bar{f}(w) \quad \text{and} \quad \bar{t}'_{ij} = \bar{t}_{ij}/\bar{f}(w)\,. \tag{1.27}$$

Equation (1.26) is the BIE in the Fourier transformed domain and provides the solution of a given dynamic problem in terms of the frequency parameter w, i.e. the response of a linear elastic solid subjected to harmonic dynamic disturbances. The time domain solution due to general dynamic disturbances can be obtained, if required, through a Fourier synthesis of a sequence of frequency dependent solutions.

For reasons similar to those mentioned in the discussion of the time domain integral representations, when the infinite space fundamental solution pair $[\bar{u}_{ij}, \bar{t}_{ij}]$ is used for the representation of a half-space, the domain of integration B in Eq. (1.26) becomes the entire free surface of the semi-infinite solid medium. This can be avoided if half-space fundamental solutions are used which incorporate the necessary boundary conditions. A variety of half-space fundamental solutions exists for the frequency domain case, e.g. Johnson [45], and Kobayashi and Nishimura [58]. In the case of foundation problems, use of these solutions reduces the domain of integration B to only the contact surface between the soil and the foundation. However, approximate solutions of SSI problems can also be obtained, with a high degree of accuracy, if the simpler infinite space fundamental solutions are used along with a truncated boundary comprising the soil–foundation contact surface and a small part of the surrounding free surface.

1.4. The two-dimensional case

There are cases in SSI analysis in which the particular geometry of a problem, e.g. long buildings on mat foundations or machine strip footings, suggests a two-dimensional (2D) approximation of the field quantities, the body forces, and the boundary and initial conditions. Thus, all of the above functions are assumed to depend upon two space variables x_η $(\eta = 1, 2)$, and one needs to consider only the plane $x_3 = 0$. For a plane region D bounded by a closed curve C and subjected to boundary displacements $u_i(x_\gamma, t)$ and boundary tractions $t_{(n)i}(x_\gamma, t)$ an integral representation identity, similar to Eq. (1.15), can be written as (de Hoop [41, 42])

$$\beta(\boldsymbol{\xi})u_k(\xi_\alpha, t) = \int_C \{v_{ik}[x_\alpha, t;\, \xi_\beta \mid t_{(n)i}(x_\gamma, t)] - \sigma_{(n)ik}[x_\alpha, t;\, \xi_\beta \mid u_i(x_\gamma, t)]\}\, \mathrm{d}C(x_\delta) \tag{1.28}$$

where zero body forces and initial conditions have been assumed and the functional operators v_{ik} and $\sigma_{(n)ik}$, corresponding to an infinite plane region, can be deduced from their 3D time domain counterparts through a spatial integration along the x_3-axis, i.e.

$$v_{ik}[x_\alpha, t; \xi_\beta \mid f(t)] = \int_{-\infty}^{\infty} u_{ik}[\boldsymbol{x}, t; \xi_\alpha \mid f(t)]\, \mathrm{d}x_3 \qquad \text{on } C \tag{1.29a}$$

$$\sigma_{(n)ik}[x_\alpha, t; \xi_\beta \mid f(t)] = \int_{-\infty}^{\infty} t_{(n)ik}[\boldsymbol{x}, t; \xi_\alpha \mid f(t)]\, \mathrm{d}x_3 \qquad \text{on } C\,. \tag{1.29b}$$

The components of the 2D time domain fundamental displacement vector appearing in Eq. (1.29a) are given by

$$\begin{aligned}
v_{3\alpha} &\equiv 0 \\
v_{33} &= \frac{H(t-(r/c_2))}{2\pi\rho c_2^2} \int_r^{c_2 t} f\left(t - \frac{\eta}{c_2}\right) \frac{\mathrm{d}\eta}{(\eta^2 - r^2)^{1/2}} \\
v_{\alpha\beta} &= \frac{1}{2\pi\rho} \left\{ \frac{\partial^2}{\partial\xi_\alpha \partial\xi_\beta} \left[H\left(t - \frac{r}{c_1}\right) \int_r^{c_1 t} \frac{\mathrm{d}\eta}{(\eta^2 - r^2)^{1/2}} \int_0^{t-(\eta/c_1)} \zeta f\left(t - \frac{\eta}{c_1} - \zeta\right) \mathrm{d}\zeta \right.\right. \\
&\quad \left. - H\left(t - \frac{r}{c_2}\right) \int_r^{c_2 t} \frac{\mathrm{d}\eta}{(\eta^2 - r^2)^{1/2}} \int_0^{t-(\eta/c_2)} \zeta f\left(t - \frac{\eta}{c_2} - \zeta\right) \mathrm{d}\zeta \right] \\
&\quad \left. + \frac{\delta_{\alpha\beta}}{c_2^2} H\left(t - \frac{r}{c_2}\right) \int_r^{c_2 t} f\left(t - \frac{\eta}{c_2}\right) \frac{\mathrm{d}\eta}{(\eta^2 - r^2)^{1/2}} \right\}.
\end{aligned} \tag{1.30}$$

The components of the 2D time domain fundamental stress tensor appearing in Eq. (1.29b) are given by

$$\begin{aligned}
&\sigma_{(n)\alpha\gamma} = \tau_{\alpha\beta\gamma} n_\beta\,, \\
&\sigma_{(n)3\alpha} = \sigma_{(n)\alpha 3} = 0\,, \qquad \sigma_{(n)33} = \tau_{\alpha 33} n_\alpha
\end{aligned} \tag{1.31}$$

where

$$\begin{aligned}
\tau_{\alpha\beta\gamma} &= -\rho(c_1^2 - 2c_2^2) \frac{\partial v_{\delta\gamma}}{\partial \xi_\delta} \delta_{\alpha\beta} - \rho c_2^2 \left(\frac{\partial v_{\alpha\gamma}}{\partial \xi_\beta} + \frac{\partial v_{\beta\gamma}}{\partial \xi_\alpha} \right) \\
\tau_{\alpha\beta 3} &= \tau_{\alpha 3\gamma} = \tau_{333} = 0 \\
\tau_{\alpha 33} &= -\rho c_2^2 \frac{\partial v_{33}}{\partial \xi_\alpha}\,, \qquad \tau_{33\alpha} = -\rho(c_1^2 - 2c_2^2) \frac{\partial v_{\alpha\beta}}{\partial \xi_\beta}\,.
\end{aligned} \tag{1.32}$$

In a manner similar to the one used to develop the 3D BIE in the frequency domain, i.e. Eq. (1.26), one can derive a 2D frequency domain integral representation identity. The necessary fundamental solutions for the displacement and traction tensors are given by Eqs. (1.22) and (1.25), respectively, where the coefficient $a = 2$, the indices i and j assume the values 1 and 2, and the functions ψ and χ are given in terms of the modified Bessel function of the second kind as

$$\psi = K_0\left(\frac{-\mathrm{i}wr}{c_2}\right) - \frac{c_2}{\mathrm{i}wr}\left[K_1\left(\frac{-\mathrm{i}wr}{c_2}\right) - \frac{c_2}{c_1} K_1\left(\frac{-\mathrm{i}wr}{c_1}\right)\right]$$

$$x = K_2\left(\frac{-\mathrm{i}wr}{c_2}\right) - \frac{c_2^2}{c_1^2} K_2\left(\frac{-\mathrm{i}wr}{c_1}\right). \qquad (1.33)$$

Half-plane fundamental solutions, in both the time and the frequency domains, are also available, e.g. Buchen [11] in the time domain, and Dravinski [24] and Kobayashi [59] in the frequency domain.

1.5. Numerical implementation

For the mixed boundary value problems encountered in SSI an analytical treatment of the boundary integral equations established in the previous sections becomes an almost impossible task, given the complicated geometries and time variations of the related functions. In this section some numerical schemes are discussed which are applicable to arbitrary foundation geometries and time variations of the forcing functions. In the static-like frequency domain case only a spatial discretization scheme is required in order to proceed with the numerical solution of the problem under consideration. The time domain case, however, appears to be more involved because in addition to the spatial discretization, a discretization scheme in time is necessary. Thus, the numerical treatment of the 3D time domain problem will be given first, followed by the frequency domain case.

1.5.1. Time domain approach

With reference to the work of the authors [46, 47], a numerical treatment of the integral equation (1.15) consists of the following two basic steps: (i) the real time axis is discretized into a number of equally spaced time intervals over each one of which a constant variation of tractions and displacements is assumed, and (ii) the boundary B is discretized into a number of rectangular elements and a constant distribution of tractions and displacements is also assumed over each of one of these boundary elements. Equation (1.15) can then be cast into a discretized relationship between boundary tractions and displacements over each rectangular element and for each time step.

In the context of the procedure outlined above, consider first the time discretization of the traction vector into a sequence of rectangular impulses as

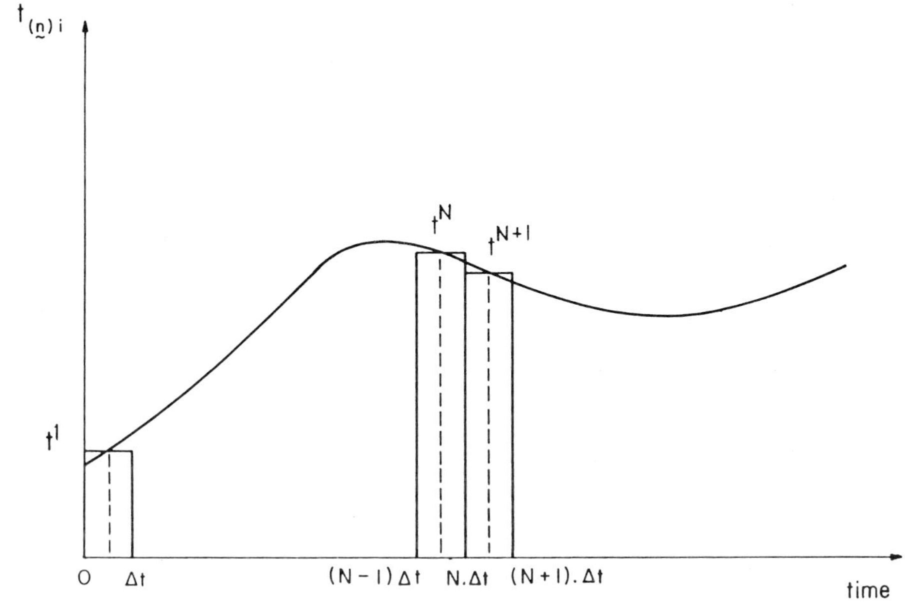

Fig. 1. Approximation of the surface traction $t_{(n)i}$ by a sequence of rectangular pulses.

shown in Fig. 1. This approximation of the continuous traction vector can be expressed, for each time interval q, i.e. $(q-1)\Delta t < t \leq q\Delta t$, as

$$t_{(n)i}(\boldsymbol{x}, t) \approx t^q_{(n)i}(\boldsymbol{x})\{H[t-(q-1)\Delta t] - H(t - q\Delta t)\}\,, \tag{1.34}$$

where $H(t-\tau)$ is the Heaviside step function, and $t^q_{(n)i}(\boldsymbol{x})$ expresses the intensity of the rectangular impulse traction at time $t=(q-0.5)\Delta t$, i.e.

$$t^q_{(n)i}(\boldsymbol{x}) = t_{(n)i}[\boldsymbol{x}, (q-0.5)\Delta t]\,. \tag{1.35}$$

Direct substitution of the forcing function $f(t)$ in Eq. (1.13) by the traction vector approximated as in Eq. (1.34) yields a time discretized expression for the Stokes displacement tensor

$$\begin{aligned} u_{ik}(\boldsymbol{x}, t; \boldsymbol{\xi} \mid t^q_{(n)k}) = \frac{1}{4\pi\rho}\Bigg\{&\left(\frac{3r_ir_k}{r^5} - \frac{\delta_{ik}}{r^3}\right)\left[H\left(t-\frac{r}{c_1}\right)F_1 - H\left(t-\frac{r}{c_2}\right)F_2\right]t^q_{(n)k}(\boldsymbol{\xi}) \\ &+ \frac{1}{c_1^2}\frac{r_ir_k}{r^3}H\left(t-\frac{r}{c_1}\right)t_{(n)k}\left(\boldsymbol{\xi}, t-\frac{r}{c_1}\right) \\ &+ \frac{1}{c_2^2}\left(\frac{\delta_{ik}}{r} - \frac{r_ir_k}{r^3}\right)H\left(t-\frac{r}{c_2}\right)t_{(n)k}\left(\boldsymbol{\xi}, t-\frac{r}{c_2}\right)\Bigg\}\,, \end{aligned} \tag{1.36}$$

where

$$F_\beta = \begin{cases} [t^2 - (r/c_\beta)^2]/2\,, & \text{if } (q-1)\Delta t < t-(r/c_\beta) \leq q\Delta t \\ [2t(\Delta t) - (\Delta t)^2]/2\,, & \text{if } q\Delta t < t-(r/c_\beta)\,,\ \beta = 1, 2\,. \end{cases} \tag{1.37}$$

A time-discretized expression for the Stokes traction tensor can be obtained in a similar way [46], i.e. by first representing the continuous displacement vector $u_i(\boldsymbol{x}, t)$ as a series of rectangular impulses, similar to Eq. (1.34), and then substituting the forcing function $f(t)$ in Eq. (1.14) by the time-discretized displacement vector. In doing so, it is obvious that the time derivatives of the time-retarded displacement function $u_i(\boldsymbol{x}, t)$ appearing in Eq. (1.14) are zero. However, the error due to this approximation can be practically eliminated for sufficiently small time increments.

The spatial discretization scheme that is used in 3D analysis is shown in Fig. 2(a). It consists of a number M of rectangular elements into which the boundary B of Eq. (1.15) has been discretized. If a constant variation of tractions and displacements is assumed over each rectangular element, and in view of Eq. (1.36) and the similar expression for the time-discretized traction tensor, the boundary integral equation (1.15), when written for every boundary element M, will yield a system of $3M$ algebraic equations for each time step N, which in matrix form can be written as

$$0.5\{u_s^{N,R}\} = \sum_{n=q}^{N} \sum_{m=1}^{M} \{[G^{n-q+1,m}]\{t_s^{N-n+q,m}\} - [T^{n-q+1,m}]\{u_s^{N-n+q,m}\}\}$$

$$R = 1, 2, \ldots, M , \qquad (1.38)$$

where $\{u_s^{N,R}\}$ is the displacement vector at time step N, $\{t_s^{k,m}\}$ is the traction vector at time step k, the superscripts R and m indicate, respectively, the "receiver" and the "source" elements, as shown in Fig. 2(b), and the subscript s

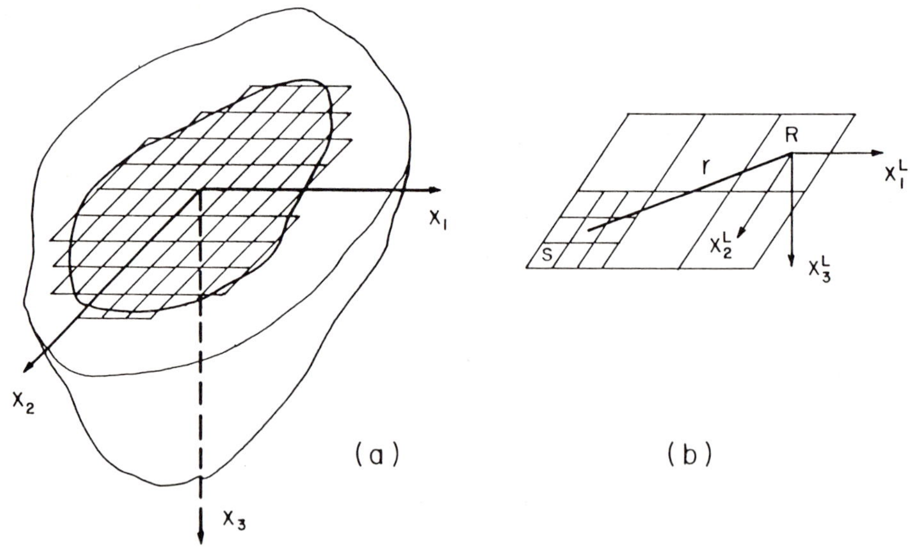

Fig. 2. Geometry and discretization of a 3D surface foundation.

is used to denote the soil medium. The displacement tensor $[G^{n,m}]$ and the traction tensor $[T^{n,m}]$ represent the time- and space-discretized equivalents of the time domain fundamental solutions, Eqs. (1.13) and (1.15), and are given by

$$G_{ik}^{n,m} = \frac{1}{4\pi\rho}\left\{B_c^n \int_{(s)} \frac{1}{r^2}\left(\frac{3r_ir_k}{r^3} - \frac{\delta_{ik}}{r}\right)\left[H\left(t-\frac{r}{c_1}\right)F_1 - H\left(t-\frac{r}{c_2}\right)F_2\right]\mathrm{d}(s)\right.$$

$$\left. + \frac{B_{c_1}^n}{c_1^2}\int_{(s)} \frac{r_ir_k}{r^3}\, d(s) + \frac{B_{c_2}^n}{c_2^2}\int_{(s)}\left(\frac{\delta_{ik}}{r} - \frac{r_ir_k}{r^3}\right)\mathrm{d}(s)\right\}, \tag{1.39}$$

$$T_{ij}^{n,m} = T_{ijk}^{n,m} e_k\,, \tag{1.40}$$

$$T_{ijk}^{n,m} = \frac{1}{4\pi}\left\{B_c^n(-6c_2^2)\int_{(s)}\left(5\frac{r_ir_jr_k}{r^7} - \frac{\delta_{ij}r_k + \delta_{ik}r_j + \delta_{jk}r_i}{r^5}\right)\right.$$

$$\times\left[H\left(t-\frac{r}{c_1}\right)F_1 - H\left(t-\frac{r}{c_2}\right)F_2\right]\mathrm{d}(s)$$

$$- B_{c_1}^n\int_{(s)}\left[2\frac{c_2^2}{c_1^2}\left(6\frac{r_ir_jr_k}{r^5} - \frac{\delta_{ij}r_k + \delta_{ik}r_j + \delta_{jk}r_i}{r^3}\right) + \left(1 - 2\frac{c_2^2}{c_1^2}\right)\frac{r_k\delta_{ij}}{r^3}\right]\mathrm{d}(s)$$

$$\left. + B_{c_2}^n\int_{(s)}\left[2\left(6\frac{r_ir_jr_k}{r^5} - \frac{\delta_{ij}r_k + \delta_{ik}r_j + \delta_{jk}r_i}{r^3}\right) - \frac{\delta_{ik}r_j + \delta_{jk}r_i}{r^3}\right]\mathrm{d}(s)\right\} \tag{1.41}$$

where (s) is the area of the source element m, and r is the distance from the center of the receiver element R to each point of the element m, as shown in Fig. 2(b). The coefficients B_c^n, $B_{c_1}^n$, and $B_{c_2}^n$ are used to specify whether or not the source element m has an influence on the receiver element R during the time step n, and they are related to the dilatational and shear wave velocities by

$$B_c^n, B_{c_\alpha}^n = \begin{cases} 1\,, & \text{if } r \in B_{n,c}, B_{n,c_\alpha} \\ 0\,, & \text{if } r \notin B_{n,c}, B_{n,c_\alpha}\,, \end{cases} \qquad \alpha = 1, 2\,. \tag{1.42}$$

where

$$\begin{aligned} B_{n,c} &= \{\boldsymbol{x};\boldsymbol{x} \in B \text{ and } c_2(n-1)\Delta t < r < c_1 n\Delta t\}\,, \\ B_{n,c_1} &= \{\boldsymbol{x};\boldsymbol{x} \in B \text{ and } c_1(n-1)\Delta t < r \leq c_1 n\Delta t\}\,, \\ B_{n,c_2} &= \{\boldsymbol{x};\boldsymbol{x} \in B \text{ and } c_2(n-1)\Delta t < r \leq c_2 n\Delta t\}\,. \end{aligned} \tag{1.43}$$

The (3×3) displacement tensor $G_{ik}^{n,m}$ defined by Eq. (1.39) represents the displacement component in the i-direction at the center of the element R during the time step n, due to a rectangular impulse traction component in the

k-direction acting over the element m during the first time step. A similar physical meaning is attached to the $T_{ij}^{n,m}$ traction tensor defined by Eq. (1.40).

At the free surface of the half-space, i.e. at $x_3 = 0$, the nine terms $G_{ik}^{n,m}$ of Eq. (1.39) reduce to five since

$$G_{13}^{n,m} \equiv G_{31}^{n,m} \equiv G_{23}^{n,m} \equiv G_{32}^{n,m} \equiv 0 \,. \tag{1.44}$$

Similarly, for any plane boundary, e.g. $x_3 = 0$, only the off-diagonal terms $T_{13}^{n,m}$, $T_{23}^{n,m}$, $T_{31}^{n,m}$, and $T_{32}^{n,m}$, of the (3×3) matrix $[T^{n,m}]$ are different from zero. These terms represent a weak coupling between the horizontal and vertical motions [87, 88]. If one chooses to neglect the influnece of these terms, i.e. relaxed boundary conditions are assumed, then a decoupling of the vertical and the horizontal motions occurs and these two distinct vibration modes can be considered independent of each other. Furthermore, if relaxed boundary conditions are assumed in the study of surface foundations, Eq. (1.38) is reduced to

$$0.5\{u_s^{N,R}\} = \sum_{n=q}^{N} \sum_{m=1}^{M} \{[G^{n-q+1,m}]\{t_s^{N-n+q,m}\} \qquad R = 1, 2, \ldots, M \,. \tag{1.45}$$

Use of Eq. (1.45) requires a discretization of only the contact surface between the soil and the foundation since the tractions on the free surface outside the contact area are zero. However, for the case of embedded foundations Eq. (1.38) must be used, which theoretically requires a discretization over the contact area as well as the entire part of the remaining free surface. For practical purposes, though, only a small zone of the free surface around the embedded foundation has to be discretized because, as it can be seen in Eqs. (1.39) and (1.41), the influence of boundary elements far away from the foundation diminishes rapidly as the distance $\boldsymbol{r}$ increases.

The spatial integrations over the area of each rectangular boundary element required by Eqs. (1.39) and (1.41) can be performed, in general, with the help of any standard numerical integration scheme. In order to accurately take into account the influence of the wave propagation effects, an improved spatial integration arrangement is proposed in Refs. [46, 47], where each "source element" is divided into a number of subelements and the integrations indicated in Eqs. (1.39) and (1.41) are performed over the area (s) of each "source subelement".

Special care should be taken, however, for the case of singular elements, i.e. when $R = m$ in Eqs. (1.38) and (1.45). It is first observed that because of the traveling character of the waves emanated at the singular element m, a singularity can occur only during the time step k, i.e. the first time step, at which the rectangular impulse traction $\{t^{k,m}\}$ is applied on the element m. If R_1 and R_2 are radii travelled by the dilatational and shear waves, respectively, from the center

$(x_1, x_2, x_3) = (0, 0, 0)$ of the singular element m during the time step k, application of Eq. (1.39) over the circular area of radius R_1 yields

$$G_{ik}^{1,m} = \frac{1}{4\pi\rho}\left\{\frac{1}{2}\left(\frac{1}{c_2^2} - \frac{1}{c_1^2}\right)\int_0^{2\pi}\int_0^{R_2}\left(\frac{3r_i r_k}{r^2} - \delta_{ik}\right) dr\, d\theta\right.$$

$$+ \frac{1}{2}\int_0^{2\pi}\int_0^{R_1}\frac{1}{r^2}\left[(\Delta t)^2 - \left(\frac{r}{c_1}\right)^2\right]\left(\frac{3r_i r_k}{r^2} - \delta_{ik}\right) dr\, d\theta$$

$$\left. + \frac{1}{c_1^2}\int_0^{2\pi}\int_0^{R_1}\frac{r_i r_k}{r^2}\, dr\, d\theta + \frac{1}{c_2^2}\int_0^{2\pi}\int_0^{R_2}\left(\delta_{ik} - \frac{r_i r_k}{r^2}\right) dr\, d\theta\right\}, \qquad (1.46)$$

where the Cartesian coordinates r_i are assumed to be substituted by their cylindrical counterparts. Equation (1.46) expresses the uniform displacement of the circular area of radius R_1 at the end of the time interval $[(k-1)\Delta t, k\Delta t]$ due to the rectangular impulse traction $\{t^{k,m}\}$, and it is apparent that presents no singularities with respect to the spatial variables. Furthermore, if plane boundary elements are used and the time increment Δt is chosen to be sufficiently small so that the circular area of radius R_1 lies within the boundaries of the singular element m, all the off-diagonal terms of the tensor $G_{ij}^{1,m}$ of Eq. (1.46) become zero. This last observation simplifies the computation of the inverse of the total influence matrix $[G^1]$ which is involved in subsequent calculations in the following sections. With regard to the singularity of the traction tensor $T_{ij}^{1,m}$ of Eq. (1.41) it is easy to prove, on the basis of the assumptions made above for plane elements and sufficiently small time increments, that for relaxed boundary conditions

$$T_{ijk}^{1,m} \equiv 0\,. \qquad (1.47)$$

A word about the appropriate selection of the time increment Δt is in order. In the work of the authors [46–52] no restriction has been applied with respect to the time increment that one should use when the above-described operations are attempted. However, a constant time increment Δt has been used throughout their work defined by

$$\Delta t = (1/c_1)[(\Delta x \Delta y)/\pi]^{1/2}\,. \qquad (1.48)$$

Equation (1.48) expresses the time required by the dilatational wave emanated at the center of a typical rectangular boundary element, with sides Δx and Δy, to cover a circular area equal to the area of the above element. This selection of the time interval eliminates interelement influences during the first time step at which the rectangular impulse traction $\{t^{k,m}\}$ is applied on the element m and thus the diagonal character of the corresponding influence matrix $[G^1]$, as it was discussed previously, is preserved. However, this observation also holds true for any time interval smaller than the one described by Eq. (1.48). For time intervals greater than the one described by Eq. (1.48) interelement influences should be considered in the computation of the influence matrix $[G^1]$, which then becomes non-diagonal, and some caution should be exercised in applying Eq. (1.46) particularly if adjacent elements do not lie in the same plane.

In view of Eqs. (1.38) and (1.45) a step-by-step time marching solution of a system of $3M$ equations can be established, i.e. first for $N = q$, then for $N = q + 1$, etc., provided that $3M$ boundary conditions are available at each time step. Although it seems that a total number of $2N(3M)^2$ discrete kernels is required when Eq. (1.38) is used, versus $N(3M)^2$ for Eq. (1.45), this number can be substantially reduced if the causality and time translation properties of the discretized fundamental solutions are properly taken into account. Furthermore, it can be shown [46] that for practical problems involving a boundary element mesh of finite dimensions, the maximum number of discrete kernels that needs to be evaluated is reduced to $2N'(3M)^2$ when Eq. (1.38) is used, or to $N'(3M)^2$ in conjunction with Eq. (1.45), where

$$N' \leq r_d/(c_2 \Delta t) , \tag{1.49}$$

with r_d being the distance between the two points of the boundary element mesh which are furthest apart from each other.

A number of time-stepping formulations for the solution of boundary integral equations in connection with various physical problems have been reported, e.g. Friedman and Shaw [32, 112] and Shaw [113] for 2D wave propagation in acoustics, Shaw [114] for transient diffusion, Mitzner [78] for 3D acoustic wave scattering, and Cole, Kosloff and Minster [14] for elastodynamic antiplane strain problems. The doubly asymptotic BEM proposed by Underwood and Geers [127] for 3D SSI problems works also in the time domain but uses a static BEM analysis to determine the stiffness matrix of the soil which subsequently is used in a standard dynamic finite element equation for the superstructure. A time-stepping BEM similar to the one described in this section, but based on a Green's function that corresponds to surface loads on a half-space, has been recently reported by Wolf and Obernhuber [140].

It is important to note that the boundary element discretization used in the above formulation, i.e. rectangular element with a constant variation of tractions and displacements, has been successfully applied by a number of researchers in the general area of elastodynamics for time domain and frequency domain formulations involving simple as well as complicated geometries, e.g. Refs. [1, 8, 16, 20–23, 87, 88, 123]. One can attribute the popularity of this boundary element discretization to its simplicity which results in a reduced numerical effort without any substantial compromise on the accuracy of the solution of most engineering problems. However, higher-order curvilinear triangular or quadrilateral elements with quadratic shape functions have also been used by Manolis [72] in the context of a 3D time-stepping BEM formulation. In this case a numerical spatial integration over both the non-singular and the singular elements is necessary

1.5.2. Frequency domain approach

The spatial discretization procedure followed in the numerical formulation of the 3D time domain case can be applied in the frequency domain as well. Thus, once again a set of M rectangular elements is used to approximate the 2D boundary of a 3D region, as shown in Fig. 2, and a uniform variation of tractions and displacements is assumed over each one of these elements. On the basis of these approximations Eq. (1.26) can be written, for elements on the boundary only, in the summation form

$$0.5\{\bar{u}_s^R\} = \sum_{m=1}^{M} \{[\bar{G}^m]\{\bar{t}_s^m\} - [\bar{T}^m]\{\bar{u}_s^m\}\} \qquad R = 1, 2, \ldots, M\,, \tag{1.50}$$

where $\{\bar{u}_s^R\}$ is the frequency transformed displacement vector of the "receiver" element R, $\{\bar{t}_s^m\}$ is the frequency transformed traction vector applied on the "source" element m, and the subscript s denotes the soil medium. The (3×3) influence matrices $[\bar{G}^m]$ and $[\bar{T}^m]$ are given by

$$\begin{aligned} \bar{G}_{ij}^m &= \int_{(s)} \bar{u}'_{ij}(\boldsymbol{x}, \boldsymbol{\xi}, w)\, \mathrm{d}s(\boldsymbol{x}) \\ \bar{T}_{ij}^m &= \int_{(s)} \bar{t}'_{ij}(\boldsymbol{x}, \boldsymbol{\xi}, w)\, \mathrm{d}s(\boldsymbol{x})\,. \end{aligned} \tag{1.51}$$

For non-singular elements m, i.e. $R \neq m$ in Eq. (1.50), standard two-dimensional numerical integration techniques can be used for the evaluation of the above integrals. However, in the case of singular elements m, i.e. $R = m$, the integrations indicated in Eq. (1.51) should be interpreted in the sense of Cauchy principal value [16]. An analytical evaluation of these integrals over the area of a rectangular singular element can be done in a straightforward manner. In doing so one should first observe that all the terms of the traction tensor $\bar{T}_{ij}^m$ as well as all the non-diagonal terms of the displacement tensor $\bar{G}_{ij}^m$ are zero [22]. The singularities of the non-zero diagonal terms of the $\bar{G}_{ij}^m$ tensor can be computed by means of a series expansion of the exponential terms appearing in formulae (1.23) and (1.24). Recently, Rizzo, Shippy and Rezayat [101, 102] proposed an alternative BIE formulation for time-harmonic problems which succeeds in eliminating the singularities of the related integrals and thus standard numerical integration schemes can be used throughout the entire computational process. Walker [131] has also discussed some general methods of avoiding singularities contained in fundamental solutions.

With regard to the coupling between the horizontal and the vertical modes of vibration, a set of observations can be made similar to those discussed earlier for the time domain formulation. Thus, if relaxed boundary conditions are used, i.e. horizontal displacements result in horizontal tractions only while vertical displacements result in vertical tractions only, a decoupling between the horizontal and the vertical motions occurs. Furthermore, for the study of surface foundations the influence of the fundamental traction vector $[\bar{T}^m]$ on the boundary $x_3 = 0$ vanishes and Eq. (1.50) can be simplified to

$$0.5\{\bar{u}_s^R\} = \sum_{m=1}^{M} [\bar{G}^m]\{\bar{t}_s^m\} , \qquad R = 1, 2, \ldots, M . \tag{1.52}$$

Nevertheless, regardless of the assumptions made for the contact conditions at the soil–foundation interface, Eq. (1.50) should be used in connection with embedded foundations. In principle, Eq. (1.50) requires a discretization of the entire infinite surface boundary. For reasons, though, similar to those mentioned in the time domain case, various authors, e.g. Dominguez [20, 21] and Dominguez, Alarcon and Abascal [23], have found that only a small area of the free field surrounding the foundation needs to be taken into account for the results to be accurate within limits satisfying practical purposes.

Equation (1.50), and similarly Eq. (1.52), represents a system of $3M$ algebraic equations requiring the specification of a total of $3M$ boundary conditions prior to its solution for some particular value of the frequency parameter w. The total number of $2(3M)^2$ discrete kernels required by Eq. (1.50), versus a number of only $(3M)^2$ discrete kernels required by Eq. (1.52), can be substantially reduced in view of the symmetry and space translation properties built-in in the fundamental solutions and preserved in their discretized form.

Similar numerical treatments of the 3D problem in the frequency domain can be found in the works of Dominguez and his co-workers [20–23], Kitamura and Sakurai [56, 57], Ottenstreuer and Schmid [87] and Sato, Kawase and Yoshida [105] for both relaxed and completely bonded boundary conditions. In reference [105] a Green's function corresponding to a surface harmonic loading on an elastic, homogeneous half-space is used as the necessary fundamental solution. Furthermore, Apsel [3] succeeded in numerically obtaining a general Green's function for layered viscoelastic media which can provide the basis for a BEM formulation of a wide range of problems involving uniform as well as layered half-spaces. Higher-order elements have also been used in 3D frequency domain analysis, e.g. curvilinear triangular or quadrilateral elements with second order shape functions in Refs. [101, 102]. The required spatial integrations over such elements are usually performed numerically for both singular and non-singular elements.

Time domain results can be obtained through a Fourier synthesis of the steady-state solutions of a sequence of frequencies provided by a Fourier domain analysis. This procedure has been applied by Niwa, Kobayashi and Azuma [83], while in Refs. [67, 101] the discrete inverse Fourier transform is used, in conjunction with the Fast Fourier Transform algorithm, to compute transient domain values. Similarly, if a Laplace domain analysis is performed first, the time domain solution can be obtained in a second step through a Laplace transform inversion which in the general case is also performed numerically [69, 70, 79, 125]. A comparative study of three BEM approaches, i.e. the time domain, the Laplace transform domain, and the Fourier transform domain, was conducted by Manolis [71] on the basis of a plane strain dynamic problem.

1.5.3. The two-dimensional case

The numerical formulation of the 2D problem in the time or the frequency domains can be treated in exactly the same way as the 3D problem. Here, the surface integrals of the 3D case are reduced to line integrals and subsequently the boundary of the region under consideration is discretized in a number of line elements. The discretized boundary integral equations (1.38) and (1.50) for the time and the frequency domains, respectively, can be viewed then as representing the 2D problem as well. Similarly, if relaxed boundary conditions are assumed at the contact surface between an infinitely long foundation and the underlying half-space, a decoupling between the horizontal and the vertical motions occurs. For the special case of surface foundations under relaxed boundary conditions Eqs. (1.45) and (1.52) are valid for the 2D case, too.

For more specific details on several numerical treatments of the 2D problem in the time domain, the interested reader should refer to the works of Niwa et al. [84], Cole et al. [14], Tzung and Lee [126], Mansur and Brebbia [73–75], Nardini and Brebbia [80, 81] and Spyrakos and Beskos [117–119]. Numerical formulations of the 2D frequency domain case have been presented by Dominguez [20, 21], Alarcon, Dominguez and del Cano [2], Manolis and Beskos [69], and Mita and Takanashi [77], among others. Higher-order 2D elements have also been used in various references, e.g. [70, 73–75, 80, 81, 125].

2. Foundation substructure

In this section the equations describing the motion of the foundation substructure will be established. Both surface and embedded massless foundations, rigid or flexible, will be studied in the frequency and the time domains. It can be shown that once this kinematic interaction problem has been solved the mass of the foundation and the superstructure can be taken into account in a straightforward manner during the solution of the inertial interaction problem [51, 53, 65, 132].

2.1. Rigid foundations

The contact surface of a foundation can be discretized into a number of M rectangular elements in the same fashion as it was done previously for the surface of the soil medium, Fig. 2. On the assumption of a constant variation of the displacements and tractions over each element, these functions can be associated with a "nodal" point at the center of each element. Then, the displacements on the contact surface of a rigid foundation can be expressed, for the time step N, as

$$\{u_f^N\} = [S]\{D^N\}, \tag{2.1}$$

where $\{u_f^N\}$ is the $(3M \times 1)$ displacement vector corresponding to the centers of the M rectangular elements on the contact surface and $\{D^N\}$ is the (6×1) rigid-body displacement vector given by

$$\{D^N\} = [\Delta_1^N \quad \Delta_2^N \quad \Delta_3^N \quad \phi_1^N \quad \phi_2^N \quad \phi_3^N]^T; \tag{2.2}$$

with Δ_i^N and ϕ_i^N being the rigid-body displacements and small rotations, respectively, with respect to a system of coordinate axes attached to the center of the foundation as shown in Fig. 2(a). The $(3M \times 6)$ transformation matrix $[S]$ is the assemblage of the M submatrices $[S^m]$ each one of which expresses the displacement vector at the center of the element m $(m = 1, 2, \ldots, M)$ due to a unit rigid-body motion, and can be written as

$$[S^m] = \begin{bmatrix} 1 & 0 & 0 & 0 & x_3^m & -x_2^m \\ 0 & 1 & 0 & -x_3^m & 0 & x_1^m \\ 0 & 0 & 1 & x_2^m & -x_1^m & 0 \end{bmatrix} \tag{2.3}$$

where x_i^m are the coordinates of the center of the element m.

The externally applied load vector $\{P^N\} = \{P_1^N, P_2^N, P_3^N, M_1^N, M_2^N, M_3^N]^T$ consisting of forces P_i^N and moments M_i^N $(i = 1, 2, 3)$ can be expressed in terms of the contact stresses $\{t_f^N\}$ applied on the foundation as

$$\{P^N\} = [K]\{t_f^N\}, \tag{2.4}$$

where the $(6 \times 3M)$ transformation matrix $[K]$ is the assemblage of a number M of (6×3) submatrices $[K^m]$ given by

$$[K^m] = A_m[S^m]^T, \tag{2.5}$$

with A_m being the area of the element m.

The frequency domain equivalents of Eqs. (2.1) and (2.4) can be obtained by substituting the time-dependent external loads, contact stresses, and displacements by their frequency domain counterparts, i.e.

$$\{\bar{u}_f\} = [S](\bar{D}\} \qquad (2.6)$$

$$\{\bar{P}\} = [K]\{\bar{t}_f\} . \qquad (2.7)$$

2.2. *Flexible foundations*

The thin plate theory, which has been found to give quite satisfactory results in the static analysis of foundations [31], is generally used to simulate the flexible foundation mat in dynamic analyses as well. Furthermore, it can be assumed that the in-plane deformations of the elastic foundation plate are negligible and the otherwise flexible foundation essentially behaves as a rigid disk in the direction of its own plane. Thus, the governing equation of a massless foundation plate can be written as

$$\frac{\partial^4 u_f}{\partial x^4} + 2\frac{\partial^4 u_f}{\partial x^2 \partial y^2} + \frac{\partial^4 u_f}{\partial y^4} - q\frac{12(1-\nu_p^2)}{E_p h^3} = 0 , \qquad (2.8)$$

where E_p, ν_p and h are the modulus of elasticity, the Poisson's ratio, and the thickness of the plate, respectively, $u_f(x, y, t)$ is the out-of-plane displacement, and $q(x, y, t)$ is the distributed transverse load. Following standard finite element procedures Eq. (2.8) can be discretized into a set of algebraic equations which for the time step N can be written as

$$[K_c]\{u_f^N\} = \{P^N\} + \{R^N\} \qquad (2.9)$$

where $[K_c]$ is the total stiffness matrix of the foundation plate after condensation, in order to eliminate the rotational degrees of freedom, $\{u_f^N\}$ is the vector of the transverse displacements at the nodal points, $\{P^N\}$ is the vector of the nodal externally applied transvere forces, and $\{R^N\}$ is the vector of the nodal transverse forces due to SSI, as shown in Fig. 3. In principle any finite element mesh can be used with Eq. (2.9), but in order to better match it with the previously established boundary element discretization of the soil medium it is obvious that the foundation plate should also be discretized into a set of rectangular elements. Among the various rectangular plate elements that are available in the literature, the authors [46, 50] chose, for simplicity, to use the non-conforming twelve degree-of-freedom plate element derived on the basis of a shape function given by

$$u_f = a_1 + a_2x + a_3y + a_4x^2 + a_5xy + a_6y^2 + a_7x^3 + a_8x^2y + a_9xy^2 + a_{10}y^3 + a_{11}x^3y + a_{12}xy^3 , \qquad (2.10)$$

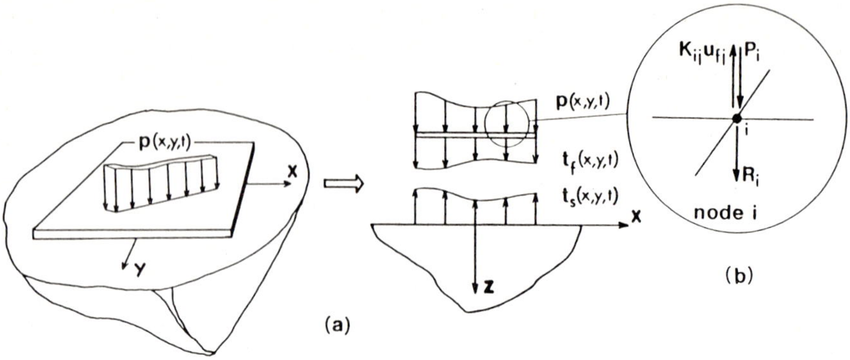

Fig. 3. System of forces applied on a flexible foundation: (a) total external forces, (b) nodal forces.

where a_i $(i = 1, \ldots, 12)$ are coefficients depending on nodal coordinates. The stiffness matrix of the above plate element is given in an explicit form in various references, e.g. [33]. Since a constant distribution of contact stresses over each boundary element has been assumed, by properly locating each nodal point of the finite element discretization to coincide with the center of an underlaying boundary elment, as shown in Fig. 4, one can write

$$\{R^N\} = [A]\{t_f^N\} \tag{2.11}$$

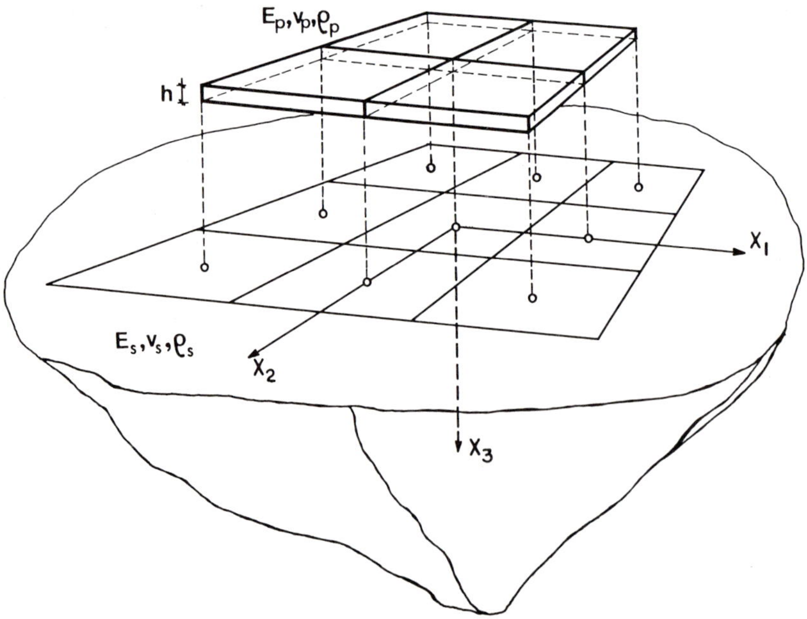

Fig. 4. Boundary and finite element discretization of the soil and the flexible foundation, respectively.

where $[A]$ is a diagonal matrix whose element A_{ii} represents the area of the element $i = 1, 2, \ldots, M$. An alternative matching of the foundation (FEM) and the soil (BEM) discretization schemes has been proposd by Iguchi and Luco [43]. However, it is possible that a more reasonable matching between the soil and the foundation would have resulted if a BEM simulation of the foundation plate had been used.

The corresponding frequency domain formulation can be derived from the above time domain case in a straightforward manner since Eq. (2.8) does not involve any time derivatives. Thus, Eq. (2.8) becomes

$$\frac{\partial^4 \bar{u}_f}{\partial x^4} + 2\frac{\partial^4 \bar{u}_f}{\partial x^2 \partial y^2} + \frac{\partial^4 \bar{u}_f}{\partial y^4} - \bar{q}\,\frac{12(1-v_p^2)}{E_p h^3} = 0\,, \tag{2.12}$$

where $\bar{u}_f = \bar{u}_f(x, y, w)$, and $\bar{q} = \bar{q}(x, y, w)$, and subsequently Eqs. (2.9) and (2.11) become, respectively,

$$[K_c]\{\bar{u}_f\} = \{\bar{P}\} + \{\bar{R}\}\,, \tag{2.13}$$

$$\{\bar{R}\} = [A]\{\bar{t}_f\}\,. \tag{2.14}$$

It should be noted at this point that the FEM formulation presented in this section can also be used for rigid foundation problems by simply increasing the stiffness of the flexible plate. However, the procedure descirbed in the previous section, which is applicable to rigid foundations only, involves far less degress of freedom than the FEM formulation and therefore is computationally advantageous.

2.3. *Two-dimensional foundations*

The modelling of 2D foundations does not present any special difficulty and follows the guidelines established previously for the 3D examples. Thus, in view of the 2D configuration shown in Fig. 5, the rigid-body displacement vector can

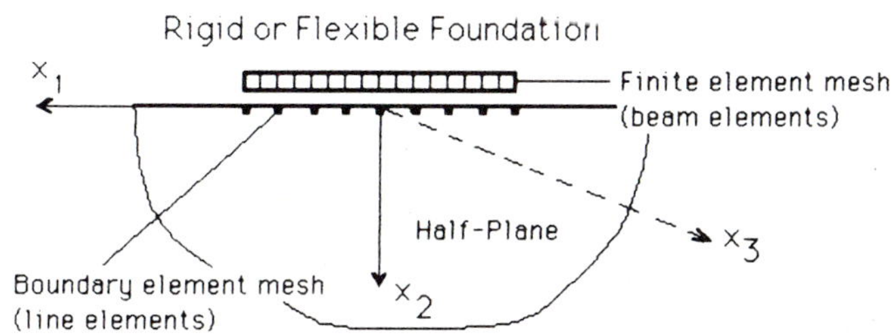

Fig. 5. Geometry and discretization of a 2D foundation.

be written as

$$\{D\} = [\Delta_1, \Delta_2, \phi_3]^{\mathrm{T}} \tag{2.15}$$

and Eq. (2.3) can be modified as

$$[S^m] = \begin{bmatrix} 1 & 0 & -x_2 \\ 0 & 1 & x_1 \end{bmatrix} . \tag{2.16}$$

Then the time domain compatibility and equilibrium equations (2.1) and (2.4), or their counterparts (2.6) and (2.7) in the frequency domain, remain in exactly the same form when they are referring to 2D rigid foundation problems.

The 2D flexible foundation problem can also be treated in the same manner as its 3D counterpart with the understanding that the plate elements which were used previously in the 3D case are now reduced to beam elements [119]. Therefore, the time domain equations (2.9) and (2.11) and the frequency domain equations (2.13) and (2.14) can also be used in the previusly shown form for 2D flexible foundation problems.

3. Foundation–soil system

3.1. Coupling equations

The soil and the foundation substructures, which have been analyzed separately in the previous sections, can be coupled by enforcing the compatibility and equilibrium conditions at their common interface. In doing so, the assumption of full continuity of stresses and displacements between the two substructures is implied, i.e. there is no separation between the soil and the foundation. The compatibility of interaction displacements at the contact surface between the soil and the foundation requires that

$$u_{\mathrm{s}}(\boldsymbol{x}, t) = u_{\mathrm{f}}(\boldsymbol{x}, t) . \tag{3.1}$$

Similarly, the equilibrium of interaction forces at the soil–foundation interface requires that

$$t_{\mathrm{s}}(\boldsymbol{x}, t) = -t_{\mathrm{f}}(\boldsymbol{x}, t) ; \tag{3.2}$$

For the purposes of the time domain formulation the above general equations can be discretized in time and space following the general procedures described previously for the soil and foundation substructures. Thus, for the time step N the discretized expressions of Eqs. (3.1) and (3.2) can be written in a matrix form as

$$\{u_s^N\} = \{u_f^N\} \tag{3.3}$$

$$\{t_s^N\} = -\{t_f^N\} \tag{3.4}$$

where the above ($3M \times 1$) displacement and traction vectors are related to a number of M boundary elements on the soil surface or to a corresponding number of M nodal points on the foundation contact surface.

The spatially discretized frequency domain equivalents of Eqs. (3.1) and (3.2) can also be written in the matrix form

$$\{\bar{u}_s\} = \{\bar{u}_f\} \tag{3.5}$$

$$\{\bar{t}_s\} = -\{\bar{t}_f\}\ ; \tag{3.6}$$

where, once again, the ($3M \times 1$) displacement and traction vectors correspond to a number of M soil boundary elements or foundation nodal points.

In the remainder of this section the dynamic response of foundations to externally applied forces or obliquely incident seismic waves is discussed.

3.2. *Externally applied dynamic loads*

In view of the discretized expressions developed in previous sections for the soil and the foundation substructures and the coupling conditions of this section, the time or frequency domain response of soil–foundation systems subjected to externally applied dynamic loads can be found as the solution of a system of linear algebraic equations with unknowns the displacements and the tractions on the soil boundary or the foundation contact surface. The few examples that are discussed next are intended to illustrate this procedure as it is applied to 3D problems. The same basic steps can be followed for 2D problems [2, 20, 21, 117–119] and thus their discussion is omitted in order to avoid repetition.

3.2.1. *Rigid foundations*

The time domain response of a rigid surface or embedded foundation can be computed, in general, from the system of equations (1.38), (2.1), (2.4), (3.3), and (3.4). A solution of this system of linear albegraic equations can be accomplished by a step-by-step time marching, i.e. for the time step $N = q$, then for $N = q + 1$, etc. A detailed discussion of the solution of this general problem can be found in Ref. [46]. However, for illustration purposes the step-by-step time marching solution of the special problem of a rigid surface foundation under relaxed boundary conditions is given here. In this case Eq. (1.38) of the above system of equations should be replaced by Eq. (1.45) which requires a discretization of only the soil surface in contact with the foundation. Then, for a sequence of rectangular impulse forces starting at time step q, the rigid-body displacements of the surface foundation at any time step $N \geq q$ is given by [46, 47]

$$\{D^N\} = 2[Q]\{\{P^N\} + [K][G^1]^{-1}\{RT^N\}\}\,, \tag{3.7}$$

where

$$[Q] = [[K][G^1]^{-1}[S]]^{-1} \tag{3.8}$$

$$\{RT^N\} = \begin{cases} 0\,, & N = q \\ \sum\limits_{n=q+1}^{N} [G^{n-q+1}]\{t_{\mathrm{a}}^{N-n+q}\}\,, & N > q \end{cases} \tag{3.9}$$

$$\{t_{\mathrm{a}}^N\} = [G^1]^{-1}[S][Q]\{P^N\}\,. \tag{3.10}$$

The frequency domain response of a rigid surface or embedded foundation can be obtained, in general, as the solution of the system of equations (1.50), (2.6), (2.7), (3.5), and (3.6). For the special case of surface foundations under relaxed boundary conditions Eq. (1.50) should be substituted by Eq. (1.52). The solution of either system of equations will result to a force–displacement relationship of the form [20, 21]

$$\{\bar{P}\} = [K(w)]\{\bar{D}\} \tag{3.11}$$

where $[K(w)]$ is the (6×6) impedance matrix of the rigid foundation.

3.2.2. Flexible foundations

Equations (1.38), (2.9), (2.11), (3.3), and (3.4) can be viewed together as a system of linear algebraic equations with unknowns being the displacements and the contact tractions at the centers of the elements of the soil boundary element mesh or at the nodal points of the foundation finite element mesh during time step N. Once again a step-by-step time marching solution is required similar to the one shown for rigid surface foundations. The time domain response of surface flexible foundations is discussed in Ref. [50], where relaxed boundary conditions are used and subsequently Eq. (1.38), of the above system of equations, has been replaced by Eq. (1.45).

The frequency domain response of flexible foundations can be given as the solution of the system of equations (1.50), (2.13), (2.14), (3.5), and (3.6). The special case of surface foundations under relaxed boundary conditions, where Eq. (1.52) is used instead of Eq. (1.50), has been studied by Ottenstreuer [88].

Savidis and Richter [106] applied a similar hybrid method to analyze the response of flexible surface foundations, while some analytic solutions for circular plates are given in Refs. [44, 61, 108]. To the authors best knowledge no results have been presented as yet for the dynamic response of 3D flexible embedded foundations by either the time or the frequency domain BEM.

3.3. Obliquely incident seismic waves

The dynamic response of an elastic half-space to a plane incident wave can be described, in the absence of any foundation, by the free field displacement vector

$$\{u_g\} = \{u_g(\boldsymbol{x}, t)\} = [u_{gx_1} \quad u_{gx_2} \quad u_{gx_3}]^T , \tag{3.12}$$

and the free field traction vector

$$\{t_g\} = \{t_g(\boldsymbol{x}, t)\} = [t_{gx_1} \quad t_{gx_2} \quad t_{gx_3}]^T , \tag{3.13}$$

where $\{u_g(\boldsymbol{x}, t)\}$, and $\{t_g(\boldsymbol{x}, t)\}$ depend upon the vertical and horizontal angles of incidences, θ_V and θ_H as they are shown in Fig. 6, and the apparent velocity $c = c_1/\cos\theta_V$ or $c = c_2/\cos\theta_V$, depending on the type of the incident wave. When a foundation is placed on the soil a scattered displacement field $\{u_s(\boldsymbol{x}, t)\}$ and a scattered traction field $\{t_s(\boldsymbol{x}, t)\}$ are generated in addition to the free field displacements and tractions. Then, the total displacement field $\{u(\boldsymbol{x}, t)\}$ in the soil medium is given by [122]

$$\{u(\boldsymbol{x}, t)\} = \{u_g(\boldsymbol{x}, t)\} + \{u_s(\boldsymbol{x}, t)\} , \tag{3.14}$$

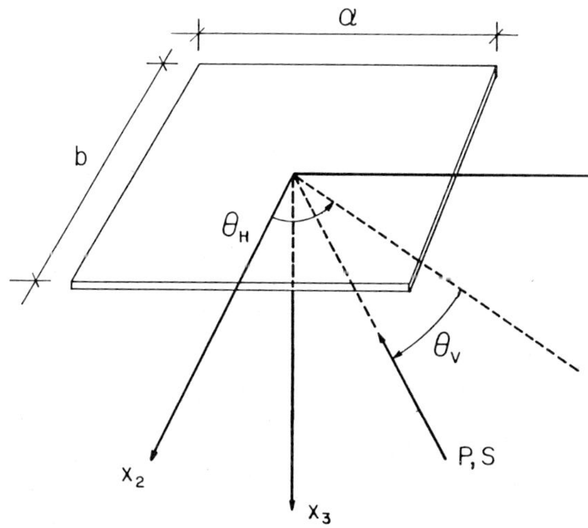

Fig. 6. Nomenclature for a foundation subjected to obliquely incident seismic waves.

and similarly the total traction field by

$$\{t(\boldsymbol{x}, t)\} = \{t_g(\boldsymbol{x}, t)\} + \{t_s(\boldsymbol{x}, t)\} . \tag{3.15}$$

Following the procedures established previously, the various quantities appearing in Eqs. (3.14) and (3.15) can be discretized in time and space. For the discretization in time the rectangular impulse approximation shown in Fig. 1 can be used again. Thus, the total displacement field, for example, can be approximated by a sequence of rectangular impulses as

$$\{u(\boldsymbol{x}, t)\} = \{u^q(\boldsymbol{x})\}[H[t-(q-1)\Delta t] - H(t-q\Delta t)] , \tag{3.16}$$

where $\{u^q(\boldsymbol{x})\}$ is the intensity of the rectangular impulse during the time interval $(q-1)\Delta t < t < q\Delta t$. On the assumption of a constant variation of the free and scattered fields over each ractangular boundary element of the discretized soil surface, Eqs. (3.14) and (3.15) can be written in the following matrix form for each time step q:

$$\{u^q\} = \{u_g^q\} + \{u_s^q\} \tag{3.17}$$

$$\{t^q\} = \{t_g^q\} + \{t_s^q\} , \tag{3.18}$$

where each of the displacement and traction vectors appearing in Eqs. (3.17) and (3.18) contains an M number of (3×1) vectors expressing the total, the free, and the scattered displacement and traction vectors at the center of each boundary element. With regard to the various qantities showing on the right hand sides of Eqs. (3.17) and (3.18) one can observe that: (i) the total displacement field $\{u^q\}$ can be expressed in terms of the rigid-body displacements by using the restraint Eq. (2.1) for the case of a rigid foundation, or the displacement vector $\{u_f^N\}$ of Eq. (2.9) for the case of a flexible foundation, (ii) the scattered displacement field $\{u_s^q\}$ is due to the motion of the foundation alone and is given in terms of the contact tractions by Eqs. (1.38) or (1.45), and (iii) the free field displacement and traction vectors, which could have arbitrary time variations, are assumed to be known for the particular problem at hand.

Equations (3.17) and (3.18) can be used along with the discretized governing equations for the soil and the foundation substructures to establish a system of linear algebraic equations the solution of which will provide the response of the soil–foundation system to a prescribed free field wave excitation. Thus, for a flexible surface foundation under relaxed boundary conditions, for example, Eqs. (1.45), (2.9), (2.11), (3.3), (3.4), (3.17), and (3.18) can be viewed together as a system of equations with unknowns the displacements and the tractions on the contact surface between the soil and the foundation. Assuming that no external forces are applied, i.e. $\{P^N\} = 0$, and observing also that for surface foundations

$\{t_g^q\} = 0$, a step-by-step time marching solution of the above system of equations can be obtained in the form [50]

$$\{u^N\} = [F]\{\{P_e^N\} + [A][G^1]^{-1}\{RT^N\}\} , \tag{3.19}$$

where

$$[F] = [(0.5)[A][G^1]^{-1} + [K_c]]^{-1} \tag{3.20}$$

$$\{P_e^N\} = (0.5)[A][G^1]^{-1}\{u_g^N\} \tag{3.21}$$

$$\{RT^N\} = \begin{cases} 0 , & N = q \\ \sum_{n=q+1}^{N} [G^{n-q+1}]\{t_a^{N-n+q}\} , & N > q \end{cases} \tag{3.22}$$

$$\{t_a^N\} = (0.5)[G^1]^{-1}\{[F]\{P_e^N\} - \{u_g^N\}\} . \tag{3.23}$$

Similar solutions are given by the authors for problems involving rigid surface foundations [46, 47] and rigid embedded foundations [52].

In the frequency domain the equivalents of Eqs. (3.17) and (3.18) can be written as

$$\{\bar{u}\} = \{\bar{u}_g\} + \{\bar{u}_s\} \tag{3.24}$$

$$\{\bar{t}\} = \{\bar{t}_g\} + \{\bar{t}_s\} . \tag{3.25}$$

Equations (3.24) and (3.25) supplemented by the spatially discretized governing equations of the soil and the foundation substructures, derived in sections 1 and 2, and the coupling equations (3.5) and (3.6), form a system of linear algebraic equations which can provide the response of a foundation–soil assemblage to a harmonic wave excitation. The solution of this system of equations for rigid foundations, in the absence of external forces, will result in a relationship of the form

$$\{\bar{D}\} = [C(w)]\{\bar{u}_g\} , \tag{3.26}$$

where $[C(w)]$ is the $(6 \times 3M)$ input motion matrix. An extensive discussion of the physical significance of the input motion matrix is given by Wong and Luco [143]

4. Numerical examples

A variety of numerical results based on the time domain and, primarily, the frequency domain formulations as described in the previous sections, is available in the literature. Due to obvious space limitations, only a few selected examples of the applicaton of the BEM in foundation dynamics are presented here. These examples refer to massless foundations, subjected to externally applied loads or seismic wave excitations. In an effort to compare the results obtained by the time domain BEM with those of the frequency domain BEM, most of the examples presented here are due to harmonic excitations only. However, some results which intend to demonstrate the capability of the time domain BEM in handling, in a straightforward manner, any time variation of the dynamic forcing function are also given.

4.1. Rigid surface foundations

The dynamic response, under relaxed boundary conditions, of a square, rigid, surface foundation subjected to an externally applied vertical force, described by

$$P_3(t) = 200\sin(wt) \quad \text{(kips)}\,, \tag{4.1}$$

is shown in Fig. 7. The amplitude of the resulting harmonic motion is plotted versus the dimensionless frequency a_0 which is defined as

$$a_0 = wb/c_2\,, \tag{4.2}$$

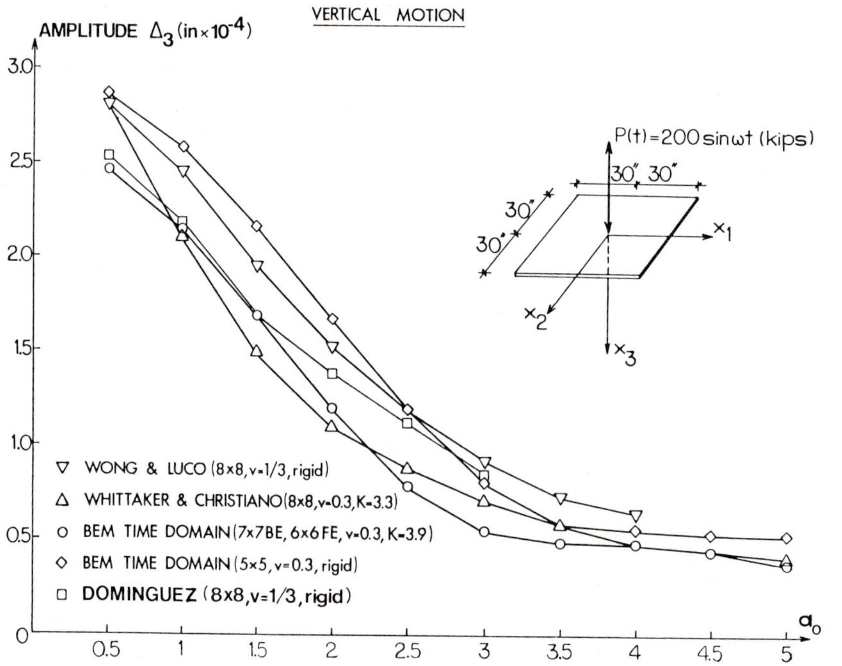

Fig. 7. Vertical response of a rigid surface foundation to a harmonic, vertical, concentrated force.

with $2b$ being the length of the side of the square foundation. The rigidity of the foundation plate is accounted for by two distinct models. The first model is based on the formulation described in section 2.1 which for this particular problem requires only one (the vertical) degree of freedom to be considered. The second model consists of a finite element idealization of the foundation plate, as the one discussed in section 2.2, and of course requires a much larger number of degrees of freedom, i.e. one at each nodal point, to be specified. In both cases the soil medium is characterized by the following elastic constants: modulus of elasticity $E_s = 8.84 \times 10^6$ psi, mass density $\rho_s = 2.82 \times 10^{-4}$ lb · s^2/in^4, and a Poisson's ratio varying among the references shown in Fig. 7 from $\nu_s = 0.3$ to 1/3. For the finite element model a material characterization of the foundation plate is also required, and in this example the following elastic constants are used: Poisson's ratio $\nu_p = 0.3$, and mass density $\rho_p = 7.34 \times 10^{-4}$ lb · s^2/in^4. The modulus of elasticity of the foundation plate can be adjusted in such a way that a relative rigidity requirement is satisfied. For all practical purposes, the foundation plate is considered to be "rigid" if the relative stiffness ratio, defined as

$$K = \frac{E_p h^3 (1 - \nu_s)}{12(1 - \nu_p^2)\mu_s b}, \tag{4.3}$$

takes values greater than 1 [133, 134]. In Eq. (4.3), h is the thickness of the foundation plate, and μ_s is the half-space shear modulus. The results obtained by the time domain BEM in conjunction with the rigid or finite element models [47, 50], and by the frequency domain BEM in conjunction with the rigid model [20], are compared with those reported by Wong and Luco [142] who used the rigid model, and by Whittaker and Christiano [133] who used the finite element model. All five methods seem to be in good agreement, but a detailed comparision of the computational costs associated with each method is not possible due to lack of appropriate information. Nevertheless, one should notice the substantially smaller number of elements required by the time domain BEM in order to produce the same level of accuracy as the other methods shown in Fig. 7.

A variety of results concerning other possible modes of vibration and 3D rigid foundation geometries have been obtained through the use of the BEM, e.g. Refs. [20, 46, 47, 87], as well as various other well established techniques, e.g. Refs. [129, 142].

In Fig. 8(a)–(c) the amplitudes of the vertical, rocking, and horizontal displacements of a square, rigid, surface foundation subjected to a harmonic SV-wave excitation are plotted versus the dimensionless freqency a_0. The soil medium of this example is characterized by elastic constants: $E_s = 1.79 \times 10^7$ psi, $\nu_s = 1/3$, and $\rho_s = 5 \times 10^{-4}$ lb · s^2/in^4. If the incident SV-wave is impinging on the foundation at a vertical angle of incidence $\theta_V = 75.5°$ in a direction parallel to the x_2-axis, i.e. $\theta_H = 0$, then the free-field displacements at the surface $x_3 = 0$ can be expressed

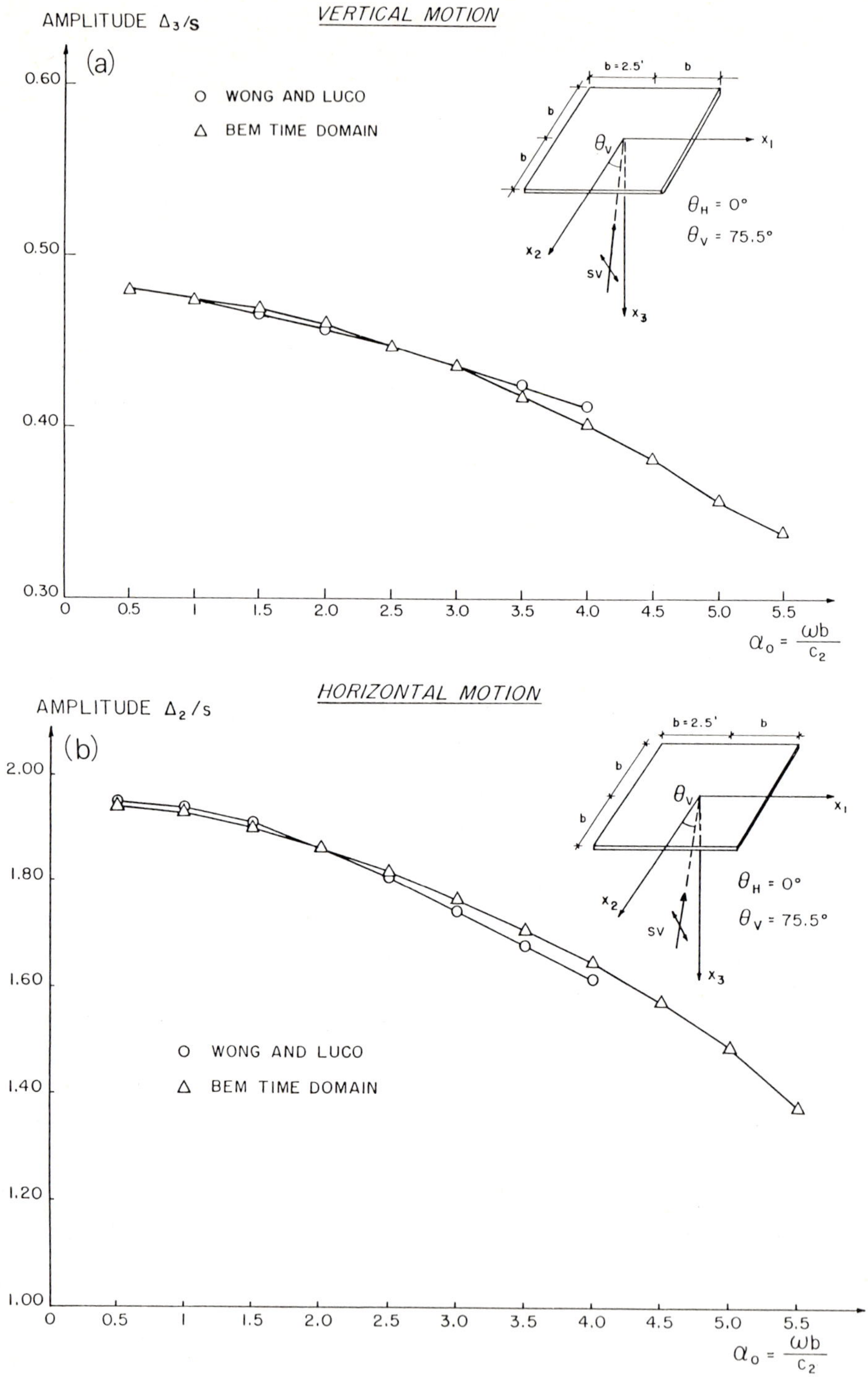

Fig. 8. Harmonic response versus frequency of a square rigid surface foundation subjected to an SV-wave excitation.

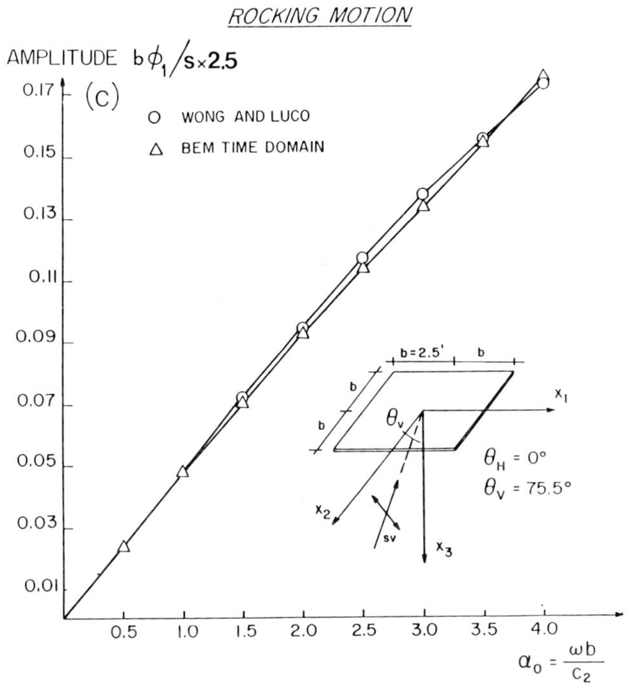

Fig. 8 (continued).

as [143]

$$
1/s \begin{Bmatrix} u_{gx_1} \\ u_{gx_2} \\ u_{gx_3} \end{Bmatrix} = \begin{Bmatrix} 0 \\ 1.95 \\ 0.48 \end{Bmatrix} \sin[w(t - x_2/c_2)]\,, \tag{4.4}
$$

where s is the amplitude of the incident wave. In Fig. 8(a)–(c) the results obtained by the time domain BEM [47], under the assumption of relaxed boundary conditions, seem to be in excellent agreement with those reported by Wong and Luco [143] who considered complete bond between the foundation and the half-space. In this case the assumption of relaxed boundary conditions allowed the computations with respect to each of the excited modes of vibration to proceed independently from one another and thus, the required computational effort was substantially reduced. A 5×5 element discretization (with a 3×3 subelement mesh per element, as explained before) is used by the time domain BEM, while an 8×8 element discretization was used in Ref. [143]. The rigidity of the foundation has been accounted for in both Refs. [47] and [143] on the basis of the model described in section 2.1.

The response of a 2D rigid surface foundation under relaxed boundary conditions when subjected to a harmonic Rayleigh wave excitation is shown in Fig. 9(a, b). The results correspond to an elastic half-plane characterized by $E_s =$

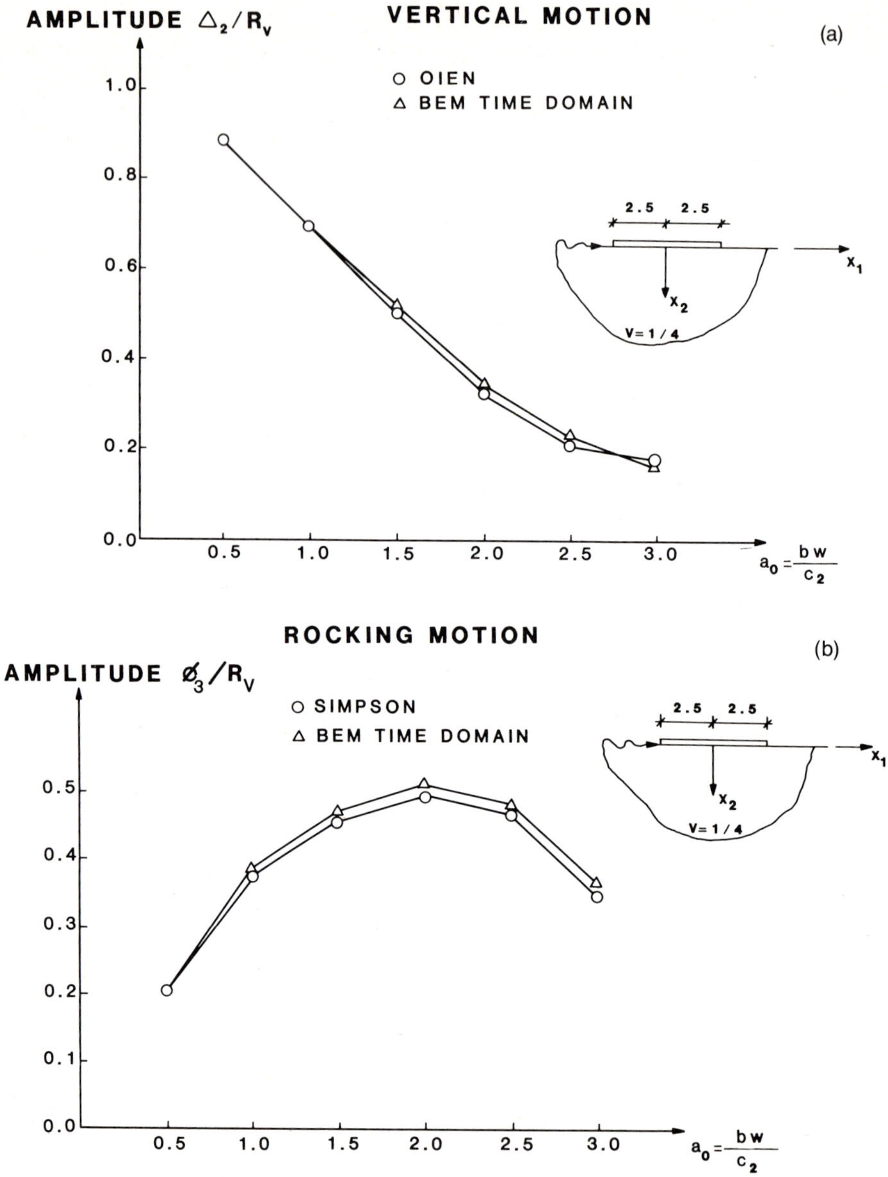

Fig. 9. Harmonic response versus frequency of a rigid strip foundation subjected to a Rayleigh wave excitation.

2.58984×10^9 lb/ft^2, $\rho_s = 10.368$ lb/s^2/ft^4, and $\nu_s = 1/3$. For these particular choices of elastic constants the amplitude of the vertical free-field surface displacement for a harmonic Rayleigh wave propagating in the x_1–x_2 plane is given by [66]

$$u_{g_3} = R_v \sin[w(t - x_1/9190.168] \tag{4.5}$$

where R_v is the amplitude of the harmonic Rayleigh wave. In Fig. 9(a, b) the results obtained by the 2D time domain BEM of Ref. [118] are plotted along with those obtained by Oien [86] and Simpson [116]. A good agreement between the various methods is observed even though the results reported in Ref. [116] correspond to complete bonding between the soil and the foundation.

Parametric studies on the problem of rigid surface foundations under the influence of various types of harmonic waves impinging on the foundation at various angles of incidence can be found in Refs. [21], and [143] among others.

4.2. Flexible surface foundations

A foundation plate is considered to be "flexible" if $0 < K < 1$, where K is the relative stiffness ratio defined by Eq. (4.3). The response of a relatively soft, square surface plate, $K = 0.004$, to an externally applied uniform load is shown in Figs. 10 and 11. The results given in both these figures correspond to the material properties used in Fig. 7 and have been obtained through an application of the

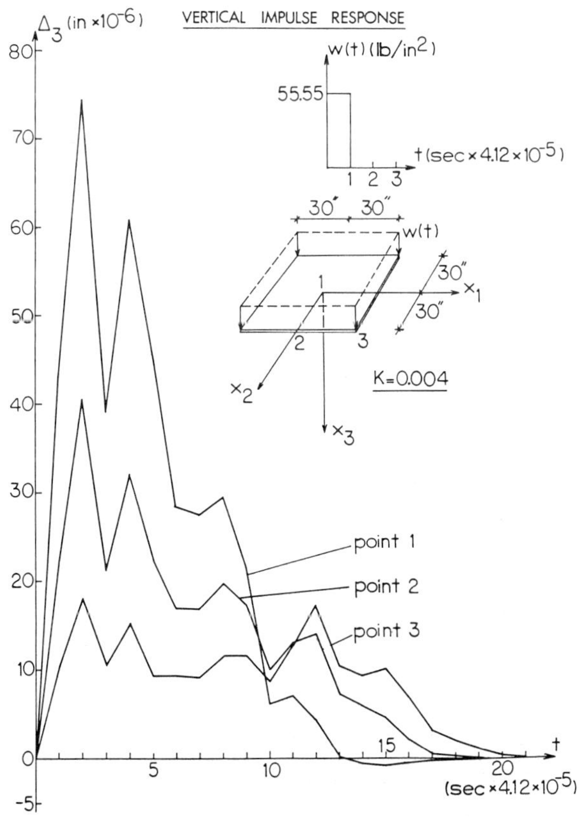

Fig. 10. Vertical response versus time of a square flexible foundation to a unifomly distributed vertical impulse load.

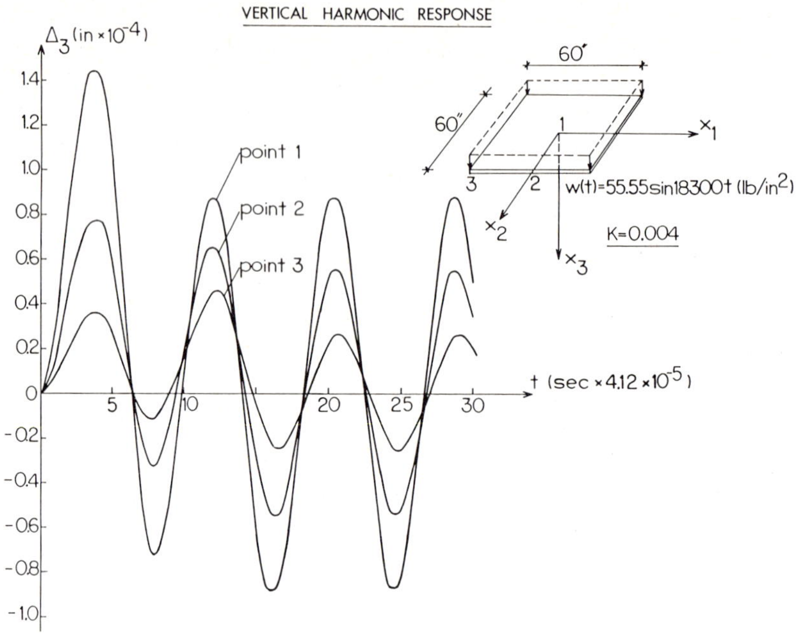

Fig. 11. Vertical response versus time of a square flexible foundation to a harmonic uniformly distributed vertical load.

time domain BEM and a finite element modelling of the flexible foundation plate as described in section 2.2.

The above-described foundation plate is first subjected to a uniformly distributed vertical impulse load defined by

$$p(t) = \begin{cases} 55.55\ (\mathrm{lb/in^2})\,, & \text{first time step} \\ 0\,, & \text{elsewhere}\,. \end{cases} \tag{4.6}$$

The vertical displacements of three characteristic points of the square plate are plotted versus time in Fig. 10. The responses of these three points show a marked similar behavior but substantial differences in magnitudes of displacements which result in the well known "dishing" of the flexible foundation. A 5×5 boundary element mesh, with a 3×3 subelement mesh per boundary element, in conjunction with a 4×4 finite element discretization scheme were used by the authors [50] to obtain the results of Fig. 10.

Next, the same flexible foundation is subjected to a harmonic, uniformly distributed vertical load defined by

$$p(t) = 55.55 \sin(18300t)(\mathrm{lb/in^2}) \tag{4.7}$$

The vertical motion of three characteristic points is again plotted versus time in

Fig. 11 [50]. With regard to Fig. 11 one should notice the transient part of the motion of the foundation which in this particular high-frequency motion causes displacements substantially larger than the ones observed during the subsequent steady-state part of the motion. The detection of transient motions preceding harmonic responses is due to the time domain formulation utilized in this example. On the contrary frequency domain formulations provide the amplitude of the steady-state part of the motion only and are unable to predict transient phenomena of this kind unless an inverse transformation into time domain is performed as a second step. The discretization scheme employed in this example is similar to the one used for Fig. 10.

In obtaining the results shown in Fig. 11 the step-by-step time marching solution developed in the previous sections was used. However, the impulse responses shown in Fig. 10 could also have been used in the general framework of a discrete convolution approach to produce the response of the above foundation to any forcing function. This last procedure, although appears to be very efficient and accurate [46], is based on the superposition principle and thus its application to the analysis of nonlinear problems is precluded. A similar convolution approach, limited to circular, rigid foundations, was proposed by Veletsos and Verbic [128].

For more detailed discussions on the subject of surface flexible foundations the interested reader is referred to Refs. [43, 44, 46, 50, 88, 133, 134].

4.3. Rigid embedded foundations

The response of a 60 in × 60 in square and 20 in deep rigid foundation subjected to externally applied vertical loads, horizontal loads, and overturning moments is given in Figs. 13(a)–(c) and 14(a) (c) [52]. In both figures the soil medium is represented by an elastic half-space with constants: $E_s = 8.84 \times 10^6$ psi, $\rho_s = 2.82 \times 10^{-4}\ \text{lb} \cdot \text{s}^2/\text{in}^4$, and $\nu_s = 0.3$. For reasons explained in previous sections, and related to the fundamental solutions used in this article, a number of boundary elements outside the contact surface between the soil and the foundation are required for the analysis of an embedded foundation, even under relaxed boundary conditions, and thus the discretization scheme shown in Fig. 12 is used.

In this example a BEM time domain analysis using the convolution approach is performed in two consecutive steps. In the first step, the response of the soil–foundation system to externally applied rectangular impulse forces is obtained. The rectangular impulse vertical force, horizontal force, and overturning moment are described, respectively, by

$$P_{03} = P_{01} = M_{02} = \begin{cases} 200\,, & \text{first time step} \\ 0\,, & \text{elsewhere}\,, \end{cases} \tag{4.8}$$

where the units for the magnitudes P_{03} and P_{01} are kips while those for M_{02} are

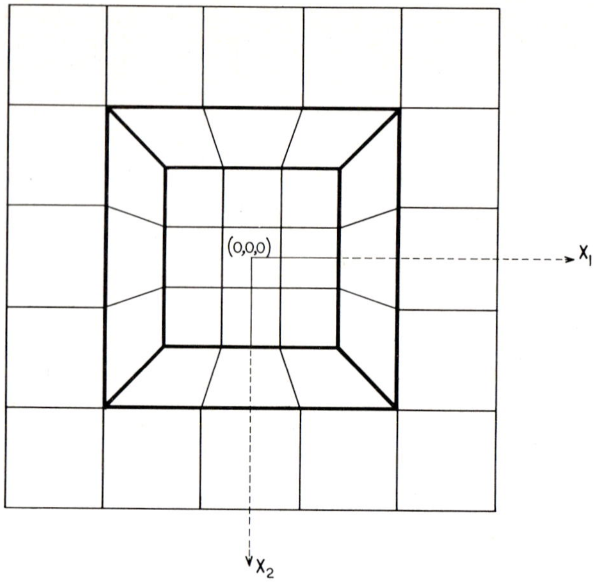

Fig. 12. Discretization for a rigid embedded foundation.

kip-in. In Figs. 13(a)–(c) the vertical, horizontal, and rocking displacements of the foundation due to the forcing functions of Eq. (4.8) are plotted versus time. Once again, relaxed boundary conditions have been assumed that permit the computations with respect to the various motions shown in Figs. 13(a)–(c) to be performed independent of each other.

In a second step, the results recorded in Figs. 13(a)–(c) can be used to obtain the response of the same embedded foundation to any forcing function. Consider, for example, a transient vertical force $P_3(t)$ applied at the center of resistance of the soil–foundation system described before. This general forcing function can be discretized into a sequence of rectangular impulses, in the manner described by Eq. (1.25), as

$$P_3(t) = P_3^n[H[t-(n-1)\Delta t] - H(t-n\Delta t)] . \tag{4.9}$$

If D_3^n represents the vertical rigid-body displacement recorded in Fig. 13(a) at time step n, then the vertical rigid-body displacement D_{t3}^N of the same foundation at any time step N due to the general force $P_3(t)$ is given by

$$D_{t3}^N = \sum_{n=1}^{N} P_3^n \cdot D_3^{N-n+1}/P_{03} \tag{4.10}$$

where P_{03} is the magnitude of the rectangular impulse force of Eq. (4.8) during the first time step. Similar expressions can be written for the horizontal forces and

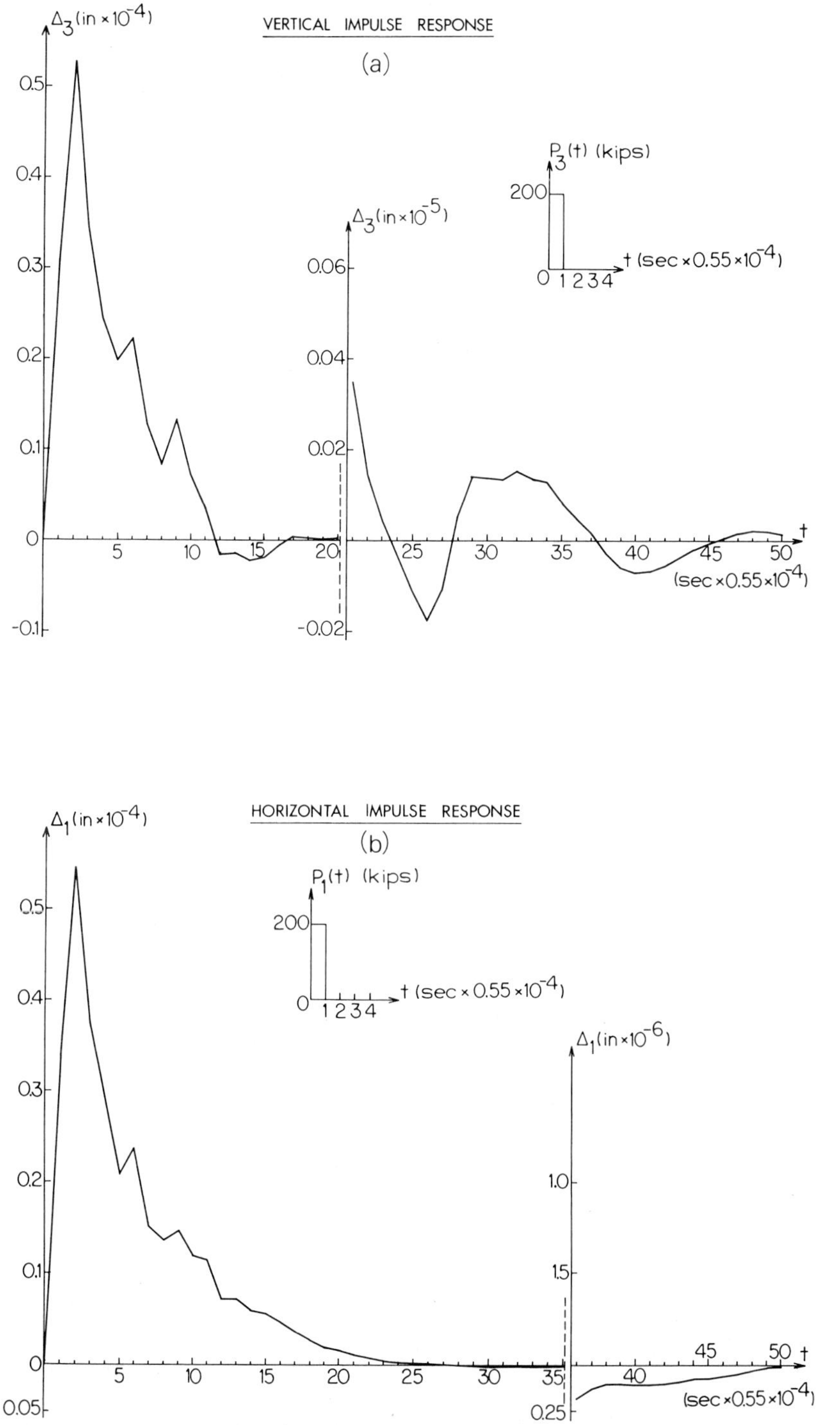

Fig. 13. Impulse force response versus time for a rigid embedded foundation.

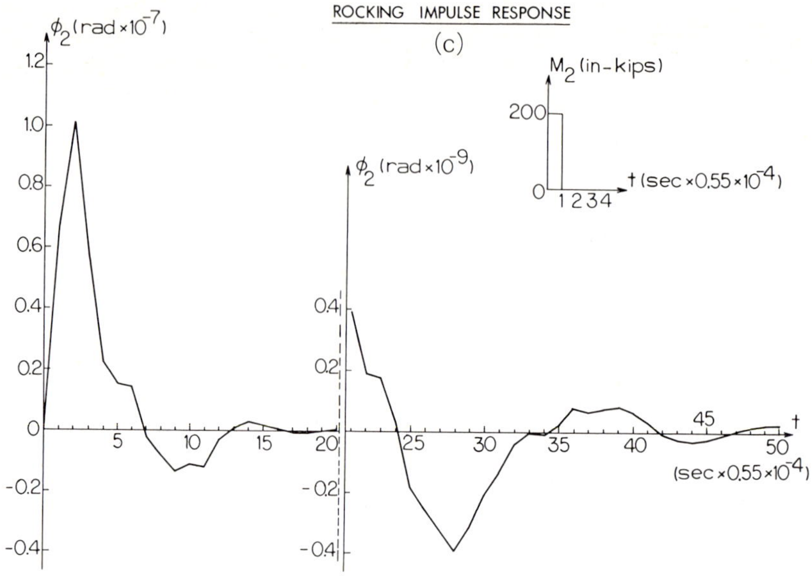

Fig. 13 (continued).

the overturning moments. In order to compare the results obtained by this time domain methodology with those reported by Dominguez [20] in the frequency domain, the response of the previously described soil–foundation system to harmonically varying externally applied forces and moment, defined as

$$\begin{aligned} P_3(t) &= 200\sin(wt) \quad \text{(kips)}\,, \\ P_1(t) &= 200\sin(wt) \quad \text{(kips)}\,, \\ M_2(t) &= 200\sin(wt) \quad \text{(kips-in)}\,, \end{aligned} \tag{4.11}$$

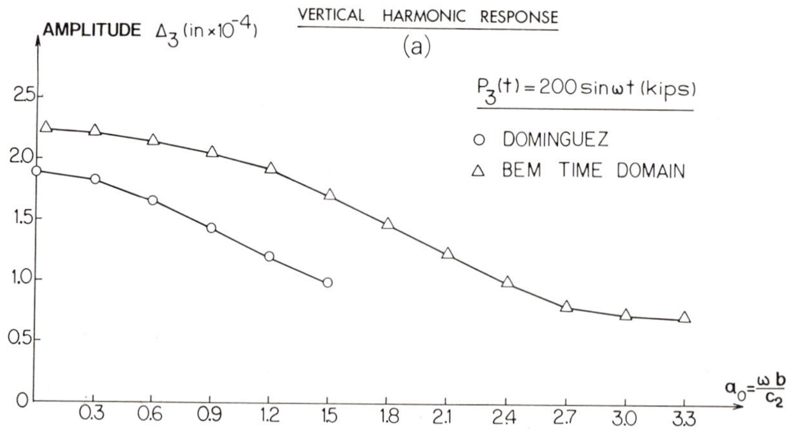

Fig. 14. Harmonic force response versus time for a rigid embedded foundation.

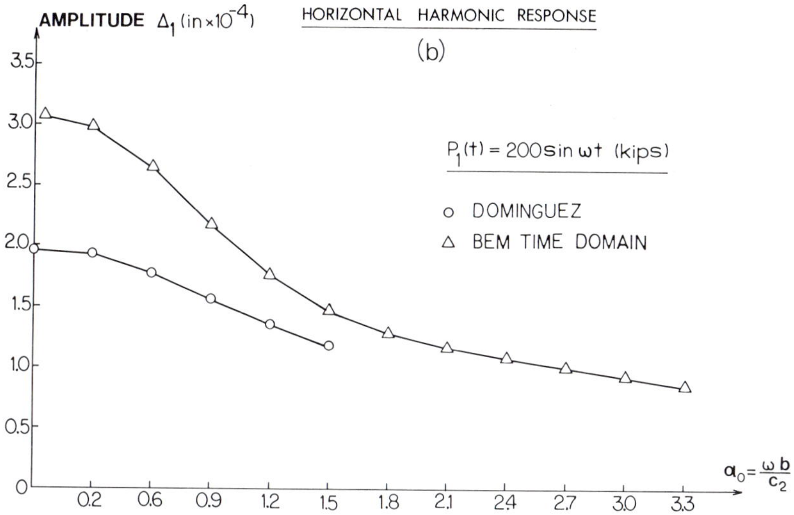

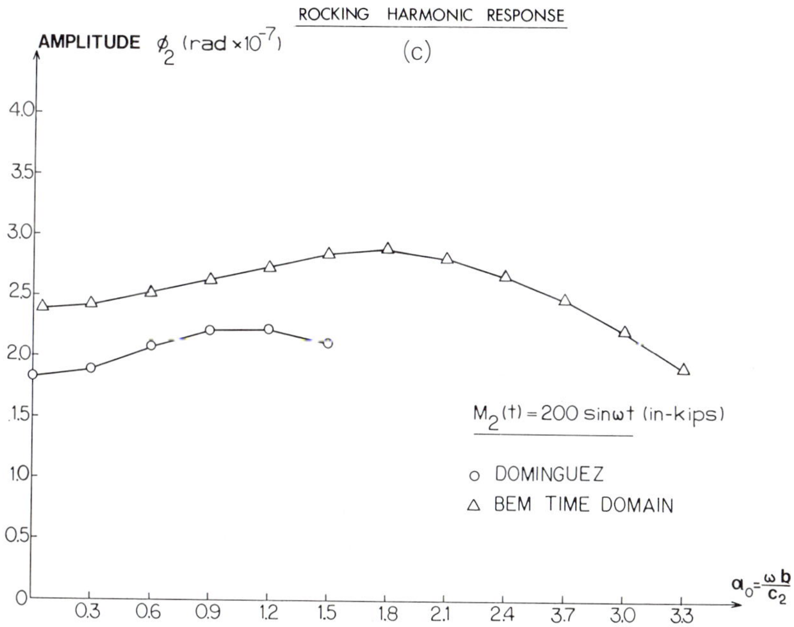

Fig. 14 (continued).

is plotted versus the dimensionless frequency a_0 in Figs. 14(a)–(c). A relative agreement of the two methods is observed, with the time domain BEM being always on the conservative side. However, due to the time domain BEM, transient phenomena, similar to those reported in Fig. 11, have been captured for the entire range of frequencies and for every mode of vibration shown in Figs. 14(a)–(c) [46, 52].

A number of numerical results, obtained through the application of the BEM, for rigid embedded foundations subjected to externally applied loads or incident wave excitations can be found in [20, 21, 46, 52] for the 3D case, and in [2, 20, 21, 117, 118] for the 2D case.

5. Special topics

In the previous sections of this chapter an attempt was made to illustrate the formulation and the basic computational aspects of the BEM as applied to a class of highly idealized SSI interaction problems. To this extent some relatively simple and well understood problems of rigid or flexible foundations resting on homogeneous, isotropic and linear elastic half-spaces were examined. However, in the short time of its existence the BEM has proven itself as a powerful and efficient numerical methodology in dealing with general problems of higher complexity and more practical substance. Some of those applications in the dynamic SSI interaction area will be briefly referred to in this section. But since most of these areas of research are currently under development the following list of subjects and references should not be considered as a complete one, but rather as a mere indication of the capabilities of the BEM and perhaps as some of the future research directions in the SSI area.

5.1. More realistic soil models

The BEM formulation of the soil substructure of the previous section was based on the simplifying assumption of homogeneous, isotropic and linear elastic half-space or half-plane. However, most of the real soil conditions cannot be approximated by such an "idealistic" model. It is well known that real soil profiles increase in rigidity with depth, as a result of overburden pressure, or consist of a number of relatively well separated arbitrary topographic conditions or soil layers each one of which exibits its own mechanical characteristics. It is also well documented through various laboratory tests that soils deform differently in different directions as a result of their anisotropic fabric, with these differences being more pronounced in the vertical and the horizontal directions. Furthermore, the linear elastic soil material assumed in the previous sections fails, by definition, to embody any internal energy-absorbing mechanism or "material damping." There is evidence, however, that the influence of the energy dissipation caused by material damping is of significant importance particularly when it approaches the same order of magnitude as the geometrical damping or when a significant portion of the radiated energy is reflected, e.g. in the case of a soil layer resting on top of a rigid base rock. The following paragraphs present, from the BEM point of view, some of the work done with: (i) viscoelastic half-spaces,

(ii) anisotropic half-spaces, and (iii) irregular site conditions and layered soil media, as related to the dynamic SSI area of research.

It should be mentioned at this point that for an inhomogeneous or anisotropic elastic solid it is possible, in principle, to formulate an integral equation method in terms of the fundamental solutions of an equivalent homogeneous and isotropic elastic material and regard the influence induced by the inhomogeneous or anisotropic character of the above solid as additional body forces included in a volume integral. Even though this method seems numerically attractive, e.g. it has been applied by Niwa, Kitahara and Hirose [85] for inhomogeneous bodies, the existence of the above-mentioned volume integrals prohibits its application to bodies of infinite extent, unless some artificially introduced boundaries can be tolerated.

5.1.1. Viscoelastic half-spaces

In stating the general problem of homogeneous, isothermal, linear viscoelastodynamics one will observe two major differences from the equivalent linear elastodynamic problem established by Eqs. (1.1)–(1.5). These are: (i) a time-dependent set of variables takes the place of the previously considered elastic constants, e.g. the Lamé set of constants, and (ii) a constitutive equation which relates the stress tensor to the displacement gradient history substitutes Hooke's law which is expressed by Eq. (1.5). In principle then, time and frequency domain BEMs similar to the ones described in previous sections should apply for the solution of the linear viscoelastodynamic problem as well. However, only a limited number of applications have been reported in the area.

The most frequently treated problem in linear visocelasticity is the quasi-static one, i.e. no inertia forces are taken into account in the equilibrium equation. The theoretical foundations of this problem as well as the basic ingredients required for a possible solution within a BEM formalism, i.e. the reciprocal integral equation and the necessary fundamental solutions, can be found in the classic works of Gurtin and Sternberg [37], Gurtin [36] and Sternberg and Al-Khozaie [120]. On the basic of these theoretical developments one can proceed to numerically formulate a BEM solution of the viscoelastic quasi-static problem in both the time and the frequency domains. Furthermore, with an appropriate use of the so-called "correspondence principle," e.g. see Refs. [13, 17, 18, 30], the BIE of the frequency transformed quasi-static problem of linear viscoelasticity can be written in a form resembling the familiar BIE in ordinary elastostatics with the understanding, of course, that the material variables depend on the frequency transform parameter under consideration. Once this static-like problem has been solved in the transformed domain, the solution of the original time domain problem can be obtained using an appropriate inversion technique [89, 107]. A 2D application of this frequency domain methodology was first reported by Rizzo and Shippy [100] and Shippy [115]. Following the same ideas the 3D problem of a spherical cavern enclosed by an elastic shell, embedded in an infinite viscoelastic medium, and subjected to an internal pressure has been solved by a BEM–FEM coupling technique by Hisatake, Ito and Ueda [40]. The correspondence principle of viscoelasticity is also valid for dynamic problems as explained, e.g., in Chirstensen [13], Manolis and Beskos [69] and Chapter 4 of this book. In the last two references BEM applications can also be found.

Robertson and Thomas [103] and Christensen [13] formulated and solved, under the assumption of constant Poisson's ratio, the complete linear viscoelastodynamic problem using a Green's function approach. Their approach was improved by Felszeghy, Sackman and Goldsmith [29]. However, the solutions provided in Refs. [13, 103, 29] are applicable to bounded bodies only, a fact that severely limits their application to SSI problems where infinite or semi-infinite soil media are encountered. A promising approach that can be used, along with the classical methods reported in Ref. [26], for establishing fundamental solutions for the complete linear viscoelastodynamic problem has been recently reported by Gaul [34].

Even though a viscoelastic model is a convenient mathematical representation of material damping in soils, it has been pointed out, e.g. Richart, Hall and Woods [98] or Hardin and Drnevich [38], that this is not a realistic assumption and the energy loss mechanism in soils is related to their general anelastic behavior. However, a widely accepted approximation, similar to the viscoelastic assumption, has been introduced according to which the elastic moduli of the problem under consideration are substituted, in the frequency domain only in accordance with the correspondence principle, by equivalent complex moduli [35, 121, 141]. Thus, the shear modulus for example is expressed in the form

$$\mu^* = \mu(1 + ig) \tag{5.1}$$

where μ is the shear modulus of the static case, and g is called the hysteretic damping coefficient and is independent of frequency. This approach has been used in BEM frequency domain formulations for the study of dynamic SSI problems, e.g. Wolf and Darbre [136–138], von Estorff and Schmid [27], and Abascal and Dominguez [1]. In addition, Wolf and Obernhuber [139] succeeded in calculating the time domain dynamic response of a rigid circular disk on a viscoelastic half-space by applying an inverse Fourier transformation to an approximate frequency domain solution.

5.1.2. *Anisotropic half-spaces*

While an isotropic material is mechanically characterized by two independent elastic constants, i.e. the two Lamé constants, a much larger number of elastic parameters, 21 in the most general case, is required for the mechanical characterization of an anisotropic material [62, 76]. As the number of elastic symmetries in the material increases, the number of elastic constants that are necessary for its mechanical characterization decreases. For the most commonly met soil deposits, formed usually by sedimentation or gravity forces, a cross-anisotropic fabric with a vertical axis of symmetry is frequently assumed, which reduces the number of required elastic constants to five. This number can be reduced even further for certain soil conditions [35].

The BEM has found a limited application in dealing with the static problem of anisotropic media, e.g. Vogel and Rizzo [130], Tomlin and Butterfield [124], Wilson and Cruse [135], and Nishimura and Kobayashi [82]. However, to the authors' best knowledge, the BEM has not been applied as yet to dynamic problems involving anisotropic half-spaces, although the basic ingredients for such an application are available. Indeed, Betti's reciprocal theorem remains valid for anisotropic media, and a variety of fundamental (Green's) solutions have been reported in the literature. For example, the equivalent Lamb's problem for an anisotropic half-space has been studied by Burridge [12] for a generally oriented free surface with respect to the crystallographic axes of the medium, while Payton in a series of papers, e.g. [90, 91, 93, 95], considered a variety of wave generating sources in 2D anisotropic media of infinite extent. Three-dimensional, time-dependent fundamental solutions for unbounded anisotropic media, i.e. extensions of Stokes' solutions, have also been reported. Thus, Payton [92] obtained closed-form solutions for a simplified three-parameter transversely isotropic elastic medium, while later the same author [94] produced closed-form solutions for the various displacement components on the axis of symmetry of a transversely isotropic medium when excited by a suddenly applied body force. Fundamental solutions for a transversely isotropic 3D half-space have also been reported by Payton [96, 97] for a buried point source, and by Ryan [104] for a surface load directed parallel to the vertical axis of material symmetry. The problem of surface loading on a general anisotropic half-space was also examined analytically and experimentally by Ricketts and Goldsmith [99]. Closed-form solutions for 3D half-spaces have not been obtained, but those reported in Refs. [96, 99, 104] can be evaluated numerically.

5.1.3. Irregular site conditions – soil layering

It is well understood today that not only the local material properties of the soil medium play an important role on the dynamic behavior of soil–structure systems but the local underground and surface topography as well. With the exception of some simple cases, e.g. see Luco [63, 64], a numerical solution of such problems is the only realistic approach in view of the complications that are imposed by arbitrary geometries and distributions of mechanical characteristics. The BEM is well suited for problems involving dynamic SSI and irregular site conditions, its application being a straightforward matter that has been studied in various works.

Consider for example the rather general problem shown in Fig. 15(a) of a foundation sitting on top of a number of soil deposits of arbitrary shape which in turn are supported by a number of soil layers. For the boundary points of every of the infinite or finite regions R_i ($i = 1, 2, \ldots, K$) a system of discretized BIEs can be written, e.g. Eq. (1.38) in the time domain or Eq. (1.50) in the transformed

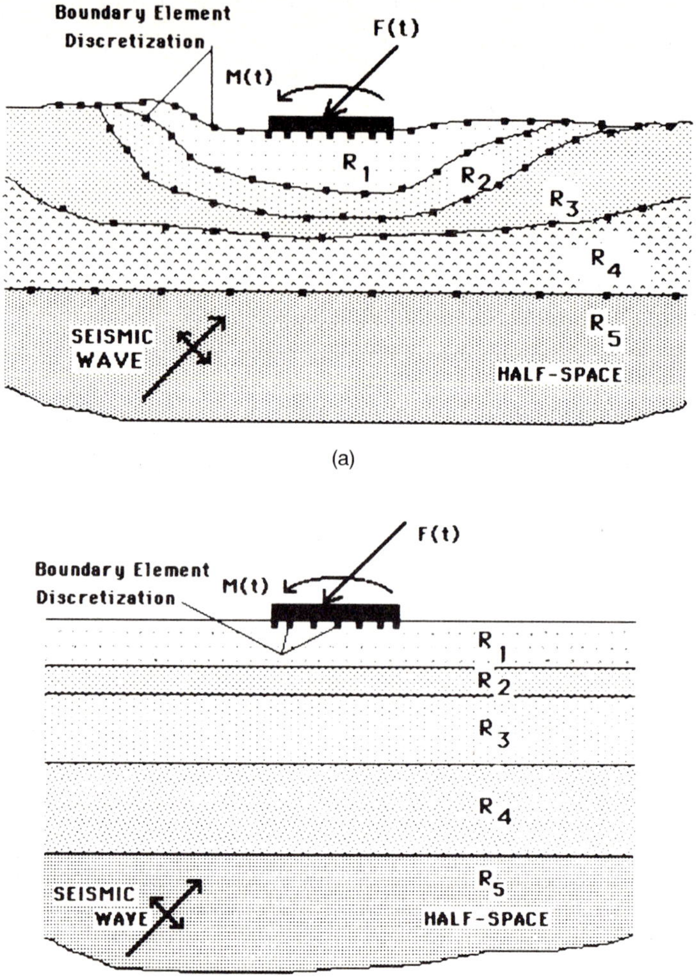

(b)

Fig. 15. Irregular site conditions and layered soil media – BEM discretizations.

domain, relating the bundary tractions and displacements of the region under consideration. This set of equations when supplemented by the equations that ensure equilibrium of stresses and continuation of displacements among the various regions as well as the boundary conditions on the free surface and the contact surface between the soil and the foundation will form a system of algebraic equations that can be solved in the manner discussed in previous sections of this article. The boundary discretization of each region shown in Fig. 15(a) is drawn on the assumption that infinite space fundamental solutions are used, but this is not always the case as it will be shown later. The procedure outlined above has been applied successfully to a number of problems similar to the one shown in Fig. 15(a), e.g. Dravinski [24], Kobori and Shinozaki [60], Toki and Sato [123], and Abascal and Dominguez [1].

A favorite topic among the researchers in the area seems to be the geometrically simpler case of a half-space that consists of a number of horizontal layers [28], Fig. 15(b). Of course the general, but straightforward, method outlined above for irregular site conditions can form the basis for a BEM application to these problems, as well. Indeed, this is the case in Ref. [1, 27], where the dynamic response of strip surface or embedded foundations on a single layer over an elastic half-plane or bedrock is studied in the frequency domain. A vibration isolation problem involving a multi-layered soil medium has been studied using layer interface discretization by Beskos, Leung and Vardoulakis [8]. Infinite-space solutions are used in Refs. [1, 8, 27] and discretizations over the infinite boundaries of each layer are theoretically necessary. However, due to the expected diminishing influence of faraway elements this problem can be avoided by appropriately truncating those infinite boundaries, Fig. 15(a). Alternatively, for this relatively simple geometry of a horizontally layered half-space a number of frequency-dependent fundamental solutions have been reported. Apparently the more general of those solutions are the ones provided by Luco and Apsel [4, 68] in a numerical form. Of engineering interest are also the analytical solutions of Bouchon [9] and particularly the solutions of Kausel and Peek [54] given in an explicit form. Recently, Wolf and Darbre [136–138] utilized the concept of the "source" line load, instead of the "source" point load that is used throughout this article, to derive frequency-dependent fundamental solutions for layered half-planes. Use of the BEM in conjunction with any of the above fundamental solutions in SSI interaction problems minimizes the amount of the required discretization as only the contact surface at the soil–foundation interface needs to be discretized [3, 136–138].

5.2. Piles

The pile structures encountered in this article are idealized as beams in the transverse direction, and columns in the axial direction. Since, to the authors' best knowledge, a time domain BEM has not been applied as yet to this area of research, only the frequency domain formulation of Banerjee and his co-workers [19, 109–111] will be considered herein.

The equations describing the dynamic response of massive piles can then be written in terms of the amplitude of the axial motion $\bar{u}_3 = \bar{u}_3(x_3, w)$ as

$$mw^2\bar{u}_3 + E_pA_p(\mathrm{d}^2\bar{u}_3/\mathrm{d}x_3^2) = -\pi D\bar{t}_3\,; \tag{5.2}$$

and in terms of the amplitude of the flexural, transverse, motion $\bar{u}_1 = \bar{u}_1(x_3, w)$ as

$$mw^2\bar{u}_1 - E_pI_p(\mathrm{d}^4\bar{u}_1/\mathrm{d}x_3^4) = D\bar{t}_1\,, \tag{5.3}$$

where E_p, A_p and I_p are, respectively, the modulus of elasticity, the area of the cross-section, and the moment of inertia of the pile, m is the mass per unit length of the pile, D is the diameter of the pile, and $\bar{t}_1 = \bar{t}_1(x_3, w)$ and $\bar{t}_3 = \bar{t}_3(x_3, w)$ are the amplitudes of the lateral and axial tractions on the pile. The general solutions of the differential Eqs. (5.2) and (5.3) can be written as

$$\bar{u}_3 = A_1 \sin(gx_3) + A_2 \cos(gx_3) \tag{5.4}$$

$$\bar{u}_1 = B_1 \sin(hx_3) + B_2 \cos(hx_3) + B_3 \sinh(hx_3) + B_4 \cosh(hx_3) , \tag{5.5}$$

where

$$g = [mw^2/E_pA_p]^{1/2} \qquad h = [mw^2/E_pI_p]^{1/4} ,$$

and the arbitrary constants can be determined from prescribed boundary conditions at the ends of the pile. Analytic expressions for the particular solutions of the differential equations (5.2) and (5.3) cannot be found in a direct way due to the complexity of the traction functions that are encountered in pile problems. However, an indirect solution to this problem is described in Refs. [109, 110]. First a point load is applied at some arbitrary location of the pile shaft and the general solutions (5.4) or (5.5) are used to describe the variation of displacements along the two unloaded segments of the pile above and below the point load. The boundary conditions at the head and the base of the pile augmented by the equilibrium and compatibility conditions at the location where the point load is applied provide the necessary equations for the calculation of the arbitrary constants. Then, as a second and final step, the response to a distributed loading can be obtained by integrating the solutions found in the first step over the length of the pile. Finally, a superposition of the solutions obtained for both the axial and the transverse motions yields

$$\{\bar{u}_p\} = [R]\{\bar{t}_p\} + \{\bar{d}_p\} , \tag{5.6}$$

where $\{\bar{d}_p\}$ is the displacement vector due to harmonic boundary displacements at the head of the pile, i.e. it corresponds to the general solutions of Eqs. (5.2) and (5.3), the matrix $[R]$ contains the constants of Eqs. (5.2) and (5.3) determined through the particular solutions, and the subscript p is generally used to denote quantities associated with the pile substructure.

When it comes to simulating the semi-infinite soil medium, the half-space frequency domain fundamental solutions of Kobayashi and Nishimura [58] have been used in Refs. [109, 110], in contrast to the infinite-space fundamental solutions which have been employed in this article in connection with foundation analysis. Thus, the soil medium needs to be discretized only at its contact surface with the pile. After a space discretization, similar to the one discussed in the previous sections, has been applied, the boundary integral equation describing the response of the soil medium can be written in the familiar form of a set of linear algebraic equations as

$$\{\bar{u}_s\} = [G]\{\bar{t}_s\} . \tag{5.7}$$

The compatibility and equilibrium conditions on the pile–soil interface are written directly in their frequency domain form as

$$\{\bar{u}_s(x_3, w)\} = \{\bar{u}_p(x_3, w)\} \tag{5.8}$$

$$\{\bar{t}_s(x_3, w)\} = -\{\bar{t}_p(x_3, w)\} \tag{5.9}$$

where the (2×1) displacement and traction vectors comprise the transverse and axial components of the pile motion, e.g. $\{\bar{u}_s\} = [\bar{u}_{s1}, \bar{u}_{s3}]^T$, etc.

Equations (5.6) and (5.7), along with the spatially discretized equivalents of Eqs. (5.8) and (5.9) form a system of equations with unknowns the displacements and tractions along the pile–soil interface. It is then elementary to find that the tractions acting on the pile are given by

$$\{\bar{t}_p\} = -[[R] + [G]]^{-1}\{\bar{d}_p\} . \tag{5.10}$$

The total axial forces, shear forces and bending moments acting on the pile at any depth x_3 can be easily computed by integrating the contact tractions and the inertia forces along the pile, i.e.

$$\begin{aligned} \bar{P}_3(x_3) &= \int_L^{x_3} [\pi D\bar{t}_3 + mw^2\bar{u}_3]\, dx_3 + A_b\bar{t}_{b3} \\ \bar{P}_1(x_3) &= \int_L^{x_3} [D\bar{t}_1 + mw^2\bar{u}_1]\, dx_3 \\ \bar{M}_2(x_3) &= \int_L^{x_3} [D\bar{t}_1 + mw^2\bar{u}_1]x_3\, dx_3 , \end{aligned} \tag{5.11}$$

where A_b and t_{b3} are the area and the vertical pressure at the base, respectively, and L the length of the pile.

The externally applied vertical force V, horizontal force H, and moment M can be expressed in terms of the displacements acting on the pile cap as

$$\{\bar{P}\} = [S]\{\bar{d}_p\} , \tag{5.12}$$

where $\{\bar{P}\} = [\bar{V} \quad \bar{H} \quad \bar{M}]^T$, and $\{\bar{d}_p\} = [\bar{u}_3 \quad \bar{u}_1 \quad \bar{\theta}_2]^T$ is the vector of the prescribed harmonic displacements of the pile cap. The (3×3) matrix $[S]$ is called the pile impedance matrix. Its elements are given in a complex form by the general formula

$$S_{ij} = K_{ij} + \mathrm{i}C_{ij} , \tag{5.13}$$

with the real part K_{ij} representing the impedance (stiffness) of the pile foundation and the imaginary part C_{ij} expressing the radiation damping as well as the material damping, if any.

The above basic formulation has also been used by Sen, Kausel and Banerjee [111], with the aid of the fundamental solutions obtained by the second of the authors [54], to study the dynamic behavior of piles and pile groups embedded in a layered soil medium.

Some results concerning the response of single piles to harmonic axial loads are shown in Fig. 16 [110]. The real and the imaginary parts of the axial impedances are plotted versus the dimensionless frequency which in this case is defined as

$$a_0 = wD/c_2 , \tag{5.14}$$

where D is the pile diameter. The results obtained by the frequency domain BEM

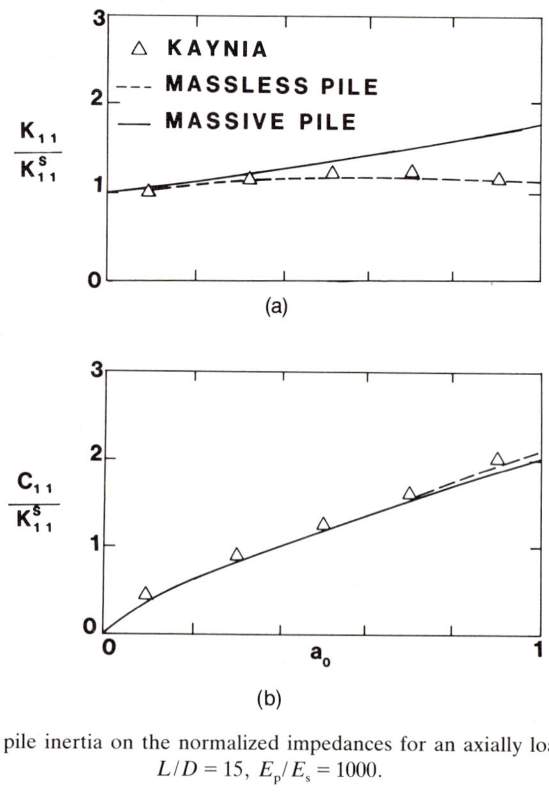

Fig. 16. Influence of pile inertia on the normalized impedances for an axially loaded single pile with $L/D = 15$, $E_p/E_s = 1000$.

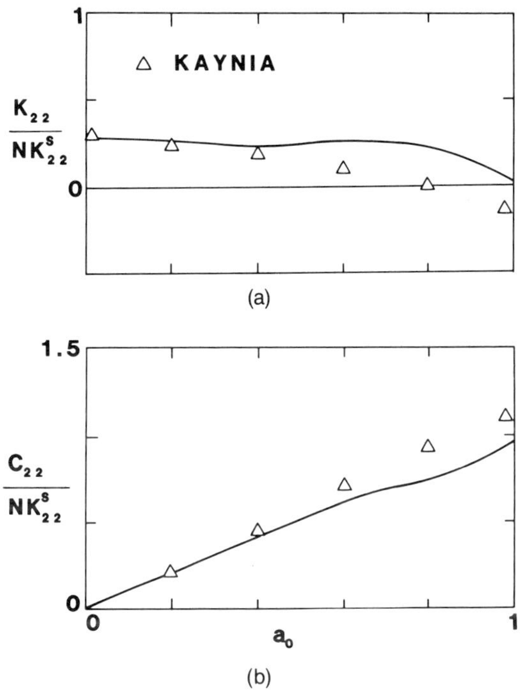

Fig. 17. Normalized impedances of laterally loaded 3 × 3 pile groups with $L/D = 15$, $E_p/E_s = 1000$, $S/D = 2$.

outlined in this article are presented with a solid line for the case of a massive pile and with a broken line for the case of a massive pile. It is obvious from Fig. 16 that the pile inertia influences substantially the real part of the axial impedance, particularly at high frequencies, while it has a minor effect on the imaginary part. In Fig. 16 a comparison with the results obtained by Kaynia [55] is also presented.

An example of the dynamic behavior of pile groups is shown in Fig. 17. The normalized impedance, i.e. the ratio of the impedance of the pile group over the single pile static stiffness times the number of piles in the group, of a laterally loaded 3 × 3 group of piles is plotted versus the dimensionless frequency for a spacing between piles equal to two pile diameters ($S/D = 2$). The results obtained by the BEM approach of Ref. [110] seem to be in a good agreement, at least for some part of the frequency range, with those obtained previously by Kaynia [55].

Examples of the application of the BEM in various other problems related to the dynamic analysis of piles and pile groups, e.g. piles under lateral loading, piles on layered media, etc., can be found in Refs. [109–111].

5.3. Vibration isolation of foundations

This is a subject of significant concern to foundation engineers as the number of sources that transmit vibrations through the ground keeps increasing in our industrialized environment. It is also a subject where the superiority of the BEM, in comparison to other numerical methods, can be fully demonstrated as far as accuracy and efficiency are concerned when dealing with media of infinite or semi-infinite extent.

A comprehensive account of the work on the subject up to 1969 can be found in the book of Richart, Hall and Woods [98] where, in addition, an extensive reference to the experimental work conducted by Woods [144] is made. A more recent report on the work in the area can be found in Emad and Manolis [25] and especially in Beskos and his co-workers [6–8] which are actually the only BEM applications on the subject. In Refs. [6–8, 25] the problem considered is that of vibration isolation by the use of open or infilled trenches appropriately located around the source of vibration (active isolation) or around the structure to be protected (passive isolation). The soil medium in Refs. [6, 7, 25] is assumed to be a linear elastic, homogeneous and isotropic half-space with a trench, and in all cases a 2D frequency domain analysis is performed using infinite plane fundamental solutions. In Ref. [8] the vibration isolation problem on a multi-layered soil medium is considered. It should be mentioned at this point that while in Ref. [25] the vibration source is just a vertical concentrated force, a rigid foundation under dynamic forces is the source of disturbance in Refs. [6–8] and thus the order of the problem is considerably increased.

The application of the BEM to the vibration isolation problem, as it is treated in Refs. [6–8, 25], is not different and actually it can be considered as a particular case to the one described in section 5.1.3 for irregular site conditions. However, in Refs. [6, 7, 25] the required computational effort seems to be far less than the one required by the problems in section 5.1.3 since in this case a surface discretization scheme is all that is needed. Indeed, Fig. 18(b) shows a possible discretization scheme employing constant boundary elements for the problem of

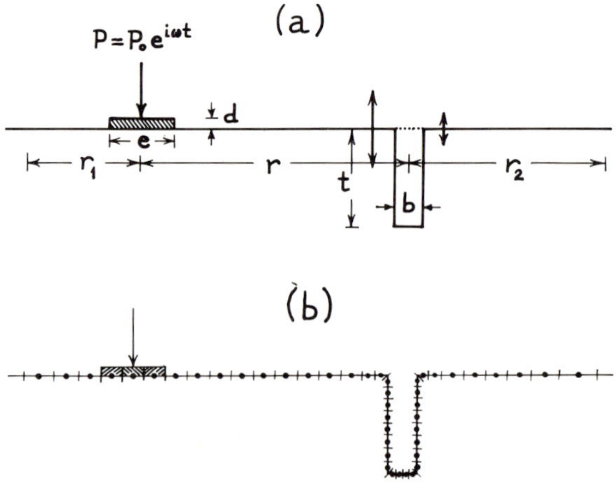

Fig. 18. Passive vibration isolation of machine foundation by open or infilled trench: (a) geometrical configuration; (b) boundary discretization.

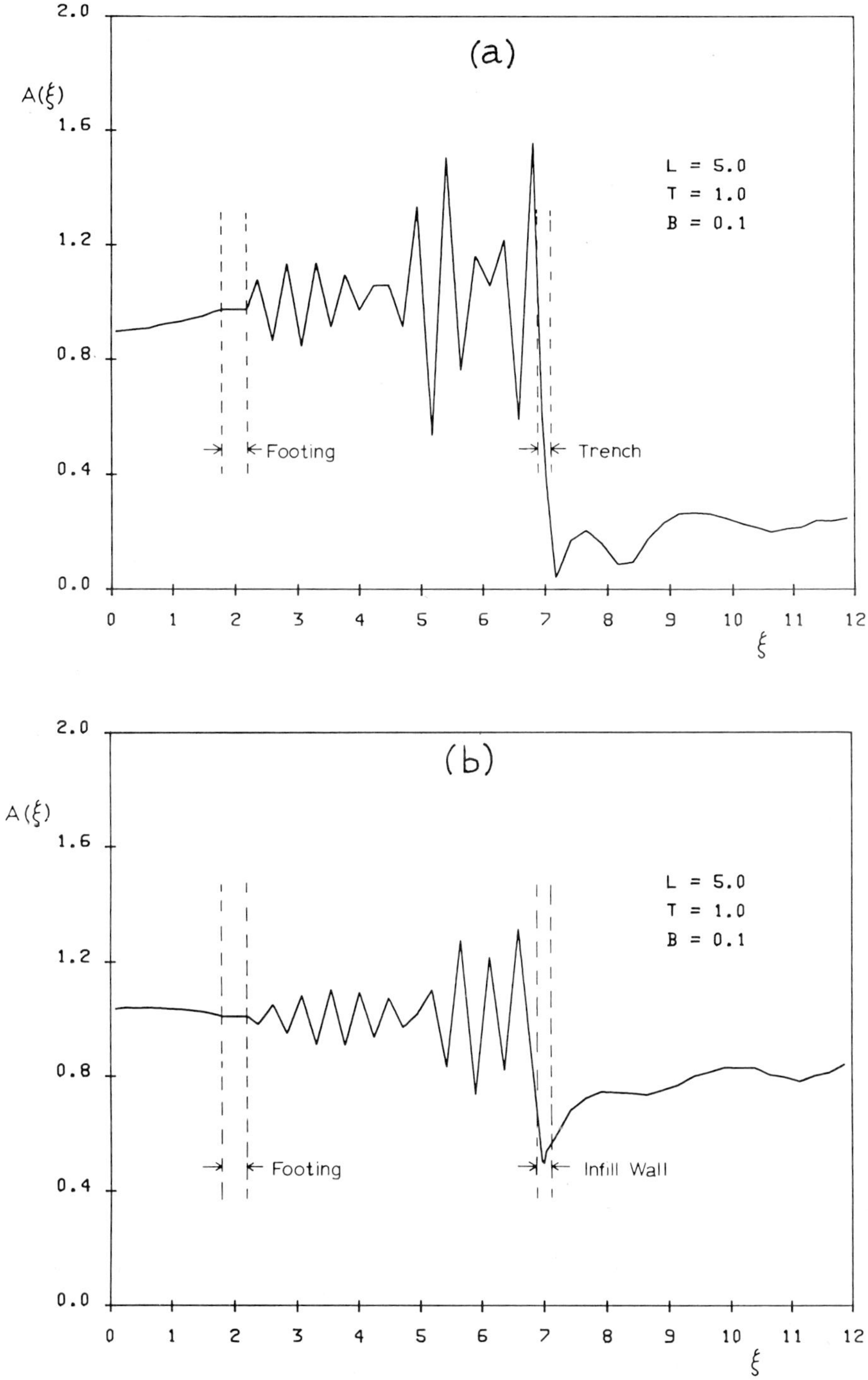

Fig. 19. Normalized amplitude $A(\xi)$ for passive vibration isolation of machine foundation by (a) open trench and (b) concrete filled trench.

passive vibration isolation in of Fig. 18(a). It is apparent that a straightforward application of the time or frequency domain BIE, in conjunction with the simple boundary conditions of the particular problem at hand, will produce the required displacements at any point on the surface of the half-space. It should, once again, be pointed out that if half-plane fundamental solutions were used a minimum discretization of the trench surface would be necessary. Nevertheless, the reader familiar with the application of the FEM in similar problems should appreciate by now the economy in data preparation, at least, that is required by the BEM even in the "adverse" case of not using the most appropriate fundamental solutions.

Some of the results obtained in Refs. [6, 7] and concerning the passive vibration isolation problem of Fig. 18 are shown next for the following numerical data: material properties of soil medium: $G_s = 132\ \mathrm{MN/m^2}$, $\nu_s = 0.25$, $\gamma_s = 17.5\ \mathrm{kN/m^2}$, $c_{Rs} = 250\ \mathrm{m/sec}$ and $\zeta_s = 6\%$; material properties of infill (concrete): $G_c = 34.29 G_s$, $\nu_c = \nu_s$, $\gamma_c = 1.37\gamma_s$, $c_{Rc} = 5c_{Rs}$ and $\zeta_c = 5\zeta_s$, where G, ν, γ, c_R and ζ stand for shear modulus, Poisson's ratio, specific weight, Rayleigh wave velocity and hysteretic damping coefficient, respectively; magnitude of applied force $P_0 = 1\ \mathrm{kN/m^2}$ and its operational frequency $w = 50\ \mathrm{Hz}$. Fig. 19(a,b) portrays the normalized vertical amplitude $A(\xi)$ at the surface of the soil as a function of the dimensionless distance $\xi = x/L_R$, where L_R is the Rayleigh wavelength for the open and the concrete infilled trench cases. $A(\xi)$ is defined as the ratio of the amplitude when the trench is present to the amplitude in the absence of the trench. In both cases the normalized dimensions of the trench are $T = t/L_R = 1.0$ and $B = b/L_R = 0.1$, the normalized dimensions of the foundation are $\pi = e/L_R = 0.5$ and $\delta = d/L_R = 0.1$, while the normalized distance between foundation and trench is $L = r/L_R = 5.0$. While it is obvious that in both cases there is a wave amplitude reduction behind the trench, this reduction is more pronounced for the case of the open trench, which proves to be, at least for the specific data utilized here, a successful wave barrier.

6. Concluding remarks

In this chapter an attempt is made to present the basic theoretical concepts of the BEM along with its numerical applications as they are related to some problems frequently encountered in SSI anaysis. Outlines of both the frequency and the time domain treatments of the problem under consideration are discussed and a brief comparison study is given in the "Numerical examples" section. The material covered in this exposition deals primarily with linear elastic, homogeneous and isotropic materials, and the use of the corresponding infinite-space fundamental solutions is emphasized. However, a brief reference to viscoelastic and anisotropic half-spaces is made where a need for further research exists. Furthermore, the benefits of utilizing certain simplifying assumptions, e.g. relaxed boundary conditions, in order to minimize the required numerical effort has been demonstrated.

Use of the BEM in the SSI area takes full advantage of the unique characteristic of the method which requires discretization of only the boundary of the domain of interest. This way the dimensions of the problem are reduced by one and the radiation conditions are automatically taken into account, thereby eliminating the need of non-reflecting boundaries which are usually employed by other volume discretization methods, i.e. the FEM or the FDM. The versatility of the BEM is also evident in view of the results presented in this chapter and other works referred to herein for various SSI problems. It is also shown that this surface gridding method can be properly combined with other volume gridding methods, such as the FEM or the FDM, to exploit the advantages of both methods.

In comparing the frequency and the time domain approaches which have been discussed in this work one should observe that the time domain BEM presents the advantages of providing the transient solution of the problem under consideration in one direct step, not two as the frequency domain methods do, and also forming the basis for an extension to nonlinear SSI problems. In general a time domain methodology seems to be a more natural way of treating problems related to wave propagation. However, the static-like character of the frequency domain BEM makes it easier to handle and get valuable insight of complicated problems at rather low numerical cost. In any case, both approaches have been reported to have excellent convergence characteristics and to possess a high degree of accuracy and efficiency.

Acknowledgements

The authors would like to express their gratitude to the National Science Foundation for supporting their work on soil–structure interaction by the BEM through the grants NSF/CEE-8024725 and NSF/CEE-8109723 to the University of Minnesota.

References

[1] R. Abascal and J. Dominguez, Dynamic Behavior of Strip Footings on Non-Homogeneous Viscoelastic Soils, in: *Dynamic Soil–Structure Interaction*, D.E. Beskos, T. Krauthammer and I. Vardoulakis, Eds. (A.A. Balkema, Rotterdam, 1984) pp. 25–35.

[2] E. Alarcón, J. Dominguez and F. del Cano, Dynamic Stiffnesses of Foundations, in: *New Developments in Boundary Element Methods*, C.A. Brebbia, Ed. (CML Publications, Southampton, 1980) pp. 264–280.

[3] R.J. Apsel, Dynamic Green's Functions for Layered Media and Applications to Boundary Value Problems, Ph.D. Thesis, University of California, San Diego, CA (1979).

[4] R.J. Apsel and J.E. Luco, On the Green's Functions for a Layered Half-Space. Part II, *Bulletin of the Seismological Society of America* **73**, 931–951 (1983).

[5] P.K. Banerjee and R. Butterfield, *Boundary Element Methods in Engineering Science* (McGraw-Hill, London, 1981).

[6] D.E. Beskos, B. Dasgupta and I.G. Vardoulakis, Vibration Isolation of Machine Foundations, in: *Vibration Problems in Geotechnical Engineering*, G. Gazetas and E.T. Selig, Eds. (ASCE, New York, 1985) pp. 138–151.

[7] D.E. Beskos, B. Dasgupta and I.G. Vardoulakis, Vibration Isolation Using Open or Filled Trenches. Part I: 2-D Homogeneous Soil, *Computational Mechanics* **1**, 43–63 (1986).

[8] D.E. Beskos, K.L. Leung and I.G. Vardoulakis, Vibration Isolation of Structures from Surface Waves in Layered Soil, in: *Recent Applications in Computational Mechanics*, D.L. Karabalis, Ed. (ASCE, New York, 1986) pp. 125–140.

[9] M. Bouchon, A Simple Method to Calculate Green's Functions for Elastic Layered Media, *Bulletin of the Seismological Society of America* **71**, 959–971 (1981).

[10] C.A. Brebbia and P. Chuang, Application of the Boundary Element Method for Solving Elastodynamic Problems, in: *Soil Dynamics and Earthquake Engineering*, A.S. Cakmak, A.M. Abdel-Ghaffar and C.A. Brebbia, Eds. (A.A. Balkema, Rotterdam, 1982) pp. 381–408.

[11] P.W. Buchen, The Elastodynamic Green's Tensor for the 2D Half-Space, *Journal of the Australian Mathematical Society* **20** (Series B), 385–400 (1978).

[12] R. Burridge, Lamb's Problem for an Anisotropic Half-Space, *Quarterly Journal of Mechanics and Applied Mathematics* **24**, 81–98 (1971).

[13] R.M. Christensen, *Theory of Viscoelasticity*, 2nd Edition (Academic Press, New York, 1982).

[14] D.M. Cole, D.D. Kosloff and J.B. Minster, A Numerical Boundary Integral Equation Method for Elastodynamics. I, *Bulletin of the Seismological Society of America* **68**, 1331–1357 (1978).

[15] T.A. Cruse and F.J. Rizzo, A Direct Formulation and Numerical Solution of the General Transient Elastodyamic Problem I, *Journal of Mathematical Analysis and Applications* **22**, 244–259 (1968).

[16] T.A. Cruse, A Direct Formulation and Numerical Solution of the General Transient Elastodynamic Problem II, *Journal of Mathematical Analysis and Applications* **22**, 341–355 (1968).

[17] G. Dasgupta and J.L. Sackman, An Alternative Representation of the Elastic-Viscoelastic Correspondence Principle for Harmonic Oscillations, *Journal of Applied Mechanics* **44**, 57–60 (1977).

[18] G. Dasgupta and J.L. Sackman, A Quadrature Representation of the Viscoelastic Analogy in the Frequency Domain, *Journal of Applied Mechanics* **45**, 955–956 (1978).

[19] T.G. Davies, R. Sen and P.K. Banerjee, Dynamic Analysis of Axially Loaded Piles, in: *Dynamic Soil–Structure Interaction*, D.E. Beskos, T. Krauthammer and I. Vardoulakis, Eds. (A.A. Balkema, Rotterdam, 1984) pp. 93–99.

[20] J. Dominguez, Dynamic Stiffness of Rectangular Foundations, Publication No. R78-20, Department of Civil Engineering, Massachussetts Institute of Technology, Cambridge, MA (1978).

[21] J. Dominguez, Response of Embedded Foundations to Travelling Waves, Publication No. R78-24, Department of Civil Engineering, Massachussetts Institute of Technology, Cambridge, MA (1978).

[22] J. Dominguez and E. Alarcón, Elastodynamics, in: *Progress in Boundary Element Methods*, Vol. 1, C.A. Brebbia, Ed. (Halsted Press, New York, 1981) pp. 213–257.

[23] J. Dominguez, E. Alarcón and R. Abascal, The Boundary Element Method in Elastodynamics, *Proc. 10th IMACS World Congress*, Concordia University, Montreal (1982).

[24] M. Dravinski, Scattering of Elastic Waves by an Alluvial Valley of Arbitrary Shape, Report No. CE 80-06, Department of Civil Engineering, University of Southern California, Los Angeles, CA (1980).

[25] K. Emad and G.D. Manolis, Shallow Trenches and Propagation of Surface Waves, *Journal of Engineering Mechanics, ASCE* **111**, 279–282 (1985).

[26] A.C. Eringen and E.S. Suhubi, *Elastodynamics* – Vol 11, Linear Theory (Academic Press, New York, 1975).

[27] O. von Estorff and G. Schmid, Application of the Boundary Element Method to the Analysis of the Vibration Behavior of Strip Foundations on a Soil Layer, in: *Dynamic Soil–Structure Interaction*, D.E. Beskos, T. Krauthammer and I. Vardoulakis, Eds. (A.A. Balkema, Rotterdam, 1984) pp. 11–17.

[28] W.M. Ewing, W.S. Jardetzky and F. Press, *Elastic Waves in Layered Media* (McGraw-Hill, New York, 1957).

[29] S.F. Felszeghy, J.L. Sackman and W. Goldsmith, Dynamic Response of a Bounded, Three-Dimensional, Viscoelastic Body, *Journal of the Acoustical Society of America* **59**, 551–556 (1976).
[30] W. Flügge, *Viscoelasticity* (Blaisdell Publishing Company, Waltham, Massachusetts, 1967).
[31] B. Frederick, Thick Rectangular Plates on an Elastic Foundation, *Transactions of ASCE* **122**, 1069–1085 (1957).
[32] M.B. Friedman and R. Shaw, Diffraction of Pulses by Cylindrical Obstacles of Arbitrary Cross Section, *Journal of Applied Mechanics* **29**, 40–46 (1962).
[33] R.H. Gallagher, *Finite Element Analysis – Fundamentals* (Prentice-Hall, Englewood Cliffs, NJ, 1975).
[34] L. Gaul, Analytical and Experimental Study on the Dynamic Interaction Between Viscoelastic Soil and Machine Foundations, in: *Dynamic Soil–Structure Interaction*, D.E. Beskos, T. Krauthammer and I. Vardoulakis, Eds. (A.A. Balkema, Rotterdam, 1984) pp. 113–123.
[35] G. Gazetas, Analysis of Machine Foundation Vibrations: State of the Art, *Soil Dynamics and Earthquake Engineering* **2**, 2–42 (1983).
[36] M.E. Gurtin, Variational Principles in the Linear Theory of Viscoelasticity, *Archive for Rational Mechanics and Analysis* **13**, 179–191 (1963).
[37] M.E. Gurtin and E. Sternberg, On the Linear Theory of Viscoelasticity, *Archive for Rational Mechanics and Analysis* **11**, 291–356 (1962).
[38] B.O. Hardin and V.P. Drnevich, Shear Modulus and Damping in Soils: Measurements and Parameter Effects, *Journal of the Soil Mechanics and Foundations Division, ASCE* **98**, 603–624 (1972).
[39] F. Hartmann, Elastostatics, in: *Progress in Boundary Element Methods*, Vol. 1, C.A. Brebbia, Ed. (Halsted Press, New York, 1981) pp. 84–167.
[40] M. Hisatake, T. Ito and H. Ueda, Three Dimensional Symmetric Coupling of Boundary and Finite Element Methods, in: *Boundary Elements*, C.A. Brebbia, T. Futagami, M. Tanaka, Eds. (Springer-Verlag, Berlin, 1983) pp. 985–994.
[41] A.T. de Hoop, Representation Theorems for the Displacement in an Elastic Solid and their Application to Elastodynamic Diffraction Theory, Thesis, Technische Hogeschool Delft, The Netherlands (1958).
[42] A.T. de Hoop, The Surface Line Source Problem, *Applied Scientific Research* (B)**8**, 349–356 (1959/1960).
[43] M. Iguchi and J.E. Luco, Dynamic Response of Flexible Rectangular Foundations on an Elastic Half-Space, *Earthquake Engineering and Structural Dynamics* **9**, 239–249 (1981).
[44] M. Iguchi and J.E. Luco, Vibration of Flexible Plate on Viscoelastic Medium, *Journal of the Engineering Mechanics Division, ASCE* **108**, 1103–1120 (1982).
[45] L.R. Johnson, Green's Function for Lamb's Problem, *Geophysical Journal of the Royal Astronomical Society* **37**, 99–131 (1974).
[46] D.L. Karabalis, Dynamic Response of Three-Dimensional Foundations, Ph.D. Thesis, University of Minnesota, Minneapolis (1984).
[47] D.L. Karabalis and D.E. Beskos, Dynamic Response of 3-D Rigid Surface Foundations by Time Domain Boundary Element Method, *Earthquake Engineering and Structural Dynamics* **12**, 73–93 (1984).
[48] D.L. Karabalis and D.E. Beskos, Earthquake Response of 3-D Foundtions by the Boundary Element Method, *Proceedings of the 8th World Conference of Earthquake Engineering*, San Francisco (1984) pp. 769–776.
[49] D.L. Karabalis, C.C. Spyrakos and D.E. Beskos, Dynamic Response of Surface Foundations by Time Domain Boundary Element Method, in: *Dynamic Soil–Structure Interaction*, D.E. Beskos, T. Krauthammer and I. Vardoulakis, Eds. (A.A. Balkema, Rotterdam, 1984) pp. 19–24.
[50] D.L. Karabalis and D.E. Beskos, Dynamic Response of 3-D Flexible Foundations by Time Domain BEM and FEM, *Soil Dynamics and Earthquake Engineering* **4**, 91–101 (1985).
[51] D.L. Karabalis and D.E. Beskos, Three-Dimensional Soil-Structure Interaction by Boundary Element Methods, in: *Topics in Boundary Element Research*, Vol. IV, C.A. Brebbia, Ed. (CML Publications, Southampton) (in print).
[52] D.L. Karabalis and D.E. Beskos, Dynamic Analysis of 3-D Rigid Embedded Foundations by

Time Domain Boundary Element Method, *Computer Methods in Applied Mechanics and Engineering* **56**, 91–120 (1986).

[53] E. Kausel, R.V. Whitman, J.P. Morray and F. Elsabee, The spring method for embedded foundations, *Nuclear Engineering and Design* **48**, 377–392 (1978).

[54] E. Kausel and R. Peek, Dynamic Loads in the Interior of a Layered Stratum: An Explicit Solution, *Bulletin of the Seismological Society of America* **72**, 1459–1481 (1982).

[55] A.M. Kaynia, Dynamic Stiffness and Seismic Response of Pile Groups, Ph.D. Thesis, Massachusetts Institute of Technology, Cambridge, MA (1982).

[56] Y. Kitamura and S. Sakurai, A Numerical Method Determining Dynamic Stiffness under Rigid Foundations Resting on an Elastic Half Space, *Transactions of JSCE* **11**, 94–95 (1979).

[57] Y. Kitamura and S. Sakurai, Dynamic Stiffness for Rectangular Rigid Foundations on a Semi-Infinite Elastic Medium, *International Journal for Numerical and Analytical Methods in Geomechanics* **3**, 159–171 (1979).

[58] S. Kobayashi and N. Nishimura, Green's Tensors for Elastic Half-spaces. – An Application of Boundary Integral Equation Method, in: *Memoirs of the Faculty of Engineering, Kyoto University* **42**, 228–241 (1980).

[59] S. Kobayashi, Some Problems of the Boundary Integral Equation Method in Elastodynamics, in: *Boundary Elements*, C.A. Brebbia, T. Futagami and M. Tanaka, Eds. (Springer-Verlag, Berlin, 1983) pp. 775–784.

[60] T. Kobori and Y. Shinozaki, Applications of the Boundary Integral Equation Method to Dynamic Soil-Structure Interaction Analysis under Topograpic Site Condition, in: *Boundary Elements*, C.A. Brebbia, T. Futagami and M. Tanaka, Eds. (Springer-Verlag, Berlin, 1983) pp. 731–740.

[61] S. Krenk and H. Schmidt, Vibration of an Elastic Circular Plate on an Elastic Half Space – A Direct Approch, *Journal of Applied Mechanics* **48**, 161–168 (1981).

[62] A.E.H. Love, *A Treatise on the Mathematical Theory of Elasticity*, 4th Edition (Dover, New York, 1944).

[63] J.E. Luco, Impedance Functions for a Rigid Foundation on a Layered Medium, *Nuclear Engineering and Design* **31**, 204–217 (1974).

[64] J.E. Luco, Vibrations of a Rigid Disc on a Layered Viscoelastic Medium, *Nuclear Engineering and Design* **36**, 325–340 (1976).

[65] J.E. Luco, Linear Soil-Structure Interaction: A Review, in: *Earthquake Ground Motion and Its Effects on Structures*, AMD-Vol. 53 (ASME, New York, 1982) pp. 41–57.

[66] J.E. Luco and H.L. Wong, Dynamic Response of Rectangular Foundations for Rayleigh Wave Excitation, in: *Proceedings of the 6th World Conference in Earthquake Engineering* (New Delhi, India, 1977) pp. 1542–1548.

[67] J.E. Luco and H.L. Wong, Response of structures to nonvertically incident seismic waves, *Bulletin of the Seismological Society of America* **72**, 275–302 (1982).

[68] J.E. Luco and R.J. Apsel, On the Green's Functions for a Layered Half-Space. Part I, *Bulletin of the Seismological Society of America* **73**, 909–929 (1983).

[69] G.D. Manolis and D.E. Beskos, Dynamic Stress Concentration Studies by Boundary Integrals and Laplace Transform, *International Journal for Numerical Methods in Engineering* **17**, 573–599 (1981).

[70] G.D. Manolis and D.E. Beskos, Dynamic Response of Lined Tunnels by an Isoparametric Boundary Element Method, *Comupter Methods in Applied Mechanics and Engineering* **36**, 291–307 (1983).

[71] G.D. Manolis, A Comparative Study on Three Boundary Element Method Approaches to Problems in Elastodynamics, *International Journal for Numerical Methods in Engineering* **19**, 73–91 (1983).

[72] G.D. Manolis, Boundary Element Method for Soil-Structure Interaction, in: *Dynamic Soil–Structure Interaction*, D.E. Beskos, T. Krauthammer and I. Vardoulakis, Eds. (A.A. Balkema, Rotterdam, 1984) pp. 85–91.

[73] W.J. Mansur and C.A. Brebbia, Numerical Implementation of the Boundary Element Method for Two Dimensional Transient Scalar Wave Propagation Problems, *Applied Mathematical Modelling* **6**, 299–306 (1982).

[74] W.J. Mansur and C.A. Brebbia, Formulation of the Boundary Element Method for Transient Problems Governed by the Scalar Wave Equation, *Applied Mathematical Modelling* **6**, 307–311 (1982).

[75] W.J. Mansur and C.A. Brebbia, Transient Elastodynamics Using a Time-Stepping Technique, in: *Boundary Elements*, C.A. Brebbia, T. Futagami and M. Tanaka, Eds. (Springer-Verlag, Berlin, 1983) pp. 677–698.

[76] W.P. Mason, *Physical Acoustics and the Properties of Solids* (D. van Nostrand, Princeton, NJ, 1958).

[77] A. Mita and W. Takanashi, Dynamic Soil-Structure Interaction Analysis by Hybrid Method, in: *Boundary Elements*, C.A. Brebbia, T. Futagami and M. Tanaka, Eds. (Springer-Verlag, Berlin, 1983) pp. 785–794.

[78] K.M. Mitzner, Numerical Solution for Transient Scattering from a Hard Surface of Arbitrary Shape-Retarded Potential Technique, *The Journal of the Acoustical Society of America* **42**, 391–397 (1967).

[79] G.V. Narayanan and D.E. Beskos, Numerical Operational Methods for Time-Dependent Linear Problems, *International Journal for Numerical Methods in Engineering* **18**, 1829–1854 (1982).

[80] D. Nardini and C.A. Brebbia, A New Approach to Free Vibration Analysis Using Boundary Elements, in: *Boundary Element Methods in Engineering*, C.A. Brebbia, Ed. (Springer-Verlag, Berlin, 1982) pp. 312–326.

[81] D. Nardini and C.A. Brebbia, Transient Dynamic Analysis by the Boundary Element Method, in: *Boundary Elements*, C.A. Brebbia, T. Futagami and M. Tanaka, Eds. (Springer-Verlag, Berlin, 1983) pp. 719–730.

[82] N. Nishimura and S. Kobayashi, A Boundary Integral Equation Formulation for Three-Dimensional Anisotropic Elastostatics, in: *Boundary Elements*, C.A. Brebbia, T. Futagami and M. Tanaka, Eds. (Springer-Verlag, Berlin, 1983) pp. 345–354.

[83] Y. Niwa, S. Kobayashi and N. Azuma, An Analysis of Transient Stresses Produced Around Cavities of Arbitrary Shape During the Passing of Traveling Waves, *Memoirs of the Faculty of Engineering, Kyoto University* **37**, 28–46 (1975).

[84] Y. Niwa, T. Fukui, S. Kato and K. Fujiki, An Application of the Integral Equation Method to Two-Dimensional Elastodynamics, *Theoretical and Applied Mechanics* **28**, 281–290 (1980).

[85] Y. Niwa, M. Kitahara and S. Hirose, Elastodynamic Problems for Inhomogeneous Bodies, in: *Boundary Elements*, C.A. Brebbia, T. Futagami and M. Tanaka, Eds. (Springer-Verlag, Berlin, 1983) pp. 751–763.

[86] M.A. Oien, Steady Motion of a Rigid Strip Bonded to an Elastic Half-Space, *Journal of Applied Mechanics* **38**, 328–334 (1971).

[87] M. Ottenstreuer and G. Schmid, Boundary Elements Applied to Soil-Foundation Interaction, in: *Boundary Elements Methods*, C.A. Brebbia, Ed. (Springer-Verlag, Berlin, 1981) pp. 293–309.

[88] M. Ottenstreuer, Frequency Dependent Dynamic Response of Footings, in: *Soil Dynamics and Earthquake Engineering*, A.S. Cakmak, A.M. Abdel-Ghaffar and C.A. Brebbia, Eds. (A.A. Balkema, Rotterdam, 1982) pp. 799–809.

[89] A. Papoulis, A New Method of Inversion of the Laplace Transform, *Quarterly of Applied Mathematics* **14**, 405–414 (1957).

[90] R.G. Payton, Two-Dimensional Anisotropic Elastic Waves Emanating from a Point Source, *Proceedings of Cambridge Philosophical Society* **70**, 191–210 (1971).

[91] R.G. Payton, Wave-Front Singularities for Two-Dimensional Anisotropic Elastic Waves, *Proceedings of Cambridge Philosophical Society* **72**, 105–116 (1972).

[92] R.G. Payton, Green's Tensor for a Constrained Transversely Isotropic Elastic Solid, *Quarterly Journal of Mechanics and Applied Mathematics* **28**, 473–481 (1975).

[93] R.G. Payton, Two-Dimensional Anisotropic Elastic Waves Emanating from a Point Source When the Associated Normal Curve Has Double Points, *International Journal of Engineering Science* **13**, 183–194 (1975).

[94] R.G. Payton, Symmetry-Axis Elastic Waves for Transversely Isotropic Media, *Quarterly of Applied Mathematics* **35**, 63–73 (1977).

[95] R.G. Payton, Plane Strain Displacement and Stress Waves Induced in a Transversely Isotropic Elastic Solid by a Line Source, *Journal of Applied Mathematics and Physics* **29**, 262–272 (1978).

[96] R.G. Payton, Epicenter Motion of a Transversely Isotropic Elastic Half-Space Due to a Suddenly Applied Buried Point Source, *International Journal of Engineering Science* **17**, 879–887 (1979).

[97] R.G. Payton, Epicenter and Epicentral-Axis Motion of a Transversely Isotropic Elastic Half-Space, *SIAM Journal of Applied Mathematics* **40**, 373–389 (1981).

[98] F.E. Richart, Jr., J.R. Hall, Jr. and R.D. Woods, *Vibrations of Soils and Foundations* (Prentice-Hall, Englewood Cliffs, NJ, 1970).

[99] T.E. Ricketts and W. Goldsmith, Wave Propagation in an Anisotropic Half-Space, *International Journal of Rock Mechanics and Mining Sciences* **9**, 493–512 (1972).

[100] F.J. Rizzo and D.J. Shippy, An Application of the Correspondence Principle of Linear Visco-Elasticity Theory, *SIAM Journal of Applied Mathematics* **21**, 321–330 (1971).

[101] F.J. Rizzo, D.J. Shippy and M. Rezayat, Boundary Integral Equation Analysis for a Class of Earth-Structure Interaction Problems, *Final Report*, NSF Research Grant No. CEE-8013461, University of Kentucky, Lexington, KY (1985).

[102] F.J. Rizzo, D.J. Shippy and M. Rezayat, A Boundary Integral Equation Method for Radiation and Scattering of Elastic Waves in Three Dimensions, *International Journal for Numerical Methods in Engineering* **21**, 115–129 (1985).

[103] S.R. Robertson and C.R. Thomas, Forced Motion of a Viscoelastic Solid, *The Journal of the Acoustical Society of America* **49**, 1673–1675 (1971).

[104] R.L. Ryan, Pulse Propagation in a Transversely Isotropic Half-Space, *Journal of Sound and Vibration* **12**, 511–524 (1971).

[105] T. Sato, H. Kawase and K. Yoshida, Dynamic Response Analysis of Rigid Foundations Subjected to Seismic Waves by Boundary Element Method, in: *Boundary Elements*, C.A. Brebbia, T. Futagami and M. Tanaka, Eds. (Springer-Verlag, Berlin, 1983) pp. 765–774.

[106] S.A. Savidis and T. Richter, Dynamic Response of Elastic Plates on the Surface of the Half-Space, *International Journal for Numerical and Analytical Methods in Geomechanics* **3**, 245–254 (1979).

[107] R.A. Schapery, Approximate Methods of Transform Inversion for Viscoelastic Stress Analysis, *Proceedings of the 4th US National Congress on Applied Mechanics*, R.M. Rosenberg, Ed. (ASME, New York, 1962) pp. 1075–1085.

[108] H. Schmidt and S. Krenk, Asymmetric Vibrations of a Circular Elastic Plate on an Elastic Half Space, *International Journal of Solids and Structures* **18**, 91–105 (1982).

[109] R. Sen, T.G. Davies and P.K. Banerjee, Dynamic Behavior of Axially and Laterally Loaded Piles and Pile Groups Embedded in Non-Homogeneous Soil, Report G7/1983/3, Department of Civil Engineering, SUNY, Buffalo, NY (1983).

[110] R. Sen, T.G. Davies and P.K. Banerjee, Dynamic Analysis of Piles and Pile Groups Embedded in Homogeneous Soils, *Earthquake Engineering and Structural Dynamics* **13**, 53–65 (1985).

[111] R. Sen, E. Kausel and P.K. Banerjee, Dynamic Analysis of Piles and Pile Groups Embedded in Non-Homogeneous Soils, *International Journal for Numerical and Analytical Methods in Geomechanics* **9**, 507–524 (1985).

[112] R.P. Shaw and M.B. Friedman, Diffraction of Pulses by Deformable Cylindrical Obstacles of Arbitrary Cross Section, in: *Proceedings of the 4th US National Congress of Applied Mechanics*, R.M. Rosenberg, Ed. (ASME, New York, 1962) pp. 371–379.

[113] R.P. Shaw, Diffraction of Acoustic Pulses by Obstacles of Arbitrary Shape with a Robin Boundary Condition, *The Journal of the Acoustical Society of America* **41**, 855–859 (1967).

[114] R.P. Shaw, An Integral Equation Approach to Diffusion, *International Journal of Heat and Mass Transfer* **17**, 693–699 (1974).

[115] D.J. Shippy, Application of the Boundary-Integral Equation Method to Transient Phenomena in Solids, in: *Boundary Integral Equation Method: Computational Applications in Applied Mechanics*, AMD-Vol. 11, T.A. Cruse and F.J. Rizzo, Eds. (ASME, New York, 1975) pp. 15–30.

[116] I.C. Simpson, On the Interaction of Rayleigh Surface Waves with Structures, *Earthquake Engineering and Structural Dynamics* **6**, 247–263 (1978).

[117] C.C. Spyrakos, Dynamic Response of Strip-Foundations by the Time Domain BEM-FEM Method, Ph.D. Thesis, University of Minnesota, Minneapolis, Minnesota, 1984.
[118] C.C. Spyrakos and D.E. Beskos, Dynamic Response of Rigid Strip Foundations by Time Domain Boundary Element Method, *International Journal for Numerical Methods in Engineering* **23**, 1547–1565 (1986).
[119] C.C. Spyrakos and D.E. Beskos, Dynamic Response of Flexible Strip Foundations by Boundary and Finite Elements, *Soil Dynamics and Earthquake Engineering* **5**, 84–96 (1986).
[120] E. Sternberg and S. Al-Khozaie, On Green's Functions and the Saint-Venant's Principle in the Linear Theory of Viscoelasticity, *Archive for Rational Mechanics and Analysis* **15**, 112–145 (1964).
[121] J.L. Tassoulas and E. Kausel, Elements for the Numerical Analysis of Wave Motion in Layered Strata, *International Journal for Numerical Methods in Engineering* **19**, 1005–1032 (1983).
[122] S.A. Thau, Radiation and Scattering from a Rigid Inclusion in an Elastic Medium, *Journal of Applied Mechanics, ASME* **34**, 509–511 (1967).
[123] K. Toki and T. Sato, Seismic Response Analyses of Ground with Irregular Profiles by the Boundary Element Method, in: *Boundary Elements*, C.A. Brebbia, T. Futagami and M. Tanaka, Eds. (Springer-Verlag, Berlin, 1983) pp. 699–709.
[124] G.R. Tomlin and R. Butterfield, Elastic Analysis of Zoned Orthotropic Continua, *Journal of the Engineering Mechanics Division, ASCE* **100**, 511–529 (1974).
[125] O. Tullberg, BEMDYN – A Boundary Element Program for Two-Dimensional Elastodynamics, in: *Boundary Elements*, C.A. Brebbia, T. Futagami and M.Tanaka, Eds. (Springer-Verlag, Berlin, 1983) pp. 835–845.
[126] F.K. Tzung and T.H. Lee, A New Transmitting Boundary Representation for Time-Domain Analysis of Soil-Structure Interaction Problems, *Proceedings of the 6th International Conference on Structural Mechanics in Reactor Technology* (K4/4), Paris (1981).
[127] P. Underwood and T.L. Geers, Doubly Asymptotic Boundary Element Analysis of Dynamic Soil-Structure Interaction, *International Journal of Solids and Structures* **17**, 687–697 (1981).
[128] A.S. Veletsos and B. Verbic, Basic Response Functions for Elastic Foundations, *Journal of the Engineering Mechanics Division, ASCE* **100**, 189–202 (1974).
[129] A.S. Veletsos and Y.T. Wei, Lateral and Rocking Vibrations of Footings, *Journal of the Soil Mechanics and Foundations Division, ASCE* **97**, 1227–1248 (1971).
[130] S.M. Vogel and F.J. Rizzo, An Integral Equation Formulation of Three Dimensional Anisotropic Elastostatics Boundary Value Problems, *Journal of Elasticity* **3**, 203–216 (1973).
[131] S. Walker, Fundamental Solutions, in: *Progress in Boundary Element Methods*, Vol. 1, C.A. Brebbia, Ed. (Halsted Press, New York 1981) pp. 13–44.
[132] R.V. Whitman and J. Bielak, Foundations, in: *Design of Earthquake Resistant Structures*, E. Rosenblueth, Ed. (John Wiley & Sons, New York, 1980) pp. 223–260.
[133] W.L. Whittaker and P. Christiano, Dynamic Response of Plate on Elastic Half-Space, *Journal of the Engineering Mechanics Division, ASCE* **108**, 133–154 (1982).
[134] W.L. Whittaker and P. Christiano, Response of a Plate and Elastic Half-Space to Harmonic Waves, *Earthquake Engineering and Structural Dynamics* **10**, 255–266 (1982).
[135] R.B. Wilson and T.A. Cruse, Efficient Implementation of Anisotropic Three Dimensional Boundary-Integral Equation Stress Analysis, *International Journal for Numerical Methods in Engineering* **12**, 1383–1397 (1978).
[136] J.P. Wolf and G.R. Darbre, Dynamic-Stiffness Matrix of Soil by the Boundary-Element Method: Conceptual Aspects, *Earthquake Engineering and Structural Dynamics* **12**, 385–400 (1984).
[137] J.P. Wolf and G.R. Darbre, Dynamic-Stiffness Matrix of Soil by the Boundry-Element Method: Embedded Foundation, *Earthquake Engineering and Structural Dynamics* **12**, 401–416 (1984).
[138] J.P. Wolf and G.R. Darbre, Dynamic-Stiffness Matrix of Embedded and Pile Foundations by Indirect Boundary Element Method, *Nuclear Engineering and Design* **80**, 331–342 (1984).
[139] J.P. Wolf and P. Obernhuber, Non-Linear Soil-Structure-Interaction Analysis Using Dynamic Stiffness or Flexibility of Soil in the Time Domain, *Earthquake Engineering and Structural Dynamics* **13**, 195–212 (1985).
[140] J.P. Wolf and P. Obernhuber, Non-Linear Soil-Structure-Interaction Analysis Using Green's

Function of Soil in the Time Domain, *Earthquake Engineering and Structural Dynamics* **13**, 213–223 (1985).
[141] J.P. Wolf, *Dynamic Soil–Structure Interaction* (Prentice-Hall, Englewood Cliffs, NJ, 1985).
[142] H.L. Wong and J.E. Luco, Dynamic response of rigid foundations of arbitrary shape, *Earthquake Engineering and Structural Dynamics* **4**, 579–587 (1976).
[143] H.L. Wong and J.E. Luco, Dynamic response of rectangular foundations to obliquely incident seismic waves, *Earthquake Engineering and Structural Dynamics* **6**, 3–16 (1978).
[144] R.D. Woods, Screening of Elastic Surface Waves by Trenches, Ph.D. Thesis, University of Michigan (1967).

CHAPTER 12

Fluid–Structure Interaction

Carlos A. BREBBIA

Computational Mechanics Institute
Wessex Institute of Technology
Ashurst Lodge, Ashurst
Southampton, SO1 2DD, United Kingdom

and

José C.F. TELLES

Department of Civil Engineering
COPPE – Federal University of Rio de Janeiro
Rio de Janeiro 21945, Brazil

Boundary Element Methods in Mechanics
Edited by D.E. Beskos

Introduction

The determination of internal pressures produced at the interface between fluid and structural elements, i.e. interaction problems, is of great importance in current engineering practice. Interaction problems are all somehow related to each other and appear in many different branches of engineering such as acoustics, hydrodynamics, electrostatics, electromagnetics and many other wave propagation problems; e.g. earthquake analysis and vibration of elastic bodies in an infinite or semi-infinite medium.

A particularly important problem which falls in this category is the study of propagation of surface waves and their interaction with fixed or floating bodies commonly found in offshore and ocean engineering applications. Until recently [5, 25], the linear wave theory was a starting assumption which leads to linearized boundary conditions applied at the mean free surface of the fluid. More recently, however, extensions to include second-order or even complete nonlinear analysis have been presented [11, 18, 20], but more effort is yet to be concentrated on this subject. A common feature in both cases, i.e. linear and nonlinear wave theory, is that the Boundary Element Method (BEM) has emerged as a powerful solution technique to the corresponding approaches.

Another important application of the BEM which also comprises recent nonlinear applications is the study of sloshing problems [1, 16, 26, 21].

In the present chapter we shall mainly concentrate on the different existing techniques to solve fluid–structure interaction problems by either using a pure boundary element procedure or a coupling analysis in which the versatility of the finite element method can be advantageous for a finite region of the problem.

1. Methods of analysis

In order to understand the way in which fluid and structure systems can be combined one has to look at the different formulations which may be used. The most widely available techniques nowadays are the finite and boundary element methods. In many cases we need to combine both solutions and to do so properly one needs to investigate the differences between the two of them. Sometimes it is possible to solve the problem in terms of boundary elements only for the solid and the fluid. In this case the problem reduces to a combination of two boundary element regions which can be done in the usual manner, i.e. satisfying continuity and equilibrium at the interface. Special care should be taken, however, to identify the way in which the variables need to be transformed before combination, i.e. velocities in the fluid may be combined with displacements in the structure and potentials with pressures through Bernoulli's equation.

The relevant analytical formulations used in the case of fluid–structure interaction studies are (i) Finite Elements; (ii) Boundary Elements; and (iii) Simplified Boundary Elements. Let us discuss them in that order.

1.1. *Finite elements*

The finite element theory is well known and will not be reviewed here. It is important, however, to mention that it is based on the reduction of the order of some derivatives and uses an error distribution approach to find the optimum approximation. The technique only requires approximate satisfaction of the natural boundary conditions of the problems and of the governing equation within the region. For the general case of a structural system the equations produced are

$$\mathbf{M}\ddot{U} + \mathbf{C}\dot{U} + \mathbf{K}U = F\ , \tag{1.1}$$

where $\mathbf{K}$ is the stiffness, $\mathbf{C}$ the damping (usually taken to be proportional to $\mathbf{K}$ or $\mathbf{M}$ or both) and $\mathbf{M}$ is the mass matrix; U are displacements and F equivalent nodal forces.

The finite element matrices obtained in this way are symmetric because they are deduced from symmetric operators using the same basis functions for the approximation and its weighting function.

Finite elements in this chapter will only be used in certain applications to define the structure. The fluid itself will always be described using boundary elements.

1.2. *Boundary elements*

While the classical finite element theory uses the same type of functions for the approximate and weighting functions, special types of functions are postulated in boundary elements for the weighting. They are such that they satisfy the equilibrium equations of the problem and usually (but not necessarily) are Green's functions with finite or semi-infinite domains.

Under these conditions the potential u at any point on the surface Γ of a boundary element region is related to the potential and fluxes ($q = \partial u/\partial n$) at all the other points by the equation [5]

$$cu + \int_\Gamma uq^*\,\mathrm{d}\Gamma = \int_\Gamma qu^*\,\mathrm{d}\Gamma\ . \tag{1.2}$$

The coefficient c depends on the type of point under consideration, i.e. it is $\frac{1}{2}$ for smooth boundaries and proportional to the solid angle for corner points. The boundary Γ is divided into two parts, one where the u are known ($u = \bar{u}$ on Γ_1) and the other where q are given ($q = \bar{q}$ on Γ_2). Notice that $\Gamma = \Gamma_1 + \Gamma_2$ and that there may be other types of boundary conditions, such as $\alpha u + \beta q = \gamma$, where α, β and γ are known parameters.

If one applies the above relationship at all points on the boundary a system of equations is obtained which, before applying the boundary conditions can be written in matrix form as

$$\mathbf{H}U = \mathbf{G}Q\,, \tag{1.3}$$

where U and Q are the vectors of the displacements u and fluxes q, respectively.

After the boundary conditions on Γ_1 and Γ_2 have been taken into consideration the columns of the above matrices can be rearranged to produce the system,

$$\mathbf{A}X = B \tag{1.4}$$

where X is the vector of unknown u and q values.

Notice that the values of the Q vector in Eq. (1.3) are the actual values of the fluxes at the nodes. In finite element analysis instead, forces or flux values are weighted and the integrated fluxes are concentrated at the nodes, i.e. in Eq. (1.1) the values of F are equivalent nodal forces. It is possible to relate the equivalent to the actual values of forces or fluxes at the nodes by defining a distribution matrix $\mathbf{N}$ whose terms represent the weighting of the boundary values of flux or tractions by the interpolation functions used in the finite element model, i.e.

$$F = \mathbf{N}Q\,. \tag{1.5}$$

In order to combine the boundary element region with the finite element one (see Fig. 1) one can proceed to apply compatibility and equilibrium at the nodes and superimpose the two systems or to transform the boundary element matrices into equivalent finite element ones.

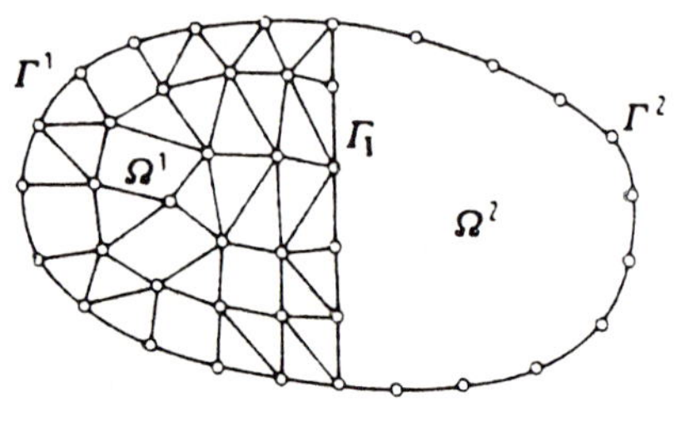

Fig. 1. Domain divided into finite element and boundary element regions.

In the case of time-independent problems such as the ones defined by Eq. (1.3) one can start by inverting $\mathbf{G}$ to obtain,

$$\mathbf{G}^{-1}(\mathbf{H}U) = Q . \qquad (1.6)$$

Next we premultiply both sides by the distribution matrix $\mathbf{N}$ to obtain the weighted flux vector F of finite elements, i.e.

$$(\mathbf{N}\mathbf{G}^{-1}\mathbf{H})U = \mathbf{N}Q , \qquad (1.7)$$

which can be represented as

$$\mathbf{K}' = \mathbf{N}\mathbf{G}^{-1}\mathbf{H} ,$$
$$F' = \mathbf{N}Q . \qquad (1.8)$$

Hence Eq. (1.7) has the following finite element form:

$$\mathbf{K}'U = F' . \qquad (1.9)$$

The major discrepancy that now arises between the above formulation and finite elements is the fact that the matrix $\mathbf{K}'$ is generally asymmetric. The asymmetry arises due to the approximation involved in the discretization process and the choice of the fundamental solution. The matrix can be made symmetric by minimizing the square of the errors in the non-symmetric off-diagonal terms as the asymmetry is usually small. This gives a new matrix whose coefficients are defined by

$$\mathbf{K} = \tfrac{1}{2}(\mathbf{K}' + \mathbf{K}'^{,\mathrm{T}}) \qquad (1.10)$$

This matrix can now be assembled with the finite element matrices to give a global system of equations.

The other possibility, which does not require any matrix inversion, is to operate directly with the finite element matrices and transform them into equivalent boundary element ones for combination [8]. In the example presented in Fig. 1 one has, for the boundary element region

$$[\mathbf{H}^2 \quad \mathbf{H}_\mathrm{I}^2]\begin{Bmatrix} U^2 \\ U_\mathrm{I}^2 \end{Bmatrix} = [\mathbf{G}^2 \quad \mathbf{G}_\mathrm{I}^2]\begin{Bmatrix} Q^2 \\ Q_\mathrm{I}^2 \end{Bmatrix} \qquad (1.11)$$

and for the finite element region

$$[\mathbf{K}^1 \quad \mathbf{K}_\mathrm{I}^1]\begin{Bmatrix} U^1 \\ U_\mathrm{I}^1 \end{Bmatrix} = [\mathbf{N}^1 \quad \mathbf{N}_\mathrm{I}^1]\begin{Bmatrix} Q^1 \\ Q_\mathrm{I}^1 \end{Bmatrix} , \qquad (1.12)$$

where the compatibility and equilibrium conditions over the interface Γ_I are simply

$$U_I^1 = U_I^2 = U_I$$

and

$$-Q_I^1 = Q_I^2 = Q_I . \tag{1.13}$$

The two systems of equations can therefore be reordered and become

$$[\mathbf{H}^2 \quad \mathbf{H}_I^2 \quad -\mathbf{G}_I^2]\begin{Bmatrix} U^2 \\ U_I \\ Q_I \end{Bmatrix} = [\mathbf{G}^2]\{Q^2\} \tag{1.14}$$

and

$$[\mathbf{K}^1 \quad \mathbf{K}_I^1 \quad \mathbf{N}_I^1]\begin{Bmatrix} U^1 \\ U_I \\ Q_I \end{Bmatrix} = [\mathbf{N}^1]\{Q^1\} . \tag{1.15}$$

Finally, systems (1.14) and (1.15) can be written together in the form

$$\begin{bmatrix} \mathbf{K}^1 & \mathbf{K}_I^1 & \mathbf{N}_I^1 & \mathbf{0} \\ \mathbf{0} & \mathbf{H}_I^2 & -\mathbf{G}_I^2 & \mathbf{H}^2 \end{bmatrix} \begin{Bmatrix} U^1 \\ U_I \\ Q_I \\ U^2 \end{Bmatrix} = \begin{bmatrix} \mathbf{N}^1 & \mathbf{0} \\ \mathbf{0} & \mathbf{G}^2 \end{bmatrix} \begin{Bmatrix} Q^1 \\ Q^2 \end{Bmatrix} . \tag{1.16}$$

Hence, the system of equations (1.16) can be solved in standard boundary element fashion after the application of the boundary conditions over the boundaries Γ_1 and Γ_2. This second procedure has been extensively applied to solve elastodynamic problems [15].

For the case of general elastodynamic problems, Nardini and Brebbia [22] have demonstrated how to obtain a system of partial differential equations in time similar to those obtained in finite elements, i.e.

$$\mathbf{H}U - \mathbf{G}P + \mathbf{M}\ddot{U} = \mathbf{0} , \tag{1.17}$$

where the vector $\boldsymbol{U}$ represents now the nodal displacements and the vector $\boldsymbol{P}$ contains the nodal tractions. This vector $\boldsymbol{P}$ is the equivalent to vector $\boldsymbol{Q}$ in potential problems.

After applying the homogeneous kinematic boundary conditions and grouping on the right-hand side all the known tractions, we obtain a smaller system of equations, which for the sake of simplicity will continue to be defined as above, i.e. as

$$\mathbf{H}U + \mathbf{M}\ddot{U} = \mathbf{G}P \tag{1.18}$$

where $\boldsymbol{P}$ is the vector of the known values of p on the boundary Γ_2 and $\boldsymbol{U}$ and $\ddot{\boldsymbol{U}}$ are the free displacements and accelerations for the structure, i.e. the ones after substituting the homogeneous boundary conditions.

Following the same ideas as before we can write

$$\mathbf{G}^{-1}\mathbf{H}U + \mathbf{G}^{-1}\mathbf{M}\ddot{U} = \boldsymbol{P} \tag{1.19}$$

and distributing

$$(\mathbf{N}\mathbf{G}^{-1}\mathbf{H})U + (\mathbf{N}\mathbf{G}^{-1}\mathbf{M})\ddot{U} = \mathbf{N}\boldsymbol{P} \tag{1.20}$$

which can now be defined as,

$$\mathbf{K}'U + \mathbf{M}'\ddot{U} = \boldsymbol{F}' . \tag{1.21}$$

The matrices $\mathbf{K}'$ and $\mathbf{M}'$ are unsymmetric. As before we can attempt to symmetricise them as follows:

$$\mathbf{K} = \tfrac{1}{2}(\mathbf{K}' + \mathbf{K}'^{,\mathrm{T}}) ; \qquad \mathbf{M} = \tfrac{1}{2}(\mathbf{M}' + \mathbf{M}'^{,\mathrm{T}}) . \tag{1.22}$$

The coupling of boundary element solutions with finite element ones or the coupling of two boundary element solutions will tend to produce densely populated matrices. Certain cases, such as problems extending to infinity, can be accurately modelled using the BEM but the system of equations is such that the equations produced by the boundary elements are all coupled and this gives a full system of equations. Because of this it is sometimes preferable to use approximate boundary elements which do not produce a fully coupled system.

1.3. Approximate boundary elements

These elements are an approximation to the boundary elements developed in the previous section but contrary to the approximate finite element formulations they only require integration over the surface rather than the domain. Because of this they are popular and if applied properly produce accurate results. The idea behind it is to use some engineering intuition and approximate Green's functions which may be of practical interest in many cases where the full fundamental solution may be difficult to obtain or cumbersome to use. The advantage of this formulation is that contrary to the boundary element treatment of the previous section, each boundary node is usually only related to its neighbors rather than to all the boundary nodes.

To illustrate the technique consider for instance the case of wave or Helmholtz equation [5, 6]

$$\nabla^2 u + \kappa^2 u = 0 \tag{1.23}$$

in the domain Ω where κ is the wave-number and u the potential. The fundamental solution for the two-dimensional case is

$$u^* = \tfrac{1}{4}\mathrm{i}H_0^{(1)}(\kappa r)\,, \tag{1.24}$$

where H is the Hankel function and r is the distance from the source point under consideration to any other field point.

As the u^* function satisfies the wave equation we can write for any source point external to the boundary the integral relation

$$\int_{\Gamma_\mathrm{I}} u^* q \,\mathrm{d}\Gamma = \int_{\Gamma_\mathrm{I}} q^* u \,\mathrm{d}\Gamma\,, \tag{1.25}$$

where Γ_I defines the interface as shown in Fig. 2. In terms of the fundamental solution, Eq. (1.25) becomes

$$\int_{\Gamma_\mathrm{I}} H_0^{(1)}(\kappa r) q \,\mathrm{d}\Gamma = \int_{\Gamma_\mathrm{I}} \frac{\partial H_0^{(1)}(\kappa r)}{\partial n}\, u \,\mathrm{d}\Gamma\,. \tag{1.26}$$

If one assumes to be sufficiently far away from the scene of perturbation to the solution, or Γ_I has a circular shape with a sufficiently large radius R so as to have $R \equiv n$, one can simplify the Hankel function and its derivative and obtain the expression

$$\left(\frac{2}{\pi\kappa}\right)^{1/2} \mathrm{e}^{-\mathrm{i}\pi/4} \int_{\Gamma_\mathrm{I}} \frac{1}{R^{1/2}} \left(\frac{\partial u}{\partial R} - \mathrm{i}\kappa u\right) \mathrm{e}^{\mathrm{i}\kappa R}\,\mathrm{d}\Gamma = 0\,. \tag{1.27}$$

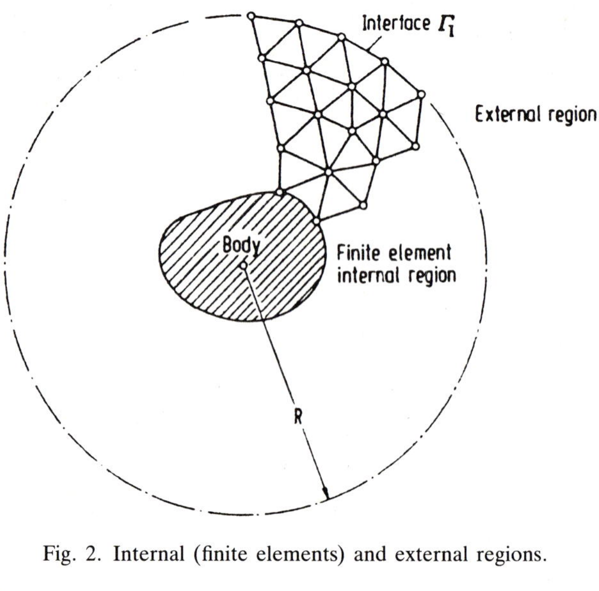

Fig. 2. Internal (finite elements) and external regions.

As R can be consided constant over Γ_{I} we can write $\mathrm{d}\Gamma = R\,\mathrm{d}\theta$ and (1.27) becomes

$$\int_0^{2\pi} R^{1/2}\left(\frac{\partial u}{\partial R} - \mathrm{i}\kappa u\right)\mathrm{d}\Gamma = 0\,. \tag{1.28}$$

For this particular case this is just another form of finding the Sommerfeld's condition which can be written on Γ_{I} as follows:

$$\frac{\partial u}{\partial R} - \mathrm{i}\kappa u = 0\,. \tag{1.29}$$

Notice that the above methodolody offers a procedure to deduce the appropriate radiation conditions for stress analysis, soil dynamics, fluid flow, acoustics and many other engineering problems.

2. Matrix procedure

In coupled problems one frequently needs to combine two or more regions together. This may be due to one of the regions being a fluid and another one a solid for instance. Consider for simplicity the case of a body divided into only two regions (Fig. 3) called Ω^1 and Ω^2. The interface between the two regions will be called Γ_{I} as usual and the external surfaces of regions Ω^1 and Ω^2 will be defined as Γ_1 and Γ_2 respectively. The potential function, its velocity and acceleration vectors over each region will be called $\boldsymbol{U}$, $\dot{\boldsymbol{U}}$ and $\ddot{\boldsymbol{U}}$, respectively. The fluxes at the nodes and their integrated values are designated by $\boldsymbol{Q}$ and $\boldsymbol{F}$. Hence $\boldsymbol{U}^1$, $\boldsymbol{Q}^1$, $\boldsymbol{F}^1$ will denote the function, fluxes and integrated fluxes on the external surface of region 1, i.e. on Γ_1 and $\boldsymbol{U}^2$, $\boldsymbol{Q}^2$, $\boldsymbol{F}^2$ the same values for the external surface of region 2, i.e. on Γ_2. $\boldsymbol{U}_{\mathrm{I}}^1$, $\boldsymbol{Q}_{\mathrm{I}}^1$ and $\boldsymbol{F}_{\mathrm{I}}^1$ are similar quantities on the interface Γ_{I} considering that

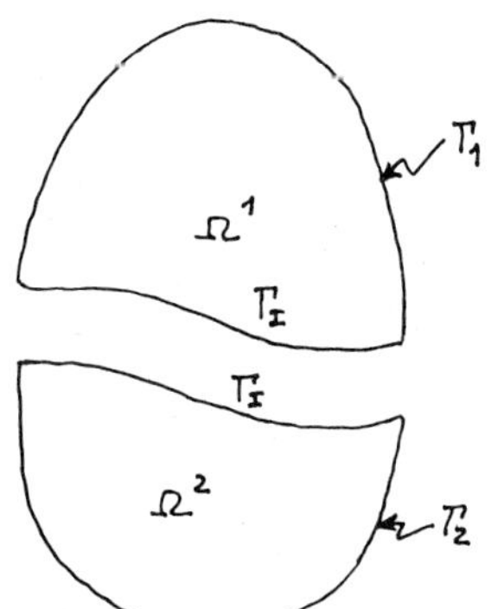

Fig. 3. Coupling of two bodies.

it belongs to Ω^1 and $\boldsymbol{U}_{\mathrm{I}}^2$, $\boldsymbol{Q}_{\mathrm{I}}^2$ and $\boldsymbol{F}_{\mathrm{I}}^2$ those values when considering that Γ_{I} belongs to Ω^2. Notice that in the other case the normals are in the opposite directions.

The conditions for coupling on the interface are compatibility and equilibrium which can be expressed as

(i) Compatibility; $\boldsymbol{U}_{\mathrm{I}} = \boldsymbol{U}_{\mathrm{I}}^1 = \boldsymbol{U}_{\mathrm{I}}^2$ on Γ_{I} with the following additional conditions in dynamic problems:

$$\left.\begin{aligned} \dot{\boldsymbol{U}}_{\mathrm{I}} = \dot{\boldsymbol{U}}_{\mathrm{I}}^1 = \dot{\boldsymbol{U}}_{\mathrm{I}}^2 \\ \ddot{\boldsymbol{U}}_{\mathrm{I}} = \ddot{\boldsymbol{U}}_{\mathrm{I}}^1 = \ddot{\boldsymbol{U}}_{\mathrm{I}}^2 \end{aligned}\right\} \quad \text{on } \Gamma_{\mathrm{I}} \tag{2.1}$$

(ii) Equilibrium, which can be expressed in terms of nodal flux values as

$$\boldsymbol{Q}_{\mathrm{I}}^1 + \boldsymbol{Q}_{\mathrm{I}}^2 = \mathbf{0} \qquad \text{on } \Gamma_{\mathrm{I}}\,, \tag{2.2}$$

or in terms of integrated values as

$$\boldsymbol{F}_{\mathrm{I}}^1 + \boldsymbol{F}_{\mathrm{I}}^2 = \mathbf{0} \qquad \text{on } \Gamma_{\mathrm{I}}\,, \tag{2.3}$$

with the first form being usually applied with boundary elements and the second one with finite elements. The systems of equations for dynamic equilibrium in the two regions can be written as follows: For region 1

$$\begin{bmatrix} \mathbf{M}_{11}^1 & \mathbf{M}_{1\mathrm{I}}^1 \\ \mathbf{M}_{\mathrm{I}1}^1 & \mathbf{M}_{\mathrm{II}}^1 \end{bmatrix} \begin{Bmatrix} \ddot{\boldsymbol{U}}^1 \\ \ddot{\boldsymbol{U}}_{\mathrm{I}}^1 \end{Bmatrix} + \begin{bmatrix} \mathbf{C}_{11}^1 & \mathbf{C}_{1\mathrm{I}}^1 \\ \mathbf{C}_{\mathrm{I}1}^1 & \mathbf{C}_{\mathrm{II}}^1 \end{bmatrix} \begin{Bmatrix} \dot{\boldsymbol{U}}^1 \\ \dot{\boldsymbol{U}}_{\mathrm{I}}^1 \end{Bmatrix} + \begin{bmatrix} \mathbf{K}_{11}^1 & \mathbf{K}_{1\mathrm{I}}^1 \\ \mathbf{K}_{\mathrm{I}1}^1 & \mathbf{K}_{\mathrm{II}}^1 \end{bmatrix} \begin{Bmatrix} \boldsymbol{U}^1 \\ \boldsymbol{U}_{\mathrm{I}}^1 \end{Bmatrix} = \begin{Bmatrix} \mathbf{F}^1 \\ \mathbf{F}_{\mathrm{I}}^1 \end{Bmatrix} \tag{2.4}$$

and for region 2,

$$\begin{bmatrix} \mathbf{M}_{22}^2 & \mathbf{M}_{2\mathrm{I}}^2 \\ \mathbf{M}_{\mathrm{I}2}^2 & \mathbf{M}_{\mathrm{II}}^2 \end{bmatrix} \begin{Bmatrix} \ddot{\boldsymbol{U}}^2 \\ \ddot{\boldsymbol{U}}_{\mathrm{I}}^2 \end{Bmatrix} + \begin{bmatrix} \mathbf{C}_{22}^2 & \mathbf{C}_{2\mathrm{I}}^2 \\ \mathbf{C}_{\mathrm{I}2}^2 & \mathbf{C}_{\mathrm{II}}^2 \end{bmatrix} \begin{Bmatrix} \dot{\boldsymbol{U}}^2 \\ \dot{\boldsymbol{U}}_{\mathrm{I}}^2 \end{Bmatrix} + \begin{bmatrix} \mathbf{K}_{22}^2 & \mathbf{K}_{2\mathrm{I}}^2 \\ \mathbf{K}_{\mathrm{I}2}^2 & \mathbf{K}_{\mathrm{II}}^2 \end{bmatrix} \begin{Bmatrix} \boldsymbol{U}^2 \\ \boldsymbol{U}_{\mathrm{I}}^2 \end{Bmatrix} = \begin{Bmatrix} \boldsymbol{F}^2 \\ \boldsymbol{F}_{\mathrm{I}}^2 \end{Bmatrix}. \tag{2.5}$$

Introducing the compatibility and equilibrium conditions (2.1) and (2.3) leads to

$$\left[\begin{array}{c|c|c} \mathbf{M}_{11}^1 & \mathbf{M}_{1\mathrm{I}}^1 & \mathbf{0} \\ \mathbf{M}_{\mathrm{I}1}^1 & \mathbf{M}_{\mathrm{II}}^1 + \mathbf{M}_{\mathrm{II}}^2 & \mathbf{M}_{\mathrm{I}2}^2 \\ \mathbf{0} & \mathbf{M}_{2\mathrm{I}}^2 & \mathbf{M}_{22}^2 \end{array}\right] \begin{Bmatrix} \ddot{\boldsymbol{U}}^1 \\ \hline \ddot{\boldsymbol{U}}_{\mathrm{I}} \\ \hline \ddot{\boldsymbol{U}}^2 \end{Bmatrix} + \left[\begin{array}{c|c|c} \mathbf{C}_{11}^1 & \mathbf{C}_{1\mathrm{I}}^1 & \mathbf{0} \\ \mathbf{C}_{\mathrm{I}1}^1 & \mathbf{C}_{\mathrm{II}}^1 + \mathbf{C}_{\mathrm{II}}^2 & \mathbf{C}_{\mathrm{I}2}^2 \\ \mathbf{0} & \mathbf{C}_{2\mathrm{I}}^2 & \mathbf{C}_{22}^2 \end{array}\right] \begin{Bmatrix} \dot{\boldsymbol{U}}^1 \\ \hline \dot{\boldsymbol{U}}_{\mathrm{I}} \\ \hline \dot{\boldsymbol{U}}^2 \end{Bmatrix}$$

$$+ \left[\begin{array}{c|c|c} \mathbf{K}_{11}^1 & \mathbf{K}_{1\mathrm{I}}^1 & \mathbf{0} \\ \mathbf{K}_{\mathrm{I}1}^1 & \mathbf{K}_{\mathrm{II}}^1 + \mathbf{K}_{\mathrm{II}}^2 & \mathbf{K}_{\mathrm{I}2}^2 \\ \mathbf{0} & \mathbf{K}_{2\mathrm{I}}^2 & \mathbf{K}_{22}^2 \end{array}\right] \begin{Bmatrix} \boldsymbol{U}^1 \\ \hline \boldsymbol{U}_{\mathrm{I}} \\ \hline \boldsymbol{U}^2 \end{Bmatrix} = \begin{Bmatrix} \boldsymbol{F}^1 \\ \hline \mathbf{0} \\ \hline \boldsymbol{F}^2 \end{Bmatrix} \tag{2.6}$$

or in general,

$$\mathbf{M}\ddot{U} + \mathbf{C}\dot{U} + \mathbf{K}U = F . \tag{2.7}$$

An important special case of these equations is that of steady-state problems, i.e. problems for which the functions and right-hand side vectors vary harmonically with time as

$$F = \hat{F}\, \mathrm{e}^{\mathrm{i}\omega t} ; \qquad U = \hat{U}\, \mathrm{e}^{\mathrm{i}\omega t} . \tag{2.8}$$

For this case Eqs. (2.7) reduce to

$$(-\omega^2\mathbf{M} + \mathrm{i}\omega\mathbf{C} + \mathbf{K})\hat{U} = \hat{F} \tag{2.9}$$

where ω is the exciting frequency. This equation governs a large range of engineering problems.

3. Interaction problems

Fluid–structure interaction problems can be divided into *external* fluid problems such as wave forces on offshore structures, floating bodies, wind acting on buildings, bridges, etc., and *internal* fluid problems such as when a fluid is contained inside a duct or vessel.

3.1. External fluid problems

External fluid problems have been analysed by dividing the fluid into a series of finite elements and extending the mesh sufficiently far from the body. The difficulty then lies in applying the right boundary conditions, but this can be achieved using boundary elements or the approximate elements previously discussed. The use of infinite elements, although popular with certain researchers, is not recommended as it entitles a domain rather than a boundary analysis. The description of the fluid using boundary elements is most attractive, especially when the boundary elements can be converted into equivalent finite matrices and combined with the finite element analysis for the structural part. Representing the fluid with finite elements does not present any special difficulties in most cases but requires working with a number of unnecessary internal unknowns which can be very large in three-dimensional problems.

3.1.1. Wave forces on submerged bodies

Consider the problem described in Fig. 4(a), i.e. a submerged cylindrical body under wave forces. The waves are assumed to be harmonic and linear – Airy

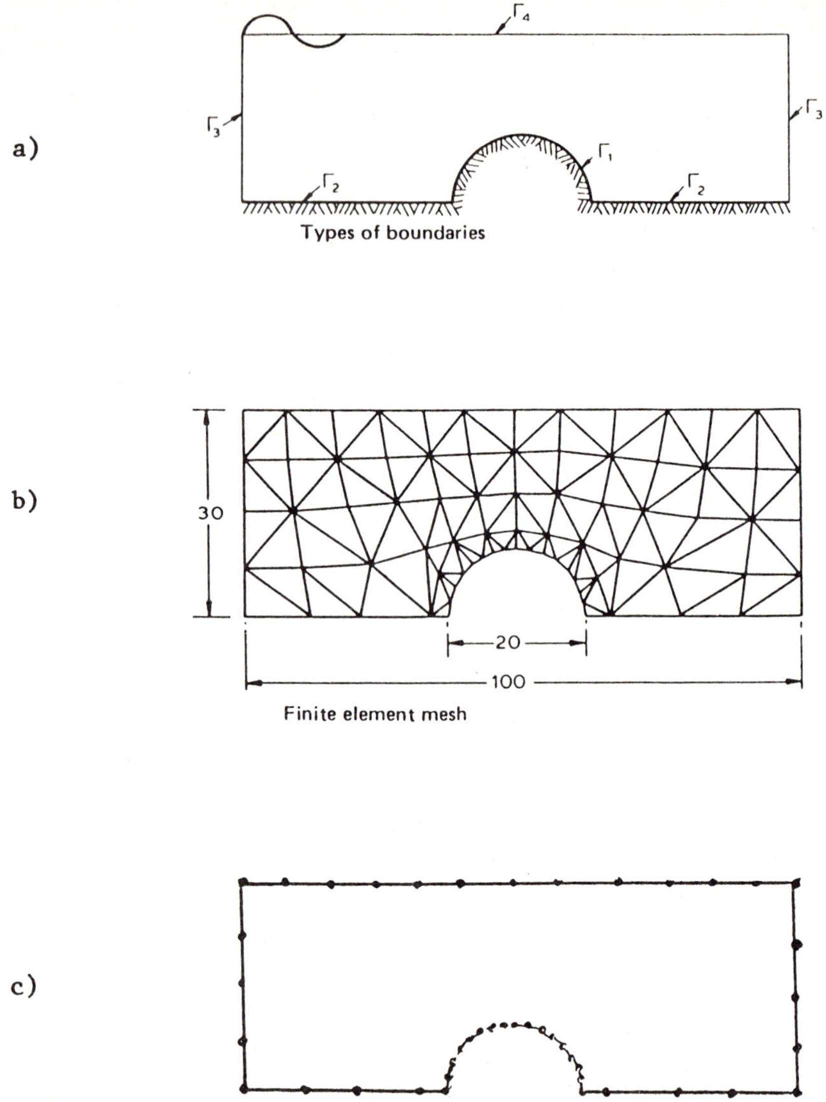

Fig. 4. Wave forces on a half cylinder: (a) type of boundaries; (b) finte element mesh; (c) boundary element mesh.

type – in which case the wave potential can be decomposed into two parts: the incident and the scattered potentials, i.e.

$$u = u_0 + u_s , \tag{3.1}$$

where, $u_s(x, y, t) = \hat{u}(x, y)\, e^{-i\omega t}$, ω is the frequency, $\kappa = \omega/c$ is the wave-number and c is the wave velocity.

Within the domain Ω the equation

$$\nabla^2 u(x, y, t) = 0 \qquad \text{in } \Omega \tag{3.2}$$

needs to be satisfied with the boundary conditions:

(i) bottom condition:

$$\frac{\partial u}{\partial n} = 0 \qquad \text{on } \Gamma_2$$

(ii) free surface condition:

$$\frac{\partial^2 u}{\partial t^2} + g \frac{\partial u}{\partial y} = 0 \qquad \text{on } \Gamma_4 \tag{3.3}$$

(iii) obstruction condition:

$$\frac{\partial u}{\partial n} = 0 \qquad \text{on } \Gamma_1$$

(iv) radiation condition in terms of scattered potential:

$$\lim_{|x| \to \infty} \left\{ \frac{\partial u_s}{\partial n} + \frac{\kappa}{\omega} \frac{\partial u_s}{\partial t} \right\} = 0 \qquad \text{on } \Gamma_3 .$$

The u_0 solution for the incident wave will satisfy the first three boundary conditions and is defined by the wave theory on Γ_3. Hence the problem can be expressed in terms of the time independent scattered field. This gives

$$\nabla^2 \hat{u}(x, y) = 0 \qquad \text{in } \Omega \tag{3.4}$$

with

(i) bottom condition:

$$\frac{\partial \hat{u}}{\partial n} = 0 \qquad \text{on } \Gamma_2$$

(ii) free surface condition:

$$\frac{\partial \hat{u}}{\partial n} - \frac{\omega^2}{g} \hat{u} = 0 \qquad \text{on } \Gamma_4$$

(iii) obstruction condition:

$$\frac{\partial \hat{u}}{\partial n} = -q_0 \qquad \text{on } \Gamma_1 , \tag{3.5}$$

where q_0 is the velocity produced by the incident field;

(iv) radiation condition – approximate:

$$\frac{\partial \hat{u}}{\partial n} - \mathrm{i}\kappa \hat{u} = 0 \qquad \text{on } \Gamma_3 .$$

The numerical solution for the region Ω can be found in terms of boundary or finite elements as shown in Figs. 4(b) and (c).

The *finite element* formulation of the problem results in the following system of equations:

$$\{\mathbf{K} + (\mathrm{i}\kappa + \omega^2/g)\mathbf{M}\}\hat{U} = \boldsymbol{F} . \tag{3.6}$$

The *boundary element* formulation is based on the single two-dimensional Green's function $u^* = (1/2\pi)\ln(1/r)$ and the integral relationship

$$c_i\hat{u}_i + \int_{\Gamma_1+\Gamma_2} \frac{\partial u^*}{\partial n}\hat{u}\,\mathrm{d}\Gamma + \int_{\Gamma_3}\left\{\frac{\partial u^*}{\partial n} - \mathrm{i}\kappa u^*\right\}\hat{u}\,\mathrm{d}\Gamma + \int_{\Gamma_4}\left\{\frac{\partial u^*}{\partial n} - \frac{\omega^2}{g}u^*\right\}\hat{u}\,\mathrm{d}\Gamma + \int_{\Gamma_1} u^* q_0\,\mathrm{d}\Gamma = 0 \tag{3.7}$$

where the constant c_i is proportional to the solid angle at the boundary point i (for smooth boundaries, $c_i = \frac{1}{2}$). We then need to propose a certain type of element to discretize the Γ surface of the domain (Fig. 4c), carry out a series of analytical or numerical interpretations and finally obtain the system of equations

$$\{\mathbf{H} - \mathrm{i}\kappa\mathbf{G}' - (\omega^2/g)\mathbf{G}''\}\hat{U} = -\mathbf{G}Q_0 , \tag{3.8}$$

where $\mathbf{G}'$ and $\mathbf{G}''$ are matrices which apply only on the Γ_3 and Γ_4 parts of the boundary, while $\mathbf{G}$ corresponds to the Γ_1 part of the boundary.

Equation (3.8) can be written in compact form as

$$\mathbf{A}\hat{U} = \boldsymbol{B} . \tag{3.9}$$

After solving Eq. (3.9) the potentials are known and then the wave forces can be computed by first calculating the pressures using Bernoulli's equation. The dynamic pressure is given by

$$p(x, y, t) = -\rho\,\mathrm{Re}\{\partial u(x, y, t)/\partial t\} , \tag{3.10}$$

where ρ is the fluid density. Once the pressures are known we can integrate them over the obstruction to find the wave forces.

The forces on cylindrical obstructions of the type studied here were computed for a depth to radius ratio of $h/a = 3.0$ and for a range of wave numbers as shown in Figs. 5 and 6. Results obtained using different types of boundary elements were compared against those produced by linear finite elements and the analytical results due to Chakrabarti and Naftzger [9, 19]. The constant boundary element results for vertical and horizontal forces obained using a mesh of 32 elements are shown in Figs. 5 and 6. To improve the results, two meshes consisting one of 46 linear elements and the other of 8 quadratic elements were employed. The results appear now to be in better agreement with those obtained in Refs. [9, 19]. In particular the quadratic elements seem to be better suited for this problem as they can represent more accurately the geometry of the half-cylinder which was not possible to do when using constant or linear elements.

Matrices obtained in this way can then be coupled to the structural matrices for the half cylinder obstruction to study the effect of wave–structure interaction on the combined system.

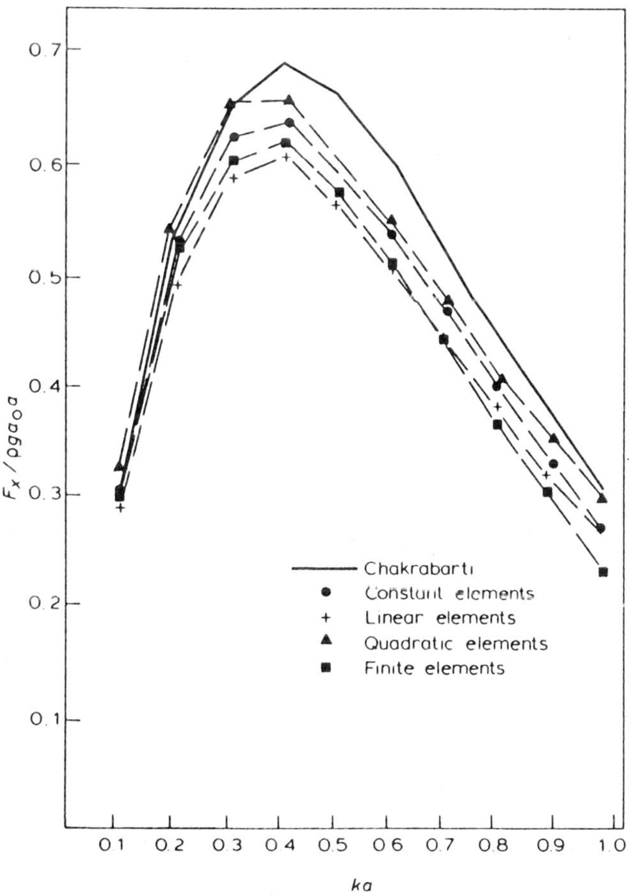

Fig. 5. Horizontal force on half-cylinder.

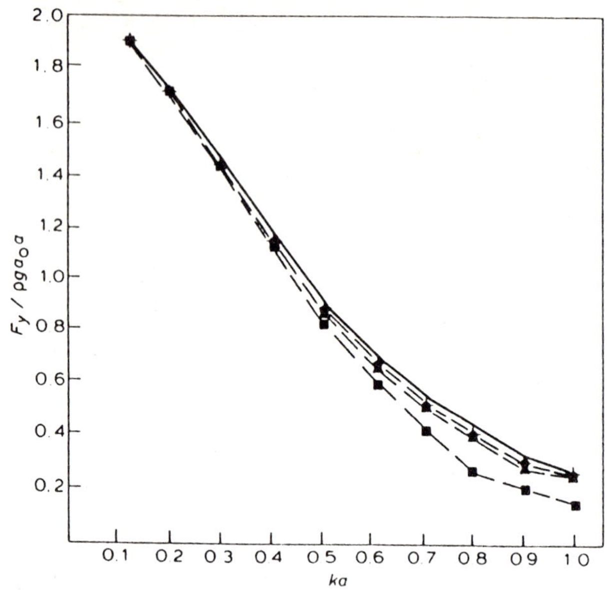

Fig. 6. Vertical force on half-cylinder. (Key as in Fig. 5.)

3.1.2. Added mass and damping coefficients

A problem with important practical applications is the determination of the added mass and damping matrices for a fluid in order to combine their effect with structural matrices. Sometimes these values can be reduced to a single coefficient when one knows the geometry of the system. In Ref. [3] results are presented for the mass and damping coefficients of a heaving half-submerged cylinder, starting from the governing equations for a floating body as shown in Fig. 7. The body is assumed to be rigid and subject to the three velocity components indicated in the figure, which can be written all together in terms of a velocity vector $\dot{X}$. The problem can now be solved numerically by subdividing the fluid using a finite element mesh as indicated in Ref. [23] and shown in Fig. 8 or by using the boundary element grid of Fig. 9 [3]. The finite element solution can be obtained

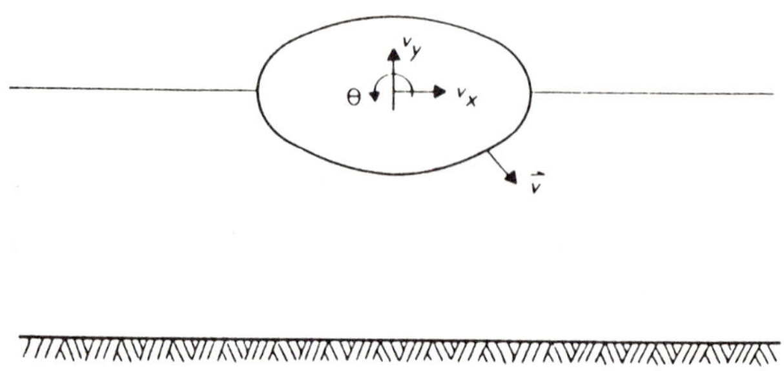

Fig. 7. Rigid 2D motion.

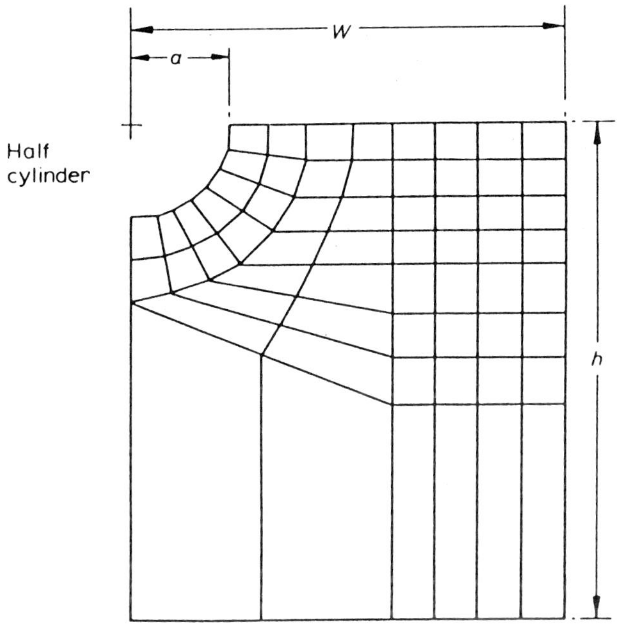

Fig. 8. Finite element mesh (62 cubic elements 359 nodes). $a = 10$; $h/a = 6$; $w/a = 5$.

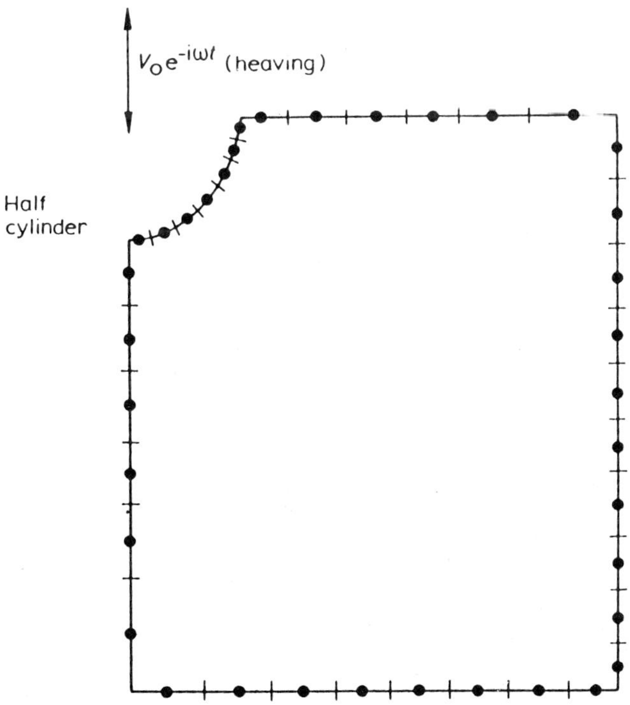

Fig. 9. Boundary element mesh (61 constant elements). $a = 10$; $h/a = 6$; $w/a = 5$.

as indicated in the other reference. Let us now assume that boundary elements are used. The starting matrix relationships can be written as

$$\mathbf{A}\hat{U} = \mathbf{G}Q_0 \,. \tag{3.11}$$

The matrix $\mathbf{A}$ is assumed to include the radiation and the surface boundary conditions as shown in Eq. (3.8). The values of Q_0 are functions of the velocities imposed on the rigid body and can be written as

$$Q_0 = \mathbf{R}\dot{X} \,, \tag{3.12}$$

where each velocity component is assumed to be harmonic, i.e. $q = q_0\,\mathrm{e}^{-\mathrm{i}\omega t}$. We can hence apply unit velocities in turn to represent surge, heave and roll and solve for the potentials. Then we compute the pressures and the forces on the body as indicated in the previous example. The forces on the floating body are represented by a vector F and once this is done we can find the added mass and damping coefficients. If the total force is represented by $\vec{F}$ we can then write

$$\begin{aligned} c_{\mathrm{m}} &= -\omega M_{\mathrm{H}}\,\mathrm{Im}\{\vec{F}\} \\ c_{\mathrm{d}} &= \omega M_{\mathrm{H}}\,\mathrm{Re}\{\vec{F}\} \end{aligned} \tag{3.13}$$

where c_{m} and c_{d} are the added mass and damping coefficients for the rigid floating body and M_{H} is the mass of water displaced by the half-cylinder.

For the case of the heaving motion of a half-submerged cylinder (Figs. 8 and 9) the boundary conditions on the symmetry axes can be taken to be $q = 0$. This is the only motion studied here for simplicity. Numerical results were obtained employing a 61 constant boundary element mesh shown in Fig. 9 and compared with those given by Newton [23] and Porter [24] and the agreement was satisfactory. Notice that the boundary element results presented in Fig. 10 achieved similar accuracy as those obtained with a much finer finite element mesh consisting of 62 cubic elements with 359 nodes.

3.1.3. Vertical cylinders

Let us now consider the problem of a train of surface waves propagating in water of constant depth which encounters one or more vertical cylinders of constant but arbitrary cross-section. Assuming that the cylinders extend from the seabed and pierce the free surface (Fig. 11), the scattered wave potential can be decomposed as

$$u_{\mathrm{s}}(x, y, z, t) = -\frac{\mathrm{i}gH}{2\omega}\,\frac{\cos[k(z+d)]}{\cosh(kd)}\,\hat{u}(x, y)\,\mathrm{e}^{-\mathrm{i}\omega t} \,, \tag{3.14}$$

where H is the wave height and the seabed and the linearized free surface conditions are directly satisfied.

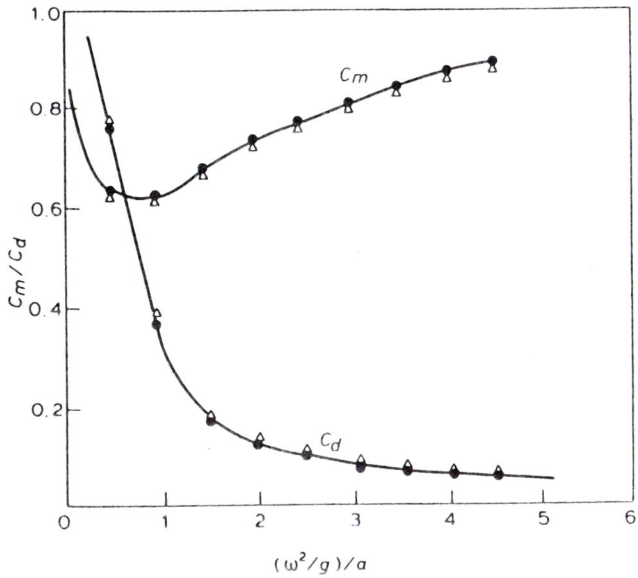

Fig. 10. Added mass and damping coefficients for a heaving half-submerged cylinder. (——), analytical curves by Porter; (●) finite element by Newton; (△), constant boundary element.

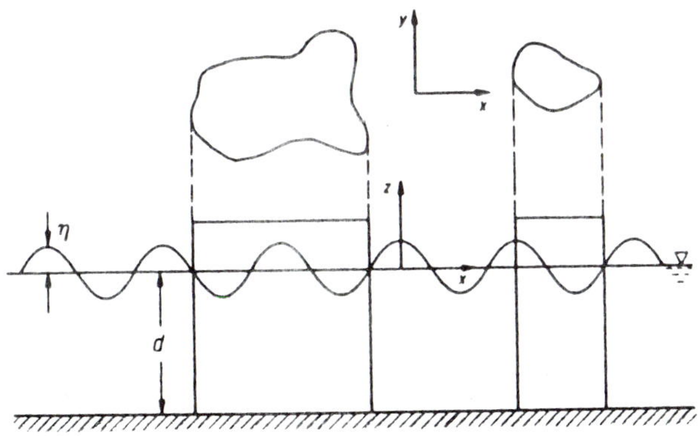

Fig. 11. Vertical cylinders of arbitrary section.

Since the fluid is assumed to be incompressible and the flow irrotational, i.e. the general problem is governed by the Laplace equation, one has that in this case the Helmholtz equation is satisfied in the form

$$\nabla^2\hat{u} + k^2\hat{u} = 0\,. \tag{3.15}$$

Thus, the actual problem is to compute the reduced potential which satisfies Eq. (3.15) with the boundary conditions (iii) and (iv) of Eqs. (3.5). Note that the

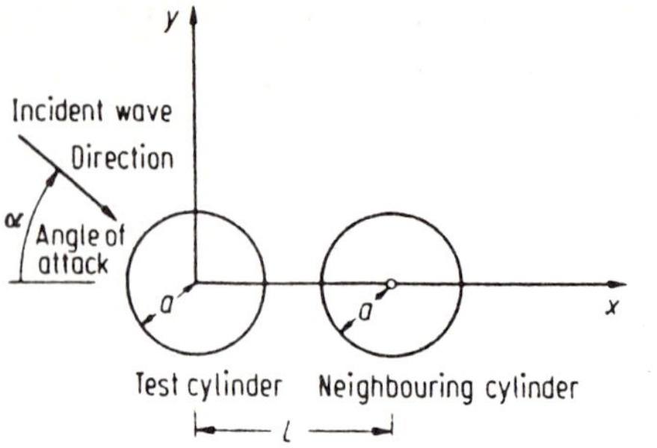

Fig. 12. Interaction between a pair of circular cylinders.

reduced incident wave potential $\hat{u}_0$ is given by

$$\hat{u}_0(x, y) = \exp[\mathrm{i}k(x \cos\alpha + y \sin\alpha)]\,, \tag{3.16}$$

where α is the angle of incidence.

A boundary integral equation equivalent to Eq. (3.7) can therefore be written with the aid of the fundamental solution given in Eq. (1.24). Some comments on the numerical evaluation of the integrals involved after discretization can be found in Refs. [4, 17].

An application of the present approach has been presented by Masetti and Wrobel [17] in which the interaction effects between two circular cylinders of equal radii are discussed (Fig. 12). The results obtained using 24 constant elements for each cylinder are depicted in Fig. 13 for three different incident angles. As can be seen, the $\alpha = 0°$ case produces an increase in the wave force of up to 50% and is the most unfavorable configuration tested.

3.2. Internal fluid problems

Consider now a type of fluid–structure interaction problem, i.e. when the fluid is enclosed within a vessel as shown in Fig. 14. The vessel may be represented using standard structural finite elements and the fluid using boundary elements. Representing the fluid using finite elements is not appropriate in most cases as we are seldom interested in the internal nodes results but only wish to construct fluid matrices which can be combined with the structural system [16].

Let us assume first that the fluid is incompressible and inviscid and that its behavior can be represented by a potential that will be called u. The boundary element matrices for this case can be be written as

$$\mathbf{H}\boldsymbol{U} = \mathbf{G}\boldsymbol{Q} \tag{3.17}$$

where $\boldsymbol{U}$ and $\boldsymbol{Q}$ are the values of nodal potentials and normal velocities.

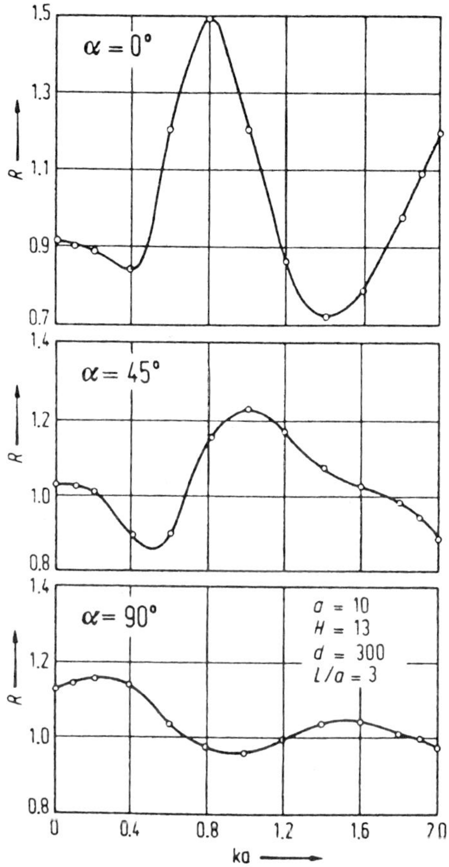

Fig. 13. Variation of force ratio for a pair of circular cylinders; R = ratio of horizontal force on test cylinder to force on isolated one.

Consider for simplicity that the movement is harmonic. Hence the velocity at any point k can be written in terms of the frequency ω and the normal displacements w_k as

$$q_k = \mathrm{i}\omega w_k \, . \tag{3.18}$$

The velocity potential can be related to the pressure p_k through Bernoulli's equation. For harmonic motion this gives

$$u_k = -\frac{1}{\mathrm{i}\omega\rho} \, p_k \, . \tag{3.19}$$

Hence we can write (3.17) as

$$\mathbf{H}\boldsymbol{P} = \omega^2\rho\mathbf{G}\boldsymbol{W} \, . \tag{3.20}$$

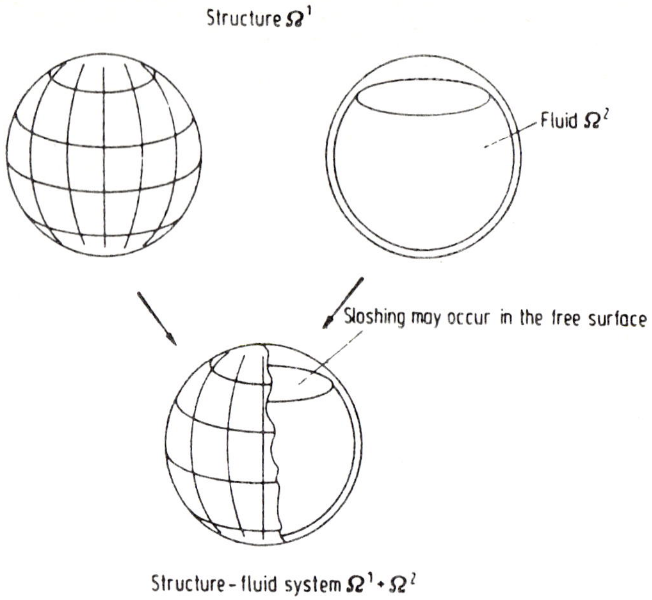

Fig. 14. Internal fluid problem.

If one restrains at least one degree of freedom by providing one reference value of the potential, one can obtain from (3.20)

$$P = \omega^2 \rho \mathbf{H}^{-1} \mathbf{G} W . \qquad (3.21)$$

In order to proceed and combine (3.21) with the usual finite element standard matrices one needs to work in terms of equivalent nodal forces. The relationship between nodal pressure and equivalent forces can be written in terms of the distribution matrix $\mathbf{N}$ such that the nodal forces are

$$F = \mathbf{N} P . \qquad (3.22)$$

Equation (3.21) now becomes

$$F = \omega^2 \rho (\mathbf{N} \mathbf{H}^{-1} \mathbf{G}) W = \mathbf{M}_F W . \qquad (3.23)$$

The form of $\mathbf{M}_F$ is the same as for a finite element formulation. Hence one can now superimpose it to the mass of the vessel and obtain the matrices for a coupled fluid–structure system. Notice that the $\mathbf{H}$ and $\mathbf{G}$ matrices of Eq. (3.23) have the same form as for a static potential problem. $\mathbf{M}_F$ will generally be asymmetric but can be symmetricized by applying the procedure shown in Eq. (1.10).

3.2.1. Free surface boundary conditions

In order to represent sloshing on the free surface of the fluid one needs to take into consideration the linearized kinematic and dynamic conditions [5, 29], i.e.

$$\frac{\partial u}{\partial z} = -\frac{1}{g}\frac{\partial^2 u}{\partial t^2} \tag{3.24}$$

where z is the normal to the free surface. Notice that $\partial u/\partial z$ is the velocity of the fluid in the vertical direction.

For harmonic problems one can write

$$q = \frac{\partial u}{\partial z} = \frac{\omega^2}{g} u . \tag{3.25}$$

This normal velocity can be written as a function of the $\boldsymbol{W}$ displacements and the potential as a function of the pressure as shown before, i.e. for a point k one finds

$$q_k = \mathrm{i}\omega w_k = -\frac{\omega}{g}\frac{1}{\mathrm{i}\rho} p_k \tag{3.26}$$

and thus

$$p_k = g\rho w_k . \tag{3.27}$$

Onece again these pressures can be converted into equivalent nodal forces using the distribution matrix $\mathbf{N}$. The equivalent nodal forces $\boldsymbol{F}_s$ on the free surface are defined as

$$\boldsymbol{F}_s = g\rho \mathbf{N} \boldsymbol{W}_s , \tag{3.28}$$

where $\boldsymbol{F}_s$ and $\boldsymbol{W}_s$ represent the forces and displacements on the nodes of the sloshing or free surface. Relationship (3.28) applies only to the nodes on the free surface and can be included in the stiffness matrix describing the structural response. The nodes on the sloshing surface can usually be eliminated by a matrix condensation technique as they are not directly related to the nodes on the structure.

3.2.2. Compressible fluids

Compressible fluid equations have important applications in engineering as well as in problems involving compressive waves inside pressure vessels, acoustics, wave propagation in general, etc. Mathematically the problem follows the lines of the wave diffraction problems already discussed. This is because the going equation for compressible potential flow has the same form as the one already seen [26], i.e.

$$\nabla^2 u = \frac{1}{c^2}\frac{\partial^2 u}{\partial t^2} \tag{3.29}$$

where c is the speed of sound in the fluid. For harmonic motion, Eq. (3.29) becomes

$$\nabla^2 u + k^2 u = 0\,. \tag{3.30}$$

References

[1] H.N. Abramson, Ed., The Dynamic Behavior of Fluids in Moving Containers, NASA-SP No. 106 (1966).

[2] S. Alliney, An Approximate Numerical Model for Radiation Problems, *Applied Mathematical Modelling* **6**, 192–196 (1982).

[6] M.C. Au and C.A. Brebbia, Numerical Prediction of Wave Forces Using the Boundary Element Method, *Applied Mathematical Modelling* **6**, 218–228 (1982).

[4] M.C. Au and C.A. Brebbia, Diffraction of Water Waves for Vertical Cylinders Using Boundary Elements, *Applied Mathematical Modelling* **7**, 106–114 (1983).

[5] C.A. Brebbia, J.C.F. Telles and L.C. Wrobel, *Boundary Element Techniques* (Springer-Verlag, Berlin, 1984).

[6] C.A. Brebbia and S. Walker, Simplified Boundary Elements for Radiation Problems, *Applied Mathematical Modelling* **2**, 135–137 (1978).

[7] C.A. Brebbia and R. Adey, Circulation Problems, in: Proceedings of Finite Element Symposium, Atlas Computer, Rutherford Laboratory, Didcot, UK (HMSO, London, 1974).

[8] C.A. Brebbia and S. Walker, *Boundary Element Techniques in Engineering* (Newnes–Butterworths, London, 1980).

[9] S.K. Chakrabarti and R.A. Naftzger, Nonlinear Wave Forces on Halfcylinder and Hemisphere, *Journal of the Waterways, Harbors and Coastal Engineering Division of ASCE* **100**, 189–204 (1974).

[10] H.S. Chen and C.C. Mei, Oscillations and Wave Force in a Man-Made Harbor in the Open Sea, in: *Proceedings of the 10th Symposium on Naval Hydrodynamics* (London, 1974) pp. 573–594.

[11] O.M. Faltinsen, Numerical Solutions of Transient Nonlinear Free-Surface Motion Outside or Inside Moving Bodies, in: *Proceedings of 2nd International Conference on Numerical Ship Hydrodynamics* (Berkeley, CA, 1977).

[12] O.M. Faltinsen and F.C. Michelsen, Motions of Large Structures in Waves at Zero Froude Number, in: *Proceedings of International Symposium on the Dynamics of Marine Vehicles and Structures in Waves* (London, 1974).

[13] C.J. Garrison and P.Y. Chow, Wave Forces on Submerged Bodies, *Journal of the Waterways, Harbors and Coastal Engineering Division of ASCE* **98**, 375–392 (1972).

[14] F. John, On the Motion of Floating Bodies II, *Communications in Pure and Applied Mathematics* **3**, 45–101 (1950).

[15] S. Kobayashi, Fundamentals of Boundary Integral Equation Methods in Elastodynamics, in: *Topics in Boundary Element Research*, Vol. 2, C.A. Brebbia, Ed. (Springer-Verlag, Berlin, 1985) pp. 1–54.

[16] T. Komatsu, Fluid–Structure Interaction, in: *Progress in Boundary Element Methods*, Vol. 2, C.A. Brebbia, Ed. (Springer-Verlag, Berlin, 1982) pp. 182–199.

[17] I.Q. Masetti and L.C. Wrobel, A Study of the Interaction Between Cylinders in Waves Using Boundary Elements, in: *Offshore Engineering*, F. Carneiro et al., Eds. (Pentech Press, London, 1984) pp. 163–185.
[18] K. Masuda and W. Kato, Hybrid BEM for Calculating Nonlinear Wave Forces on Three-Dimensional Bodies, in: *Boundary Elements*, C.A. Brebbia, T. Futagami and M. Tanaka, Eds. (Springer-Verlag, Berlin, 1983) pp. 209–215.
[19] R.A. Naftzger and S.K. Chakrabarti, Scattering of Waves by Two-Dimensional Circular Obstacles in Finite Water Depth, *Journal of Ship Research* **23**, 32–42 (1979).
[20] T. Nakayama, Boundary Element Analysis of Nonlinear Water Wave Problems, *International Journal for Numerical Methods in Engineering* **19**, 953–970 (1983).
[21] T. Nakayama and K. Washizu, The Boundary Element Method Applied to the Analysis of Two-Dimensional Nonlinear Sloshing Problems, *International Journal for Numerical Methods in Engineering* **17**, 1631–1646 (1981).
[22] D. Nardini and C.A. Brebbia, Boundary Integral Formulation of Mass Matrices for Dynamic Analysis, in: *Topics in Boundary Element Research*, Vol. 2, C.A. Brebbia, Ed. (Springer-Verlag, Berlin, 1985) pp. 191–208.
[23] R.E. Newton, Finite Element Analysis of Two-Dimensional Added Mass and Damping, in: *Finite Elements in Fluids*, Vol. 1, R.H. Gallagher et al., Eds. (John Wiley and Sons, New York, 1975) pp. 219–232.
[24] W.R. Porter, Univeristy of California, Berkeley, Institute of Engineering Research, Report Series 82, Issue 16 (1960).
[25] T. Sarpkaya and M. Isaacson, *Mechanics of Wave Forces on Offshore Structures* (Van Nostrand Reinhold, New York, 1981).
[26] S. Walker, Boundary Elements in Fluid-Structure Interaction Problems, in: *New Developments in Boundary Element Methods*, C.A. Brebbia, Ed. (CML Publications, Southampton, 1980) pp. 347–364.
[27] S. Walker and C.A. Brebbia, Harbour Resonance Problems Using Finite Elements, *Advances in Water Resources* **1**, 205–211 (1978).
[28] J.V. Wehausen and E.V. Laitone, Surface Waves, in: *Handbuch der Physik*, Vol. 9, S. Flügge and C. Truesdell, Eds. (Springer-Verlag, Berlin, 1960) pp. 446–778.
[29] L.C. Wrobel, S.H. Sphaier and P.T.T. Esperanca, Propagation of Surface Waves, in: *Topics in Boundary Element Research*, Vol. 2, C.A. Brebbia, Ed. (Springer-Verlag, Berlin, 1985) pp. 156–190.

Author Index

Subject Index